NATO ASI Series

Advanced Science Institutes Series

A series presenting the results of activities sponsored by the NATO Science Committee, which aims at the dissemination of advanced scientific and technological knowledge, with a view to strengthening links between scientific communities.

The Series is published by an international board of publishers in conjunction with the NATO Scientific Affairs Division

A Life Sciences B Physics	Plenum Publishing Corporation London and New York
C Mathematical and Physical Sciences D Behavioural and Social Sciences E Applied Sciences	Kluwer Academic Publishers Dordrecht, Boston and London
F Computer and Systems Sciences G Ecological Sciences H Cell Biology I Global Environmental Change	Springer-Verlag Berlin Heidelberg New York London Paris Tokyo Hong Kong Barcelona Budapest

NATO-PCO DATABASE

The electronic index to the NATO ASI Series provides full bibliographical references (with keywords and/or abstracts) to more than 30000 contributions from international scientists published in all sections of the NATO ASI Series. Access to the NATO-PCO DATABASE compiled by the NATO Publication Coordination Office is possible in two ways:

- via online FILE 128 (NATO-PCO DATABASE) hosted by ESRIN, Via Galileo Galilei, I-00044 Frascati, Italy.
- via CD-ROM "NATO Science & Technology Disk" with user-friendly retrieval software in English, French and German (© WTV GmbH and DATAWARE Technologies Inc. 1992).

The CD-ROM can be ordered through any member of the Board of Publishers or through NATO-PCO, Overijse, Belgium.

Series I: Global Environmental Change, Vol. 17

The ASI Series Books Published as a Result of Activities of the Special Programme on Global Environmental Change

This book contains the proceedings of a NATO Advanced Research Workshop held within the activities of the NATO Special Programme on Global Environmental Change, which started in 1991 under the auspices of the NATO Science Committee.

The volumes published as a result of the activities of the Special Programme are:

Vol. 1: **Global Environmental Change.** Edited by R. W. Corell and P. A. Anderson. 1991.
Vol. 2: **The Last Deglaciation: Absolute and Radiocarbon Chronologies.** Edited by E. Bard and W. S. Broecker. 1992.
Vol. 3: **Start of a Glacial.** Edited by G. J. Kukla and E. Went. 1992.
Vol. 4: **Interactions of C, N, P and S Biogeochemical Cycles and Global Change.** Edited by R. Wollast, F. T. Mackenzie and L. Chou. 1993.
Vol. 5: **Energy and Water Cycles in the Climate System.** Edited by E. Raschke and D. Jacob. 1993.
Vol. 6: **Prediction of Interannual Climate Variations.** Edited by J. Shukla. 1993.
Vol. 7: **The Tropospheric Chemistry of Ozone in the Polar Regions.** Edited by H. Niki and K. H. Becker. 1993.
Vol. 8: **The Role of the Stratosphere in Global Change.** Edited by M.-L. Chanin. 1993.
Vol. 9: **High Spectral Resolution Infrared Remote Sensing for Earth's Weather and Climate Studies.** Edited by A. Chedin, M.T. Chahine and N.A. Scott. 1993.
Vol. 10: **Towards a Model of Ocean Biogeochemical Processes.** Edited by G. T. Evans and M. J. R. Fasham. 1993.
Vol. 11: **Modelling Oceanic Climate Interactions.** Edited by J. Willebrand and D. L. T. Anderson. 1993.
Vol. 12: **Ice in the Climate System.** Edited by W. Richard Peltier. 1993.
Vol. 13: **Atmospheric Methane: Sources, Sinks, and Role in Global Change.** Edited by M. A. K. Khalil. 1993.
Vol. 14: **The Role of Regional Organizations in the Context of Climate Change.** Edited by M. H. Glantz. 1993.
Vol. 15: **The Global Carbon Cycle.** Edited by M. Heimann. 1993.
Vol. 16: **Interacting Stresses on Plants in a Changing Climate.** Edited by M. B. Jackson and C. R. Black. 1993.
Vol. 17: **Carbon Cycling in the Glacial Ocean: Constraints on the Ocean's Role in Global Change.** Edited by R. Zahn, T. F. Pedersen, M. A. Kaminski and L. Labeyrie. 1994.

Carbon Cycling in the Glacial Ocean: Constraints on the Ocean's Role in Global Change

Quantitative Approaches in Paleoceanography

Edited by

Rainer Zahn

GEOMAR
Forschungszentrum für Marine Geowissenschaften
Christian-Albrechts-Universität
Wischhofstraße 1-3, D-24148 Kiel

Thomas F. Pedersen

Department of Oceanography
University of British Columbia
Vancouver, B.C. Canada V6T 1W5

Michael A. Kaminski

Micropalaeontology Unit
Department of Geological Sciences
University College London
Gower Street, London, WC1E 6BT, U.K.

Laurent Labeyrie

Centre des Faibles Radioactivités
Laboratoire Mixte CNRS-CEA
Avenue de la Terrasse, F-91190 Gif-sur-Yvette

Springer-Verlag
Berlin Heidelberg New York London Paris Tokyo
Hong Kong Barcelona Budapest
Published in cooperation with NATO Scientific Affairs Division

Proceedings of the NATO Advanced Research Workshop on Carbon Cycling in the Glacial Ocean: Constraints on the Ocean's Role in Global Change, held at Fellhorst, Germany, September 17–19, 1992

Library of Congress Cataloging-in-Publication Data
Carbon cycling in the glacial ocean: constraints on the ocean's role in global change: quantitative approaches in paleoceanography / edited by Rainer Zahn ... [et al.]. p. cm. – (NATO ASI series. Series I, Global environmental change ; vol. 17)
"Proceedings of the NATO Advanced Research Workshop on Carbon Cycling in the Glacial Ocean: Constraints on the Ocean's Role in Global Change, held at Fellhorst, Germany, September 17-19, 1992" – P. iv. Includes index.

ISBN-13: 978-3-642-78739-3 e-ISBN-13: 978-3-642-78737-9
DOI: 10.1007/978-3-642-78737-9

1. Ocean-atmosphere interaction–Congresses. 2. Carbon cycle (Biogeochemistry)–Congresses. 3. Glacial epoch–Congresses. I. Zahn, Rainer, 1956- . II. NATO Advance Research Workshop on Carbon Cycling in the Glacial Ocean–Constraints on the Ocean's Role in Global Change (1992: Fellhorst, Germany) III. Series. GC190.2.C37 1994 551.46'8–dc20 93-49420

Softcover reprint of the hardcover 1st edition 1994

Typesetting: Camera ready by authors
31/3145 - 5 4 3 2 1 0 - Printed on acid-free paper

FOREWORD

Growing interest in global climatic fluctuations and the need for more precise modeling of past and future climatic change has increased the demand for a better understanding of how the ocean's carbon cycle has changed during the past. Glacial-interglacial changes in atmospheric CO_2 recorded in polar ice cores clearly point to the importance of ocean processes in determining the atmosphere's chemistry. Precise knowledge about these processes becomes especially important in view of man's impact on atmospheric CO_2 due to fossil fuel burning and deforestation. The possibility that increasing atmospheric concentrations of CO_2 and other greenhouse gases may significantly alter Earth's climate underscores the need to better constrain the reconstruction of biological and chemical cycles in relation to the dramatic changes of atmospheric CO_2 during the past.

A wealth of proxy-data has been published over the past decade which documents past changes of the oceanic carbon cycle. Despite this, quantitative estimates of the fluxes between the different carbon reservoirs are not well enough constrained to serve as reliable controls, for example, for calibrating predictive climate models. Uncertainties involved in quantifying the glacial marine carbon cycle and evaluating its relevance for the state of global climate stem from the fact that spatial and temporal variations of a single paleoceanographic proxy are not uniquely tied to one specific environmental parameter, but usually document the combined variations of several components of the oceanic carbon system.

The past several years have witnessed a rapid development of new tools and concepts that are used for reconstructing water mass properties, primary productivity, and chemical cycling in the world ocean. The traditional boundaries between the major subdisciplines sedimentology, micropaleontology, marine geochemistry, and stratigraphy, which were originally independent fields of research, have become increasingly perforated. Different methods and tools have been developed, often quite independently within these subdisciplines, which serve as proxies for the same parameters in the ocean system. For example, glacial-to-interglacial changes in paleoproductivity and bottom water chemistry can be estimated using statistical analysis of foraminiferal faunal community structures, organic carbon content in the sediments, carbon isotope composition, the cadmium-to-calcium ratio in the shells of foraminifera, and inorganic geochemical oxygen and nutrient tracers. Considerable advances have been made in analysing the stable isotope and elemental composition of the sedimentary organic fraction which have resulted in the development of new tracers and markers for the past biological and geochemical operation of the ocean.

Some of the tracers show coherent patterns of variation through time, others apparently contradict each other. For instance, carbon isotope data from planktonic foraminifera argue that Southern Ocean surface water nutrients increased during glacial periods, whereas planktonic cadmium/calcium data from the same region suggest little or no change. In view of the potential importance of the high-latitude ocean as a control for atmospheric CO_2 this apparent discrepancy warrants the application of other tools which promise to shed brighter light on polar nutrient scenarios. Furthermore, inconsistencies have been observed between the variability of the ice core CO_2 record and the marine carbon isotope record which seem to require several, possibly independent mechanisms to account for variations in the ocean-atmosphere carbon exchange. If true, these mechanisms can be separated only by applying multi-proxy approaches that combine the analysis of several tracers which are independently tied to the marine carbon system. Potentially helpful tracers such as nitrogen isotopes, for example are just appearing on the horizon, and should bolster the collection of "standard" tools in paleoceanography.

In view of these new developments, there exists a need to: (1) evaluate the uncertainties involved in the different analytical strategies; (2) correlate results obtained by the using the independent approaches; and (3) resolve contradictory trends between different proxies. Our motivation for organizing the NATO Advanced Research Workshop was to bring recognized scientists from the various subdisciplines together who have independently developed tools and methods to determine quantitatively various parameters involved in the marine carbon cycle. Some of the speakers at the workshop applied similar analytical strategies but presented stimulating and controversial interpretations of proxy data. Our intention was not only to promote interdisciplinary research but also to ignite discussion amongst scientists with controversial viewpoints regarding the implications of their proxy-data with respect to carbon cycling and paleoclimate.

The workshop has been made possible through a grant by the Scientific Affairs Division of the North Atlantic Treaty Organization. We wish to thank Dr. L. da Cunha (director SAD-NATO) and his staff for their patience, and T. Wagner and D. Nürnberg - both at GEOMAR/Kiel - for their help in running the logistics during the meeting.

Much of the work on this book was done by Angela Keir who could not see this volume come to completion. Angela died unexpectedly briefly before this book was finished. She supported us with great humor in organizing the workshop and kept track of all the loose ends before, during, and after the workshop. We are deeply indebted to her and will always remember her. During the final stage of getting the book together, J. Schönfeld (GEOMAR/Kiel) helped us sorting through the files and preparing the final layout of the book.

We hope that this book will stand as a comprehensive progress report in the collaborative, multi-disciplinary and international effort to understand the role of the ocean in mediating climate change.

GEOMAR/Kiel, UC/London, CFR/Gif-Sur-Yvette, UBC/Vancouver
November 1993 The Editors

TABLE OF CONTENTS

FOREWORD

LIST OF AUTHORS AND PARTICIPANTS

I - OPERATION OF THE OCEAN-ATMOSPHERE CARBON CYCLE

II - VARIATIONS OF THE OCEAN'S CARBON RESERVOIR: FAUNAL VERSUS GEOCHEMICAL RECORDS

IV - GEOCHEMICAL AND MICROPALEONTOLOGICAL INDEXES OF PALEO-PRODUCTIVITY

LIST OF AUTHORS AND PARTICIPANTS

Fatima Abrantes, Serviços Geologicos de Portugal, Nucleo de Geologia Marinha, Rua da Academia das Ciencias, 19-20, 1200 Lisboa, Portugal; ph +351-1-3463-915, fax +351-1-3424-609

Ahuva Almogi-Labin, Geological Survey of Israel, 30, Malkhei Yisrael St. Jerusalem 95501, Israel; ph. +972-2-314211, fax +972-2-380688

Mark Altabet, Woods Hole Oceanographic Institution, Chemistry Dept., Clark Lab, Woods Hole, MA 02543, U.S.A.; ph +1-508-457-2000 (ext. 2890), fax +1-508-457-2193, E-mail maltabet@cliff.whoi.edu

Alexander Altenbach, Geologisch-Paläontologisches Institut, Christian-Albrechts-Universität Kiel, Olshausenstrasse 40-60, D-24118 Kiel, Germany; ph +49-431-880-2939, fax +49-431-880-4376

Edouard Bard, Université d'Aix-Marseille III, URA CNRS 132/Case 431, Avenue Escadrille Normandie-Niemen, F-13397 Marseille, Cedex 13, France; ph +33-91-2880-37, fax +33-91-9892-60

Jean-Marc Barnola, Laboratoire de Glaciologie et Geophyique de l'Environment, BP 96, F-38402 St. Martin d'Heres Cedex, France; ph +33-7682-4200, fax +33-7682-4201

Wolfgang Berger, Scripps Institution of Oceanography, University of Carlifornia, San Diego, La Jolla, CA 92093, U.S.A.; ph +1-619-452-2750, fax +1-619-5340784

Jelle Bjima, Alfred-Wegener-Institut, für Polar and Marine Research, Postfach 120161, Columbusstrasse, D-27570 Bremerhaven, Germany; ph +49-471-4831-0, fax +49-471-4831-149

Ann Boersma, Microclimates Research Consultants, 540 Gate Hill, Stoney Point, N.Y. 10980, U.S.A.; ph +1-914-942-1394

Edward Boyle, Massachusetts Insitute of Technology, Dept. of Earth, Atmospheric, and Planetary Science, Room E34- 258, Cambridge, MA 02139, U.S.A.; ph. +1-617-253-3388, fax +1-617-253-8630, E-mail eaboyle@mit.edu

Barrie Dale, Dept. of Geology, University of Oslo, P.B. 1047 Blindern, N-0316 Oslo 3, Norway; ph +47-2-856656, fax +47-2-854215

Gert J. De Lange, University of Utrecht, Institute of Earth Sciences, Dept. of Geochemistry, Section Marine Earth Sciences, P.O. Box 80.021, NL-3508 TA Utrecht, The Netherlands; ph +31-30-535034, fax +31-30-535030, E-mail gdelange@earth.ruu.nl

Gerald Ganssen, Free University Amsterdam, Geomarine Center, Institute of Marine Sciences, P.O. Box 7161, NL-1007 MC Amsterdam, The Netherlands; ph +31-20-548-5586, fax +31-20-646-2457

Christoph Heinze, Max-Planck-Institut für Meteorologie, Bundesstraße 55, D-20146 Hamburg, Germany; ph. +49-40-41173-245 fax +49-40-41173-298, E-mail Heinze@dkrz-hamburg.d400.de

Christoph Hemleben, Geologisch-Palaeontologisches Institut, Sigwartstr. 10, D-72076 Tübingen, Germany; ph +49-7071-292496, fax +49-7071-296-990, E-mail epiot0@compserv.zdv.uni-tuebingen.de

Juan Carlos Herguera, Scripps Institution of Oceanography, UCSD, Mail Code 0215, La Jolla, CA 92093, U.S.A.; now at: Centro de Investigación Científica y de Enseñanza Superior de

Ensenada, Diviśión de Oceanología, Carretera Transpeninsular Km. 107, Ensenada, Baja California; México

Eystein Jansen, University of Bergen, Department of Geology, Section B, Allégaten 41, 5007 Bergen, Norway; ph +47-5-213491, fax +47-5-324416, E-mail jansen@geol.uib.no

John P. Jasper, Biogeochemical Laboratories, Department of Geological Sciences and Department of Chemistry, Indiana University, Bloomington, IN 47405-1403, U.S.A., ph. +1-812-855-1382, fax +812-855-7961, E-mail JasperJ@UCS.Indiana.edu

Andrew M. Johnston, University of Dundee, Department of Biological Sciences, Dundee DD1 4HN, United Kingdom, ph. +44-382-23-181, fax +44-382-22-318, E-mail bso4@uk.ac.dundee.primeb

Michael Kaminski, University College London, Micropalaeonotology Unit, Department of Geological Sciences, Gower Street, London, WC1E 6BT, United Kingdom; ph +44-71-387-7050 (ext. 2432), fax +44-71-388-7614, E-mail ucfbmak@ucl.ac.uk

Robin S. Keir, GEOMAR, Research Center for Marine Geosciences, Wischhofstrasse 1-3, D-24148 Kiel, Germany; ph +49-431-7202-281, fax +49-431-7202-293

Eddy Keppens, Vrjie University-Fakulteit van deWetenschappen, Eenheid Geochronologie, Pleinlan 2, B-1050, Brussels, Belgium, ph. +32-2-641-33 95 or -80, fax +32-2-641-33-91

Laurent Labeyrie, Centre des Faibles Radioactives, Laboratoire Mixte CNRS-CEA, Avenue de la Terrasse, F-91190 Gif sur Yvette, Cedex France ph +33-1-6982-3536, fax: +33-1-6982-3568, E-mail labeyrie@eole.cfr.cnrs-gif.fr

Ingeborg Levin, Institut für Umweltphysik, Im Neuenheimer Feld 366, D-69120 Heidelberg, Germany; ph +49-6221-563350, fax +49-6221-563405, E-mail LV@svr1.uphysn.uni-heidelberg.de

Andreas Mackensen, Alfred-Wegener-Institut, für Polar and Marine Research, Postfach 120161, Columbusstrasse, D-27570 Bremerhaven, Germany; ph +49-471-4831-221, fax +49 471-4831-149

Augusto Mangini, Institut für Umweltphysik, Im Neuenheimer Feld 366, D-69120 Heidelberg, Germany; ph +49-6221-56-33-08, fax +49-6221-563405

Daniel McCorkle, Woods Hole Oceanographic Institution, Department of Geology & Geophysics, 100 C McLean, Woods Hole, MA 02543, U.S.A.; ph +1-508-548-1400 (ext. 2949), fax +1-508-457-2183, E-mail dmccorkl@cliff.whoi.edu

Elisabeth Michel, Centre des Faibles Radioactives, Laboratoire Mixte CNRS-CEA, Avenue de la Terrasse, F-91198 Gif sur Yvette Cedex, France; ph +33-1-6982-3534, fax: +33-1-6982-3568

Alan Mix, Oregon State University, College of Oceanography, Oceanography Admin. Bldg. 104, Corvallis, OR 97331-5503, U.S.A., ph +1-503-737-2296, fax +1-503-737-2064, E-mail mix@oce.orst.edu

Joseph Montoya, Harvard University, Biological Laboratories 179, 16 Divinity Ave., Cambridge, MA 02138, U.S.A.; ph +1-617-496-8537, fax +1-617-496-5854, E-mail montoya2@husc.harvard.edu

Guy Munhoven, Institut d'Astrophysique, Université de Liège, Avenue de Cointe, 5, B-4000 Liège, Belgium; ph +32-41-529-980, fax +32-41-527-474

Jean-Robert Petit, Laboratoire de Glaciologie et Geophyique de l'Environment, BP 96, F-38402 St. Martin d'Heres Cedex, France; ph +33-7682-4244, fax +33-7682-4201

Greg Rau, University of California Santa Cruz, NASA Ames Research Center, P.O. Box 1000, Moffett Field, CA 94035-1000, U.S.A.; ph +1-415-604-6061, fax +1-415-604-1088, E-mail greg_rau@qmgate.arc.nasa.gov

Robert Ross, Geologisch-Paläontologisches Institut, Christian-Albrechts-Universität Kiel, Olshausenstrasse 40-60, D-24118 Kiel, Germany; now at: Institute of Geosciences, Shizuoka University, 836 Oya, Shizuoka 422, Japan; ph +81-54-237-1111, fax +81-54-238-0491

Paul Saager, Free University, Department of Earth Sciences, De Boelelaan 1085, 1081 HV Amsterdam, The Netherlands; ph +31-20-548-2431 or -3585, fax +31-20-646-2457, E-mail saap@geo.vu.nl

Constance Sancetta, National Science Foundation, Ocean Sciences Division, Room 609, 1800 G St. NW, Washington, DC 20550, U.S.A.; ph +1-202-357-7312, fax +1-202-357-7621

Michael Sarnthein, Geologisch-Paläontologisches Institut, Christian-Albrechts-Universität Kiel, Olshausenstrasse 40-60, D-24118 Kiel, Germany; ph +49-431-880-2882, fax +49-431-880-4376

Ralph Schneider, Universität Bremen, Fachbereich 5, Geowissenschaften, Klagenfurter Straße, D-28359 Bremen, Germany; ph +49-421-218-3579, fax +49-421-218-3116

Detmar Schnitker, University of Maine, Darling Marine Center, Walpole, ME 04573, U.S.A.; ph +1-207-563-3146, fax +1-203-536-3119

Hans Schrader, University of Bergen, Department of Geology, Section B, Allégaten 41, N-5007 Bergen, Norway; ph +47-5-213533, fax +47-05-324416

Graham Shimmield, Grant Institute of Geology and Geophysics, University of Edinburgh, West Mains Road, Edinburgh EH9 3JW, United Kingdom; ph +44-31-650-8502, fax +44-31-668-3184, E-mail G.Shimmield@ed.ac.uk

Daniel Sigman, Woods Hole Oceanographic Institution, Dept. of Geology & Geophysics, 100 C McLean, Woods Hole, MA 02543, U.S.A.; ph +1-508-548-1400 (ext. 2949), fax +1-508-457-2183

Kozo Takahashi, Woods Hole Oceanographic Institution, Chemistry Dept., Clark Lab, Woods Hole, MA 02543, U.S.A.; now at: Hokkaido Tokai University, Department of Marine Sciences and Technology, School of Engeneering, 5-1-1-1 Minamisawa, Minami-ku, Sapporo 005, Japan; ph +81-011-571-5111 (ext. 615), fax +81-11-571-7879

Carlos Vale, Ministério do Mar, Secretaria de Estado das Pescas, Instituto National de Investigação das Pescas, Av. de Brasília, 1400 Lisboa, Portugal; ph +351-1-301-08-14, fax +351-1-301-59-48

Rainer Zahn, GEOMAR, Research Center for Marine Geosciences, Wischhofstrasse 1-3, D-24148 Kiel, Germany; ph +49-431-7202-265, fax +49-431-725391, E-mail rzahn@geomar.uni-kiel.d400.de.

This volume is dedicated to the memory of Angela Keir, who passed away suddenly following minor surgery on September 7, 1993, at a Kiel hospital. All of the participants of the Fellhorst Conference remember Angela for her cheerful disposition and unselfish dedication in organizing the meeting and ensuring that the conference agenda ran smoothly. She was the single most instrumental person in handling the logistics of both the conference itself as well as the initial compilation of this volume. All the contributors of this volume corresponded with Angela over a period of two years. Her sudden departure leaves a void in the hearts of her friends and co-workers on two continents which will not soon be filled.

PART I:

OPERATION OF THE OCEAN-ATMOSPHERE CARBON CYCLE

THE RECENT STATE OF CARBON CYCLING THROUGH THE ATMOSPHERE

Ingeborg Levin
Institut für Umweltphysik
University of Heidelberg
Im Neuenheimer Feld 366
D-69120 Heidelberg
Germany

ABSTRACT. A summary of our knowledge about the recent carbon cycle is presented from the perspective of atmospheric observations. Particular emphasis is given to carbon isotopes as a tool to determine global carbon sources and sinks as well as their recent anthropogenic changes. In the case of atmospheric carbon dioxide, the yearly anthropogenic emissions are small if compared to the gross exchange between the most important reservoirs: biosphere and ocean. In order to derive net fluxes caused by climatic or anthropogenic perturbations, the gross fluxes have to be known as accurately as possible. One isotopic tool is to trace the penetration of bomb ^{14}C into the ocean surface water so as to derive the atmosphere/ocean gas-exchange rate. Moreover, the distribution of $\delta^{13}C$ in atmospheric CO_2 can help to constrain the amount of excess CO_2 taken up (or released) by the global biosphere. As for atmospheric methane, anthropogenic emissions today exceed natural release rates by nearly a factor of two. Although the total yearly methane emissions to the atmosphere are believed to be known to within 10-20%, large uncertainties still remain in the estimates for individual sources, in particular for natural ecosystems like wetlands and the oceans. Isotope observations provide strong constraints also for the global atmospheric methane budget because different source types (biogenic vs. thermogenic) produce their own characteristic isotopic fractionations during methane production. These isotopic signatures can be used further to predict the future development of atmospheric methane concentrations and to understand better the drastic changes observed in the past under different conditions.

NATO ASI Series, Vol. I 17
Carbon Cycling in the Glacial Ocean:
Constraints on the Ocean's Role in Global Change
Edited by R. Zahn et al.

INTRODUCTION

Climate on Earth strongly depends on the radiative balance of its atmosphere, and, thus, on the abundance of radiatively active trace gases. Largely due to human activities, since the Industrial Revolution the chemical composition of the atmosphere has changed dramatically. Direct measurements during the last decades clearly show an increasing burden of many greenhouse gases in the atmosphere. The atmospheric carbon dioxide concentration is recently increasing by about 0.5 % per year (Keeling et al., 1989a). The recent methane increase, although slowing down in the last several years, shows an increase of about 0.6 - 0.7 % per year (Steele et al., 1992). Moreover, analysis of ancient air trapped in ice from polar regions allows insight into the change of these trace gas concentrations in the past. From a presumably "undisturbed" Holocene situation several hundred years ago until today atmospheric CO_2 concentration increased by more than 25 % from a pre-industrial level of about 280 ppmv to about 355 ppmv extrapolated for 1991 (Neftel et al., 1985; Friedli et al., 1986; Keeling et al., 1989a). The change of atmospheric methane concentration has been even more dramatical, from a pre-industrial level of about 0.8 ppmv to 1.7 ppmv at the end of 1990 (Etheridge et al., 1988; Steele et al., 1992).

Any predictions of the anthropogenically influenced future climate on Earth, or the understanding of past climatic changes, e.g. the glacial/interglacial cycles, strongly rely on an accurate knowledge of the naturally or anthropogenically influenced trace gas concentrations in the present day atmosphere. Strictly speaking, this implies we have to quantitatively understand the recent geophysical cycles, e.g. the carbon cycle in the case of methane and carbon dioxide. This also means that we have to deal with large perturbations caused by man's impact.

Firstly, our current picture of the recent CO_2 and CH_4 cycles will be outlined. In attempting to disentangle the complexity of their geophysical cycles, from the perspective of the atmosphere, emphasis is given to isotope studies as an experimental tool being used most profitably where the detailed information on the spatial and temporal variability of sources and sinks is missing. The known parameters influencing the levels of the atmospheric trace gases carbon dioxide and methane will be discussed in the perspective of their variations in the past.

ATMOSPHERIC CARBON DIOXIDE

The recent state of the atmospheric carbon dioxide cycle is summarized in Figure 1 (Watson et al., 1990). Large amounts of carbon dioxide (about 15% of the atmospheric inventory) are annually exchanged between the atmosphere and the biosphere and theatmosphere and the ocean surface waters, respectively. Compared to this large 'steady state' carbon transfer between reservoirs, the total net yearly input of CO_2 into the atmosphere from the burning of fossil fuels and from deforestation is smaller by more than an order of magnitude. Keeping in mind that the errors of the exchange rates between the reservoirs are in the order of at least ±20%, immediately illustrates the difficulty to univocally determine the fate of anthropogenic CO_2 emissions today and in future times.

Due to the large diversity of the terrestrial biosphere, and a large temporal and spatial variability of dissolved inorganic carbon in the surface ocean, reliable net changes in the sizes of these reservoirs are difficult to obtain by direct observations. We, thus, atmosphere and the

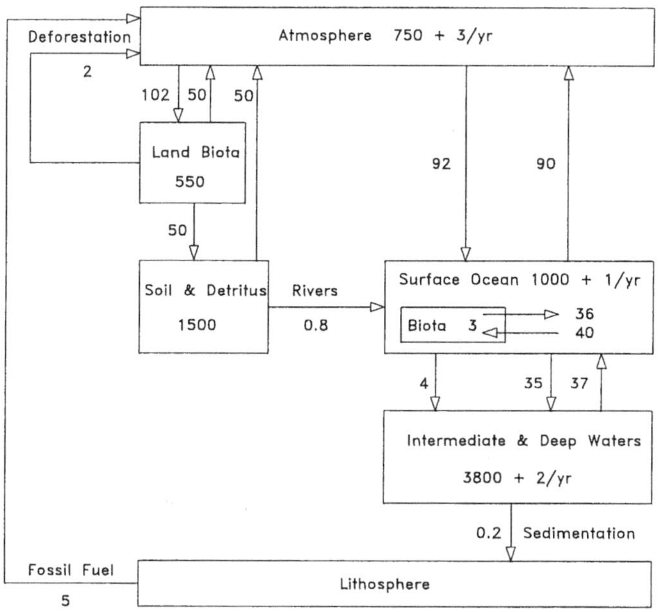

Fig. 1: Global carbon reservoirs and fluxes. The numbers apply to the present day situation. Fluxes, e.g. between atmosphere and surface ocean are gross annual exchanges. Units in gigatons of carbon (GtC; 1 Gt = 10^9 metric tons = 10^{12}kg) for reservoir sizes and GtC yr^{-1} for fluxes (adopted from Watson et al., 1990).

ocean surface waters, respectively. Compared to this large "steady state" carbon transfer between reservoirs, the total net yearly input of CO_2 into the atmosphere from the burning of fossil fuels and from deforestation is smaller by more than an order of magnitude. Keeping in mind that the errors of the exchange rates between the reservoirs are in the order of at least ±20%, immediately illustrates the difficulty to univocally determine the fate of anthropogenic CO_2 emissions today and in future times.

Due to the large diversity of the terrestrial biosphere, and a large temporal and spatial variability of dissolved inorganic carbon in the surface ocean, reliable net changes in the sizes of these reservoirs are difficult to obtain by direct observations. We, thus, basically have to rely on concentration measurements of CO_2 in the atmosphere, the latter integrating the fluxes from all sources and sinks. At present, the atmosphere is the best-known of all carbon reservoirs, and a lot of information about the CO_2 cycle has been derived from the interpretation of the observed spatial and temporal variability of atmospheric concentration by model simulations. These models have been constrained by the changing stable and radioactive carbon isotope patterns observed in the atmosphere as well as in the ocean. Two examples of how to handle this isotopic tool are illustrated below.

DETERMINING THE AIR-SEA CO_2 EXCHANGE RATE WITH BOMB RADIOCARBON

As a consequence of the atmospheric nuclear weapon tests in the fifties and early sixties large amounts of radiocarbon have been artificially introduced into the atmosphere, leading to a global increase of the $^{14}C/^{12}C$ ratio in atmospheric CO_2 of nearly a factor of two (Levin et al., 1992), and, thus, to a substantial disequilibrium of ^{14}C between atmosphere, biosphere and surface ocean water. The gas exchange rates between atmosphere and ocean can be derived from the observed total excess ^{14}C inventory in the ocean at a given time if compared to the pre-bomb level (Stuiver, 1980). From the GEOSECS 1972/3 data in the Atlantic, the 1973/4 data in the Pacific and the 1976 data in the Indian oceans, a global mean air-sea CO_2 exchange flux of 20±3 mol m^{-2} yr^{-1} has been estimated for the 10-year period of 1962 to 1972 (Stuiver et al., 1981; Broecker et al., 1985; 1986). Assuming an unchanged transfer velocity for the air-sea exchange in 1990, and taking into account the 10% increase of atmospheric CO_2 from 1972 to 1990, leads to the total gas exchange rate in 1990 of about 90 GtC yr^{-1} (compare Figure 1).

THE FATE OF ANTHROPOGENIC CO_2

The estimate of the net uptake of anthropogenic CO_2 by the oceans of about 2.0±0.8 GtC yr^{-1} given in Figure 1 for the period of 1980 - 1989 has been derived from ocean circulation models which were also calibrated as to reproduce the distribution of bomb ^{14}C in the ocean (e.g. Siegenthaler and Oeschger, 1987). These model estimates of the yearly net oceanic CO_2 uptake are, however, in contradiction with results from a recent study by Tans et al. (1990). When using the global distribution of ΔpCO_2 between atmosphere and surface ocean water, the *^{14}C calibrated* gas-exchange rate (see above), and the fairly well known mainly northern hemispheric input of fossil fuel CO_2 into the atmosphere, Tans et al. (1990) predict through three dimensional atmospheric model estimates a north/south CO_2 concentration gradient in the atmosphere to be about a factor of 2 larger than actually observed. They, thus, came to the conclusion, that a large portion of the anthropogenic excess CO_2 (ca. 2 GtC yr^{-1}) must be taken up in the northern hemisphere. The larger CO_2 uptake, when assigned to the northern oceans, would, however, require larger air-sea pCO_2 differences than actually observed in the North Atlantic and North Pacific oceans. Tans et al. (1990), therefore, attributed the northern hemispheric sink to net uptake by the biosphere. They adjusted the global distributions of the net CO_2 uptake by the biosphere, and came to an estimate of the total global uptake of excess CO_2 by the world oceans in the range of only 0.3 to 0.8 GtC yr^{-1}, a number, considerably lower than the value of about 2 GtC yr^{-1} derived from ocean models.

A possibility to solve this problem and to identify the nature of the strong northern hemispheric sink, again comes from isotope observations: Keeling et al. (1989b) with their 3-D model came to a similar estimate as Tans et al. (1990) of a total northern hemispheric sink of about 2 GtC yr^{-1}. They, however, constrained their estimates with the observed north-south gradient of the stable isotope ratio $^{13}C/^{12}C$ in atmospheric CO_2, and identified the northern *oceans* as the dominant CO_2 sink. Net *biospheric* CO_2 uptake, due to strong fractionation during photosynthesis, would largely increase the $^{13}C/^{12}C$ north-south gradient in contradiction to the observations.

In conclusion, atmospheric observations and observations in the ocean surface water give two different answers for the fate of fossil fuels. For an accurate prediction of the future development of the atmospheric CO_2 concentration, however, it is important to know which fraction of the anthropogenic emissions enters the biosphere and which fraction enters the oceans because these two reservoirs behave very differently. If the CO_2 is once stored in the oceans it is unlikely that it will re-enter the atmosphere soon, whereas CO_2 fluxes associated with the biospheric reservoir are very sensitive to changes, e.g. due to human activities, nutrients, water, or climatic change.

ATMOSPHERE - BIOSPHERE EXCHANGE

The second reservoir exchanging substantial amounts of CO_2 with the atmosphere is the terrestrial biosphere. The large annual variations of atmospheric CO_2 concentration in the northern hemisphere have been attributed to biospheric activity. A strong constraint for this assumption is the correlation observed between the $^{13}C/^{12}C$ ratio and the atmospheric CO_2 concentration. Substantial isotope fractionation of about 18‰ for $^{13}C/^{12}C$ in CO_2 occurs mainly during plant assimilation (Vogel, 1980), the respired CO_2 from plants and soils is, thus, also depleted in ^{13}C by this amount. During gas exchange between atmosphere and ocean, only very little fractionation occurs (Siegenthaler and Münnich, 1981). $\delta^{13}C$ measurements performed in parallel to atmospheric CO_2 concentration observations, therefore, allow to distinguish between these two sources resp. sinks. Heimann and Keeling (1989) used the CO_2 concentration and $\delta^{13}C$ observations of the Scripps Institution of Oceanography global network (Keeling et al., 1989a) to verify their 3D model simulations of the global atmospheric CO_2 cycle. They found that the temporal and spatial distribution of the exchange fluxes between the atmosphere and the biosphere resp. the oceans (summarized in Figure 1) are consistent with these observations. However, this ^{13}C constraint has one problem: CO_2 from the burning of fossil fuels, being also of biogenic origin has a very similar ^{13}C signature as soil or plant respiration CO_2. Part of the seasonality in atmospheric CO_2 concentration, particularly in mid northern latitudes may, thus, come from a seasonally varying admixture of fossil fuel CO_2, even at oceanic clean air sites. Fossil fuel CO_2 is distinguishable from biogenic CO_2 only via its $^{14}C/^{12}C$ ratio: fossil fuel CO_2 is free of radiocarbon.

$^{14}CO_2$ observations at the oceanic baseline station Izaña at Tenerife indeed show a seasonality with an about 10‰ lower $\Delta^{14}C$ during late winter, *i.e.* at the maximum atmospheric concentration, than during mid to late summer, when the seasonal concentration cycle is at its minimum (Levin et al., 1992). Only a small part of about 2-3‰ of the $\Delta^{14}C$ seasonality can be attributed to natural processes such as fractionation effects during atmosphere-biosphere exchange or a seasonally varying input of natural $^{14}CO_2$ from the stratosphere into the troposphere. The latter has been estimated to be of the order of 1-2‰. The remaining ^{14}C seasonality, if caused by variable contributions of fossil fuel CO_2 at the measurement site, corresponds to a maximum fossil fuel admixture at Izaña in April/May of about 1-2 ppmv, more than 10% of the total observed seasonal CO_2 amplitude of about 8 ppmv at this site (Navascues and Rus, 1991).

The above examples show that our knowledge on the gross and net fluxes of CO_2 between the dominant carbon reservoirs is still limited, at least in view of the atmosphere and its recent CO_2 concentration changes. On the one hand, the anthropogenic perturbations partly mask natural atmospheric signals used to deconvolve the steady state source and sink strengths. On the other hand, also the processes, e.g. at the interface between the atmosphere and the largest carbon reservoir, the oceans, as well as vertical mixing within the oceans, which is important in controlling atmospheric CO_2 concentration levels, are far from being well understood. This is certainly because of lack of good temporal and spatial coverage of oceanic observational data (e.g. pCO_2), but also because the processes controlling e.g. gas exchange at the air/sea interface still require a more reliable understanding.

ATMOSPHERIC METHANE

Atmospheric methane has its principal sources at the Earth's surface. The most important sink for methane is the oxidation in the troposphere mainly through reaction with the OH radical. About 5-10 % of the tropospheric methane are oxidized in aerated soils (Dörr et al., 1993). Table 1 summarizes the most important sources and destruction processes of atmospheric methane (Cicerone and Oremland, 1988; Watson et al., 1990). The sources of methane can be divided into two groups: biogenic sources like wetlands, rice paddies, cows, etc., and thermogenic sources like natural gas, coal associated gas, and biomass burning.

The total yearly methane source of today about 500 Tg CH_4 yr^{-1} is believed to be known to within ±20%. This global source strength has mainly been derived from balancing the fairly well known main sink by reaction with OH in the atmosphere (Crutzen, 1991), the uptake by soils, and the observed yearly increase in the atmosphere. The variability in the estimates of individual source types, particularly in natural emissions, is, however, much larger. For example, the role of the global ocean as a source (or, perhaps, a sink) of atmospheric methane remains still unclear. The emission rates from individual sources are derived from direct flux measurements in the relevant source areas and extrapolation to the global scale. Again, important additional information to determine the relative contributions from individual sources come from isotopic measurements on source methane and in the atmosphere (Cicerone and Oremland, 1988). The global mean $\delta^{13}C$ of all methane sources can be calculated from the global mean atmospheric $\delta^{13}C$ and the fractionation factors for methane oxidation with OH in the atmosphere (Cantrell et al., 1990) and by soil microbial activity (King et al., 1989; Levin et al., 1993). On the basis of the fractionation factors cited above, we derive a $\delta^{13}C$ value of the global mean methane source of -52.3±1.6‰ in good agreement with the weighted mean of all sources in Table 1.

The so-called *best estimate* of the recent global methane budget given in Table 1 is mainly derived from an observational network of methane concentration (Steele et al., 1992) and isotopes (Quay et al., 1991) in the atmosphere in combination with 3D atmospheric modelling of the methane cycle (Fung et al., 1991). From the summary in Table 1 it is obvious that present day anthropogenic emissions exceed natural release rates by nearly a factor of two. This, and the possibility that the oxidizing capacity of the atmosphere may have changed since pre-industrial times as a result of increasing concentrations of carbon monoxide, non-methane hydrocarbons, and methane itself (Sze, 1977), makes it extremely difficult to establish the natural budget of atmospheric CH_4, *i.e.* for the Holocene period.

TABLE 1: Recent budget of atmospheric methane: sources, sinks, and $\delta^{13}C$ values

Sources	Production Rate ($1\ Tg = 10^{12}\ g = 10^{-3}\ Gt$)		
	Range[1] ($Tg\ CH_4\ yr^{-1}$)	Best estimate[2] ($Tg\ CH_4\ yr^{-1}$)	$\delta^{13}C$[3,4] (‰ PDB)
Bogs and Tundra (50-70°N)		35	-64±3
Swamps and Alluvial (< 50°N)	100-200	80	-59±5
Termites	10-100	20	-64±5
Oceans	5-20	10	-40±10
Gas Hydrates	0-100	5	-40±10
Animals	65-100	80	-62±3
Rice	25-170	100	-63±3
Landfills	20-70	40	-51±2
Biomass Burning	20-80	55	-25±3
Natural Gas	25-50	40	-40±2
Coal Mining	19-50	35	-35±3
Total		500	-52.7
Atmospheric Increase	40-48	44	?
AtmosphericBurden 1990 ($Tg\ CH_4$)[6]		5050	-47.1±0.2
Sinks			Fractionation α
Reaction with OH	400-600	430	1.0054±0.0009[7]
Removal by Soils	15-45	30	1.019±0.003[5,8]

[1]Watson et al., 1990
[2]Fung et al., 1991
[3]Stevens and Engelkemeier, 1988
[4]Quay et al., 1991
[5]Levin et al., 1993
[6]Steele et al., 1992
[7]Cantrell et al., 1990
[8]King et al., 1989

SOME FACTORS INFLUENCING NATURAL CH_4 EMISSIONS AND DESTRUCTION RATES

In many environments, *e.g.* in wetland areas or lake and ocean sediments, only part of the methane produced in the anaerobic zone is actually emitted into the atmosphere. Reeburgh et al. (1993) estimated that more than 50% of the global gross methane production are oxidized before entering the atmosphere. For example, the concentration distribution of methane in tundra soils suggests an aerobic surface zone of rapid methane oxidation and an anaerobic subsurface zone, *i.e.* below the water table, of methane production (Whalen and Reeburgh, 1992). In the case of a predicted future warming, temperature change in tundra soils will be largest in the upper soil layers, namely in the aerobic zone. This will cause a higher ratio of CH_4 oxidation to CH_4 production, and, thus, smaller emissions and a negative feedback, opposite in sign to the often proposed positive feedback when only accounting for the temperature dependency of methane production by microorganisms. Whalen and Reeburgh (1992) emphasize that thc extent of actual CH_4 oxidation is very sensitive to water table changes as this immediately effects the depth of the oxidized surface zone. Climate induced water table changes in regions where this water table is below the soil surface are, thus, an important parameter influencing the ratio between methane production and emission. For example, a higher precipitation rate as predicted by climate models for higher latitudes when associated with a water table rise, would provide a positive feedback for a trace gas induced warmer climate in the future. The feedback associated with water table changes is most important for northern wetland regions as the water table here is often below the soil surface (Whalen and Reeburgh, 1992) but may also be of relevance in the tropics under changed hydrological conditions.

Microbial oxidation of methane in moist aerated soils is a small sink of atmospheric methane today (about 7% of the total destruction by OH (Dörr et al., 1993)). The issue of microbial oxidation of methane in aerated soils as a sink for atmospheric methane has been investigated in several studies in recent years (Keller et al., 1983; Steudler et al., 1989; Born et al., 1990; Whalen and Reeburgh, 1990; Dörr et al., 1993). Dörr et al. found a strong correlation between soil permeability (soil texture class) and the methane uptake rate by soils at various sites in Central Europe. Methane consumption by soils is basically controlled by the gas transport resistance within the soil, soil temperature has been found to be of minor influence. As the soil permeability strongly depends on soil humidity, in a postulated drier climate during glacial periods the total global uptake of atmospheric methane by soils may have been larger than in the present day climate.

Bacterial oxidation by microbes is associated with strong isotope fractionation. In the case of a larger soil sink we would expect an isotope enrichment of atmospheric CH_4. A higher oxidation/production ratio, for example during glacial periods, would cause a higher $\delta^{13}C$ value of the global methane source mix, and, thus, also a higher $\delta^{13}C$ value of atmospheric methane. Isotope measurements on methane from ice cores of Holocene and glacial times would, thus, help to understand the reasons for the large glacial/interglacial concentration changes of atmospheric methane already observed in the glacier archive (Barnola et al., 1987; Chappellaz et al., 1990).

CONCLUSIONS

From our recent knowledge on the geophysical cycles of the most important atmospheric greenhouse gases CO_2 and CH_4, it seems clear that the changes of atmospheric methane are mainly driven by continental sources and the oxidizing capacity of the atmosphere. The atmospheric CO_2 concentration levels - to a large extend - seem to be controlled by ocean properties, although the terrestrial biospheric carbon reservoir must have been subject to large changes between glacial and interglacial situations. In order to understand, *e.g.* this transition between glacial to interglacial warmings where CO_2 and CH_4 increases are roughly in phase with the climatic change (Raynaud and Siegenthaler, 1993), we, therefore, still have to find the coupling mechanisms between synchronous changes on the continents and in the oceans. Possibly, this unknown coupling is connected to a very sensitive, yet unknown, mechanism affecting both, the oceanic and the continental biosphere ?

ACKNOWLEDGEMENTS

This work has partly been funded by the German Minister of Science and Technology, Bonn, under contract No. 07KFT 110, and by the European Community, Brussels, under Contract No. EPOC-CT90-0017. I wish to thank Dietmar Wagenbach for many helpful discussions.

REFERENCES

Barnola JM, Raynaud D, Korotkevich YS, Lorius C (1987) Vostok ice core: A 160,000 year record of atmospheric CO_2. *Nature 329*: 408-414

Born M, Dörr H, Levin I (1990) Methane consumption in aerated soils in the temperate zone. *Tellus 42B*: 2-8

Broecker WS, Peng T-H, Östlund G, Stuiver M (1985) The distribution of bomb radiocarbon in the ocean. *J Geophys Res 90:* 6953-6970

Broecker WS, Ledwell JR, Takahashi T, Weiss R, Merlivat L, Memery L, Peng T-H, JÑhne B, Münnich KO (1986) Isotopic versus micrometeorological ocean CO_2 fluxes: a serious conflict. *J Geophys Res 91, C9:* 10 517-10 527

Cantrell CA, Shetter RE, McDaniel AH, Calvert JG, Davidson JA, Lowe DC, Tyler SC, Cicerone RJ, Greenberg JP (1990) Carbon kinetic isotope effect in the oxidation of methane by hydroxyl radical. *J Geophys Res 95*: 22 455 - 22 462

Cicerone RJ, Oremland RS (1988) Biogeochemical aspects of atmospheric methane. *Global Biogeochemical Cycles 2*: 299-327

Chappellaz J, Barnola JM, Raynaud D, Korotkevich YS, Lorius C (1990) Ice-core record of atmospheric methane over the past 160,000 years. *Nature 345*: 127-131

Crutzen PJ (1991) Methane's sinks and sources. Nature 350: 380-381

Dörr H, Katruff L, Levin I (1993) Soil texture parameterization of the methane uptake in aerated soils. *Chemosphere* 26: 709-725

Etheridge DM, Pearman GI, de Silva F (1988) Atmospheric trace gas variations as revealed by air trapped in an ice-core from Law Dome, Antarctica. *Ann Glaciol 10*: 28-33

Friedli H, Loetscher H, Oeschger H, Siegenthaler U, Stauffer B (1986) Ice-core record of the $^{13}C/^{12}C$ record of atmospheric CO_2 in the past two centuries. *Nature 324*: 237-238

Fung IY, John J, Lerner J, Matthews E, Prather M, Steele LP, Fraser PJ (1991) Three dimensional model synthesis of the global methane cycle *J Geophys Res 96, D7*: 13 033-13 065.

Heimann M, Keeling CD (1989) A three-dimensional model of atmospheric CO_2 transport based on observed winds: 3. Seasonal cycle and synoptic time scale variations. *Geophysical Monograph 55*: 277-303

Keeling CD, Bacastow RB, Carter AF, Piper SC, Whorf TP, Heimann M, Mook WG, Roeloffzen H (1989a) A three-dimensional model of atmospheric CO_2 transport based on observed winds: 1. Analysis of observational data. *Geophysical Monograph 55*: 165-236

Keeling CD, Piper SC, Heimann M (1989b) A three-dimensional model of atmospheric CO_2

transport based on observed winds: 4. Mean annual gradients and interannual variations. *Geophysical Monograph 55*: 305-363
Keller M, Goreau TJ, Wofsy SC, Kaplan WA, McElroy MB (1983) Production of nitrous oxide and consumption of methane by forest soils. *J Geophys Res 91, D11*: 11 791-11 802
King SL, Quay PL, Lansdown JM (1989) The $^{13}C/^{12}C$ kinetic isotope effect for soil oxidation of methane at ambient atmospheric concentrations. *J Geophys Res 94, D15*: 18273-18277
Levin I, Bösinger R, Bonani G, Francey R, Kromer B, Münnich KO, Suter M, Trivett NBA, Wölfli W (1992) Radiocarbon in atmospheric carbon dioxide and methane: global distribution and trends. In: RE Taylor, A Long and RS Kra (eds), *Radiocarbon After Four Decades: an Interdisciplinary Perspective*. Springer, Berlin Heidelberg New York, p 503
Levin I, Bergamaschi P, Dörr H, Trapp D (1993) Stable isotopic signature of methane from different sources in Germany. *Chemosphere* 26: 163-179
Navascues B, Rus C (1991) Carbon dioxide observations at Izaña baseline station, Tenerife (Canary Islands): 1984-1988. *Tellus 43B*: 118-125
Neftel A, Moor E, Oeschger H, Stauffer B (1985) Evidence from polar ice-cores for the increase in atmospheric CO_2 in the past two centuries. *Nature 314*: 611-613
Quay PD, King SL, Stutsman J, Steele LP, Fung IY, Gammon RH, Brown TA, Farwell GW, Grootes PM, Schmidt FH (1991) Carbon isotopic composition of atmospheric CH_4: fossil and biomass burning source strength. *Global Biogeochemical Cycles 5*: 25-47
Raynaud D, Siegenthaler U (1993) Role of trace gases: the problem of lead and lag. In: Eddy JA, Oeschger H (eds) *Global Changes in the Perspective of the Past*. Dahlem Workshop Report ES 12. John Wiley and Sons, Chichester (in press)
Reeburgh WS, Whalen SC, Alperin MJ (1993) The role of methylotrophy in the global methane budget. In: JC Murrell and DC Kelley (eds) *Microbial Growth on C-1 Compounds*. Intercept, Andover, UK, p 1
Siegenthaler U, Münnich KO (1981) $^{13}C/^{12}C$ Fractionation during CO_2 transfer from air to sea. In: Bolin B (ed) *SCOPE 16, Carbon cycle modelling*. Wiley, Chichester New York Brisbane Toronto, p 249
Siegenthaler U, Oeschger H (1987) Biospheric CO_2 emissions during the past 200 years reconstructed by deconvolution of ice core data. *Tellus 39B*: 140-154
Steele LP, Dlugokencky EJ, Lang PM, Tans PP, Martin RC, Masarie KA (1992) Slowing down of the global accumulation of atmospheric methane during 1980s. *Nature 358*: 313-316
Stevens MC, Engelkemeier A (1988) Stable carbon isotopic composition of methane from some natural and anthropogenic sources. *J. Geophys. Res.*: 725-733
Steudler PA, Bowden RD, Mellilo JM, Aber JD (1989) Influence of nitrogen fertilization on methane uptake in temperate forest soils. *Nature 341*: 314-316
Stuiver M (1980) ^{14}C distribution in the Atlantik Ocean. *J Geophys Res 85, C5*: 2711-2718
Stuiver M, Oestlund HG, McConnaughey TA (1981) GEOSECS Atlantic and Pacific ^{14}C distribution. In: Bolin B (ed) *SCOPE 16, Carbon cycle modelling*. Wiley, Chichester New York Brisbane Toronto, p 201
Sze ND (1977) Anthropogenic CO emissions: Implictions for the atmospheric CO-OH-CH_4 cycle. *Science 195*: 5151-5172
Tans PP, Fung IY, Takahashi T (1990) Observational constraints on the global atmospheric CO_2 budget. *Science 247*: 1431-1438
Vogel JC (1980) Fractionation of the carbon isotopes during photosynthesis. *Sitzungsberichte der Heidelberger Akademie der Wissenschaften, Mathematisch-naturwissenschaftliche Klasse*, 3. Springer, Heidelberg
Watson RT, Rodhe H, Oeschger H, Siegenthaler U (1990) Greenhouse gases and aerosols, In: Houghton JT, Jenkins GJ, Ephraums JJ (eds) *Climate Change, The IPCC Scientific Assessment*. Cambridge University Press, Cambridge New York Port Chester Melbourne Sydney, p 1
Whalen SC, Reeburgh WD (1990) Consumption of atmospheric methane by tundra soils. *Nature 346*: 406-409
Whalen SC, Reeburgh WS (1992) Interannual variations in tundra methane emission: A 4-year time series at fixed sites. *Global Biogeochemical Cycles 6*: 139-159

GLACIAL OCEAN CARBON CYCLE MODELING

Christoph Heinze[1]
Institut für Meereskunde der Universität Hamburg
D-22529 Hamburg
Germany

ABSTRACT. The Hamburg Oceanic Carbon Cycle Circulation Model as a global tracer model suited for glacial ocean studies is briefly described. Experimental strategies for paleoclimatic model experiments are outlined. In sensitivity experiments the effect of a hydrography shift due to a change in sea level on the marine carbon cycle is investigated.

INTRODUCTION

The glacial carbon cycle offers a unique opportunity for scientists of all disciplines to study the behaviour of a world under distinctly different climatic conditions than today's. While often the principle of actualism is applied to problems of the Earth's history, consideration of the last glaciation may help us to understand the climatic system today and the potential environmental changes we may have to face in the near future. The basis of our understanding of the last glaciation is data from ice cores and marine sediment cores. These records provide a clear climatic signal; unfortunately these data are usually proxy records, i.e., are only approximate analogs of the real data we are measuring today. Any physical modeling of the glacial climate system has to be validated against this data base. Therefore, geochemical models that try to reproduce the proxy records are necessary and have to be used parallel to any physical modeling effort.

This paper is focused on modeling paleoclimatic tracers with the Hamburg Oceanic Carbon Cycle Model. A brief description of the model and associated modeling strategies is presented, that should enable any experimental scientist to find appropriate interfaçes to bring in her or his data and to make suggestions for experiments. As an example for the procedure that

[1] Present address: Max-Planck-Institut für Meteorologie, Bundesstraße 55, D-20146 Hamburg, Germany

NATO ASI Series, Vol. I 17
Carbon Cycling in the Glacial Ocean:
Constraints on the Ocean's Role in Global Change
Edited by R. Zahn et al.

has to be carried out for a sensitivity experiment, one aspect of sea level lowering is investigated.

MODEL DESCRIPTION

The carbon cycle model was developed by Maier-Reimer at the Max-Planck-Institute for Meteorology with contributions of Bacastow (Maier-Reimer and Hasselmann, 1987; Bacastow and Maier-Reimer, 1990; Maier-Reimer and Bacastow, 1990). Descriptions and supplementary information are given also in Heinze et al. (1991) and Heinze and Maier-Reimer (1992). Like the geochemical box models of Broecker and Peng (1986), Keir (1988), and Michel (1991), the Hamburg Carbon Cycle Model can be considered as a box model. The differences between previous box models and the Hamburg Carbon Cycle Model are: (a) the number of boxes is much higher, (b) exchange rates between all boxes are constrained by the Navier-Stokes-Equations, thermodynamic laws, and realistic forcing fields, (c) the model can directly be used for testing ocean general circulation models. Advantages of this model are: (a) comparatively good horizontal and vertical resolution of tracer distributions that may allow quantitatively more correct results, (b) no calibration has to be carried out with respect to water transport exchange rates between the boxes. Disadvantages are: (a) the model is more expensive, slower and less easy to handle than small box models, (b) results may be more difficult to interpret.

Basic parts of the Hamburg Carbon Cycle Model are geometry, reservoirs, transport rates by oceanic currents, and the representation of geochemical interactions. The Hamburg Carbon Cycle Model is based on the grid and model geometry of the Hamburg Large Scale Geostrophic Ocean Circulation Model (Maier-Reimer et al. , 1992). Both models use an E-grid (Arakawa and Lamb, 1977) with 72x72 grid points in each horizontal layer. This corresponds to an approximate 3.5°x3.5° degree horizontal resolution (ca. 380 km grid point spacing). Here the 11 layer version is considered with the model layers centered at 25, 75, 150, 250, 450, 700, 1000, 2000, 3000, 4000, and 5000 m water depth. A 15 layer version with higher vertical resolution in the upper 300 m of the model ocean and a monthly time step is also available (Kurz and Maier-Reimer, 1992; Maier-Reimer, accepted for publ.). The model topography follows realistic contours. The respective model bottom layer can be of varying depth corresponding to the actual averaged bottom depth for one grid cell. Due to the coarse discretization of the model domain, not all oceanic provinces can be represented perfectly, the model favors pelagic oceanic areas. For any comparison with observed data, it should be kept in mind that each grid cell of the model gives an average value for the whole area covered by that cell.

The model includes the following reservoirs: troposphere, ocean, and sea floor surface sediments. The oceanic carbon cycle model can also be coupled to a terrestrial biosphere model (Esser, 1987). The velocity and thermohaline fields (water temperature and salinity) are com-

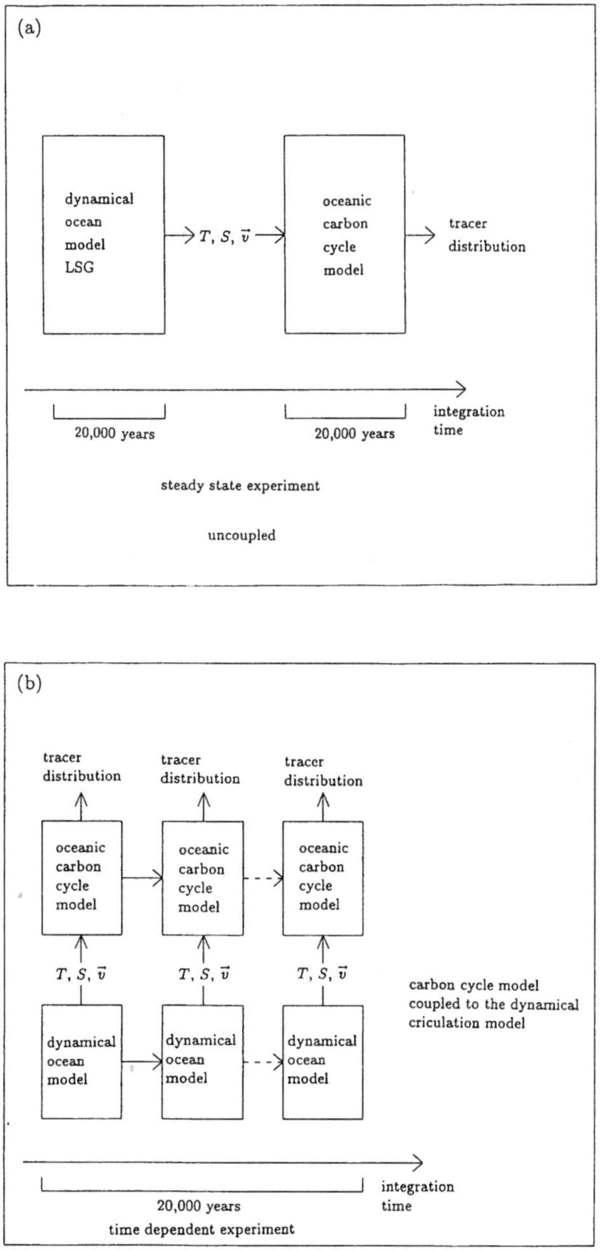

Figure 1: Different strategies for the use of velocity fields for the carbon cycle model. (a) "Frozen" fields. (b) Permanent update.

puted separately from the carbon cycle model with a dynamic ocean circulation model, namely the Hamburg Large Scale Geostrophic Ocean Circulation Model (Maier-Reimer et al. , 1992). This model is based on the Navier-Stokes-equation (Newtons second law transformed for hydrodynamics and balance of forces), the continuity equation and transport equations for heat and salt. Mixing due to eddy motion, velocity shear and static instability of the water column are parameterized by the advection scheme providing increasing eddy diffusion with increasing current velocities and by a convective adjustment algorithm (exchange of thermohaline properties between two overlying grid cells, if the water in the upper cell becomes denser than the water in the underlying cell, while retaining the heat and salt budgets). The circulation model is integrated with a monthly time step. Initial thermohaline fields are taken from Levitus (1982). Wind forcing is taken from the Hellerman and Rosenstein (1983) monthly mean wind stress data set. Thermal forcing is performed by relaxation to air temperatures from the COADS data set (Woodruff et al., 1987). A simple sea ice model (thermodynamic with simplified rheology) is included that provides salt brine ejection during freezing processes and a corresponding change in static stability of the water column. For the standard case used here, the circulation model was integrated for 15,000 model years to quasi-equilibrium (Maier-Reimer et al., 1992). For paleoclimatic purposes, annual mean values of velocity, temperature, and salinity are used as input data for the carbon cycle model. The convective adjustment values are used as annual sums of convective overturning events, thus conserving the seasonal effect of deep water production approximately also in the annually averaged case.

Figure 1 shows the different strategies for supplying the carbon cycle model with velocity, temperature, and salinity values: either a single 'frozen' velocity field is used (quasi-equilibrium conditions), or, for time dependent applications, the circulation model and the carbon cycle model are run parallel, and the oceanic current and thermohaline arrays are updated for the carbon cycle model at each time step (e.g., Winguth, 1992).

The geochemical processes included in the carbon cycle model are:meridional diffusive transport of CO_2 in a simplified zonally averaged one layer model atmosphere, gas exchange between atmosphere and ocean surface layer, primary new production of particulate organic matter and biogenic calcite particles, particle flux through the water column, remineralization of particulate organic matter, re-dissolution of calcite particles, and exchange with sediment reservoirs. Conceptually the model follows the principles outlined in Broecker and Peng (1982, p. 11). The tracers considered are: atmospheric CO_2, dissolved inorganic carbon ΣCO_2, total alkalinity, dissolved oxygen, dissolved phosphate (as nutrient component), POC (particulate organic carbon), particulate $CaCO_3$, organic sediment (though quantitatively almost negligible here), and availability of $CaCO_3$ bearing sediment. For all carbon tracers the three isotopes ^{12}C, ^{13}C, and ^{14}C are considered throughout. In order to facilitate the implementation on the computer PO_4 is used in 'per carbon' units by the Redfield ratio value P:C=1:122. The rare carbon isotopes are treated separately from each other. At the end of each model run these values are computed relative to the averaged model atmosphere value and hereby calibrated and subsequently transformed to the usual geochemical notations as $\delta^{13}C$ (relative to the PDB standard for an atmospheric value of $\delta^{13}C$=-6.5‰) and $\Delta^{14}C$ with fractionation correction (Stuiver and Polach, 1977). 'Master tracers' are ΣCO_2 and total alkalinity. The other carbon system values like ionic carbonate and the pH-value are computed from this pair by a two-dimensional Newton-Raphson-algorithm. The entire model conserves the total molar content of matter for each chemical element which is considered. The radiocarbon budget is closed by an atmospheric ^{14}C source balancing the radioactively decaying amount in the total model system. Very precise matter conservation is essential, e.g., for any reliable computation of the CO_2 airborne fraction and calcite lysocline depth level values in low gradient areas.

The basic equation for the carbon cycle model is the continuity equation (tracer conservation equation). One equation must be solved for each tracer at each grid point:

$$\frac{\partial c}{\partial t} = -div\,\vec{j} + q \qquad (1)$$

where
c is the tracer concentration, t is time, div is the divergence operator $\left(\frac{\partial}{\partial x} + \frac{\partial}{\partial y} + \frac{\partial}{\partial z}\right)$,
$\vec{j}$ is the flux density vector, $\vec{j} = \vec{v} \cdot c$ ($\vec{v}$ is the current velocity), and
q is the term for sources (>0) and sinks (<0)

For constant fluid volume (incompressibility) this equationcan be written as

$$\frac{\partial c}{\partial t} + u\frac{\partial c}{\partial x} + v\frac{\partial c}{\partial y} + w\frac{\partial c}{\partial z} = q \qquad (2)$$

where u, v, and w are the velocity components in the zonal (x), meridional (y) and vertical (z) directions. This equation is also called advection equation. In the model, by approximation of the differentials by finite differences, the integral form of equation (2) is established at each grid point for the finite box volume represented by that grid point:

$$V \cdot \frac{\partial c}{\partial t} \;=\; \sum_{i=1}^{N}\left[(c - c_i) \cdot A_i \cdot v_i\right] \;+\; q \cdot V \qquad (3)$$

where V is the box volume, c is the tracer concentration in the box, c_i is the tracer concentration in the ith neighboring box, N is the number of wet boxes surrounding the central box, A_i is the area between the central box and the ith neighboring box, v_i is the velocity component directing from the central box to the ith neighboring box

This equation is solved numerically for finite time steps (here 1 year) by an implicit upwind scheme for the tracer concentration (the prognostic variable)at the new time level:

$$\frac{c^{t+\Delta t} - c^{t}}{\Delta t} + \sum_{i=1}^{N} v_i \frac{c_i^{t+\Delta t} - c^{t+\Delta t}}{\Delta s_i} \;=\; q \qquad (4)$$

where $\Delta s_i = V/A_i$ is a length scale, Δt is the time step and the index t and $t+\Delta t$ indicates the previous and new time level.

The geochemical interactions are formulated in the source and sink term q. The global sum of q for each chemical element must be zero.

CO_2 gas exchange flux between atmosphere and surface layer is given by a bulk formula:

$$S = \lambda \left[(pCO_2)_{ocean} - (pCO_2)_{atmosphere} \right] \tag{5}$$

with a bulk coefficient of λ=19 mol m^{-2} a^{-1} at 270 ppm atmospheric CO_2 partial pressure.

Biological primary production of organic matter is modelled following Michaelis-Menten kinetics according to the nutrient supply available for consumption (e.g., Parsons and Takahashi, 1973):

$$P = I_\phi \cdot \frac{V_{max} \cdot [PO_4]^2}{K_s + [PO_4]} \tag{6}$$

where I_ϕ is a latitudinal function for the parameterization of growth conditions according to the meridional profile of incoming solar radiation entering the ocean surface layer (Bacastow and Maier-Reimer, 1990). V_{max} and K_s are adjustable parameters representing the maximum uptake rate of nutrients by phytoplankton and the half saturation constant (the nutrient concentration where the nutrient uptake rate is half of the maximum limit). Respective changes in nutrient and dissolved carbon and particulate organic carbon are coupled by a fixed Redfield ratio of P:C=1:122 for the standard run. Biogenic $CaCO_3$ production is coupled to the POC production by a rain ratio of 1:4 for C atoms bound to calcite or organic matter with a strong reduction of calcite production for temperatures lower than 5°C. Oxygen is set to solubility equilibrium in the surface model layer after each time step.

The vertical flux of newly produced biogenic organic carbon and $CaCO_3$ particles through the water column is parameterized by prescribing a vertical redistribution profile (for details see Heinze et al., 1991), i.e., a fixed percentage of the biogenic matter extracted from the surface layer is added to deeper layers instantaneously. $CaCO_3$ penetration is relatively deeper than that of POC. Remineralization is performed by a given rate under oxygen consumption according to a Redfield ratio of P:-O_2=1:172. The remineralization rate is independent of O_2 until O_2 is depleted. No negative oxygen concentrations are allowed, i.e., remineralization stops as long as the entire oxygen in one box is consumed. Dissolution of calcite is formulated proportional to CO_3^{2-} undersaturation. A prescribed percentage of particulate calcite is allowed to reenter the dissolved form in any case. Both adjustable parameters are tuned to result in a satisfactory reproduction of the present CO_3^{2-} distribution.

The equations for the sediment boxes have no advection term and consist only of source and sink terms. A fixed percentage of particulate matter in the respective ocean model bottom layer is allowed to enter the sediment reservoirs underlying each model water column. The cycle of matter is closed by assuming that a given portion of the accumulated particulate matter in the sediment is re-transferred to the lowest wet model layer by bioturbation. No real stratification of sediment is modelled, but a certain pool size. For organic carbon the total sediment pool size is almost negligible. For $CaCO_3$ the parameters are adjusted to result in a plausible pool size for available $CaCO_3$ that can in principle be redissolved into the bottom water (Broecker and Takahashi, 1977). Alternatively a more comprehensive $CaCO_3$ sediment model was coupled to the ocean carbon cycle model by Archer and Maier-Reimer (accepted for publ.).

Fractionation factors for ^{13}C versus ^{12}C are taken from Mook (1986) for evaporative sea/air transfer and set to 0.98 for uptake during plankton growth. The solubility of CO_2 in sea water is computed after Weiss (1974). The chemical constants for the inorganic carbon chemistry are taken from Mehrbach et al. (1973), Edmond and Gieskes (1970), and Culberson and Pytkowicz (1968). The solubility product of $CaCO_3$ is computed according to Ingle (1975) with pressure correction after Edmond and Gieskes (1970). Additional borate buffering is included. The change in nitrogen concentration during plankton growth and remineralization is considered for total alkalinity changes by use of a Redfield ratio of C:N=1:16 though the nitrogen cycle is not modelled explicitly. For comparison with paleoclimatic $CaCO_3$ records, calcite lysocline depth levels are computed as depth intercepts of the modelled vertical CO_3^{2-} profiles with the critical CO_3^{2-} concentration profile given by Broecker and Takahashi (1978).

To integrate the model from an initial state to equilibrium approximately 20,000 years of integration time are necessary corresponding to 20 CPU hours on one processor of a Cray YMP. The per time step CPU demand decreases with approaching equilibrium conditions. After 10,000 years integration time more than 95% of the adjustment has taken place.

EXPERIMENTAL STRATEGIES FOR PALEO-CLIMATE STUDIES

Basic strategy for paleoclimatic model experiments is to carry out sensitivity experiments relative to a control run (reference or standard run) (Figure 2). This control run usually represents the late Holocene or recent ocean carbon cycle system. This run has a specific setting of carbon cycle model parameters and a specific tracer distribution characteristic for the late Holocene. Examples for the standard run of the Hamburg Carbon Cycle Model are given in Figures 3 - 8 for annual primary organic carbon new production, biogenic calcite new production as well as $\delta^{13}C$ and CO_3^{2-}. In the CO_3^{2-} sections, the lysocline depth levels correspond to the critical CO_3^{2-}concentration (Broecker and Takahashi, 1978) and are indicated by dotted lines. The results are not considered as the best possible since fine tuning of the parameters is

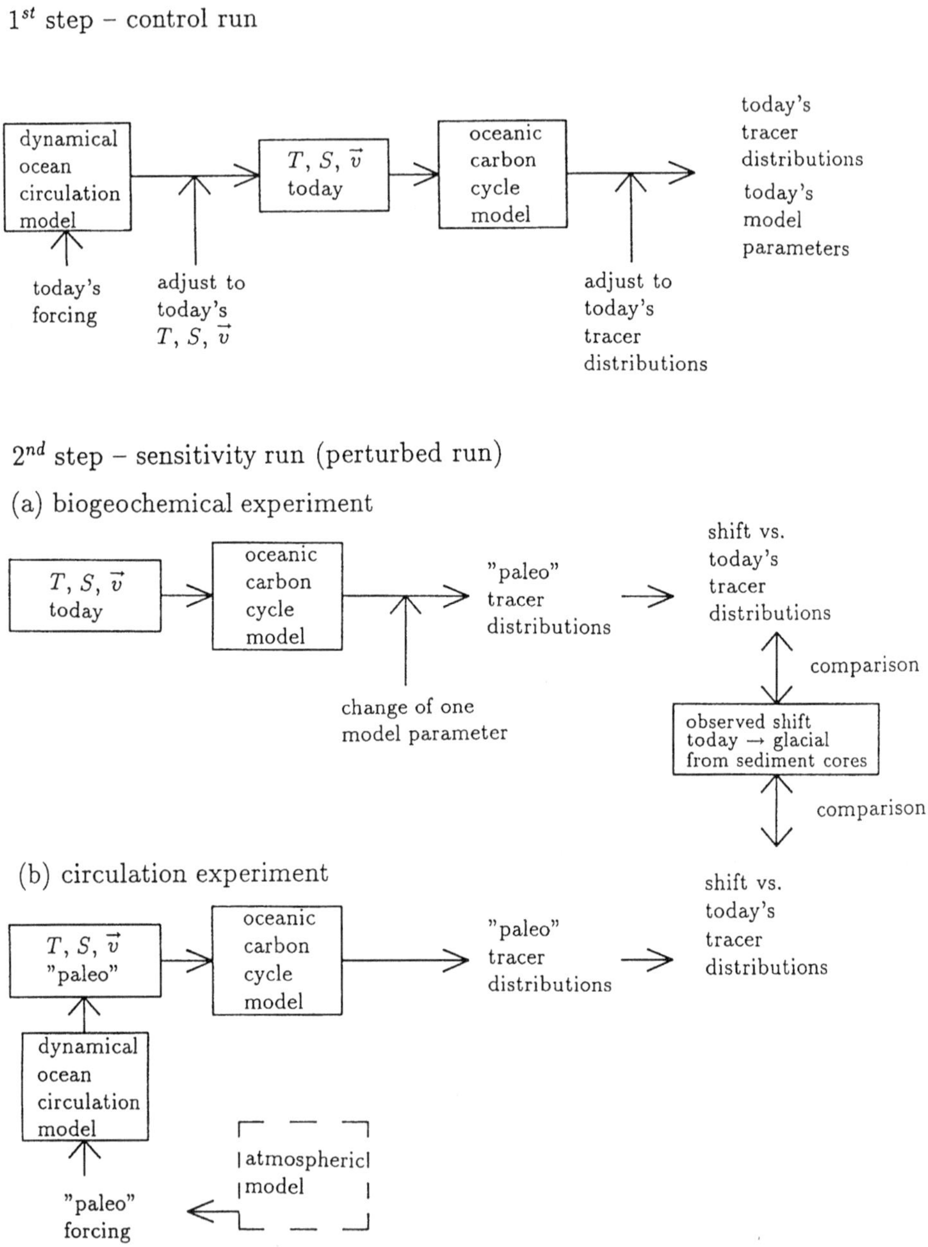

Figure 2: Experimental strategies: Sensitivity experiments.

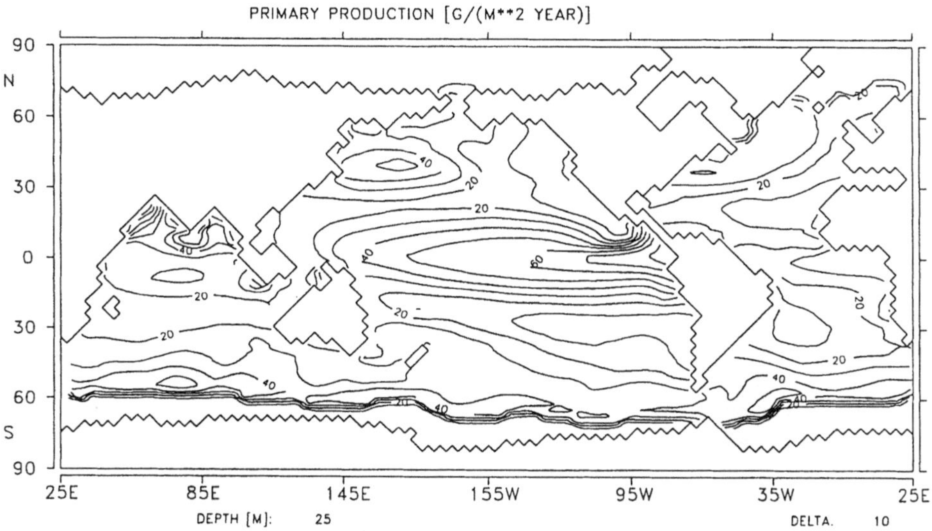

Figure 3: Organic matter primary new production in grams carbon per m^2 and year for the standard case.

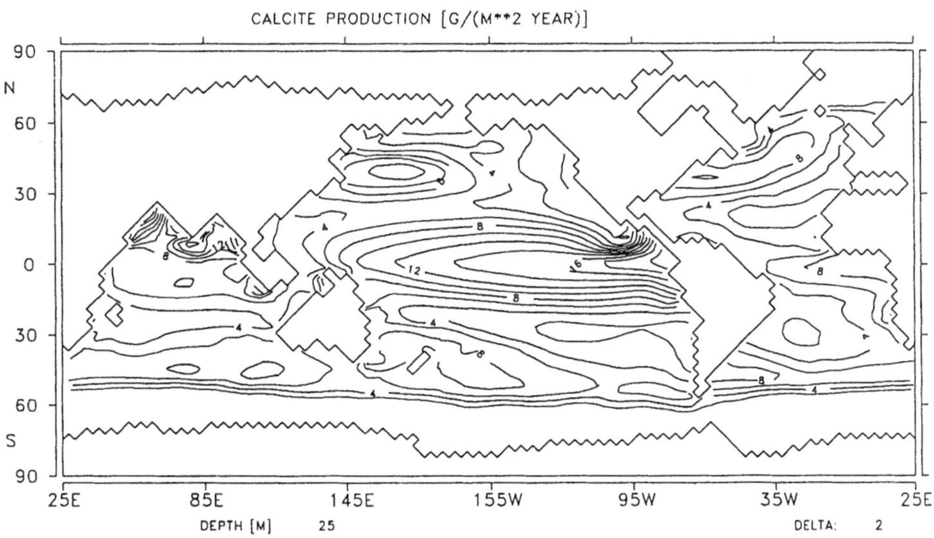

Figure 4: Biogenic $CaCO_3$ primary new production in grams carbon per m^2 and year for the standard case.

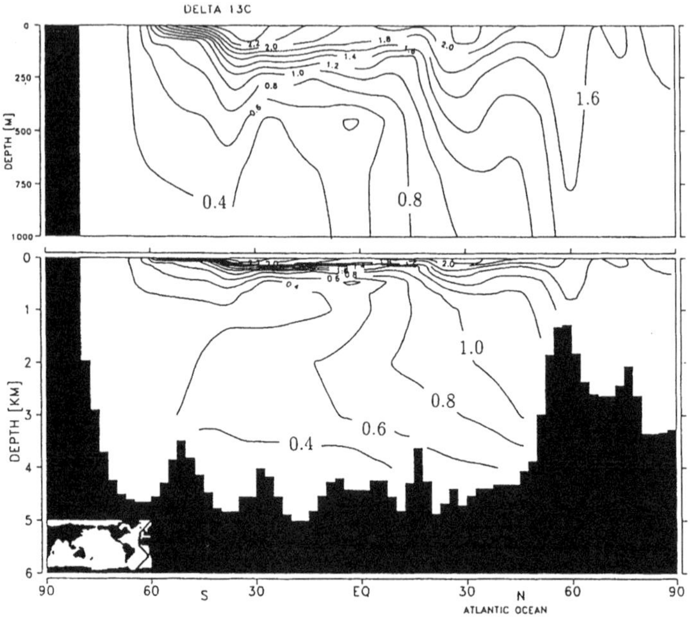

Figure 5: $\delta^{13}C$ cross section along the eastern Atlantic Ocean for the standard case (contour interval = 0.2 permil)

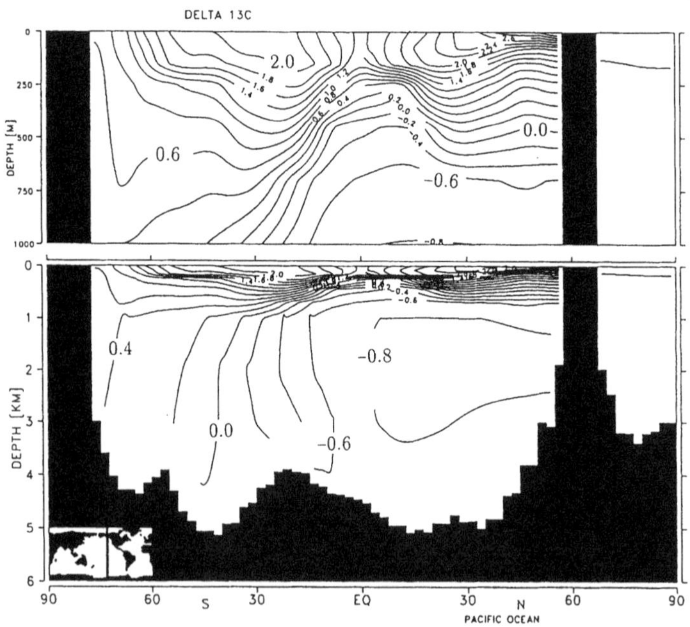

Figure 6: $\delta^{13}C$ cross section along the eastern Pacific Ocean for the standard case (contour interval = 0.2 permil)

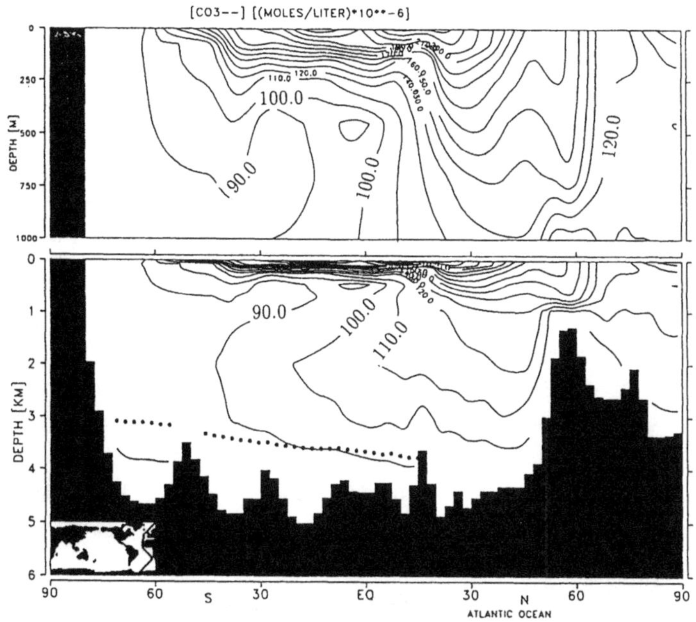

Figure 7: $[CO_3^{2-}]$ cross section along the eastern Atlantic Ocean for the standard case. (Dotted line: Model $CaCO_3$ lysocline horizon.)

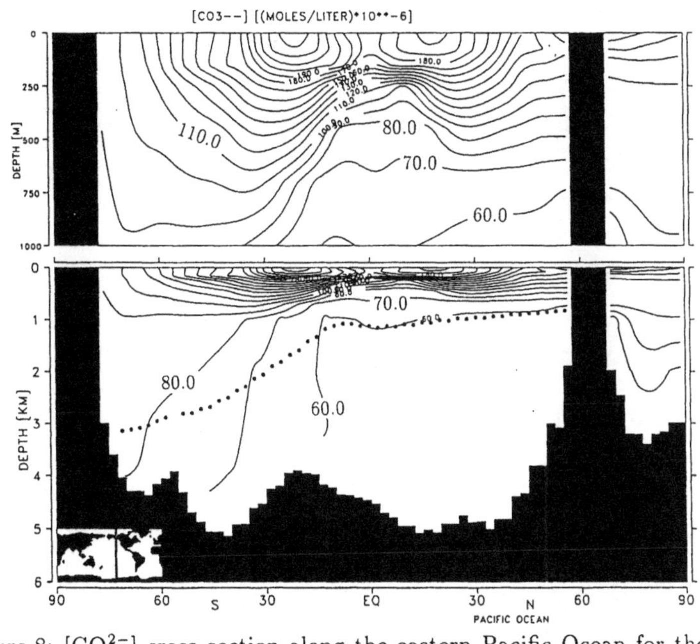

Figure 8: $[CO_3^{2-}]$ cross section along the eastern Pacific Ocean for the standard case. (Dotted line: Model $CaCO_3$ lysocline horizon.)

not yet finished. However, this run (Heinze and Maier-Reimer,1992) can serve as a reference for further improvements.

In sensitivity experiments the change of the model carbon cycle system is tested for changes in one (or more) model parameters. With the prescribed model parameter shift, the model is integrated from the late Holocene tracer distribution as initial condition again up to quasi-equilibrium (Figure 2). The resulting tracer distribution from this sensitivity test is then compared with the control run data set. The differences between these two model data sets can then be compared to observed tracer changes between the late Holocene and, e.g., the Last Glacial Maximum (appr. 18,000--20,000 years ago). This type of model strategy was applied in Heinze et al. (1991) for testing different hypotheses on the glacial pCO_2 reduction by oceanic processes. In this study purely biogeochemical experiments and circulation experiments (cf. Figure 2) were carried out on the basis of the following thinking models: increased biological productivity at high latitudes (Siegenthaler and Wenk, 1984; Sarmiento and Toggweiler, 1984; Knox and McElroy, 1984); increased Redfield ratio C:P (Broecker, 1982); decreased calcite production (rain ratio change, Berger and Keir, 1984); inventory changes of PO_4, $CaCO_3$, and POC (Broecker, 1982; Berger and Keir, 1984); and increased CO_2 gas solubility in the ocean surface layer (Broecker and Peng,1986, 1987). In addition two different circulation fields were tested: one with only half the current velocities and one with a second stable mode of ocean circulation (Mikloajewicz, pers. comm., see Heinze et al., 1991) under present day forcing conditions (but starting from initial conditions for salinity which are different from those used for the control run). The modelled tracer-shifts from the control run values to the perturbed run were directly compared to sediment core $\Delta\delta^{13}C_{planktonic-benthic}$ and $\Delta\delta^{13}C_{benthic}$ and $CaCO_3$ lysocline depths. None of the sensitivity runs resulted in satisfactory reproduction of the observed tracer shifts though considerable changes in the tracer concentrations and distribution patterns could be achieved (see the summary in Figure 18 of Heinze et al., 1991). The study was encouraging in that we have now some confidence to be in principle able to model the glacial ocean tracer distributions once a model parameter set has been found which corresponds to the real mechanisms.

A first attempt to model the glacial ocean including its atmospheric forcing was carried out by Lautenschlager et al. (1993). At first the ECHAM T21 Atmospheric General Circulation Model was used to provide fresh water and wind stress data for glacial conditions. The resulting anomalies with respect to the present day control atmosphere were added to the standard forcing of the Large Scale Geostrophic Ocean Circulation Model and the model was integrated again up to quasi-equilibrium. The model output fields for current velocity, T, and S were taken as input fields for a subsequent integration of the carbon cycle model. The resulting tracer distribution shows a strong change in that the Atlantic and Pacific Oceans practically switch their roles and tracer patterns. Though not resulting in a $\delta^{13}C$ distribution that is sufficiently consistent with the glacial $\delta^{13}C$ pattern obtained from the sedimentary record,the study demonstrates the importance of a correct paleo-forcing for any further attempt to reliably model the glacial ocean taking into account the extreme sensitivity of the ocean circulation to changes in the fresh water flux to the surface layer.

The experimental approach described above uses observed data only for comparison. Other methods have been developed in order to use the observed data quantitatively by directly incorporating them into the modelling procedure. These methods are called inverse approaches or assimilation techniques (Figure 9). For instance, Heinze and Hasselmann (accepted for publ.) superimpose the full non-linear carbon cycle model on a simple linear model on the basis of the tracer changes resulting from the fixed parameter perturbations prescribed for the sensitivity experiments discussed in Heinze et al. (1991). This simple model is fitted to observed data and demonstrates the great potential to reproduce the observed changes by combining changes of parameters. This method has the advantage that the reduced simple linear model is much less time and resource consuming than the full non-linear model while enabling the multiple estimate of simultaneous changes in several model parameters. On the other hand, the linear model gets increasingly unreliable the greater the parameter perturbations become.

The most comprehensive procedure in combining observed data and models is the assimilation of observed data and thereby correcting the parameters of the original full model (e.g., Kalman, 1960; Long and Thacker, 1989). Quasi-equilibrium and time dependent cases can be considered. The basic strategy is to integrate the full model for a certain time, then compare the modelled tracer changes with observations, estimate the combination of changes in model parameters that are necessary for minimising the difference between observed and modelled tracer distributions and then repeat the integration with the corrected setting of model parameters. As neither the models nor the data are perfect and because data are available only for discrete locations and times no perfect agreement of data and model results can be achieved. To obtain reliable results one has to apply a correct weighting of the observed (input) data with respect to their reliability. By allowing the observed data to vary slightly from their given values inherent systematic errors of the records can also be identified by the assimilation procedure. Adjoint methods allow to fit entire model state trajectories to observed data time series (e.g., Giering, 1989). Assimilation methods at present are used mainly for short and medium range weather forecasts (e.g., Hollingsworth, 1988) but - in principle - can (and will) be used for paleoclimatic studies with almost unlimited options for future investigations.

THE EFFECT OF SEA LEVEL LOW STANDS ON THE CARBON SYSTEM

None of the model experiments carried out so far give an adequate explanation for the marine glacial tracer record and the findings for lower glacial atmospheric pCO_2 as recorded in

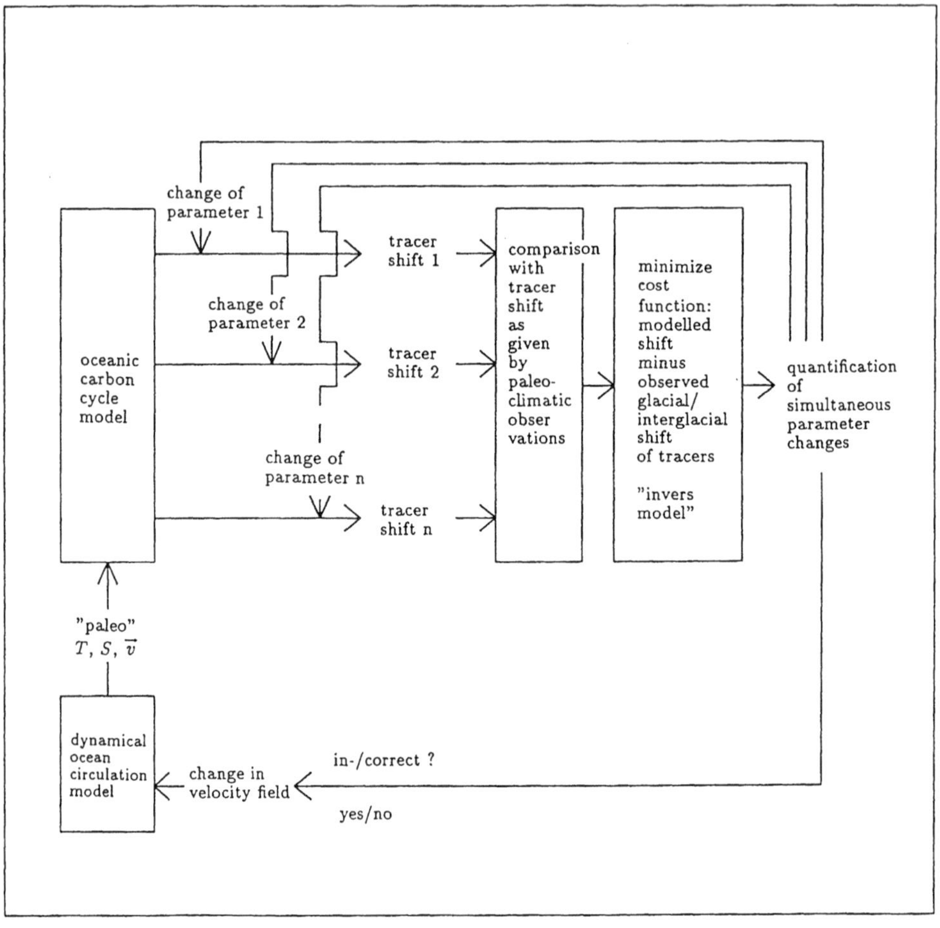

Figure 9: Experimental strategies: Inverse approach.

of the glacial/interglacial tracer shifts. At present, however, it cannot be ruled out that shifts in the biogeochemical machinery occurred on interglacial-glacial time scales. The processes associated with such potential shifts include many 'vital effects' that are not yet understood and hence cannot be modelled adequately. Based on the sedimentary record it has been speculated that changes in the chemical contrast between the ocean's intermediate and deep waters could be one of the most important options for explaining the global interglacial-glacial carbon re-distribution (Boyle, 1988; deMenocal et al., 1992; Mix et al., 1991). Until a plausible reproduction of a corresponding ocean velocity field is available it is worthwhile to ask what other mechanisms could have contributed to the glacial-interglacial carbon system changes.

We ask here what the effect of glacial sea level lowering could have been on the carbon cycle and corresponding tracer distributions. The maximum glacial sea level lowering is estimated to be about 130 m (Chappell and Shackleton, 1986; Fairbanks, 1989). The average increase of salinity in the world ocean which results from the extraction of fresh water associated with the sea level lowering is approximately 4 % of its present value(i.e. $\Delta S \approx +1‰$; Broecker and Peng, 1986) while the average cooling of the deep ocean water masses probably has not exceeded 2°C (Labeyrie et al., 1987). Hydrographic changes in the ocean's surface layer may have deviated substantially from the deep ocean shifts with regionally stronger cooling and freshening (Duplessy et al., 1992; Pichon et al., 1992).

The effect of those hydrographic changes on the carbon system were tested in several sensitivity experiments. These experiments are unrealistic in the sense that the velocity fields used are not compatible to the assumed hydrographic shift, for example the change in surface salinity was not allowed to change the ocean circulation. Assuming that the basic gross world ocean circulation pattern was not too far away from the actual present ocean the approach may be justified.

In the first experiment HO (Hydrography Only)the control run velocity field was used and all chemical constants (solubilities, solubility products, mass action constants, rain ratio) computed as if all ocean water temperatures were 2°C lower and salinities would be 1‰ higher than today. In this run nutrients and the carbon cycle tracers had the same initial inventories as in the control run. In the second experiment HI (Hydrography and Inventories) the nutrient and carbon cycle tracer inventories (including alkalinity) were increased parallel to the hydrography shift also by 4% of their present value corresponding to the salinity increase. To test additionally a strong freshening of the ocean surface layer experiment HS (Hydrography Surface Salinity) was carried out (salt pump experiment, cf. Broecker and Peng, 1987). The

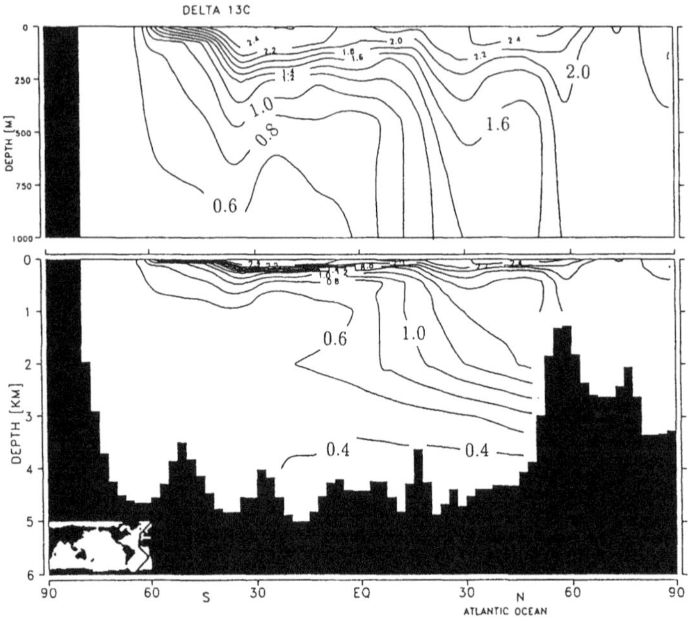

Figure 10 $\delta^{13}C$ cross section along the eastern Atlantic Ocean for scenario HIC (contour interval = 0.2 permil)

parameter perturbation is identical to HO with the difference that the surface salinity was decreased by 4% relative to the control run value. Finally, in experiment HIC (Hydrography, Inventories and Circulation) the second Holocene mode circulation (see Heinze et al., 1991) was used in addition to the changes implemented in run HI. Every run re-started from the control run distributions and was integrated for 20,000 years (20,000 time steps) until equilibrium.

The results are summarized in Tables 1-3 for atmospheric pCO_2, organic matter new production, calcite new production, $\Delta\delta^{13}C_{intermediate\text{-}deep}$ for the Atlantic and Pacific Oceans, and $CaCO_3$ lysocline depth variations in the Indo-Pacific and the eastern and western North Atlantic. $\Delta\delta^{13}C_{intermediate\text{-}deep}$ was computed from grid points corresponding to the locations of DSDP Site 502B (intermediate Atlantic; deMeneocal et al., 1992), V19-27 (intermediate Pacific; Mix et al., 1991), M13519 (deep Atlantic, Sarnthein et al.; 1984) and RC13-110 (deep Pacific; Mix et al., 1991). Lysocline depths were calculated at grid points representing the cores investigated by Crowley (1983; eastern North Atlantic), Balsam (1983; western North Atlantic), Peterson and Prell (1985; Indian Ocean), and Farrell and Prell (1989; Pacific Ocean). The observed records were averaged over the period 18,000-65,000 years before present.

experiment	atm. pCO_2 [ppm]	POC new production [GtC/year]	$CaCO_3$ new production [GtC/year]
control	278.5	8.99	1.97
HO	258.5	8.99	1.86
HS	240.8	8.99	1.86
HI	262.7	9.34	1.94
HIC	251.9	9.31	1.94

Table 1: Results of the sensitivity experiments for atmospheric pCO_2, POC new production and $CaCO_3$ new production.

Scenarios HO and HS which are purely hydrography-driven as well as experiment HI show practically no $\Delta\delta^{13}C_{intermediate\text{-}deep}$ changes. They reveal, however, different responses of the lysocline depth level. For run HO and HS lysocline depths change only little with a general trend towards slightly greater depths except for the Pacific value in run HS where the lysocline shallows slightly. This shoaling can be explained by the decreased outgassing of upwelled CO_2 due to the prescribed drastic decrease in the surface salinity. Changes of lysocline depths are remarkably stronger in the case HI with additional inventory increases (Figures 10 and 11, Table 3). In this case the shift predicted for the Pacific approaches the magnitude of the observed interglacial-glacial deepening. The general deepening of the lysocline in the North Atlantic in the three experiments HO, HS and HI contradicts the strong shallowing in the western part and the slight shallowing in the eastern part of the Atlantic which is implied by sediment data. The combined scenario HIC with the additional circulation change (Figures 12 and 13) leads to a further deepening of the lysocline in the Indo-Pacific. While the only subtle decrease in Pacific Ocean $\Delta\delta^{13}C_{intermediate\text{-}deep}$ is inconsistent with the observed interglacial/glacial shift of about +0.4‰, both the $\Delta\delta^{13}C_{intermediate\text{-}deep}$ change (Figure 14) and the $CaCO_3$ lysocline depth level changes predicted for the Atlantic now agree with the observed trend. The effect of the circulation change postulated for the Atlantic Ocean overrides the effect of the assumed inventory shift (associated with the fresh water extraction) on the deepening of the lysocline there. The change in Atlantic $\Delta\delta^{13}C_{intermediate\text{-}deep}$ is still only about 20% of the observed increase.

In all experiments the global new production values change only slightly (Table 1). The reduction in pCO_2 associated with experiments HO and HS is in part due to a slight decrease of calcite production in the ocean surface layer due to the temperature dependence of the modelled rain ratio.

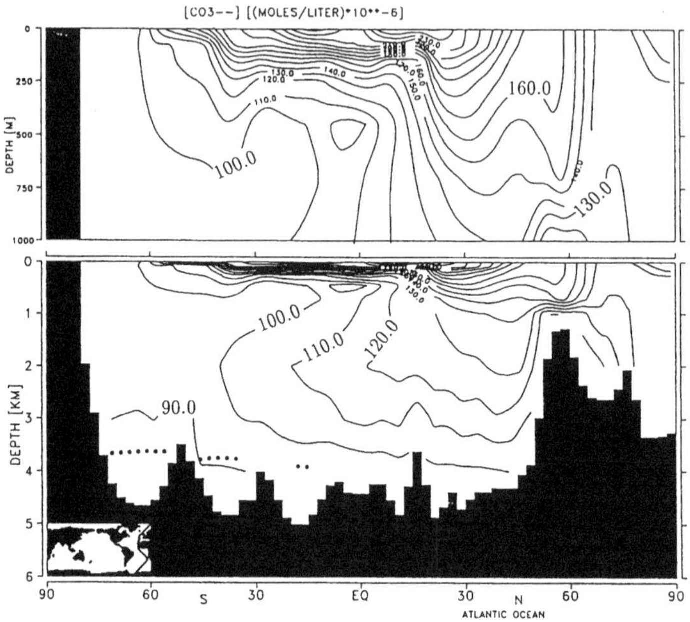

Figure 10: $[CO_3^{2-}]$ cross section along the eastern Atlantic Ocean for scenario HI. (Dotted line: Model $CaCO_3$ lysocline horizon.)

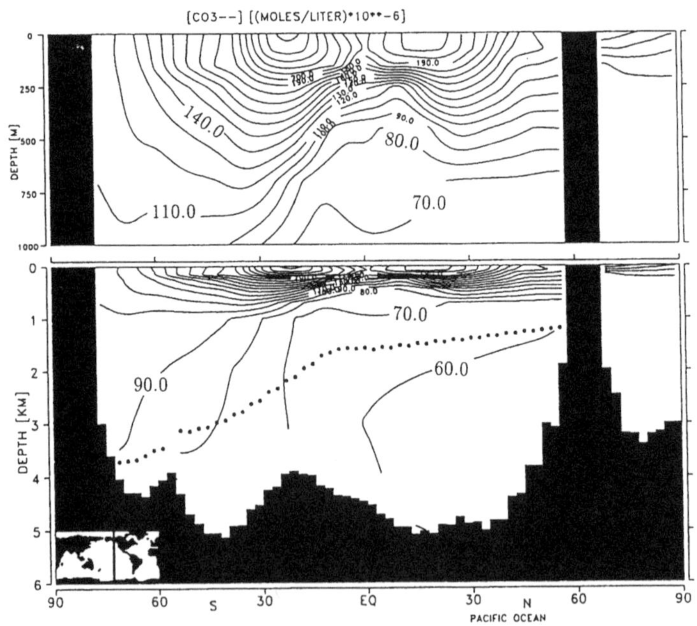

Figure 11: $[CO_3^{2-}]$ cross section along the eastern Pacific Ocean for scenario HI. (Dotted line: Model $CaCO_3$ lysocline horizon.)

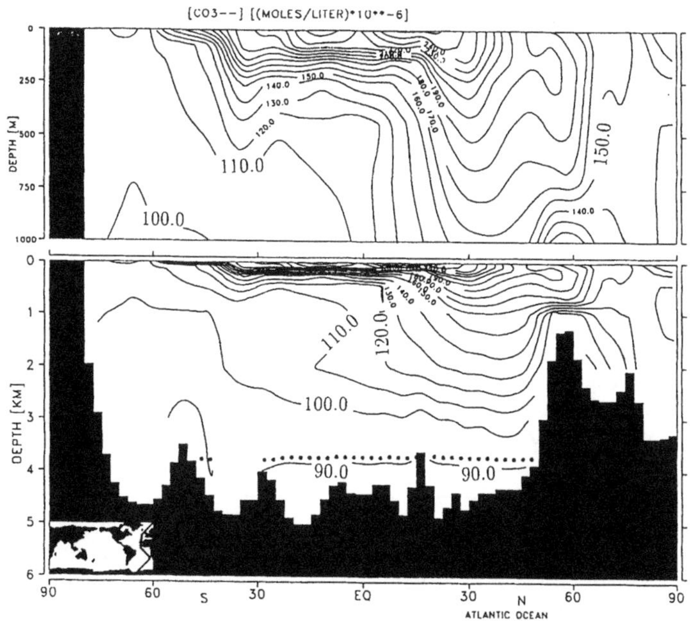

Figure 12: $[CO_3^{2-}]$ cross section along the eastern Atlantic Ocean for scenario HIC. (Dotted line: Model $CaCO_3$ lysocline horizon.)

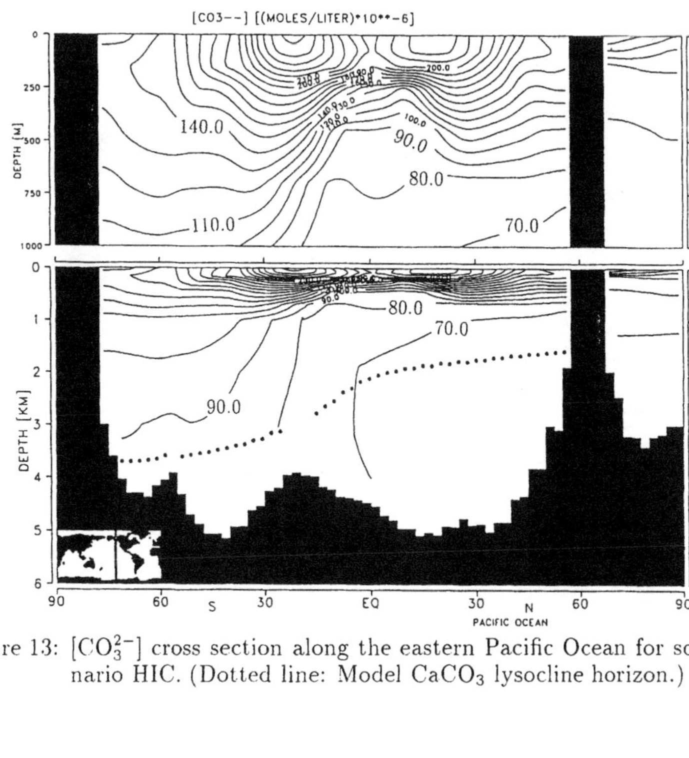

Figure 13: $[CO_3^{2-}]$ cross section along the eastern Pacific Ocean for scenario HIC. (Dotted line: Model $CaCO_3$ lysocline horizon.)

experiment	$\Delta\delta^{13}C$ intermediate-deep change [permil] Atlantic Ocean	$\Delta\delta^{13}C$ intermediate-deep change [permil] Pacific Ocean
HO	0.00	- 0.01
HS	0.00	- 0.01
HI	0.00	0.00
HIC	0.17	- 0.02
observation	0.90	0.44

Table 2: Results of the sensitivity experiments for $\Delta\delta^{13}C_{intermediate\text{-}deep}$ differences relative to the control run values.

experiment	lysocline depth level change [m] W-Atlantic	lysocline depth level change [m] E-Atlantic	lysocline depth level change [m] Indian	lysocline depth level change [m] Pacific
HO	>bottom	137	156	15
HS	>bottom	131	145	-102
HI	>bottom	385	527	463
HIC	- 869	- 297	911	1016
observation	- 1589	- 155	130	696

Table 3: Results of the sensitivity experiments for $CaCO_3$ lysocline depth level changes relative to the control run (negative values indicate a shoaling, positive values a deepening)

SUMMARY

Up to perhaps half the observed glacial reduction of atmospheric pCO_2 can be explained by a decrease of surface salinity in combination with a decrease in sea surface temperature. The additional change in the tracer inventories has only a minor impact on the CO_2 airborne fraction but a considerable influence on the depth level of the calcite lysocline. Variations in sea level can well account for a substantial part of the Indo-Pacific $CaCO_3$ lysocline deepening during the glaciation. All these processes leave no imprints on the $\Delta\delta^{13}C_{intermediate\text{-}deep}$ distribution. The most promising approach for further modeling studies is a reproduction of the correct glacial nutrient distribution with an appropriate velocity field as the result of a correct glacial surface forcing. Incorporation of observed paleoclimatic tracer distributions including oxygen isotopes, ^{14}C, nutrient tracers ($\delta^{13}C$, Cd/Ca), paleoproductivity, and $CaCO_3$ saturation indices in modeling procedures with inversion techniques will help to identify the weaknesses in the data base, the models, and our understanding of the carbon cycle, and to quantify its interactions. At present we are able to model the actual ocean large scale carbon cycle tracer distributions satisfactorily. Any serious forecast of the sensitivity of the system to changes in external and internal parameters need long term cooperation of measuring and modeling scientists.

ACKNOWLEDGEMENTS

I am grateful for the discussions with my colleagues at the Max-Planck-Institut für Meteorologie and the University of Hamburg and for the helpful review of the manuscript by David Archer.

REFERENCES

Arakawa A, and VR Lamb (1977) Computational design of the basic dynamical process of the UCLA general circulation model. Meth. Com. Phys. 16: 173-283.
Archer D, and E Maier-Reimer (accepted for publication). Deep sea $CaCO_3$ burial and atmospheric pCO_2. Nature.
Bacastow RB and E Maier-Reimer (1990) Circulation model of the oceanic carbon cycle. Climate Dyn. 4: 95-125.
Balsam WL (1983) Carbonate dissolution on the Muir Seamount (Western North Atlantic): Interglacial/glacial changes. J. Sed. Pet. 53, 719-731.
Berger WH, and RS Keir (1984) Glacial-Holocene changes in atmospheric CO_2 and the deep-sea record. In "Climate processes and climate sensitivity", JE Hansen and T Takahashi, editors, pp. 337-351. American Geophysical Union, Geophysical Monograph 29, Washington D.C..
Boyle EA (1988) The role of vertical chemical fractionation in controlling late Quaternary atmospheric carbon dioxide. J. Geophys. Res. 93: 15,701-15,714.
Broecker WS (1982) Ocean chemistry during glacial time. Geochim. Cosmochim. Acta 46: 1689-1705.
Broecker WS, and TH Peng (1982) Tracers in the Sea. ELDIGIO Press, Lamont-Doherty Geological Observatory, Columbia University, Palisades, New York, 690 pp..
Broecker WS, and TH Peng (1986) Carbon cycle: 1985 - Glacial to interglacial changes in the operation of the global carbon cycle. Radiocarbon 28: 309-327.
Broecker WS, and TH Peng (1987) The oceanic salt pump: Does it contribute to the glacial--interglacial difference in atmospheric CO_2 content ? Global Biogeochem. Cyc. 1: 251-259.
Broecker WS, and T Takahashi (1977) Neutralization of fossil fuel CO_2 by marine calcium carbonate. In: The fate of fossil fuel CO_2 in the oceans, NR Anderson and A Malahoff, editors, Plenum Press, New York, 213-241.
Broecker WS, and T Takahashi (1978) The relationship between lysocline depth and in situ carbonate ion concentration. Deep-Sea Res. 25: 65-95.
Chappell J, and NJ Shackleton (1986) Oxygen isotopes and sea level. Nature 324: 137-140.
Crowley TJ (1983) Calcium-carbonate preservation patterns in the central North Atlantic during the last 150,000 years. Mar. Geol. 51: 1-14.

Culberson CH, and RM Pytkowicz (1968) Effect of pressure on carbonic acid, boric acid and the pH in sea water, Limnol. Oceanogr. 13: 403-417.
deMenocal PB, DW Oppo, RG Fairbanks, and WL Prell (1992) Pleistocene $\delta^{13}C$ variability of North Atlantic intermediate water. Paleoceanography 7: 229-250.
Duplessy JC, L Labeyrie, M Arnold, M Paterne, J Duprat, and TCE van Weering (1992) Changes in surface salinity of the North Atlantic Ocean during the last deglaciation. Nature 358: 485-488.
Edmond JM, and TM Gieskes (1970) On the calculation of the degree of saturation of sea water with respect to calcium carbonate under in situ conditions. Geochim. Cosmochim. Acta 34: 1,261-1,291.
Esser G (1987) Sensitivity of global carbon pools and fluxes to human and potential climatic impacts. Tellus 39B: 245-260.
Fairbanks RG (1989) A 17,000-year glacio-eustatic sea level record: influence of glacial melting rates on the Younger Dryas event and deep-ocean circulation. Nature 342: 637-642.
Farrell JW, and WL Prell (1989) Climatic change and $CaCO_3$ preservation: An 800,000 year bathymetric reconstruction from the central equatorial Pacific Ocean. Paleoceanography 4: 447-466.
Giering, R (1989) Assimilation von Satellitendaten in ein dreidimensionales numerisches Modell der atlantischen Zirkulation. Diplomarbeit. Max-Planck-Institut für Meteorologie, Hamburg, 94 p..
Heinze C, and K Hasselmann (accepted for publication) Inverse multi-parameter modeling of paleo-climate carbon cycle indices. Quart. Res.
Heinze C, and E Maier-Reimer (1992). The Hamburg Oceanic Carbon Cycle Circulation Model (Cycle 1). Technical Report No.\ 5, Deutsches Klimarechenzentrum, Hamburg 31 pp..
Heinze C, E Maier-Reimer and K Winn (1991). Glacial pCO_2 reduction by the World Ocean: Experiments with the Hamburg Carbon Cycle Model. Paleoceanography 6: 395-430.
Hellerman S, and M Rosenstein (1983) Normal monthly wind stress over the world ocean with error estimates, J. Phys. Oceanogr. 13: 1,093--1,104.
Hollingsworth A (1988) The role of real-time-four-dimensional data assimilation in the quality control, interpretation, and synthesis of climate data. In: Oceanic circulation models: Combining data and dynamics. DLT Anderson and J Willebrand, editors, NATO ASI series, Vol.\ 284, Kluwer, Dordrecht, 304-343.
Ingle SE (1975) Solubility of calcite in the ocean, Marine Chemistry, 3, 301-319.
Kalman R (1960) A new approach to linear filtering and prediction problems. J.Basic Eng.Trans.ASME, Series D, 35-45.
Keir, R (1988) On the late Pleistocene ocean geochemistry and circulation. Paleoceanography 3: 413-446.
Knox F, and MB McElroy (1984) Changes in atmospheric CO_2: Influence of the marine biota at high latitude. J. Geophys. Res. 89: 4629-4637.
Kurz KD, and E Maier-Reimer (1992) Iron fertilization of the Austral Ocean -- the Hamburg Model assessment. Global Biogeochem. Cyc. 7: 229-244.
Labeyrie LD, JC Duplessy, and PL Blanc (1987) Variations in mode of formation and temperature of oceanic deep waters over the past 125,000 years. Nature 327: 477-482.
Lautenschlager M, U Mikolajewicz, E Maier-Reimer, and C Heinze (1993) Application of ocean models for the interpretation of atmospheric general circulation model experiments on the climate of the last glacial maximum. Paleoceanography 7: 769--782.
Levitus S (1982) Climatological Atlas of the World Ocean, NOAA Professional Paper 13, Rockville Md..
Long R, and W Thacker (1989) Data assimilation into a numerical equatorial ocean model. Part I: the model and assimilation algorithm. Dyn. Atmos. Oceans 13: 413-440.
Maier-Reimer, E. (accepted for publication) Geochemical cycles in an ocean general circulation model. Preindustrial tracer distributions. Global Biogeochem. Cycl.
Maier-Reimer E, and R Bacastow (1990) Modeling of geochemical tracers in the ocean. In 'Climate-Ocean Interaction', M E Schlesinger, editor, pp. 233-267. Kluwer, Dordrecht.
Maier-Reimer E, and K Hasselmann (1987) Transport and storage of CO_2 in the ocean - an inorganic ocean-circulation carbon cycle model. Climate Dyn. 2: 63-90.

Maier-Reimer E, K Hasselmann, and U Mikolajewicz (1993) Mean circulation of the Hamburg LSG OGCM and its sensitivity to the thermohaline surface forcing. J. Phys. Oceanogr. 23: 731--757.

Mehrbach C, CH Culberson, JE Hawley, and RM Pytkowicz (1973) Measurement of the apparent dissociation constants of carbonic acid in seawater at atmospheric pressure. Limnol. Oceanogr. 18: 897-907.

Michel E (1991) L'Ocean au dernier maximum glaciaire: Le cycle du carbone et la circulation. Constraintes isotopiques et modelisation. PhD thesis. Universit\'{e} de Paris-Sud, 191 pp..

Mix AC, NG Pisias, R Zahn, W Rugh, C Lopez, and K Nelson (1991) Carbon 13 in Pacific deep and intermediate waters, 0-370 KA: Implications for ocean circulation and Pleistocene CO_2. Paleoceanography 6: 205-226.

Mook WG (1986) ^{13}C in atmospheric CO_2. Netherlands Journal of Sea Research 20: 211-223.

Parsons TR, and M Takahashi (1973) Biological Oceanographic Processes. Pergamon Press. 186 pp.

Peterson LC, and WL Prell (1985) Carbonate preservation and rates of climatic change: An 800 kyr record from the Indian Ocean. In: The Carbon Cycle and Atmospheric CO_2: Natural Variations Archean to Present, E T Sundquist and W S Broecker, editors, pp. 251-269. American Geophysical Union, Geophysical Monograph 32, Washington D.C..

Pichon J-J, LD Labeyrie, G Bareille, M Labracherie, J Duprat, and J Jouzel (1992) Surface water temperature changes in the high latitudes of the southern hemisphere over the last glacial-interglacial cycle. Paleoceanography 7: 289-318.

Sarmiento JL, and JR Toggweiler (1984) A new model for the role of the oceans in determining atmospheric pCO_2. Nature 308, 620-624.

Sarnthein M, H Erlenkeuser, R von Grafenstein and C Schröder (1984) Stable-isotope stratigraphy for the last 750,000 years: "Meteor" core 13519 from the eastern Atlantic. Meteor Forsch.-Ergebn., Reihe C, 38: 9-24.

Siegenthaler U, and T Wenk (1984) Rapid atmospheric CO_2 variations and ocean circulation. Nature 308: 624-626.

Stuiver M, and HD Polach (1977) Discussion reporting of ^{14}C data. Radiocarbon 19: 355-363.

Weiss RF (1974) Carbon dioxide in water and sea water: the solubility of a non-ideal gas, Mar. Chem. 2: 203-215.

Winguth AME (1992) Windinduzierte interannuale Variabilität in der Warmwassersphäre von 1981 bis 1987 Teil II: Fluktuationen im Kohlenstoffkreislauf. Max-Planck-Institut für Meteorologie, Examensarbeit Nr. 14, Hamburg, 158 pp..

Woodruff SD, RJ Slutz, RL Jenne, and PM Steurer (1987) A comprehensive ocean-atmosphere data set, Bulletin of the American Meteorological Society 68: 1,239-1, 250.

Zahn R, K Winn, and M Sarnthein (1986) Benthic foraminiferal $\delta^{13}C$ and accumulation rates of organic carbon: *Uvigerina peregrina* group and *Cibicidoides wuellerstorfi*.. Paleoceanography 1, 27-42.

GLACIAL-INTERGLACIAL CHANGES IN CONTINENTAL WEATHERING: POSSIBLE IMPLICATIONS FOR ATMOSPHERIC CO_2

Guy Munhoven and Louis M François
Institut d'Astrophysique
Université de Liège
Avenue de Cointe, 5
B–4000 Liège
Belgium

ABSTRACT. An eleven-box model of the ocean-atmosphere subsystem of the global carbon cycle is developed to study the potential contribution of continental rock weathering and oceanic sedimentation to variations of atmospheric CO_2 pressure over glacial-interglacial timescales. The model is capable of reproducing the distribution of total dissolved inorganic carbon, total alkalinity, phosphate, $\delta^{13}C$, and $\Delta^{14}C$ between the various ocean basins today, as well as the partial pressure of atmospheric CO_2. A simple sedimentation scheme at 20 different depth levels drives carbonate deposition and dissolution as a function of the depths of carbonate and aragonite lysoclines in each ocean basins considered (Atlantic, Antarctic and Indo-Pacific). The coral-reef erosion-deposition cycle is also taken into account. Furthermore, a simple cycle of oceanic strontium isotopes has been added to this model to take advantage of the $^{87}Sr/^{86}Sr$ data recently published by Dia et al. (1992) for the last 300,000 years. These data emphasize the importance of weathering of continental silicate rocks at glacial-interglacial timescales. They are used to construct several scenarios of changes of continental weathering over the last glacial cycles. They suggest that the flux of alkalinity delivered to the ocean from continental silicate weathering may have been substantially larger during glacial times than today. We show that such variations of continental weathering may explain at least in part the observed changes of the partial pressure of atmospheric CO_2 between glacial and interglacial periods.

NATO ASI Series, Vol. I 17
Carbon Cycling in the Glacial Ocean:
Constraints on the Ocean's Role in Global Change
Edited by R. Zahn et al.

INTRODUCTION

Analyses of air bubbles trapped in Antarctic (Vostok) ice sheets have shown that the concentration of atmospheric CO_2 has fluctuated between 180 and 280 ppm during the last glacial-interglacial cycle (Barnola et al., 1987). The reason for this variation of atmospheric CO_2 over timescales of 10-100 kyr is not yet understood, but many authors have speculated on an oceanic cause, invoking changes in the dynamics, the chemistry or the biology of the world ocean (e. g. Broecker, 1982; Knox and McElroy, 1984; Sarmiento and Toggweiler, 1984; Broecker and Peng, 1989; Martin, 1990). The oceanic circulation has a characteristic time of $\approx$ 1000 yr and involves carbon and alkalinity fluxes of rather important magnitude. Changes in the rates of dynamical and chemical oceanic processes have thus the ability to modify substantially the atmospheric CO_2 content over timescales of several thousand years.

Simple box models of the glacial ocean have been developed and widely used to test the numerous mechanisms hypothesized to explain the CO_2 changes. However, the authors of these models have often considered the global ocean as a closed system, neglecting the long-term cycle of rock weathering on the continents followed by river transport and sedimentary deposition on the seafloor. Nevertheless Keir and Berger (1983) were the first to explicitly take into account the interactions with shelf and deep-sea sediments in a two-box model of the oceanic carbon cycle used to test the phosphate-extraction mechanism proposed by Broecker (1982). Further Berger and Killingley (1982) and Berger and Keir (1984) proposed that large atmospheric CO_2 variations could be driven by coral-reef growth accompanying sea level rise. The resulting drop in surface alkalinity propagates to the deep sea where dissolution of seafloor carbonates occurs. Dissolution spikes are indeed observed in deep seafloor carbonate sequences and Farrell and Prell (1989) have summarized the past changes of the calcite lysocline and compensation depth over the last million years in the Central Equatorial Pacific. The time lag between coral-reef buildup and deep seafloor carbonate dissolution is critical to generate a significant atmospheric CO_2 signal. The so-called coral-reef hypothesis has been recently reconsidered by Opdyke and Walker (1992a) who used a 2-box model of the oceanic carbon and alkalinity cycles.

In this paper, we analyse the importance of chemical weathering of both silicate and carbonate minerals for glacial-interglacial changes in atmospheric CO_2. For this purpose, we use a 10-box model of the world ocean taking into account river input as well as seafloor sedimentation. Our study considers recent observations which are possible indicators of past weathering intensities.

$^{87}Sr/^{86}Sr$: INDICATOR OF PAST WEATHERING ?

Recently, Dia et al. (1992) have published data from planktonic foraminifera and corals in the Pacific recording the evolution of the $^{87}Sr/^{86}Sr$ ratio of sea-water over the last 250 kyr, i. e. over the last two glacial cycles. The strontium isotope data show cyclic variations with a periodicity of ≈ 100 kyr. A good correlation with $\delta^{18}O$ in the same core is observed. The amplitude of the signal is rather important with the $^{87}Sr/^{86}Sr$ ratio varying between 0.709155 and 0.709175. In view of the rather long residence time of strontium in the ocean (2-4 Myr) compared with the 100 kyr period of the cyclicity, such a variation of the isotopic signal suggests that important changes in the river flux of oceanic strontium have occured over glacial-interglacial times. River strontium comes from the weathering of continental silicates and carbonates. The weathering of carbonate rocks is not of major importance because carbonate rocks have been precipitated from ancient sea-water and thus have a mean isotopic ratio relatively close to the oceanic value. Igneous silicate rocks in contrast show a wide range of isotopic compositions with $^{87}Sr/^{86}Sr$ ratios generally higher (at least for the old granitic shield) than the oceanic value (Holland, 1984). A change in silicate rock weathering, either in the global flux or in the distribution over the continents, can significantly alter the strontium isotopic ratio of sea-water. The isotopic fluctuations recorded over the last two glacial cycles thus possibly signify an important change in the global rate of silicate weathering during the same period. Such drastic changes in weathering intensity over glacial cycles are supported by Ge/Si data recently reported by Froelich et al. (1992), suggesting a variation in the mean river concentration of dissolved silica originating from weathering by a factor of 2-3.

The strontium isotopic record of ancient sea-water over Cenozoic and even Phanerozoic timescales has been known for many years (Peterman et al., 1970; Veizer and Compston, 1974; Burke et al., 1980; Koepnick et al., 1988). For these time periods, which are much longer than the residence time of strontium in the ocean, the isotopic changes are large and easily measured. The interpretation of the Cenozoic and Phanerozoic records is still much debated. Some authors (e. g. Brass, 1976; Berner and Rye, 1992) call upon variations of the isotopic composition of the material from igneous rock weathering entering the oceans due to changes in the relative importance of the continental shields (characterized by a high $^{87}Sr/^{86}Sr$ ratio) versus young volcanic rocks as sources of strontium. Edmond (1992) provides a plausible mechanism for this hypothesis : taking the Himalaya as an example, he shows that rock metamorphism during collisions between continental plates may transfer radiogenic strontium from K-rich minerals into Ca- and Na-rich ones, which, when uplifted, weather more rapidly and thus become a source of ^{87}Sr. Several other studies (Raymo et al., 1988; Raymo, 1991) have proposed that the isotopic signal is an index of past intensities of continental weathering. François and Walker (1992) and François et al. (1992) have shown that, on the basis of this hypothesis, it is possible to reconstruct in a numerical simulation an atmospheric CO_2 history consistent with the timing of Phanerozoic glaciations, as well as CCD and carbonate deposition rates compatible with the Cenozoic records.

Over timescales of several million years, it is usually assumed that atmospheric carbon dioxide is controlled primarily by the balance between the input from volcanoes and the loss in weathering reactions of continental silicate rocks (Walker et al., 1981; Berner et al., 1983; Lasaga et al., 1985). The role of silicate weathering in the glacial-interglacial carbon budget of the ocean-atmosphere system has nevertheless often been neglected, because of the rather long residence time of oceanic carbon with respect to continental weathering and sedimentation ($\approx$ 150 kyr) or more specifically with respect to silicate weathering followed by carbonate deposition on the seafloor ($\approx$ 400 kyr). However this simple argument regarding the oceanic residence time of carbon is not really relevant as far as atmospheric CO_2 is concerned, because of the high non-linearity of CO_2 dissolution and carbonate reactions in sea-water. The long-term mean release of CO_2 from volcanoes (and hence the global weathering rate of continental silicates) is not well-known but estimates generally fall in the range of $5\text{-}10\times10^{12}$ mol/yr (Walker, 1977; Walker et al., 1981; Berner et al., 1983; Meybeck, 1987). Following François and Walker (1992), let us assume an intermediate value of 7×10^{12} moles for the amount of Ca-Mg originating from continental silicate weathering and entering the ocean each year. This flux corresponds to an alkalinity input of 14×10^{12} eq/yr. With a volume of 1.35×10^{18} m^3 for the world ocean, a flux of this magnitude, if not compensated by carbonate deposition on the seafloor, has the ability to raise the average alkalinity of sea-water by 0.2 meq/kg in only 20 kyr. Such a drop in global alkalinity, if reflected in surface water, can induce a drop in atmospheric CO_2 by $\approx$ 130 ppm, a change larger than the amplitude of the glacial-interglacial signal. Of course, in the real system, the rise in alkalinity would be accompanied by a rise in carbonate ion concentration and thus by an increase in the deposition rate of seafloor carbonates, an increase which compensates at least partly for the initial rise in alkalinity.

As mentioned above, the isotopic data of Dia et al. (1992) show a cyclic change of 28 ppm in the $^{87}Sr/^{86}Sr$ ratio of sea-water over the last two glacial cycles. What is the change in the weathering rate required to sustain such an isotopic shift, if we assume that weathering intensity is the only important factor that has varied ? Is this needed change significant with respect to that postulated in the simple example discussed previously ? The effect of a variation ΔF_{riv} of the river flux of strontium on the $^{87}Sr/^{86}Sr$ isotopic ratio of ocean water r_o can be measured by the quantity $(r_{riv}-r_o).(\Delta F_{riv}/q_o).(9.43+r_o)/(9.43+r_{riv})$ which is numerically equal to the time derivative of r_o induced by the flux variation (the factor

$(9.43+r_o)/(9.43+r_{riv})$ is close to unity and it comes from the consideration of not only ^{87}Sr and ^{86}Sr but also ^{84}Sr and ^{88}Sr in the strontium budget; for details see Holland (1984)). In this expression, r_{riv} is the river strontium isotopic ratio which today is close to 0.7119 (Wadleigh et al., 1985) and q_o is the amount of strontium in the oceanic reservoir (0.12×10^{18} moles today). The data of Dia et al. (1992) record changes of r_o between 0.709155 and 0.709175 over typical time intervals of 40 kyr. The necessary variation of the strontium river flux required to sustain such an isotopic shift is thus $\Delta F_{riv} = 2.2\times10^{10}$ mol/yr, which is about 70% of the present-day river flux of strontium. It is clear that some fraction of the isotopic shift can be induced by a variation in r_{riv} in response to a change in the type of rock weathered, due for example to a change in the distribution of precipitation and runoff worldwide. However, a change in the type of rock weathered, if due to a totally independent factor (such as runoff distribution) can act with an equal probability towards an increase or a decrease in r_{riv} the latter possibility leading to an even higher estimate of ΔF_{riv}. Note also that metamorphic effects (invoked by Edmond (1992)) have not the ability to change substantially the value of r_{riv} on the geologically short timescale of the glacial-interglacial cycle. We can thus conclude that the isotopic data suggest variations of the weathering intensity over the continents of the order of 70% of the present-day flux. As shown before, if silicate weathering has varied by this amount, then atmospheric CO_2 must have been strongly affected by these changes.

In the following sections, we will discuss these topics by constructing a coupled model of the carbon and strontium geochemical cycles to be applied over glacial-interglacial timescales. We will focus on the description of the long-term exchanges of carbon between the continental crust, the ocean-atmosphere system and seafloor sediments. We believe that the changes in atmospheric CO_2 recorded at Vostok are certainly not due solely to changes in continental weathering of rocks. Presumably many other factors are at work, such as the much discussed effect of a redistribution of oceanic carbon in response to dynamical changes. Thus we do not aim at an explanation for the Vostok record, but the available strontium isotopic record suggests that silicate weathering has changed dramatically over the last two glacial cycles and we will show that these changes cannot be neglected as far as atmospheric carbon dioxide is concerned.

MODEL CHARACTERISTICS

The model of the ocean-atmosphere subsystem of the global carbon cycle which we have been using for this study is a low resolution box model with one single atmospheric box and ten oceanic boxes: deep Atlantic, Indo-Pacific, and Antarctic, temperate Atlantic and Indo-

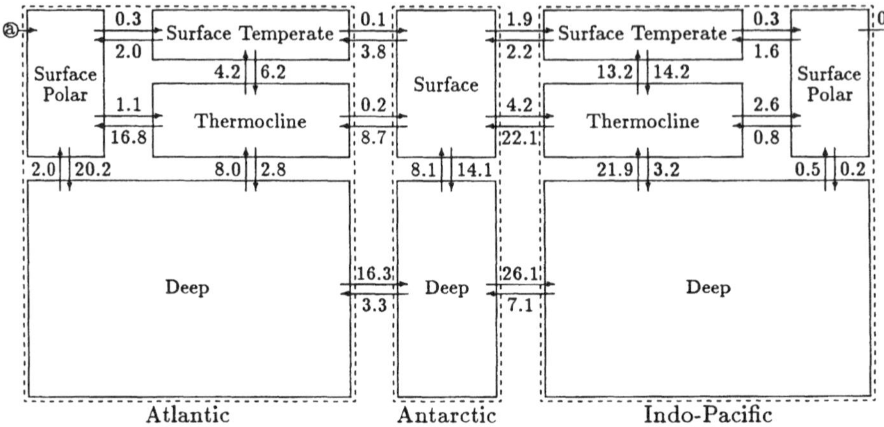

Figure 1. Ocean circulation. Arrows indicate water fluxes in Sverdrups (i. e. 10^6 m^3/yr).

Pacific thermoclines and surfaces, high-latitude Atlantic and Pacific surfaces and Antarctic surface (Figure 1). The volumes of these different reservoirs are calculated as a function of sea level, which is allowed to vary in time. For this purpose we use five depth profiles (one per surface reservoir) obtained by sampling the 1°×1° bathymetric data of the Scripps Institution of Oceanography at 400 m depth intervals and by linear interpolation between these sampling points. Furthermore, we have associated a series of twenty sedimentary layers with each one of these profiles (each layer thus covering a depth interval of 400 m).

A set of 264 ordinary differential equations is used to determine the following tracers : total dissolved inorganic carbon, total alkalinity, phosphate and dissolved oxygen for the oceanic reservoirs, the partial pressure of CO_2 in the atmosphere, the $\delta^{13}C$ and $\Delta^{14}C$ for all of the preceding carbon species and finally the carbonate content and the $^{87}Sr/^{86}Sr$ ratio of the sedimentary reservoirs. The water exchange is fixed and has been determined from the annual mean velocity distribution of the Large Scale Geostrophic Oceanic General Circulation Model of the Max-Planck-Institut für Meteorologie in Hamburg, FRG (Maier-Reimer and Mikolajewicz, 1991). This velocity field has been integrated over the surfaces separating the different boxes and then been decomposed into an advective and a mixing component. To reproduce the observed $\Delta^{14}C$ distribution of the real ocean we must take only 15% of the mixing part of the Hamburg model into account. This is due to the very large differences in the resolutions of the two models. Lumping many small boxes together into a single large homogeneous reservoir is equivalent to increasing the mixing rates between them *ad infinitum*. This is exactly what is done when moving from the large scale Hamburg GCM to our box model. To keep the overall ventilation times of the different ocean basins (i. e. of the reservoirs of the box model) right, it is thus necessary to reduce the mixing through the separating surfaces.

The resulting circulation is presented in Figure 1. When comparing our transports to those of earlier box models of equivalent complexity (Bolin et al. (1983), Broecker and Peng's (1986, 1989) PANDORA and Keir's (1988) CYCLOPS – see Keir (1988) for a summary), we find some common features as well as some striking differences. With the notable exception of North Atlantic Deep Water (NADW) formation which ranges in the vicinity of 20 Sv (1 Sv = 10^6 m^3/yr) for all of the four models compared here, our transport rates, like those of PANDORA and CYCLOPS, are in general much lower than those calculated by Bolin et al. (1983). We will thus focus on PANDORA and CYCLOPS for this discussion. Our upper ocean exchange is similar to that of CYCLOPS in the Atlantic and to that of PANDORA in the Pacific. Like PANDORA, our model derives 30% of its NADW source waters from upwelling in the temperate Atlantic, which is not the case for CYCLOPS. It presents also nearly the same exchange between the deep Atlantic and Antarctic reservoirs as PANDORA, which is about half as important as that in CYCLOPS. The most important differences between our model and the other two can be found in the Antarctic-Pacific interaction and in the intra-Pacific circulation. These two are of course not independent. Whereas the deep Antarctic-deep Pacific exchange is dominated by mixing in both CYCLOPS and PANDORA, our model shows a well pronounced advection (19 Sv) from the Antarctic into the Pacific. This advective flux is followed by a strong upwelling in the temperate Pacific and a return flow from the Pacific intermediate layer to the surface Antarctic reservoir. Finally, our model presents a net formation of deep water inside the Antarctic region, whereas CYCLOPS and PANDORA present a net upwelling (although their fluxes differ by a factor of two).

Besides this advective transport by the marine circulation, we have considered the transfer of organic and inorganic matter from the surface to the thermocline and deep water : a fixed fraction of the phosphate entering each surface reservoir is used for organic matter production and the inorganic (calcium carbonate) production is supposed to be proportional to the organic production. We take 30% of the inorganic production as aragonite, the rest as calcite. The ratio $CaCO_3/C_{org}$ varies from 0.05 in the Northern Pacific reservoir to 0.4 in the Equatorial Indo-Pacific. These ratios have been adjusted to get for total alkalinity the best possible agreement with GEOSECS data. The organic matter is entirely recycled in the thermocline and the deep ocean. Production as well as recycling of organic matter are constrained by the Redfield ratio C/P/N = 106/1/16. One part of the calcium carbonate is transferred to sediments and the remaining part is dissolved in the deep ocean reservoirs, depending on the depth of the calcite and aragonite lysoclines (i. e. the saturation depths with respect to each of these minerals). Carbonate sediments are allowed to redissolve when these lysoclines become shallower.

Sea level variations over the period of interest for our problem are assumed to follow linearly the mean sea-water $\delta^{18}O$. Furthermore, we suppose that sea level variations by themselves do not change the total amounts of dissolved inorganic carbon, of total alkalinity and of phosphate, as well as of salt in the oceans. This means that when sea level is falling oceanic compounds become more concentrated, and *vice versa*. The temperature evolution of the different water masses is taken from the reconstruction of Labeyrie et al. (1987). These reconstructions only deal with deep basins and as no data were available to us for the surface reservoirs, the surface temperature variations have been given the same amplitude as their deep counterparts (3° in the Atlantic and 2° in the Pacific) with present-day values of 2°C in polar and 19°C in temperate regions.

MODEL RUNS AND DISCUSSION

Among the numerous model experiments performed, four have been selected to illustrate the sensitivity of the system to changes in continental weathering. The results of these simulations are summarized in Figures 2 and 4 to 7. The first experiment (Run #1) shows the response of the model ocean-atmosphere to the glacial-interglacial variations in sea level and temperature, including the associated change in sea-water salinity. When sea level is falling, the volume of oceanic water is reduced but no exchange of carbon and alkalinity with crustal pools (either through river runoff or carbonate sedimentation) is allowed, meaning that total inorganic carbon, alkalinity and phosphate are slightly more concentrated, with potential effects on the atmospheric CO_2 pressure. The opposite applies when sea level rises. Temperature and salinity changes also influence atmospheric CO_2 through the dependence of carbonate speciation and CO_2 solubility on these factors. Finally, ocean productivity can react to the slight change in phosphate concentration. The overall response of the model atmospheric CO_2 pressure to the combination of these effects is given by curve #1 of Figure 2. This curve may be considered as the response of a *closed* ocean to the temperature and salinity changes, without any change in circulation. The amplitude of the CO_2 signal is about 30 ppm. Note however that a CO_2 peak is obtained in the model near 5 kyr BP. This peak, not observed in the Vostok core, is due to a similar peak in the Atlantic temperature history (≈ 2°C warmer today than 5 kyr BP) reconstructed by Labeyrie et al. (1987): when the Atlantic temperature peak is removed from the model data, the CO_2 peak disappears. If this peak at 5 kyr BP is not considered, then the amplitude of the signal is only 20 ppm, a value

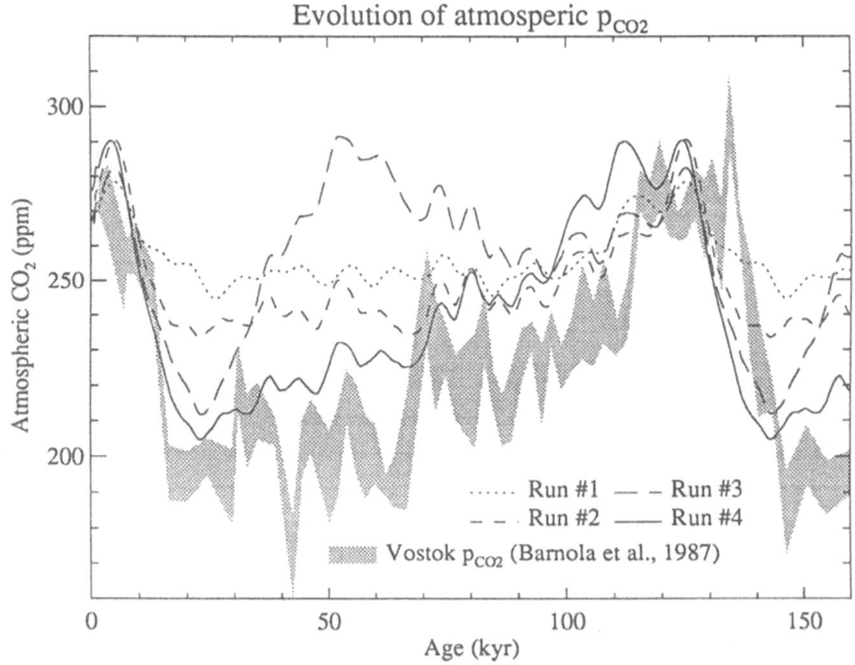

Figure 2. Model-generated variations of atmospheric CO_2 pressure compared with the observations from ice cores (Vostok). Four runs have been performed with the following characteristics (a more detailed description of model runs is given in the text):
Run #1 : sea level, salinity and temperature changes only, no weathering-sedimentation cycle;
Run #2 : sea level, salinity and temperature changes, with cycle of coral-reef erosion and deposition, but with constant continental weathering rates;
Run #3 : sea level, salinity and temperature changes, with cycle of coral-reef erosion and deposition, and variations of the weathering rate of continental silicate rocks from Figure 6 (curve #3), but fixed isotopic composition of the weathering products;
Run #4 : sea level, salinity and temperature changes, with cycle of coral-reef erosion and deposition, both the weathering rate of continental silicate rocks (Figure 6, curve #4) and the isotopic composition of the weathering products (Figure 5, curve #4) vary. Note the large difference in the silicate weathering rate with respect to run #3 (see Figure 5).

slightly higher than the rough estimate of 12 ppm made recently by Broecker (1992). It must be emphasized however that our result contains internal oceanic feedbacks which are not considered in simple estimates of the direct effect of temperature and salinity.

Model run #2 illustrates the importance of considering the ocean-atmosphere as an *open* system. In this experiment, the ocean is fed by *constant* river fluxes of, first, carbon and alkalinity (in a 1:2 ratio, as in $CaCO_3$) originating from weathering of "ancient" (as opposed to freshly deposited coral-reefs) continental carbonate rocks, and, second, of alkalinity from continental silicate weathering. Because silicate weathering adds alkalinity alone (i. e. without carbon) to sea-water and because $CaCO_3$ (i. e. carbon and alkalinity in a 1:2 ratio) deposition is the only flux out of the model ocean-atmosphere, an additional input of carbon to the system equal to half the source of alkalinity from silicate weathering is needed to balance the carbon and alkalinity budgets. This additional carbon input is provided by the release of CO_2 from volcanic origin (hydrothermal and metamorphic), a flux which is also assumed to be *constant* with time.

Coral reefs build up on continental shelves during sea level high stands and are eroded due to exposure when sea level is low. This simple coral-reef buildup and erosion scheme is important since the coral-reef flux to limestones has represented almost 11×10^{12} mol $CaCO_3$ per year over the last 5000 yrs (Kinsey and Hopley, 1991), compared to a total net oceanic sink of about 30×10^{12} mol $CaCO_3$ per year. The growth of coral-reefs in our model is determined by the combined effect of the following factors. (1) They are supposed to grow only in water shallower than 100 m on the continental shelf. The favourable area S_{100} thus changes in time as sea level rises and falls. (2) Furthermore, and similar to Opdyke and Walker (1992b) and Walker and Opdyke (submitted; and personal communication), we suppose that their growth depends on the rate of sea level variation : coral-reefs can neither build up during sea level drops, nor during very rapid rises. The corresponding limiting function L has the shape of a trapezoid. It is equal to 0 for sea level variations lower than −5 mm/yr (i. e. for sea level drops faster than 5 mm/yr) or higher than +25 mm/yr (i. e. for sea level rises faster than 25 mm/yr) and is equal to 1 between +5 mm/yr and +15 mm/yr. Between −5 mm/yr and +5 mm/yr it varies linearily from 0 to 1, and between +15 mm/yr and +25 mm/yr, from 1 to 0. (3) Finally, we assume that coral-reef growth is enhanced when the degree of supersaturation Ω_{CaCO3} of sea-water with respect to $CaCO_3$ increases. The total precipitation G of $CaCO_3$ by coral-reefs (see Figure 3) is then $G = k \times S_{100} \times L \times (\Omega_{CaCO3} -1)$, where we have chosen $k = 0.7$ mol/m^2/yr to produce a mean $CaCO_3$ production of 11×10^{12} mol/yr over the last 5000 years (Kinsey and Hopley, 1991). This approach differs from that of Opdyke and Walker (1992a) who have forced their model with a hypothetical net variation, including as well precipitation as erosion. The above scheme only drives net precipitation. In our model, erosion of coral-reefs is proportional to the exposed area, with a rate of the same order of magnitude as the growth rate, in terms of $CaCO_3$ fluxes per unit area. The $CaCO_3$ flux originating from coral-reef (i. e. "freshly" deposited carbonates) erosion is added to that originating from weathering of "ancient" continental carbonate rocks.

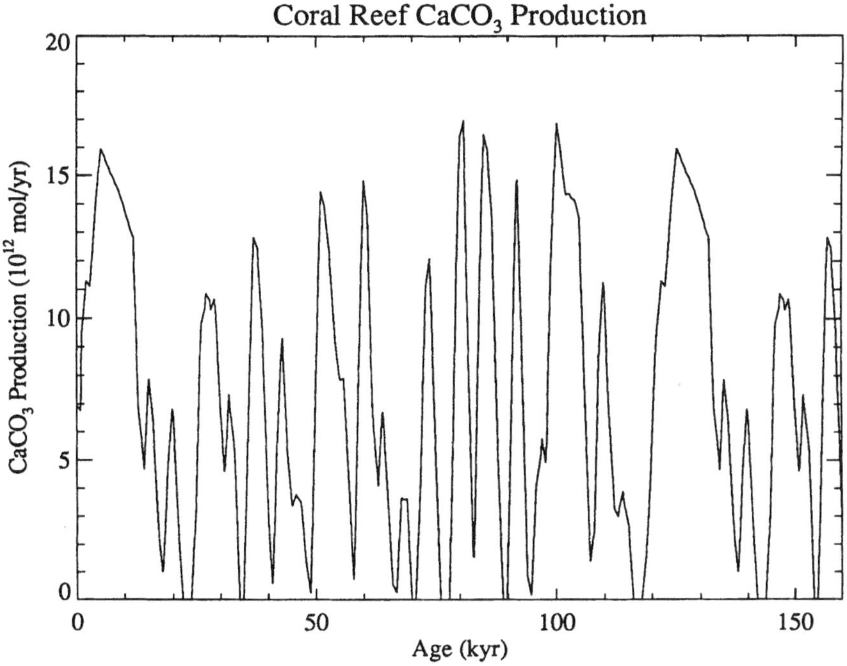

Figure 3. Model-generated coral-reef growth evolution.

Planktonic carbonates deposited on the ocean bottom above the lysocline are kept in sedimentary reservoirs until a rise of the lysocline exposes them to sufficiently corrosive waters so that they can be redissolved. Above the lysocline, there is no dissolution and below, we take it as proportional to $([CO_3^{2-}]_{loc}-[CO_3^{2-}]_{sat})^{4.5}$, following Keir (1980), where $[CO_3^{2-}]_{loc}$ is the local carbonate ion concentration given by the model and $[CO_3^{2-}]_{sat}$ is the carbonate ion concentration at saturation calculated from the relationship of Broecker and Takahashi (1978). We have integrated this law over the top surface of each sedimentary reservoir. The dissolution constant has been fixed at 10^4 mol/m^2/yr, a value consistent with the transition zone thicknesses (CCD minus lysocline) of the Atlantic ($\approx$ 800 m) and the Pacific ($\approx$ 600 m).

The variation of the atmospheric CO_2 pressure (p_{CO2}) in this model run #2 is also presented in Figure 2. Compared with run #1, the opening of the ocean-atmosphere system, which comprises the addition of river inputs as well as the cycles of coral-reef buildup-erosion and of pelagic carbonate sedimentation-dissolution, has increased substantially the amplitude of the CO_2 signal, which is now of ≈ 55 ppm. The overall shapes of the two signals also differ slightly. The peak at 5 kyr BP mentioned previously and due to the Atlantic temperature history has also been strengthened, so that the difference between pre-industrial and glacial minimum p_{CO2} is only ≈ 35 ppm. This value is too small by a factor of more than 2, compared with the corresponding difference in the Vostok record (≈ 80 ppm). This means that, in our model, the Vostok signal cannot be explained by the coral-reef hypothesis alone, despite the rather important coral-reef fluxes involved. Presumably, the reason is the inclusion of negative feedbacks such as the partial compensation of the coral-reef signal by lysocline shifts.

The previous discussion of runs #1 and #2 has overlooked the data on oceanic strontium isotopic composition recently published by Dia et al. (1992). These data are shown as the shaded area (detrended data including error bars) in Figure 4. As discussed earlier these data support major changes of continental weathering through time, the rate of which has precisely been held constant in run #2. Before a comparison of model results with the observed strontium isotopic signal can be performed, we need to know the isotopic composition of the weathered material from various types of rocks, so that the effect of continental weathering on the $^{87}Sr/^{86}Sr$ ratio of sea-water can be estimated. Whereas the Sr/(Ca+Mg) ratio of 0.0015 and the $^{87}Sr/^{86}Sr$ ($r_{carb} = 0.708$) for carbonate rocks are quite well-defined (Holland, 1984), this is not the case for silicate (igneous) material. Rivers draining continental silicate rocks show $^{87}Sr/^{86}Sr$ ratios of 0.704-0.706 for young volcanics and of 0.710-0.730 for old shield crustal rocks (Holland, 1984). When using a ratio of 13:17 (Wadleigh et al., 1985) between the present-day riverine fluxes of Sr coming respectively from silicate (igneous) rocks and from carbonates, we need to assume a typical $^{87}Sr/^{86}Sr$ ratio $r_{sil} = 0.717$ for the mean silicate (igneous) material weathered today to get the observed riverine ratio $r_{riv} = 0.7119$ (Wadleigh et al., 1985). To get rid of any long-term trend, the strontium isotopic budget equation of the ocean has to be equilibrated by the Sr exchange with oceanic basalts presenting a ratio $r_{bas} = 0.704$, a mean for the interactions with mid-ocean ridge basalts at high temperature (0.703) and with pelagic crustal material at low temperature (0.705).

The most probable candidate for the cause of the variation of the strontium isotopic ratio of sea-water shown in Figure 4 is silicate weathering. Indeed it is unlikely that the oceanic basaltic flux or its isotopic composition r_{bas} show changes with a periodicity close to that of the glacial-interglacial cycle. The connection between climatic cycles and continental weathering is much more direct, through changes in runoff or in the rate of mechanical and chemical breakdown of rocks. Of the continental components, carbonate weathering is also a poor candidate, because its isotopic ratio $r_{carb} \approx 0.708$ is too close to the mean oceanic value

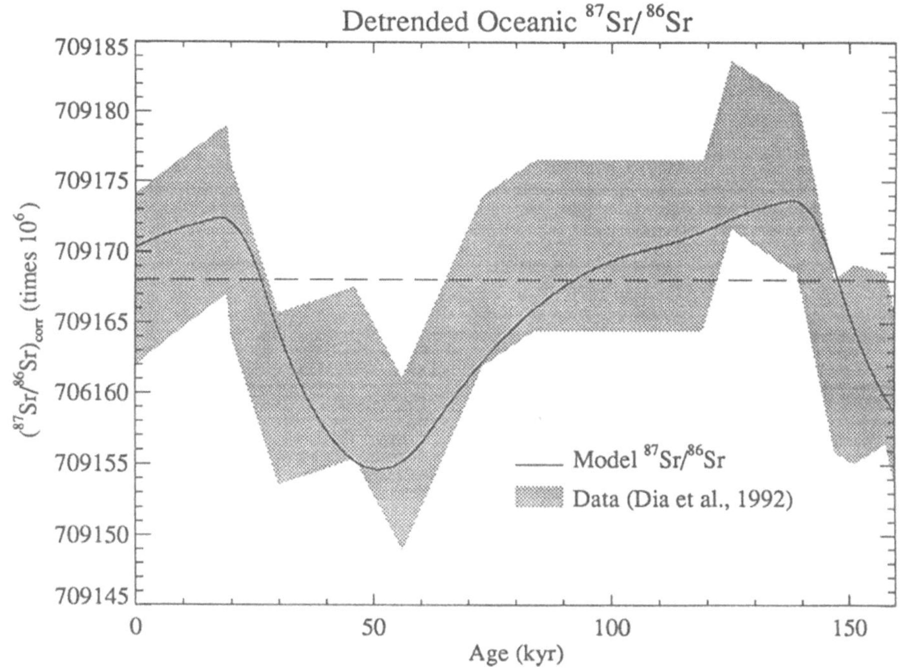

Figure 4. Variation (solid line) of the $^{87}Sr/^{86}Sr$ ratio of sea-water obtained in runs #2, #3 and #4 when the model silicate weathering follows the isotopic compositions and rates of Figures 5 and 6 respectively. The model signal is compared with the observed range (shaded area).

$r_o = 0.70916$: the present-day flux should be enhanced by a factor of at least four to explain the observed isotopic shifts. By contrast, silicate rocks have a mean isotopic ratio r_{sil} much more different from that of sea-water, with a very wide spectrum of values. Two end-member scenarios thus come to mind : *either* the mean isotopic ratio r_{sil} of the silicate weathering has changed through time (due to a worldwide change in the distribution of precipitation, runoff and erosion) while its global intensity F_{sil} remained constant, *or* its global intensity F_{sil} has varied without any changes of the mean isotopic ratio.

For the first of these two scenarios – constant global rate of silicate weathering F_{sil}, but varying mean isotopic composition r_{sil} – the variation in r_{sil} necessary to reproduce the observed glacial-interglacial change of the isotopic ratio of sea-water is shown by the short-dashed line labelled 'Run #2' in Figure 5. The amplitude of the signal is relatively important with r_{sil} varying between 0.714 and 0.724. Such a variation is not precluded, since it falls

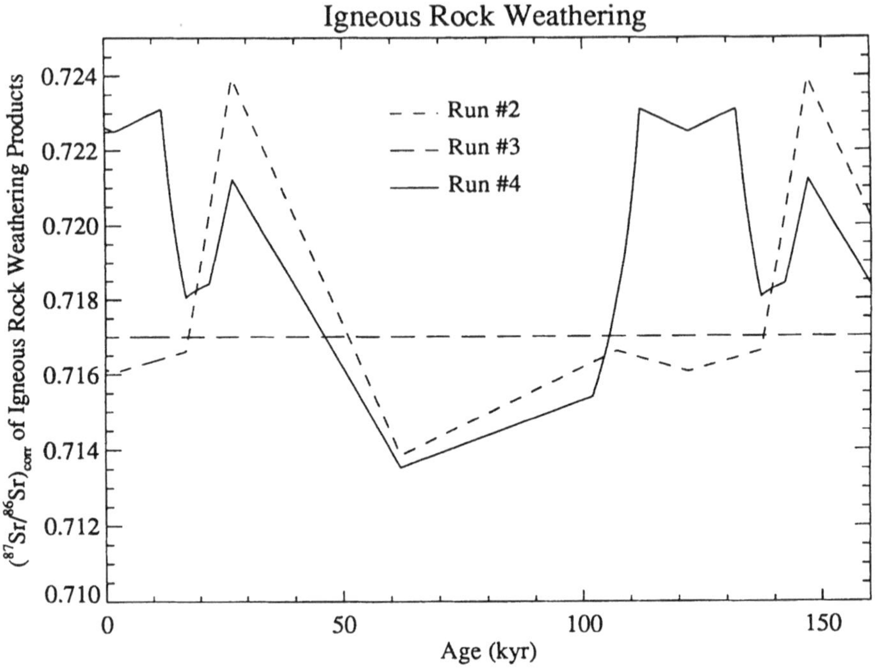

Figure 5. Isotopic compositions of the weathering products from continental (igneous) silicate rocks adopted in model runs #2, #3 and #4.

within the observed range of $^{87}Sr/^{86}Sr$ ratios of rivers draining different types of igneous rocks (0.704-0.730), but we might question whether or not it is compatible with the expected change in the distribution over the continents of precipitation, runoff and weathering during glacial cycles. This question is outside the scope of this paper, because it would require an atmospheric general circulation model to be run and coupled with a model of rock weathering taking into account the detailed lithology of the earth's surface. Of course, because the rate of silicate weathering is held constant in this scenario, the variation of the atmospheric CO_2 pressure is that of run #2 presented earlier : the shape of the signal is approximately correct, but not the amplitude.

The second of these two scenarios is investigated in run #3, where the (igneous) silicate weathering input F_{sil} has been adjusted empirically until a best fit of the model sea-water isotopic composition with the observed data is obtained, with r_{sil} fixed at 0.717. The shape of the F_{sil} signal (Figure 6, 'Run #3') is the same as that of r_{sil} in the previous scenario (Figure 5) 'Run #2') because in both experiments the product $(r_{sil}-r_o)\times F_{sil}$ must be essentially the same. As can be seen in Figure 4, the isotopic ratio r_o of model sea-water (solid line) falls within the range of observed values. For this best fit experiment, the input of Ca+Mg (proportional to that of Sr, F_{sil}) from silicate weathering has the glacial-interglacial variation shown by the long-dashed line of Figure 6. It presents a very marked peak toward high weathering rates centered near 30 kyr BP and a small depression toward low values between 100 and 50 kyr BP. When the carbon cycle model of run #2 – which includes variations of temperature and salinity, as well as the coral-reef and pelagic deposition cycles – is forced with this alkalinity input, the atmospheric CO_2 pressure varies as shown by the curve labelled 'Run #3' in Figure 2. The peak in silicate weathering centered near 30 kyr results in a sharp minimum of atmospheric CO_2 at ≈ 22 kyr BP, when the Vostok signal is also at minimum. The sharp rise in CO_2 during the last deglaciation is also approximately reproduced. However, if we go further back in time, we meet a distinct maximum in model CO_2 near 60-50 kyr with peak values as high as 290 ppm (i. e. characteristic of an interglacial), in sharp conflict with the observations. The reason for this maximum is the small depression in silicate weathering between 100 and 60 kyr. If the observed strontium isotopic ratio of sea-water is to be reproduced without changing r_{sil}, this depression and hence the increased CO_2 cannot be avoided, because of its necessity for yielding decreasing r_o values in the period 90-60 kyr BP.

None of the two end-member weathering scenarios adopted to reproduce the sea-water strontium isotopic cycle observed by Dia et al. (1992) is entirely successful as far as atmospheric CO_2 is concerned. In each case some features of the Vostok CO_2 are captured,but major disagreements remain. Of course, to keep the model as simple as possible, we ignored some mechanisms (such as oceanic dynamical response or biological fertilization) of potential importance for atmospheric CO_2 variations. Such mechanisms could be invoked to solve the apparent disagreement between observed Sr isotopic ratios and atmospheric CO_2.

It is however possible to reproduce *simultaneously* both observed signals in the same run, provided that F_{sil} and r_{sil} are both allowed to vary. Indeed it is relatively easy to create "something like" the observed atmospheric p_{CO_2} if it is assumed that the rate of silicate weathering has been higher during glacial times than today. The physical basis for such an assumption is that during glacial times mechanical breakdown of rocks was probably more extensive (because continental areas subject to seasonal freezing and unfreezing can be expected to have been larger) and that enhanced mechanical breakdown should lead to higher rates of chemical weathering in response to larger areas of rock surface available to the

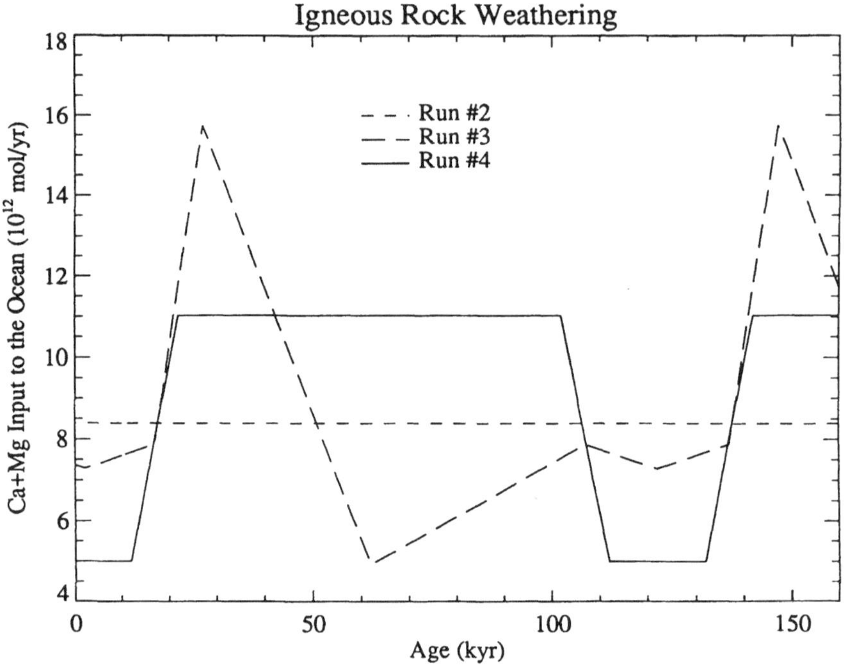

Figure 6. Rates of continental weathering of (igneous) silicate rocks adopted in model runs #2, #3 and #4 (in terms of Ca+Mg input to the ocean).

weathering reactions. Ge/Si data (Froelich et al., 1992) further support glacial weathering rates higher than today. In run #4, we have thus adopted a two-step function for the rate of silicate weathering F_{sil} (Figure 6) through time, with glacial rates higher than interglacial ones by a factor of 2.2 (in the range of values suggested by Ge/Si data). The resulting pCO_2 (Figure 2) resembles the Vostok signal, with a correct amplitude and a good general trend. The history of the mean isotopic composition r_{sil} of weathered silicate rocks which must be assumed to reproduce oceanic strontium isotopic data is shown in Figure 5. During interglacial times, r_{sil} reaches a value of ≈ 0.723, while during the early phase of the glacial period (100-60 kyr BP) it decreases to values between 0.714 and 0.716.

Finally, Figure 7 shows the history of the Atlantic, Antarctic and Indo-Pacific lysocline depths generated by this model run #4. The model Indo-Pacific lysocline is about 600 m deeper during glacial times than at present with a major drop occuring near 80 kyr and a rise

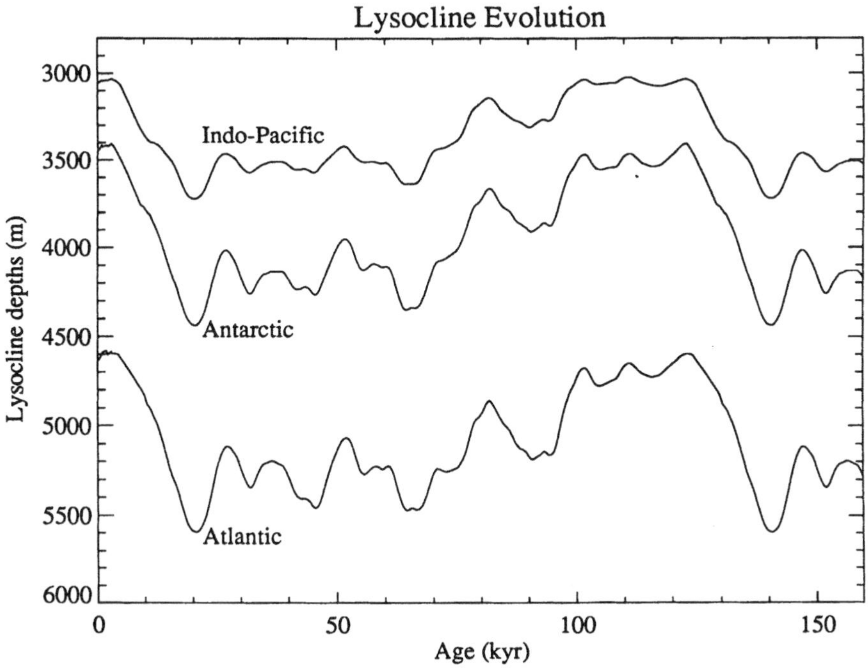

Figure 7. Histories of the lysocline depths of the model ocean basins calculated in run #4.

near 20 kyr BP, in good agreement with the data of Farrell and Prell (1989) for the equatorial Pacific. The behaviour of the observed Pacific lysocline has the biggest chance to be of global significance, because of the large size of the basin. The lysoclines of other basins are certainly more affected by possible circulation changes which have been ignored here. Because of the constancy of the oceanic circulation, model Atlantic and Antarctic lysoclines changes are parallel to those of the Indo-Pacific basin.

CONCLUSIONS

In this paper we have analysed the contribution of continental rock weathering to the glacial-interglacial cycles of atmospheric CO_2. For this purpose, a 10-box model of the oceanic carbon cycle has been built, in which sedimentary processes have been particularly emphasized. The model ocean has been fed by weathering fluxes. Of these fluxes, specific attention has been given to the weathering of silicate rocks. Indeed cyclic variations in the strontium isotopic ratio of sea-water have been observed recently by Dia et al. (1992) at glacial-interglacial timescales. Although large error bars exist on these data, they suggest that continental weathering rates, and more specifically silicate weathering, have changed markedly throughout glacial-interglacial cycles. These observations are strengthened by recent data on the Ge/Si ratio of Pleistocene sea-water reported by Froelich et al. (1992). The interpretation of the strontium isotopic signal remains speculative, because the mean isotopic composition of the weathering products may have changed with time, and no simple estimate of its history can be derived. However our model shows that it is important to take weathering into account in studies of the glacial-interglacial oceanic carbon cycle and that the contribution of continental weathering cannot be neglected. It is even possible to construct a scenario in which both the rate of silicate weathering and the isotopic ratio of its products change through time, and for which the model predicts correctly the observed variation of the strontium isotopic ratio of sea-water as well as that of the atmospheric CO_2 pressure. The glacial-interglacial change in atmospheric CO_2 in this scenario is due in large part to silicate weathering, with the rest originating from variations in temperature and salinity as well as from coral-reef erosion-deposition cycles. The scenario looks *a priori* fully plausible, but would need to be tested with a detailed global model of continental weathering, a model still to be constructed. Anyway it is likely that other oceanic processes not considered here, such as circulation which for simplicity was held constant, are also at work. Future studies should include these topics, but this simplification does not alter the main conclusion reached here that the contribution of silicate weathering to the glacial-interglacial cycles of atmospheric CO_2 was not negligible.

ACKNOWLEDGMENTS

We would like to thank Ján Veizer and Robin Keir for their reviews which substantially improved this paper. Support for this work was provided by the Commission of the European Communities under Contract No. EPOC-CT90-0017 (G Munhoven) and by the Belgian National Fund for Scientific Research (G Munhoven and LM François). Additionnal support came from the Belgian National Impulse Programmes "Information Technology - Supercomputers" (contract No. IT/SC/30) and "Global Change" (contract No. GC/12/017).

REFERENCES

Barnola JM, Raynaud D, Korotkevich YS, Lorius C (1987) Vostok ice core provides 160,000 year record of atmospheric CO_2. Nature 329: 408-414

Berger WH, Keir RS (1984) Glacial-Holocene changes in atmospheric CO_2 and the deep-sea record. In: Hansen JE, Takahashi T (eds) Climate Processing and Climate Sensitivity. American Geophysical Union, Washington DC, (Geophysical Monograph Series, vol 29, p 337)

Berger WH, Killingley JS (1982) The Worthington effect and the origin of the Younger Dryas. J Mar Res 40:27-38

Berner RA, Lasaga AC, Garrels RM (1983) The carbonate-silicate geochemical cycle and its effect on the atmospheric carbon dioxide over the past 100 million years. Am J Sci 283:641-683

Berner RA, Rye DM (1992) Calculation of the Phanerozoic strontium isotope record of the oceans from a carbon cycle model. Am J Sci 292:136-148

Bolin B, Björkström A, Holmen K, Moore B (1983) The simultaneous use of tracers for ocean circulation studies. Tellus 35B:206-236

Brass GW (1976) The variation of the marine $^{87}Sr/^{86}Sr$ ratio during Phanerozoic time : interpretation using a flux model. Geochim Cosmochim Acta 40: 721-730

Broecker WS (1982) Ocean chemistry during glacial time. Geochim Cosmochim Acta 46:1689-1705

Broecker WS (1992) The Glacial World According to Wally. Lamont-Doherty Geological Observatory of Columbia University, Palisades NY

Broecker WS, Peng TH (1986) Carbon cycle, 1985 : Glacial to interglacial changes in the operation of the global carbon cycle. Radiocarbon 28:309-327

Broecker WS, Peng TH (1989) The cause of the glacial to interglacial atmospheric CO_2 change : a polar alkalinity hypothesis. Global Biogeochemical Cycles 3:215-239

Broecker WS, Takahashi T (1978) The relationship between lysocline depth and in situ carbonate ion concentration. Deep-Sea Res 25:65-95

Burke WH, Denison RE, Hetherington EA, Koepnick RB, Nelson HF, Otto JB (1980) Variation of seawater $^{87}Sr/^{86}Sr$ throughout Phanerozoic time. Geology 10:516-519

Dia AN, Cohen AS, O'Nions RK, Shackleton NJ (1992) Seawater Sr isotope variation over the past 300 kyr and influence of global climate cycles. Nature 356:786-788

Edmond JM (1992) Himalayan tectonics, weathering processes and the strontium isotope record in marine limestones. Science 258:1594-1597

Farrell JW and Prell WL (1989) Climatic change and $CaCO_3$ preservation : an 800,000 year bathymetric reconstruction from the central equatorial Pacific Ocean. Paleoceanography 4:447-466

François LM, Walker JCG (1992) Modelling the Phanerozoic carbon cycle and climate : constraints from the $^{87}Sr/^{86}Sr$ isotopic ratio of seawater. Am J Sci 292:81-135

François LM, Walker JCG, Opdyke BN (1993) The history of global weathering and the chemical evolution of the ocean-atmosphere system. In: Takahashi E, Jeanloz R, Rubie DC (eds) Chemical Evolution of the Earth and Planets, American Geophysical Union, Washington DC (Geophysical Monograph Series, vol 74, IUGG Series, vol 14)

Froelich PN, Blanc V, Mortlock RA, Chillrud SL, Dunstan W, Udomkit A, Peng TH (1992) River fluxes of dissolved silica to the ocean were higher during glacials : Ge/Si in diatoms, rivers and oceans. Paleoceanography 7:739-767

Holland HD (1984) The Chemical Evolution of the Atmosphere and Oceans. Princeton University Press, Princeton NJ

Keir RS (1980) The dissolution kinetics of biogenic calcium carbonates in seawater. Geochim Cosmochim Acta 44:241-252

Keir RS (1988) On the Late Pleistocene ocean geochemistry and circulation. Paleoceanography 3:413-445

Keir RS, Berger WH (1983) Atmospheric CO_2 content in the last 120,000 years : the phosphate-extraction model. J Geophys Res 88:6027-6038

Kinsey D, Hopley D (1991) The significance of coral-reefs as global carbon sinks - Response to greenhouse. Palaeogeog Palaeocl Palaeoecol 89:1-15
Knox F, McElroy MB (1984) Changes in atmospheric CO_2 : influence of the marine biota at high latitude, J Geophys Res 89:4629-4637
Koepnick RB, Denison RE, Dahl DA (1988) The Cenozoic seawater $^{87}Sr/^{86}Sr$ curve : Data review and implications for correlation of marine strata. Paleoceanography 3:743-756
Labeyrie LD, Duplessy JC, Blanc PL (1987) Variations in mode of formation and temperature of oceanic deep waters over the past 125,000 years. Nature 327:477-482
Lasaga AC, Berner RA, Garrels RM (1985) An improved geochemical model of atmospheric CO_2 fluctuations over the past 100 million years. In: Sundquist ET, Broecker WS (eds) The Carbon Cycle and Atmospheric CO_2 : Natural Variations Archean to Present. American Geophysical Union, Washington DC (Geophysical Monograph Series vol 32, p 397-411)
Maier-Reimer E, Mikolajewicz U (1991) The Hamburg Large Scale Geostrophic Ocean General Circulation Model (Cycle 1). Technical Report No 2. Deutsches Klimarechen-zentrum, Hamburg
Martin JH (1990) Glacial-interglacial CO_2 change: the iron hypothesis. Paleoceanography 5:1-13
Meybeck M (1987) Global chemical weathering of surficial rocks estimated from river dissolved loads. Am J Sci 287:401-428
Opdyke BN, Walker JCG (1992a) Return of the coral-reef hypothesis : Basin to shelf partitioning of $CaCO_3$ and its effect on atmospheric CO_2. Geology 20:733-736
Opdyke BN, Walker JCG (1992b) Glacial to interglacial, basin to shelf partitioning of $CaCO_3$ and its effect on atmospheric CO_2. In: Sarnthein M, Thiede J, Zahn R (eds) Fourth International Conference on Paleoceanography ICP IV, Short- and Long-Term Global Change: Records and Modelling 21-25 Sept 1992. GEOMAR, Kiel (GEOMAR REPORT 15)
Peterman ZE, Hedge CE, Tourtelot HA (1970) Isotopic composition of strontium in sea-water throughout Phanerozoic time. Geochim Cosmochim Acta 34:105-120
Raymo ME (1991) Geochemical evidence supporting T.C. Chamberlin's theory of glaciation. Geology 19:344-347
Raymo ME, Ruddiman WF, Froelich PN (1988) Influence of late Cenozoic mountain building on ocean geochemical cycles. Geology 16:649-653
Sarmiento JL, Toggweiler JR (1984) A new model for the role ofthe oceans in determining atmospheric pCO_2. Nature 308:621-624
Veizer J, Compston W (1974) $^{87}Sr/^{86}Sr$ composition of seawater during the Phanerozoic. Geochim Cosmochim Acta 38:1461-1484
Wadleigh MA, Veizer J, Brooks W (1985) Strontium and its isotopes in Canadian rivers : fluxes and global implications. Geochim Cosmochim Acta 49:1727-1736
Walker JCG (1977) Evolution of the atmosphere. MacMillan, New York
Walker JCG, Opdyke BN (submitted) The influence of variable rates of shelf carbonate deposition on atmospheric carbon dioxide and pelagic sediments. Paleoceanography
Walker JCG, Hays PB, Kasting JF (1981) A negative feedback mechanism for the long-term stabilization of Earth's surface temperature. J Geophys Res 86:9776-9782

PART II:

VARIATIONS OF THE OCEAN'S CARBON RESERVOIR: FAUNAL VERSUS GEOCHEMICAL RECORDS

THE RELATIONSHIP BETWEEN SURFACE WATER MASSES, OCEANOGRAPHIC FRONTS AND PALEOCLIMATIC PROXIES IN SURFACE SEDIMENTS OF THE GREENLAND, ICELAND, NORWEGIAN SEAS.

Truls Johannessen[1], Eystein Jansen, Astrid Flatøy, Ana Christina Ravelo[2]
Department of Geology
University of Bergen
Allégaten 41
N-5007 Bergen, Norway.

ABSTRACT

A detailed study of paleoclimatic proxy data (stable isotopes, planktonic foraminiferal census data, carbonate content, and Ice Rafted Detritus (IRD)) in the surface sediments of the Greenland, Iceland and Norwegian Seas (GIN-seas) shows that different proxies are closely related to the surface water masses, to the position of oceanic fronts and to the sea ice extent. Both stable isotopes, foraminifers and sedimentological data differentiate between Polar water with extensive sea ice cover, Arctic water with only seasonal sea ice cover, and warm Atlantic water. The fronts that border these surface water masses are also well defined. Polar water is characterized by lower carbon and oxygen isotope values than Arctic water, and a slightly lower content of *Neogloboquadrina pachyderma* sinistral in the Polar Front region. Carbonate content is low and IRD input is high in Polar waters. Arctic water has highest carbon and oxygen isotope values, and is completely dominated by *N. pachyderma* sin. The Arctic Front is reflected by a clear isotopic gradient and by a strong switch from *N. pachyderma* sin. dominance to *Globigerina quinqueloba* dominance. Atlantic Water is defined by lower carbon and oxygen isotope values and by dominance of *N. pachyderma* dextral and increased amounts of *Globigerina bulloides*. The results have implications for paleoceanographic reconstructions of cold environments and point to the possibility of better defining sea ice margins and convective regions as well as frontal positions in past high latitude oceans. Applying these results to the Last Glacial Maximum and the Younger Dryas indicates more dynamic and less sea ice covered surface conditions in the GIN-seas than in earlier reconstructions.

---

[1] Present address: Centre for Studies of Environment and Resources, University of Bergen, Thormøhlensgate 20, N-5020 Bergen, Norway.
[2] Marine Science Board, University of California, Sta. Cruz, CA 95064, USA.

NATO ASI Series, Vol. I 17
Carbon Cycling in the Glacial Ocean:
Constraints on the Ocean's Role in Global Change
Edited by R. Zahn et al.

INTRODUCTION

The high latitude areas of the North Atlantic region which includes the Greenland, Iceland, Norwegian Seas (GIN-Seas) are important for natural climate change. They are proximal to the major Quaternary ice sheets, they may trigger climate change on Milankovitch and sub-Milankovitch time scales, and the thermohaline circulation in the area constrains the global ocean ventilation, ocean chemistry and the variability of the carbon cycle. Several theories which aim at explaining causes of natural climate change, point specificly at processes in the high latitude North Atlantic and the GIN-Seas which drives or amplifies climate change (Boyle, 1988; Broecker and Denton, 1989; Imbrie et al., 1992). Imbrie et al. (1992) document that the climate system responds linearly to orbital forcing at Milankovitch frequencies, and propose, based on phase relationships between orbital forcing and the climatic response patterns, that the prime forcing function for Milankovitch scale climate change is high northern latitude summer insolation. The initial responder to this insolation forcing is suggested to be the surface ocean of the GIN-seas which transmits this response to the global climate system via ice-albedo feedbacks and via effects on the thermohaline circulation.

A critical test for this theory lies in identifying initial responses to insolation changes in the GIN-sea surface circulation and sea ice cover. So far, this has been difficult, and a refinement of the proxy methods used in reconstructions of past high latitude environments is necessary. Such a refinement is also needed in order to understand how sensitive the circulation in the region is to other forcing mechanisms, such as meltwater and fresh-water forcing and wind field forcing. A fundamental problem in explaining causal mechanisms for high amplitude, rapid climate change such as the Younger Dryas cold event and the extreme cooling at the Last Glacial Maximum (LGM) lies in the sensitivity and resolution of the paleoclimate proxies at their cold end member. Further progress in reconstructing and modelling such events depends to a large degree on the ability to accurately place ocean front boundaries, sea ice limits and on the accuracy of sea surface temperature (SST) and sea surface salinity (SSS) estimates.

The GIN-seas and the high latitude North Atlantic are the locus of modern deep sea ventilation and the surface waters in the region are source regions for global deep-water renewal and essential for thermohaline circulation. It is consequently also important for understanding variations in the global carbon cycle that the source regions of deep water, i.e. the surface waters can be accurately described.

In an attempt to improve high latitude paleoceanographic proxy methods, we conducted an extensive mapping of sedimentological parameters, planktonic foraminiferal distribution, and stable isotopes; based on surface sediments obtained from high quality box-core samples and sea water samples in all the major water masses of the GIN-seas. This study covers the major thermal gradient between warm Atlantic environments and cold Polar environments between Norway and Greenland. This includes the Arctic Front, the Polar Front and the seasonal sea ice margins. Here we report some findings from these investigations. We focus here on the geographical distribution of the different parameters and a qualitative comparison with modern water masses, front locations and sea ice boundaries. A more quantitative treatment of the data is and will be performed in separate publications. The isotope results are discussed in detail by Johannessen et al. (subm.) and Ravelo et al. (in prep). The data on foraminiferal and sediment distributions are presented here for the first time.

PRESENT OCEANOGRAPHY

The oceanographic conditions of the study area are determined by the northward flowing warm, saline surface waters of the Norwegian Atlantic Current in the east, which is a continuation of the North Atlantic Drift, and the southward flowing cold, less saline surface waters of the East Greenland Current (Figure 1). A branch of the Norwegian Current flows northwestward toward Jan Mayen, bringing warmer waters into the western part of the study area, whereas a branch of the East Greenland Current flows eastward as the East Iceland Current in the Iceland Sea between Iceland and Jan Mayen. This creates a complex zone of mixing of cold and temperate waters in the central and western parts of the study area and a gyre in the Iceland Sea. These surface waters, named Arctic waters, convect during winters, and a major component of the overflowing water that drives the formation of North Atlantic Deep Water (NADW) is formed in this area (Aagaard et al., 1985). The Arctic water is therefore a very important water mass to reconstruct in order to assess where convection has occurred in the past. As will be seen in the following chapters, the Arctic water mass has several specific characteristics in terms of its accociated faunal and isotopic distributions which make this possible.

The surface water masses define two major fronts (Figure 1): The *Polar Front* is the boundary between low saline Polar waters and cold, higher salinity Arctic waters, the *Arctic Front* is the boundary between cold Arctic waters and warm Atlantic waters (Helland-Hansen and Nansen, 1909; Alekseev and Istochin, 1956; Dietrich, 1969; Stefánsson, 1962; Hansen and Meincke, 1979; Swift and Aagaard, 1981; Hopkins, 1991). The maximum winter sea ice limit lies slightly west of the Arctic Front, while the summer sea ice limit lies west of the Polar Front. The area of seasonal sea ice cover basically comprises Arctic Water masses, i.e. the water masses where winter time convection occurs.

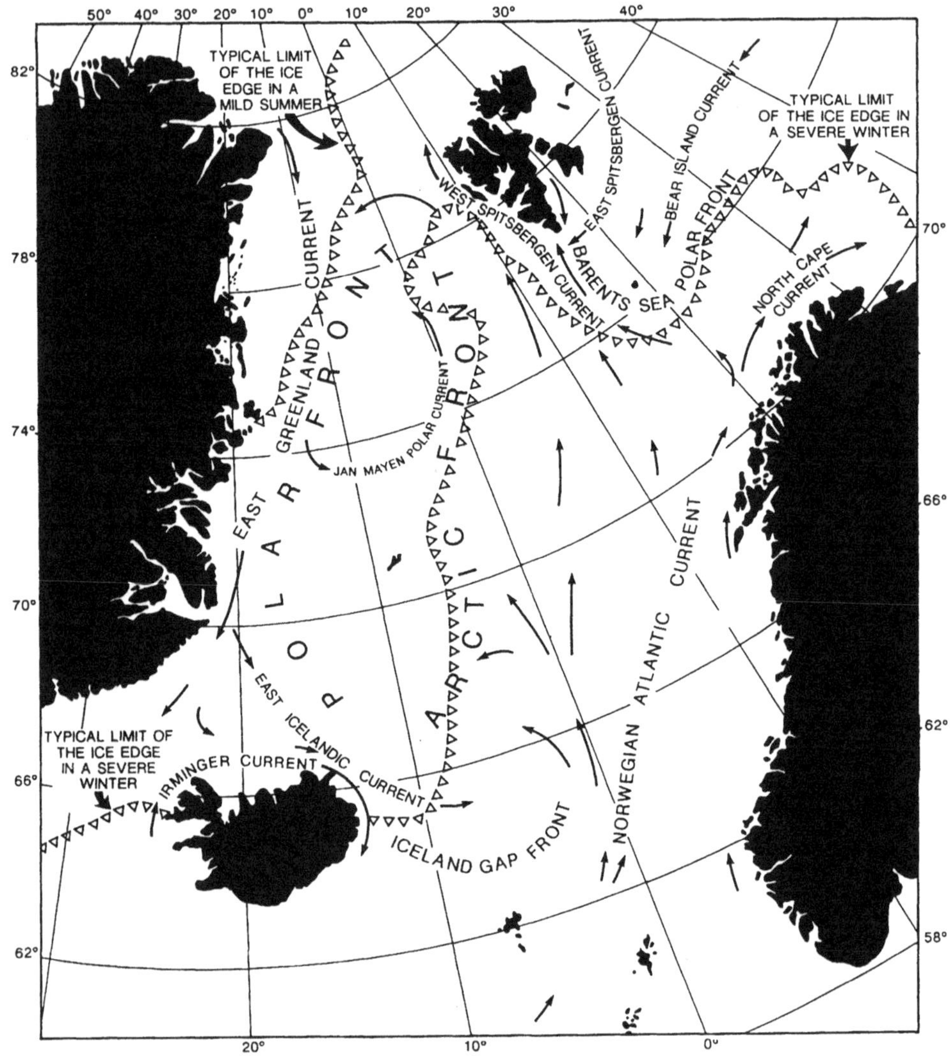

Figure 1. Overview map of the Greenland, Iceland and Norwegian Seas (GIN-Seas) showing the main surface circulation, oceanic fronts and sea ice boundaries (Modified from Hurdle 1986).

MATERIAL AND METHODS

Undisturbed surface sediment samples were collected with a vented box corer. The upper 0.5 cm of the sediment was scraped for use in this study. Additional checks that the samples reflect the actual sediment surface as closely as possible, were done by visual inspection of the surface of the box-core sample and by investigating the content of living benthic foraminifera by using the rose bengal staining method (Lutze, 1964). Samples which contained few stained benthic foraminifera and/or appeared disturbed by visual inspection were not used. Although some disturbance of the uppermost sediments normally take place during coring and handling of the coring device on ship, we consider the samples to closely represent the actual sediment surface.

The samples were stored in rose bengal stained alcohol, wet sieved on a 125 μm and 63 μm screen and dried. Foraminiferal counts and IRD-estimates were performed on the >125 μm fraction. IRD is defined as grains of quarz and feldspar in the >125 μm fraction. Samples from the upper slope and shelf areas where current winnowing of glacial sediments occur were not included. Calcium carbonate analyses were performed on a LECO carbon analyzer. Before isotope analyses, monospecific samples of planktonic foraminifera were ultrasonically cleaned in methanol and roasted in vacuo for 40 minutes at 380 °C. Isotope measurements were performed on a Finnigan MAT 251 mass spectrometer fitted on line to a reaction line following the description of Shackleton et al. (1983). Repetitive measurements of an internal carbonate standard gave a reproducibility of ± 0.06 ‰ for $\delta^{13}C$ and ± 0.07 ‰ for $\delta^{18}O$, reported at 1σ.

Although the samples are believed to be representative of the sediment surface, the actual age of the sediments is probably variable. Due to bioturbation and variable sedimentation rates they contain an integrated record of sediments deposited over a variable amount of time. There are no direct studies of bioturbation mixing rates in the GIN-Seas. Studies from other areas indicate that the bioturbated mixed layer vary in size from 2 to 8 cm (Berger and Heath, 1968; Peng et al., 1979; Jones and Ruddiman, 1982; Bard et al., 1987). Since the sedimentation rates in the studied area may vary from 2-3 cm ky^{-1} in the pelagic realm to 10-15 cm ky^{-1} in the hemipelagic sediments on the Norwegian continental margin (Kellogg et al., 1978; Jansen et al., 1983; Johannessen, 1987), a tentative mixed layer of 4 cm would indicate that the surface sediments vary in age between 400 and 2000 years. The result of this process is that our measurements represent an integrated record of some hundred to up to 2000 years, depending on the core location. The youngest sediments are probably found in the eastern part of the study area at depths between 600 and 1000m where high sedimentation rates occur (Jansen et al., 1983).

Some isotopic and temperature variability of the upper water masses over the last hundred to thousand years may be responsible for inhomogeneity in the sediments. Although the Holocene sea-surface environment in the area has been relatively stable during the last 9,000 years (Jansen and Bjørklund, 1985), there have been important shifts in the positioning of the oceanic fronts as shown by Lamb (1977) for the Little Ice Age, and by Karpuz (1993) for the Holocene Climate Optimum 8-6 kyrs ago. The exact amount of hydrographic variance is variable, but variations on the order of a few degrees near frontal boundaries have been identified in the Holocene (Karpuz 1993).

PLANKTONIC FORAMINIFER DISTRIBUTION

The sample locations where foraminiferal counts were performed are marked on the distribution maps (Figs. 2-5). They cover the main water masses and fronts of the study area, and provide enough geographical control to identify details of distribution patterns related to the fronts. Faunal counts are presented as maps showing percentage distribution relative to the total planktonic foraminifer content. Also marked are fronts and sea ice limits. The maps were constructed using the UNIX based GMT software package which contours the data with the blockmean method.

Neogloboquadrina pachyderma sinistral

Maximum relative abundace of this species is found in the western part of the area North and Northeast of Iceland, where more than 90% of the total planktonic foraminiferal fauna consists of this species (Fig. 2). The abundance pattern closely follows the position of the Arctic Front, which lies close to the 80% isoline. A slight minimum is observed close to the Polar Front, where a slight increase in the relative percentage of *Globigerina quinqueloba* is observed (see below).

This pattern has some important consequences for paleoceanographic reconstructions: Strictly speaking, high percentages of *N. pachyderma* sin. is a proxy for the Arctic Front and Arctic water masses, rather than the Polar Front, which appears connected with somewhat lower percentages. Therefore, as often practiced, using *N. pachyderma* sin. as a Polar Front proxy, makes the interpretation miss the important Arctic water mass. Hence, high percentages of *N. pachyderma* sin. is not only a proxy for Polar water masses with perennial sea ice cover, but is a proxy for Arctic water masses with only seasonal sea ice cover as well. The highest percentages are closely linked with the mixed water masses of the Arctic realm where convection occurs in the modern ocean. In terms of their foraminiferal distribution the main

Table 1. Average foraminiferal content in surface sediments under different water masses in GIN-seas.

Polar Water

	N. pach sin	N. pach dex	G. quinq.	Total
Minimum	5.00	0.00	0.00	6.00
Maximum	9283.00	226.00	347.00	9766.00
Points	12.00	12.00	12.00	12.00
Mean	1674.92	43.83	113.58	1833.00
Std Deviation	2533.62	61.88	111.75	2666.61

Arctic Water

	N. pach sin	N. pach dex	G. quinq	Total
Minimum	2261.00	21.00	0.00	2388.00
Maximum	66642.00	2578.00	29552.00	99169.00
Points	19.00	19.00	19.00	19.00
Mean	32448.68	917.58	4937.37	38363.16
Std Deviation	19007.05	824.52	8377.43	26377.78

Arctic Front Area

	N. pach sin	N. pach dex	G. quinq	Total
Minimum	2389.00	183.00	2933.00	5555.00
Maximum	56242.00	4455.00	61321.00	88991.00
Points	11.00	11.00	11.00	11.00
Mean	23526.36	1774.55	20709.18	46417.91
Std Deviation	13679.74	1165.68	16236.35	24454.46

North Atlantic Water

	N. pach sin	N. pach dex	G. quinq	Total
Minimum	357.00	2916.00	188.00	4104.00
Maximum	36400.00	15996.00	45416.00	89830.00
Points	9.00	9.00	9.00	9.00
Mean	9796.22	5293.33	10979.22	27665.00
Std Deviation	11678.93	4084.72	15404.96	27174.22

Norwegian Sea Coastal Water

	N. pach sin	N. pach dex	G. quinq	Total
Minimum	4.00	181.00	7.00	217.00
Maximum	4579.00	5456.00	9939.00	18071.00
Points	8.00	8.00	8.00	8.00
Mean	1543.75	2589.50	1718.87	6407.87
Std Deviation	1911.86	1619.93	3349.73	5823.89

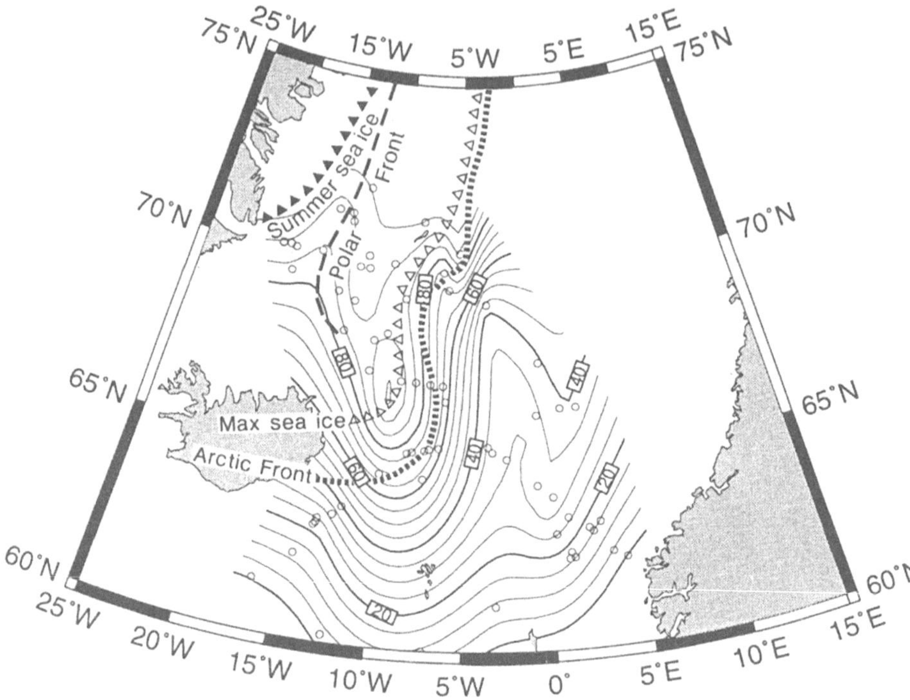

Figure 2. Percentage distribution of *N. pachyderma* sin. in surface sediments of the GIN-seas. Symbols indicate sample locations.

difference between Arctic and Polar surface waters is their foraminiferal fluxes (Table 1). As documented in Table 1, the total flux of planktonic foraminifers as well as the flux of *N. pachyderma* sin., drops dramatically as one enters Polar Surface waters, due to the lower productivity under the sea ice. The fluxes were calculated by using available sedimentation rates for late Holocene sediments from representative sediment cores from the study area (Flatøy, 1991). Due to the large variability at the boundaries between water masses, the ranges for the water masses overlap to some extent. Both mean and maximum values differ, however, very clearly.

For future studies it is important to link percentage data with flux estimates to be able to precisely define the kind of surface water masses that were present in former times. The

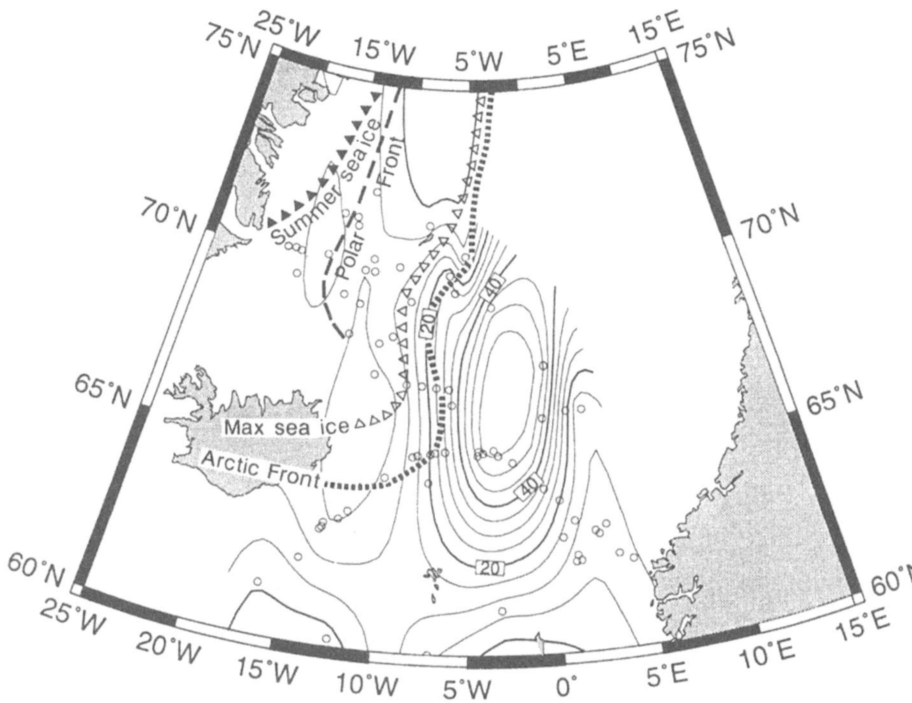

Figure 3. Percentage distribution of *G. quinqueloba* in surface sediments of the GIN-seas.

confluence of seasonal sea ice cover, deep convection and high percentages and fluxes of *N. pachyderma* sin. in the Arctic Water makes it potentially possible to define past similar convective regions. As shown below with stable isotopes, these areas can be further defined by the aid of stable isotopes.

Globigerina quinqueloba

The relative abundance pattern of *G. quinqueloba* (Fig. 3) shows a major maximum which reach 70%, East of the Arctic Front, covering the main thermal gradient towards colder waters on the western border of the warm Atlantic Water. A similar, but much weaker maximum has been noted in the North Atlantic along the front between cold and warm waters south of Iceland (Loubere, 1981), and can be traced in some of the results of Kellogg (1975) from the GIN-

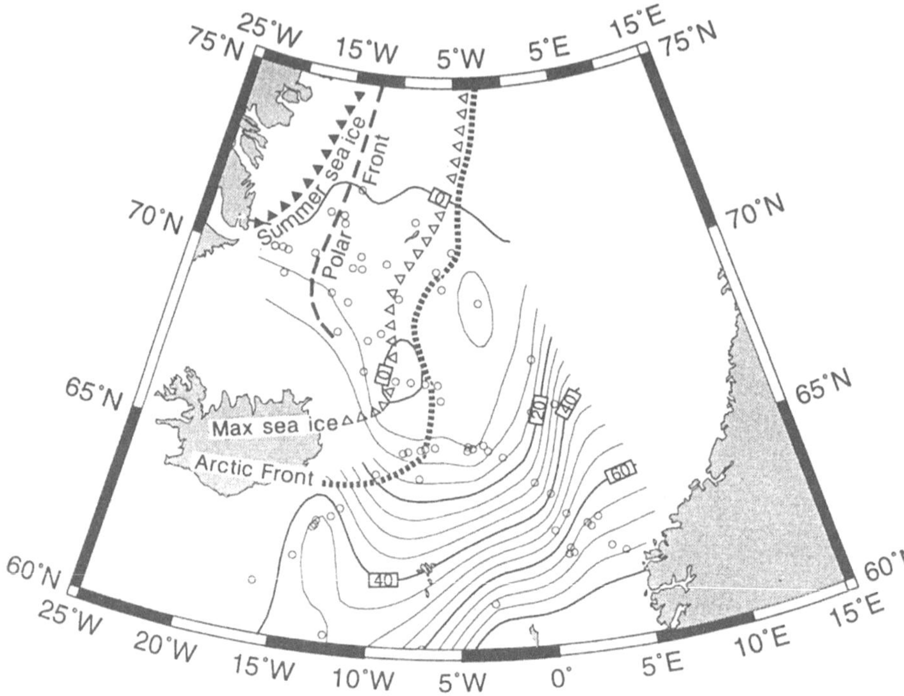

Figure 4. Percentage distribution of *N. pachyderma* dex. in surface sediments of the GIN-seas.

seas. These studies do not include the same dense sample locations and probably miss parts of the relatively narrowly defined maximum. A small maximum is also present near the Polar Front, where the relative abundance of *G. quinqueloba* is higher than 10% (Fig. 3). This small maximum in the west is, however, characterized by much smaller specimens than the larger maximum in the central parts of the study area.

Since *G. quinqueloba* is a rather small species, the size fraction used for counting may influence the results. We performed our counts in the >125μm fraction, whereas the >150μm fraction is more comonly used by the CLIMAP group and others for constructing paleotemperature transfer functions (CLIMAP project members 1976,1981; Kellogg, 1975). We tested whether this distribution pattern also is observed in the larger >150μm fraction. The large *G. quinqueloba* maximum east of the Arctic Front is dominated by specimens that are

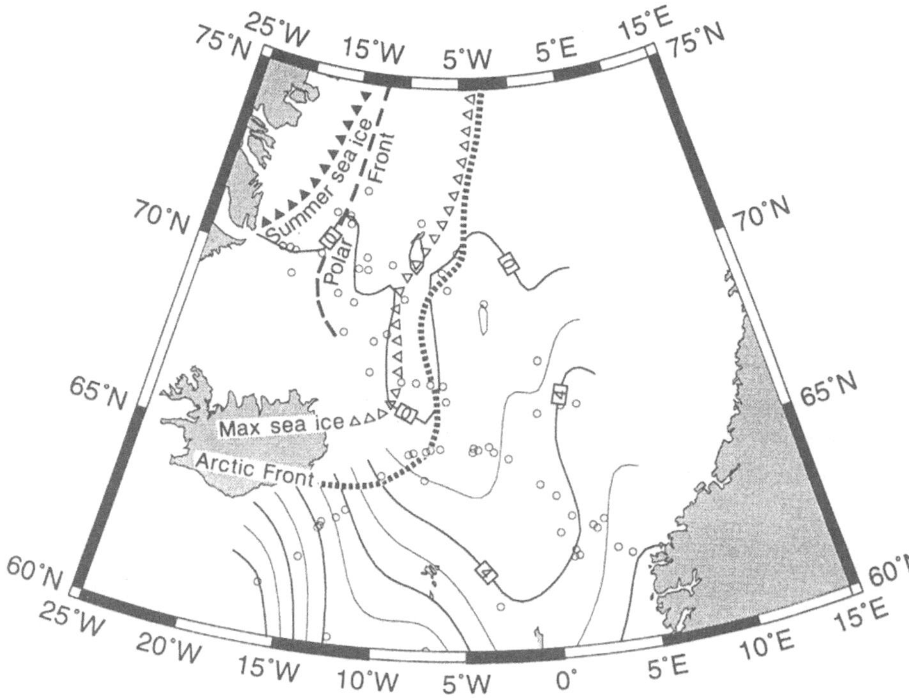

Figure 5. Percentage distribution of *G. bulloides* in surface sediments of the GIN-seas.

larger than 150µm, and almost the same percentages are recorded in the >150µm fraction as in the >125µm fraction (Johannessen, 1987). The specimens at the Polar Front are, however, so small that they mostly will fall under the >150µm fraction and will be less noted in studies using this size fraction. For paleotemperature calculations, this larger fraction may be preferable, since the bimodal distribution will obscure the temperature relation for the species. The bimodality and the maxima related to the fronts indicate that this species is related to turbulence and perhaps more productive waters at the oceanic fronts.

Our conclusion is that in areas with high foraminiferal flux, the boundary between high *G. quinqueloba* percentages and high percentages of *N. pachyderma* sin. closely defines the position of the Arctic front, whereas smaller percentages of small sized *G. quinqueloba* are related to the Polar Front.

N. pachyderma dextral

The percentage distribution of the dextrally coiled variety of *N. pachyderma* is shown in Figure 4. High percentages, more than 60%, are found in the East and are closely linked with the warm North Atlantic water in the Eastern part of the study area. Therefore the species serves as a good proxy for strong presence of inflow of Atlantic water into the Norwegian Sea.

Globigerina bulloides

Figure 5 shows the percentage distribution of *G. bulloides*. *G. bulloides* is never very abundant in the surface sediments of the GIN-seas. The highest relative abundances are found in the Southeastern part of the Norwegian Sea and in the North Atlantic. The species follows the warmest part of the North Atlantic Current, and presence of more than a few percent *G. bulloides* indicates strong warm water inflow.

In sum, detailed investigations of the present surface sediment foraminiferal distribution reveal a more diverse distribution pattern than earlier studies. This allows a distinction between Atlantic, Arctic and Polar surface water masses which are important in determining surface environments and potential convective regions. Also a more precise definition of the Polar and Arctic Fronts and areas of permanent and seasonal sea ice cover appears possible with the new data. The presence of a large *G. quinqueloba* maximum documents three major faunal assemblages in the study area. This is in contrast to only two major assemblages documented in previous investigations of high latitude planktonic foraminiferal faunas (Kellogg 1975). This provides additional resolution to paleotemperature estimates (Johannessen 1987).

OXYGEN ISOTOPES

The stable oxygen isotope distributions were measured on *Neogloboquadrina pachyderma* sinistral and dextral coiling, *Globigerina bulloides* and *Globigerina quinqueloba* (Johannessen et al. subm). The general trends are similar for the different species, and inter-specific differences can be attributed to differences in vital-effects and to season of calcification (Johannessen et al. subm.). We here only present results for *N. pachyderma* sin. since this species covers the whole study area.

The $\delta^{18}O$ record of planktonic foraminifers generally describes the surface current system, as well as the different water masses and fronts situated within the GIN-Seas (Figure 6). The high $\delta^{18}O$ values from the Arctic Water in the Iceland Sea (Figure 6) reflect the cold Arctic Water

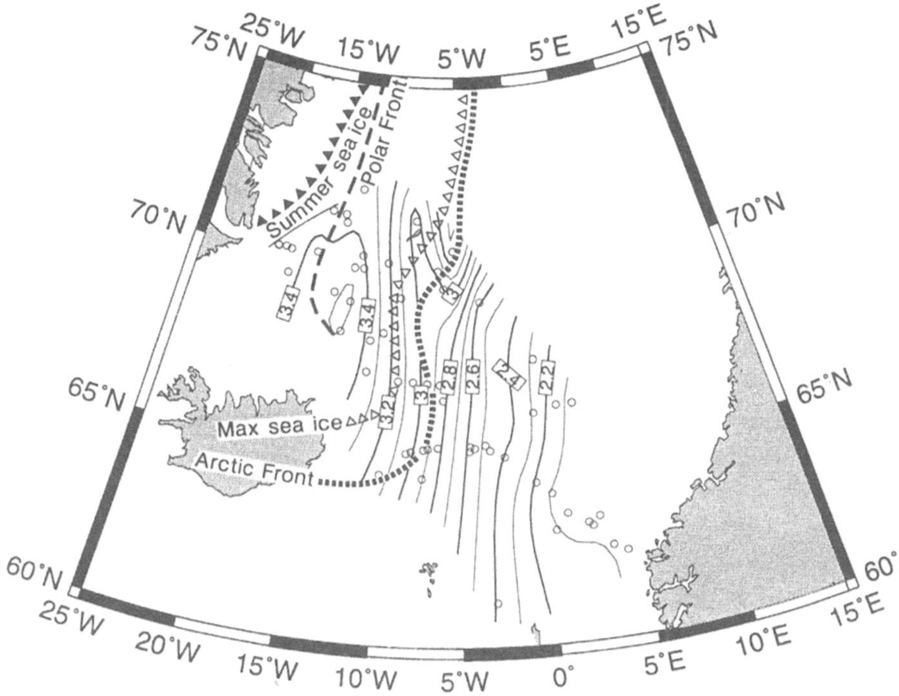

Figure 6. Oxygen isotopic distribution of *N. pachyderma* sin. in surface sediments of the GIN-seas. (From Johannessen et al. subm.).

which is a mixing product of salty North Atlantic water with cold, less salty Polar Water with relatively low oxygen isotopic values of the East Greenland Current. In the western part of the area the lighter oxygen isotopic values of *N. pachyderma* sinistral, relative to Arctic Water, reflect the low $\delta^{18}O$-values of less saline Polar Water. The lightest isotopic values are from the North Atlantic Water in the southeastern part of the area where the Norwegian Atlantic Current is warmest and saltiest, reflecting the warmer temperatures in this water mass.

The trend towards lighter $\delta^{18}O$ in the westernmost part of the area (Figure 6) defines the transition from the Arctic Surface Water to Polar Surface Water, i.e. the Polar Front. It is worth mentioning that this light isotopic excursion in $\delta^{18}O$ is recorded by *N. pachyderma* sin. despite the lower temperatures in Polar waters (as low as -0.7 ^{o}C close to the permanent sea ice cover), which works in the opposite direction.

The Arctic Front is well defined by the $\delta^{18}O$-distribution. The strong gradient in $\delta^{18}O$ across the transition between the Arctic Water and the North Atlantic Water marks the area were the Arctic Front is situated in the present Norwegian Sea (Figure 1). This points to a good potential for reconstructing the different water masses and associated fronts based on $\delta^{18}O$, particularly in conjunction with foraminiferal census counts and flux estimates.

Arctic Surface Water has a $\delta^{18}O$ of sea water close to 0 ‰ (Johannessen et al. subm.). This is rather constant due to constant salinities in Arctic water, and the range in Arctic Water $\delta^{18}O$ thus reflects the temperature range. Based on calculations of predicted equilibrium calcite $\delta^{18}O$ (Johannessen et al. subm.), planktonic foraminifera living in this water should have the heaviest $\delta^{18}O$ of the entire region. As expected, the heaviest isotopic values for *N. pachyderma* sinistral are found in connection with this water mass. The important Arctic water mass is hence uniquely defined by gradients in $\delta^{18}O$ on both the warm (Atlantic) and cold (Polar) side.

A comparison between calculated $\delta^{18}O$ equilibrium values and measured foraminiferal $\delta^{18}O$ show that all the major species of the GIN-sea predominantly calcify within the uppermost 30m of the water column in the study area. Hence their $\delta^{18}O$-values reflect the combined isotopic effects of the salinity and temperature of the surface layer (Johannessen et al. subm.). A clear separation in $\delta^{18}O$-values between sinsitral and dextrally coiling *N. pachyderma* in the warm waters, is best explained by seasonal separation of calcification, whereby the dextrally coiling form calcifies during summer, and the sinistral form calcifies during spring (May/June) and/or autumn (Johannessen et al. subm.).

CARBON ISOTOPES

We focus here on the carbon isotopic composition of *N. pachyderma* sin., and refer to Johannessen et al. (subm.) and Ravelo et al. (in prep.) for more detailed information and interpretations.

The carbon isotopic distribution of *N. pachyderma* sin. is shown in Figure 7. Two distinct $\delta^{13}C$ trends are found across the study area. One trend, from low to high $\delta^{13}C$ values, marks the transition zone from Polar Water to Arctic Water, across the Polar front area. The second trend, from high to low $\delta^{13}C$ values, marks the transition from the Arctic Water to the North Atlantic Water, crossing the Arctic front. As for $\delta^{18}O$, the heaviest values are confined to the Arctic water mass, and the major water masses and the fronts that delimit them can be clearly

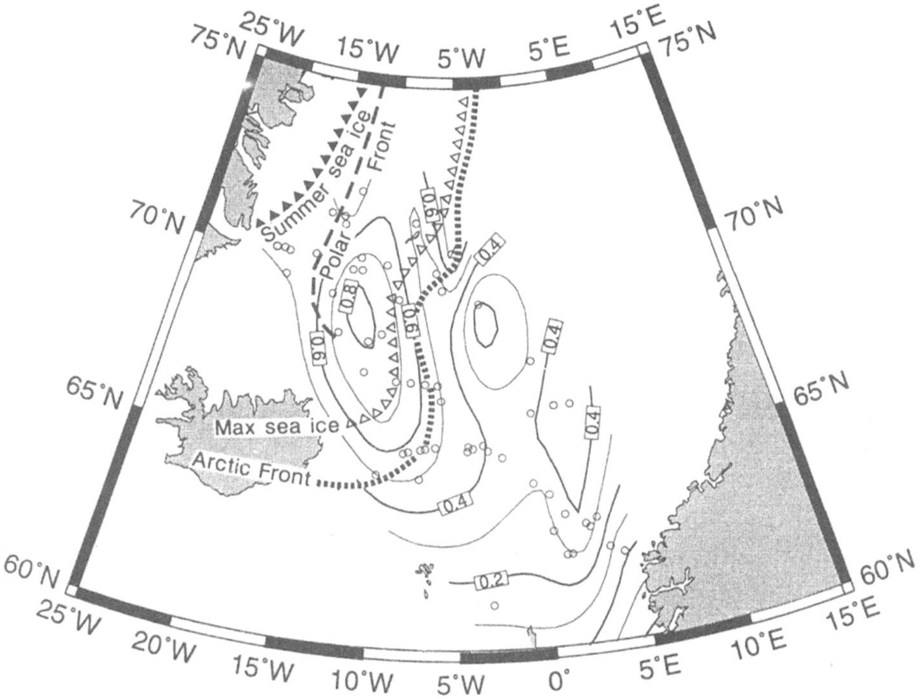

Figure 7. Carbon isotopic distribution of *N. pachyderma* sin. in surface sediments of the GIN-seas. (From Johannessen et al. subm.).

depicted by the $\delta^{13}C$ of planktonic foraminifers.

The different common species all record the $\delta^{13}C$ gradient between Atlantic and Arctic waters (Johannessen et al. subm.), and it appears as if these isotopic gradients are a robust phenomenon of the surface waters in the region. In order to link the foraminiferal $\delta^{13}C$ trends from surface sediments to the chemistry of the surface ocean, we have performed an extensive survey of carbon isotope variability in the water column in the GIN-seas (Johannessen et al. subm., Ravelo et al. in prep.). As seen in Figure 8, we also observe similar $\delta^{13}C$-gradients as those measured in foraminiferal calcite, in the $\delta^{13}C$ of total dissolved inorganic carbon (ΣCO_2) in the surface sea water. The heaviest $\delta^{13}C$-values are observed in the ASW both in foraminiferal calcite and in the water column. The causes for these gradients are discussed in

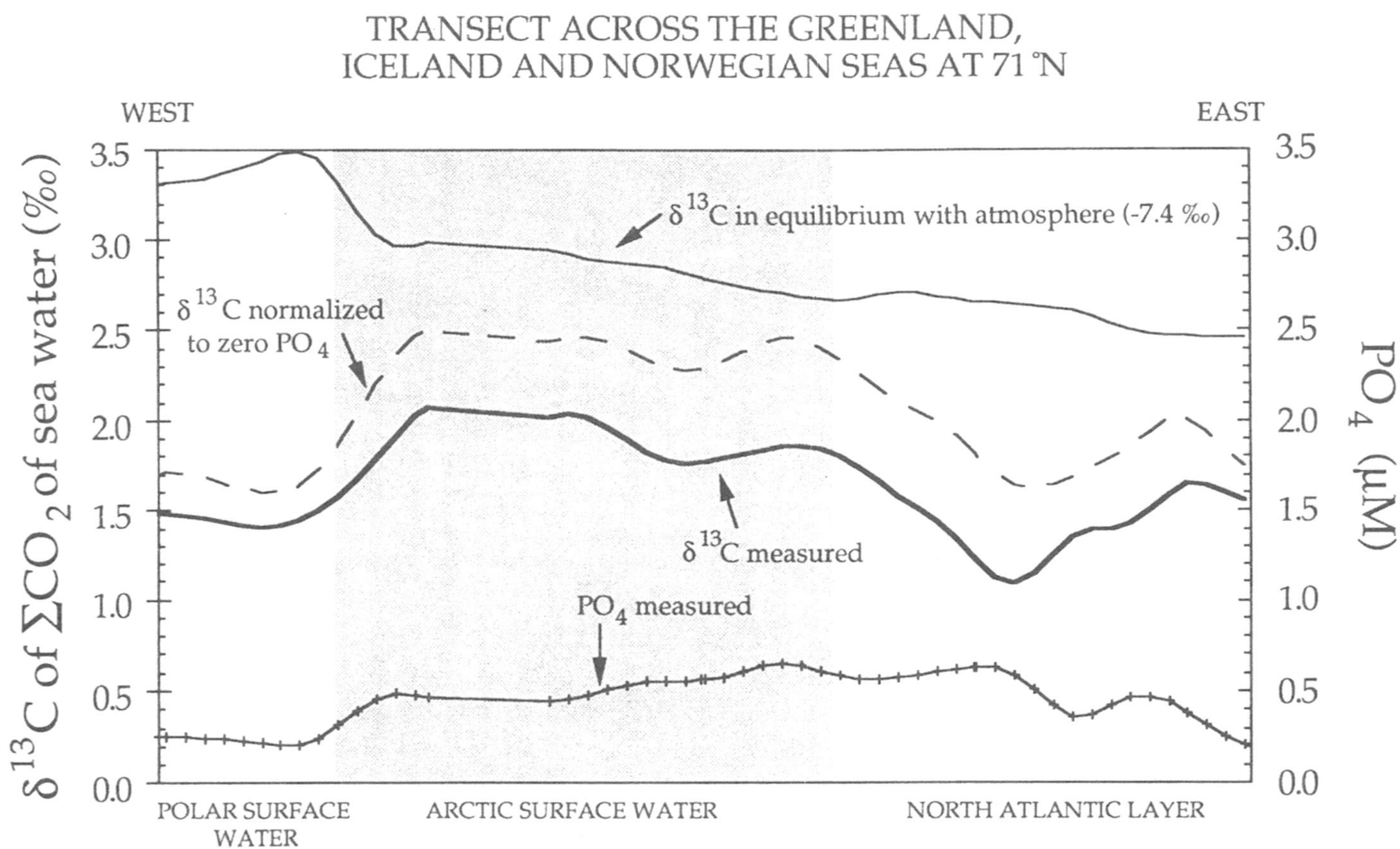

Figure 8. Phosphate content of the sea surface water along transect A-A', plotted together with $\delta^{13}C$ of ΣCO_2, $\delta^{13}C$ phosphate normalized (adjusted to zero phosphate) and $\delta^{13}C$ predicted. The predicted $\delta^{13}C$ values for surface water are calculated by using the empirical fractionation between CO_2 (aq) - HCO_3^- determined by Mook [1974] and a $\delta^{13}C$ of -7.4 ‰ (year 1982) for the atmosphere [Keeling and Revelle, 1986]. (From Johannessen et al. subm.).

detail by Johannessen et al. (subm.), and Ravelo et al. (in prep.), and we refer to these papers for more detailed documentation and discussion. Our conclusion is that the gradients cannot be solely explained by variations in nutrient content (i.e. biological production) because the gradients in surface water $\delta^{13}C$ are decoupled from variations in the nutrient content of surface waters (Fig. 8) (Johannessen et al. subm., Ravelo et al. in prep.). Thus, to a large degree, the gradient between Atlantic and Arctic waters reflects temperature controlled isotopic fractionation due to partial isotopic equilibration between atmospheric CO_2 and the bicarbonate of surface waters. This process would increase the $\delta^{13}C$ of the colder water. The decrease in $\delta^{13}C$ in Polar waters appears related to less isotopic equilibration, due to the reduced gas exchange in the perennially ice covered Polar waters and/or a different "non-Redfield" combination of nutrients and $\delta^{13}C$ in Polar waters (Ravelo et al. in prep).

CALCIUM CARBONATE DISTRIBUTION

A compilation of the carbonate distribution in surface sediments of the GIN-seas is shown in Figure 9. The highest percentages are found in the central basin of the Norwegian Sea. This is also the deepest part of the area, which indicates that carbonate dissolution is not a prominent feature in the main parts of the GIN-seas, except on some shelf and slope areas, where strong local dissolution has been described (P.I. Steinsund, pers. comm. 1990, Flatøy 1991). High sedimentation rates in hemipelagic environments lead to dilution and masking of carbonate in these environments and are responsible for the low carbonate percentages in the eastern part of the study area, off Norway. On the other hand, if the data are recalculated as carbonate flux, there are similar carbonate accumulation rates in these areas as in the central basin.

Along the Arctic Front, where dilution is unimportant, the carbonate values drop off from more than 40% to between 20 and 40% clearly showing that high carbonate percentages follow the warmer water masses. As seen in Figure 1, there is a branch of Atlantic water penetrating into the northern part of the Iceland Sea close to Jan Mayen Iceland. This forms the northern part of the Iceland Sea gyre, which is closely mimicked by the enhanced carbonate values in the northern part of the Iceland Sea.

Further west, the carbonate content is further reduced at the Polar Front, and Polar waters have carbonate contents less than 10%. Because carbonate content is influenced by other factors than surface water productivity (dilution, dissolution) carbonate content is less accurate than other proxies in reconstructing the past environments of the region. In a qualitative way, however, the results document that carbonate content is strongly related to surface water masses. Dilution can be compensated for in records which have a good time control which allows calculation of

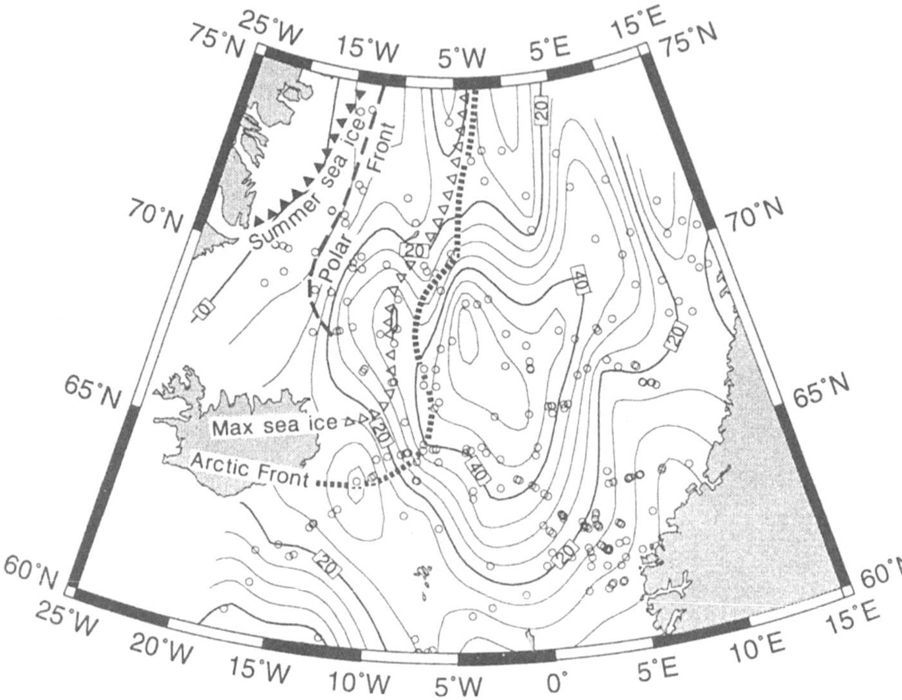

Figure 9. Distribution of calcium carbonate in surface sediments of the GIN-seas.

carbonate fluxes. This will give a more accurate picture than percentage data, in particular in high latitude areas where dilution by Ice Rafted Detritus (IRD) is common.

ICE RAFTED DETRITUS (IRD)

Sedimentation of Ice Rafted Detritus is at present confined to the Northwestern part of the study area, as documented by the distribution pattern in Figure 10. The main increase appears related to the Polar Front, while the Arctic Front and Arctic waters appear to be low in IRD, indicating that icebergs in general follow the East Greenland Current moving along the Greenland margin instead of drifting out to the more central parts of the study area. A second maxium is, however, observed in a few samples along the Arctic Front in the North (Fig. 10). This is probably related to some iceberg drift following the Greenland Sea gyre North of Jan Mayen Island. The Greenland Sea gyre therefore appears more prone to disperse icebergs eastward than the Iceland Sea gyre. In general, sedimentation of IRD reflects Polar water masses. This applies, however, to the present situation with only minor iceberg calving from the Greenland Ice Sheet and in Arctic Canada, and the situation may be different in past environments with more extensive calving and also larger icebergs which will drift further away from their sources before melting.

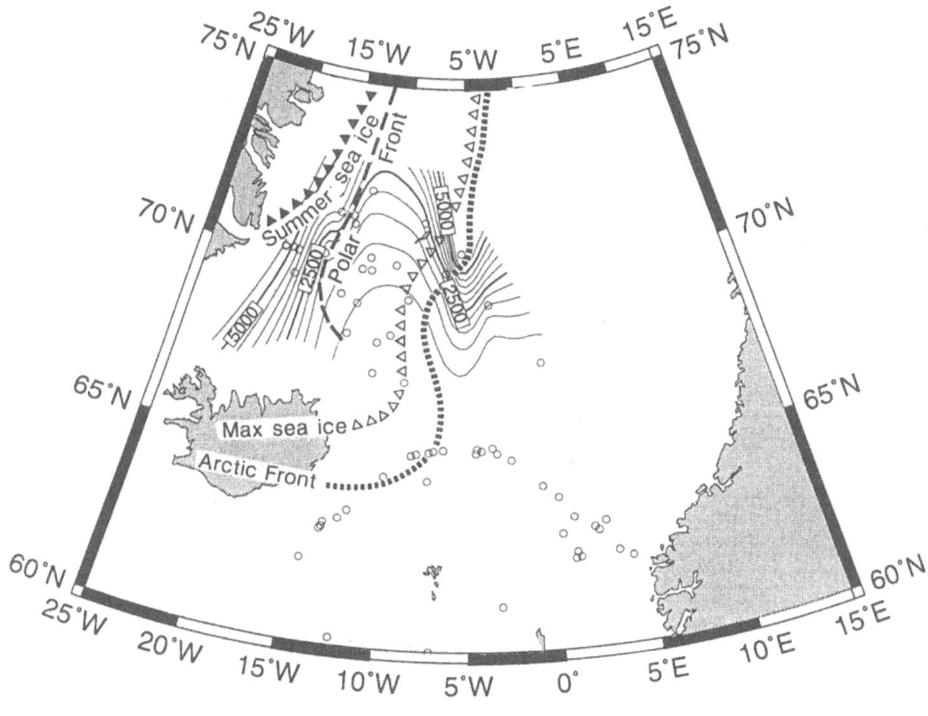

Figure 10. Distribution of IRD in surface sediments of the GIN-seas. Units are number of grains per gram of dry sediment.

DISCUSSION AND IMPLICATIONS

The detailed mapping of paleoceanographic proxies in surface sediments of the GIN-seas shows that fronts and surface water masses can be defined in more detail than before. This is the result of both the large number of samples in a relatively small, but critical area, and by using a multi-parameter approach which provide several criteria for defining a water mass uniquely and distinguish it from other water masses.

Combining foraminiferal $\delta^{18}O$ and $\delta^{13}C$ data shows that the major surface water masses in the area can be uniquely defined isotopically (Fig. 11). Combining these with foraminiferal census data enables a precise definition of both the Arctic and Polar fronts, as well as Polar, Arctic and Atlantic surface water masses. Of particular importance is the ability to define the Arctic water mass and distinguish this from the Polar water mass. Cold water areas with perennial sea ice cover can thus be distinguished from those with minor and only seasonal sea ice cover. Equally important is the combined use of $\delta^{18}O$ and $\delta^{13}C$ for reconstructing potential convective regions such as those in Arctic waters in the present Iceland Sea. These improvements may be particularly useful for synoptic map reconstructions of past environments where geographical gradients which provide evidence of fronts and water mass boundaries similar to those described in our study, may be evaluated and interpreted in similar ways.

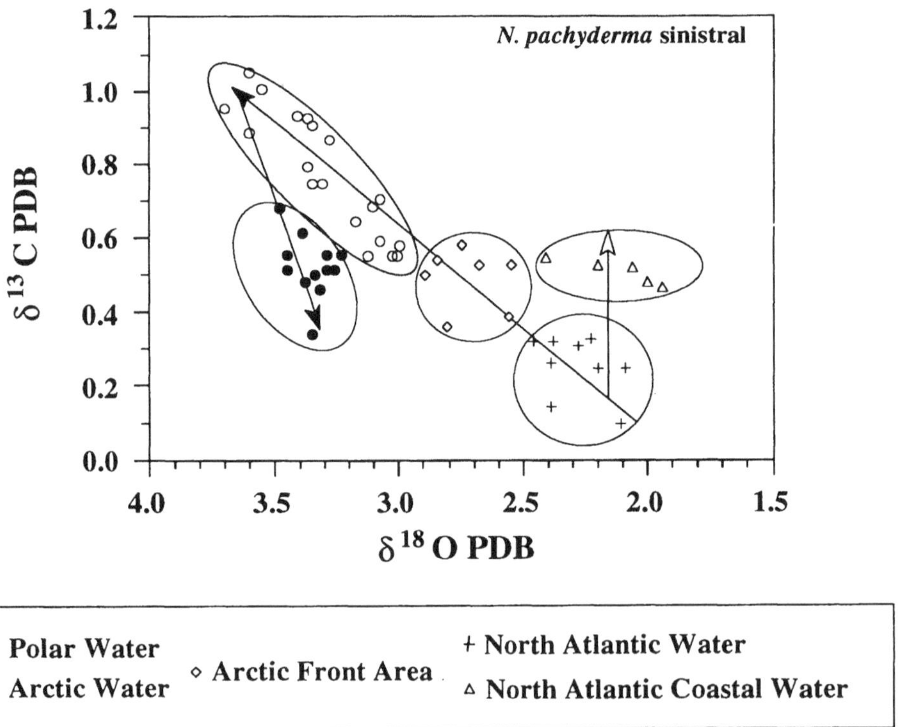

Figure 11. Isotopic zonation based on the $\delta^{18}O$ and $\delta^{13}C$ of *N. pachyderma* sin. in surface sediments of the GIN-seas.

Table 2 summarizes how the proxies we studied relate to water masses and fronts in the present GIN-seas. The internal variability in specific water masses is sometimes large, and this will lead to mean values that are not so clearly distinguishable. Hence, using the maps, to track gradients and trends, in addition to these criteria will aid the interpretations.

The Polar Front and Polar waters which have reduced salinities and long ice-covered seasons are characterized by low foraminiferal flux, less than 10% carbonate, often lower $\delta^{18}O$ and most times lower $\delta^{13}C$ than in more ice free waters, and may have a contribution of small specimens of *G. quinqueloba*, in addition to the dominant *N. pachyderma* sin.

The Arctic Front and Arctic waters which are a mixture of Polar and Atlantic waters and convect to intermediate depths during winter, are characterized by relatively high foraminiferal flux, 20-40% carbonate, very high $\delta^{18}O$ and $\delta^{13}C$ values and almost complete dominance of *N. pachyderma* sin. On the warm side of the Arctic Front there is a very strong maximum in the relative abundance of *G. quinqueloba*.

Atlantic waters are characterized by high foraminiferal flux, high (more than 40%) carbonate content in pelagic environments, relatively low $\delta^{18}O$ and $\delta^{13}C$ values, and dominance of *N. pachyderma* dex., in addition to significant presence of *G.bulloides*.

As an example of the implications of our results, we use these criteria on two published paleoclimatic time slices, the Last Glacial Maximum (LGM) and the Younger Dryas:

Time slice reconstructions of $\delta^{18}O$ and $\delta^{13}C$ during the LGM (Keigwin & Boyle 1989, Sarnthein et al. 1992) document strong gradients between the Northeastern North Atlantic and the GIN-seas. The highest $\delta^{18}O$ and $\delta^{13}C$ -values are recorded in a rather extensive area in the Norwegian Sea. Gradients in $\delta^{18}O$ are related to temperature and salinity variations. Today, the highest values are found in seasonally open waters of the Arctic type, and we may postulate that similar gradients at the LGM may indicate presence of a front similar to the Arctic Front. This argument is further strengthened by the carbon isotopic gradient. It is on seasonally open waters that we today find the highest values of $\delta^{13}C$, and these high values are partially dependant upon a high enough gas exchange rate between the atmosphere and the surface ocean to allow a partial isotopic equilibration. It is unlikely that similar gradients with a clear $\delta^{13}C$ maximum could occur if the waters in the GIN-seas were perennially sea ice covered at the LGM, as proposed in the CLIMAP reconstruction (CLIMAP project members 1976,1981). The isotopic results hence indicate an inflow of surface waters to the Southeastern Norwegian Sea and seasonally open waters akin to present day Arctic waters in the Norwegian Sea during the LGM. Relatively high foraminiferal flux is described from the eastern Norwegian Sea

Table 2. Summary of stable isotope, foraminifer and IRD-data for surface sediments under different water masses in the GIN-seas.

Parameters	Polar Water	Arctic Water	Arctic Front Area	North Atlantic Water	Norwegian Sea Coastal Water
% N. pachyderma sin.	84.35-96.13	81.53-99.5	22.3-76.1	2.9-47.6	1.8-44.6
% N. pachyderma dex	0.32-4.84	0.5-4.4	1.1-15.7	5.95-63	33.8-83.5
% G. quingueloba	1.66-13.61	0.0-15.76	22.6-71	3.2-51.8	3.2-14
% G. bulloides	0.00-0.00	0.0-3.46	0.0-4.5	0.4-13.5	2.5-9.9
$\delta^{18}O$ N. pachyderma sin.	2.95-3.45	3.07-3.69	2.41-3.1	2.11-2.39	1.94-2.20
$\delta^{13}C$.N. pachyderma sin.	0.34-0.61	0.65-1.05	0.16-069	0.10-0.33	0.25-0.53
No/gram N. pachyderma sin.	5-9992	4085-58560	4030-66642	736-36400	4-4579
No/gram N. pachyderma dex.	0-230	21-1946	496-4455	2916-15996	181-5456
No/gram G. quinqueloba	0-347	0-11323	7159-61321	619-45416	7-1314
No/gram Total	6-10394	4106-71829	18071-99169	8486-89830	217-10266
No/gram Ice rafted detrius	221-10090	0-511	0-0	0-0	0-0
No/cm*2/ka N. pachyderma sin.	7-29976	5719-175680	5642-199926	2208-22394	6-13737
No/cm*2/ka N. pachyderma dex.	0-690	29-5838	694-13365	8748-63582	253-16368
No/cm*2/ka G. quinqueloba	0-1041	0-33969	10023-183963	867-136248	10-3942
No/cm*2/ka Total	8-31182	5748-215487	25299-297507	11880-269490	304-30798
No/cm*2/ka Ice rafted detrius	309-30270	1-1533	0-0	0-0	0-0

Remark: The flux estimates are based on the variability recorded in cores in the GIN-Seas and varies between 1.4-3 no/cm2/ka.

(Ramm 1989, T.Dokken pers. comm. 1993), but only in a narrow belt close to the continental margin. A major portion of the GIN-seas apparently experienced much denser sea ice cover than the present Iceland Sea (Veum et al., 1992). The implication of this situation is that surface waters in parts of the GIN seas apparently could be potential sources for deep waters, and that convection may have taken place in this area, accounting for the high benthic carbon isotope values from the Norwegian sea during the LGM (Veum et al. 1992).

During the Younger Dryas, this picture is more clearly developed. High $\delta^{18}O$ and $\delta^{13}C$ values are typical of large areas of the GIN-seas (Sarnthein et al., 1992; Johannessen, 1992; Johannessen, 1993; Veum et al. 1992), and high foraminiferal flux is recorded in a well developed band in the eastern part of the Norwegian Sea, documenting environments which are very close to those in the present Iceland Sea (Veum et al., 1992; E. Jansen unpubl. results). The diatom floras, however, indicate more sea ice influence than at present in Arctic waters (Karpuz & Jansen 1992; Karpuz, 1993). The distinct isotopic gradients between the Norwegian Sea and the North Atlantic points to the location of the Arctic front between the two regions. A large area covered by Arctic-like water masses could potentially lead to strong convection and deep water formation. GIN-sea benthic isotope records (Veum et al., 1992) as well as records from the North and South Atlantic (Jansen & Veum, 1991; Charles and Fairbanks, 1992) indicate that ventilated deep waters formed in the high latitude North Atlantic region during the Younger Dryas. The Arctic type of surface water masses indicated by our results are fully compatible with this situation, and point to the GIN-seas as being an important site for convection during the Younger Dryas cold phase.

ACKNOWLEDGEMENTS

Sturla Rolfsen, Odd Hansen, and Rune Søraas operated the mass spectrometer, and Rune Søraas and Karl Johan Karlsen prepared samples. Jane Ellingsen helped finalizing figures and Nalan Koc Karpuz provided us with Figure 1. Paal Wessel gave us access to the GMT software package. This study was supported by grants from the Norwegian Science Foundation (NAVF) to the POC-project and a visiting scientist grant to ACR. TJ was supported by VISTA, the Nordic Council of Ministers and Bergen University. All persons and institutions are heartily thanked.

REFERENCES

Aagaard K, Swift JH, Carmack EC (1985) Thermohaline Circulation in the Arctic Mediterranean Seas, *Journal of Geophysical Research, 90*, 4833-4846.

Alekseev AP, Istochin BV 1(956) Chart and constant current in the Norwegian and Greenland Seas, *Trans. Knipovich Polar Sci. Inst., 9*, 62-68.

Bard E, Arnold M, Duprat J, Moyes J, Duplessy, JC (1987) Reconstruction of the last deglaciation: Deconvolved records of $\delta^{18}O$ profiles, micropaleontological variations and accelerator mass spectrometric ^{14}C dating, *Climate Dynamics, 1*, 101-112.

Berger WH, Heath GR (1968)Vertical mixing in pelagic sediments, *Journal of Marine Research, 2*, 134-143.

Boyle EA (1988) Vertical oceanic nutrient fractionation and glacial/interglacial CO_2 cycles, *Nature*, 331, 55.56.

Broecker WS, Denton, GH (1989) The role of ocean-atmosphere reorganizations in glacial cycles, *Geochimica et Cosmochimica Acta*, 53, 2465-2501.

Charles CD, Fairbanks, RG (1992) Evidence from Southern Ocean sediments for the effect of North Atlantic deep-water flux on climate, *Nature*, 355, 416-419.

CLIMAP Project Members (1976) The surface of the ice-age earth, *Science* 191, 1131-1144.

CLIMAP Project Members (1981) Seasonal reconstruction of the earth's surface at the last glacial maximum, *Geological Society of America Map and Chart Series, MC-36.*

Dietrich G (1969) Atlas of the hydrography of the North Atlantic Ocean. Based on the Polar survey of the International Geophysical Year, winter and summer 1958, *ICES Outside Series.*

Flatøy A (1991) Rekonstruksjon av vannmasser, fronter og isforhold i GIN-havet. Unpubl. Masters Thesis, University of Bergen, 1991.

Hansen B, Meincke, J (1979) Eddies and meanders in the Iceland-Færø area, *Deep-Sea Research, 26*, 1067-1082.

Helland-Hansen B, Nansen F. (1909) The Norwegian Sea. Its physical oceanography based upon the Norwegian researches 1900-1904, *Reports on Norwegian Fishery and Marine Investigations, 2*, 1-359.

Hopkins TS (1991) The GIN Sea - A synthesis of its physical oceanography and literature review 1972-1985, *Earth-Science Reviews, 30*, 175-318.

Imbrie J, Boyle EA, Clemens S, Duffy A, Howard WR, Kukla G, Kutzbach J, Martinson DG, Morley JJ, Peterson LC, Pisias NG, Prell WL, Raymo ME, Shackleton NJ, Toggweiler, RJ (1992) On the structure and origin of major glaciation cycles: 1. Linear responses to Milankovitch Forcing, *Paleoceanography* 7, 701-738

Jansen E, Sejrup HP, Fjæran T, Hald, M, Holtedahl H, Skarbø O (1983) Late Weichselian paleoceanography of the southeastern Norwegian Sea, *Norsk Geologisk Tidsskrift, 2-3*, 118-146.

Jansen E, Bjørklund KR (1985), Surface water circulation in the Norwegian Sea 15 000 B. P. to present, *Boreas 14*, 243-257.

Jansen E, Veum T (1991) Evidence for two step deglaciation and its impact on North Atlantic deep-water circulation, *Nature* 343, 612-616.

Johannessen, G (1993) Paleoseanografiske tidsbilder fra GIN-havet de siste 15000 år, Unpubl. Masters Thesis, Univ. of Bergen.

Johannessen, T (1987) Resente planktoniske foraminiferer fra the Norskehavet, Islanshavet og Nordatlanteren. Taksonomi, faunafordeling og stabil-isotopsammensetning, Unpubl. Masters Thesis, University of Bergen.

Johannessen, T (1992) Stable isotopes as climatic indicators in ocean and lake sediments, Unpubl. Dr. scient. thesis, Univ. of Bergen.
Johannessen T, Jansen E, Ravelo AC (subm.), Distribution of carbon isotopes in recent subpolar and polar planktonic foraminifera: Relationship to water masses, nutrients and circulation, Submitted to Paleoceanography 1993.
Jones G, Ruddiman WR (1982) Assessing the global meltwater spike, *Quaternary Research, 17*, 148-172.
Karpuz NK (1993) Application of marine diatoms for paleoceanographic reconstructions of the Greenland, Iceland and Norwegian Seas through the last 14 ka. Dr. scient. thesis, Bergen Univ.
Karpuz NK, Jansen E (1992) A high resolution diatom record of the last deglaciation from the SE Norwegian Sea: documentation of rapid climatic changes. *Paleoceanography 7*, 499-520.
Keigwin LD, Boyle EA (1989) Late Quaternary Paleochemistry of high-latitude surface waters, *Palaeogeography, Palaeoclimatology, Palaeoecology*, 73, 85-106.
Kellogg TB (1975) Late Quaternary climate changes in the Norwegian and Greenland Seas. In: Bowling, S.A. and G. Weller (eds.), *Climate of the Arctic*, Univ. of Alaska, 3-36.
Kellogg T B, Duplessy JC, Shackleton NJ (1978) Planktonic foraminiferal and oxygen isotopic stratigraphy and paleoclimatology of Norwegian Sea deep-sea cores, *Boreas, 7*, 61-73.
Lamb WH (1977) Climate variations and Changes in the Wind and Ocean circulation: The little Ice Age in the Northeast Atlantic. *Quarternary Research, 11*, 1-20, 1977.
Loubere P (1981) Oceanographic parameters reflected in the seabed distribution of planktic foraminifera from the North Atlantic and Mediterranean sea, *J. Foram. Res.* 11, 137-158.
Lutze GF (1964) Zum Färben rezenter Foraminiferen, *Meymiana, 14*, 43-47.
Peng TH, Broecker WS, Berger WH (1979) Rates of benthic mixing in deep-sea sediments as determined by radio-active tracers, *Quaternary Research, 11*, 141-149.
Ramm M (1989) Late Quaternary carbonate sedimentation and paleoceanography in the eastern Norwegian Sea, *Boreas* 18, 255-272.
Ravelo AC, Johannessen T, Jansen E (in prep.) The Modern Distribution of ^{13}C of ΣCO_2 of the Sea Water in the GIN-Seas, manuscript in. prep.
Sarnthein ME, Jansen E, Arnold M, Duplessy JC, Erlenkeuser H, Flatøy A, Veum T, Vogelsang E, Weinelt MS (1992) $\delta^{18}O$ time-slice reconstruction of meltwater anomalies at termination 1 in the north Atlantic between 50 and 80°N, *The last deglaciation: absolute and radiocarbon chronologies*, E. Bard and W. S. Broecker (eds.), NATO ASI Series, Vol. I2, Springer-Verlag Berlin Heidelberg.
Shackleton NJ, Imbrie J, Hall MA (1983) Oxygen and carbon isotope record of The East Pacific core V19-30: Implications for the formation of deep water in the late Pleistocene North Atlantic, *Earth Planetary Science Letter 65*, 233-244.
Stefánsson U (1962) North Icelandic Waters, *Rit Fiskideildar, 3*, 269.
Swift JH, Aagaard (1981) Seasonal transitions and water mass formation in the Iceland and Greenland seas, *Deep-Sea Research, 28A*, 1107-1129.
Veum T, Jansen E, Arnold M, Beyer I, Duplessy JC (1992) Water mass exchange in the North Atlantic and the Norwegian Sea during the past 28,000 years, *Nature*, 356, 783-785.

IS THERE A RELATIONSHIP BETWEEN ATMOSPHERIC CO_2 AND MANGANESE IN THE OCEAN?

Augusto Mangini, Hans-Jörg Rutsch, Martin Frank, Anton Eisenhauer, Jörg-Detlef Eckhardt[2]
Heidelberger Akademie der Wissenschaften
Im Neuenheimer Feld 366
69120 Heidelberg
Germany

ABSTRACT. Trace element and $^{230}Th_{excess}$ profiles of sediment cores from two strategic localities yield further experimental arguments to our Mn-CO_2-hypothesis. At the Galapagos Microplate, which is supplied with hydrothermal Mn, we observe low Mn concentrations (as well as low Mn/Al and low Mn/Fe ratios) during isotopic stages 2 and 3 suggesting a longer residence time of Mn in the water column and enhanced export of Mn from this area to the glacial ocean. In the Angola Basin, west of the Congo Fan, peaks of Mn (and of Mn/Fe) coincide with peaks of $^{230}Th_{excess}$, which exceed the production flux at Terminations II and I. As ^{230}Th is immobile in the sediments, the peaks suggest that the chemistry in the water column has changed significantly during the last part of glacial stages 6 and 2. Assuming a build-up of Mn in the water column during these periods, we infer that peaks of Mn in the sediments originate from rapid precipitation of MnO_2 at the beginning of interglacials. We estimate the impact of Mn deposition on the alkalinity in the range from 3 to 10 μequ./l. This could explain a part of the rapid increase of atmospheric CO_2 observed in the record of ice cores at the beginning of interglacial stages.

INTRODUCTION

In earlier publications we argued that the periodic recycling of Mn from pelagic sediments into the water column may have played an important role for the atmospheric CO_2-budget in the Late Pleistocene (Mangini et al., 1988, 1990 and 1991). This hypothesis assumes that during (or rather at the end of glacial periods) Mn had a longer residence time and as a consequence a higher concentration in the water column. The beginning of the interglacial should then be

[2] Institut für Petrographie und Geochemie der Universität Karlsruhe, Kaiserstraße 12, D-76131 Karlsruhe, Germany

NATO ASI Series, Vol. I 17
Carbon Cycling in the Glacial Ocean:
Constraints on the Ocean's Role in Global Change
Edited by R. Zahn et al.

characterized by rapid precipitation of Mn from the water column. This precipitation is driven by the progressive ventilation of the deep ocean and occurs on a time scale of hundreds of years. Precipitation as Mn-oxide lowers the alkalinity of the ocean because of the conversion of Mn^{2+} (dissolved) to Mn^{4+} (solid). The most interesting aspect of this idea is that this process is significantly faster than the response time of carbonate in the ocean, estimated to be 2,000 to 6,000 years (Broecker and Peng, 1987; Boyle, 1988). The rapid increase of atmospheric CO_2 (as observed in ice cores) at the beginning of the interglacial stages is proportional to an increase of CO_2 in the ocean and thus nearly proportional to a decrease of carbonate (and thus of alkalinity) in the water column. The "alkalinity pump" was introduced by Keir and Berger (1983). However, the long response time of the carbonate in the deep ocean makes it difficult to explain the rapid response of atmospheric CO_2 by the ocean chemistry of carbonate alone. One possible mechanism to solve this problem was suggested by Berger (1982) (and later Berger and Keir, 1984), who assumed that build-up of shelf carbonates during deglacial transgression may deliver a considerable amount of CO_2 to the atmosphere, through decreasing the alkalinity of upper ocean waters.

Another mechanism that was proposed to explain at least a part of the rapid CO_2 increase, is recycling of Mn back and forth from the sediments. Our present knowledge about the geochemistry of Mn raises serious questions about this hypothesis, mainly because it is generally believed that the glacial ocean was not anoxic, and "Black Sea-type" conditions did not occur, at least in the major oceanic deep basins. In the earlier papers we derived our hypothesis from peaks of Mn observed at the glacial-interglacial transitions in pelagic sediments. Our model was based on the assumption that these peaks originated from precipitation of Mn^{4+} at the end of the glacials. We assumed in our hypothesis that Mn^{2+} is mobilised from the sediments during periods of lower oxygen concentration in the water column and that it then builds up in the water column. It is precipitated as Mn^{4+} when the ventilation rate of the ocean becomes faster again at the beginning of interglacials. This assumption contradicts the model proposed by other authors, who do not allow a build-up of Mn^{2+} in the water column, and who attribute peaks of Mn in the sediments to reprecipitation of Mn^{4+} at the redoxcline, which was remobilised as Mn^{2+} in the deeper sections of the sediments (c.f. Finney et al., 1988).

Because we do not fully understand the processes driving the chemistry of Mn in the water column yet, which are probably controlled by a steady state between reduction and oxidation, we focussed on sediment cores from strategic localities, for example in the neighbourhood of hydrothermal areas, which are the main sources of Mn to the ocean. Under present-day conditions Mn precipitates close to its hydrothermal sources due to its short residence time in the water column of 50 years (Weiss, 1977). In these metalliferous sediments Mn attains concentrations of up to several percent, whereas open-ocean sediments display concentrations of Mn which are often 1 or 2 orders of magnitude lower (c.f. Marchig and

Gundlach, 1982). The latter may not be true universally, e.g. the North Pacific pelagic clay province hosts abundant Mn in surface sediments which persists at high concentration to considerable depths at some sites. Following our hypothesis, Mn in the water column should survive horizontal transport to more distant areas during glacial periods. If so, we expect a more homogeneous distribution throughout the ocean and a lower flux of Mn to the sediments close to the hydrothermal sources during glacial periods. This is an assumption we can test by investigating sediment cores from hydrothermal areas.

Another strategy that we follow is the measurement of trace element profiles at high resolution in sediment cores with a good oxygen isotope stratigraphy. The aim is to find out if peaks of Mn are synchronous with peaks of other components and radionuclides, such as ^{230}Th. ^{230}Th is of special interest because it is produced by decay of uranium in the ocean at a constant rate, and therefore the flux of ^{230}Th to the sediments should be constant throughout time. We previously reported peaks of $^{230}Th_{excess}$ at the climatic transitions 6/5 and 8/7 in deep sea sediments that cannot be explained by a constant ^{230}Th flux model (Eisenhauer et al., 1990; Mangini et al., 1990). If we can discard variability of compaction of the sediments or of dilution of the $^{230}Th_{excess}$ concentration with carbonate as the cause of the peaks and, furthermore, if peaks of $^{230}Th_{excess}$ occur synchronously with the peaks of Mn, we have a strong argument for our hypothesis that the explanation for the peaks of Mn and $^{230}Th_{excess}$ has to be sought in the water column and not in the sediments. The reason is that ^{230}Th, once deposited, is immobile in the sediments, while reduced Mn is not. Thus, the $^{230}Th_{excess}$ record may trace variations of the chemistry of the ocean's water column in the past.

In this paper we present a summary of results of sediment cores that are described in more detail in publications elsewhere (Eckhardt, 1992; Schneider, 1991; Schneider and Müller, 1993; Frank et al., 1993)

$^{230}Th_{excess}$- AND Mn-PROFILES OF CORE KLH 093, GALAPAGOS MICRO-PLATE, EASTERN EQUATORIAL PACIFIC

Core KLH 093 (Figs. 1 and 2) is located in the southern part of the Galapagos Microplate (1°29,97'N, 102°03,81'W, water depth: 3259m), only a few km away from the active spreading centres and the southern triple junction of the Microplate. The sediments of this core should therefore be influenced by hydrothermal activity (for detailed information see Frank et al., 1993). The core is 290 cm long and consists of a foraminiferal ooze with a carbonate content varying between 60 and 80%. A $\delta^{18}O$ stratigraphy of the planktonic foraminiferal species *N. dutertrei* was established and is shown in Fig. 1c together with the isotope stage transitions as defined by Martinson et al. (1987). From this stratigraphy we deduced an age of

95 kyr at the core bottom and sedimentation rates for the different isotope stages that range from 2.7 to 4.9 cm/ky.

The $^{230}Th_{excess}$, as well as the Mn, Al and Fe concentrations were measured at a sampling resolution of 5 cm. The Mn concentrations range from 0.02 to 0.46% (Fig. 1a). They are in general quite low compared to the concentrations measured in metalliferous sediments, which probably reflects a higher carbonate dilution resulting from the overlying equatorial productivity zone. The Mn profile shows high concentrations in the core sections corresponding to isotope stages 1, 4 and 5 and low concentrations during stages 2 and 3. In the latter sections the Mn/Al ratio profile shows values close to the average of 0.06 derived for modern pelagic sediments (Chester and Aston, 1976). Compared to this value the enrichment of Mn in the isotope stages 1, (parts of) 4, and 5 is 6 to 12 fold. The Fe/Al ratio oscillates at levels between 2 and 3 throughout the core. Assuming the Fe/Al ratio of the lithogenous pelagic clay to about 0.47 (Chester and Aston, 1976), an only slightly variable supply of Fe from the hydrothermal centres can be deduced. The enrichment of Fe ranges between 4 and 7 and is similar to the enrichment of Mn in stages 1 and 5. Finally, the Mn/Fe ratio profile shows the same variations as the profile of Mn, thus excluding dilution effects. The average fluxes of Mn to the sediments during isotope stages 1, 4 and 5 amount to about 100 $\mu mol/cm^2 kyr$ which is about on order of magnitude higher than in an average pelagic sediment. During the stages 2 and 3 the average flux of about 10 $\mu mol/cm^2 kyr$ is comparable to that of average pelagic sediment (c.f. Mangini et al., 1985).

It was concluded that during the isotope stages 5, 4, and to a lesser extent during Stage 1 the supply of hydrothermal Mn to these sediments was higher than during the isotope Stages 2 and 3, when most of the Mn appears not to have been deposited at the core site. This missing amount of Mn may have been transported elsewhere due to a change in the bottom-water current regime, or alternatively diminished deep water oxygenation may have prevented the oxidation of Mn^{2+} to Mn^{4+}. In any event, the drastic changes coinciding with the isotope stage transitions suggest a climatically-controlled deposition of Mn. The marked peak at the beginning of Stage 1 and some amount of the increased concentration of Mn during this climatic stage must be ascribed to remobilisation and reprecipitation at the redoxcline (Berger et al., 1983). If we assume an average content of 0.2% Mn for the case of complete deposition of the hydrothermal Mn supply at this location, the standing crop missing in the sections corresponding to the isotope Stages 2 and 3 amounts to 3100 $\mu mol/cm^2$. The standing crop that can be calculated from an integration over the Mn-peak at the beginning of Stage 1 amounts to about 350 $\mu mol/cm^2$. It is clear from the comparison of the two values that this locality may have been a source of Mn for the open ocean during climatic Stages 3 and 2 (Fig. 1a). The decay-corrected $^{230}Th_{excess}$ concentration profile shows increased values in the sections corresponding to isotope stage 4 and 5, but not during Stage 1.

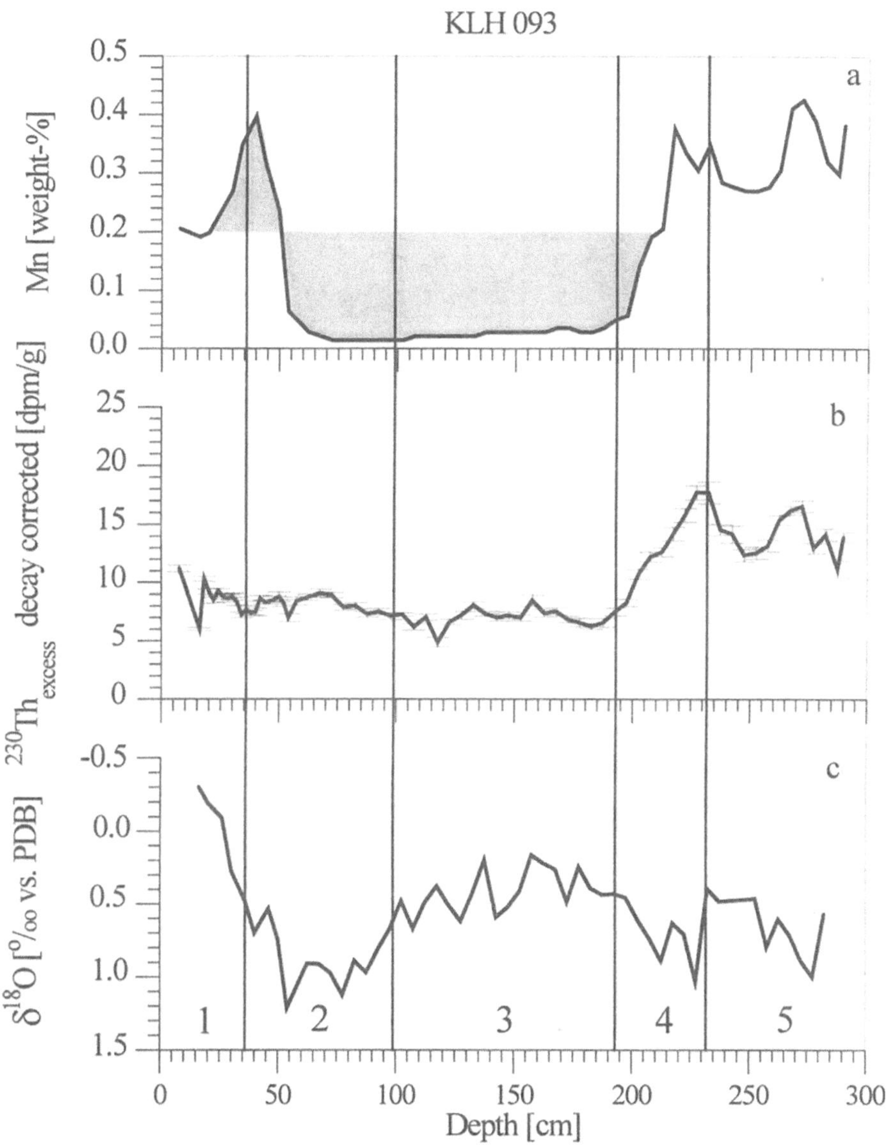

Figure 1: Profiles of Mn (a), decay-corrected $^{230}Th_{excess}$ (b) and of $\delta^{18}O$, *N. dutertrei* (c) against depth of core KLH 093 from the Galapagos Microplate locality, which is influenced by hydrothermal activity (Frank et al., 1993). The difference between the two shaded areas in the Mn profile represents the net loss of Mn during glacial stages 2 and 3. The vertical lines represent the transitions of the isotope stages following Martinson et al. (1987)

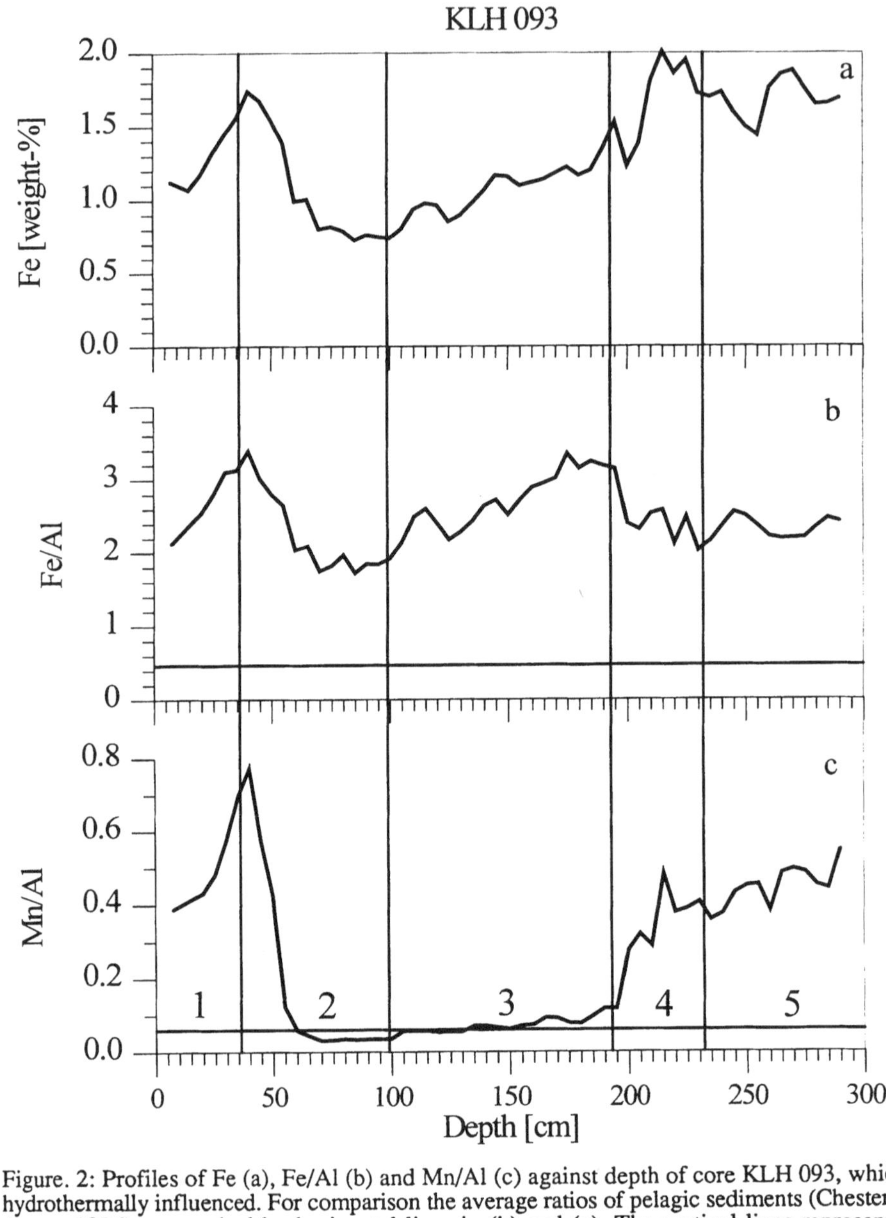

Figure. 2: Profiles of Fe (a), Fe/Al (b) and Mn/Al (c) against depth of core KLH 093, which is hydrothermally influenced. For comparison the average ratios of pelagic sediments (Chester and Aston, 1976) are marked by horizontal lines in (b) and (c). The vertical lines represent the transitions of the isotope stages following Martinson et al. (1987). The ratios of Fe/Al and of Mn/Al are significantly higher than those of the average sediments, with the exception of the Mn/Al ratio during isotope Stages 2, 3, and part of 4. In this section the ratio corresponds to the average level.

$^{230}Th_{excess}$ AND Mn PROFILES OF SEDIMENT CORE GEOB 1008-3 FROM THE ANGOLA BASIN, OFF WEST AFRICA

Core GeoB 1008-3 (Figs. 3 and 4) is located in the Angola Basin on the continental slope west of the Congo fan (6°34.9'S 10°18.8'E, water depth: 3124m). The core was intensively analyzed at Bremen University in a study of the variation of paleoproductivity in the Angola Basin (Schneider, 1991; Schneider and Müller, 1993). Sedimentation of biogenic components has been influenced by the periodic variation of the intensity of trade winds, which induce coastal upwelling of deep water rich in nutrients. The sediment accumulation rate of this core is high due to the supply of detritic material by the Congo River. A very good oxygen isotope stratigraphy of *G. ruber* was established. Together with the sedimentation rate it yields an extremely high time-resolution (Fig. 3c). The transition from stage 6 to 5 was determined at 840 cm depth, resulting in an average sedimentation rate of 6.5 cm/kyr. Accumulation rates between the substages of the oxygen isotope profile in the section from 0 - 840 cm range from 4.2 to 16.7 cm/kyr. The average $CaCO_3$ content is lower than 6% throughout the last 130 kyr, but displays two distinct peaks of up to 16% in the section representing oxygen isotope stage 6. The average biogenic opal content is 13.7%, with maximum values of up to 25% in stage 3, and a minimum value of 3.4% in stage 5.5. The average C_{org} content amounts to 2.5%, with a modulation similar to the one of opal and values of up to 4% in stage 3, and the lowest value of 1.3% in stage 5.5. Biogenic opal as well as C_{org} contents reflect the high bioproductivity at this site. The $\delta^{18}O$ profile indicates a section with disturbed sedimentation around stage 4.

We analyzed continuous sections (strips of 25 cm length) for ^{230}Th and Mn to a depth of 1200 cm. No $^{230}Th_{excess}$ was found in the section from 490 to 570 cm suggesting a higher age of the probably reworked material in this disturbed section. Uranium has an average concentration of 4.8 ppm. Enrichment of authigenic U (maximum: 3 ppm) is observed near but preceding both terminations (Fig. 4). Excess ^{230}Th for each sample is determined as:

$$^{230}Th_{excess} = {}^{230}Th_{total} - {}^{234}U,$$

when $^{238}U \leq 4.8$ ppm,

and as

$$^{230}Th_{excess} = {}^{230}Th_{total} - {}^{234}U_{average} - ((^{234}U - {}^{234}U_{average}) * (1-\exp(-\lambda * t)),$$

when $^{238}U > 4.8$ ppm), $\lambda = 9.24 * 10^{-6}\ yr^{-1}$.

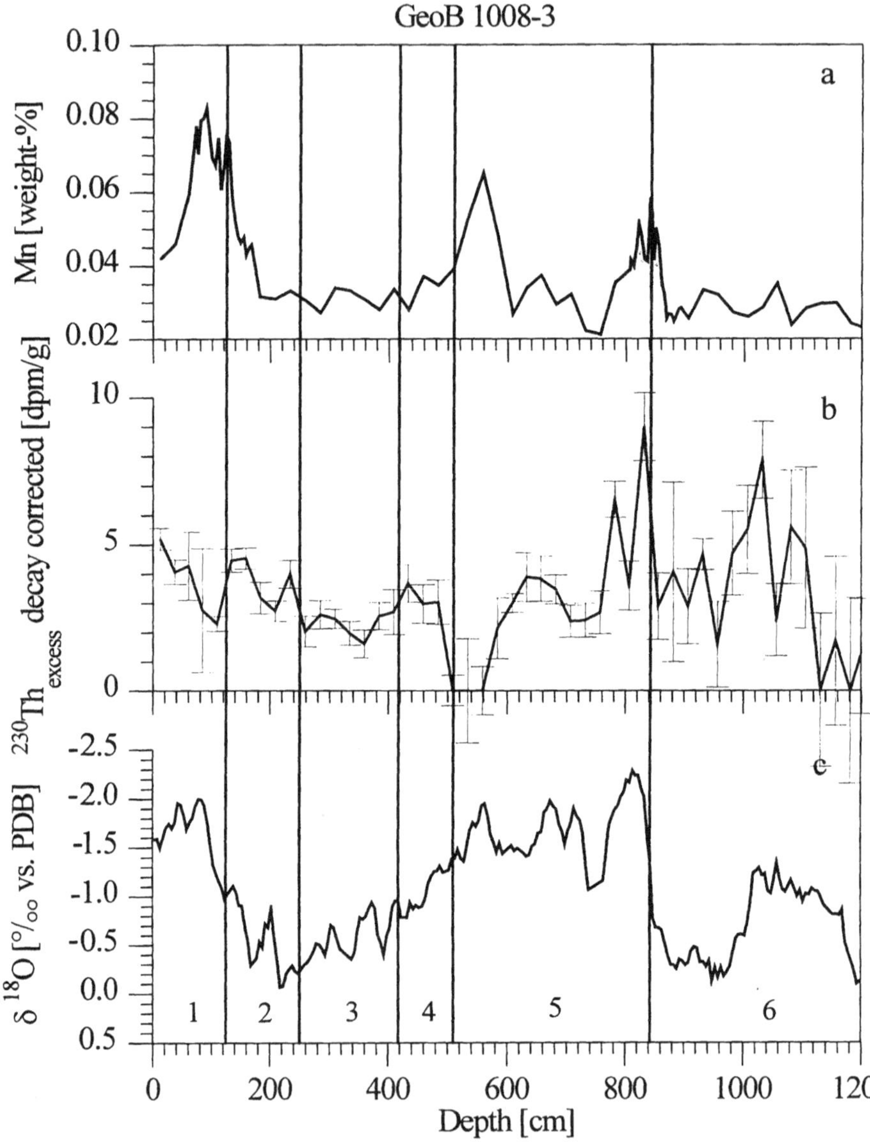

Figure 3: Profiles of Mn (a), decay-corrected $^{230}Th_{excess}$ (b) and of $\delta^{18}O$, *G. ruber* (three point running average, from Schneider (1991)) (c) against depth of core GeoB 1008-3 from the Angola Basin. The vertical lines represent the transitions of the isotope stages following Martinson et al. (1987). Outstanding peaks of decay-corrected $^{230}Th_{excess}$ coincide with peaks of Mn at the Terminations II and I. The shaded areas represent the standing crops of the peaks of Mn.

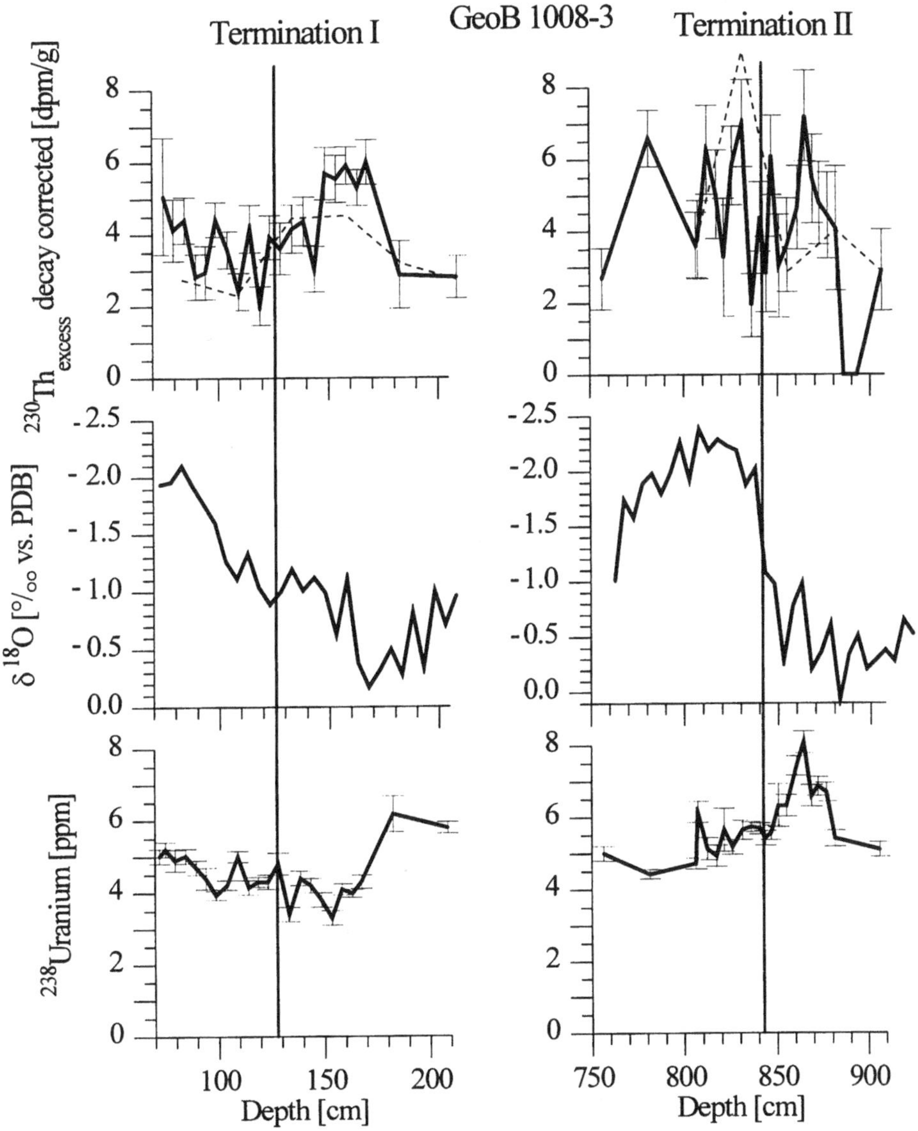

Figure 4: High resolution sections of the peaks of $^{230}Th_{excess}$ (a), of the $\delta^{18}O$ curve (b) and of the ^{238}U content (c) of the Terminations I and II in core GeoB 1008-3. The dashed curves represent the low resolution profiles. The $^{230}Th_{excess}$ peak at Termination I clearly precedes the oxygen isotope signal. The vertical lines represent the depths of the terminations following Martinson et al. (1987), that were derived from the $\delta^{18}O$ stratigraphy (Schneider, 1991).

The profile of decay-corrected $^{230}Th_{excess}$ shows a large peak in the section representing the transition from stage 6 to 5 and a smaller one near the Termination I (Fig. 3b). The core sections at the Terminations were sampled once again later and analyzed at a higher sampling resolution of 5 cm (discontinuous samples). Fig. 4 shows that the high resolution profile reproduces well the earlier profile in Termination I. However, at Termination II the high resolution profile does not reproduce in detail the pattern of the large peak measured at low resolution. This pattern suggests that the peak extends over a longer section of the core. Despite this mismatch, the low-resolution peak of ^{230}Th shown in Fig. 3 seems reliable because we observed a peak of similar magnitude in core M 16772 from the Guinea Basin (Mangini et al., 1993). Fig. 4 shows that the concentration of $^{230}Th_{excess}$ begins to rise at 170 cm (15,400 yr BP), thus preceding the Termination I $\delta^{18}O$ signal by approx. 3,000 yr. This age at 170 cm represents a non-corrected ^{14}C-age derived via comparison of the oxygen isotope curve of core GeoB 1008-3 with those from sediment cores at similar localities (Schneider, 1991). At Termination II the maximum of the low resolution signal seems to coincide with the transition from stage 6.1 to 5.5 (130,000 yr BP). The high resolution profile, however, suggests a rise of $^{230}Th_{excess}$ at the time of event 6.2 (135,000 yr BP) following Imbrie et al. (1984).

The increased sediment accumulation rate of about 15 cm/ky at the beginning of the Holocene dilutes the $^{230}Th_{excess}$ concentration to a stronger extent than at the transition from Stage 6 to 5. The fluxes of ^{230}Th (determined from the $^{230}Th_{excess}$ concentration x sediment accumulation rate x dry bulk density) corroborate the peaks. The standing crops of ^{230}Th in excess of the expected fluxes amount to 16 dpm/cm^2 (Termination I) and 27 dpm/cm^2 (Termination II), respectively. These standing crops represent the average of the peaks of the high and the low resolution profiles.

The concentration of Mn shows distinct peaks in the sections representing Terminations I and II (Fig. 3a). The standing crop of Mn in these peaks (in excess of a base line fixed at 350 ppm and 300 ppm) amounts to 160 µmol/cm^2 and 540 µmol/cm^2, respectively. The single-point spike at 560 cm is related to the reworked material in this section. Its origin is unclear and we shall omit consideration of this peak in the following discussion.

DISCUSSION AND CONCLUSIONS

It is well documented that Mn undergoes diagenetic redistribution within sediments, and there is a huge redox-based literature on this process (c.f. Froelich et al., 1979; Klinkhammer, 1980; Thomson et al., 1984; Finney et al., 1988). Our model requires remobilisation of Mn in the sediments but also the migration of Mn into the water column. A possible glacial scenario was discussed extensively in an earlier paper (Mangini et al., 1990). An increase of the Mn concentration in the water column can be achieved either when deep water becomes anoxic

(Black-Sea type conditions), or when the flux of oxygen from an oxic or suboxic water column into the sediments is less than the demand established by aerobic bacteria as they degrade organic matter. In the second case, Mn will be continuously recycled between the sediment and the water column, and the level of Mn in the water column will be a function of the flux of Mn from the sediments and of the kinetics of oxidation. This process will support the transport of (hydrothermal) Mn to areas more distant from its sources at the ridges than one would expect from the relaxation length of Mn undergoing horizontal transport in an oxic ocean.

Close to the spreading centres of the Galapagos Microplate, which probably are a source of hydrothermal Mn, we found lower Mn concentrations (and Mn/Fe and Mn/Al ratios) during stages 2 and 3, suggesting a period of export of Mn to the open ocean. This finding suggests a longer residence time of Mn in the water column of the equatorial Pacific during glacial periods, as is predicted in our model. However, we cannot evaluate how long the residence time of Mn in the water column might have been. More data from other hydrothermal areas are required to check the validity of these conclusions on a larger scale. Unpublished data of sediment cores collected near the EPR at 20°S and dated by oxygen isotope stratigraphy show maxima of the Mn concentration in interglacial stages and minima in glacial stages (Riech, pers. comm.). Such apparent climatically-controlled periodicity supports our results from core KLH 093. Similar features were observed in the Mn profiles of the Vulcan cores from the northern equatorial Pacific presented by Finney et al. (1988). These authors ascribed the peaks of Mn in the sediments to diagenetic remobilisation of Mn in the sediments and to reprecipitation of MnO_2 at the redoxcline only. If our model is correct, a major part of Mn in the peaks in open ocean sediments far from hydrothermal sources originates from precipitation of Mn from the water column, triggered either by better ventilation of the water column, or by a smaller flux of organic matter into the sediments. At Termination I there is direct evidence supporting the view that the Panama Basin was not anoxic during the last glacial. This conclusion was derived from Iodine/C_{org} ratio profiles (Pedersen et al., 1988) and is also corroborated by modelled O_2 levels derived from the $\Delta\delta^{13}C$ of benthic and planktonic foraminifera (Lyle et al., 1988). Therefore one may assume that in this area it was the flux of organic matter that controlled the diagenetic remobilisation of Mn.

The sediment core GeoB 1008-3 from the Angola Basin shows peaks of Mn, $^{230}Th_{excess}$ and of other components at the climatic transitions 2/1 and 6/5. While the peaks of Mn are "easy" to explain, because they can be ascribed to diagenetic remobilisation, the peaks of $^{230}Th_{excess}$ are not, because ^{230}Th is assumed to be immobile in the sediments. Possible explanations are that: 1) the peaks are a local phenomenon in the Angola Basin; 2) the peaks are generated by compaction of the sediment (via dissolution of $CaCO_3$) or by 3) extremely low sedimentation rates or by 4) a lithogenic component enriched in ^{230}Th; 5) peaks of remobilised authigenic uranium could have generated additional ^{230}Th in the sediments; and finally, 6) the peaks might reflect periods of enrichment of ^{230}Th in the water column. We shall consider each possibility in turn:

1) *(the peaks are a local phenomenon)*: The results of core GeoB 1008-3 are supported by data from other cores. Peaks of $^{230}Th_{excess}$ of similar magnitude (and standing crop) were observed in high resolution profiles of eight other sediment cores at the Termination II throughout the Pacific and Atlantic Ocean (Mangini et al., 1993). For example core M16772 located in the Guinea Basin displays a very similar profile of $^{230}Th_{excess}$ as that of core GeoB 1008-3. Generally, $^{230}Th_{excess}$ peaks at Termination I are significantly smaller than those observed at Termination II. They were detected in cores from the Atlantic (e.g. core 23235, Eisenhauer et al., 1990; core GPC-5, Bacon and Rosholt, 1982) but can hardly be seen in the Pacific. They are not detected in the core KLH 093. This observation seems in agreement with the results of Yang et al. (1990), who did not find anomalies of the ^{230}Th flux at the 2/1 transition in sediment cores from the eastern equatorial Pacific. The fact that peaks of Mn are seen at both Terminations, whereas the peaks of ^{230}Th are not is a major unsolved question. If one assumes that the size of the $^{230}Th_{excess}$ peaks corresponds to the duration of the events of Mn build-up (as discussed below), the smaller peaks at Termination I suggest that periods of Mn build-up were shorter and did not occur on an ocean-wide scale.

2) and 3) *(the peaks are due to dissolution of $CaCO_3$ or extremely low sedimentation rates)*: we have observed the peaks in cores with very low or no $CaCO_3$ content (e.g. Site 580, Eisenhauer et al., 1987; and core GeoB 1008-3), which excludes compaction due to carbonate dissolution as the reason for the $^{230}Th_{excess}$ enrichment. Furthermore, normalisation with ^{10}Be clearly shows the so called "^{230}Th-anomalies" (Eisenhauer et al., 1990) which originate from abnormally high fluxes of ^{230}Th relative to ^{10}Be. Because one would expect ^{230}Th to be diluted to a similar extent as ^{10}Be (assuming that variations of the production rate of ^{10}Be do not have a strong effect on the flux of this isotope to the sediments) due to its relatively long residence time in the water column of the ocean of about 400 to 1,000 years (c.f. Kusakabe et al., 1991) we can also exclude periods of extremely low sedimentation rates as the cause of these peaks.

4) *(the peaks are generated by a lithogenic component enriched in ^{230}Th)*: The concentrations of the peaks where the average $^{230}Th_{excess}$ content is exceeded amount to at least 3 dpm/g in core GeoB 1008-3, and were found to be as high as 25 dpm/g in core 169 KL from the Clarion Clipperton Fracture Zone (Mangini et al., 1993). Such a high $^{230}Th_{excess}$ content does not seem justifiable for detritus distributed to deep sea sediments on a rather global scale. Normalisation to ^{232}Th which is of lithogenic origin renders the peaks of $^{230}Th_{excess}$ even more prominent.

5) *(peaks are generated by remobilisation of authigenic uranium)*: An alternative model to explain the peaks of $^{230}Th_{excess}$ would require that authigenic uranium was incorporated in the top sections of reduced sediments at an early stage of sediment formation, followed by later release of U to the water column or to deeper sections that are still anoxic. The facts that the

peaks of $^{230}Th_{excess}$ coincide with climatic boundaries and that an enrichment of U was observed in Atlantic sediments of the last glacial stage just underneath the oxic-suboxic boundary (Thomson et al., 1990) (core GeoB 1008-3 corroborates their results, Fig. 4) warrants discussion. Assuming the standing crop of authigenic U at a level of 7 ppm (the highest value observed by Thomson et al., (1990)), and assuming a deepening of the redoxcline (and thus loss of uranium) after 10 kyr due to burn-down of organic matter, the amount of the ^{230}Th produced from authigenic U would be no more than 0.5 dpm/g. This value probably even overestimates "authigenic" ^{230}Th because U should "come and go" from the layers at the redoxcline within a period of time shorter than 10 kyr. To explain the peaks of $^{230}Th_{excess}$ (3 dpm/g in core GeoB 1008-3) one has to postulate levels of U in the sediments at least one order of magnitude higher than observed, which seems unrealistic. The mismatch between ^{230}Th and U becomes even larger at the Termination II in cores from the Pacific, where the sediments show no significant authigenic U enrichment at all. Thus, the contribution from authigenic U to the peaks of $^{230}Th_{excess}$ (if any) should be rather small.

Of the possibilities listed above, our favourite explanation at present, although speculative and controversial, is

6) *(the peaks may reflect periods of enrichment of ^{230}Th in the water column)*: We ascribe the observed peaks to a rapid deposition of ^{230}Th from the water column when the oceanic system changed from a glacial to an interglacial circulation pattern. This, however, requires that the water column (or probably the deepest layer) was loaded with ^{230}Th to a much higher concentration than today. We derive this conclusion by assuming that the peaks of $^{230}Th_{excess}$ were not caused by mobilisation in the sediment. The standing crop of "excess-$^{230}Th_{excess}$" can then be applied to hindcast the level of ^{230}Th in the glacial ocean and its glacial residence time. From the two peaks of $^{230}Th_{excess}$ in core GeoB 1008-3 we evaluate the residence time at 2,000 and 3,300 yrs, respectively. These residence times were derived from the standing crop in the peaks of $^{230}Th_{excess}$ (exceeding the average $^{230}Th_{excess}$ content) of 16 and 27 dpm/cm^2, respectively, divided by the production flux of ^{230}Th in the water column of 8.1 dpm/cm^{2}ky. Obviously, such a long residence time of ^{230}Th is controversial when compared to measured residence times as short as 20-40 yrs in the present-day ocean. However, as discussed above, other possible explanations also exhibit weaknesses. Despite this caveat, we suggest based on our data that the transfer of ^{230}Th into the sediments of the glacial ocean (especially during isotope stage 6) was not as efficient as in the present day ocean. A model showing how the fluxes, and how the concentration of ^{230}Th in sediments could have developed, when the ocean changed from glacial to interglacial mode, is shown in Fig. 5. The model suggests periods of build-up of ^{230}Th in the deep water column of 1,000 to 5,000 yr duration within glacials, followed by rapid precipitation at the climatic transitions. As shown in this figure this hypothesis predicts ^{230}Th fluxes into sediments nearly the same as that predicted by the

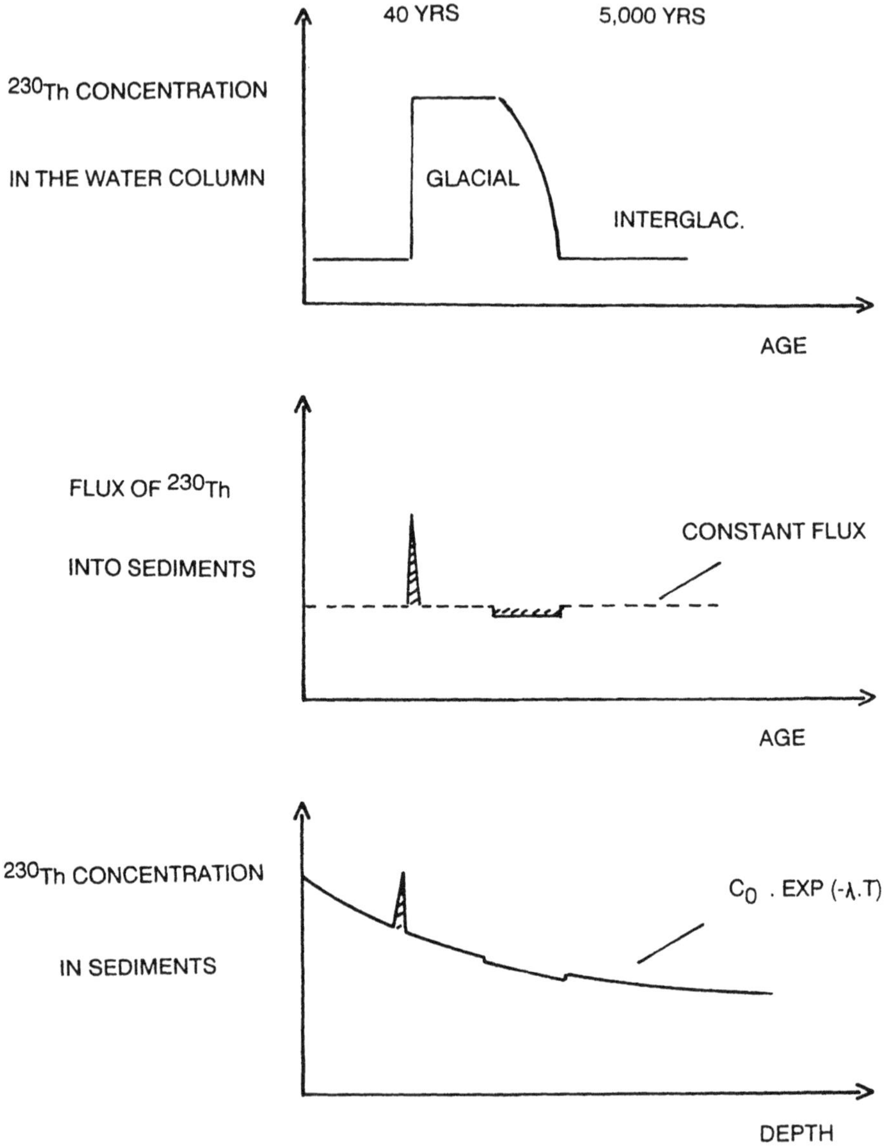

Figure 5: Scheme suggested for the development of the concentration of ^{230}Th in the water column, of the ^{230}Th flux into sediments and of the depth distribution of $^{230}Th_{excess}$ in sediments, during glacials, interglacials and at the transition from glacial to interglacials in the open ocean (from Mangini et al., 1993).

constant flux model (Francois et al., 1990), because the mean lifetime of radioactive ^{230}Th has been much longer than its residence time in the water column at any time in the past.

The rate of transfer of ^{230}Th from the water column into the sediments is assumed to be mainly controlled by adsorption to Mn- and Fe-oxyhydroxide coated surfaces of settling particles. If our assumption of a longer glacial residence time of ^{230}Th in the water column is correct, then the most unsophisticated model to explain this is to assume a longer residence time of Mn in the water column as well. To explain the build-up of the concentration of Mn in the water column we then postulate short periods of anoxia in the water column at the end of glacial stages. Their duration may be evaluated from the longer glacial residence time of ^{230}Th at 1,000 to 6,000 years.

This conclusion is in contradiction to the general opinion, that the glacial ocean was never anoxic. However, we doubt that such short periods of anoxia can be recorded (and be resolved) in sediments accumulating at rates below 1 cm/kyr. Bioturbation together with burn-down of organic matter may cancel the record. At ODP site 658 (off North West Africa) significant depletions in high resolution profiles of $\delta^{13}C$ at the Terminations I to VI were determined (Sarnthein and Tiedemann, 1990), which can be interpreted to represent short periods of extremely low oxygen concentration in the water column. We suggest that during these periods a release of Mn from the sediments occurred as a consequence of the low oxygen content in the deep ocean.

If excess Mn in the peaks originates from the water column, we may then estimate its concentration in the glacial water column at 1-3 µmol/l, based on the standing crop of Mn in the peaks (Mangini et al., 1990). Assuming that at the end of glacial periods this oxidation and precipitation of MnO_2 occurred, could this have had an impact on the atmospheric CO_2? The answer is positive because precipitation of MnO_2 leads to a decrease of the alkalinity by 2 µeq/l per µmol of Mn precipitated. Thus, a precipitation of 3 µmol Mn/l causes a drop of alkalinity of 6 µeq/l.

As modelled by Broecker and Peng (1987) and in more detail by Boyle (1988), half the increase of atmospheric CO_2 at the end of glacials can be ascribed to a rapid export of carbon from the ocean system due to a slower biological pump, while the other half is accomplished by a reorganisation of the carbonate system and by precipitation of calcium carbonate ("carbonate compensation", Broecker and Peng, 1987). Boyle evaluated that a drop of atmospheric CO_2 by 54 ppmV requires an increase in alkalinity of 61 µeq/l. Correspondingly a drop in alkalinity in the ocean leads to a proportional increase of atmospheric CO_2. In our hypothesis we suggest that a part of the drop of the alkalinity, when atmospheric CO_2 was raised by about 80 ppmV at the end of glacial stages, could have been accomplished by precipitation of Mn (and of other trace elements) from the water column. The interesting point about the Mn system is that its response is at least one order of magnitude faster than the one of the carbonate system. One can conclude that Mn adds the smaller high frequency noise (in the range of 10 ppmV) to the overall pattern of atmospheric CO_2 recorded in the ice cores, which is driven by the carbonate system. Arbitrarily limiting the maximum concentration of dissolved Mn to 5 µmol/l would constrain the maximum contribution to the atmospheric CO_2 change to about 9 ppmV.

Finally, one may ask how the peaks of Mn and of $^{230}Th_{excess}$ in GeoB 1008-3 are related to spikes of aragonite preservation recorded at the Termination I in the E.N. Atlantic, in the W.E. Pacific, in the Caribbean and on the western continental slope of India (Berger, 1977). These aragonite "preservation events" are associated with an adjustment of the carbonate in the ocean and with a smaller flux of organic carbon to the deep sea. This question cannot be answered yet, because the timing is crucial here. Maximum preservation of aragonite coincides with the maximum change of the oxygen isotope signal. As it can be seen in Fig. 4, the $^{230}Th_{excess}$ peak at Termination I apparently precedes the maximum change of $\delta^{18}O$ suggesting that ^{230}Th (and thus its carrier phase) reacted faster than carbonate to the change of chemistry in the water column. This seems reasonable, because the compensation of the carbonate system is supposed to proceed with a time constant of at least several thousands of years. In contrast, precipitation of Mn could occur on a time scale of hundreds of years. However, at Termination II ^{230}Th apparently coincides with the transition. More data are required before further conclusions regarding these relationships can be drawn.

SUMMARY

Our model suggests that during or rather at the end of glacial periods manganese had a longer residence time in the water column, and that at the beginning of interglacials precipitation of Mn from the water column ocurred. These periodic events enable transfer of Mn from the hydrothermal sources to more distant areas causing a more homogeneous distribution of Mn throughout the oceans. If this model is correct, precipitation of Mn-oxide corresponds to a lowering of alkalinity of the ocean and may have had an impact on atmospheric CO_2.

Further, we speculate that the recycling behaviour of Mn may influence the efficiency of transfer of the radionuclide ^{230}Th from the water column into sediments. We relate peaks of $^{230}Th_{excess}$ in the sediments to periods of enhanced Mn deposition at the end of glacials. However, this would require a build-up of ^{230}Th together with Mn in the glacial water column to values that contradict our present knowledge of the geochemistry of Th in the ocean. Finding an answer to this intriguing contradiction may lead to a better understanding of glacial to interglacial changes of ocean chemistry.

REFERENCES

Bacon, M.P., Rosholt, J.N. (1982) Accumulation rates of ^{230}Th, ^{231}Pa and some transition metals on the Bermuda Rise, Geochim. Cosmochim. Acta 46: 651-666

Berger, W.H. (1977) Deep-sea carbonate and the deglaciation preservation spike in pteropods and foraminifera. Nature 269: 301-304

Berger, W.H. (1982) Increase of carbon dioxide in the atmosphere during deglaciation: the coral reef hypothesis. Naturwissenschaften 69: 87-88

Berger W.H., Finkel, R.C., Killingley, J.S. and Marchig, V. (1983) Glacial-Holocene transition in deep sea sediments: manganese spike in the east-equatorial Pacific. Nature 303: 231-233

Berger, W.H. and Keir, R.S. (1984) Glacial-Holocene changes in the atmospheric CO_2 and the deep-sea record. In: Hansen, J.E. and Takahashi, T. (eds.) Climate processes and climate sensitivity. Geophys Monogr. 29, Maurice Ewing Series, vol. 5, American Geophys. Union, Washington, D.C.: 337-351

Broecker, W.S. and Peng, T.-H. (1987) The role of $CaCO_3$ compensation in the glacial to interglacial CO_2 change. Glob. Biogeochem. Cycles 1: 15-30

Boyle E.A. (1988) The role of vertical chemical fractionation in controlling Late Quaternary atmospheric carbon dioxide. J. Geophys. Res. 93: 701-714

Chester, R., and Aston, S.R. (1976) The geochemistry of deep-sea sediments. In: Riley, J.P. and Chester, R. (eds.) Chemical Oceanography vol. 6, New York, Academic Press

Eckhardt, D.J. (1992) Geochemische Untersuchungen an jungen Sedimenten von der Galapagos Mikroplatte: Hydrothermale und stratigraphisch signifikante Signale. Karlsruher Geochem. H. 3: 157 pp.

Eisenhauer, A., Mangini, A., Segl, M., Beer, J., Bonani, G., Suter, M., Hofmann, H.J., Wölfli, W. (1987) High resolution ^{10}Be and ^{230}Th in DSDP Site 580. Nucl. Instr. Methods Phys. Res. B29: 326-334

Eisenhauer, A., Mangini, Botz, R., Walter, P., Beer, J., Bonani, G., Suter, M., Hofmann, H.J., Wölfli, W. (1990) High resolution ^{230}Th and ^{10}Be stratigraphy of polar core M23235-2. In: Bleil, U. and Thiede, J. (eds.) Geologic History of the Polar Oceans: Artic versus Antartic. Kluwer Dordrecht: 475-487

Finney, B. P., Lyle, M. W. and Heath, G.R. (1988) Sedimentation at MANOP site H (Eastern Equatorial Pacific) over the past 400,000 years: climatically induced redox variations and their effects on transition metal cycling. Paleoceanography 3: 169-189

Francois, R., Bacon, M.P. and Suman, D.O. (1990) ^{230}Th profiling in deep sea sediments: high resolution record of flux and dissolution of carbonate in the Equatorial Pacific during the last 24,000 years. Paleoceanography 5: 761-788

Frank, M., Eckhardt, D., Eisenhauer, A., Kubik, P.W., Dittrich-Hannen, B., Segl, M., and Mangini, A. (1993) ^{10}Be, ^{230}Th and ^{231}Pa in Galapagos Microplate sediments: implications for hydrothermal activity and paleoproductivity changes during the last 100,000 years. submitted to Paleoceanography

Froelich, P.N., Klinkhammer, G.P., Bender, M.L., Luedtke, N.A., Heath, G.R., Cullen, D., Dauphin, P., Hammond, D., Hartman, B., and Maynard, V. (1979) Early oxidation of organic matter in pelagic sediments of the eastern equatorial Atlantic: suboxic diagenesis. Geochim. Cosmochim. Acta 43: 1075-1090

Imbrie, J., Hays, J.D., Martinson, D.G., McIntyre, A., Mix, A.C., Morley, J.J., Pisias, N.G., Prell, W.L. and Shackleton, N. (1984) The orbital theory of Pleistocene climate: support from a revised chronology of the marine δ^{18}O record. In: Kukla, G. and Saltzmann, B. (eds.) Milankovitch and Climate 1. Reidel Publ. Co.: 269-305

Keir, R.S., and Berger, W.H. (1983) Atmospheric CO_2 content in the last 120,000 years: the phosphate extraction model. J. Geophys. Res. 88: 6027-6083

Klinkhammer, G.P. (1980) Early diagenesis in sediments from the eastern equatorial Pacific, II. Pore water metal results. Earth. Planet. Sci. Lett. 49: 81-101

Kusakabe, L., Ku, T.L., Southon, J.R., Liu, S., Vogel, J.S., Nelson, D.E., Nakaya, S. and Cusinmano, G.L. (1991) Be isotopes in rivers/estuaries and their oceanic budgets. Earth. Planet. Sci. Lett. 102: 265-276

Lyle, M., Murray, D.W., Finney, B.P., Dymond, J., Robbins, J.M., Brooksforce, K. (1988) The record of the Late Pleistocene biogenic sedimentation in the eastern tropical Pacific Ocean. Paleoceanography 3: 39-59

Mangini, A., Halbach, P., Puteanus, D. and Segl, M. (1985) Chemistry and growth history of Central Pacific Mn-crusts and their economic importance. In: Teleki, P.G. (ed.), Marine Minerals: resource assessment strategies. Reidel Publ. Co.: 205-220

Mangini, A., Eisenhauer, A. and Walter, P. (1988) The relevance of manganese in the oceans for the climatic cycles in the Quaternary, Sitzungsberichte der Heidelberger Akademie der Wissenschaften, Jahrgang 1990, 4. Abhandlung, Springer Heidelberg: 263-289

Mangini, A., Eisenhauer, A., Walter, P. (1990) Response of Manganese in the ocean to the climatic cycles in the Quaternary. Paleoceanography 5: 811-821

Mangini, A., Eisenhauer, A., Walter, P. (1991) A spike of CO_2 in the atmosphere at glacial-interglacial boundaries induced by rapid deposition of manganese in the oceans. Tellus 43B: 97-105

Mangini, A., Eisenhauer, A., Wilhelm, A. (1993) The residence time of ^{230}Th in the glacial ocean. submitted to Earth Planet. Sci. Lett

Marchig, V. and Gundlach, H. (1982) Iron-rich metalliferous sediments on the East Pacific Rise: prototype of undifferentiated metalliferous sediments on divergent plate boundaries. Earth Planet. Sci. Lett. 58: 361-382

Martinson, D.G., Pisias, N.J., Hays, J.D., Imbrie, J., Moore jr., T.C. and Shackleton, N.J. (1987) Age dating and the orbital theory of the ice ages: development of a high-resolution 0 to 300.000-year chronostratigraphy. Quaternary Research 27: 1-29

Pedersen, T. F., Pickering, M., Vogel, J.S., Southon, J.N., and Nelson, D.E. (1988) The response of benthic foraminfera to productivity cycles in the eastern equatorial Pacific: faunal and geochemical constraints on glacial bottom water oxygen levels. Paleoceanography 3: 157-168

Sarnthein, M., Tiedemann, R. (1990) Younger Dryas-style cooling events at glacial terminations I-VI at ODP Site 658: associated benthic δ^{13}C anomalies constrain meltwater hypothesis. Paleoceanography 5: 1041-1055

Schneider, R. (1991) Spätquartäre Produktivitätsänderungen im östlichen Angola-Becken: Reaktionen auf Variationen im Passat-Monsun-Windsystem und in der Advektion des Benguela Küstenstroms. Berichte Fachbereich Geowissenschaften. Univ. Bremen Nr. 21: 198 pp.

Schneider, R., Müller, P.J., and Wefer, G. (1993) Late-Quaternary paleoproductivity changes off the Congo deduced from stable carbon isotopes of planktonic foraminifera. submitted to Palaeogeogr., Palaeoclimatol., Palaeoecol.

Thomson, J., Wilson, T.R.S., Culkin, F. and Hydes, D.J. (1984) Non-steady state diagenetic record in eastern equatorial Atlantic sediments. Earth. Planet. Sci. Lett. 71: 23-33

Thomson, J., Wallace, H.E., Colley, S., and Toole, J. (1990) Authigenic uranium in Atlantic sediments of the last glacial stage - a diagenetic phenomenon. Earth. Planet. Sci. Lett. 98: 222-232

Weiss, R.F. (1977) Hydrothermal manganese in the deep sea: scavenging residence time and Mn/^{3}He relationships. Earth. Planet. Sci. Lett. 37: 257-262

Yang, Y.L., Elderfield, H., Ivanovich, M. (1990) Glacial to Holocene changes in carbonate and clay sedimentation in the equatorial Pacific Ocean, estimated from ^{230}Th profiles. Paleoceanography 5: 789-809

BENTHIC FORAMINIFERAL ASSEMBLAGES AND THE $\delta^{13}C$-SIGNAL IN THE ATLANTIC SECTOR OF THE SOUTHERN OCEAN: GLACIAL-TO-INTERGLACIAL CONTRASTS

A. Mackensen, H. Grobe, H.-W. Hubberten[1], G. Kuhn
Alfred Wegener Institute for Polar and Marine Research
Columbusstraße
27515 Bremerhaven, Germany

ABSTRACT. We use benthic foraminiferal assemblages and benthic $\delta^{13}C$ to interpret glacial-to-interglacial contrasts in two gravity cores from the lower bathyal Antarctic continental margin at 69°S, and the abyssal Agulhas Basin at 43°S. As a recent analogue, sediment surface samples from an eastern Atlantic Ocean and Weddell Sea transect between 20°-70°S are discussed. In the investigated area, benthic foraminiferal assemblages reflect both ocean circulation and surface productivity. Also at most stations from a belt with seasonally high surface productivity between 48°S-55°S, the $\delta^{13}C$ values of epibenthic *Cibicidoides* spp., including *F. wuellerstorfi* are depleted relative to the bottom water $\delta^{13}C_{\Sigma CO2}$ and hence do not follow the 1:1 relationship established for more northern areas. This bears implications for the interpretation of large glacial-interglacial $\delta^{13}C$ shifts from the Southern Ocean: Significant parts of this shift can be caused by a northward migration of high productivity belts associated with the Polar Front and the winter sea-ice limit rather than indicating nutrient-rich glacial Southern Ocean deep and bottom water.

During interglacial climatic optima, seasonally open surface water accompanied by relatively high opal and very low carbonate accumulation characterises the Antarctic continental margin environment. The recent benthic foraminiferal fauna indicates moderate productivity, but during peak warm periods (oxygen isotope Stages: 11, 9, 7.5, 7.3, 5.5 and 1.1) very low numbers of benthic foraminifera are inferred to represent maximum organic matter fluxes with severe calcite dissolution on the sea floor. Equally high $\delta^{13}C$ values in surface and bottom water as inferred from planktic and benthic foraminifera, may indicate deep convection and bottom water formation during interglacials. In contrast, during glacials, very low opal accumulation, moderate carbonate accumulation, a benthic fauna that is presently associated with low productivity, as well as different benthic and planktic $\delta^{13}C$ values are consistent with both a reduced primary productivity and a stratified water column, suggesting suppressed bottom water generation.

NATO ASI Series, Vol. I 17
Carbon Cycling in the Glacial Ocean:
Constraints on the Ocean's Role in Global Change
Edited by R. Zahn et al.

In the Agulhas Basin high carbonate and low organic carbon (C_{org}) accumulation reflect the late Holocene position of the site investigated well north of the present-day Polar Front. Low Holocene $\delta^{13}C$ values of 0.3‰ and a benthic foraminiferal fauna that indicates a southern bottom water mass which is corrosive to carbonate is in agreement with the injection of North Atlantic Deep Water into Circumpolar Deep Water at intermediate depths, which does not affect bottom waters of this basin. During glacial periods, a specific southern fauna, associated with high productivity today, low carbonate, high sediment and C_{org} accumulation, and by 1.1‰ lower $\delta^{13}C$ values indicate a bottom water mass of southern origin, a northward shift of the high productivity belt by 7°, and strongly diminished injection of NADW into the Southern Ocean.

INTRODUCTION

Glacial-interglacial changes in atmospheric CO_2 recorded in polar ice cores are intimately connected with processes that redistribute and partition carbon within the world's ocean (Broecker and Peng, 1986, 1989). These processes include changes of the global ocean circulation and nutrient inventory, first suggested and documented by quantitative micropaleontology and subsequently inferred and proven by marine geochemistry (Streeter, 1973; Schnitker, 1974; Lohmann, 1978; Streeter and Shackleton, 1978; Douglas and Woodruff, 1981; Broecker, 1982; Curry and Lohmann, 1982; Curry et al., 1988; Duplessy et al., 1988; Boyle and Keigwin, 1982, 1987; Boyle, 1988). Although different methods and tools have been applied to reconstruct glacial nutrient distribution and circulation patterns of the ocean, quantitative estimates regarding exchange fluxes between the main basins or between surface, intermediate and deep water masses have not been well constrained. Moreover, as more data have become available it is evident that glacial water mass properties, primary productivity and nutrient cycling reconstructed by different proxies do not always show coherent variation through time. Instead the data apparently contradict each other in some locations. Since the southern high-latitude deep ocean is of special importance for the ocean's nutrient distribution and atmospheric CO_2 (cf. Keir, 1988; Broecker and Peng, 1989; Martin, 1990; Boyle, 1992), it is of particular interest that there appears to be a discordance between the glacial to interglacial changes in the $\delta^{13}C$ values and the cadmium content of benthic foraminifera from the deep Southern Ocean (Boyle, 1992; Broecker, 1993).

In this paper we present recent benthic foraminiferal faunal distributions and benthic foraminiferal $\delta^{13}C$ from various sites in the Southern Ocean (Mackensen et al., 1990, 1993, in press). The data are used as an analogue for interglacial conditions and compared to down-core changes in three South Atlantic Ocean and Weddell Sea cores. We use the data further to reconstruct glacial-interglacial contrasts in paleoproductivity and deep and bottom water circulation.

New benthic foraminiferal and stable isotope data from two cores spanning the last 400 kyr are presented.

MATERIAL AND METHODS

The investigated surface sediment and water samples are from a traverse between 20°-70°S through the eastern Atlantic Ocean and Weddell Sea. The profile runs from Walvis Ridge to the north, well off the African continental margin through Cape Basin, across the Agulhas Ridge and through the Agulhas Basin; it crosses Meteor Rise and the Atlantic-Indian Ridge close to Bouvet Island, runs through the Weddell Sea Basin and terminates on the Antarctic continental margin to the South (Figure 1). For exact sample positions as well as sedimentological and micropaleontological treatment of sediment samples, benthic foraminiferal counting procedures and stable isotope analyses see Mackensen et al. (1990, 1993, in press) and Grobe and Mackensen (1992). Only standard laboratory procedures are used. Samples from Walvis Ridge are from Schmiedl (unpublished data).

Living benthic foraminiferal faunas were differentiated from dead assemblages by staining with Rose Bengal (Mackensen et al., 1985). To facilitate analogy with late Quaternary core material, potential fossil assemblages (dead fauna reduced by non-resistant agglutinants) were calculated using a Q-mode Principal Component Analysis with subsequent Varimax rotation (Mackensen et al., 1990, 1993). The carbon isotope composition of the total dissolved inorganic carbon ($\delta^{13}C_{\Sigma CO2}$) was measured on water samples from hydrographic stations and bottom water from surface sediment samples taken with a multiple corer. These measurements were compared to $\delta^{13}C$ values determined on Rose Bengal-stained specimens of the benthic foraminifer *Anomalina wuellerstorfi* Schwager 1866, more recently assigned to *Planulina*, *Cibicidoides* or *Cibicides* and now to *Fontbotia* (Loeblich and Tappan, 1988). Because *F. wuellerstorfi* (Schwager) *sensu strictu* reaches its environmentally-controlled southern distribution limits close to the polar front, standing stock values in the investigated area are very low (Mackensen et al., 1990, 1993). Therefore, we also analysed the closely related genera *Cibicidoides kullenbergi*, *Cibicidoides* sp. cf. *C. kullenbergi*, *Cibicidoides* sp. cf. *F. wuellerstorfi*, *Cibicidoides* sp. cf. *C. robertsonianus*, and *Cibicidoides* sp. indet..

Two of the gravity cores used to compare glacial and interglacial benthic foraminiferal faunas and $\delta^{13}C$ values (Core PS1388: 69°02'S, 5°55'W; Core PS1506: 68°45'S, 5°53'W) were from the eastern Antarctic continental margin at water depths of 2531 m and 2405 m (Figure 1). The third core used in this study (Core PS2082: 43°13'S, 11°45'E) was from the Agulhas Basin in the eastern South Atlantic Ocean at a water depth of 4661 m (Figure 1). R-mode Principal Component Analysis was used to determine down-core changes in benthic foraminiferal assemblages (Woodruff, 1985).

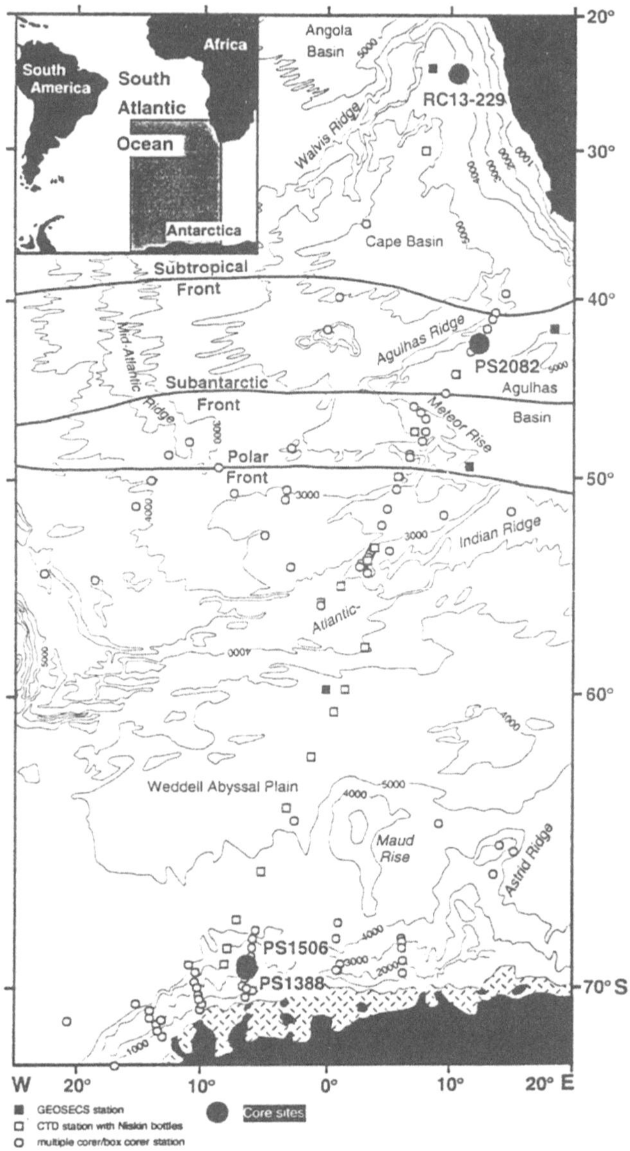

Figure 1: Sample locations and bathymetry of the eastern South Atlantic Ocean and Weddell Sea. Open circles indicate sediment surface samples, squares oceanographic water rosette sampling stations, including GEOSECS stations (Kroopnick, 1980), and large solid circles give positions of sediment cores. Bathymetric contours (given in 1000 m intervals) are from General Bathymetric Chart of the Oceans. Positions of the oceanographic fronts are after Peterson and Stramma (1991). For detailed positions of sediment surface and water samples see Mackensen et al. (1990, 1993, in press).

The $\delta^{13}C$ composition of the epibenthic phytodetritus feeding foraminifera *Epistominella exigua* (Gooday, 1988) and the planktic foraminifera *Neogloboquadrina pachyderma* were determined in both cores from the Antarctic continental margin (Table 1, data of Core PS1388 in Mackensen et al., 1989). In addition, at site PS1506 the isotopic composition of the probably endobenthic living *Oridorsalis umbonatus* (cf. Belanger et al., 1981; Mackensen et al., 1985) was measured (Table 1). At site PS2082, the stable isotope composition of preferentially epibenthic *Cibicidoides* spp., including *F. wuellerstorfi* (Lutze and Thiel, 1987) as well as planktic polar *N. pachyderma* and subpolar *Globigerina bulloides* were determined (Table 2).

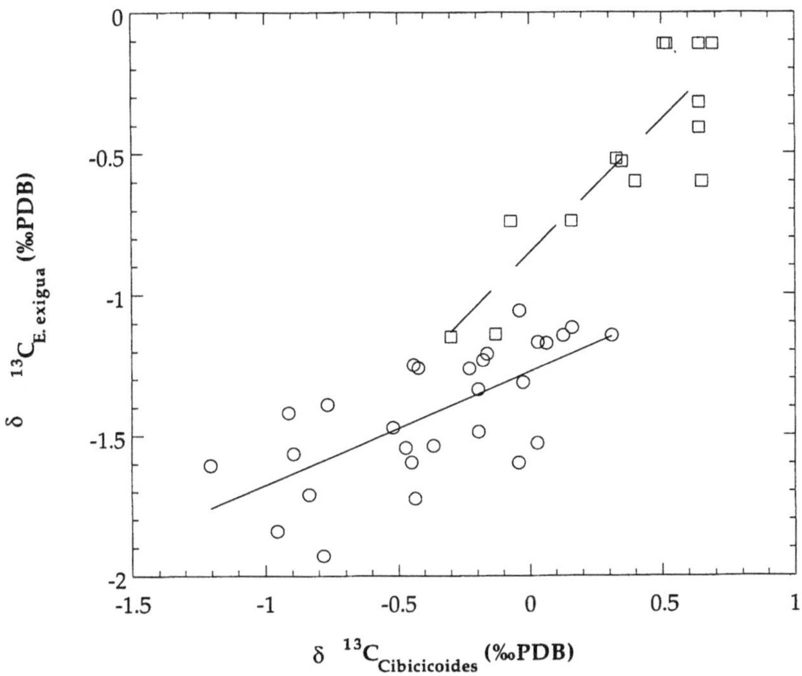

Figure 2: $\delta^{13}C$ comparison of *Epistominella exigua* versus *Cibcicidoides* spp. (including *Fontbotia wuellerstorfi*.). Squares = stained (living) species from the Southern Ocean between 40°-70°S (offset = -0.85‰). Circles = down-core samples from Core PS2082 (offset = -1.26‰).

The $\delta^{13}C$ values of *E. exigua* were calibrated versus the *Cibicidoides* scale by paired measurements of *E. exigua* and *Cibicidoides* spp. from stained surface sediment samples (-0.85, R=0.87) and from Pleistocene core material of PS2082 (-1.26‰, R=0.65) (Figure 2). Although calculation of a linear regression of all paired analyses of *Cibicidoides* and *E. exigua* from the Weddell Sea, regardless whether down-core samples or surface samples, results in an offset of -1.01‰, for comparison between cores we used an offset of *E. exigua* relative to *Cibicidoides* of -1.2‰. This value agrees with the calculated offset based on 14 paired analyses of interglacial sediments in PS2082 (-1.18‰, R=0.69), and is close to -1.15‰ given by Woodruff et al. (1980). This is the most reliable offset, because there was consistently low productivity during the Holocene at this site and hence there is no productivity induced change of either *E. exigua* or *Cibicidoides* $\delta^{13}C$ values.

AGE MODELS

Age models were constructed by correlation of benthic and planktic foraminiferal $\delta^{18}O$ records with the SPECMAP record (Imbrie et al., 1984). The graphical correlations were aided by a computer program of Paillard et al. (1992). Because of the paucity of biogenic carbonate in Antarctic continental margin Cores PS1388 and PS1506, the oxygen isotope stratigraphy was also correlated with several lithological parameters (Grobe et al., 1990; Grobe and Mackensen, 1992), the opal and Ba content (Shimmield et al., this volume), the benthic foraminiferal fauna (Mackensen et al., 1989) and the $\delta^{13}C$ records of infaunal benthic foraminifera (Table 3). Core PS1506 was sampled with a resolution of approximately 500 years during peak interglacials and 3000 years during glacials.

During short sequences of severe calcite dissolution the age model of Core PS2082 was corroborated by a detailed *Cycladophora davisiana* record (U. Brathauer, pers. comm., 1993) (Table 4). Core PS2082 was sampled with a resolution of 4000 years during interglacials and less than 1000 years during glacial periods. Our age model gives similar interglacial sedimentation rates, but two-fold glacial sedimentation rates compared to what Oppo et al. (1990) have proposed for Core RC13-229 from the northern Cape Basin at 25°30'S and 11°20'E (Figures 3, 4).

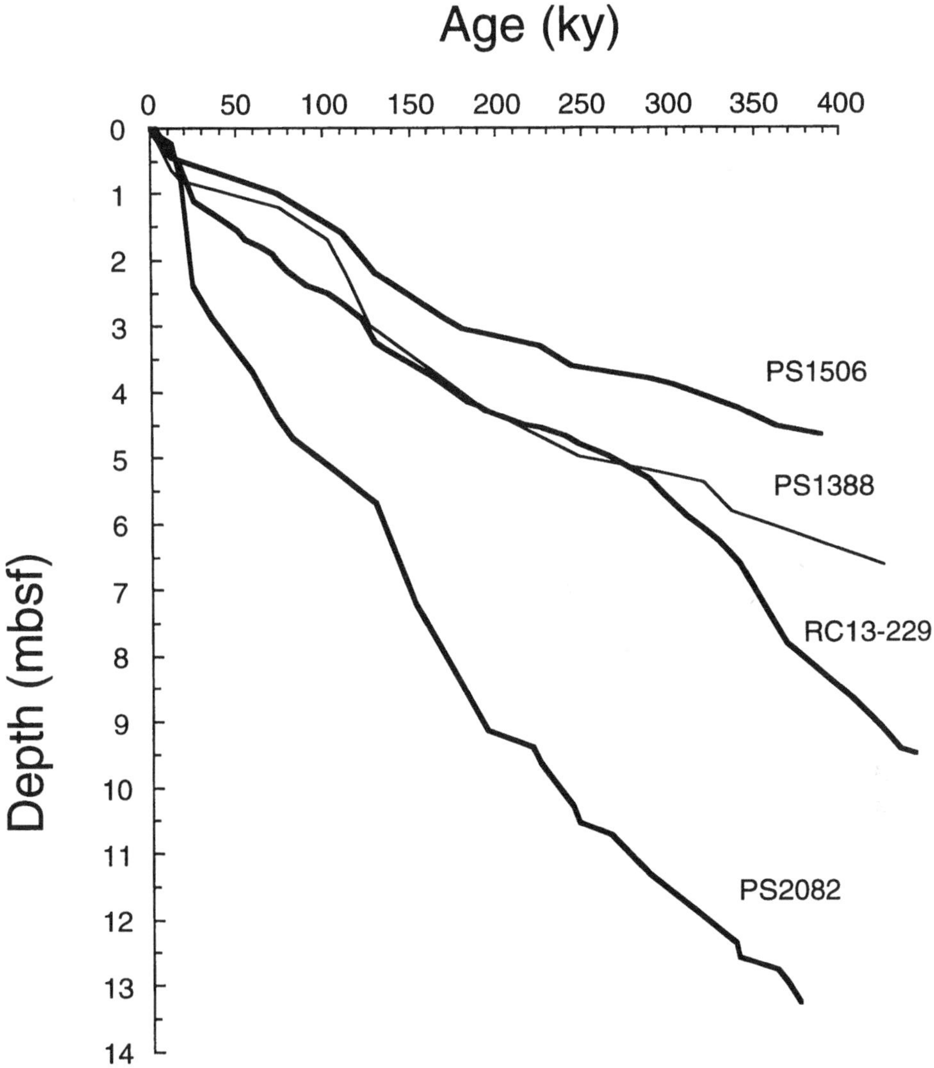

Figure 3: Age models of cores discussed in this paper. Age model of RC13-229 is from Oppo et al. (1990).

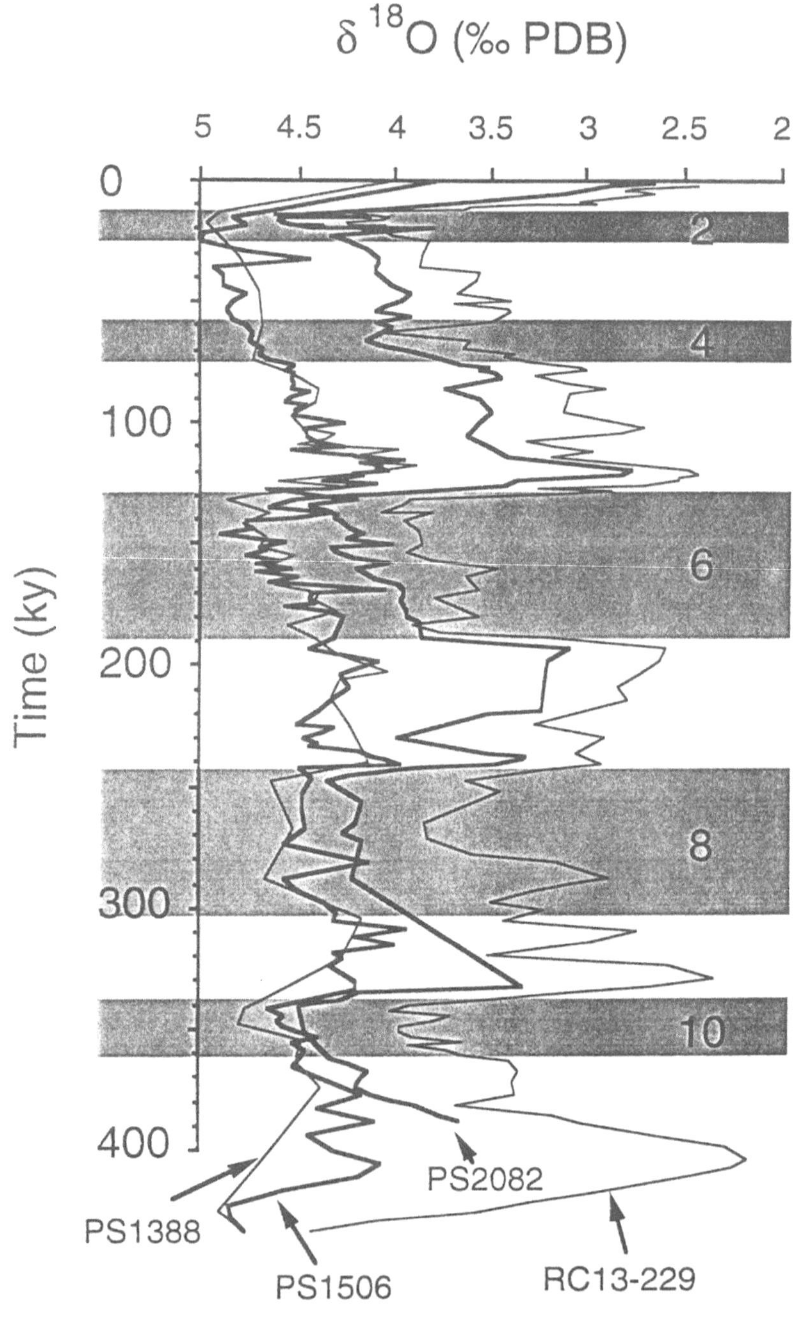

Figure 4: Benthic $\delta^{18}O$ records of PS1388, PS1506, PS2082 and RC13-229 according to the age models used in this study.

INTERGLACIAL RECENT CIRCULATION AND PRODUCTIVITY

Hydrography

Waters from the North Atlantic, the Antarctic Circumpolar Current (ACC) and the Weddell Sea enter the South Atlantic and are fed into a circulation system which is driven primarily by winds and thermohaline processes. The near-surface, wind-driven circulation in the southern Atlantic and the Weddell Sea is dominated by a system of gyres and the circumpolar current systems (Peterson and Stramma, 1991). The cyclonic Weddell Gyre to the south extends to the bottom (Reid, 1989) and is separated from the anticyclonic Subtropical Gyre in the North by the ACC. The ACC consists of several zonal jets and bands of relative quiescence (Peterson and Whitworth, 1989). The Polar Front (PF) and the Subantarctic Front (SAF), subdivide the ACC in (1) the Antarctic Zone south of the PF, (2) the Polar Frontal Zone between the PF and the SAF, and (3) the Subantarctic Zone north of the SAF (Peterson and Stramma, 1991) (Figures 1, 5).

The deep and bottom water masses south of the Atlantic-Indian Ridge are the Weddell Sea Bottom Water (WSBW) with potential temperatures of <-0.7°C, overlain by the Antarctic Bottom Water (AABW, 0>q>-0.7) (Reid, 1989; Foldvik and Gammelsrød, 1988) (Figure 5). AABW bathes parts of the Agulhas Basin below 4000 m water depth, but most of the deeper parts of this basin are filled by Circum Polar Deep Water (CPDW), as is the Cape Basin. Parts of the CPDW in the eastern South Atlantic cyclonically turns around with the Weddell Gyre. This provides a layer of warm, saline, nutrient-rich and oxygen-poor Weddell Sea Deep Water (WDW) at intermediate depths along the eastern Weddell Sea continental margin (Reid, 1989). North of about 55°S the CPDW encounters the warm, saline, oxygen-rich and nutrient-poor North Atlantic Deep Water (NADW) at intermediate depths and is divided into an upper (Upper CPDW) and a lower layer (Lower CPDW). The nutrient-poor and oxygen-rich Antarctic Intermediate Water (AAIW) forms between the PF and the SAF and spreads northward above the UCPDW.

Sediments

The surface sediments on the eastern continental margin consist of silty and clayey muds with varying amounts of coarse sand and gravely drop stones. Above 1600 m water depth the sand size fraction exceeds 40 %, and below 1400 m the mud content (<63 mm) exceeds 50 %, of the bulk dry sediment. On the shelf carbonate contents reach up to 9.5 % of the bulk dry sediment, whereas at the shelf break and on the upper slope where the core of WDW resides, they never exceed 1 %. On the slope between 2500-3500 m, carbonate contents range from 3 % through 13 %. The Calcite Compensation Depth (CCD) is at 4000 m. C_{org} contents of the

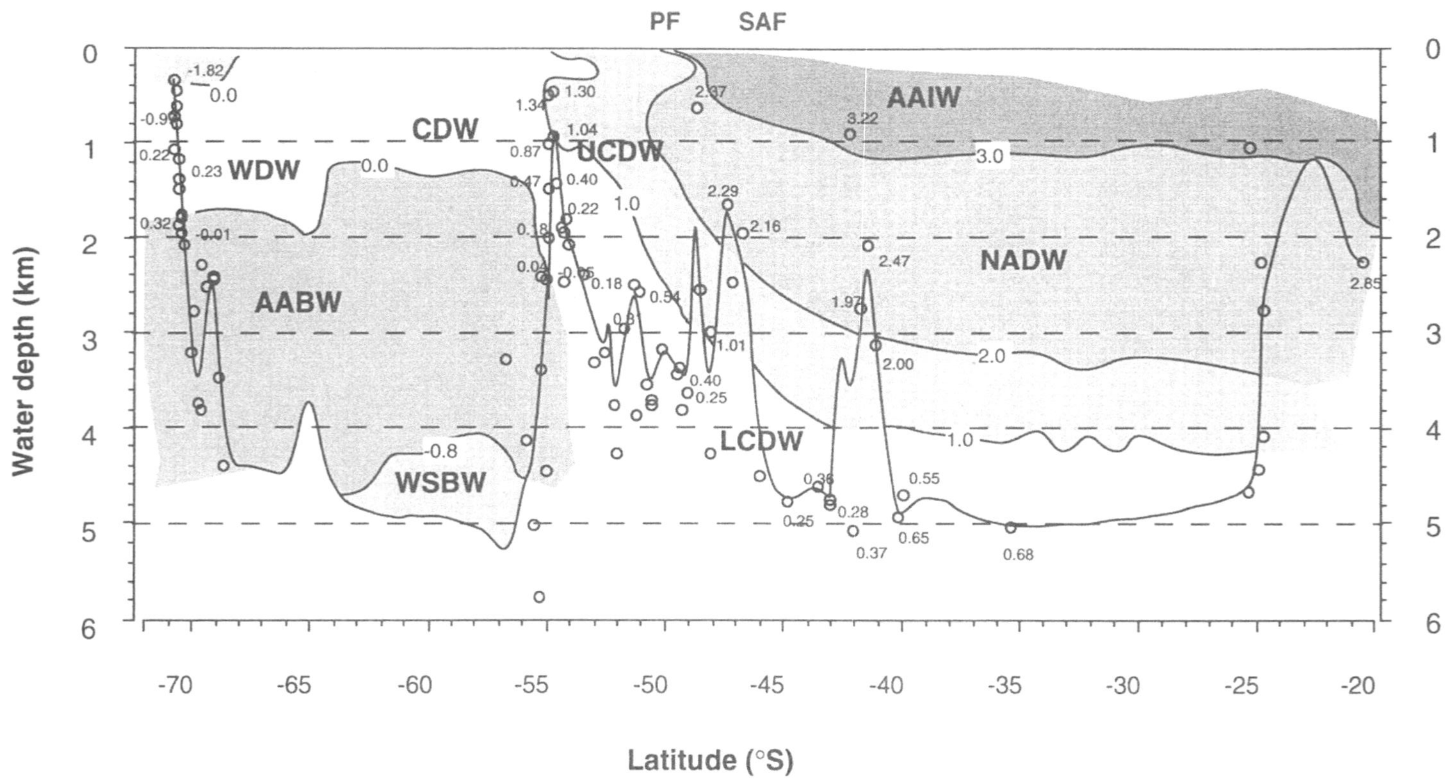

Figure 5: Distribution of potential bottom water temperatures as measured at the time of surface sediment sampling in the eastern South Atlantic and Weddell Sea. Sample latitude and water depth is indicated as well as an idealized bathymetry. Areas between isotherms are shaded for comparison. Positions of surface oceanographic fronts (after Peterson and Stramma, 1991) and the deep and bottom water mass distribution (after Reid, 1989) are indicated without defining oceanographic boundaries. (PF=Polar Front, SAF=Subantarctic Front; CPDW=Circumpolar Deep Water, UCPDW= Upper CPDW, LCPDW=Lower CPDW, AABW=Antarctic Bottom Water; NADW=North Atlantic Deep Water).

surface sediments are low (< 0.9 %), with lowest values between 500-1800 m and at abyssal depths (Mackensen et al., 1990; Grobe and Mackensen, 1992).

In the eastern South Atlantic, in a belt South of the PF and north of the average winter sea ice limit, high primary productivity results in high accumulation rates of biogenic opal (Demaster, 1981; Defelice and Wise, 1981; Wefer and Fischer, 1991). This is reflected by the grain size distribution of the sediments (Mackensen et al., 1993). South of the PF, diatomaceous ooze and muddy diatomaceous ooze are the most common surface sediments. Only in shallow areas on top of the Atlantic-Indian Ridge, winnowing and an increased supply of ice-rafted debris cause a dominance of the sand-sized and coarser fractions. North of the PF, the dominance of sand-sized grains is due to the increased flux of planktic foraminifera. There, the CCD is below 5000 m depthand the depth of the foraminiferal lysocline is at 4000 m, as defined by decreasing sand (i.e., planktic foraminifera) content and increasing fragmentation of planktic foraminifera below 3000 m water depth. South of the PF at water depths < 4000 m and on the Atlantic-Indian Ridge, C_{org} contents > 0.5% of the bulk sediment are found in sediments with high biogenic silica content and rapid deposition rates. Due to the decay of organic matter and CO_2 formation at the sediment surface, most of the foraminifera dissolve, forcing the lysocline up to only a few hundred meter water depth.

40 % of C_{org} exported from the photic zone of the Southern Ocean into the deep sea is synthezised in the coastal zones near and on the continental shelves and in the Polar Front Zone, despite the fact that these zones occupy only 10% of the surface ocean area (Wefer and Fischer, 1991). The annual primary production in the eastern Atlantic Polar Frontal Zone is 83 g C m^{-2}, compared to an average annual primary production in the Southern Ocean of around 26 g C m^{-2} (Wefer and Fischer, 1991).

Benthic Foraminiferal Assemblages

The distribution of recent benthic foraminiferal assemblages defined by Q-mode principal component analysis mirrors the hydrographic and the trophic conditions in the eastern Weddell Sea and South Atlantic, i.e. ocean circulation and productivity (Mackensen et al., 1990, 1993). In the eastern Weddell Sea and South Atlantic Ocean, two Southern Component Bottom Water faunas dominate the sea-floor bathed by WSBW, AABW and LCPDW and reflect the northward flow of AABW and LCPDW: one assemblage characterized by the arenaceous *Multifidella nodulosa* below the CCD, and a second *Nuttallides umbonifer* dominated assemblage above the CCD (Figure 6). In the region of highest bio-siliceous primary productivity, south of the polar front and north of the winter sea ice limit, these faunas are overprinted by a Southern High-productivity Fauna, dominated by infaunal species such as *Pullenia bulloides*, *Melonis pompilioides* and *M. zaandamae* (Figure 7).

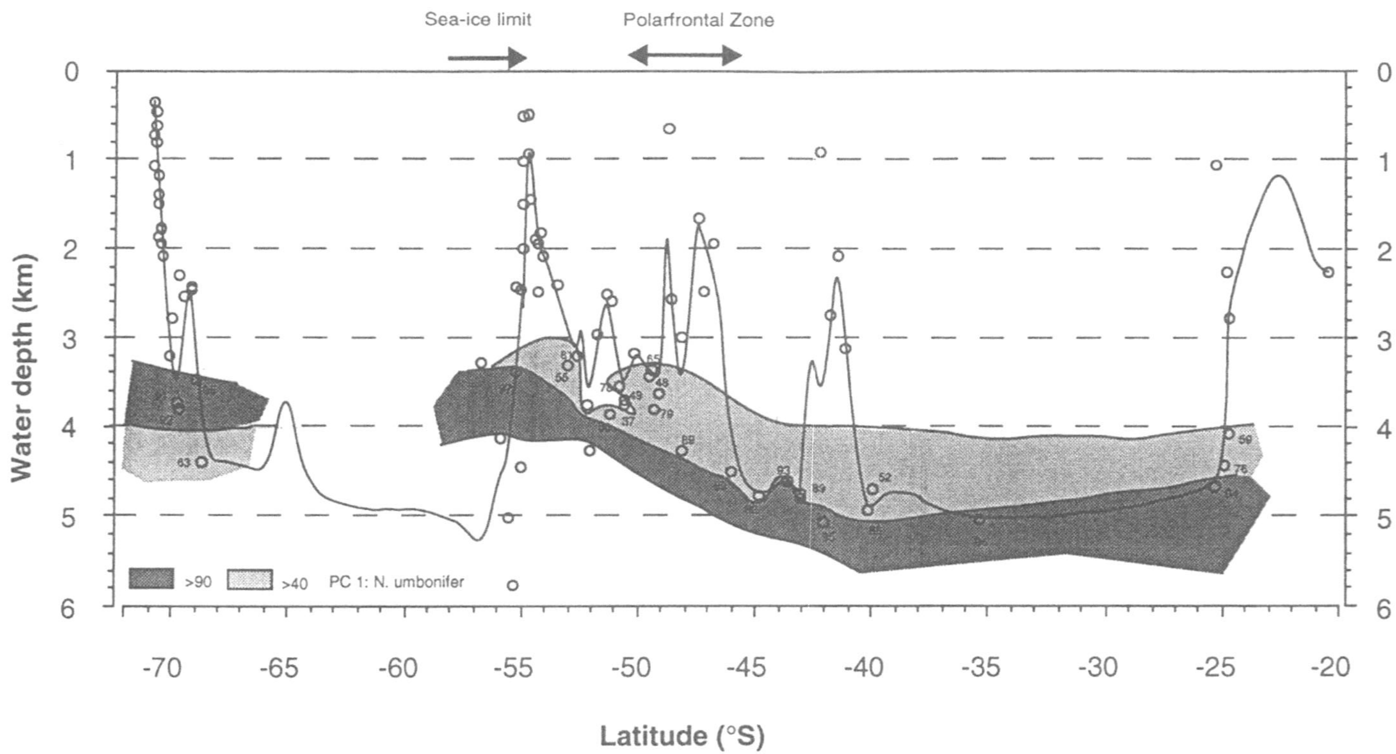

Figure 6: Distribution of one of the Southern Component Bottom Water Faunas, dominated by *Nuttallides umbonifer*. Contours are 100x Varimax Component Loadings. For detailed positions of sediment surface samples see Mackensen et al. (1990, 1993).

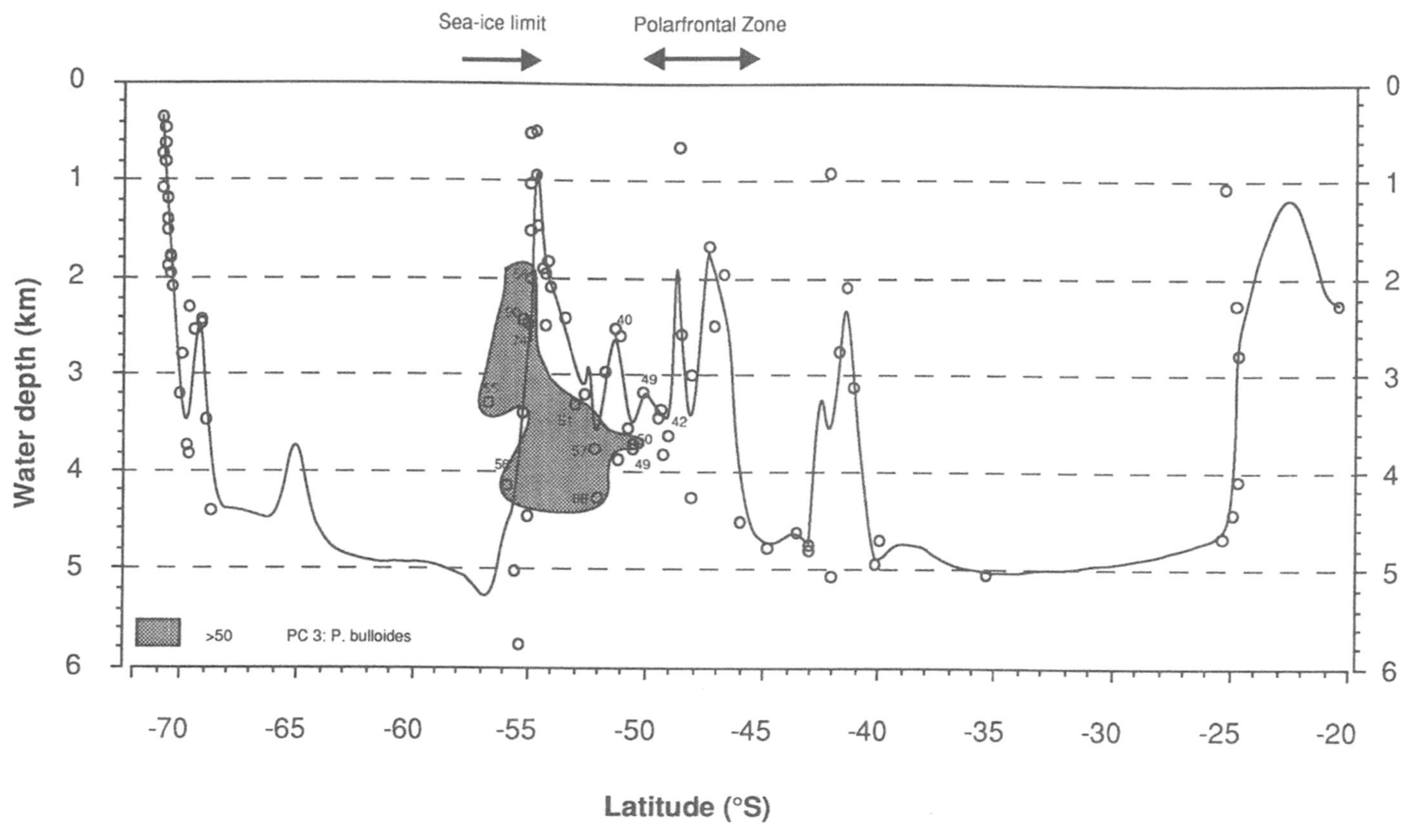

Figure 7: Distribution of Southern High-productivity Fauna. Contours are 100x Varimax Component Loadings. For detailed positions of sediment surface samples see Mackensen et al. (1990, 1993).

NADW mixes with CPDW at intermediate depth, as reflected in the distribution of the Northern Component Deep Water Fauna, here dominated by *Epistominella exigua* (Figure 8). The Northern High-Productivity Assemblage overprinting this fauna is dominated by the shallow endobenthic *Bulimina aculeata*. This fauna inhabits highly productive areas in the Antarctic and Polar Frontal Zone which are bathed in CPDW, UCPDW and NADW. Consequently *B. aculeata* also dominates on the Antarctic upper continental margin down to 2000 m water depth, where the core of warm (> 0 °C) WDW bathes the slope and fine, organic-rich sediments are deposited (Figure 9).

Total Dissolved Inorganic, Foraminiferal and Sedimentary C_{org}

In the eastern South Atlantic Ocean and Weddell Sea, the general $\delta^{13}C_{\Sigma CO2}$ distribution is in close agreement with the oxygen content and the nutrient distribution (Kroopnick, 1980; Whitworth and Nowlin, 1987; Reid, 1989) (Figure 10a). Very low $\delta^{13}C$ values of UCPDW (<0.3‰ PDB) in the PF region and the eastern limb of the Weddell Gyre coincide with nutrient maxima. However, a significant decoupling of the dissolved phosphate signal from the $\delta^{13}C_{\Sigma CO2}$ signal is indicated in the abyssal Weddell Sea (Figure 10b). We attribute this to temperature-dependent fractionation processes during gas exchange of surface waters with the atmosphere at sites of bottom-water formation (Charles and Fairbanks, 1990; Broecker, 1993; Mackensen et al., in press; Zahn and Keir, this volume). To quantify decoupling of WSBW $\delta^{13}C$ values from phosphate contents, we reduced measured $\delta^{13}C$ values by calculated $\delta^{13}C$ values using the global relation between phosphate and $\delta^{13}C$ of Kroopnick (1985). The remaining $\delta^{13}C$ values are due to the air-sea exchange of CO_2 at the sites of bottom water formation in the southern Weddell Sea (Figure 10b).

Multiple corer water from the sediment-water interface is slightly depleted in $\delta^{13}C$ relative to bottom-water $\delta^{13}C_{\Sigma CO2}$ as sampled several meters above the sea floor (Figure 11). Most of this depletion is probably due to contamination of the near-bottom water with pore waters because of disturbance and decompressing during sampling. Surface sediment $\delta^{13}C_{org}$ is 3-4‰ lower south of the polar front than north of it, and the $\delta^{13}C_{org}$ in freshly accumulated phytodetritus is 3 to 4‰ lower than surface sediment $\delta^{13}C_{org}$ (Figure 12). Antarctic plankton $\delta^{13}C_{org}$ values are lower than low-latitude plankton values (Rau et al., 1989), but although plankton isotope values change during transport through the water column, these different $\delta^{13}C_{org}$ values obviously are still present in sediments collected north and south of the polar front (Figure 12) (Fischer, 1991). We attribute the strong $\delta^{13}C_{org}$ enrichment at the sediment-water interface to high benthic respiration rates in the fluff-layer which produces CO_2 depleted in ^{13}C, and residual organic matter in the sediment, which is enriched in ^{13}C relative to the supplied phytodetritus. Comparison of live *F. wuellerstorfi* $\delta^{13}C$ and related genera with bottom-water $\delta^{13}C_{\Sigma CO2}$ exhibits a significant lowering of foraminiferal $\delta^{13}C$ values at most stations between the

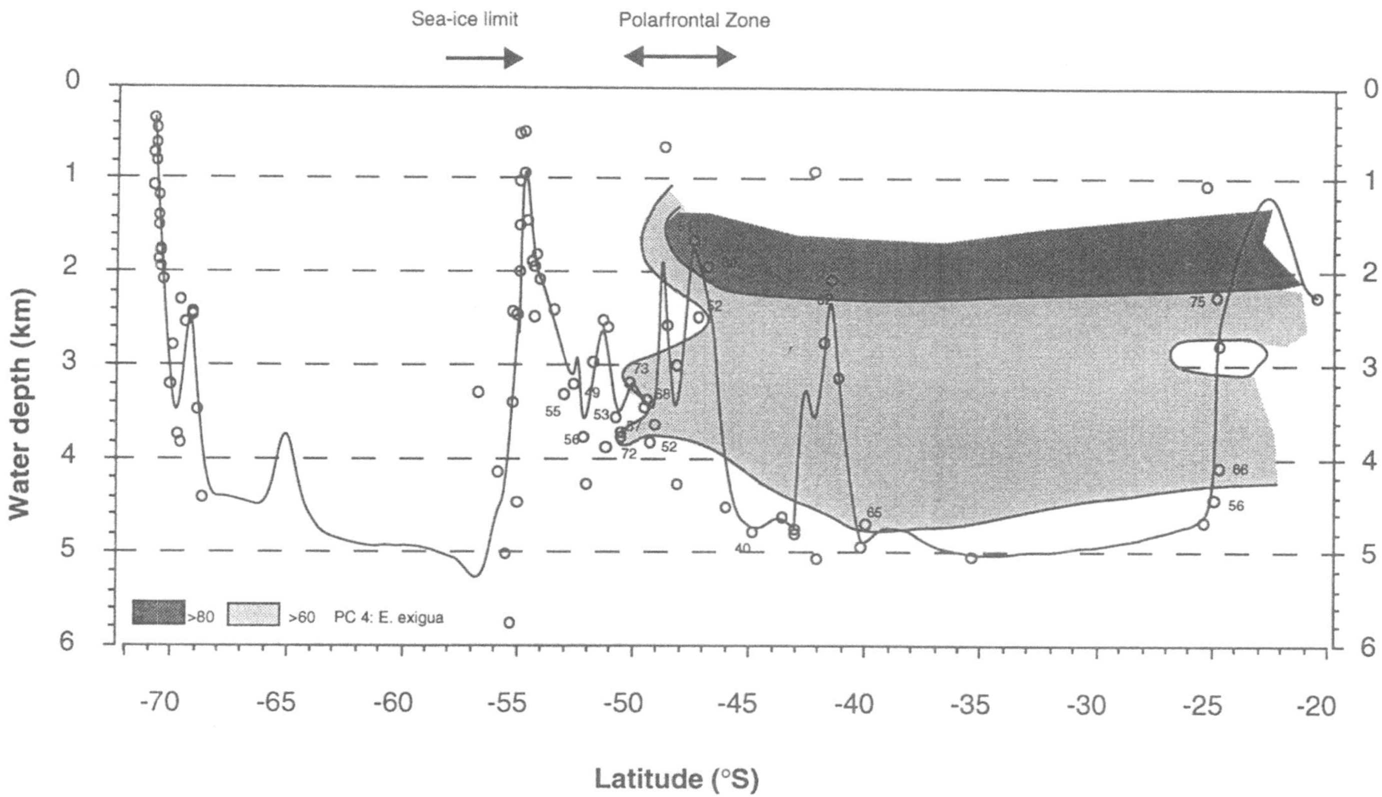

Figure 8: Distribution of Northern Component Deep Water Fauna. Contours are 100x Varimax Component Loadings. For detailed positions of sediment surface samples see Mackensen et al. (1990, 1993).

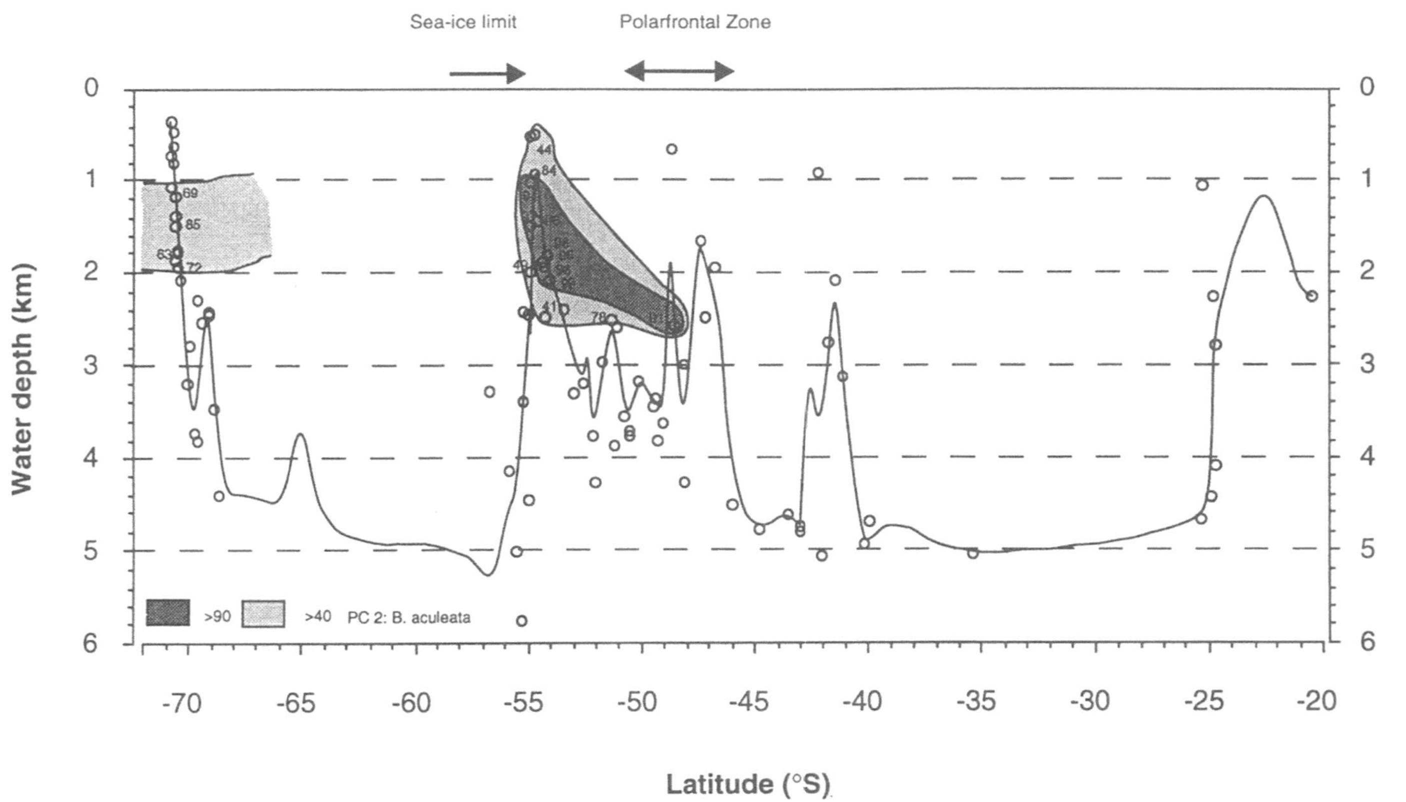

Figure 9: Distribution of Northern High-productivity Fauna. Contours are 100x Varimax Component Loadings. For detailed positions of sediment surface samples see Mackensen et al. (1990, 1993).

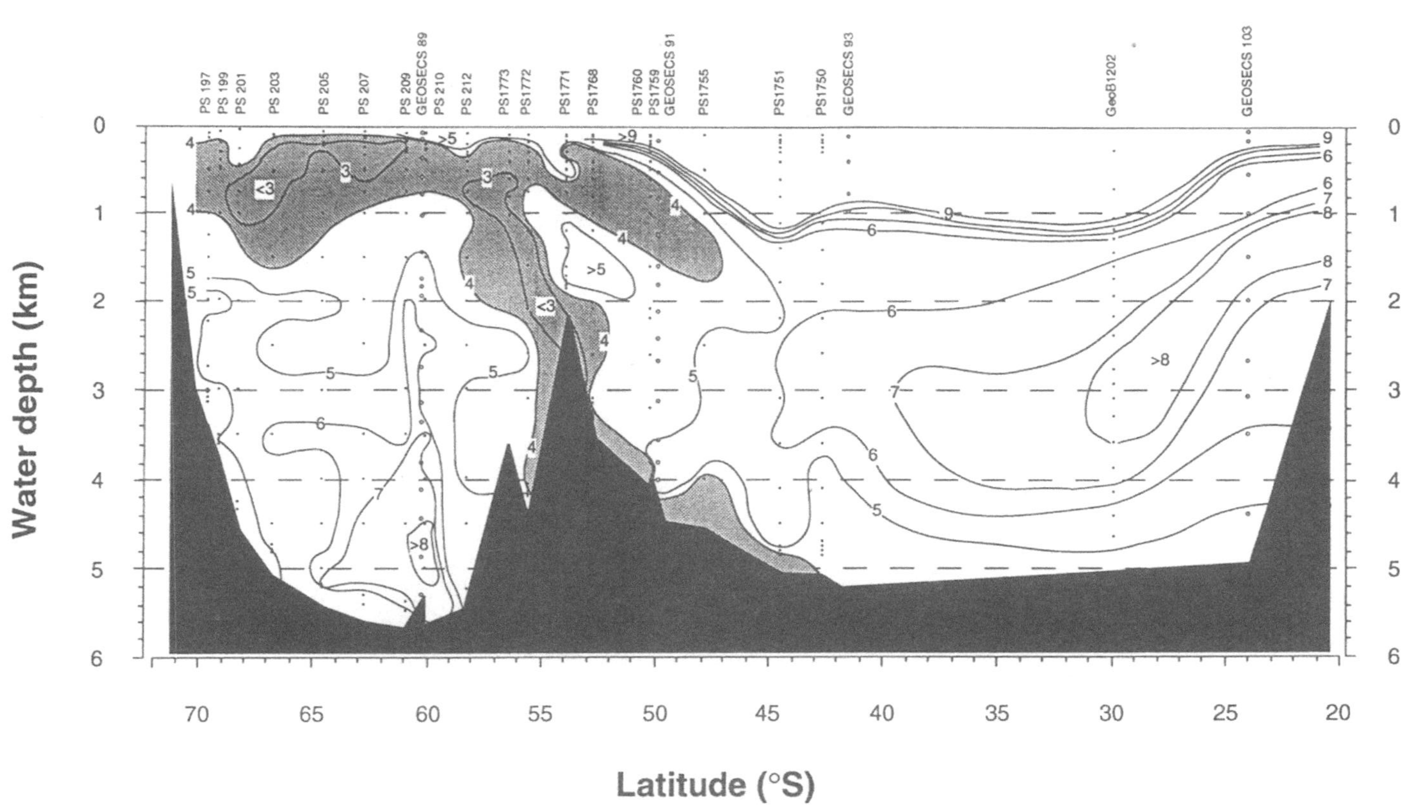

Figure 10a: Latitudinal distribution of $\delta^{13}C_{\Sigma CO2}$ versus depth between 20 and 70°S in the eastern South Atlantic Ocean and Weddell Sea. Station identifications are indicated across the top. The contour interval is 0.1‰ and the decimal point is dropped on the contours. The shading highlights the $\delta^{13}C$ minimum (<0.4‰) and nutrient maximum. Note the $\delta^{13}C$ maximum in the abyssal Weddell Sea, not coupled with a comparable nutrient depletion. From Mackensen et al. (in press).

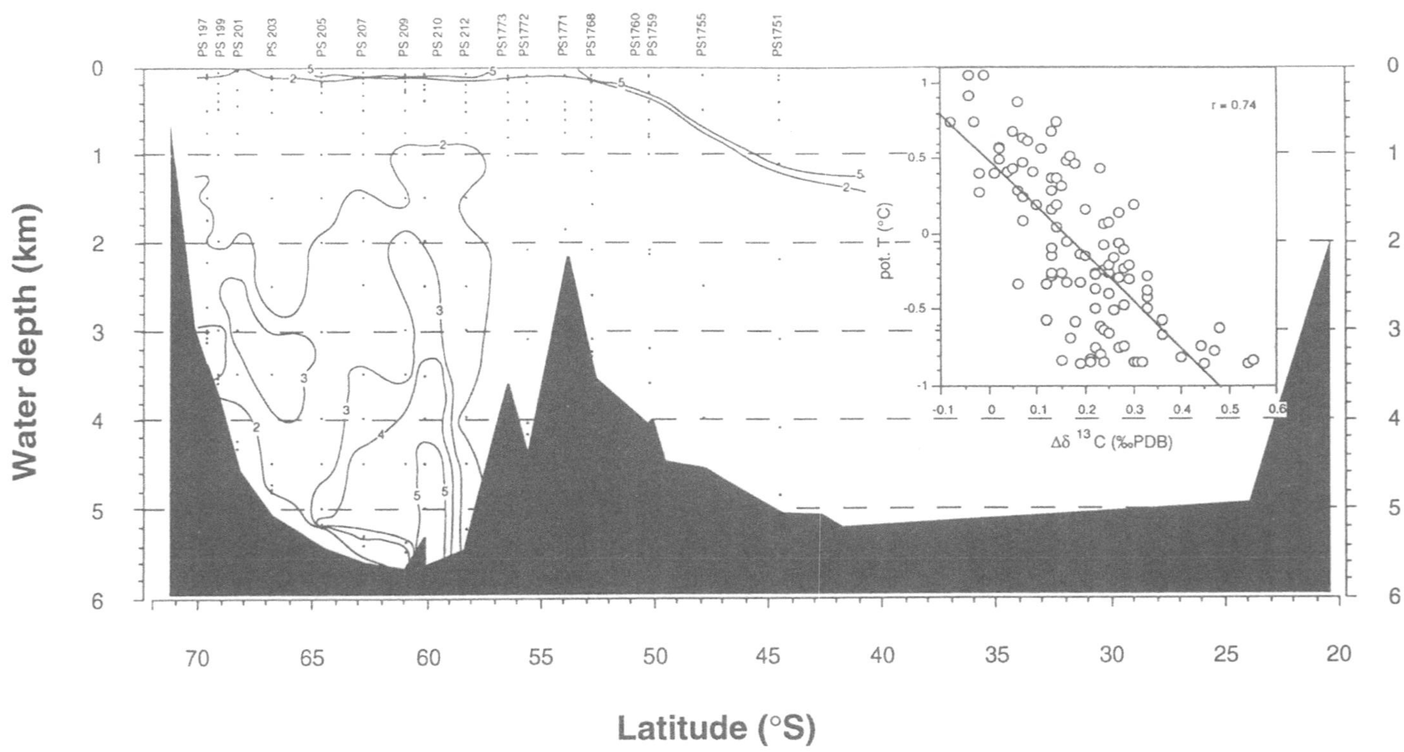

Figure 10b: Latitudinal distribution of $Dd^{13}C_{SCO2}$ (measured $\delta^{13}C$ reduced by $\delta^{13}C$ as calculated by using measured phosphate concentrations from the same Niskin bottle and the global regression of Kroopnick [1985]). Station identifications are indicated across the top. The contour interval is 0.1‰, and the decimal point is dropped on the contours. High values in the Weddell Sea and north of 50°S indicate high amounts of newly formed Weddell Sea Bottom Water and Antarctic Intermediate Water, respectively. Inset figure gives correlation between potential temperature and $Dd^{13}C_{SCO2}$ in deepwater and bottom-water masses of the Weddell Sea to illustrate conservative mixing between newly formed bottom water and older deep water. From Mackensen et al. (in press).

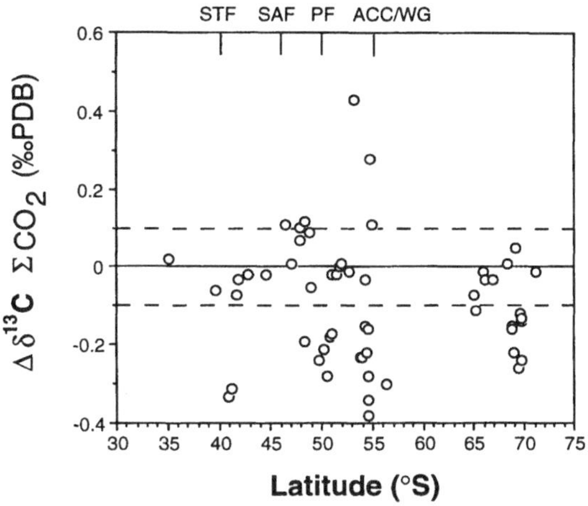

Figure 11. $\delta^{13}C_{\Sigma CO2}$ difference between bottom water taken from multiple corer and regional bottom waters (Niskin bottles). STF = Subtropical Front, SAF = Subantarctic Front, PF = Polar Front, ACC/WG = Antarctic Circumpolar Current/Weddell gyre boundary. After Mackensen et al. (in press).

Subtropical Front (≈41°S) and the southern boundary of the Antarctic Circumpolar Current (≈55°S) (Figure 13) (Mackensen et al., in press). This is different from the more northern South Atlantic Ocean (Bickert, 1992) and we relate this benthic foraminiferal $\delta^{13}C$ depletion to strong seasonal surface ocean productivity with rapidly-deposited organic matter and the development of a phytodetritus layer at the sea floor. High amounts of isotopically very light CO_2 released into the fluffy layer probably alter the isotopic composition of the $\delta^{13}C_{\Sigma CO2}$ therein and hence influence the $\delta^{13}C$ of *F. wuellerstorfi* and related genera, which reproduce and calcify just at or shortly after times of high food supply.

GLACIAL-TO-INTERGLACIAL CONTRASTS

Core PS2082: 43°S, North of the Present-Day Polar Front Zone

Core PS2082 was chosen from a suite of gravity cores crossing the eastern Atlantic Polar Front region because of its undisturbed calcareous record of the last 400,000 years, with the exception of some dissolution during glacial periods. According to the authors' knowledge, with the exception of Core MD84-527 (Labracherie et al., 1989), Core PS2082 is the south

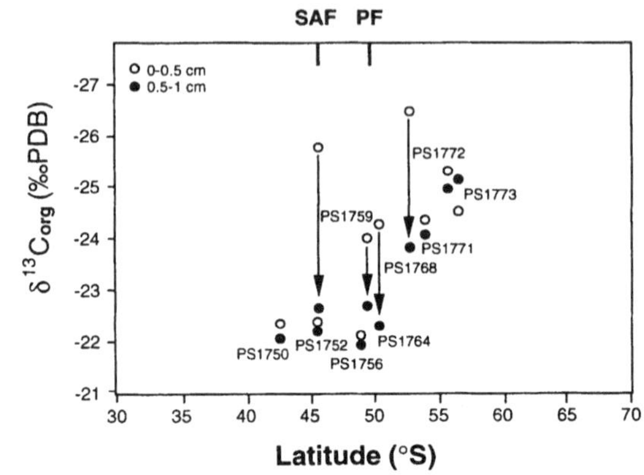

Figure 12: Regional pattern of surface sediment $\delta^{13}C_{org}$ values. Open circles = 0-0.5 cm, full dots = 0.5-1.0 cm depth in sediment. Significant ^{13}C enrichments (arrows) are observed close to the Subantarctic and Polar Front where freshly accumulated phytodetritus was present. Isotopically light CO_2 is released into the fluff-layer and influences the $\delta^{13}C$ record of *F. wuellerstorfi* (see also Figure 13). From Mackensen et al. (in press).

ernmost core with an almost complete *Cibicidoides* $\delta^{13}C$ record. Along this core, benthic and planktic $\delta^{13}C$ records vary between approximately 1.2‰ and -1.3‰ during the last 390,000 years (Table 2, Figure 14) such that low values coincide with glacial periods. An overall trend of increasing values with decreasing age is clearly visible. The $\delta^{13}C$ record of *Cibicidoides* shows a shift of ca 1.1‰ between glacials and interglacials; this is about the largest amplitude observed so far.

Down-core changes in the benthic foraminiferal composition are indicated by R-mode principal components PC 1 and PC 2 explaining 16 and 10% of the total variance, respectively (Figure 14, center). PC 1 on its positive half is similar to the recent Southern Component Bottom Water Fauna (Mackensen et al., 1993). The most prominent components include *Nuttallides umbonifer* and *F. wuellerstorfi*. This fauna reaches maximum values during peak interglacial periods 9, 7.5, 7.3-7.1, 5.5 and 1. The second end-member of PC 1 corresponds to the recent Southern High-Productivity Fauna characterized by deep endobenthic species, such as *Steinforthia davisi*, *Melonis zaandami*, *M. pompilioides*, *Pullenia subcarinata* and others. This fauna dominates during glacial stages 8, 6, 4 and 2. The negative end-member of PC 2 corresponds to the recent Northern Component Deep Water Fauna, characterized by *E. exigua* and *O. umbonatus*. This fauna is dominant during interglacial oxygen isotope stages 7, 5 and 3, but not during peak interglacial stages 9, 7.5, 7.3, 5.5 and 1.1. Positive values of PC 2 do

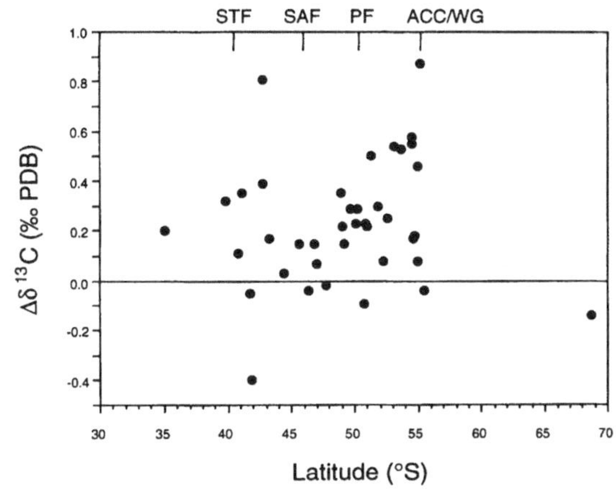

Figure 13. Carbon isotope difference between regional bottom water and benthic foraminifers. STF = Subtropical Front, SAF = Subantarctic Front, PF = Polar Front, ACC/WG = Antarctic Circumpolar Current/Weddell gyre boundary. After Mackensen et al. (in press).

not correspond to any recent analogue fauna, but seem to indicate a transition between a Southern Component Bottom Water Fauna and a Southern High-Productivity Fauna.

To further document changes in surface ocean productivity in Core PS2082 we calculated paleoproductivity rates (PP in Figure 14, right-hand side) applying the empirical equations of Müller and Suess (1979) and Stein (1986). Both formulas imply primary productivities of 40 g C $m^{-2}a^{-1}$ during interglacials and 80-160 g C $m^{-2}a^{-1}$ during glacials (Table 2). Differences between curves during glacials are caused by different weighing of sedimentation rates and considerations of water depth as important factors influencing C_{org} accumulation rates (Stein (1986). The recent values of 60 g C $m^{-2}a^{-1}$ agree well with estimates of Berger (1989). Glacial values of >80 g C $m^{-2}a^{-1}$ agree with sediment trap data from the Polar Front Zone (Wefer and Fischer, 1991).

Core PS1506: 69°S, Eastern Weddell Sea Continental Margin

Core PS1506 was chosen from a suite of gravity cores from the Weddell Sea continental margin because of relatively high sedimentation rates (1-4 cm kyr^{-1}) and carbonate contents (up to 23%). Usually, on the Antarctic continental margin, enhanced sedimentation is coupled with low biogenic carbonate contents (Grobe and Mackensen, 1992). During the last approximately 400,000 years the $\delta^{13}C$ records of *N. pachyderma*, *E. exigua* and *O. umbonatus* exhibit the characteristic glacial-to-interglacial cycles with low $\delta^{13}C$ values during glacials (Figure 15,

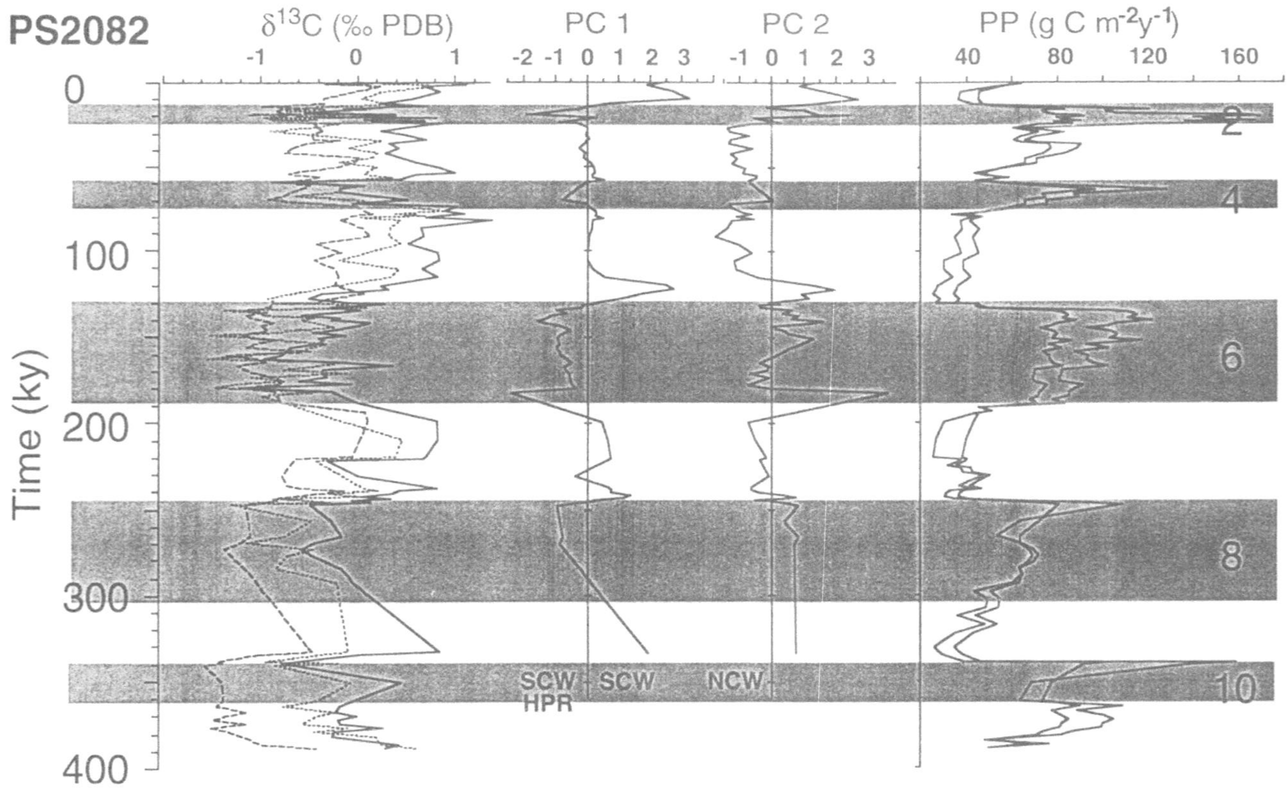

Figure 14: $\delta^{13}C$ records of *Cibicidoides* (stippled),*N. pachyderma* (dotted) and *G. bulloides* (full), as well as benthic foraminiferal assemblages (PC1 and PC2) and paleoproductivity rates (Müller and Suess, 1979; Stein, 1986) in PS2082 from the abyssal Agulhas Basin (SCW=Southern Component Bottom Water Fauna; HPR=Southern High Productivity Fauna; NCW=Northern Component Deep Water Fauna).

Table 1). The glacial-interglacial $\delta^{13}C$ amplitude in the planktic *N. pachyderma* record and the endobenthic *O. umbonatus* record is >1.0‰, whereas that of the epibenthic, phyto-detritus feeding *E. exigua* increases is only 0.6‰. After a correction of +1.2‰ is applied for the *E. exigua* $\delta^{13}C$ values to account for possible vital effects the interglacial values are similar to the *N. pachyderma* record but glacial values in the benthic $\delta^{13}C$ record are more possitive than the planktic values.

The benthic foraminiferal fauna at the mid-depth eastern Antarctic continental slope does not change much between glacials and interglacials. Only during times of rapid climatic amelioration does the number of benthic foraminifera (BFN) drastically decrease so that the last four climatic optima are very clearly depicted by low benthic counts (Figure 15). Based on these low numbers we infer severe calcite dissolution during times of enhanced bio-siliceous productivity. This contention is supported by the opal, carbonate and barium records (Grobe and Mackensen, 1992; Shimmield et al., this volume). When plotted on a log-scale, the BFN indicates severe calcite dissolution during stages 11 and 9. Both climatic optima have been previously reported to represent the warmest interglacial conditions in the Southern Ocean (Oppo et a., 1990; Hodell, 1993). The negative end-member of Principal Component 1 (PC1) is characterized by infaunal species such as *Melonis pompilioides* and *Pullenia* spp. PC1 accounts for just 10% of the variance but it clearly shows that infaunal species are characteristic members of the interglacial fauna before oxygen isotope stage 12, indicating enhanced productivity then.

DISCUSSION

Our results suggest that the benthic $\delta^{13}C$ signal in Core PS2082 from the Agulhas Basin at 43°S and 4.6 km water depth reflects drastic changes in local or regional particulate organic matter fluxes which are superimposed on the global carbon inventory signal of 0.32-0.46‰ (Duplessy et al. ,1988; Curry et al., 1988) and the signal of varying NADW advection into the Southern Ocean. The changes in organic matter supply were caused by latitudinal fluctuations of the high productivity belt associated with the polar front and the northern sea-ice limit. Benthic foraminiferal assemblages and inferred paleoproductivity rates suggest a northward shift of the polar front by some 7° and an increase of primary productivity by at least a factor of two during glacial periods 2, 4, 6, 8, 10 and parts of stage 3 (Figure 14). This is in agreement with findings of Defelice and Wise (1981) and with results from the southern Indian Ocean (Pichon et al., 1992; Francois et al., 1993). It also agrees with conclusions of Mortlock et al. (1991) and Charles et al. (1991) who found high glacial productivity north of the polar front but low productivity south of the Front, where present-day productivity reaches a maximum. On the other hand, benthic faunas indicate that the present-day situation where the site is bathed in LCPDW is found only during peak warm interglacial periods 5.5, 7.3, 7.5 and 9. During

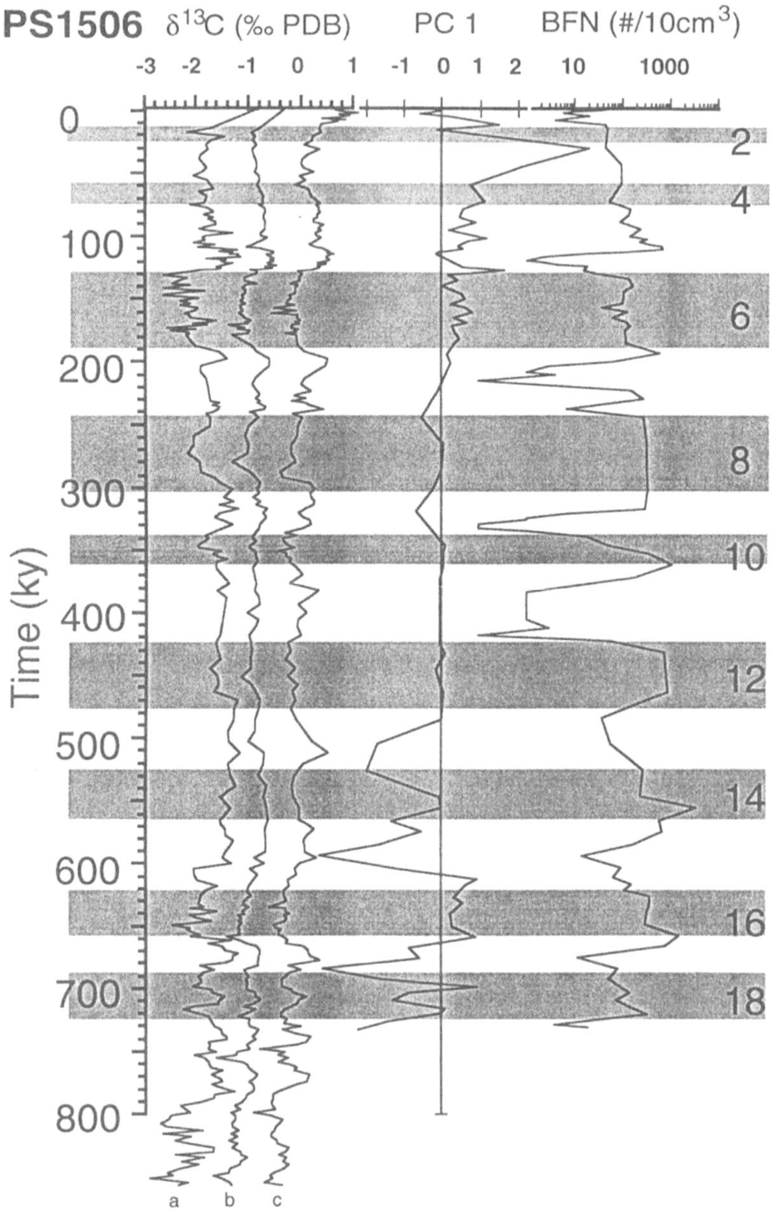

Fig 15: $\delta^{13}C$ records of *O. umbonatus* (a), *E. exigua* (b) and *N. pachyderma* (c) in PS1506, as well as benthic foraminiferal assemblages (PC1) and the Benthic Foraminiferal Number (BFN) in closeby PS1388 from the Antarctic continental margin.

stages 3, 5.1 through 5.3, 7.1 and 7.4. (probably not in stage 9) the Agulhas Basin was filled with NADW or an equivalent water mass. This micropaleontological evidence is further supported by slightly increasing $\delta^{13}C$ values during interglacial periods that may reflect increasing NADW fluxes, or more decreasing AABW production, after the climatic optima.

Comparison with Core RC13-229 from the northernmost Cape Basin (Oppo et al., 1990) reveals similar $\delta^{13}C$ fluctuations during the last 400,000 years, although Core PS2082 seems to exhibit a trend towards lower values down to Stages 8 and 10; values from the Agulhas Basin are consistently lower in stages 2, 8, and 10 than in the Cape Basin (Figure 16). Core-top values in the Cape and in the Agulhas Basin are 0.39 and 0.31‰, respectively. Both sites are bathed by LCPDW; the Agulhas Basin, however, is influenced by AABW and the Cape Basin by some 20% of NADW. The shift during the last deglaciation is 0.8‰ in the Cape Basin and 1.1‰ in the Agulhas Basin. Based on the present-day relation between *Cibicidoides* $\delta^{13}C$ values and seasonal high productivity (Mackensen et al., in press), we suggest that glacial $\delta^{13}C$ values in Core PS2082 include a productivity-induced lowering of about 0.4‰. Core PS2082 is situated close to the polar front region, which experienced highly different surface ocean productivity during glacial-interglacial times. There is no evidence in support of a productivity belt associated with the polar front as far north as 26°S during glacials, and thus Core RC13-229 should show similar $\delta^{13}C$ values during interglacial times and consistently higher $\delta^{13}C$ values during glacial times when compared to Core PS2082 (Figure 16). This obviously does not hold true for Stages 4 and 6 when $\delta^{13}C$ values are lower at the northern Core RC13-229 which may infer increased glacial upwelling west of Namibia and Angola and a shift in plankton quality (Oberhänsli, 1991). A mean Southern Ocean glacial-interglacial $\delta^{13}C$ amplitude of 0.81‰ (Curry et al., 1988) and 0.99‰ (Mackensen et al., in press) was calculated based on $\delta^{13}C$ records from the area. This amplitude is greater than in all other regions. It implies that during the Last Glacial Maximum Southern Ocean nutrient contents were higher than e.g. in the eastern equatorial Pacific Ocean (Duplessy et al., 1988; Curry et al., 1988; Oppo et al., 1990; Raymo et al., 1990). However, most of the cores used for this calculation may have been substantially influenced by a northward migration the polar front and/or Subantarctic Front system, with possible effects of changing plankton assemblages and primary productivity on the foraminiferal $\delta^{13}C$ records (Hays et al., 1976; Defelice and Wise, 1981).

Assuming that *N. pachyderma* (Charles and Fairbanks, 1990) and *E. exigua* (the latter adjusted by +1.2‰) reliably record near-surface water and bottom-water $\delta^{13}C$ then should deep convection and bottom water formation be documented by similar and positive $\delta^{13}C$ values on the order of +1‰ in both the planktic and benthic records from Core PS1506 at the mid-slope Antarctic continental margin (Figure 16). Because of a lower glacial-interglacial $\delta^{13}C$ amplitude in the benthic record the benthic values are significantly higher during glacial periods than the planktic values. This implies that either the value of 1.2‰ applied for *E. exigua* to correct for possible vital effects changes on glacial-interglacial time scales or that during cold glacial peri-

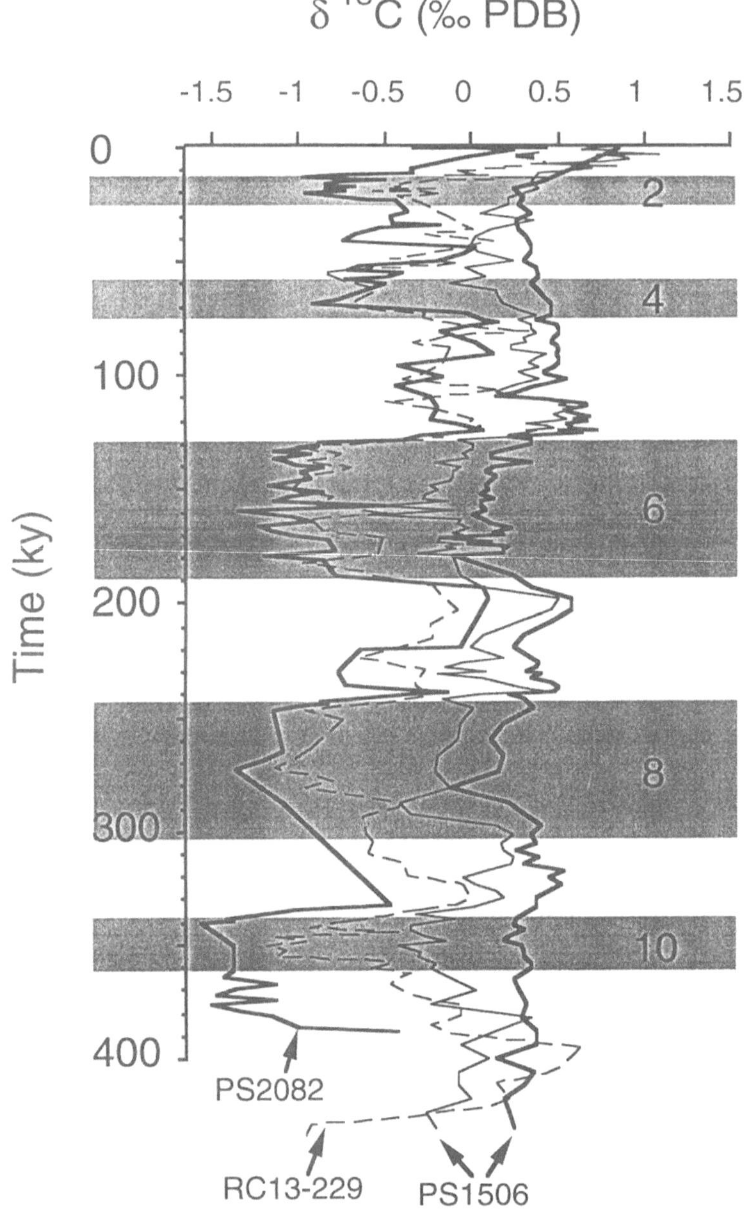

Figure 16: $\delta^{13}C$ records of *Cibicidoides* from PS2082 and RC13-229 with *E. exigua* record of PS1506 (bold graph) corrected to the *Cibicidoides* scale and the *N. pachyderma* record of PS1506 (thin graph).

ods deep convection and bottom water formation was suppressed and a stratified water column existed such that planktic values became lighter than the benthic values. The latter case would imply almost no surface productivity to keep the $\delta^{13}C_{\Sigma CO2}$ of the surface water at its equilibrium level with the atmosphere (Figure 16). Nevertheless, either case calls for reduced surface productivity during glacial periods. Discarding the possibility of varying vital effects for *E. exigua*, the reduced $\delta^{13}C$ amplitude of *E. exigua* may just reflect the global carbon inventory shift combined with a reduced flux of NADW into the Southern Ocean.

Although the benthic foraminiferal assemblage does not vary much during the last 400,000 years in Core PS1506, the benthic foraminiferal number, along with carbonate and opal contents, indicates intensive calcite dissolution for the interglacial stages 11 and 9 and peak warm events 7.5, 7.3, 5.5 and 1.1 occurred (cf. Mackensen et al., 1989; Grobe and Mackensen, 1992). Together with high opal and Ba contents during the same periods (Grobe et al., 1990; Shimmield et al., this volume) this further underscores that primary productivity must have been enhanced at the Antarctic continental margin during the climatic optima of the last four climatic cycles. Prior to Stage 12 glacial-to-interglacial contrasts were not as pronounced and calcite dissolution apparently did not occur. During this earlier time the negative end member of PC 1 represents a fauna closely related to the recent Southern High Productivity Fauna which is an important constituent of the benthic foraminiferal assemblage during interglacial stages 17, 15, and 13 (Figure 15). The benthic foraminiferal fauna together with benthic and planktic $\delta^{13}C$ data and carbonate and opal records from the Antarctic continental margin all consistently call for enhanced primary productivity and bottom water formation during peak interglacial periods and low production with reduced bottom water formation during moderate interglacials and glacials. A lowered sea-level conceivably resulted in reduced bottom water formation because extensive ice shelves and large seasonal polynyas like today did not exist (Grobe and Mackensen, 1992). This would also explain reduced primary production during glacial periods.

CONCLUSIONS

Benthic $\delta^{13}C$ records from core sites to the north and south of the Southern Ocean Polar Front are offset by more than 0.4‰ during glacials when compared to interglacials. We use this enhanced offset to infer increased glacial productivity at the northern core sites which was presumably brought about by a northward shift of the polar front with its high-productivity belt. In addition, biological productivity off W-Africa would have been increased due to intensified coastal upwelling whereas productivity in the Weddell Sea remained roughly constant. This scenario is supported by systematic incursions of a southern High-Productivity Fauna to the northern Core PS2082 during glacial periods indicating a northward migration of the polar front from presently 55-50°S to a glacial position at 43°S.

The benthic foraminiferal faunas indicate that southern component bottom waters entered the northern Agulhas Basin only during peak interglacial periods (Stages 9, 7.5, 7.3, 5.5 and 1.1). We speculate that AABW production was most extensive during deglaciations and climatic optima. This is also implied by the $\Delta\delta^{13}C_{benthic-planktic}$ record from the Antarctic continental margin. The advection of northern component deep water to the eastern Atlantic sector of the Southern Ocean was strongest during moderate interglacial stages 5.3 through 5.1 and 3.

ACKNOWLEDGEMENTS

We acknowledge the assistance of R.V. *Polarstern* crews and masters during numerous cruises into the Southern Ocean. We are grateful to M. Heyn and H. Röben who skilfully carried out all of the on-shore laboratory preparations. Finally, we thank R. Zahn and M. Kaminski for inviting A.M. to this NATO workshop. This is Contribution 67 of the Sonderforschungsbereich 261 and publication No. 691 of the Alfred Wegener Institute for Polar and Marine Research.

REFERENCES

Belanger, P.E., Curry, W.B. and Matthews, R.K., 1981. Core-top evaluation of benthic foraminiferal isotopic ratios for paleoceanographic interpretations. Palaeogeogr., Palaeoclimatol., Palaeoecol., 33: 205-220.

Berger, W.H., 1989. Global maps of ocean productivity. In: Productivity of the Ocean: Present and Past. W.H. Berger, V.S. Smetacek and G. Wefer (eds.). Dahlem Workshop Reports, J. Wiley and Sons, 429-455.

Bickert, T., 1992. Rekonstruktion der spätquartären Bodenwasserzirkulation im östlichen Südatlantik über stabile Isotope benthischer Foraminiferen, Rep., 27, 205 pp., Fachbereich Geowissenschaften, Univ. Bremen, Bremen, Germany.

Boyle, E.A., 1988. Cadmium: chemical tracer of deepwater paleoceanography. Paleoceanography, 3(4): 471-489.

Boyle, E.A., 1992. Cadmium and $\delta^{13}C$ paleochemical ocean distributions during the stage 2 glacial maximum. Ann. Rev. Earth Planet. Sci., 20: 245-287.

Boyle, E.A. and Keigwin, L.D., 1982. Deep circulation of the North Atlantic over last 200,000 yeras: geochemical evidence. Science, 218: 784-787.

Boyle, E.A. and Keigwin, L., 1987. North Atlantic thermohaline circulation during the past 20,000 years linked to high-latitude surface temperature. Nature, 330: 35-40.

Broecker, W.S., 1982. Glacial to interglacial changes in ocean chemistry. Prog. Oceanog., 11: 151-197.

Broecker, W.S., 1993. An oceanographic explanation for the apparent carbon isotope-cadmium discordancy in the glacial Antarctic? Paleoceanography, 8(2):137-139.

Broecker, W.S. and Peng, T.-H-, 1986. Global carbon cycle: 1985. Radiocarbon, 28:309-327.

Broecker, W.S. and Peng, T.-H-, 1989. The cause of the glacial to interglacial atmospheric CO_2 change: A polar alkalinity hypothesis. Global Biogeochem. Cycles, 3: 215-239.
Charles, C.D. and Fairbanks, R.G., 1990. Glacial to interglacial changes in the isotopic gradients of Southern Ocean surface water. In: Geological History of Polar Oceans: Arctic versus Antarctic. U. Bleil and J. Thiede (eds.), NATO/ASI Ser. C: 519-538.
Charles, C.D., Froelich, P.N., Zibello, M.A., Mortlock, R.A., Morley, J.J., 1991. Biogenic opal in Southern Ocean sediments over last 450,000 years: Implications for surface water chemistry and circulation. Paleoceanography, 6: 697-728.
Curry, W. B., and Lohmann, G.P., 1982. Carbon isotopic changes in benthic foraminifera from the western South Atlantic: Reconstruction of glacial abyssal circulation patterns, Quat. Res.,18: 218-235.
Curry, W.B., Duplessy, J.C., Labeyrie, L., and Shackleton, N.J., 1988. Changes in the distribution of $\delta^{13}C$ of deep water SCO_2 between the last glaciation and the Holocene. Paleoceanography, 3(3): 317-341.
Defelice, D.R. and Wise, S.W., Jr., 1981. Surface lithofacies , biofacies, and diatom diversity patterns as modesls for delineation of climatic change in the southeast Atlantic Ocean. Mar. Microplaeontol., 6: 29-70.
Demaster, D.J., 1981. The supply and accumulation of silica in the marine environment. Geochimica et Cosmochimica Acta, 45: 1715-1732.
Douglas, R.G. and Woodruff, F., 1981. Deep-sea benthic foraminifera. The Oceanic Lithosphere, The Sea, C. Emiliani (ed.), vol. 7, 1233-1327. Wiley-Interscience, New York, N.Y.
Duplessy, J.-C., Shackleton, N.J., Fairbanks, R.G., Labeyrie, L., Oppo, D., and Kalllel, N., 1988. Deepwater source variations during the last climatic cycle and their impact on the global deepwater circulation. Paleoceanography, 3(3): 343-360.
Fischer, G., 1991. Stable carbon isotope ratios of plankton carbon and sinking organic matter from the Atlantic sector of the Southern Ocean. Mar. Chem., 35:581-596.
Foldvik, A. and Gammelsrød, T., 1988. Notes on Southern Ocean hydrography, sea-ice and bottom water formation. Palaeogeogr., Palaeoclimatol., Palaeoecol., 67: 3-17.
Francois, R., Altabet, M.A. and Burckle, L.H., 1993. Glacial to interglacial changes in surface nitrate utilization in the Indian sector of the Southern Ocean as recorded by sediment $d^{15}N$. Paleoceanography, 7(5): 589-606.
Gooday, A.J., 1988. A benthic foraminiferal response to the deposition of phytodetritus in the deep sea. Nature, 322: 70-73.
Grobe, H. and Mackensen, A., 1992. Late Quaternary climatic cycles as recorded in sediments from the Antarctic continental margin. In: The Antarctic Paleoenvironment: A Perspective on Global Change, J.P. Kennett and D.A. Warnke (eds.). Antarctic Res. Ser., 56(1):349-376.
Grobe, H., Mackensen, A., Hubberten, H.-W., Spiess, V. and Fütterer, D.K., 1990. Stabel isotope record and late Quaternary sedimentation rates at the Antarctic continental margin. In: Geological History of Polar Oceans: Arctic versus Antarctic. U. Bleil and J. Thiede (eds.), NATO/ASI Ser. C: 539-572.
Hays, J.D., Lozano, J.A., Shackleton, N., amd Irving, D., 1976. Reconstruction of the Atlantic and Western Indian Ocean sectors of the 18,000 B.P. Antarctic Ocean. In: R.M. Cline and J.D. Hays (eds.), Investigation of Late Quaternary paleoceanography and paleoclimatology. Geol. Soc. Am. Mem., 145: 337-373.
Hodell, D.A., 1993. Late Pleistocene Paleoceanography of the south Atlantic sector of the Southern Ocean: Ocean Drilling Program Hole 704A. Paleoceanography, 8(1): 47-67.
Imbrie, J., Hays, J.D., Martinson, D.G., MacIntyre, A., Mix, A.C., Morley, N., Pisias, N.G., Prell, W.L., and Shaxckleton, N.J., 1984. The orbital theory of Pleistocene climate: Support from a revised chronology of the marine $\delta^{18}O$ record. In: Milankovitch and Climate. A. Berger, J. Imbrie, J. Hays, G. Kukla, and B. Saltzmann (eds.), pp. 269-305, D. Reidel, Hingham, Mass.
Keir, R.S., 1988. On the late Pleistocene ocean geochemistry and criculation. Paleoceanography, 3: 413-445.
Kroopnick, P., 1980. The distribution of ^{13}C in the Atlantic Ocean. Earth and Planetary Science Letters, 49: 469-484.

Labracherie, M., Labeyrie, L.D., Duprat, J., Bard, E., Arnold, M., Pichon, J.-J. and Duplessy, J.C., 1989. The last deglaciation in the Southern Ocean. Paleoceanography, 4(6): 629-638.
Loeblich, A.R., Tappan, H., 1988. Foraminiferal genera and their classification, Vol. 1 and 2, Van Nostrand Reinhold, New York.
Lohmann, G.P., 1978. Abyssal benthonic foraminifera as hydrographic indicators in the western South Atlantic Ocean. J. Foraminiferal Res., 8: 6-34.
Lutze, G.F. and Thiel, H., 1987.Cibicidoides wuellerstorfi and Planulina ariminensis, elevated epibenthic Foraminifera, in Beobachtungen an Benthos-Foraminiferen (Teilprojekt A3), Rep. Sonderforschungsbereich 313, vol. 6, edited by A. V. Altenbach, G. F. Lutze and P. Weinholz, pp. 17-30, Univ. Kiel, Kiel, Germany, 1987.
Mackensen, A., Sejrup, H.P., and Jansen, E., 1985. The distribution of living benthic foraminifera on the continental slope and rise off southwest Norway. Mar. Micropaleontol., 9:275-306.
Mackensen, A., Grobe, H, Hubberten, H.-W., Spiess, V., Fütterer, D.K., 1989. Stable isotope stratigraphy from the Antarctic continental margin during the last one million years. Mar. Geol., 87: 315-321.
Mackensen, A., Grobe, H., Kuhn, G., and Fütterer, D.K., 1990. Benthic foraminiferal assemblages from the eastern Weddell Sea between 68 and 73°S: distribution, ecology and fossilization potential. Mar. Micropaleontol., 16: 241-283.
Mackensen, A., Fütterer, D.K., Grobe, H., and Schmiedl, G., 1993. Benthic foraminiferal assemblages from the eastern south Atlantic Polar Front region between 35 and 57°S: distribution, ecology and fossilization potential. Mar. Micropaleontol., 22: 33-69.
Mackensen, A., Hubberten, H.-W., Bickert, T., Fischer G., and Fütterer, D.K., in press. The $\delta^{13}C$ in benthic foraminiferal tests of *Fontbotia wuellerstorfi* (Schwager) relative to the $\delta^{13}C$ of dissolved inorganic carbon in the Southern Ocean deep water: Implications for glacial ocean circulation models. Paleoceanography, 8(5): 587-610.
Martin, J.H., 1990. Glacial-interglacial CO_2 change: The iron hypothesis. Paleoceanography, 5: 1-13.
Mortlock, R.A., Charles, C.D., Froelich, P.N., Zibello, M.A., Saltzman, J., Hays, J.D., and Burckle, L.H., 1991. Evidence for lower prodcutivity in the Antarctic Ocean during the last glaciation. Nature, 351: 220-222.
Müller, P.J. and Suess, E., 1979. Productivity, sedimentation rate, and sedimentary organic matter in the oceans. I.-Organic matter preservation. Deep-Sea Res., 26: 1347-1362.
Oberhänsli, H., 1991. Upwelling signals at the northeastern Walvis Ridge during the past 500,000 years. Paleoceanography, 6(1): 53-71.
Oppo, D.W., Fairbanks, R.G., Gordon, A.L., and Shackleton, N.J., 1990. Late Pleistocene Southern Ocean $\delta^{13}C$ variability. Paleoceanography, 5(1): 43-54.
Paillard, D., Lecoat, F., Chol, E., and Jodet, L., 1992. Analyseries. Computer Program for graphical correlation on a Macintosh. Preliminary version.
Peterson, R.G. and Stramma, L., 1991. Upper-level circulation in the South Atlantic Ocean. Prog. Oceanog., 26: 1-73.
Peterson, R.G. and Whitworth, T., III., 1989. The Subantarctic and Polar Fronts in relation to deep water masses through the southwestern Atlantic. J. Geophys. Res., 94(C8): 10817-10838.
Pichon, J.-J., Labeyrie, L.D., Bareille, G., Labracherie, M., Duprat, J., and Jouzel, J., 1992. Surface water temperaturr changes in the high latitudes of the Southern Hemisphere over the last glacial-interglacial cycle. Paleoceanography, 7(3): 289-318.
Raymo, M.E., Ruddiman, W.F., Shackleton, N.J., & Oppo, D.W., 1990. Evolution of Atlantic-Pacific $\delta^{13}C$ gradients over the last 2.5 m.y. Earth and Planetary Science Letters, 97: 353-368.
Rau, G. H., T. Takahashi, and D. J. des Marais, , 1989. Latitudinal variations in plankton $\delta^{13}C$: Implications for CO_2 and productivity in past oceans, Nature, 341, 516-518.
Reid, J.L., 1989. On the total geostrophic circulation of the South Atlantic Ocean: Flow patterns, tracers, and transports. Prog. Oceanog., 23: 149-244.
Schnitker, D., 1974. West Atlantic abyssal circulation during the past 120,000 years. Nature, 248: 385-387.

Stein, R., 1986. Organic carbon and sedimentation rate - Further evidence for anoxic deep-water conditions in the Cenomanian/Turonian Atlantic Ocean. Mar. Geol., 72:199-209.
Streeter, S.S., 1973. Bottom water and benthonic foraminifera in the North Atlantic glacial-interglacial contrasts. Quat. Res., 3: 131-141.
Streeter, S.S. and Shackleton, N.J., 1979. Paleocirculation of the deep North Atlantic: 150,000 yr record of benthic foraminifera and oxygen-18. Science, 203: 168-171.
Wefer, G. and Fischer, G., 1991. Annual primary prodcution and export flux in the Southern Ocean from sediment trap data. Marine Chemistry, 35(1-4): 597-613.
Whitworth, T., III., and Nowlin, W.D., Jr., 1987. Water masses and currents of the Southern Ocean at the Greenwich meridian, J. Geophys. Res., 92(C6): 6462-6476.
Woodruff, F., 1985. Changes in Miocene deep-sea benthic foraminiferal distribution in the Pacific Ocean: Relationship to paleoceanography. Geol. Soc. Am. Mem., 163:131-175.
Woodruff, F., Savin, S., and Dourglas, R.G., 1980. Biological fractionation of oxygen and carbon isotopes by Recent foraminifera. Mar. Micropaleontol., 5: 3-11.

Appendix: Tables 1-4

Table 1: Changes in benthic (*E. exigua, O. umbonatus*) and planktic (N. pachyderma) foraminiferal $\delta^{18}O$ and $\delta^{13}C$ in Core PS1506 from the Antarctic continental margin.

Depth (m)	CaCO3 (%)	E. exigua $\delta^{18}O$	E. exigua $\delta^{13}C$	O. umbonatus $\delta^{18}O$	O. umbonatus $\delta^{13}C$	N. pachyderma $\delta^{18}O$	N. pachyderma $\delta^{13}C$
0.00		3.66	-0.41	3.42	-1.14	3.51	0.91
0.00		3.81	-0.32	3.69	-0.84		
0.01	0.89			3.41	-1.03		
0.02				3.71	-0.84		
0.03	0.67	3.78	-0.34	3.80	-0.88	3.69	0.69
0.06	0.68					4.08	0.64
0.09	0.02					3.96	0.79
0.12	0.05					3.74	0.82
0.15	0.05					3.80	1.09
0.18	0.00					3.70	0.72
0.21	0.46					3.64	0.89
0.24	0.46					3.76	0.92
0.27	0.00					3.72	0.76
0.30	0.00					4.55	0.40
0.33	0.01					3.76	0.84
0.36	0.33					3.98	0.59
0.39	0.00					3.81	0.70
0.42	0.43					4.08	0.58
0.45	0.27					4.31	0.39
0.48	1.08	4.83	-0.80	4.70	-1.71	4.70	0.35
0.51	1.39	4.78	-0.95	4.62	-2.16	4.74	0.39
0.54	2.64	4.98	-0.89	4.64	-1.45	4.86	0.22
0.57	2.14	5.00	-0.93	4.67	-1.71	4.87	0.21
0.60	1.85	4.71	-0.80	4.79	-1.80	4.94	0.06
0.60		4.82	-0.92				
0.63	2.07	4.23	-1.03	4.18	-1.75	4.42	0.36
0.63		4.65	-0.84				
0.66	2.64	4.95	-0.93	4.65	-1.89	5.03	0.12
0.66		4.93	-0.91				
0.69	2.43	4.90	-0.89	4.74	-1.93	4.89	0.03
0.69		4.88	-0.87				
0.72	2.66	4.88	-0.87	4.80	-1.85	5.00	0.00
0.72		4.90	-0.84				
0.75	2.63	4.79	-0.85			4.90	0.04
0.75		4.75	-0.87	4.72	-1.78		
0.78	2.60	4.83	-0.90	4.65	-1.86	4.90	0.24
0.81	2.80	4.87	-0.82	4.53	-2.04	4.84	-0.03
0.84	2.54	4.84	-0.81	4.69	-1.77	4.82	0.09
0.87	2.92	4.78	-0.85	4.53	-1.72	4.77	-0.12
0.90	2.75	4.74	-0.81	4.52	-1.65	4.85	0.14
0.93	4.26	4.76	-0.78	4.48	-1.92	4.78	0.16
0.96	3.15	4.69	-0.73	4.48	-1.78	4.57	0.19
0.99	3.22	4.70	-0.74	4.36	-1.80	4.78	0.29
1.02	3.65	4.65	-0.73	4.38	-1.83	4.63	0.37
1.05	3.58	4.53	-0.82	4.20	-2.10	4.62	0.28
1.08	4.71	4.57	-0.71	4.32	-1.78	4.55	0.35
1.11	7.20	4.52	-0.69	4.23	-1.69	4.54	0.34
1.14	5.48	4.54	-0.69	4.42	-1.65	4.52	0.37
1.17	4.71	4.52	-0.72	4.15	-1.83	4.52	0.30
1.20	5.23	4.54	-0.73	4.34	-1.66	4.45	0.32
1.23	6.33	4.45	-0.71	4.20	-1.75	4.53	0.23
1.26	7.83	4.54	-0.69	4.31	-1.62	4.56	0.23

Table 1, Continued

Depth (m)	CaCO3 (%)	E. exigua $\delta^{18}O$	E. exigua $\delta^{13}C$	O. umbonatus $\delta^{18}O$	O. umbonatus $\delta^{13}C$	N. pachyderma $\delta^{18}O$	N. pachyderma $\delta^{13}C$
1.29	7.15	4.59	-0.72	4.23	-1.61	4.49	0.44
1.29		4.55	-0.69				
1.32	6.14	4.47	-0.70	4.28	-1.73	4.34	0.38
1.35	8.77	4.53	-0.70	4.14	-1.91	4.45	0.33
1.38	7.11	4.53	-0.72	4.12	-1.79	4.51	0.39
1.41	8.21	4.43	-0.78	4.39	-1.30	4.49	0.27
1.44	7.99	4.27	-0.77	4.17	-1.79	4.32	0.30
1.47	11.08	4.48	-0.65	4.18	-1.86	4.36	0.39
1.50	11.58	4.46	-0.74	4.20	-1.81	4.43	0.30
1.53	12.99	4.47	-0.86	4.13	-2.03	4.57	0.22
1.56	8.76	4.41	-0.92	4.30	-1.63	4.48	0.15
1.59	7.86	4.27	-1.05	4.03	-1.95	4.28	0.28
1.62	8.71	4.54	-0.79	4.25	-1.55	4.31	0.42
1.65	19.87	4.41	-0.69	4.21	-1.22	4.36	0.47
1.68	26.88	4.27	-0.55	3.91	-1.54	4.24	0.55
1.71	20.32	4.24	-0.53	4.02	-1.31	4.04	0.50
1.74	15.50	4.18	-0.58	3.77	-1.36	3.97	0.62
1.77	12.73	3.95	-0.67	3.98	-1.20	3.87	0.57
1.80	12.45	4.19	-0.63	3.98	-1.23	4.02	0.52
1.83	10.50	4.07	-0.58	3.95	-1.15	4.05	0.48
1.86	7.74	4.10	-0.52	3.88	-1.79	3.88	0.50
1.89	1.51	4.04	-0.60	3.91	-1.42	4.14	0.55
1.92	1.98	4.22	-0.53	3.83	-1.50	4.18	0.48
1.95	0.44	4.17	-0.65	4.02	-1.41	4.12	0.32
1.98	0.02	4.18	-0.62	3.91	-1.30	4.24	0.43
2.01	0.08	4.26	-0.63			4.01	0.48
2.04	0.08	4.38	-0.48			4.38	0.32
2.07	0.24	4.10	-0.70	3.89	-1.60	4.38	0.37
2.10	1.24	4.23	-0.56	3.86	-1.51	4.33	0.32
2.13	0.26	4.35	-0.80	4.10	-1.61	4.38	0.21
2.16	0.80	4.68	-0.98	4.64	-1.84	4.60	0.24
2.19	1.07	4.54	-0.85			4.59	-0.07
2.22	2.14	4.37	-0.86			4.84	0.02
2.25	1.86	4.52	-0.97	4.20	-2.62	4.87	0.02
2.28	1.81	4.62	-1.08	4.73	-2.16	4.87	-0.04
2.31	1.54	4.65	-1.08	4.31	-2.32	4.99	-0.01
2.34	1.04	4.33	-0.86	4.40	-2.45	4.77	-0.20
2.37	3.86	4.52	-1.07	4.20	-2.10	4.91	-0.06
2.40	5.39	4.78	-1.11	4.41	-2.61	4.83	-0.11
2.43	6.85	4.77	-1.06	4.06	-1.81	4.92	-0.11
2.46	8.60	4.74	-1.08	4.12	-2.48	4.68	-0.12
2.49	11.33	4.91	-1.12	4.30	-2.62	4.78	-0.23
2.52	12.46	4.71	-1.10	4.58	-1.94	4.76	-0.19
2.55	3.15	4.58	-1.15	4.38	-2.30	4.78	-0.26
2.58	4.90	4.70	-1.07	4.46	-2.10	4.30	-0.28
2.61	4.11	4.68	-1.15	4.44	-2.28	4.67	-0.13
2.64	2.28	4.77	-1.15	4.39	-1.95	3.97	-0.52
2.67	5.47	4.61	-1.11	4.54	-2.01	4.66	-0.10
2.70	4.31	4.73	-1.18	4.41	-2.20	4.01	-0.57
2.73	2.92	4.55	-1.13	4.22	-2.11	4.63	-0.06
2.76	2.92	4.72	-1.08	4.50	-2.09	4.59	-0.08
2.79	6.97	4.51	-0.97	4.19	-1.61	4.60	-0.01
2.82	6.34	4.65	-1.07	4.39	-1.74	4.43	-0.11
2.85	9.08	4.44	-1.01	4.28	-2.04	4.42	-0.04
2.88	8.08	4.06	-1.37	4.35	-1.77	4.35	-0.17
2.91	6.86	4.43	-1.08	4.22	-2.55	4.08	-0.19
2.94	6.30	4.45	-1.21	4.37	-1.99	4.45	-0.05
2.97	6.50	4.42	-0.99	4.28	-2.48	4.62	-0.14

Table 1, Continued

Depth (m)	CaCO3 (%)	E. exigua $\delta^{18}O$	E. exigua $\delta^{13}C$	O. umbonatus $\delta^{18}O$	O. umbonatus $\delta^{13}C$	N. pachyderma $\delta^{18}O$	N. pachyderma $\delta^{13}C$
3.00	4.41	4.58	-1.01	4.43	-2.17	4.54	-0.09
3.03	5.54	4.35	-1.01	4.16	-2.11	4.43	-0.30
3.06	4.51	4.26	-1.31	4.29	-2.17	4.38	-0.10
3.09	7.49	4.31	-0.95	4.17	-2.00	4.19	-0.06
3.12	6.16	4.45	-0.86	4.12	-1.56	3.78	0.07
3.15	12.68	4.09	-0.62	4.07	-1.41	3.86	0.51
3.18	9.32	4.29	-0.62	3.65	-1.89	4.02	0.47
3.21	2.33	4.24	-0.74	3.87	-1.90	3.97	0.28
3.24	1.16	4.31	-0.89	4.14	-1.80	4.43	0.07
3.27	0.00	4.40	-0.95	3.78	-1.77	4.44	0.00
3.30	0.00	4.51	-0.85			4.33	0.19
3.33	0.00	4.31	-0.81			4.30	-0.05
3.36	0.00	4.38	-0.83			4.55	-0.14
3.39	0.00	4.48	-0.88			4.19	0.09
3.42	2.10	4.40	-0.80	4.23	-1.72	3.85	-0.19
3.45	6.36	4.44	-0.85	4.33	-1.56	4.36	0.07
3.48	10.18	4.19	-0.74	4.28	-1.68	4.12	0.17
3.51	12.94	4.12	-0.69	4.20	-1.55	4.09	0.27
3.54	3.59	4.05	-0.72	4.09	-1.58	4.04	0.43
3.57	5.06	3.98	-0.99	4.08	-1.67	4.45	0.05
3.60	5.64	4.50	-0.90	4.44	-1.80	4.40	-0.16
3.63	6.74	4.43	-0.84	4.32	-1.72	4.63	0.02
3.66	6.50	4.48	-1.01	4.55	-1.73	4.51	-0.06
3.69	7.64	4.48	-1.08	4.29	-2.04	4.55	-0.08
3.72	6.86	4.46	-1.01	4.33	-2.04	4.51	-0.19
3.75	8.22	4.57	-1.04	4.40	-2.15	4.62	-0.20
3.78	8.28	4.14	-1.32	4.23	-1.93	4.34	-0.12
3.81	10.49	4.58	-0.99	4.36	-1.90	4.42	-0.41
3.84	12.61	4.50	-0.89	4.29	-1.66	3.80	-0.35
3.87	15.77	4.37	-0.80	4.35	-1.53	4.35	0.15
3.90	17.45	4.30	-0.86	4.19	-1.33	4.07	0.25
3.93	9.07	4.31	-0.83	4.20	-1.40	4.26	0.17
3.96	7.01	3.94	-0.94	3.99	-1.48	4.12	0.18
3.99	1.22	4.20	-0.81	4.10	-1.31	4.06	0.25
4.02	0.92	4.01	-0.89	3.91	-1.57	4.20	0.23
4.05	0.00	4.31	-0.67	3.98	-1.22	4.29	0.08
4.08	0.00	4.27	-0.75	4.11	-1.55	4.04	-0.05
4.11	0.00	4.33	-0.68	4.03	-1.65	4.33	0.03
4.14	0.00					4.39	0.15
4.17	0.00	4.20	-0.82	3.52	-1.33	4.34	0.18
4.20	0.00	4.20	-0.83			4.28	0.02
4.23	0.00					4.59	-0.32
4.26	0.67	4.53	-0.94	4.39	-1.82	4.71	-0.09
4.29	3.81	4.65	-0.96	4.31	-1.76	4.64	-0.35
4.32	12.32	4.57	-0.89	4.36	-1.83	4.48	-0.35
4.35	9.63	4.61	-0.98	4.21	-1.97	4.73	-0.26
4.38	5.33	4.58	-1.01	4.16	-1.95	4.53	-0.14
4.41	10.71	4.49	-0.91	4.30	-1.53	4.62	-0.43
4.44	7.02	4.39	-0.89	4.14	-1.67	4.55	-0.29
4.47	5.50	4.52	-0.90	4.29	-1.59	4.62	-0.30
4.47				4.13	-1.60		
4.50	0.00	4.46	-0.86	4.05	-1.42	4.44	-0.20
4.53	0.00	4.48	-0.86	4.12	-1.62	4.57	-0.23
4.56	0.00	4.52	-0.95	4.24	-1.63	4.31	-0.10
4.59	0.00	4.37	-0.92	3.98	-1.52	4.42	0.03
4.62	0.00	4.17	-0.88	3.76	-1.39	4.41	-0.23
4.65	0.00	4.40	-0.91	4.07	-1.61	3.71	0.34
4.68	0.00	4.10	-0.83	3.38	-1.44	4.09	0.15

Table 1, Continued

Depth (m)	$CaCO_3$ (%)	E. exigua $\delta^{18}O$	E. exigua $\delta^{13}C$	O. umbonatus $\delta^{18}O$	O. umbonatus $\delta^{13}C$	N. pachyderma $\delta^{18}O$	N. pachyderma $\delta^{13}C$
4.71	0.00	4.44	-0.82			4.38	-0.06
4.74	0.00	4.31	-1.05			4.25	0.10
4.77	0.00	4.07	-0.84			4.45	-0.07
4.80	0.00	4.19	-0.90			4.39	-0.08
4.83	0.00	4.58	-1.01	4.50	-1.56	4.31	0.00
4.86	0.83	4.85	-0.98	4.61	-1.63	4.83	-0.26
4.89	3.75	4.84	-0.96	4.68	-1.59	4.75	-0.21
4.92	7.50	4.77	-0.99	4.55	-1.55	4.74	-0.31
4.95	9.16	4.69	-0.98	4.57	-1.73	4.62	-0.13
4.98	6.66	4.39	-1.15	4.34	-1.60	4.54	-0.26
5.01	9.16	4.64	-0.97	4.44	-1.58	4.70	-0.06
5.04	6.95	4.62	-0.89	4.42	-1.66	4.31	-0.17
5.07	4.53	4.55	-0.80	4.34	-1.23	4.55	-0.11
5.10	5.80	4.58	-0.83	4.47	-1.29	4.57	-0.20
5.13	10.73	4.59	-0.90	4.00	-1.30	4.42	-0.09
5.16	9.62	4.62	-0.90	4.13	-1.40	4.51	0.07
5.19	7.86	4.52	-1.02	4.10	-1.39	4.28	0.21
5.22	8.63	4.47	-0.71	4.19	-1.18	4.24	0.52
5.25	10.95	4.33	-0.73	4.00	-1.40	4.48	0.04
5.28	11.16	4.33	-0.83	4.09	-1.36	4.24	-0.13
5.31	12.76	4.44	-0.67	4.30	-1.26	4.35	-0.18
5.34	16.05	4.41	-0.72	4.32	-1.54	4.32	-0.15
5.37	17.40	4.28	-0.66	4.20	-1.35	4.29	-0.02
5.40	17.06	4.38	-0.64	4.26	-1.54	4.45	-0.04
5.43	11.28	4.32	-0.70	3.79	-1.53	3.92	0.11
5.43		4.13	-0.62				
5.46	10.37	4.20	-0.70	3.88	-1.45	4.01	0.25
5.49	12.65			3.91	-1.45	3.89	0.12
5.52	9.15	4.29	-0.68	4.10	-1.31	4.16	0.05
5.55	6.96	4.31	-0.68	3.97	-1.37	4.02	0.07
5.58	8.26	4.08	-0.69	3.95	-1.45	4.25	0.07
5.61	3.86	4.17	-0.91	4.01	-1.50	4.10	0.29
5.64	4.21	4.35	-0.71	4.24	-1.34	4.57	-0.02
5.67	2.97	4.27	-0.99	4.21	-2.05	4.63	-0.13
5.70	2.83	4.45	-0.85	4.15	-2.03	4.90	-0.09
5.73	1.75	4.47	-0.93	4.07	-1.98	4.98	-0.26
5.76	1.74	4.56	-1.01	4.30	-1.60	5.07	-0.23
5.79	2.00	4.68	-1.04	4.46	-1.81	4.77	-0.33
5.82	1.67	4.88	-0.98	4.44	-1.90	4.72	-0.33
5.85	2.67	4.94	-1.10	4.83	-1.94	4.76	-0.29
5.88	3.92	4.78	-1.12	4.70	-1.96	4.79	-0.45
5.91	4.33	4.61	-1.02	4.79	-1.84	4.50	-0.62
5.94	2.92	4.57	-1.05	4.46	-2.10	4.94	-0.30
5.97	3.17	4.67	-1.13	4.46	-1.61	4.96	-0.30
6.00	2.92	4.81	-1.10	4.50	-1.87	4.76	-0.22
6.03	4.03	4.81	-1.22	4.70	-2.12	4.84	-0.33
6.06	5.83	4.83	-1.15	4.43	-1.96	4.81	-0.28
6.09	8.78	4.78	-1.21	4.44	-1.71	4.76	-0.30
6.12	8.50	4.72	-1.15	4.26	-1.88	4.61	-0.35
6.15	9.66	4.68	-1.19	4.27	-2.46	4.71	-0.54
6.18	12.58	4.77	-1.22	4.48	-2.12	4.52	-0.50
6.21	14.41	4.69	-1.20	3.88	-2.17	4.57	-0.39
6.24	18.69	4.68	-1.21	3.88	-2.08	4.50	-0.43
6.27	16.34	4.54	-1.19	4.05	-1.92	4.58	-0.43
6.30	20.94	4.36	-1.43	3.90	-2.10	4.57	-0.35
6.33	14.74	4.40	-1.19	3.93	-1.35	4.41	-0.26
6.36	14.62	4.52	-1.06	3.90	-1.77	4.20	-0.30
6.39	17.74	4.32	-0.87	4.00	-1.52	4.25	0.04

Table 1, Continued

Depth (m)	CaCO3 (%)	E. exigua $\delta^{18}O$	E. exigua $\delta^{13}C$	O. umbonatus $\delta^{18}O$	O. umbonatus $\delta^{13}C$	N. pachyderma $\delta^{18}O$	N. pachyderma $\delta^{13}C$
6.42	14.43	4.29	-0.80	3.77	-1.45	4.06	0.13
6.45	8.16	4.08	-0.81	3.90	-1.23	4.01	0.14
6.48	4.02	4.15	-0.83	3.98	-1.32	3.82	0.17
6.51	4.16	4.37	-0.84	4.09	-1.33	3.98	0.36
6.54	6.00	4.33	-0.85	4.00	-1.45	4.16	0.08
6.57	4.66	4.29	-0.84	3.91	-1.91	4.39	-0.24
6.60	3.00	4.26	-1.10	4.20	-1.66	4.23	0.00
6.63	3.17	4.36	-1.08	4.29	-1.86	4.44	-0.22
6.66	2.67	4.55	-1.01	3.88	-1.96	4.56	-0.30
6.69	2.08	4.55	-1.07	4.40	-1.90	4.69	-0.38
6.72	2.42	4.64	-1.04	4.47	-1.89	4.48	-0.37
6.75	2.72	4.68	-1.06	4.05	-2.06	4.70	-0.37
6.78	3.36	4.64	-1.05	4.28	-1.71	4.58	-0.20
6.81	4.20	4.59	-0.90	4.28	-1.70	4.59	-0.17
6.84	12.91	4.54	-0.88	4.40	-1.47	4.28	0.10
6.87	14.94	4.45	-0.79	4.26	-1.56	4.22	-0.03
6.90	14.86	4.40	-0.92	4.20	-1.59	4.46	-0.09
6.93	12.49	4.19	-1.11	4.18	-2.06	4.35	-0.19
6.96	6.46	4.40	-0.89	4.09	-2.22	4.48	-0.31
6.99	6.41	4.42	-1.07	4.12	-1.95	4.03	-0.39
7.02	5.00	4.28	-1.23	4.39	-1.77	4.33	-0.39
7.05	7.93	4.51	-1.00	4.19	-1.74	4.32	-0.27
7.08	19.91	4.39	-1.02	4.17	-1.69	3.95	-0.33
7.11	16.74	4.08	-1.00	4.01	-1.66	4.19	0.00
7.14	9.16	4.14	-0.83	4.28	-1.53	3.94	-0.36
7.17	14.49	4.30	-0.89	4.09	-1.52	4.18	-0.09
7.20	5.83	4.16	-0.95	3.90	-1.44	4.21	0.18
7.23	3.50	4.20	-0.88	4.16	-1.43	4.32	0.13
7.26	3.65	4.28	-0.98	4.08	-1.77	4.34	0.10
7.29	3.90	4.33	-1.06	4.23	-1.74	4.42	-0.10
7.32	1.25	4.33	-1.00	4.11	-1.77	3.70	-0.80
7.35	1.83	4.50	-1.25	4.27	-1.63	4.65	-0.31
7.38	2.83	4.47	-1.26	4.41	-2.03	4.65	-0.45
7.41	3.00	3.96	-1.63	4.47	-1.84	4.52	-0.24
7.44	3.33	4.63	-1.14	4.21	-1.76	4.51	-0.45
7.47	3.58	4.54	-1.04	3.97	-1.75	4.25	-0.20
7.50	5.58	4.47	-1.02	4.32	-1.72	4.16	-0.08
7.53	11.30	4.30	-0.97	3.98	-1.51	3.60	-0.14
7.56	22.51	4.13	-0.85	3.95	-1.72	3.91	0.16
7.60	17.83	4.03	-0.92	4.03	-1.54	4.03	0.12
7.63	18.21	4.07	-1.01	3.95	-1.52	3.94	0.11
7.66	14.99	4.01	-0.94	3.66	-1.71	4.19	-0.08
7.70	4.58	4.06	-0.91	3.98	-1.48	4.35	-0.30
7.73	7.37	4.03	-0.94	3.94	-1.29	4.50	-0.44
7.76	4.00	4.35	-1.05	4.03	-1.66	4.42	-0.55
7.80	4.16	4.55	-1.08	3.89	-1.67	4.56	-0.54
7.83	1.92	4.44	-1.15	4.25	-1.94	4.60	-0.57
7.86	2.50	4.36	-1.29	4.37	-2.30	4.66	-0.56
7.90	1.75	4.10	-1.18	4.20	-2.19	4.68	-0.50
7.93	1.17	4.57	-1.36	4.13	-2.11	4.06	-0.93
7.96	1.83	4.47	-1.28	4.18	-2.47	4.59	-0.60
8.00	0.83	4.57	-1.28	4.12	-2.53	4.54	-0.35
8.03	2.25	4.45	-1.34	4.17	-2.69	4.54	-0.42
8.06	1.08	4.47	-1.33	3.80	-2.56	4.64	-0.49
8.10	0.83	4.44	-1.22	4.09	-1.87	4.58	-0.52
8.13	0.83	4.48	-1.36	4.12	-2.64	4.45	-0.58
8.16	1.17	4.59	-1.24	4.22	-2.10	4.59	-0.59
8.20	3.33	4.60	-1.40	4.06	-2.41	4.63	-0.47

Table 1, Continued

Depth (m)	CaCO3 (%)	E. exigua $\delta^{18}O$	E. exigua $\delta^{13}C$	O. umbonatus $\delta^{18}O$	O. umbonatus $\delta^{13}C$	N. pachyderma $\delta^{18}O$	N. pachyderma $\delta^{13}C$
8.23	2.00	4.59	-1.24	4.08	-2.15	4.59	-0.47
8.26	2.75	4.28	-1.34	4.28	-1.67	4.26	-0.24
8.30	9.58	4.57	-1.18	4.41	-1.68	4.41	-0.30
8.33	4.25	4.29	-1.18	3.92	-2.40	4.07	-0.43
8.36	9.16	4.49	-1.05	4.20	-1.85	4.28	-0.29
8.40	8.41	4.30	-1.19	4.03	-2.27	4.27	-0.35
8.43	9.50	4.33	-1.20	4.19	-1.96	4.38	-0.24
8.46	8.41	4.11	-1.39	4.29	-2.31	4.30	-0.34
8.50	3.33	4.29	-1.40	4.03	-2.55	4.31	-0.54
8.53	4.33	4.25	-1.69	4.06	-2.44	4.56	-0.61
8.56	4.41	4.41	-1.53	3.99	-2.90	4.44	-0.56
8.60	4.50	4.49	-1.39	4.38	-2.16	4.34	-0.72
8.63	1.42	4.45	-1.35	4.27	-2.35	4.31	-0.37
8.66	2.25	4.29	-1.34	4.26	-1.97	4.21	-0.35
8.70	2.92	4.33	-1.10	4.20	-1.86	4.14	-0.29
8.73	5.50	4.17	-1.07	3.95	-2.02	3.25	-0.42
8.76	7.23	4.15	-0.99	4.16	-1.73	4.03	-0.07
8.80	8.41	4.07	-0.87	3.73	-1.73	3.93	0.11
8.83	9.36	3.68	-1.06	3.62	-1.45	4.05	-0.03
8.86	3.45	3.61	-0.98	3.68	-1.57	4.17	-0.17
8.90	3.75	3.98	-0.83	3.92	-1.64	4.40	-0.13
8.93	0.58	4.28	-0.95	4.11	-1.82	4.20	-0.01

Table 2. Core PS2082: benthic (*F. wuellerstorfi* and *Cibicidoides* spp.) and planktic (*N. pachyderma*, *G. bulloides*) foraminiferal $\delta^{18}O$ and $\delta^{13}C$, benthic census counts and primary productivity.

Depth (m)	C_{org} (%)	Carbonat (%)	Cibcidoides spp. $\delta^{13}C$	Cibcidoides spp. $\delta^{18}O$	G. bulloides $\delta^{13}C$	G. bulloides $\delta^{18}O$	N. pachyderma $\delta^{13}C$	N. pachyderma $\delta^{18}O$	PC 1	PC 2	PP 1 (gC/m2/y)	PP 2 (gC/m2/y)
0.00			0.31	2.80	1.04	2.28			2.502	1.276		
0.01			0.31	2.85	1.09	2.27	0.30	1.32				
0.02	0.36	42.52	0.44	2.70	1.19	2.49	0.02	1.39			64	64
0.03			-0.33	2.88	0.97	2.28	0.46	1.59				
0.04			0.03	2.66	1.11	2.42	0.62	1.31				
0.06	0.25	52.23	0.24	2.89	0.67	2.22	0.81	2.46	1.891	0.865	55	52
0.12	0.15	68.49	-0.03	3.13	0.85	2.35	0.25	2.66	2.693	1.694	46	37
0.20	0.15	63.74	-0.35	3.75	0.56	3.02	0.04	2.73	3.216	2.639	45	36
0.28	0.18	41.57	-0.34	4.09	0.24	3.17	0.30	3.11	0.832	1.239	47	43
0.37	0.22	21.65	-0.79	4.28	0.72	3.43	-0.04	3.13	0.360	0.313	52	63
0.45	0.34	7.89	-0.98	4.48	0.32	3.21	-0.31	3.38	-0.032	-0.148	59	74
0.50	0.51	4.71	-0.63	4.63	0.20	3.33	-0.54	3.39			74	102
0.56	0.53	16.28	-0.49	4.13	-0.14	3.23	-0.36	2.79	-0.336	0.587	83	120
0.60	0.53	8.65	-0.77	4.60	0.08	3.20	-0.34	3.05	-0.468	0.989	73	101
0.70	0.55	7.34	-0.84	4.60	-0.37	3.73	-0.73	3.16			74	101
0.80	0.55	6.14	-0.78	4.57	-0.52	3.66	-0.04	3.89			76	107
0.90	0.57	8.77	-0.66	4.59	0.38	3.91	-0.24	3.37	-1.704	1.437	80	140
1.00	0.65	5.83	-0.85	4.52	0.10	3.60	-1.02	3.35	-1.919	2.361	87	158
1.10	0.63	3.75	-0.78	4.38	0.38	3.77	-1.11	2.94			90	166
1.20	0.61	1.18	-0.80	4.52	0.20	3.96	-0.14	2.89			87	159
1.30	0.66	0.71	-0.73	4.45	0.09	3.74	-0.34	2.91			92	169
1.40	0.58	1.02			-0.19	3.46	-0.46	2.98			80	140
1.50	0.63	1.48	-0.96	3.81	0.80	2.76	-0.15	3.06	0.291	-0.530	85	153
1.60	0.59	1.12			0.24	3.58	-0.82	2.88			81	144
1.70	0.62	1.78			0.16	3.64	-0.97	2.93			82	145
1.70							-0.83	2.98				
1.80	0.62	1.15			0.22	3.29	-0.79	3.04			87	159
1.80							-0.38	3.31				

Table 2, Continued

Depth (m)	C_{org} (%)	Carbonat (%)	Cibcidoides spp. $\delta^{13}C$	$\delta^{18}O$	G. bulloides $\delta^{13}C$	$\delta^{18}O$	N. pachyderma $\delta^{13}C$	$\delta^{18}O$	PC 1	PC 2 (gC/m2/y)	PP 1	PP 2
1.90	0.60	1.14			0.08	3.34	-0.42	3.30			90	167
2.00	0.50	0.93			0.69	3.61	-0.56	3.14			78	136
2.10	0.54	0.81									85	151
2.20	0.58	1.26			0.15	3.43	-0.86	2.78			85	153
2.30	0.56	1.92	-0.60	4.33	0.30	3.76	-0.69	3.00	-0.876	0.297	86	157
2.40	0.43	7.74	-0.44	4.29	0.86	3.65	-0.17	3.10	-0.368	-0.579	72	120
2.50	0.37	24.85	-0.41	4.22	0.50	3.47	-0.41	2.93	-0.012	-1.368	60	65
2.60	0.39	14.52	-0.37	4.15	0.24	3.64	-0.91	2.87	-0.009	-1.281	72	83
2.70	0.30	40.24	-0.47	4.12	0.48	3.52			-0.053	-0.679	64	70
2.80	0.28	31.83	-0.45	4.08	0.70	3.73	0.08	3.17	0.045	-1.260	61	66
2.84	0.32	18.47	-0.13	4.08			0.25	3.70	-0.052	-1.307	64	71
2.84			-0.23	4.11								
2.90	0.41	31.00	-0.52	4.11	0.32	3.42			-0.087	-1.322	76	90
3.00	0.40	23.45	-0.67	4.11	0.42	3.20	-0.09	3.02	-0.269	-0.573	77	88
3.10	0.39	20.22	-0.75	4.05	0.28	3.26	0.20	3.14	-0.084	-1.274	74	84
3.20	0.31	28.85	0.03	4.03			0.15	2.87	-0.135	-1.056	66	71
3.30	0.30	24.32	-0.04	3.93	0.59	3.58	-0.19	3.10	0.080	-1.222	66	71
3.40	0.23	30.25	-0.19	3.97	0.82	3.42	0.20	2.86	0.188	-0.824	55	55
3.50	0.19	24.19	-0.64	4.06	1.00	3.31	0.08	2.91	0.222	-0.875	47	44
3.55	0.23	27.82	-0.71	4.11	0.60	2.96	0.07	2.83	0.125	-0.456	50	49
3.60	0.26	25.85	-0.39	4.00	0.50	3.42	0.35	3.03	0.143	-0.464	57	58
3.65	0.28	-0.90	-0.49	3.93	0.43	3.40	-0.02	2.63	0.525	-0.689	47	44
3.70	0.32	5.56	-0.73	4.11	-0.88	2.96	-0.58	2.54	-0.062	-0.653	54	54
3.80			-0.49	4.03	0.07	3.35	-0.13	2.71				
3.90	0.66	0.01			-0.17	3.49	-0.17	3.47			96	128
4.00	0.47	1.03			-0.18	3.41	-0.19	3.28			73	87
4.10	0.50	4.24	-0.77	4.16	0.17	3.63	-0.63	2.98			75	90
4.20	0.43	3.41	-0.91	4.06	0.51	3.28	-0.08	3.14	-0.779	0.037	66	75
4.30	0.41	9.61	-0.44	3.92	-0.05	3.46	0.28	3.16	-0.055	-1.297	66	75
4.40	0.32	16.16	-0.03	3.81	1.02	3.32	0.42	2.95	0.064	-1.232	59	64
4.50	0.19	38.48	0.06	3.66	0.89	3.11	0.87	2.41	0.214	-0.952	47	45
4.55	0.12	54.11	0.16	3.52	1.14	3.00	0.36	2.28	0.271	-0.613	38	33
4.55					1.04	2.77						
4.60	0.15	52.42	-0.04	3.57	0.84	2.81	0.69	2.30	0.226	-1.085	47	44
4.65	0.15	48.85	0.16	3.45	0.70	2.85	0.01	2.22	0.502	-0.545	43	40
4.65			-0.34	3.50								
4.70	0.18	23.25	-0.18	3.45	1.37	3.07	0.43	2.42	0.149	-1.234	41	37
4.80	0.20	28.86	0.03	3.74	0.64	2.60	0.33	2.49	0.140	-1.389	45	38
4.90	0.19	32.47	0.13	3.57	0.68	3.07	0.31	2.65	0.054	-1.747	43	35
5.00	0.16	33.89	-0.42	3.50	0.51	2.93	0.43	2.22	0.029	-0.989	44	36
5.10	0.19	30.53	-0.16	3.56	0.81	3.03			-0.015	-0.607	45	38
5.20	0.20	10.72	-0.44	3.63	0.84	2.99	-0.19	2.29	-0.014	-1.145	39	30
5.30	0.18	18.07	-0.23	3.52	0.66	3.01	0.42	2.74	0.153	-1.090	38	30
5.40	0.19	21.56	-0.20	3.42	0.81	3.08	0.33	2.63	0.552	-0.398	42	34
5.50	0.12	46.70	-0.24	2.78	0.24	2.58	-0.25	2.27	2.452	1.317	36	27
5.55	0.11	53.36	-0.08	2.82	0.31	2.10			2.721	1.928	35	27
5.60	0.09	62.97	0.07	3.36	-0.36	2.38			1.788	0.936	34	25
5.65	0.10	66.86	-0.29	3.44	-0.50	2.94	-0.92	2.39	1.103	1.141	37	28
5.70	0.10	60.94	-0.40	4.09	-0.07	3.22	-0.84	2.26	0.282	-0.073	35	26
5.75	0.12	57.02	-0.89	4.06	0.28	3.22	-0.31	2.84	0.123	0.260	44	45
5.80	0.13	47.76	-0.89	4.38	0.00	3.22	0.00	3.03	-0.172	-0.362	44	46
5.90	0.17	48.14	-0.90	4.19	-0.27	3.34	0.09	3.39	-0.226	-0.353	46	48
6.00	0.40	35.00	-1.15	4.44	0.03	3.71	-1.38	3.07	-0.984	0.618	79	103
6.10	0.44	28.88	-0.93	4.43	-0.19	3.70	-1.03	3.44	-0.669	0.334	82	112
6.20	0.47	21.31	-1.15	4.36	-0.57	3.12	-0.97	3.13	-1.190	0.823	85	115
6.30	0.49	11.34	-1.09	4.29	-0.18	3.70	-0.44	3.15	-1.293	0.632	83	111
6.40	0.57	7.50	-1.02	4.31	-0.26	3.65	-0.57	3.20	-1.583	1.596	88	122
6.50	0.51	4.59	-0.86	4.28	0.12	3.74			-1.335	1.100	80	106

Table 2, Continued

Depth (m)	C_{org} (%)	Carbonat (%)	Cibcidoides spp. $\delta^{13}C$	$\delta^{18}O$	G. bulloides $\delta^{13}C$	$\delta^{18}O$	N. pachyderma $\delta^{13}C$	$\delta^{18}O$	PC 1	PC 2 (gC/m2/y)	PP 1	PP 2
6.60	0.47	3.51	-0.99	4.25	0.01	3.66	-0.31	3.22	-0.904	0.052	77	101
6.70	0.42	11.85	-0.97	4.24			-0.23	3.09	-0.666	0.140	73	93
6.80	0.42	16.35	-0.93	4.18	-0.25	3.59	-0.72	2.97	-0.550	0.439	78	102
6.90	0.50	11.07	-1.02	4.21	-0.46	3.55	-0.78	3.01	-0.759	0.488	80	106
7.00	0.48	3.62	-1.18	4.25	-0.12	3.58	-1.50	2.90	-0.900	0.876	79	106
7.10	0.51	1.46	-1.08	4.02	-0.40	3.51	-0.97	2.98	-0.961	1.271	78	103
7.20	0.56	1.81	-0.86	4.34	-0.34	3.68	-0.84	3.30			86	117
7.30	0.53	2.08	-1.06	4.30	-0.39	3.50	-0.31	3.34	-0.905	0.856	77	93
7.40	0.58	5.65	-1.14	4.22	-0.69	3.01			-0.990	0.562	82	102
7.50	0.48	0.98	-0.49	4.03			-0.62	2.78			74	88
7.60	0.55	2.13	-1.35	4.20	-0.78	2.92	-1.14	2.99	-0.845	0.144	77	94
7.70	0.55	0.48	-1.04	4.15	-0.96	2.92	-1.54	2.52			77	93
7.80	0.52	1.21	-0.92	4.10	-0.25	2.71	-0.13	3.40	-0.497	-0.369	78	96
7.90	0.50	5.69	-1.23	3.99	0.35	3.57	-0.41	3.02	-0.736	-0.084	82	102
8.00	0.41	7.08	-1.16	3.96	-0.35	3.20	-0.76	2.77	-0.697	-0.132	70	82
8.10	0.45	7.54	-1.05	3.96	-0.36	3.33	-0.10	3.03	-0.599	-0.653	70	82
8.20	0.42	4.06	-0.84	3.97	-0.32	3.14	-1.10	2.44	-0.573	-0.108	69	81
8.30	0.43	0.77			-0.59	3.28	-0.76	2.96			70	81
8.40	0.45	4.30	-0.78	3.93	-0.07	3.62	-1.24	2.75	-0.505	-0.726	75	91
8.50	0.45	0.77	-1.20	3.93	-1.22	2.54	-1.47	2.42	-0.369	-0.066	74	87
8.60	0.47	1.23	-0.85	3.88	-0.26	3.48	-0.41	2.97	-2.455	3.567	72	85
8.70	0.45	0.28					-0.99	2.98			72	85
8.80	0.40	0.25									68	78
8.90	0.45	0.94	-0.79	3.85	-0.02	2.87					71	83
9.00	0.26	0.18									46	45
9.10	0.30	0.18									50	51
9.15	0.21	11.22	0.08	3.09							45	44
9.20	0.23	10.84	0.10	3.20	0.80	2.81			0.461	-0.728	43	30
9.30	0.21	6.24			0.81	2.67	0.45	2.77	0.621	-0.513	40	27
9.40	0.16	26.89	-0.06	3.25	0.69	2.87	0.38	2.68	0.713	-0.170	38	25
9.45	0.14	23.16	-0.65	3.52	0.66	2.85	-0.10	2.79	0.641	-0.367	40	38
9.50	0.17	9.87			-0.44	3.53	-0.57	2.95			39	37
9.50					-0.18	3.54	-0.29	2.98				
9.60	0.17	0.72									36	33
9.65	0.16	1.50									35	32
9.70	0.21	0.60									42	37
9.78	0.22	0.63									43	38
9.85	0.27	1.09	-0.77	4.00	0.01	3.58			-0.418	-0.094	50	48
9.95	0.23	0.36									45	41
10.05	0.21	10.41	-0.73	3.52	0.62	2.97			0.633	-0.631	41	37
10.10	0.18	36.38	-0.57	3.32	0.80	2.74	0.06	2.05	0.733	-0.641	46	43
10.15	0.11	57.08	-0.13	3.33	0.42	2.70	0.20	1.90	0.721	-0.379	38	33
10.20	0.11	52.77	-0.37	3.48	0.35	2.77	-0.30	1.89	1.342	0.347	37	31
10.25	0.14	28.23	-0.68	3.85	0.03	2.47	0.04	2.55	1.122	0.736	36	30
10.30	0.18	20.34	-0.87	4.01	0.33	3.15	-0.78	2.63	0.176	-0.494	40	35
10.35	0.21	11.10	-0.83	4.10	-0.06	3.24			0.200	-0.058	44	46
10.35												
10.45	0.44	5.97	-0.98	4.24	0.13	2.41	-0.28	2.94	-0.482	0.520	70	91
10.55	0.47	8.01	-1.15	4.34	-0.49	3.60	-1.29	2.93	-0.976	0.808	81	109
10.65	0.44	17.65	-1.12	4.17	-0.27	3.73	-0.48	3.28	-0.901	0.403	73	64
10.75	0.39	7.45	-1.10	4.20	-0.17	3.36	-0.75	3.06	-0.811	0.821	64	54
10.85	0.38	2.91	-1.20	4.27	-0.44	3.29	-1.15	2.99	-0.911	0.700	63	64
10.95	0.42	0.83	-1.36	4.17	-0.58	3.17	-0.45	3.13			67	69
11.05	0.43	0.76			-0.47	3.46	-0.71	2.71			69	72
11.15	0.39	0.67									67	69
11.25	0.37	0.80			-0.18	3.50	-0.84	3.14			63	63
11.35	0.38	0.78	-1.10	4.21							64	65
11.35			-1.09	4.21	-0.11	3.37	-0.56	3.21				

Table 2, Continued

Depth (m)	C_{org} (%)	Carbonat (%)	Cibcidoides spp. $\delta^{13}C$	Cibcidoides spp. $\delta^{18}O$	G. bulloides $\delta^{13}C$	G. bulloides $\delta^{18}O$	N. pachyderma $\delta^{13}C$	N. pachyderma $\delta^{18}O$	PC 1	PC 2	PP 1 (gC/m2/y)	PP 2
11.45	0.37	0.90			-0.05	3.12	-0.22	3.10			63	60
11.55	0.28	0.32									50	44
11.65	0.29	0.26									55	49
11.75	0.31	0.36									54	48
11.85	0.22	-0.07									44	36
11.95	0.31	0.97									54	48
12.05	0.22	0.08									43	35
12.15	0.20	0.18									38	30
12.23	0.17	2.68									34	26
12.28	0.13	35.73	-0.47	3.33	0.82	2.71	-0.10	1.54	1.894	0.731	36	27
12.32	0.13	43.23	-1.02	4.19	0.02	2.89	-0.39	1.57			38	29
12.40	0.23	16.53	-1.48	4.22			-0.94	3.17			46	38
12.40			-1.38	4.36								
12.50	0.66	8.57	-1.39	4.42	-0.78	3.62	-0.19	3.54			101	159
12.60	0.57	9.68	-1.57	4.50			-0.84	3.22			91	139
12.70	0.58	10.37	-1.39	4.47	0.43	3.88	-0.07	3.45			78	69
12.80	0.51	18.06	-1.38	4.34			-0.42	3.19			73	63
12.90	0.54	22.93	-1.44	4.24	-0.19	3.74	-0.77	3.18			90	108
13.00	0.44	19.23	-1.13	4.14	-0.23	3.44	-0.25	3.07			78	88
13.10	0.51	7.58	-1.36	4.16	-0.02	3.44	-0.42	3.01			82	100
13.20	0.54	5.08	-1.49	4.19	-0.21	3.68	-0.49	3.31			85	105
13.30	0.52	2.11	-1.14	4.20	-0.18	3.38	-0.56	3.13			82	100
13.40	0.54	1.47	-1.51	4.12	0.25	3.49	-0.11	3.40			81	99
13.50	0.46	2.51	-1.33	4.01	-0.23	3.40	-0.45	3.05			74	87
13.60	0.44	0.89	-1.17	3.89	-0.26	2.99	0.18	3.02			71	82
13.70	0.29	1.23									49	48
13.80	0.39	3.38	-1.00	3.76	0.41	2.98	0.23	2.70			68	76
13.89	0.24	9.35	-0.42	3.67	0.27	3.04	0.59	2.71			50	50

Primary productivity calculated using the equations after Stein (1986; PP1) and Müller &Suess (1979; PP2)

FORAMINIFERAL POPULATION DYNAMICS AND STABLE CARBON ISOTOPES

Christoph Hemleben and Jelle Bijma[1]
Institut und Museum für Geologie und Paläontologie
Universität Tübingen
Sigwartstraße 10
D-72076 Tübingen

ABSTRACT. Most planktic foraminifera live within the photic zone and exhibit a life style tied to the lunar cycle. They migrate between the reproductive depth (thermocline and/or the chlorophyll maximum) and the uppermost part of the photic zone. This ontogenetic migration pattern sets the initial $\delta^{13}C$ of the foraminiferal shell. On top of that, biological fractionation processes (vital effects) modify the signal. These processes include photosynthetic activity of the symbionts and respiration of the host/symbiont complex. *Globigerinoides sacculifer* (Brady) was chosen to model ontogenetic changes in the $\delta^{13}C$ of the shell as a function of depth migration with and without vital effects.

INTRODUCTION

Reconstructions of marine paleo-environments are frequently based on stable isotope measurements. General assumptions are that a foraminifer has a rather steady life habitat throughout ontogeny and that the $\delta^{13}C$ of ΣCO_2 is constant in the mixed layer. However, because most planktic foraminifera, if not all, change their life horizon during ontogeny and because the $\delta^{13}C$ of ΣCO_2 is not necessarily constant in the mixed layer, the isotopic composition of the foraminiferal shell is not only dependent on water-mass properties (e.g. oligotrophic vs. eutrophic) and the geographic and climatic setting (e.g. upwelling, monsoon, subarctic etc.) but to an important extent on the life history of the foraminifer. Consequently, the disequilibrium precipitation of the foraminiferal shell (so called "vital effect") reported in the literature (e.g. Berger *et al.*, 1978) can often be explained in terms of ontogenetic migration and biological fractionation. In this paper we shall discuss the potential effect of

[1] Present address: Alfred Wegener Institut für Polar- und Meeresforschung, Columbusstraße, D-27570 Bremerhaven, Germany

NATO ASI Series, Vol. I 17
Carbon Cycling in the Glacial Ocean:
Constraints on the Ocean's Role in Global Change
Edited by R. Zahn et al.

some of these processes on the stable carbon isotope composition of the shell of *G. sacculifer*.

POPULATION DYNAMICS

Despite all the useful information on biostratigraphy, stable isotopes and biology of planktic foraminifera that has been compiled (for references see, e.g., Vincent and Berger, 1981) little knowledge exists on the population dynamics of planktic foraminifers (see Hemleben *et al.*, 1989). Only a few papers deal with the small scale temporal and spatial events that occur in populations.

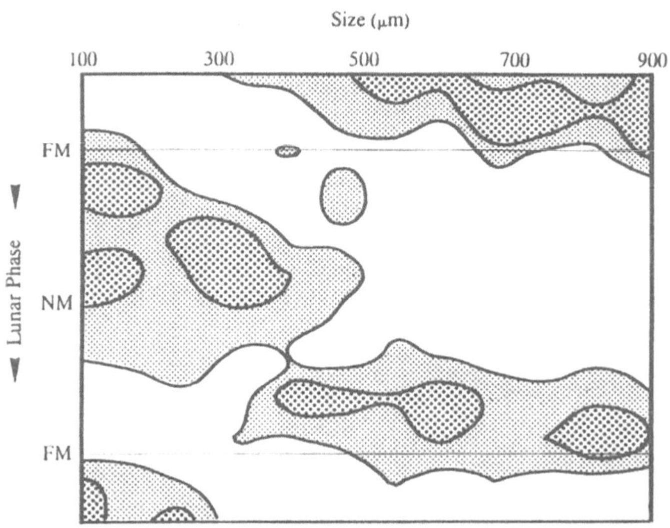

Figure 1: Lunar cycle of *G. sacculifer*. Contour plot of residual values in a size-time frame, showing that the larger size-fractions are most frequent just before full moon (FM) and that the smaller size-fractions are most frequent just after FM. Spindler *et al.* (1979).

Spindler *et al.* (1979) were the first to document that reproduction in the planktic foraminifer *Hastigerina pelagica* (d'Orbigny) is coupled to the synodic lunar cycle. Almogi-Labin (1984) has indicated that *Globigerinoides sacculifer* (Brady) reproduces at full moon in the Gulf of Elat/Aqaba. Based on a time series of plankton-net tows, a synodic lunar reproductive cycle in *G. sacculifer* (Figure 1) has been shown by Bijma *et al.* (1990) and subsequently confirmed by Erez *et al.* (1991).

Time series samples collected in the Red Sea (METEOR Cruise 5) and the northeast Atlantic Ocean (METEOR Cruise 10) demonstrate the concept of lunar cyclicity in the life cycle of other spinose planktic species, e.g. the symbiont-free, spinose species *Globigerina*

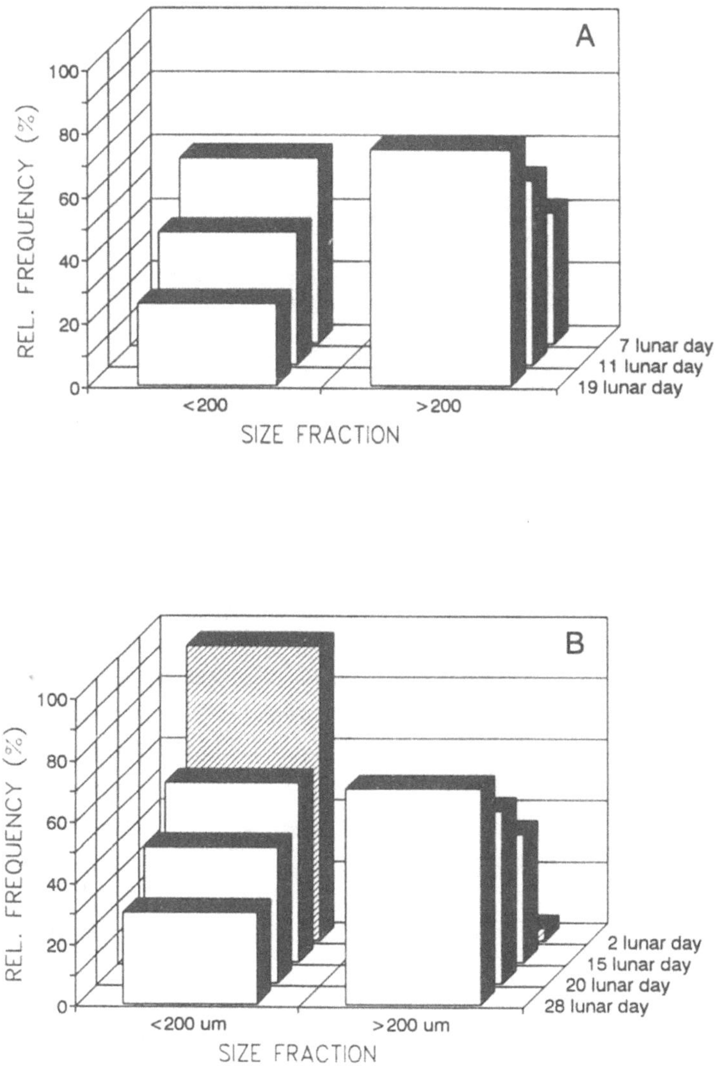

Figure 2: Relative frequency and size fraction of *G. bulloides* in respect to full moon (1. lunar day=full moon) . A: in the productive zone (0-60 m). B: in the flux zone (300-500 m).

bulloides and the symbiont bearing spinose species *Orbulina universa* (figs. 2-3). On the other hand, the deep living non-spinose foraminifers *Globorotalia scitula, G. hirsuta,* and *G. truncatulinoides* do not respond to the lunar cycle (Figure 4) and probably have a yearly life-cycle (Hemleben *et al.*, 1989). Because most spinose planktic foraminifers are associated with symbiotic algae, their main habitat is the euphotic zone. For instance, the number of *G. sacculifer* decreases drastically with depth. On the basis of absolute abundance, 95% of the shells collected in the upper 100 m of the water column live in the upper 80 m (Figure 5A). *Orbulina universa* d'Orbigny shows a subsurface maximum; and even the symbiont-barren *Globigerina bulloides* lives almost exclusively in the euphotic zone (Figure 5B-C).

The depth migration of *G. sacculifer* during ontogeny has recently been documented (Bijma and Hemleben, 1993). It was shown that this species does not inhabit a specific depth, but migrates up and down in the water column in response to its life cycle while growing (Figure 6). Thus, not only the absolute abundance changes with depth but also the size distribution changes with depth (Figure 6A). The small fraction (<300 μm) dominates the water column but their relative frequency decreases with depth until, between 60 and 80 m depth, the larger fractions (>300 μm) become dominant. Below 80 m depth, however, the small size fraction starts to dominate again. The cumulative plot of the relative frequency of mature (>366 μm) and immature specimens (<366 μm) versus water-depth (Figure 6A) demonstrates a bimodal pattern with a breakpoint between 60 to 80 m depth. The maximum of mature specimens coincides with the depth range 60 to 80 m. The relative number of immature specimens increases towards shallower and deeper environments. Apparently, reproduction in the central Red Sea takes place at approximately 60 to 80 m depth. Recruitment is shown by the ascent of immature specimens to the surface. High mortality rates in early ontogeny are probably responsible for the increase of immature specimens below the reproduction depth. The latter fraction is part of the shell flux to the sea floor.

A contour plot of the distribution of *G. sacculifer* in the productive zone demonstrates sizedependent depth preferences (Figure 6B). The plot demonstrates that the upper 20 m are the preferred habitat for immature specimens between 100 and 300 μm. The depth range between 20 and 40 m is favored by specimens from 300 to 500 μm. This is the depth where maturation takes place. Specimens between 500 and 700 μm prefer a depth range between 40 to 60 m, and specimens larger than 700 μm are found between 60 and 80 m depth, just above the reproduction depth (Figure 6B).

The depth habitat is probably dependent on the hydrological conditions and may thus differ for different water bodies. At present, we believe that the thermocline and/or the deep chlorophyll maximum determine the reproduction depth.

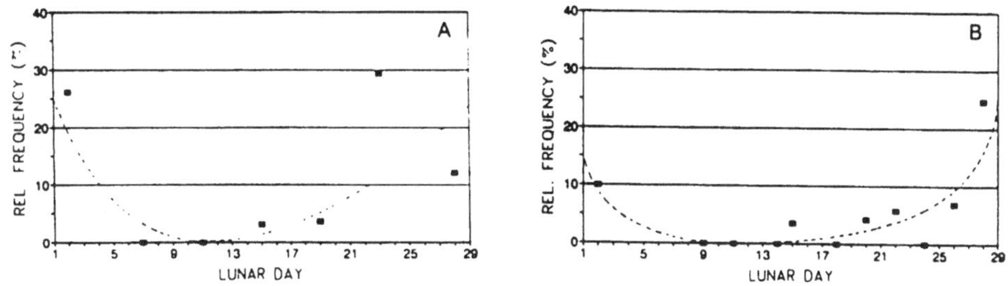

Figure 3: Relative frequency of *O. universa* in respect to the lunar period. A: in the productive zone (0-60 m). B: in the flux zone (300-500 m).

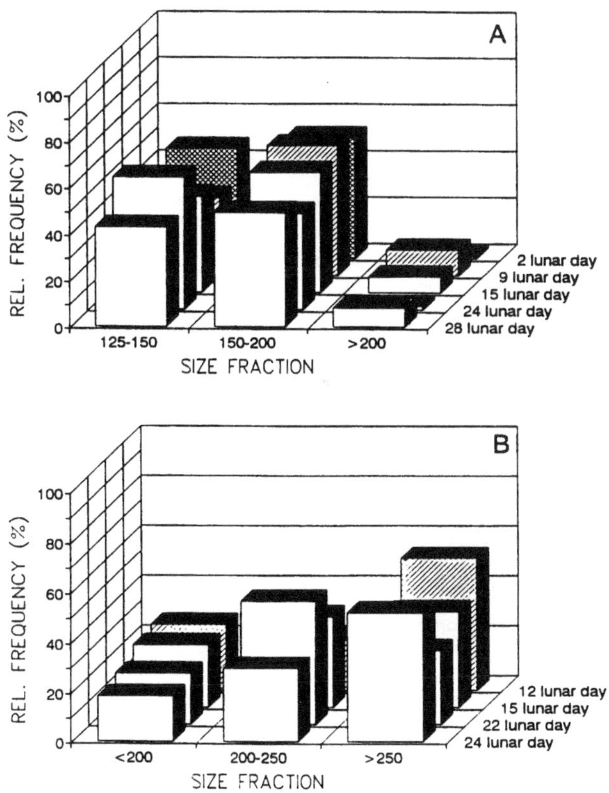

Figure 4: Non-spinose species do not respond to the lunar cycle. Samples were collected between 300-500 m depth. A: *G. scitula*. B: *G. hirsuta*.

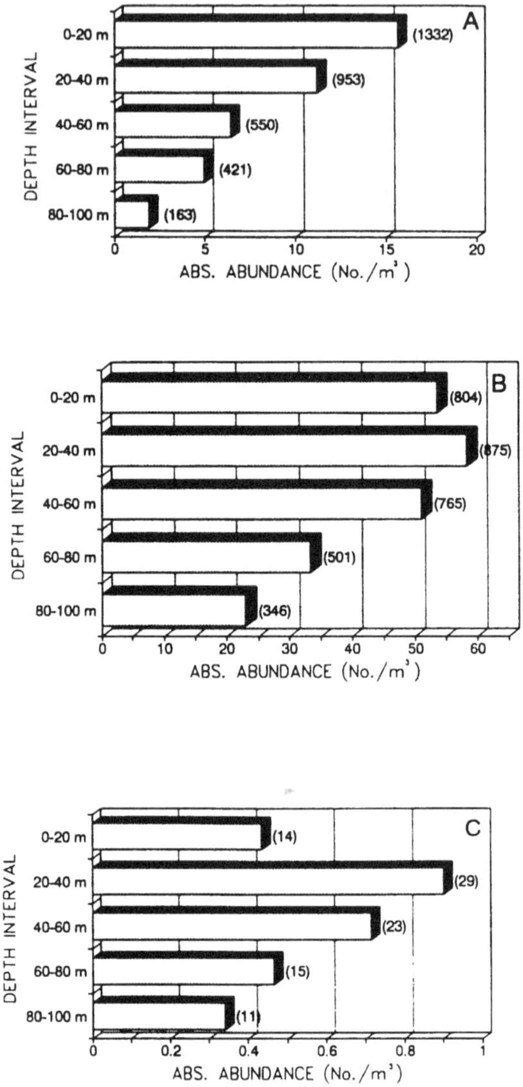

Figure 5: Mean abundance (No./m^3) of planktic foraminifers in the upper 100 m of the water column. A: *Globigerinoides sacculifer* >100 µm in the Central Red Sea. B: *Globigerina bulloides* >125 µm from the North Atlantic. C: Spherical *Orbulina universa* >125 µm from the North Atlantic. The total number of specimens are bracketed.

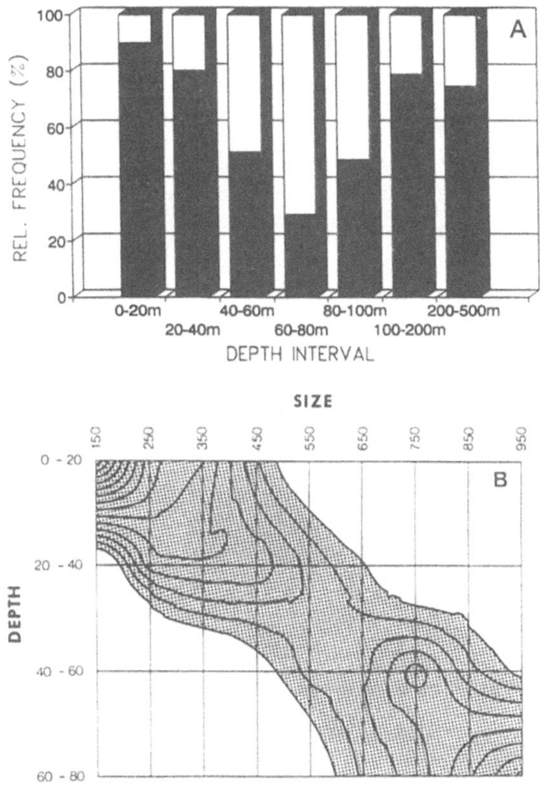

Figure 6: A: The frequency (%) of mature (>366 μm; white) and immature (<366 μm; black) *G. sacculifer* show a characteristic break between 60 and 80 m depth. B: Contour plot of residual values in a size-depth frame, showing that larger size fractions prefer deeper depth-habitats. The cross-sections of the grid represent the data points. The centers of the nine size factions (100 μm wide) and the depth intervals are listed. The shaded area contains positive residual values, indicating a larger than average presence of a size fraction. The isolines start with 0 and are drawn at 0.05 intervals. The blank area contains negative residual values, indicating a smaller than average presence of a size fraction.

$\delta^{13}C$ OF ΣCO_2 AS A FUNCTION OF DEPTH

The CO_2-fixing enzyme of most marine algae (C_3-plants), RuBP-carboxylase (ribulose-1,5bisphosphate carboxylaseoxygenase), has a higher affinity for $^{12}CO_2$ than for $^{13}CO_2$. As a result, the marine photic zone is enriched in ^{13}C relative to equilibrium with atmospheric CO_2. Simultaneously, the photosynthetically fixed carbon, which is enriched in isotopically light carbon, is decomposed by bacteria resulting in lower $\delta^{13}C$ values with depth. As a result of oxygen producing photosynthesis in the surface waters and oxygen consuming bacterial decomposition in deeper waters an empirical relationship between the $\delta^{13}C$ of the total CO_2

(ΣCO_2) and the apparent oxygen utilization (AOU) was established (Williams *et al.*, 1977). Because nutrients are utilized during photosynthesis and released by bacterial decomposition, a strong negative correlation can be seen also between the $\delta^{13}C$ of ΣCO_2 and nutrients. However, the correlation between $\delta^{13}C$ and nutrients may be disrupted (note that the surface $[PO_4]$ is less than the $[PO_4]$ at 100 m depth, yet both depth have similar $\delta^{13}C$ of ΣCO_2) due to the temperature dependence of the carbon isotope fractionation between atmospheric CO_2 and surface ocean ΣCO_2. The extent of the disruption depends on the relative strength of mixing within the sea and CO_2 exchange between ocean and atmosphere (Broecker and Maier-Reimer, 1992).

It is often assumed (e.g. Spero and Williams, 1989) that the $\delta^{13}C$ of ΣCO_2 is constant with depth in the mixed layer. This, however, is not necessarily true: In regions with a seasonal thermocline, the $\delta^{13}C$-value of ΣCO_2 is dependent on the gross assimilation rate (GAR) in the photic zone. For a station in the North Atlantic at 35°25'N, 29°30'W a subsurface $\delta^{13}C$ and oxygen maximum could be correlated to a nutrient minimum (Figure 7A-B; data courtesy Ganssen and Brummer, VU, Amsterdam). Apparently, the $\delta^{13}C$ of ΣCO_2 in the "mixed layer" cannot be assumed constant. As long as the rate of $\delta^{13}C$ enrichment through photosynthesis is faster than the rate of mixing, a $\delta^{13}C$-gradient will develop, independent of water stratification.

PREDICTING $\delta^{13}C$ OF THE FORAMINIFERAL SHELL

Already in 1978, Berger *et al.* demonstrated that $\delta^{13}C$ is dependent on the growth stage. They correlated the changing isotopic composition with environmental and metabolic signals, which are indeed the main influencing parameters. To predict the $\delta^{13}C$ of the shell of *G. sacculifer* at each ontogenetic stage under equilibrium conditions (i.e. no vital effects) we used several growth statistics. Foraminifers do not grow continuously, but by addition of separate chambers (accretionary growth). Consequently, the effect of newly formed chambers on the final isotope composition of the test can be calculated by using a mass balance equation:

$$\delta^{13}C_n = \left(\frac{W_{n-1}}{W_n}\right)\delta^{13}C_{n-1} + \left(\frac{W_n - W_{n-1}}{W_n}\right)\delta^{13}C_{water} \qquad (1)$$

where W stands for weight and n and n-1 are the nth and the (n-1)th chamber-stage. In addition, several other relationships must be known: 1) A growth relationship was derived from Figure 8 and is based on culture experiments with *G. sacculifer*,

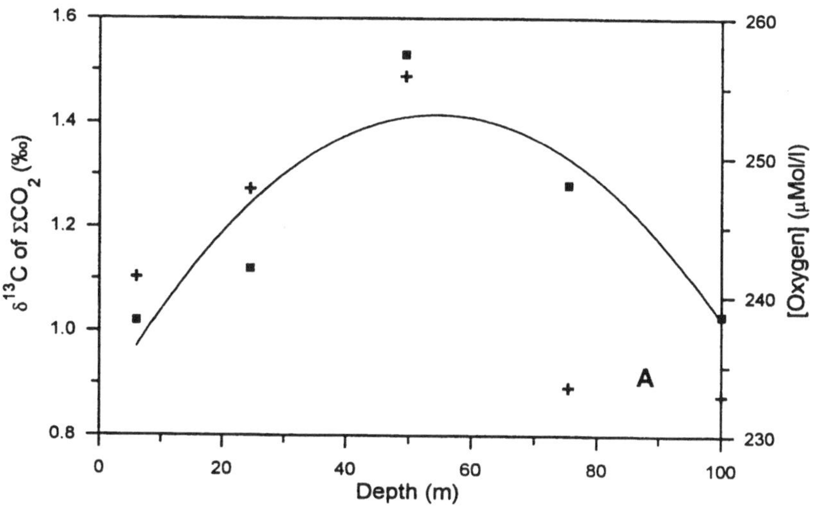

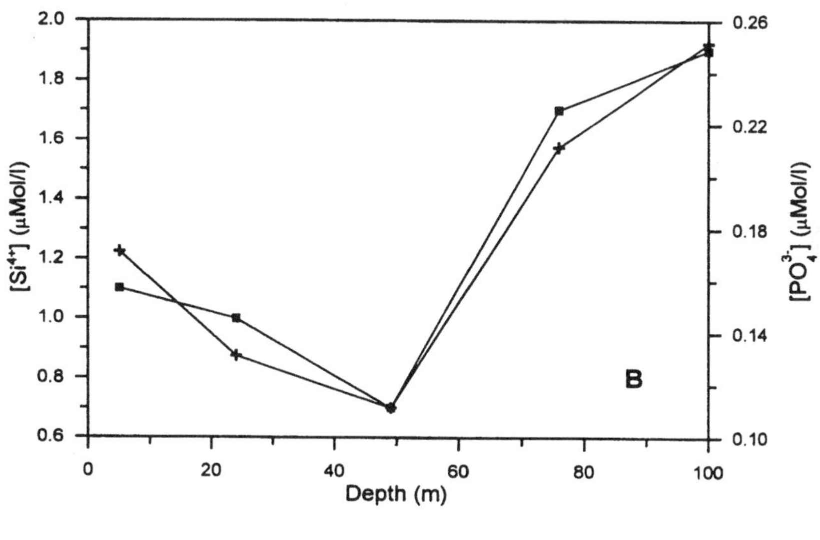

Figure 7: A: The $\delta^{13}C$ of the total dissolved CO_2 (•) and the oxygen concentration (+) at 35°25'N, 29°30'W as a function of depth. The polynom for $\delta^{13}C$ is $0.000189*depth^2 + 0.0206*depth + 0.853$. B: The silicium (•) and the phosphate concentration (+) at the same station (data courtesy of G. Ganssen and G.J.A. Brummer, Free University of Amsterdam).

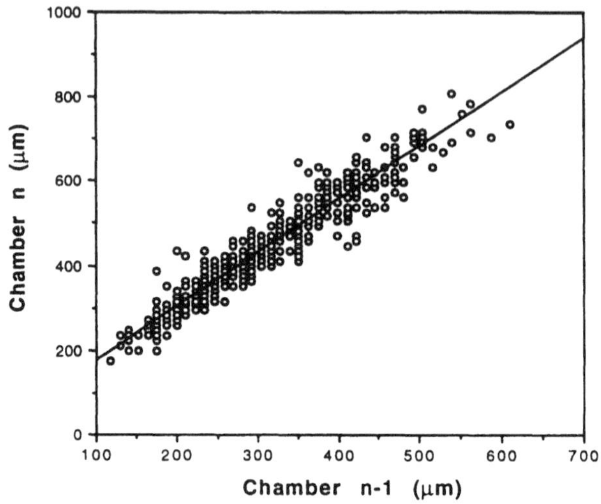

Figure 8: Growth curve of *G. sacculifer*. Best fit regression line is: $D_n = 1.27D_{n-1} + 40$ (for specimens ≥ 100 μm). For explanation see text. Growth below 100 μm was based on measurements from Brummer *et al.* 1987.

$$D_n = 1.27\, D_{n-1} + 48 \text{ (for specimens} \geq 100\ \mu\text{m)} \qquad (2)$$

where D_n and D_{n-1} are the diameters for the nth and the (n-1)th chamber-stage respectively. Growth below 100 μm was based on measurements from Brummer *et al.* 1987. 2) The sizeweight function of Anderson and Faber (1984):

$$\text{mass } (\mu g) = 2.89 - 0.0238 * \text{size} + 0.0001 * (\text{size})^2 \qquad (3)$$

is used for size classes ≥ 100 μm. For specimens < 100 μm weight we assume a linear relationship between size and weight. (e.g. a specimen of 40 μm weighs 40% of the weight at 100 μm). 3) The $\delta^{13}C$-depth relationship is based on the data of Figure 7A and, 4) the ontogenetic depth-migration relationships are extracted from Figure 6B:

$$\text{depth (m)} = -\ 0.833 * \text{size} + 83.33 \text{ (for specimens} < 100\ \mu\text{m)} \qquad (4a)$$

$$\text{depth (m)} = 0.073 * \text{size} - 4 \text{ (for specimens} \geq 100\ \mu\text{m)} \qquad (4b)$$

Using a mass-balance equation (1), the growth relationship (2), the size-weight relationship (3) and the depth migration pattern (4a and 4b), the final $\delta^{13}C$ may be calculated

(Table 1, Figure 9).

To a size of approximately 100 μm, the pre-adult specimens ascend to the surface water, thereby passing the $\delta^{13}C$-maximum at the depth of maximum GAR (gross assimilation rate). After passing the level of maximum GAR, the $\delta^{13}C$ of the test decreases (Figure 9A). During the second part of their life cycle they slowly descend to the reproduction depth, passing the $\delta^{13}C$ maximum again. During the descending phase the $\delta^{13}C$-value initially becomes lighter as long as the $\delta^{13}C$ of ΣCO_2 is still lighter than the test. When they reach a size of ca. 250 μm (at approx. 15 m depth or little deeper) the $\delta^{13}C$ of ΣCO_2 becomes heavier than the value of the test i.e. from there onwards, the test becomes heavier. When the $\delta^{13}C$ maximum in the water column is passed, at an adult test size, the tests become lighter again.

Table 1: Equilibrium $\delta^{13}C$ values (i.e. assuming no biological effect) calculated with a mass balance equation on the basis of ontogenetic migration and a changing ambient $\delta^{13}C_{water}$ with depth.

chamber (#)	shell size (μm)	water depth	$\delta^{13}C$ water	shell weight	chamber weight	$\delta^{13}C$ shell
1	16	70	1.37	0.24	0.24	1.37
2	25	63	1.40	0.38	0.14	1.38
3	28	60	1.41	0.41	0.04	1.38
4	30	58	1.41	0.45	0.04	1.39
5	34	55	1.42	0.51	0.06	1.39
6	37	53	1.41	0.56	0.04	1.39
7	40	50	1.41	0.60	0.05	1.39
8	45	46	1.40	0.68	0.08	1.39
9	52	40	1.38	0.78	0.11	1.39
10	58	35	1.34	0.87	0.09	1.39
11	69	26	1.26	1.04	0.17	1.37
12	75	21	1.20	1.13	0.09	1.35
13	85	13	1.08	1.28	0.15	1.32
14	100	3	0.92	1.51	0.24	1.26
15	175	9	1.02	1.79	0.28	1.22
16	270	16	1.13	3.76	1.97	1.17
17	391	25	1.25	8.88	5.12	1.22
18	545	36	1.35	19.61	10.72	1.29
19	740	50	1.41	40.03	20.42	1.35
20	988	68	1.38	76.95	36.91	1.37

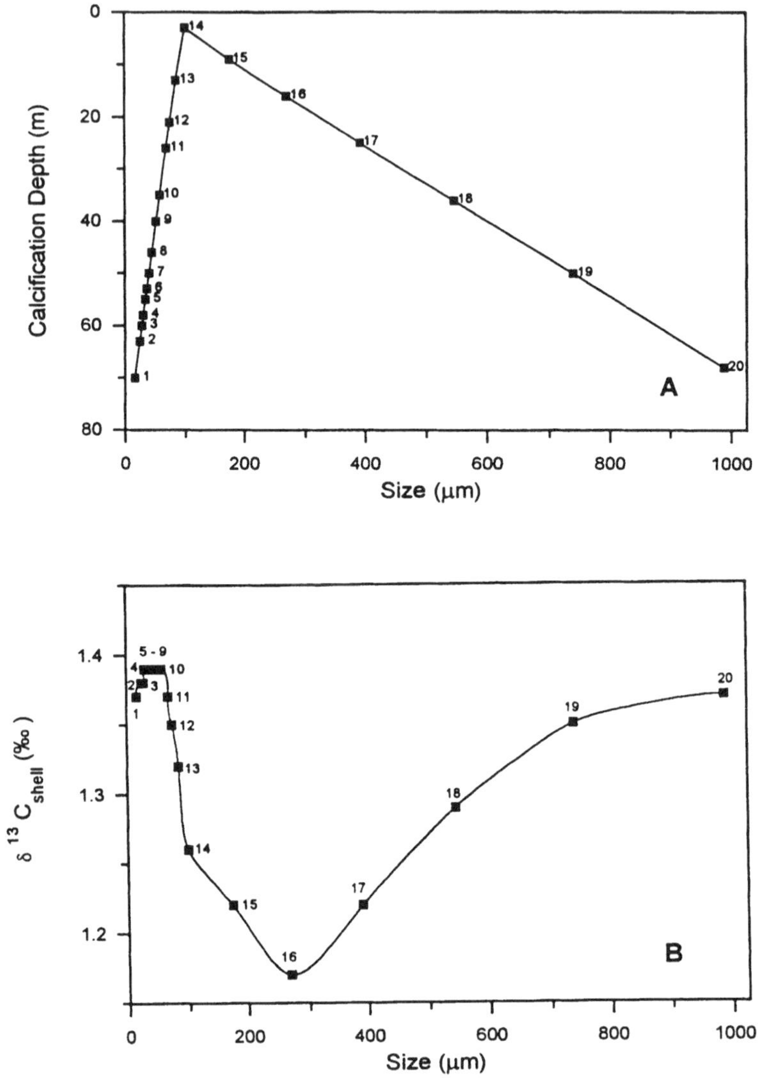

Figure 9: A: The migration of *G. sacculifer* during ontogeny between the surface and approximately 70 m depth. The numbers indicate chamber formation. B: The simulated carbon isotope ratio of the shell of *G. sacculifer* under the assumptions that 1) the $\delta^{13}C$-values of ΣCO_2 of the mixed layer change with depth (see Figure 7A), 2) that *G. sacculifer* migrates up and down in the water column in response to ontogeny, and that 3) chambers are formed in equilibrium with the ambient water (i.e. no vital effect). Chamber numbers are indicated.

DISCUSSION

For comparison, data from Oppo and Fairbanks (1989) were used. On the small size end we added some own measurements from the Red Sea. (Figure 10). This graph demonstrates that the $\delta^{13}C$ of *G. sacculifer* increase with size but that the $\delta^{13}C$ range is much larger than the equilibrium simulation and does not show an increase at smaller sizes (Figure 9A). Apparently, the shells of real foraminifers are not secreted in equilibrium with the ambient seawater and metabolic processes may control the carbon isotope signal and even obscure the ambient value. It has been suggested that $\delta^{13}C$ disequilibrium precipitation of planktic

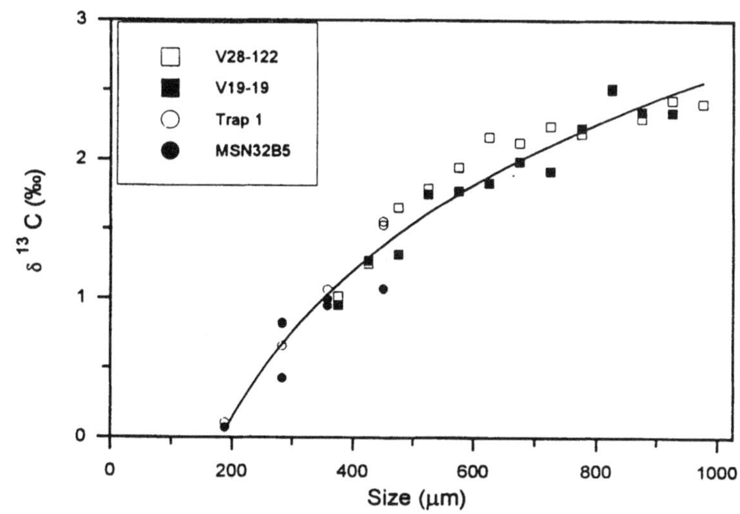

Figure 10: The carbon isotope ratio of *G. sacculifer*. Data from specimens from the Red Sea, collected with multiple o/c net and sediment trap off Barbados, and data from Oppo and Fairbanks (1989) are combined. Best fit is with $\delta^{13}C = -7.96 + 3.52*LOG(size)$.

foraminiferal calcite is controlled by both symbiotic photosynthesis (e.g. Spero and DeNiro, 1987) and foraminiferal respiration (e.g. Berger *et al.*, 1978). The primary sources of HCO_3 for calcification are controlled by diffusion processes between the aqueous environment and the cell (Bé *et al.*, 1981; Anderson and Faber, 1984; Spero and Williams, 1988; Spero *et al.*, 1992). The ΣCO_2 of the seawater between the spines, which may serve as a carbonate pool for calcification, becomes enriched with ^{13}C due to symbiont photosynthesis. As the number of symbionts and/or irradiance levels increase (Figure 11A), the ambient environment will become progressively more enriched. Respiration counteracts

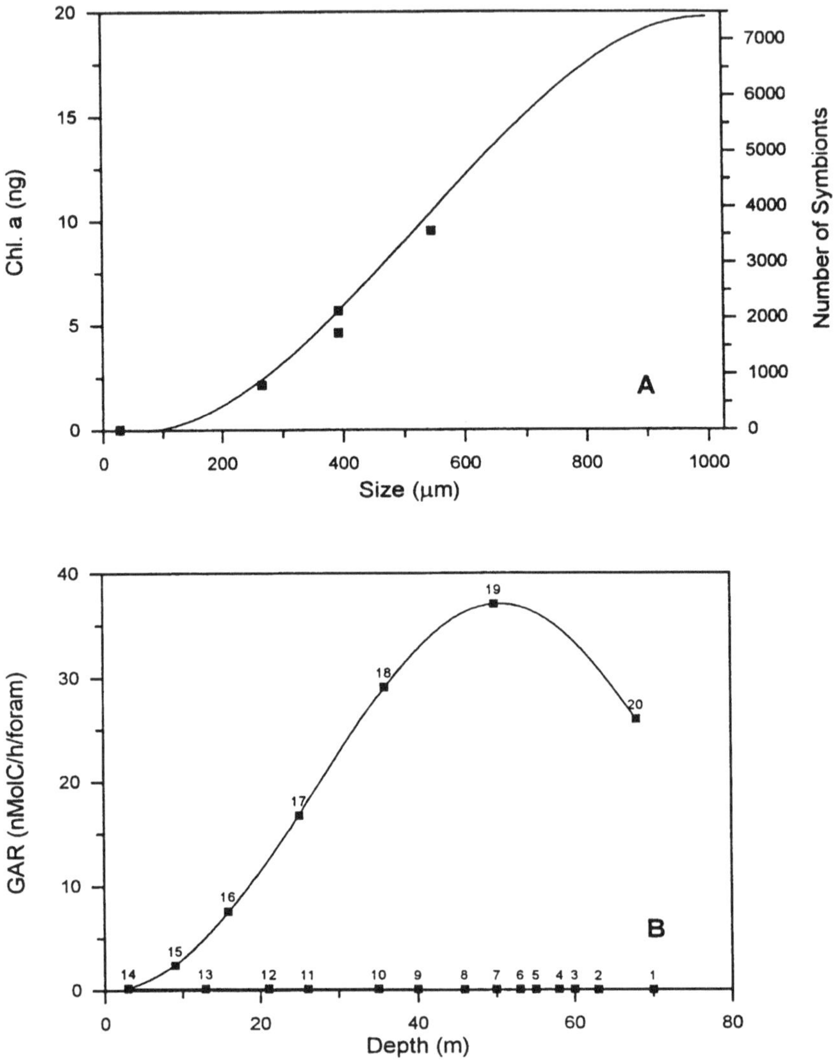

Figure 11: A: Chlorophyll a (square) concentration of *G. sacculifer* as a function of size: Chl. a = 0.191-8E-3*size+7.15E-5*size^2-4.52E-8*size^3. We assume that a symbiont (triangle) contains 2.5 pgChl a/cell (Falkowski and Dubinsky, 1981; McCloskey and Muscatine, 1984) and that 4 symbionts are sequestered at the 3-chamber stage (28 µm). The number of symbionts is plotted at the secondary y-axis. B: The gross assimilation rate of *G. sacculifer* as a function of the ontogenetic migration pattern.

the enrichment by releasing CO_2 derived from light organic compounds to the ambient environment (Spero *et al.*, 1992). ^{13}C-depleted CO_2 may also be released to an internal carbon pool which is used for calcification (Bijma, 1986).

The $\delta^{13}C$ value of the newly formed chamber thus depends on the gross assimilation rate (GAR) and on the respiration rate. The logarithmic trend of the $\delta^{13}C$-size curve (Figure 10) could be explained by (1) a reduction of the photosynthetic activity of the symbionts caused by the change to a deeper habitat in late ontogeny or (2) by an increase of the effect of respiration with size. In order to test this assumption, we modelled the GAR of the symbionts and the respiration of the host-symbiont complex as a function of ontogeny.

GROSS ASSIMILATION RATE.

The GAR is dependent on the number of symbionts and on the light intensity. During the ascending stage of the life cycle, the $\delta^{13}C$ of the test increases rapidly because both, the number of symbionts and the light intensity increase. During the descending phase from early adult stage onwards, the GAR reaches a maximum after which it decreases. Below the depth of maximum GAR, the decrease of the light intensity controls the GAR. Using the sizechlorophyll a plot (Figure 11A), the PI-curve of *G. sacculifer* (Jørgensen *et al.*, 1985) and applying the mass-balance equation, the GAR as a function of depth can be calculated:

$$GAR = P_{max} \cdot \frac{\frac{I_d}{I_k}}{\sqrt{1+\left(\frac{I_d}{I_k}\right)^2}} \qquad (5)$$

where P_{max} is the maximum photosynthetic rate of the symbionts, I_k and I_d are the saturation light intensity of the symbionts and the light intensity at depth d, respectively. The decrease in light intensity with depth is described by the Lambert-Beer equation:

$$I_d = I_0 * d^{(depth * k * z)} \qquad (6)$$

where I_0 is the surface light intensity, k is the extinction coefficient for clear (oligotrophic) open ocean water and z is a factor that compensates for the elevation of the sun. If we assume that the maximum photosynthetic rate (P_{max}) equals 8.17 pMolO_2/h,symbiont and that the saturation light intensity (I_k) equals 160 $\mu Em^{-2}sec^{-1}$ (Jørgensen *et al.*, 1985), a surface light intensity of 2000 $\mu Em^{-2}sec^{-1}$, an extinction coefficient of 0.04 (Steemann Nielsen, 1975), and z = 1.2, *G. sacculifer* reaches a maximum GAR of ca. 37 nMolC/h at approximately 50 m depth (Table 2; Figure 11B). Consequently, the $\delta^{13}C$ value of the test should decrease when the shell continues to calcify below approximately 50 m depth.

RESPIRATION RATE

The $\delta^{13}C$ of oceanic plankton is dependent on their food source. Although consumers reflect the $\delta^{13}C$ of their diet, they are normally slightly enriched (DeNiro and Epstein, 1978). Thus the $\delta^{13}C$ value generally increases towards higher trophic levels in a process which is termed cumulative fractionation (McConnaughey and McRoy, 1979a; 1979b). Depending on the region, copepods were roughly between 2‰ and 5‰ enriched compared the POC (McConnaughey and McRoy, 1979a; Mills *et al.*, 1983; Checkley and Entzeroth, 1985). Nonspinose, herbivorous foraminifers are therefore generally lighter than spinose, carnivorous species (cf. Figure 17 in Vincent and Berger, 1981). Unfortunately, in most reports on the isotopic composition of multiple species from the same environment, the analysis were carried out on different size fractions, making such a comparison invalid. Similarly, the change from juveniles stage in spinose species, feeding on detritus and phytoplankton, to neanic and older stages which have mostly a carnivorous diet, should be accompanied by an increase in the $\delta^{13}C$. Depending on the species, the transition from juvenile to neanic occurs roughly between 60 µm to 80 µm (Brummer *et al.*, 1986, Hemleben *et al.*, 1989). To our knowledge there exist no monospecific isotopic measurements from such small size fractions (pre-neanic), so that this early ontogenetic transition is not documented with isotope data.

Feeding rate also influences the $\delta^{13}C$ of the test. It was shown that the rate of chamber formation and the final size of spinose planktic foraminifers increased proportionally to the feeding frequency (Bé *et al.*, 1981; Hemleben *et al.*, 1987; Bijma *et al.*, 1992). In other words, at a constant temperature and equal size, the respiration rate or metabolic rate is increased at higher feeding rates, which in turn decreases the $\delta^{13}C$-value of the test. Bijma and Hemleben (unpubl. results) found that by doubling the feeding rate of *G. siphonifera*, the $\delta^{13}C$ of the test decreased by approximately 1‰. In contrast, Spero and Lea (in press.) calculated that the respiratory CO_2 accounts for less than 3% of the shell carbon in *G. sacculifer*.

Respiration rate is further dependent on temperature and biomass. We assume a constant temperature, but with increasing biomass, respiration of the foraminifer increases according to the formula of Fenchel and Finlay (1983):

$$\text{respiration (ngC/hr)} = 10^{0.75 * \text{LOG(volume)} - 4.99} \qquad (7)$$

Table 2: The gross assimilation rate and the respiration rate (both in nMolC/hr,foram) are calculated in dependence of number of symbionts, depth migration and size. The difference between these two is expressed as the vital effect (%). The vital effect is maximally 1 ‰.

chamber (#)	size (μm)	depth (m)	LI (μE/m², s)	Chl. a (ng)	symb. (#)	GAR (nMolC,hr)	volume (μm³)	Resp. (nMolC,hr)	GAR Resp. (nMolC,hr)	Vital effect (%)	unfract. signal (‰)	fract. signal (‰)
1	16	70	69	0.00	0	0.00	8.09	0.00	0.00	0.00	1.37	1.37
2	25	63	100	0.00	0	0.00	3.20	0.00	0.00	0.00	1.38	1.38
3	28	60	110	0.01	4	0.02	4.26	0.01	0.01	0.04	1.38	1.38
4	30	58	122	0.01	4	0.02	5.54	0.01	0.01	0.04	1.39	1.39
5	34	55	143	0.01	5	0.03	8.06	0.01	0.01	0.06	1.39	1.39
6	37	53	161	0.02	6	0.03	1.04	0.02	0.02	0.08	1.39	1.39
7	40	50	181	0.02	7	0.04	1.31	0.02	0.02	0.11	1.39	1.39
8	45	46	221	0.02	8	0.05	1.87	0.02	0.03	0.13	1.39	1.40
9	52	40	293	0.03	10	0.07	2.88	0.03	0.04	0.18	1.39	1.39
10	58	35	372	0.03	12	0.09	4.00	0.04	0.05	0.22	1.39	1.39
11	69	26	578	0.04	15	0.12	6.73	0.06	0.06	0.27	1.37	1.37
12	75	21	735	0.05	18	0.14	8.65	0.07	0.07	0.33	1.35	1.36
13	85	13	1096	0.05	20	0.16	1.26	0.08	0.07	0.33	1.32	1.32
14	100	3	1707	0.06	24	0.20	2.05	0.12	0.08	0.34	1.26	1.26
15	175	9	1313	0.74	295	2.40	1.10	0.59	1.81	7.79	1.22	1.30
16	270	16	940	2.36	943	7.60	4.05	1.69	5.91	25.46	1.17	1.43
17	391	25	615	5.30	2119	16.76	1.23	3.85	12.90	55.64	1.22	1.77
18	545	36	359	9.75	3899	29.10	3.32	7.66	21.44	92.47	1.29	2.22
19	740	50	181	15.11	6046	37.03	8.31	13.84	23.19	100.00	1.35	2.36
20	988	68	76	18.51	7403	25.97	1.98	23.10	2.87	12.40	1.37	1.49

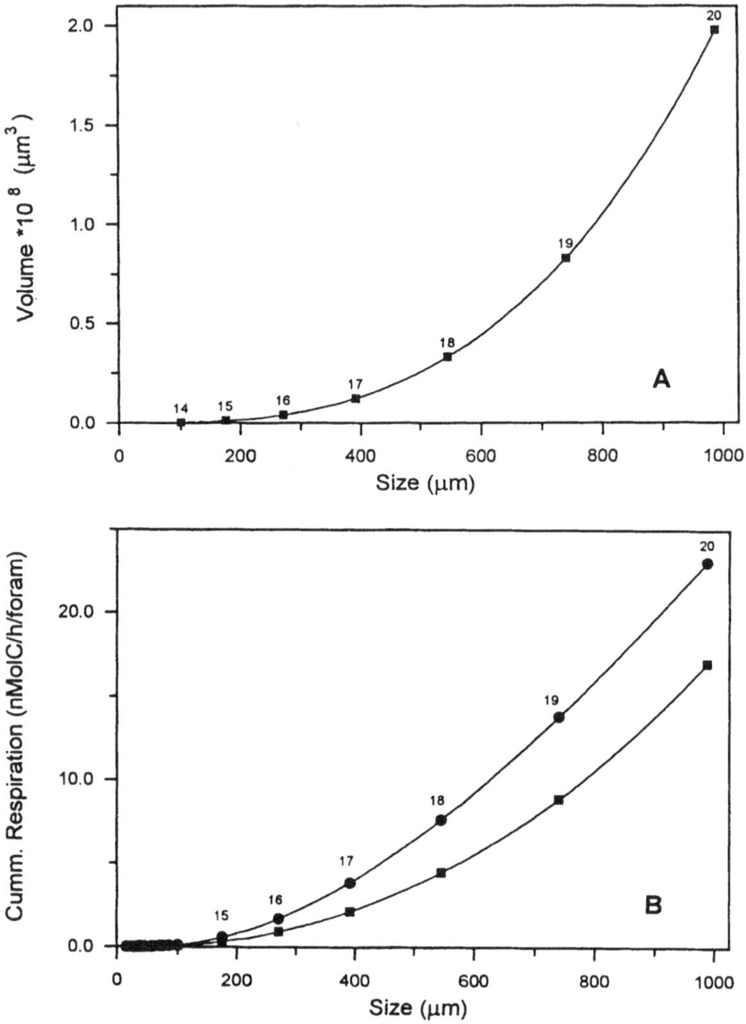

Figure 12: A: The volume of *G. sacculifer* calculated with the method of Ott *et al.* (1992): $V = 0.205 * size^3$. B: The cumulative respiration rate (square: host respiration; circle: symbiont respiration) as a function of size. The equation of Fenchel and Finlay (1983) was modified to fit *G.sacculifer*. The numbers denote chambers.

The volume increases with increasing size (Figure 12A). As a rule of thumb, respiration of the symbionts is usually considered to dissipate some 10% of the gross primary production achieved at saturation (e.g. Barnes and Hughes, 1982). In late ontogeny, the total respiration reaches the same order of magnitude as the photosynthesis of the symbionts (table 2). As the total respiration increases, the internal carbon pool and the ambient environment receive more ^{12}C and the $\delta^{13}C$-value of the test decreases.

THE OVERALL VITAL EFFECT

The effect of photosynthesis raises and the effect of respiration lowers the $\delta^{13}C$-size curve (Figure 6B). The biochemical pathways that deliver the carbon may be different as well. For instance, if carbon from the internal carbon pool (depleted by respiration) is the dominant source for calcification, the effect of respiration is important. If, on the other hand, epitactic growth is the predominant process, symbiont photosynthesis may be the most important process because carbon from the ambient environment (enriched by photosynthesis) will be used. We assume that both processes affect the carbon isotope value of the shell to an equal extent, i.e. in relation to their absolute carbon turnover rates. Consequently, we can estimate the vital effect by subtracting the absolute carbon release by respiration from the absolute carbon fixation by the symbionts. The relative vital effect at each ontogenetic stage can now be calculated by setting the maximum absolute change in carbon equal to 100 % vital effect (Figure 13A). Apparently, the vital effect is negligible in early ontogeny. Afterwards it tends to increase the $\delta^{13}C$-value of the shell. It reaches a maximum between the 18th and 19th-chamber stage. At the end of the life cycle the effect of the vital processes is to reduce the $\delta^{13}C$ of the shell. In the present simulation experiment we allow a vital effect for the carbon isotopes that is two times as strong as that for oxygen isotopes (ca. 0.5‰ e.g. Hemleben *et al.*, 1989). An increase between 1 and 2‰, in the carbon isotope composition due to light intensity alone was reported by Spero (1988) and Spero and Deniro (1987) respectively. A decrease of about 1‰, was found at higher feeding frequency in *G. siphonifera* (Bijma *et al.*, unpub. results). In other words, a vital effect of about 1‰, is very realistic for carbon isotopes. The resulting change of the $\delta^{13}C$ of the shell with ontogeny can be calculated (Figure 13B). Thus the vital effect tends to amplify the shape of the equilibrium signal that results from the migratory pattern (cf. Figure 13B with Figure 9B). Depending on local conditions, the shapes of the curve may of course vary, just as the amplitude.

Several arguments, indicate that photosynthesis of the symbionts may not be such an important controlling mechanism for carbon fractionation as we assume here: 1) *G. siphonifera* cultured in continuous darkness show the same trend as specimens cultured under light, although the slope may be slightly less inclined (Bijma and Hemleben, unpublished results). 2) Spinose non-symbiotic species such as e.g. *G. bulloides* also show the same general pattern of figure 10. 3) Chamber formation (calcification) in *G. sacculifer* predominantly occurs between 6 p.m. and 12 p.m. (Bijma, 1986).

If only host-symbiont respiration is considered the shape of the curve becomes even further away from reality. We therefore conclude that more information on the mechanisms that control foraminiferal $\delta^{13}C$ is needed before the carbon isotope black box can really be opened.

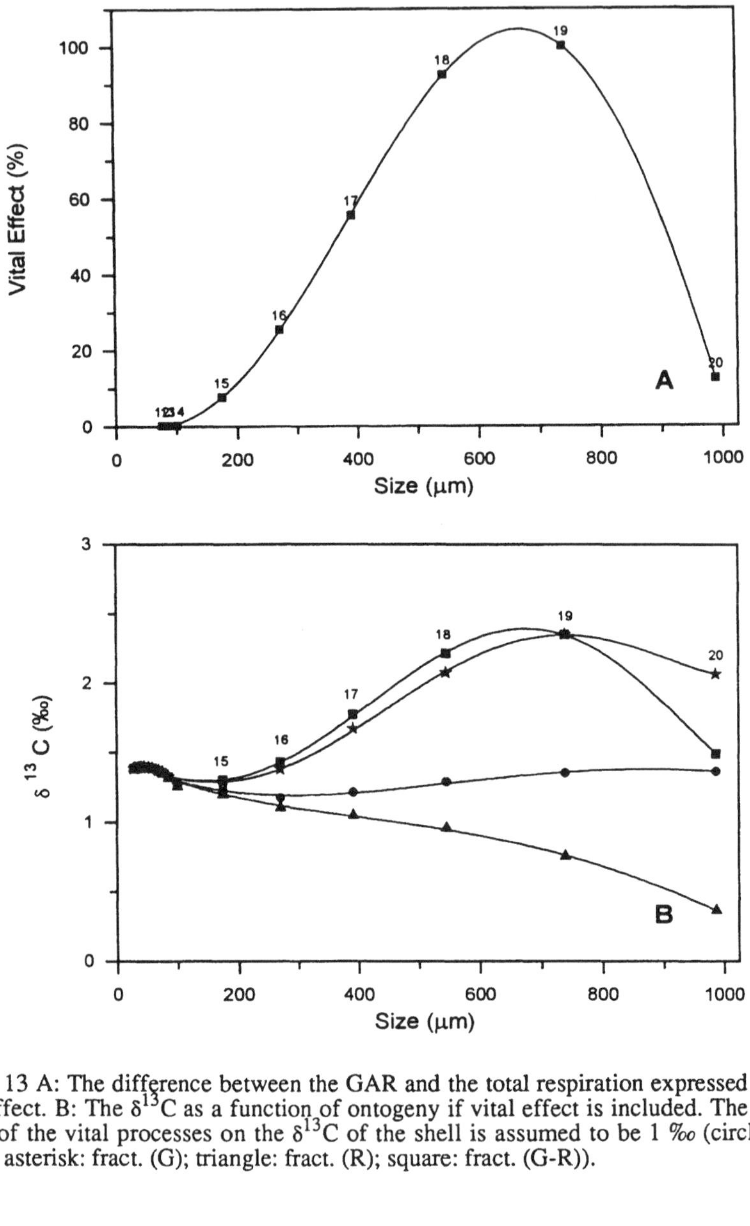

Figure 13 A: The difference between the GAR and the total respiration expressed as percent vital effect. B: The $\delta^{13}C$ as a function of ontogeny if vital effect is included. The maximum effect of the vital processes on the $\delta^{13}C$ of the shell is assumed to be 1 ‰ (circle: original signal; asterisk: fract. (G); triangle: fract. (R); square: fract. (G-R)).

ACKNOWLEDGEMENTS

We thank Ingrid Breitinger and Rolf Ott for their helpful assistance and Howard Spero and Fred Banner for critical remarks that improved the manuscript. The authors acknowledge the editor's engagement in both the meeting and assembling the volume.

REFERENCES

ANDERSON, O.R., FABER, W.W., Jr., 1984. An estimation of calcium carbonate deposition rate in planktonic foraminifera *Globigerinoides sacculifer* using 45-Ca as a tracer: a recommended procedure for improved accuracy. J. Foram. Res., 14, 303-308.

ALMOGI-LABIN, A., 1984, Population dynamics of planktic Foraminifera and Pteropoda - Gulf of Aqaba, Red Sea: Proc. Koninkl. Nederlandse Akad. Wetensch., Ser. B, 87, 4, 481-511.

BARNES, RSK, HUGHES, RN, 1982. Introduction to Marine Ecology, Blackwell Scientific Publications, 339 p.

BÉ, A.W.H. CARON, D.A., ANDERSON, O.R., 1981, Effects of feeding frequency on life processes of the planktonic foraminifer Globigerinoides sacculifer in laboratory culture: J. Mar. Biol. Assoc. U.K., 61, 257-277.

BERGER, W.H. KILLINGLEY, J.S., VINCENT, E., 1978, Stable isotopes in deep-sea carbonates: Box Core ERDC-92 west equatorial Pacific: Oceanol. Acta, 1, 2, 203-216.

BIJMA, J., 1986, Observations on the life history and carbon cycling of planktonic foraminifera from the Gulf of Elat/Aqaba. Unpubl. Master Thesis, Univ. Groningen, The Netherlands, 120 p.

BIJMA, J., J. EREZ, HEMLEBEN Ch., 1990, Lunar and semi-lunar reproductive cycles in some spinose planktonic foraminifers. Journal of Foraminiferal Research, 20 (2), 117-127.

BIJMA, J., HEMLEBEN Ch., 1993, Population dynamics of *G. sacculifer* (Brady) from the Central Red Sea: Deep Sea Res.

BIJMA, J., Ch. HEMLEBEN, H. OBERHÄNSLI, SPINDLER M., 1992. The effects of increased water fertility on tropical spinose planktonic foraminifers in laboratory cultures: J. Foram. Res., 22,3, 242-256.

BRUMMER, G.J., HEMLEBEN Ch., SPINDLER M., 1987. Ontogeny of extant spinose planktonic foraminifera (Globigerinidae): A concept exemplified by *G. sacculifer* (Brady) and *G. ruber* (d'Orbigny). Mar. Micropal., 2, 357-381.

BROECKER, W.S., MAIER-REIMER E, 1992. The influence of air and sea exchange on the carbon isotope distribution in the sea: Global Biogeochemical Cycles, 8(3), 315-320.

CHECKLEY, D.M., ENTZEROTH L.C., 1985. Elemental and isotopic fractionation of carbon and nitrogen by marine, planktonic copepods and implications to the marine nitrogen cycle: J. Plankt. Res., v. 7(4), 553-568.

DENIRO, M.J., EPSTEIN, S , 1978. Influence of diet on the distribution of carbon isotopes in animals: Geochim. and Cosmochim. Acta, v. 42, 495-506.

EREZ, J., A. ALMOGI-LABIN, AVRAHAM S., 1991, Lunar Reproduction Cycle in *Globigerinoides sacculifer* (Brady). Paleoceanography, 6 (3), 295-306.

FALKOWSKI, P.G., DUBINSKY Z., 1981. Light-shade adaptation of *Stylophora pistilata*, a hermatypic coral from the Gulf of Eilat. Nature, v. 289, p. 172-174.

FENCHEL, T., FINLAY B.J., 1983, Respiration rates in heterotrophic, free-living protozoa. A review. Microb. Ecol., v. 9, 99-122.

HEMLEBEN, CH., M. SPINDLER, ANDERSON O.R., 1989, Modern Planktonic Foraminifera, Springer Verlag, 363 pp.

HEMLEBEN, CH. SPINDLER, M. BREITINGER, I., OTT, R., 1987, Morphological and physiological responses of Globigerinoides sacculifer (Brady) under varying laboratory

conditions: Marine Micropal., 12, 305-324
JØRGENSEN, B.B. EREZ, J. REVSBECH, N.P., COHEN, Y., 1985, Symbiotic photosynthesis in a planktonic foraminiferan, Globigerinoides sacculifer (Brady), studied with microelectrodes: Limnol. Oceanogr., 30, 1253-1267
McCLOSKEY, L.R., MUSCATINEL., 1984. Production and respiration in the Red Sea coral Stylophora pistillata as a function of depth. Proc. of the Royal Soc, London, B222, 215.230.
McCONNAUGHEY, T., MCROY, C.P., 1979a. Foodweb structure and the fractionation of carbon isotopes in the Bering Sea. Mar. Biol., 53, 257-262.
McCONNAUGHEY, T., MCROY, C.P., 1979b. ^{13}C label identifies eelgrass (*Zostera marina*) carbon in the alaskan estuari foodweb. Mar. Biol., 53, 263-269.
MILLS, E.L., K. PITTMAN and F.C. TAN, 1983. Food web structure on the Scotian shelf, eastern Canada. study using ^{13}C as a food chain tracer. ICES. Rapp. Proc. Verb. Reun., 183, 111-118.
OTT, R., J. BIJMA, Ch. HEMLEBEN and M. SIGNES, 1992, A computer method for estimating volumes and surface areas of complex structures consisting of overlapping spheres. Math. Comput. Modelling, 16(12), 83-98.
REISS, Z., and L. HOTTINGER (1984) The Gulf of Aqaba. Ecological Micropaleontology. Ecological studies 50. Springer-Verlag, Berlin, Heidelberg, New York and Tokyo, 354p.
SPERO, H.J. and M.J. DENIRO, 1987, The influence of symbiont photosynthesis on the δ^{13}O and δ^{13}C values of planktonic foraminiferal shell calcite. Symbiosis, 4, 213-228.
SPERO, H.J. and D.W. LEA, in press., Intraspecific stable isotope variability in the planktonic foraminifera *Globigerinoides sacculifer*: Results from laboratory experiments. Mar. Micropal.
SPERO, H.J. and D.F. WILLIAMS, 1988. Extracting environmental information from planktonic foraminiferal δ^{13}C data. Nature, 335, 717-719.
SPERO, H.J. and D.F. WILLIAMS, 1989. Opening the carbon isotope "vital effect" black box, I: Seasonal temperatures in the euphotic zone. Paleoceanography, 4, 593-601.
SPERO, H.J., LERCHE, I. and D.F. WILLIAMS, 1992. Opening the carbon isotope "vital effect" black box, 2; Quantitative model for interpreting foraminiferal carbon isotope data. Paleoceanography, 6, 639-655.
STEEMANN NIELSEN, E., 1975. Marine photosynthesis. With special emphasis on the ecological aspects. Elsevier Scientific Publishing company, Amsterdam, 141 p.
SPINDLER, M., Ch. HEMLEBEN, U. BAYER, A.W.H. BÉ and O.R. ANDERSON, 1979, Lunar Periodicity of Reproduction in the Planktonic Foraminifer *Hastigerina pelagica*. Marine Ecology, Progress Series 1, 1, 61-64.
VINCENT, E., and BERGER, W.H. 1981. Planktonic foraminifera and their use in paleooceanography. In: The oceanic lithosphere, The Sea. C. Emiliani (ed.), v. 7, 1025-1119.
WILLIAMS, D.F. SOMMER, M.A. and BENDER, M.L., 1977, Carbon isotopic compositions of Recent planktonic Foraminifera of the Indian Ocean: Earth Planet. Sci. Lett., v. 36, 391-403.

A COMPARISON OF CARBON ISOTOPES AND CADMIUM IN THE MODERN AND GLACIAL MAXIMUM OCEAN: CAN WE ACCOUNT FOR THE DISCREPANCIES?

Edward A. Boyle
Department of Earth, Atmospheric, and Planetary Sciences
Room E34-258
Massachusetts Institute of Technology
77 Mass. Ave.
Cambridge MA 02139, U.S.A.

ABSTRACT. Oceanic cadmium, carbon isotopes, phosphorus, and dissolved carbon dioxide are correlated on a global scale because of the dominance of the biological cycling of these elements. Some differences between the distributions of these properties are seen in the modern ocean, although these differences can be rationalized. In the glacial maximum (18K) ocean, the differences between benthic foraminiferal $\delta^{13}C$ and Cd/Ca are much larger than in the modern ocean, suggesting quite different scenarios for the carbon system chemistry of the Southern Ocean, the northwest Pacific Ocean, and the intermediate waters of the Pacific. A re-examination of the modern distribution of $\delta^{13}C$ and Cd suggests that one major factor which has not been addressed in detail is the role of gas exchange on $\delta^{13}C$, but it is not clear that this factor can reconcile the differences between the tracers during the glacial maximum. The modern oceanic Cd-P relationship is reassessed in light of three new data sets and found to be consistent with previous interpretations of deep ocean data. The issue of foraminiferal accuracy and precision for $\delta^{13}C$ and Cd/Ca reconstructions is considered: more effort on living benthic foraminifera is clearly needed to resolve the uncertainties concerning foraminiferal response to bottom water properties.

INTRODUCTION

In order to examine problems such as the causes of glacial/interglacial atmospheric carbon dioxide cycles and the role of deep ocean circulation in large-scale global climate change, we need to know how oceanic distributions of carbon-system parameters have varied on glacial-interglacial time scales. The chemical properties we require for this task (e.g. total

NATO ASI Series, Vol. I 17
Carbon Cycling in the Glacial Ocean:
Constraints on the Ocean's Role in Global Change
Edited by R. Zahn et al.

dissolved carbon dioxide, alkalinity, phosphorus, oxygen) must be reconstructed indirectly from measurements on foraminiferal shells (e.g. $\delta^{13}C$, Cd/Ca). Although the carbon system properties are not derived in such direct fashion as the CO_2 content of bubbles in ice, with proper interpretation, they may be estimated reasonably well.

Reconstructions of oceanic $\delta^{13}C$ and Cd distribution histories have two distinct applications. One application is the examination of past ocean circulation patterns using $\delta^{13}C$ and Cd as non-conservative tracers of water masses. Evidence from oceanographic expeditions during the past 70 years (e.g. Bainbridge, 1980; Craig et al., 1982; Weiss et al., 1982; Wust, 1935, 1978) shows that these tracers provide information that can be used to great advantage. Non-conservative tracers are of even greater utility in evaluating past ocean circulation patterns because our ability to estimate conservative paleotracers is limited (temperature and salinity estimation in past oceans is possible (Duplessy et al., 1991) but relatively imprecise compared to the range of these tracers in deep waters). For the purpose of water mass tracing, one non-conservative tracer may be just as good as any other, given sufficient primary differences in source water composition and adequate understanding of the tracer's non-conservative characteristics. For this purpose, the precision and constancy of correlations between $\delta^{13}C$ or Cd with ΣCO_2 or P are relatively unimportant.

The other application of foraminiferal $\delta^{13}C$ and Cd is to enable estimation of carbon system properties that are of prime importance for understanding atmospheric CO_2 variability. This task has been undertaken in most cases by assuming that the modern correlations between these properties apply in the past (Figure 1 Model 1). It is obvious that this assumption is neither necessary nor is it likely to be justifiable in detail (e.g. Boyle, 1988; Boyle et al., 1985; Boyle et al., 1981; Broecker, 1982; Broecker and Maier-Reimer, 1992). Reconstruction of the oceanic carbon system does not in fact demand that the proxy tracers precisely represent the key carbon-system properties through a simple linear correlation. Instead, it is understood that one only need estimate the key properties through a mathematical mapping of the properties that we can measure onto the space of those that we desire to know. For example, if we had a complete model for the biogeochemical cycling of CO_2, P, $\delta^{13}C$, and Cd in the modern ocean, we could run this model in "inverse" mode and use the past distributions of $\delta^{13}C$ and Cd to estimate the past distributions of CO_2 and P (Figure 1, Model 2). This concept should not be taken to construe a precise uniqueness of the mathematical mapping (which cannot be justified in terms of our present knowledge of ocean chemistry). The intent here is not to emphasize literal generation of unique fields of CO_2 and P from $\delta^{13}C$ and Cd through inversion; rather, this strategy is a reminder that paleo-carbon-system estimations need not be restricted to simple linear analogies to modern correlations between the tracers.

Both of the paleotracer applications require that we can accurately assess past oceanic distributions of $\delta^{13}C$ and Cd from measurements in foraminifera. Certainly foraminiferal tests reflect these properties of the water overlying the seafloor where these organisms grow (at least

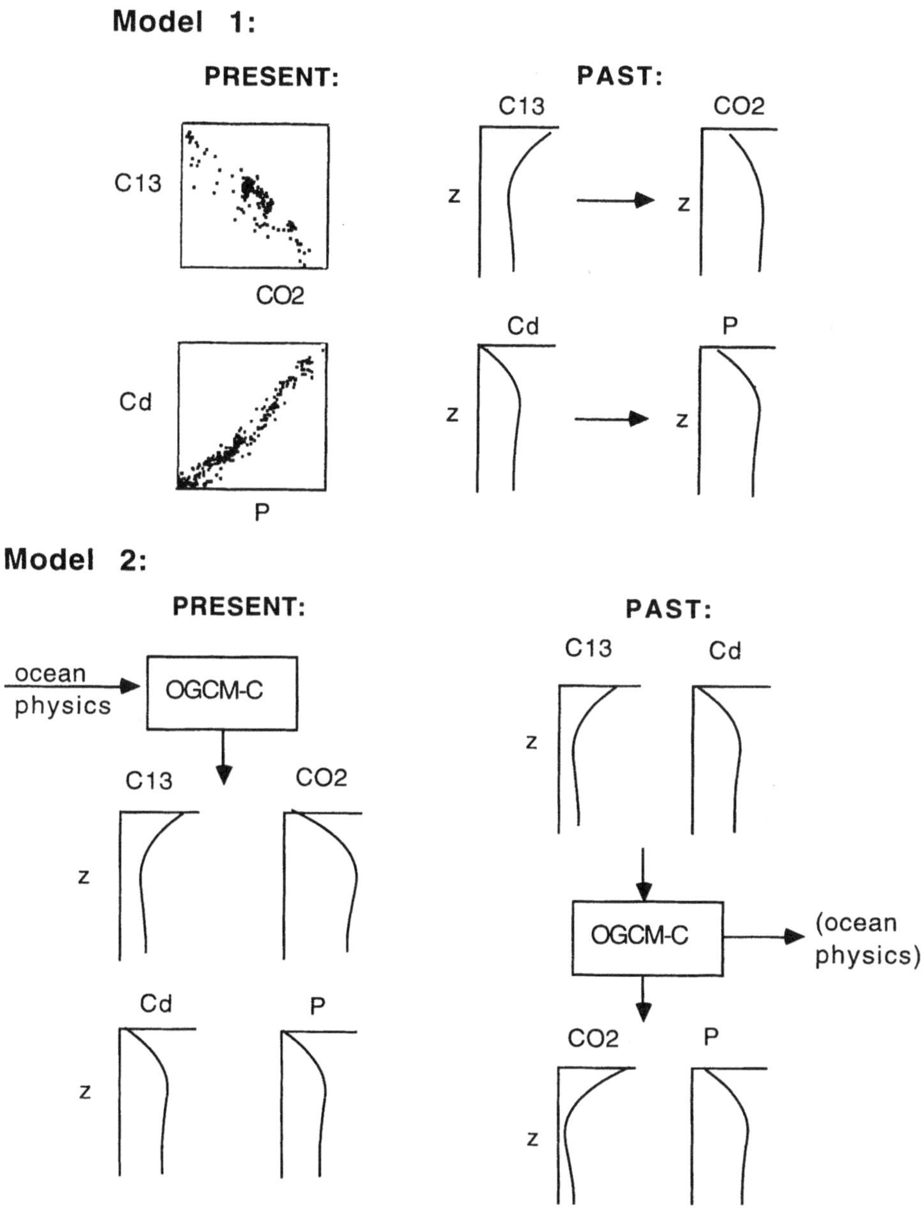

Figure 1: Oceanic $\delta^{13}C_{\Sigma CO2}$ and Cd-P interconversion models. Model 1 uses modern correlation between the tracers to estimate ΣCO_2 from $\delta^{13}C$ and P from Cd. Note that the plots are based on real data: the $\delta^{13}C_{\Sigma CO2}$ plot is the SIO data obtained during GEOSECS, and the Cd-P plot is data shown in Figure 6. Model 2 assumes that an adequate ocean general circulation model with embedded chemistry (OGCM-C) exists; then the model is used in an inversion mode to estimate distributions of ΣCO_2 and P which are consistent with $\delta^{13}C$ and Cd data.

in the large-scale aspects of differences between Atlantic and Pacific, or surface-to-deep chemical fractionations). Nonetheless, valid concerns remain concerning the absolute accuracy and precision of these foraminiferal measurements, particularly for the sake of evaluating more subtle chemical changes. This concern is heightened by the recent discovery that during the most recent glacial maximum, $\delta^{13}C$ and Cd sometimes appear to contradict one another (Boyle, 1992). Could it be that these tracers sometime mislead us about past bottom water chemistry in some circumstances? Or is there a problem in appearance only, resulting from viewing these tracers through an oversimplified conceptual model? In the latter case, the apparent "contradictions" may instead provide "information". Certainly, this situation has occurred before in paleoceanography. For example, oxygen isotope evidence for sea level change once appeared to conflict shoreline evidence for sea level change. Once reliable sea level records became established, we came to see the contradiction as information about past changes in ocean temperature. Perhaps with some luck, these apparent contradictions between $\delta^{13}C$ and Cd may serve us as well.

The purpose of this paper is to explore the nature of differences between modern and past $\delta^{13}C$ and Cd distributions in order to set the stage for evaluation of the cause of the differences. Although the current state of understanding is insufficient to immediately answer all of these questions, a discussion can distinguish real concerns from red herrings and clarify what will be necessary to resolve the uncertainties.

CARBON ISOTOPES AND CADMIUM: THE SIMPLE VIEW

The simplest possible view of carbon and cadmium is that both are incorporated into biological tissues in a strict "Redfield" ratio ^{13}C:^{12}C:Cd:P and that their variability throughout the ocean is created only by differing degrees of uptake into biological tissue, decay of that biological tissue from sinking particles, and the effects of oceanic advection and diffusion. Although this view is oversimplified, it does account for the basic correlation between properties on an ocean-wide scale (Figure 1), and provides the basic intuition with which one can interpret paleoceanographic data. The principle complications to this simple view are:

(1) Gas exchange influences $^{13}CO_2$ and $^{12}CO_2$, with no effect on P and Cd. The role of this process has perhaps been overly minimized in previous treatments of past $\delta^{13}C$ distributions. The effect of "thermodynamic" gas exchange is probably comparable to the effect of biological uptake and decay on $\delta^{13}C$ (Broecker and Maier-Reimer, 1992).

(2) The ^{13}C:^{12}C:Cd:P ratio in biological material is not precisely constant; there are regional variations that affect their relative uptake and recycling.

Although these complications are significant, they are not sufficient to overwhelm the ocean-wide correlation in the modern ocean, and it would seem at least a plausible point of departure to assume that on a large-scale basis these correlations have been similar in the past. This point of view will be adopted temporarily in this discussion, and then reconsidered in view of the new modern and sedimentary observations.

The second premise of this simplest-level interpretation is that $\delta^{13}C$ and Cd are incorporated into the shells of planktonic and benthic foraminifera in simple linear fashion, at least within a single species. This assumption is consistent with data on the global-scale relationship between benthic foraminiferal $\delta^{13}C$ and Cd from core tops and that of the bottom water in the deep ocean. Nonetheless, there is evidence that some species may shift their shell $\delta^{13}C$ in response to changing levels of organic carbon flux to the sediment (Mackensen, in press; Zahn et al., 1986), and that the Cd response of calcitic benthic foraminifera is depth-dependent above about 2500 meters (Boyle, 1992). At the outset, this discussion will adopt the assumption that large-scale global signals are recorded reasonably well by calibrated core-top benthic foraminifera. Later, this assumption will be reassessed in light of emerging sedimentary data.

PALEOCHEMICAL DISTRIBUTIONS IN CORE TOPS AND THE LAST GLACIAL MAXIMUM: A COMPARISON OF CADMIUM AND CARBON ISOTOPE EVIDENCE

This section briefly reviews data presented by Boyle (1992). Figures 2 and 3 (top) show core top data (as translated through the calibration functions) from 6 regions (note that these regions are broad; for example, "Northwest Pacific" will refer to cores from the equatorial to subpolar zones of the western North Pacific). Although the agreement between core top and modern data is not perfect, the foraminiferal core top data do show the expected ocean-wide trends. Figures 2 and 3 (bottom) show the glacial maximum data from the same suite of cores. Cd and $\delta^{13}C$ data agree (at least qualitatively) on some aspects of the glacial distribution, notably (1) the reduction (but not elimination) of $\delta^{13}C$ and Cd gradients along the "conveyor belt" from the North Atlantic to the Eastern Tropical Pacific and (2) the relatively low Cd/positive $\delta^{13}C$ in the upper waters of the Atlantic and Northern Indian Ocean relative to the deeper waters in those basins. But there are some notable discordancies as well: (1) Antarctic $\delta^{13}C$ data would indicate glacial maximum nutrient contents as high or even higher than the eastern tropical Pacific (as originally noted by Curry et al. (1988), but Cd indicates relatively little change in the nutrient contents, (2) $\delta^{13}C$ data seem to indicate relatively nutrient-depleted

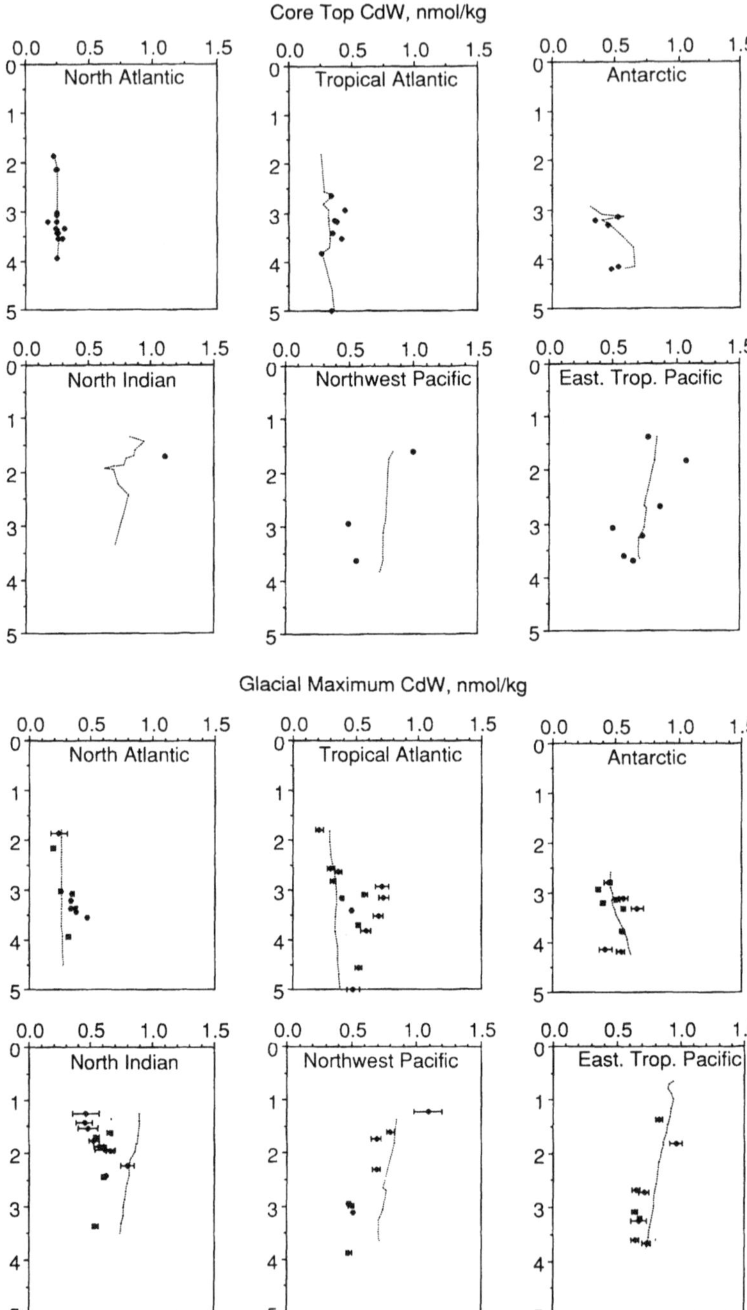

Figure 2a and 2b: Core top and glacial maximum Cd profiles (points) from 6 ocean basins compared to modern Cd (dotted line). This Figure and Figure 3 are scaled so that changes in CdW and $\delta^{13}C$ compared to present-day Atlantic-Pacific contrast are graphically equivalent.

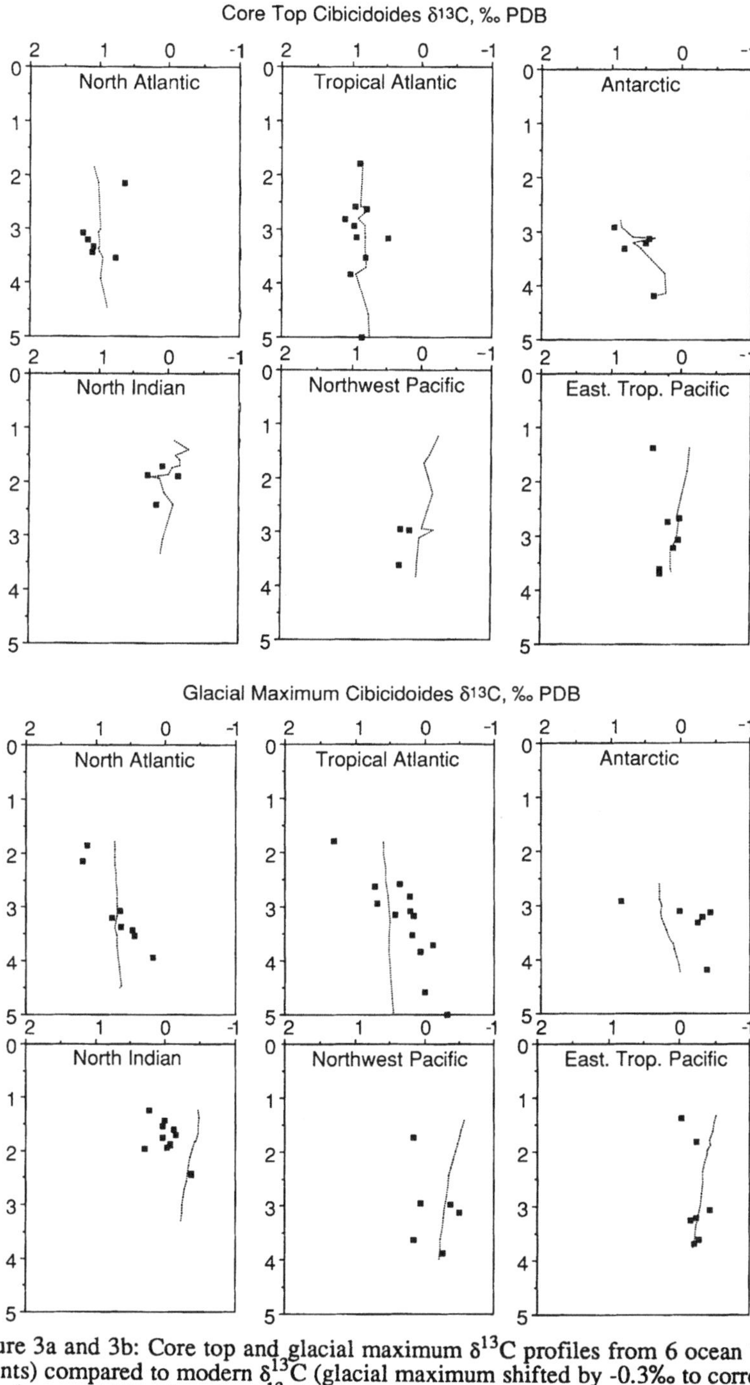

Figure 3a and 3b: Core top and glacial maximum $\delta^{13}C$ profiles from 6 ocean basins (points) compared to modern $\delta^{13}C$ (glacial maximum shifted by -0.3‰ to correct for global mean shift). Note that $\delta^{13}C$ scale is reversed so that changes in CdW and $\delta^{13}C$ have the same sense.

waters in shallower waters of the glacial maximum eastern tropical Pacific (although note that Mix et al. (1990) also observed that the $\delta^{13}C$ between shallower and deeper waters didn't seem to change much between glacial and interglacial periods), although Cd data suggest that nutrient contents of the upper waters of the Pacific have remained consistently high, (3) Cd data indicate low Cd concentrations in the glacial maximum northwest Pacific; $\delta^{13}C$ data from some tropical sites seem to agree with this, but data from the more northerly sites clearly disagree with the Cd evidence, and (4) the vertical nutrient gradient in the Atlantic, although in the same sense for both tracers, is more extreme for $\delta^{13}C$ than it is for Cd.

Another way of presenting the results is to use the analogy between properties and nutrients to create a map of the deepwater distribution. This analogy is presented for Cd in the modern and glacial maximum ocean in the depth range 2400-3800m (Figure 4) using the assumption that the Cd-P relationship in the glacial ocean is similar to that in the modern ocean (using the two-line "kink" model discussed later and shown in Figure 6). It is important to note that the stability assumption is not likely to be exactly correct, so this manner of presentation is mainly directed at making the Cd distribution more accessible to the larger audience familiar with the quantitative oceanic phosphorus distribution.

The most serious disagreements between these tracers (according to the simplest assumptions adopted above) are those in the Antarctic and in the Northwest Pacific ocean. If we demand the simplest interpretation of the tracers, it is difficult to reconcile these differences. Either the simple linear-correlation tracer model is wrong, or the calibration of the organisms for one or both properties is variable (or mis-assessed). So the next task is to re-evaluate the modern data for $\delta^{13}C$ and Cd in the ocean in order to better understand their chemical oceanography and to re-evaluate the accuracy and precision of foraminiferal $\delta^{13}C$ and Cd uptake.

$\delta^{13}C$ IN THE MODERN OCEAN: THE ONGOING REASSESSMENT

Other papers in this volume and elsewhere will present more detailed evaluation of the oceanic $\delta^{13}C$ distributions, and so it is only necessary to summarize some of their conclusions relevant to the present discussion, and then move on to discussion of the oceanic Cd distribution.

(1) Data Quality: The accuracy of published $\delta^{13}C$ data set has been examined and found adequate in some cases but in need of reevaluation in other instances. Some of the apparent conflicts between core top data and water column data may be due to inaccurate water column data (McCorkle and Keigwin, in review).

(2) Data Editing: Given evidence for some data consistency problems in the GEOSECS data, a certain amount of editing was unavoidable. Although this editing was done honestly (i.e., the

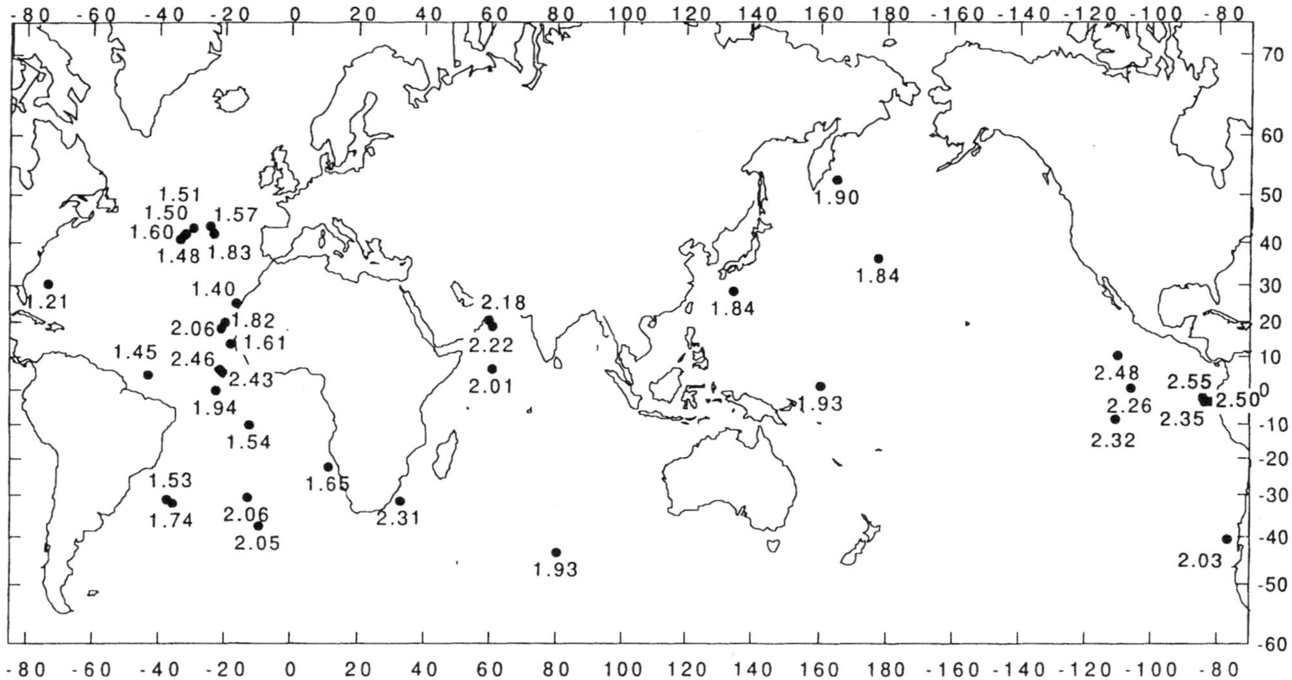

Figure 4: Distribution of analog "phosphorus" in deep waters (2400-3800m) of the glacial maximum (18K) ocean as inferred from measurements of benthic foraminiferal Cd/Ca and translated into "phosphorus" units by using the kink model for the modern Cd-P relationship.

raw data were public), editing always requires a conceptual framework for the evaluation of what is reasonable. It appears that a guiding principle in the data editing was that a certain measure of correlation between dissolved oxygen and the nutrients phosphorus and nitrate were expected, resulting in removal of some "anomalous" points in the initial editing and neglect of others during section contouring.

(3) New Data and Interpretations: One example of the effect of this data selection is given by Mackensen (in press) who obtained additional data from the Atlantic sector of the Southern Ocean and established that bottom water in the Weddell Sea has $\delta^{13}C$ values in excess of +0.8‰ - both in their new data and in the GEOSECS data. These high $\delta^{13}C$ values are correlated with high oxygen values (M. Sarnthein, personal communication), but not with low phosphorus or nitrate concentrations.

On the basis of these observations it is possible that the carbon isotope distribution is more complicated than previously supposed. Recent modeling efforts indicate that the "thermodynamic" effects of gas exchange on $\delta^{13}C$ are important (Broecker and Maier-Reimer, 1992; Keir, 1991) . The most plausible explanation for more positive $\delta^{13}C$ in some regions (e.g., in Mackensen's Weddell Sea data), in the absence of notable phosphorus or nitrate depletions, is that extensive gas exchange at low temperatures produces higher $\delta^{13}C$ values in waters that feed into the oceanic interior.

Finally, it should be remarked that variations in the $\delta^{13}C$ of organic matter from -15‰ to -28‰ have been observed (Rau et al., 1982); these differences can possibly produce measurable differences in dissolved $\delta^{13}C$. Although it is not considered likely that major $\delta^{13}C$ signals from this process occur in the modern ocean, perhaps this effect may have been more important at other times.

(4) Consequences for benthic foraminiferal $\delta^{13}C$ calibrations and paleo-$\delta^{13}C$ interpretations: It is obvious that calibrations of core tops that are based on incorrect estimates of $\delta^{13}C$ in the bottom water will produce incorrect estimates of the calibration of benthic species, and that some priority should be directed towards addressing this issue by a systematic program of new water column measurements. It is perhaps equally obvious that the new appreciation of the role of gas exchange on oceanic $\delta^{13}C$ will provoke a rethinking of the meaning of past changes in foraminiferal $\delta^{13}C$. Whatever else, future analysis will require that $\delta^{13}C$ not be considered as a simple indicator of the oceanic phosphorus distribution. It is not so for the present ocean, and only improbably so for past oceans.

CADMIUM IN THE MODERN OCEAN: PRELIMINARY ASSESSMENT OF EMERGING DATA

The global database on cadmium in the ocean (whether dissolved in seawater, attached to oceanic particles, or as vertical fluxes) is slim. More data should be gathered for oceanic Cd to determine whether its distribution has surprises comparable to those recently uncovered for $\delta^{13}C$. However, contrary to some assertions, as of yet no published data require major reinterpretation of paleoceanographic interpretations based on Cd. Because paleoceanographers are not generally familiar with aspects of oceanic Cd that have been known for over a decade, some discussion of this data is necessary.

Although it is widely known that there is a remarkable correlation of oceanic Cd and P, there is less appreciation of the characteristics of that correlation and the extent to which individual data sets deviate from perfect correlation. First, it should be remembered that prior to 1976, there were no more than a few data points on this element in seawater that are not now considered incorrect as a result of analytical problems and contamination during sampling. Following two publications in that year that established the correlation of that element with phosphorus in the ocean (one of which noted that this correlation may be global in scope), several more followed that established these characteristics in more detail. But even in the period following 1976, incorrect Cd data continued to be published. As a result of this situation, it has been difficult to assess the situation (outside of the few laboratories that measure Cd and know how to assess data quality) .

Boyle (1988) summarized the characteristics of deep ocean Cd as they pertain to paleoceanography (Figure 5 shows station locations; data shown in Figure 6). The essential characteristics that were attributed to Cd were: (1) in the deep waters of the ocean, available data suggested that the Cd-P (cadmium-phosphorus) correlation is remarkably uniform, sufficient to allow estimation of Cd from P in the deep ocean to within about 7% of the range of observed values, (2) the correlation is not a simple line; the slope of the global correelation changes from 0.20 to 0.40 (nmol Cd per μmol of P) above P levels of about 1.3 μmol/kg, and (3) highly significant deviations from the deepwater correlation are seen in near-surface waters, coastal waters, and marginal basins such as the Mediterranean Sea, so local correlations should be applied to estimate Cd from P (or estimate P from paleo-Cd) in those regions. This summary remains relevant today, but with the addition of a few new data sets, it is worth expanding and clarifying these points here.

First, the issue of data precision and accuracy as it contributes to the scatter of the global correlation should be addressed. Intercalibration errors, some contamination problems, and real oceanic differences are embedded within the scatter of the oceanic Cd-P plot (Figure 6). It is sometimes hard to distinguish real differences from artifacts. It is easy to buy or prepare standards for Cd that are accurate to better than 1%, so with proper analytical protocol it

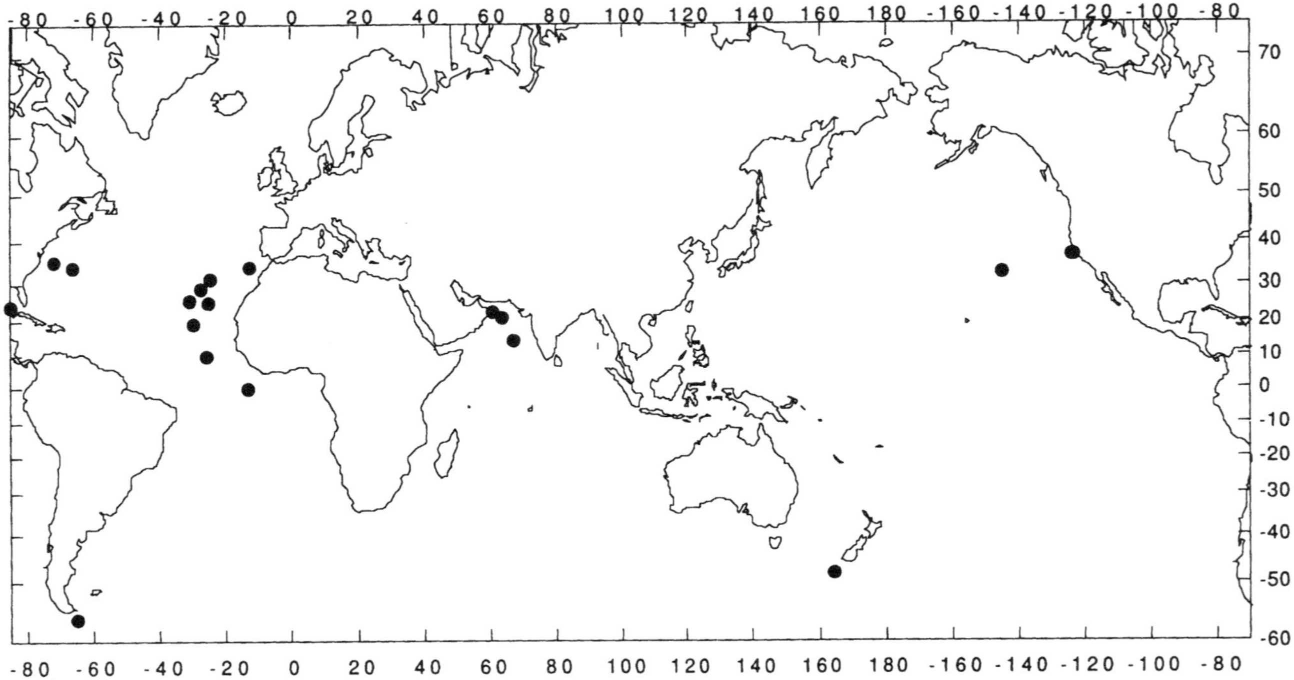

Figure 5: Global deepwater P station map. Note that in Figure 5 (and Boyle, 1988), all stations are included <u>except</u> those in the Southern Ocean and northern Indian Ocean (discussed below). Note also that an Arctic Ocean data set included in the global Cd-P plots is not shown on map.

should be possible for data sets to be calibrated as well as the overall instrumental and procedural precision allows. Almost all of the published data is undertaken by graphite furnace atomic absorption spectroscopy (GFAAS). By that technique, one can obtain precise replicate analyses within a single day better than 2%. Three factors prevent most labs from maintaining this level of precision for samples analyzed over many years: (1) it is difficult to maintain stable low-level Cd standards because Cd is strongly adsorptive and low-level standards are not stable if they are stored in glass or nitric-acid-leached polyethylene containers or left unacidified, (2) estimation of recovery efficiency

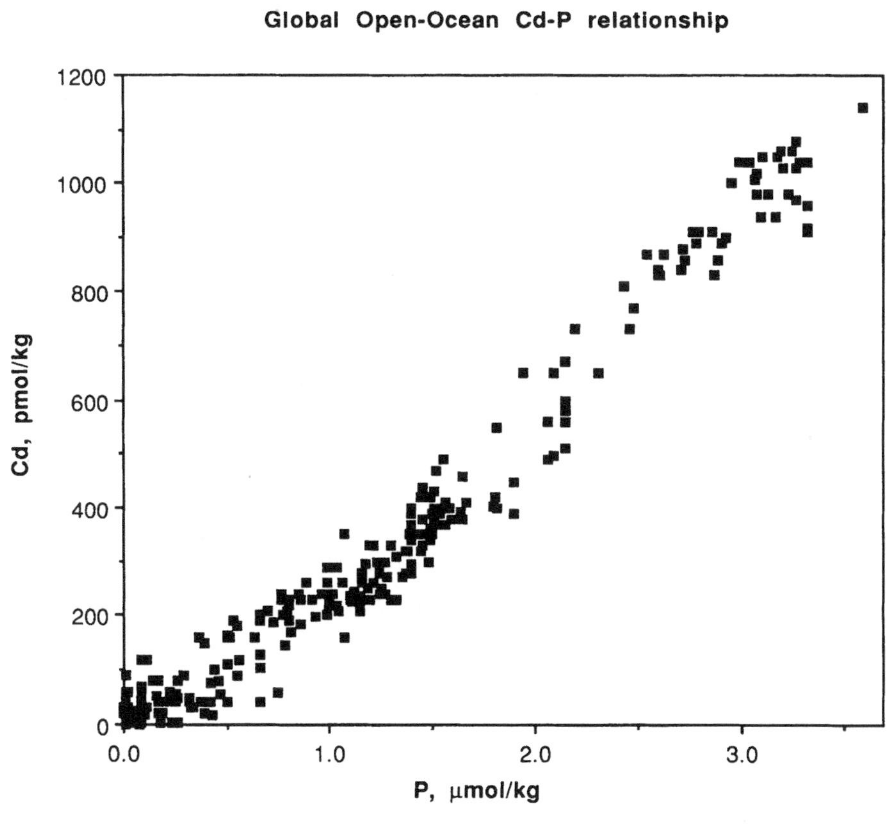

Figure 6: Boyle (1988) global deep water Cd-P plot

during preconcentration procedures introduces an additional uncertainty in converting from instrumental analyses to actual seawater concentrations, and (3) GFAAS is subject to matrix interferences, and any mismatch between standards and samples may result in a systematic offset. It should not be expected that most data sets are accurate to better than 5% (a subjective judgment which has not been tested explicitly). The accuracy of some data sets that are internally precise to this level may deviate more in accuracy. Despite these problems, the characteristics attributed to the global Cd-P relationship above are robust. However, there are likely to be some real characteristics of the ocean that are glossed over in that description; that is, some of the scatter in the Cd-P relationship represents real deviations from a simple correlation in the ocean, much as plots of nitrate-phosphorus relationships show scatter that represents real differences in the behavior of nitrogen and phosphorus.

Second, the meaning of the kink in the global Cd-P relationship should be clarified. Dividing the data set into two groups, two different regression lines result for $P<1.3$ and $P>1.3$ µmol/kg (these differences are highly significant). For both of these data groups there is little justification for any more complicated function to describe the data. These statements are true even though: (1) there are real differences from linear relationships within each group of data, (2) geographically distinct sub-sets of each group may be described by slightly different regressions, and (3) the kink is not sharp and the value chosen for the dividing point is not necessarily of exact significance within the oceanic Cd-P system. Describing the global Cd-P relationship by two straight line segments joined at a kink is only a convenient presentation of the result of numerous processes fractionating Cd from P in the ocean. This manner of presentation of the global oceanic Cd-P is mainly intended to provide an estimation of bottom-water Cd from estimates of P near a core site (given that there are many measurements of oceanic P throughout the ocean, and very few of Cd) and secondarily to draw some attention to the non-linear nature of the global Cd-P relationship.

Since that 1988 summary, some new data sets have been published. In some cases it has been asserted that these new data raise concerns about the validity of the global oceanic Cd-P relationship as described above. Two of these data sets are from the high latitude Southern Ocean (Frew and Hunter, 1992; Martin et al., 1990). These data sets are plotted on top of the field of the earlier data in Figure 7, with shallower samples indicated. Overall, the new data describe a trajectory that differs from the global ensemble. However, note that the trajectory joins the lower envelope of the global relationship for waters deeper than about 400 meters; that is, the deepwater portions of these two new data sets do not deviate from the ensemble average as seen earlier by more than the original scatter. What these data do demonstrate is that near-surface waters in portions of the Southern Ocean follow a different Cd-P relationship than that of the deep ocean. This observation is interesting and of significance in evaluating the mechanisms fractionating Cd from P in the ocean. Presumably in this region, Cd is preferentially removed from surface waters relative to P. As suggested by Frew and Hunter

(1992), this observation also may be of significance in explaining the kink in the global Cd-P relationship. The data are also of interest in attempts to use Cd in planktonic foraminifera in the Southern Ocean as a constraint on the evolution of surface water carbon chemistry. However, these observations at two stations - one in the Drake Passage and one south of New Zealand - should not be taken as representative of the Cd-P relationship of the whole Southern Ocean. Unpublished data from the author's laboratory for surface waters throughout the Atlantic sector of the Southern Ocean demonstrate that low Cd relative to P is limited in geographic extent to a narrow band; south of that zone, higher Cd values are found that are near or even higher than the global deepwater Cd-P trend. This matter will require more extensive documentation than space will allow here, so this discussion should be taken as a reminder that substantially more data from the Southern Ocean will be necessary before the nature of surface Cd-P anomalies in this region is understood.

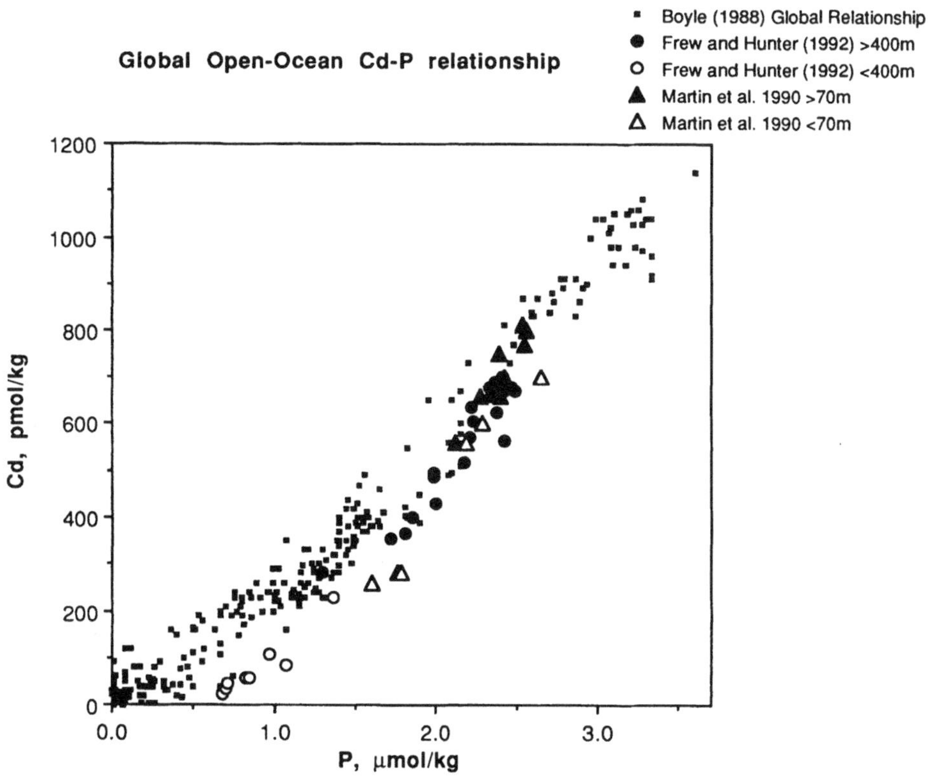

Figure 7: New data from Southern Ocean (Martin et al., 1990), triangles, and Frew and Hunter (1992), circles, highlighted and superimposed on Boyle (1988) global deepwater Cd-P plot. Data shallower than 400 m in New Zealand data and shallower than 60 m in Drake Passage are indicated by open symbols.

Another newly published data set comes from a pair of R/V Discovery stations in the Northwestern Indian Ocean (Saager et al., 1992). In the original abstract presentation of data from these station (DeBaar et al., 1987), it was asserted that this data set was dramatically different from the Cd-P relationship observed elsewhere. By the time the data was published, the reported Cd measurements had decreased by nearly a factor of two relative to the first presentation, but categorical statements about an anomalous Cd-P relationship were retained. Figure 8 shows the position and phosphorus depth profiles of these stations and the nearest GEOSECS stations; notice the large systematic differences in P between the Discovery stations (even from station to station) and the nearby GEOSECS stations, as well as other P data from this region (Wyrtki, 1971). Hence, variations in Cd-P relationships may be created by systematic analytical offsets between P data sets as well as by analytical and real differences in Cd. Putting aside the issue of the accuracy of the P data, the relevance of this new data set to the global Cd-P relationship can be put in perspective by overlaying this Indian Ocean data on the global data set (with Antarctic data from >400 meter now being included within the global deepwater set) (Figure 9). It is evident that even though the data follow a different trajectory from surface into deep waters, most samples fall within the scatter of the global Cd-P relationship. The most deviant data come from a single station very close to the continental margin. It is not clear that such large changes in Cd-P relationships are reasonable even for two nearby stations, so the possibility of internal data quality artifacts must be considered. But following the premise that near-shore waters can be anomalous (Boyle, 1988; Boyle et al., 1981; Boyle et al., 1984; Bruland and Franks, 1983; van Geen et al., 1991), this station can be excluded from the deepwater data set. When the open-ocean station (stations 5 and 7) are included within the global array without highlighting (Figure 10) it would be difficult to discern this revised Figure from the original 1988 presentation of the global Cd-P relationship (Figure 6).

On the basis of this evaluation of the new published data sets, it is evident that the global deepwater Cd-P relationship still can be adequately represented by the two-segment kink model described in 1988. Nonetheless, the discovery of significant near-surface Cd-P anomalies in the Southern Ocean emphasizes the largely uninvestigated nature of the global Cd-P relationship, and should stimulate further efforts to unravel the nature of the global Cd-P fractionation. From the perspective of paleoceanographic interpretations, the available data give no cause for concern in the use of Cd as a tracer of ocean circulation patterns. But as before, it should be emphasized that Cd is not an isotope of P and that real differences exist between these tracers; the simple-minded "Redfield" ^{13}C:^{12}C:Cd:P model should be used only in zero-order thinking, and more complicated representations (e.g., the Cd-P kink model) should be used in data analysis.

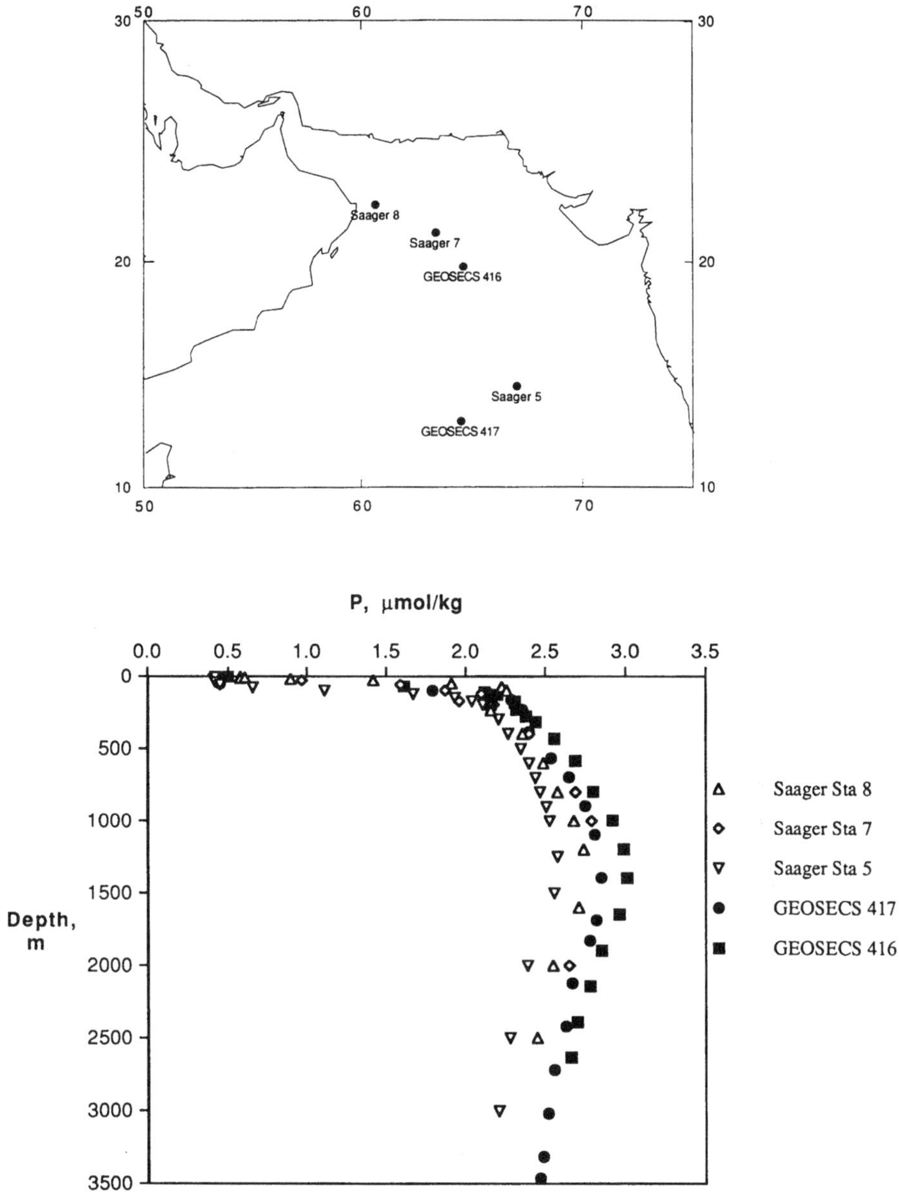

Figure 8: Station sites and phosphorus profiles from new R/V Discovery data from northwest Indian Ocean (Saager et al., 1992) compared to sites and P profiles from nearest GEOSECS stations. Note that Discovery P data (especially station 5) is low compared to nearby GEOSECS stations, as well as other Indian Ocean P data (Wyrtki, 1971).

DOES THIS NEW PERSPECTIVE ON $\delta^{13}C$ AND Cd DATA EXPLAIN GLACIAL ANOMALIES?

With this new data on oceanic $\delta^{13}C$ and Cd, are we in any better position to understand the discordancies between these properties in the glacial ocean? Certainly a rapprochement of the two types of data is thinkable; Broecker (1993) has outlined a diagrammatic conceptual model that reconciles the difference between $\delta^{13}C$ and Cd in the Southern Ocean as a consequence of more efficient gas exchange in the high latitude northern hemisphere (hence making the N.H. end-members relatively more positive than the S.H. end-members) and altered circulation patterns that turn the Southern Ocean into more of the "end of the ocean" than at present. As of yet, no one has described a quantitative model that is consistent with the

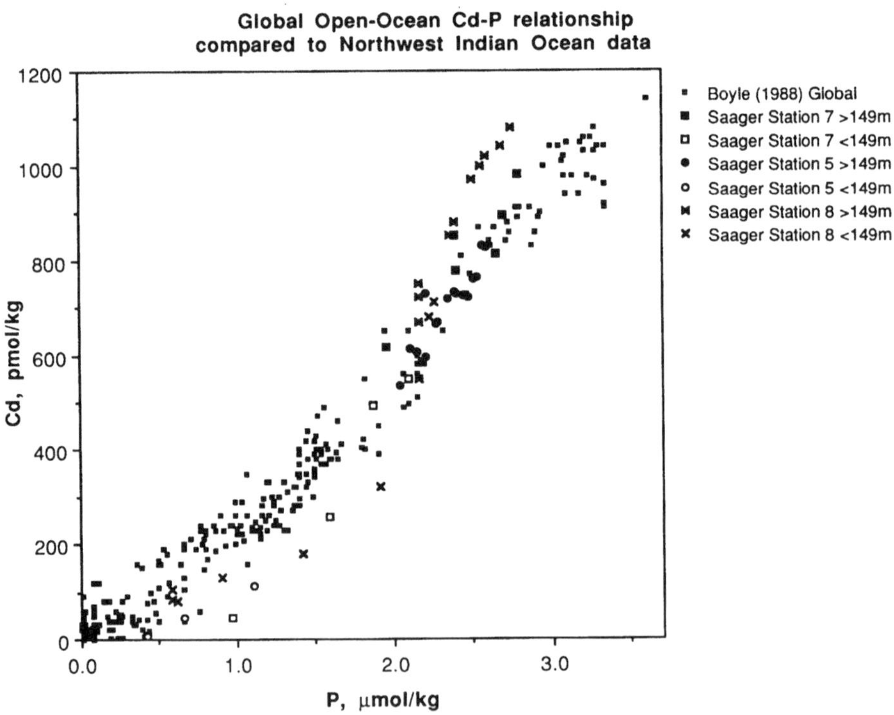

Figure 9: New data from northwest Indian Ocean (Saager et al., 1992) highlighted and superimposed on Boyle (1988) global deepwater Cd-P plot. Data shallower than 149 m are indicated by open symbols.

glacial maximum $\delta^{13}C$ and Cd distributions, and it may prove difficult to do so while obeying the other constraints such as maintaining ^{14}C ventilation of the deep ocean. Broecker's conceptual model does not attempt to address the apparent discrepancy between $\delta^{13}C$ and Cd in the deep Northwest Pacific ocean during glacial times, where Cd indicates nutrient-depleted waters while $\delta^{13}C$ doesn't.

Although the evolving perspective on $\delta^{13}C$ and Cd in the modern ocean does not provide a *simple* answer to differences between the tracers in the glacial maximum ocean, it does provide a basis for re-examining the old simplifying assumptions used previously.

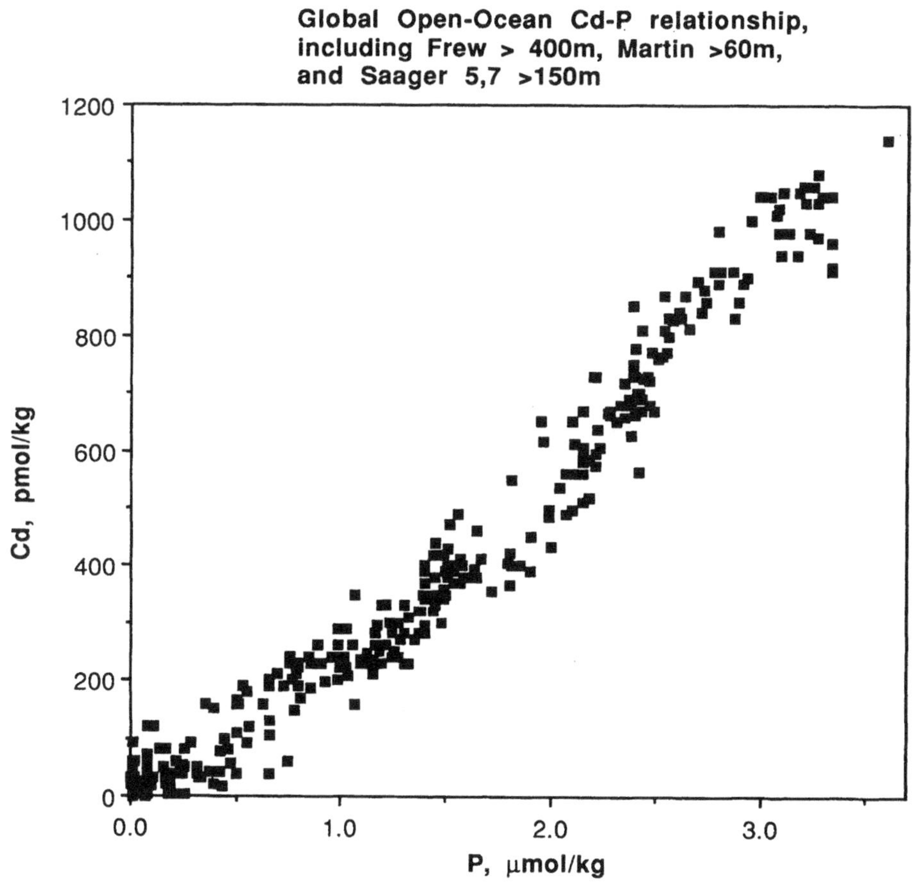

Figure 10: New global deepwater Cd-P plot, beginning with the Boyle (1988) data compilation and adding the deepwater data from the Southern Ocean and northwest Indian Ocean. Note the inconsequential differences from Figure 5.

HOW WELL DO FORAMINIFERAL $\delta^{13}C$ AND Cd REFLECT BOTTOM WATER?

Most of this discussion has been directed at understanding what reliable recorders of deepwater $\delta^{13}C$ and Cd can tell us about the ocean; it is also necessary to examine how reliable benthic foraminifera are as recorders of those properties. In order to uncover the record of deep ocean chemical changes in a timely manner, less effort overall has been directed towards establishing the accuracy and precision of benthic foraminifera as recorders of deep water properties. To date, the response of benthic foraminifera to bottom water properties has been established by comparing modern bottom water properties (often extrapolated from relatively distant stations) with (dead) foraminifera shells extracted from core top surface sediments. Although these studies show that foraminifera do respond to bottom water properties, the scatter in the calibration is an order of magnitude above the precision of the measurements (Boyle, 1988; Duplessy et al., 1984)(Figure 11). Most geochemists who work with this data believe that the scatter can be explained as follows:

Working hypothesis:

Scatter in the benthic foram core top calibration is due to the imperfection of surface sediments rather than to the imperfection of benthic foraminifera as geochemical recorders. Given biological sediment mixing of the upper centimeters of sediment and typically low sedimentation rates (order of centimeters per 1000 years), fossil foraminifera found in typical core top samples will average several thousand years in age, with a significant probability that a few individuals will have been reworked upwards from glacial times. Given the nature of most core tops (relatively low sedimentation rate; piston cores rather than box cores; etc.), one can't expect that the core top fossil calibrations should be as precise as the laboratory analytical precision.

The imperfection of core tops is undeniable, but this argument leaves the true precision and accuracy of benthic foraminifera undetermined. Shackleton et al. (1983) argued that precision could be established by downcore studies of adjacent samples of benthic foraminifera in high sedimentation rate cores. They found that the scatter between adjacent samples (separated in time by a few thousand years) was only a bit larger than analytical error, and postulated that downcore scatter in high sedimentation cores was the best estimator of the precision of benthic foraminiferal $\delta^{13}C$ response to bottom water $\delta^{13}C$.

Most geochemists and paleoclimatologists have been satisfied with Shackleton's argument, and hence they have concentrated efforts upon expanding the deep ocean climate record rather than upon testing and improving the calibration of benthic foraminifera. However,

Zahn et al. (1986) demonstrated that the incorporation of $\delta^{13}C$ in at least one genus of benthic foraminifera (Uvigerina) appeared to be sensitive to the organic carbon flux to the sediment, with benthic foraminifera found in regions of high benthic carbon flux showing a greater offset from bottom waters than those found in regions of low carbon flux. If this carbon-flux-sensitivity is accepted, then Shackleton's precision argument must be reconsidered because the carbon flux at a given site for a period of a few thousand years might show far less variability than the carbon flux between sites or at the same site on longer time spans. In that case, benthic foraminifera may indeed respond precisely to bottom water composition at any one site and time but have a variable (and possibly unknown) offset from a calibration established from the large-scale ocean core top correlation.

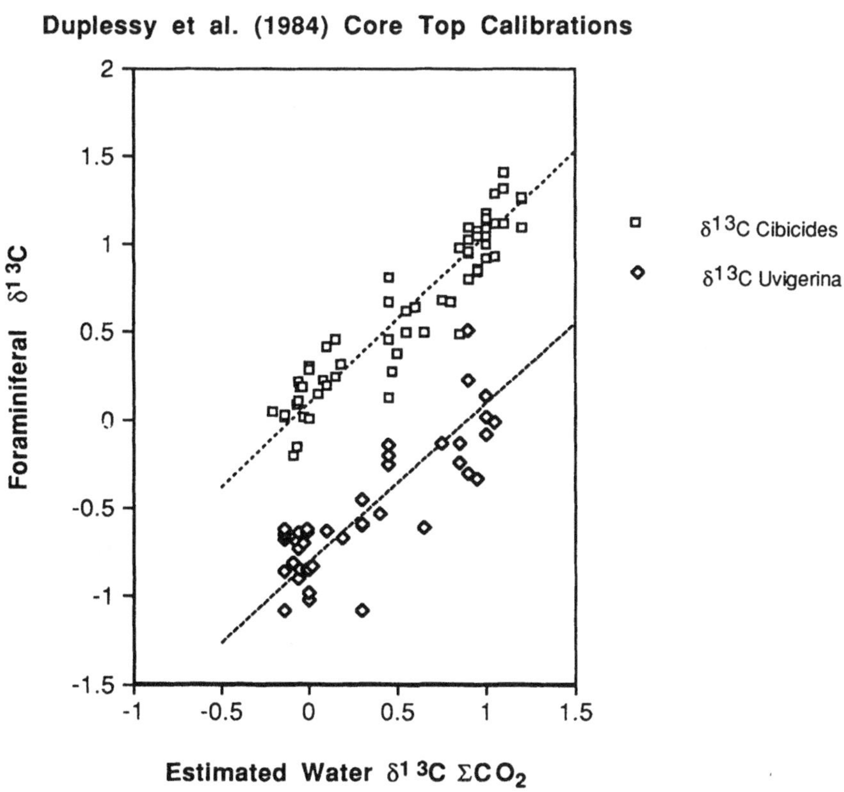

Figure 11a. Duplessy et al. (1984) $\delta^{13}C$ calibrations for Cibicides and Uvigerina. Regression lines are drawn through the data for each species.

The reason for the variable $\delta^{13}C$ offset of Uvigerina with respect to bottom water is being debated. A substantial portion of the community believes that Uvigerina has a shallow infaunal habitat (i.e., living within rather than on top of the sediment) (e.g. (Corliss, 1991; Zahn et al., 1986)). They further believe that because the pore waters of marine sediments are known to be offset from bottom water $\delta^{13}C$ because of the in-situ degradation of isotopically light organic carbon, the variable $\delta^{13}C$ offset of Uvigerina is due to variations in the depth of

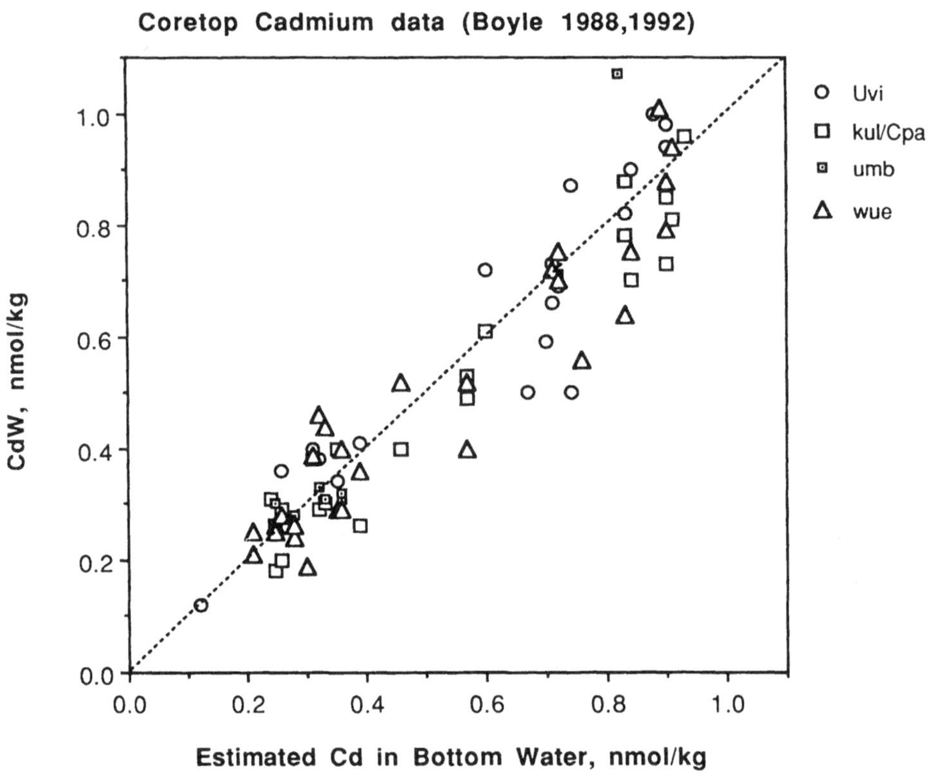

Figure 11b. Boyle (1988, 1992) Cd calibrations. Note that CdW is estimated using the simple depth-dependent estimation suggested by Boyle (1992). Dashed line is the 1:1 line of perfect agreement. Data from 5 out of the 86 samples in that data set were not plotted because they fell far off the average trend: CHN 82.001 P. ariminensis; MW 88.011 and MW 88.010, C. wuellerstorfi, and MW 88.004, C. wuellerstorfi and N. umbonifera.

growth and the intensity of organic carbon decomposition (McCorkle et al., 1990). Furthermore, Lutze and Thiel (1989) reported that they observed the species Cibicidoides wuellerstorfi living in an epifaunal environment (i.e., living on objects that protrude above the sediment). Because this species also shows essentially no average offset from bottom water $\delta^{13}C$, it is often preferred as a measure of bottom water $\delta^{13}C$ on the grounds that it usually will not be influenced by isotopically light pore waters (McCorkle et al., 1990). But there are still some problems with these widely-held assumptions: (1) No published evidence establishes that most Uvigerina *actually* live within the sediments. (2) The upper few cm of pore waters are also enriched in Cd and Ba, but there is no evidence for large offsets in Cd and Ba uptake between coexisting fossil C. wuellerstorfi and Uvigerina (Boyle, 1988; Lea and Boyle, 1989). Although there have been some ingenious rationalizations for these differences (McCorkle and Klinkhammer, 1991), there may well be other reasons for the offset, including the classically-cited "vital effect" in which organisms actively fractionate isotopes relative to their sources. For example, C. wuellerstorfi and Uvigerina are offset from one another in $\delta^{18}O$ by about 0.6 ‰ (Duplessy et al., 1970). This difference *cannot* be explained away by differences in the pore water composition and *must* be due to a $\delta^{18}O$ vital effect for at least one of these species. If organisms can differentially fractionate $\delta^{18}O$, perhaps so also may they also fractionate $\delta^{13}C$.

The preferred use of C. wuellerstorfi for $\delta^{13}C$ estimation is not the end of the story, because C. wuellerstorfi shows nearly as much scatter in its core top calibration as does Uvigerina (Figure 11a). The question of calibration accuracy and precision of paleoceanographic $\delta^{13}C$ estimates therefore is still open. There are occasional reports of surprisingly large deviations from typical calibration factors. Sarnthein et al. (1988) reported relatively negative Cibicidoides $\delta^{13}C$ in core tops underlying high productivity environments. Mackensen (this volume and Mackensen, in press) studied living benthic foraminifera (i.e. Rose Bengal stained, taken to mean living or recently living) from the South Atlantic sector of the Southern Ocean and found them to be offset from the modern bottom water by more than -0.5 ‰ in several samples. Keigwin et al. (1991) reported rapid >0.7‰ swings in Holocene $\delta^{13}C$ in C. wuellerstorfi that are not seen in Cd/Ca in the same species in the same core and that are also not seen in N. umbonifera in the same samples. These observations may not yet be generally acknowledged as proving that there are environmental effects that alter Cibicidoides $\delta^{13}C$. Some of these observations may result from the lack of appropriate bottom water data. But it behooves the geochemical tracer community to check into these instances to determine whether Cibicidoides may itself have an environmentally variable carbon isotope offset as seen in the study of Mackensen (this volume).

It should be emphasized that these problems of calibration are not confined to isotopic studies. *All* tracers show this behavior of larger scatter of core top calibrations relative to analytical precision; it is seen in ^{14}C, Cd/Ca, and Ba/Ca studies (Boyle, 1988; Broecker et al., 1990; Lea and Boyle, 1989). It should also be emphasized that these apparent discrepancies do not invalidate the geochemical tracers but rather raise questions about what accuracy and precision should be assigned to these data. These questions must be addressed to establish how much confidence we can have in subtle but nonetheless potentially important features of the data.

Most of the attention in this discussion has been focused on $\delta^{13}C$, but the accuracy and precision issue exists for Cd as well. From the perspective of foraminiferal Cd, the good news is that the various depth-dependent calibrated calcitic foraminifera - Uvigerina, C. kullenbergi, C. wuellerstorfi, and N. umbonifera do not show any calibration differences, and their downcore variability is confirmed by studies of aragonitic Hoeglundina, (which is not depth dependent: Boyle and Labeyrie, 1992). However, core top calibration scatter for Cd is still comparable to that for $\delta^{13}C$. A further problem for foraminiferal Cd is that core top calcitic foraminifera with Holocene $\delta^{18}O$ in the northwest Pacific have too little Cd compared to modern bottom water (Figure 2a). Either bottom water has changed composition since these fossil foraminifera grew, or the incorporation of Cd in this area is different than in other regions of the ocean. Although concerns such as these do not invalidate the use of foraminiferal Cd as a paleoceanographic tracer, they require further investigation. If modern northwest Pacific bottom waters are significantly different from waters that have existed in this region at other times during the Holocene (or in any other oceanic region), it will be difficult to improve our understanding of the situation by more core top studies, except in rare high deposition rate environments. Once again, the importance of examining living or recently living specimens in relation to direct measurements of Cd in the co-existing bottom water is of paramount importance, and must be undertaken in order to make progress in understanding the past variability of bottom water chemical composition.

Finally, another means for exploring the calibration reliability of foraminifera is the direct experimental manipulation of metal uptake by foraminifera, following the pioneering efforts of Delaney (Delaney, 1989; Delaney et al., 1985). Little of this work has been done yet, and most of that which has been done is on planktonic foraminifera or shallow-water benthic species. Although these studies can be of help in understanding mechanisms controlling shell composition, studies of species other than those used in deep-ocean paleochemical reconstructions will be useful more by analogy and may not be definitive regarding the calibration accuracy issue of species that are used in deep-sea reconstructions. To be truly useful the design of these experiments is critical. Because most studies follow the calcification of living specimens collected from the environment rather than specimens raised in continuous culture, there is a discontinuity in the chemical environment that may introduce artifacts. These artifacts are likely to be particularly severe for Cd, because cell protoplasm contains far more Cd relative to calcium than do their calcium carbonate shells. Hence it may take longer for the transferred specimen to adapt to the Cd tracer levels in the new environment than it will for Ca (or similar elements such as Sr and Ba). Furthermore, in the natural environment, the food that foraminifera eat will contain Cd levels appropriate to the ambient sedimentary environment, whereas in culture the foraminifera are fed brine shrimp that are typically transferred to the foraminifera from a separate container that does not contain Cd tracer. Hence in these experiments an artificial flux of Cd out of the foram (created by the digestion and elimination of food) will counter the flux of tracer Cd into the foram from the culture water. All of these concerns are not meant to disparage the use of culture experiments to unravel metal incorporation by foraminifera, but are instead offered as a plea that experiments be properly designed.

SUMMARY

The oceanic behavior of $\delta^{13}C$ and Cd are similar but not identical; in particular gas exchange is important for $\delta^{13}C$ but not for Cd. Cd is of course not identical to P in its oceanic behavior, but it is sufficiently similar that a relatively simple Cd-P relationship can capture most of the variability of one tracer from data on the other (in the modern ocean). These known processes may account for some of the differences seen in glacial maximum distributions, but a simple quantitative accounting for the observed differences has not yet been achieved. In view of the observation of anomalies and scatter in core-top calibrations of foraminifera, it is still possible that variable foraminiferal response to bottom water chemistry may account for some of the differences. The only clear way to establish whether this scatter is a matter for paleoceanographic concern is to study the chemistry of living and recently-living benthic foraminifera in relation to direct measurements of modern bottom water composition.

ACKNOWLEDGEMENTS

Thanks to Lloyd Keigwin, Dan McCorkle, and Andreas Mackensen for discussions and preprints of their recent work. This research was sponsored by NSF grant OCE 9018490. Travel and lodging to participate in the NATO meeting was sponsored by NATO.

REFERENCES

Bainbridge AE (1980) GEOSECS Atlantic Ocean Expedition, Vol. 2, Sections and Profiles U.S. Government Printing Office, Washington D.C.

Boyle EA (1988) Cadmium: chemical tracer of deep-water paleoceanography, Paleoceanography 3: 471-489.

Boyle EA. (1992) Cd and C13 paleochemical ocean distributions during the stage 2 glacial maximum. In Ann. Rev. Earth Planet. Sci., 20: 245-287.

Boyle EA, Huested SS, and Jones SP (1981) On the distribution of Cu, Ni, and Cd in the surface waters of the North Atlantic and North Pacific Ocean, J. Geophys. Res. 86: 8048-8066.
Boyle EA, Reid DF, Huested SS and Hering J (1984) Trace metals and radium in the Gulf of Mexico: an evaluation of river and continental shelf sources, Earth Planet. Sci. Lett. 69: 69-87.
Boyle EA, Chapnick SD, Bai XX and Spivack A (1985) Trace metal enrichments in the Mediterranean Sea, Earth Planet. Sci. Lett. 74: 405-419.
Boyle EA and Labeyrie L (1992) Cadmium in Hoeglundina and glacial nutrient depletion in the intermediate waters of the Northern Indian Ocean, Trans. Am. Geophys. Union 73: 275.
Broecker WS (1982) Ocean chemistry during glacial time, Geochim. Cosmochim. Acta 46: 1689-1705.
Broecker WS (1993) An oceanographic explanation for the apparent carbon isotope-cadmium discordance in the glacial Antarctic?, Paleoceanogr. 8: 137-140.
Broecker WS and Maier-Reimer E (1992) The influence of air and sea exchange on the carbon isotope distribution in the sea, Glob. Biogeochem. Cycles 6: 315-320.
Broecker WS, Peng TH, Trumbore S, Bonani G and Wolfli W (1990) The distribution of radiocarbon in the glacial ocean, Global Biogeochem. Cycles 4: 103-117.
Bruland KW and Franks RP. (1983) Mn, Ni, Cu, Zn, and Cd in the western North Atlantic. In Trace Metals in Seawater , ed. C.S. Wong, E. Boyle K.W. Bruland, J.D. Burton, and E.D. Goldberg (Eds), Plenum, N.Y.,.395-414.
Corliss BH (1991) Morphology and microhabitat preferences of benthic foraminifera from the northwest Atlantic Ocean, Mar. Micropal. 17: 195-236.
Craig H, Broecker WS, and Spencer DW (1982) GEOSECS Pacific Ocean Expedition, Vol. 4, Sections and Profiles, U.S. Government Printing Office, Washington, D.C.
Curry WB, Duplessy J-C, Labeyrie LD and Shackleton NJ (1988) Changes in the distribution of $\delta^{13}C$ of deep water CO_2 between the last glaciation and the Holocene, Paleoceanogr. 3: 317-342.
DeBaar H, Saager PM, and Howland RJ (1987) Nutrient type distributions of Ni, Cu, Zn, and Cd in the Northwest Indian Ocean (NWIO), Trans. Am. Geophys. Union 68: 66.
Delaney ML (1989) Uptake of cadmium into calcite shells by planktonic foraminifera, Chem. Geol. 78: 159-165.
Delaney ML, Be AWH, and Boyle EA (1985) Li, Sr, Mg, and Na in foraminiferal calcite shells from laboratory culture, sediment traps, and sediment cores, Geochim. Cosmochim. Acta 49: 1327-1341.
Duplessy J-C, Lalou C, and Vinot AC (1970) Differential isotopic fractionation in benthic foraminifera and paleotemperatures re-assessed, Science 168: 250-251.
Duplessy J-C, Matthews RK, Prell W, Ruddiman WF, Caralp M and Hendy CH (1984) C-13 record of benthic forminifera in the last interglacial ocean: implications for the carbon cycle and global deep water circulation, Quat. Res. 21: 225-243.
Duplessy J-C, Labeyrie L, Juillet-LeClerc A, Maitre F, Duprat J and Sarnthein M (1991) Surface salinity reconstruction of the North Atlantic Ocean during the last glacial maximum, Oceanologica Acta 14: 311-324.
Frew RD and Hunter KA (1992) Influence of Southern Ocean waters on the cadmium-phosphate properties of the global ocean, Nature 360: 144-146.
Keigwin LD, Jones GA, Lehman SJ and Boyle EA (1991) Deglacial meltwater discharge, North Atlantic Deep Circulation, and Abrupt Climate Change, J. Geophys. Res. 96: 16811-16826.
Keir RS (1991) The effect of vertical nutrient redistribution of surface ocean $\delta^{13}C$, Glob. Biogeochem. Cycles 5: 351-358.
Lea D and Boyle EA (1989) Barium content of benthic foraminifera controlled by bottom-water composition, Nature 338: 751-753.
Lutze GF and Thiel H (1989) Epibenthic foraminifera from elevated microhabitats: Cibicidoides wuellerstorfi and Planulina ariminensis, J. Foram. Res. 19: 153-158.
Mackensen A (in press) $\delta^{13}C$ in benthic foraminferal tests of Fontbotia wuellerstorfi (Schwager) relative to $\delta^{13}C$ of dissolved inorganic carbon in Southern Ocean deep water: implications for glacial ocean circulation models, Paleoceanogr.

Martin JH, Gordon RM, and Fitzwater SE (1990) Iron in Antarctic Waters, Nature 345: 156-158.

McCorkle DC and Keigwin LD (in review) Depth profiles of $\delta^{13}C$ in bottom water and core-top C. wuellerstorfi on the Ontong-Java Plateau and Emporer Seamounts, Paleoceanogr.

McCorkle DC, Keigwin LD, Corliss BH and Emerson SR (1990) The influence of microhabitats on the carbon isotopic composition of deep sea benthic foraminifera, Paleoceanogr. 5: 161-186.

McCorkle DC and Klinkhammer GP (1991) Porewater cadmium geochemistry and the porewater cadmium:$\delta^{13}C$ relationship, Geochim. Cosmochim. Acta 55: 161-168.

Mix AC, Pisias NG, Zahn R, Rugh W and Lopez C (1990) Carbon-13 in Pacific deep and intermediate waters, 0-370 kyr bp.: implications for ocean circulation and Pleistocene CO_2, Paleoceanogr. 6: 205-226.

Rau GH, Sweeney RW, and Kaplan IR (1982) Plankton C13/C12 ratio changes with latitude: differences between northern and southern oceans, Deep-Sea Res. 289: 1035-1039.

Saager PM, DeBaar HJW, and Howland RJ (1992) Cd, Zn, Ni, and Cu in the Indian Ocean, Deep-Sea Res. 39: 9-35.

Sarnthein M, Winn K, Duplessy J-C and Fontugne MR (1988) Global variations of surface ocean productivity in low- and mid-latitudes : influence on CO_2 reservoir of the deep ocean and atmosphere during the last 21,000 years, Paleoceanogr. 3: 361-399.

Shackleton NJ, Imbrie J, and Hall MA (1983) Oxygen and carbon isotope record of East Pacific core V19-30:implications for the formation of deep water in the late Pleistocene North Atlantic, Earth Planet. Sci. Lett. 65: 233-244.

van Geen A, Boyle EA, and Moore WS (1991) Trace metal enrichments in the coastal waters of the Gulf of Cadiz (Spain), Geochim. Cosmochim. Acta 55: 2173-2192.

Weiss RF, Broecker WS, Craig H and Spencer DW (1982) GEOSECS Indian Ocean Expedition, Vol. 6, Sections and Profiles U.S. Government Printing Office, Washington, D.C.

Wust G (1935, 1978) The Stratosphere of the Atlantic Ocean In: Emery W (ed) Amerind Press, New Delhi, 112 p.

Wyrtki K (1971) Oceanographic Atlas of the International Indian Ocean Expedition U.S. Government Printing Office, Washington, D.C., p. 442

Zahn R, Winn K, and Sarnthein M (1986) Benthic foraminiferal 13C and accumulation rates of organic carbon: Uvigerina peregrina group and Cibicidoides wuellerstorfi, Paleoceanogr. 1: 27-42.

TRACER-NUTRIENT CORRELATIONS IN THE UPPER OCEAN: OBSERVATIONAL AND BOX MODEL CONSTRAINTS ON THE USE OF BENTHIC FORAMINIFERAL $\delta^{13}C$ AND Cd/Ca AS PALEO-PROXIES FOR THE INTERMEDIATE-DEPTH OCEAN

Rainer Zahn and Robin Keir
GEOMAR, Research Center for Marine Geosciences
Wischhofstraße 1-3
D-24148 Kiel, Germany

ABSTRACT. Deviations of $\delta^{13}C:PO_4$ and $Cd:PO_4$ ratios from the global trend are observed within the density range $26.8 > \sigma_\Theta > 27.7$ close to where the isopycnal surfaces outcrop at the sea surface. In these areas tracer concentrations seem to be decoupled from nutrient concentrations presumably because of the effects of gas exchange and varying recycling depths for tracers and nutrients. Temperature exerts strong control on the surface water $\delta^{13}C$ anomalies in high latitudes but results from an ocean-atmosphere $\delta^{13}C$-simulating box model imply that the anomalies still exist if air-sea $\delta^{13}C$-fractionation is held constant for the warm and cold surface ocean. The warm and northern Atlantic surface waters acquire a lower $\delta^{13}C$ than expected from biological fractionation alone, whereas the Antarctic surface $\delta^{13}C$ is somewhat more positive. A dynamic balance is achieved where CO_2 evading from Antarctic waters to the atmosphere has a lower $^{13}C/^{12}C$ ratio than that invading, while the isotope ratio evading from the warm ocean is greater than that of the CO_2 return flux.

Box model simulations of Cd and PO_4 distribution suggest that differential uptake of Cd in Antarctic waters affects the ocean Cd distribution globally because these waters are entrained into intermediate waters which spread about the global ocean. Stripping Cd from Antarctic surface waters produces the greatest intermediate water depletion in the Atlantic, deeper recycling produces the greatest intermediate water Cd depletion in the North Pacific with noticeable lower $Cd:PO_4$ ratios than those of the deep waters in that ocean. Because of the polar mediterranean nature of the circulation in the northern Atlantic differential uptake of Cd in boreal Atlantic surface waters would only be detected locally. In the rest of the ocean that is capped by other surface waters, a linear correlation of Cd and PO_4 would be observed throughout the water column. Changing parameters related to biological cycling in Antarctic surface waters exerts by far the largest control on the distribution of Cd and PO_4 in the world ocean.

Because $\delta^{13}C$ and Cd are not uniquely tied to phosphorus a reliable interpretation of foraminiferal $\delta^{13}C$ and Cd/Ca records is difficult, especially if they come from shallow-ocean

NATO ASI Series, Vol. I 17
Carbon Cycling in the Glacial Ocean:
Constraints on the Ocean's Role in Global Change
Edited by R. Zahn et al.

core sites at high latitudes or other areas of oceanic upwelling. When estimating nutrient concentrations from benthic $\delta^{13}C$ and/or Cd/Ca ratios we need to know if we are on the low-nutrient or high-nutrient side of the break in the global regression lines. It may be possible to obtain this information by measuring a number of different proxies with different boundary conditions. Subtle differences between independent proxy records may tell us more about past ocean chemical cycling than coherent paleo-proxy records.

INTRODUCTION

Until recently, it appeared that the $\delta^{13}C_{\Sigma CO2}$-phosphate and cadmium- phosphate relationships in the ocean exhibited single, nearly linear correlations (Kroopnick, 1985; Boyle, 1988). At closer look, however, the data show that deviations exist from this ideal data distribution: (i) mixed-layer $\delta^{13}C_{\Sigma CO2}$ inventories are decoupled from phosphate inventories at high latitudes where gas exchange with the atmosphere exerts primary control on carbon isotope fractionation (e.g., Charles and Fairbanks, 1990; Broecker and Maier-Reimer, 1992); (ii) cadmium concentrations are lower in a significant part of the Southern Ocean than expected from the main $Cd:PO_4$ trend found elsewhere (Martin et al., 1990; Nolting et al., 1991; Frew and Hunter, 1992). Potential candidates to explain the discordant $\delta^{13}C_{\Sigma CO2}$-phosphate distribution at high latitudes are (i) gas exchange between surface waters and the overlying atmosphere with a higher carbon isotope fractionation factor at low temperatures and (ii) differential photosynthetic carbon isotope fractionation factors which vary as a function of surface water molecular CO_2 (Rau et al., 1991 and this volume) and thus determine the carbon isotope signal of ΣCO_2 released to the water column during decay of organic matter. The result of any combination of these factors is an increase of mixed-layer $\delta^{13}C_{\Sigma CO2}$ at high latitudes to nearly constant values around +1.3‰ to +1.5‰ (PDB) regardless of nutrient levels. For the discordant cadmium-phosphate pattern Frew and Hunter (1992) have suggested that preferential uptake of cadmium relative to phosphate in Antarctic surface waters during biological production reduces the Cd/PO_4 ratio of Antarctic Intermediate Water (AAIW). They further propose that a relatively strong source of AAIW into the Atlantic decreases the cadmium concentration that otherwise would be found in that ocean, and that this produces the kink observed in the main $Cd\text{-}PO_4$ correlation.

We use the GEOSECS Shorebased Dataset (Östlund et al., 1987) to map the $\delta^{13}C_{\Sigma CO2}$-phosphate distribution at high latitudes in order to examine the extent to which high-latitude processes determine the global tracer-nutrient pattern. The ocean-atmosphere $\delta^{13}C$-simulating box model of Keir (1988) is used to separate the effects of temperature and bio-fractionation on the disparate $\delta^{13}C\text{-}PO_4$ pattern at high latitudes. Later we show the results of an ocean box model which was designed to coarsely simulate the distribution of phosphate as well as other tracers. The model results generally support the hypothesis of productivity-driven variations in the cadmium-nutrient relation, and further suggest that cadmium tends to be recycled slightly shallower in the water column than is phosphate.

THE $\delta^{13}C_{\Sigma CO2}$-PHOSPHATE DISCREPANCY BETWEEN HIGH NORTHERN AND SOUTHERN LATITUDES

DATA REDUCTION

The GEOSECS Shorebased Data (Östlund et al., 1987) were extracted from magnetic tape supplied by the Physical and Chemical Oceanographic Data Facility at Scripps Institution of Oceanography, San Diego. All $\delta^{13}C_{\Sigma CO2}$ data were corrected using the correction factors listed in Kroopnick (1985) (Table 1); potential density (σ_Θ) was computed using the equations of Bryden (1973) and Fofonoff (1977), and applying the International Equation of State for seawater (UNESCO, 1981). Figure 1 shows the GEOSECS stations which were used to calculate the source signals at high latitudes (i.e., preformed $\delta^{13}C_{\Sigma CO2}$ and phosphate[1]) and to map the distribution of $\delta^{13}C_{\Sigma CO2}$ and phosphate throughout the mid-depth ocean. For the Pacific and Indian Oceans we used only stations in the western basin. In both oceans exist east-west gradients both in hydrography and chemistry. Those gradients reflect differences in ventilation due to geostrophic gradients. In order to extract oceanographically consistent tracer gradients we used data from stations along single ventilation paths. The absolute numbers (e.g. µmol PO_4 kg^{-1}) are different for the eastern and western basins, but the trend would have not changed much had we chosen the eastern stations.

GLOBAL PHOSPHATE TRANSECTS, "INTERMEDIATE WATER" VERSUS "MID-DEPTH WATER"

Salient phosphate maxima exist in the South Atlantic (>2.0 µmol PO_4 kg^{-1}), Northern Indian (>2.75 µmol PO_4 kg^{-1}) and Northern Pacific Oceans (>3.0 µmol PO_4 kg^{-1}) (Figure 2). All maxima lie at potential densities of $26.8>\sigma_\Theta>27.7$ which is at water depths generally shallower than 2000 m. The South Atlantic maximum is associated with the northward propagation of Antarctic Intermediate Water (AAIW) which convects in the region north of the Antarctic Polar Front where surface salinities are just high enough to allow surface waters to sink to mid-depths between 800-1200 m. AAIW also spreads northward in the South Indian and South

---

[1] 'preformed' will be used here to describe the observed nutrient and proxy-signals of a water mass by the time convection occurs; i.e., no correction is applied, the 'preformed' concentrations are true data.

Table 1. GEOSECS Stations, Months of Occupation and $\delta^{13}C_{\Sigma CO2}$ Correction factors Used to Calculate Preformed Tracer and Nutrient Values.

Station	Occupation	$\delta^{13}C_{\Sigma CO2}$correction[a]
North Atlantic		
5	Jul/Aug 1972	+0.2
11	Aug 1972	+0.2
15	Aug 1972	+0.1
17	Aug 1972	+0.1
18	Aug 1972	-
19	Aug 1972	-
23	Aug 1972	+0.1
South Atlantic		
74	Dec 1972	-
76	Dec/Jan 1972/73	-
78	Jan 1972	-
82	Jan 1972	-
89	Jan 1972	+0.1
South Indian		
430	Feb 1978	-
431	Feb 1978	-
433	Feb 1978	-
South Pacific		
287	Feb 1974	+0.1
290	Feb 1974	+0.2 (casts 2, 6)

[a] Kroopnick, 1985 (Table 1 on p. 59)

Pacific Oceans. However, deep water nutrient values there are higher (2.0-2.5 μmol PO_4 kg^{-1}) than in the South Atlantic (1.5-2.0 μmol PO_4 kg^{-1}) so that no isolated nutrient maxima exist along the flow path of AAIW into the southern portions of the Indian and Pacific Ocean basins.

The two northern hemisphere nutrient maxima (Indian and Pacific Oceans, Figure 2) are not related to any type water mass but rather document the cul-du-sac type circulation in these regions: there are no sources for mid-depth or deep-water ventilation in the North Indian and North Pacific Oceans and thus, the water masses there are old and oxygen depleted, and have accumulated large amounts of biologically-cycled nutrients (e.g. Kroopnick, 1985).

Some evidence has been documented in support of a North Pacific Intermediate Water (NPIW) which constitutes a salinity minimum in the mixed water region between the Kuroshio and Oyashio Extensions (Hasunuma, 1978; Yuan and Talley, 1992; Talley, 1993). Its mean densities of $26.5<\sigma_\Theta<26.8$ are considerably lower than those found in the core layer of the North Pacific nutrient maximum ($\sigma_\Theta \sim 27.4$) so that NPIW does not contribute to the mid-depth

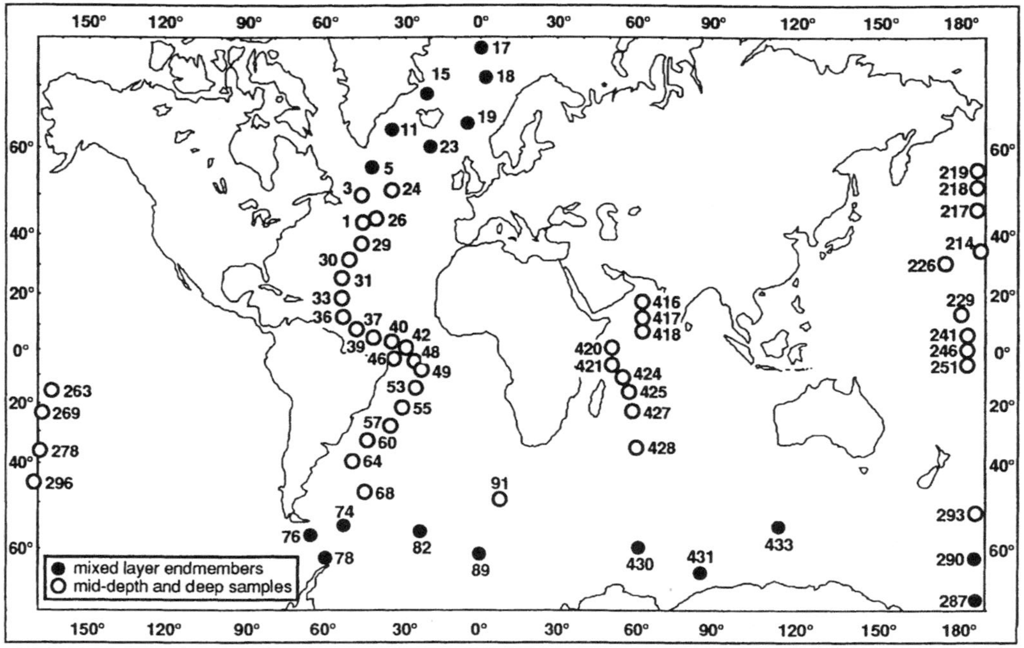

Figure. 1 GEOSECS stations used to map the $\delta^{13}C_{\Sigma CO2}$-phosphate distribution at mid-depth. Closed symbols mark those stations which were used to calculate preformed values (see Table 2).

Table 2. Preformed Values and Slopes of the $\delta^{13}C_{\Sigma CO2}$: PO_4 Regressions in Outcrop Regions of the Intermediate Water Isopycnals [a].

$\delta^{13}C_{\Sigma CO2}$ (‰ PDB)	error	PO_4 (μmol kg^{-1})	error	observations	$\Delta\delta^{13}C_{\Sigma CO2}$: ΔPO_4	error	observations
North Atlantic (GEOSECS Stations 5,11,15,17,18,19,23)							
1.56	0.36	0.48	0.26	13	-0.995	0.09	29
South Atlantic (South of 50°S; GEOSECS Stations 74,76,78,82,89)							
1.36	0.27	1.71	0.20	13	-1.415	0.13	32
South Indian (South of 53°S; GEOSECS Stations 430, 431, 433)							
1.37	0.28	1.83	0.16	9	-1.777	0.11	32
South Pacific (South of 58°S; GEOSECS Stations 287, 290)							
1.23	0.42	1.76	0.20	6	-1.544	0.20	19
"Southern Ocean" (South Atlantic + South Indian + South Pacific)							
1.33	0.30	1.76	0.19	28	-1.518	0.09	83

[a] The preformed $\delta^{13}C_{\Sigma CO2}$ and PO_4 values were calculated using data points at water depths < 100 m in the outcrop regions. Slopes were calculated using all data within the $26.8 > \sigma_\theta > 27.7$ isopycnal range in the outcrop regions. 'Preformed' is used here to describe the observed nutrient and proxy-signals of a water mass close to the surface; i.e., no correction is applied, the 'preformed' concentrations are true data.

nutrient maximum found in the subarctic North Pacific. The nutrient maximum in the South Atlantic, therefore, is the only one which is directly associated with a true intermediate-water-mass whereas the ones in the North Indian and North Pacific Oceans are the result of restricted mid-depth ventilation. We therefore prefer the term "mid-depth water" for these two maxima over the term "intermediate water" which is strictly defined on the basis of physical water mass properties (T-S).

$\delta^{13}C_{\Sigma CO_2}$:$PO_4$ REGRESSIONS FOR THE HIGH-LATITUDE OUTCROP REGIONS ($26.8<\sigma_\Theta<27.7$)

Constant input of biologically-cycled nutrients and ^{12}C-enriched ΣCO_2 to the global ocean's deep and bottom waters makes the $\delta^{13}C_{\Sigma CO_2}$-$PO_4$ relationship for the deep-ocean very stable (Kroopnick, 1985). Kroopnick determined the relationship between $\delta^{13}C_{\Sigma CO_2}$ and other nutrient-related tracers (AOU=Apparent Oxygen Utilization, PO_4) by using only those GEOSECS data which have been measured on water samples with potential densities >27.7. Densities in this range are found throughout the world ocean at water depths greater than 1800 m, except for some high-latitude outcrop regions where the 27.7 isopycnal rises to near the sea surface (Figure 2). Kroopnick's reasoning for reducing the global data set to σ_Θ >27.7 was that air-sea gas exchange would influence $\delta^{13}C_{\Sigma CO_2}$ at shallower water depth and distort the $\delta^{13}C_{\Sigma CO_2}$-AOU-$PO_4$ relationship by adding noise to an otherwise well-defined regression.

As is shown in Figure 2, the 27.7-σ_Θ cut-off line coincides with the lower horizontal boundary of the mid-depth nutrient maxima in the South Atlantic and North Indian and North Pacific Oceans. The $\delta^{13}C_{\Sigma CO_2}$-$PO_4$ data within the density range $26.8<\sigma_\Theta<27.7$ indeed show an incoherent distribution pattern as indicated by Kroopnick (1985) in that no single regression line describes all data (Figure 3). Instead, separate regression lines approximate data from northern and southern hemisphere outcrop regions. Within the density range $26.8<\sigma_\Theta<27.7$, the mixed-layer $\delta^{13}C_{\Sigma CO_2}$ values in the polar regions remain high (>1.3‰ PDB) regardless of nutrient concentrations, and the $\delta^{13}C_{\Sigma CO_2}$:$PO_4$ slope for data within this density range (i.e. surface and subsurface water) in the Southern Ocean outcrop area is higher than in the North Atlantic outcrop where nutrient concentrations are low (Table 2). Yet, comparing mixed-layer $\delta^{13}C_{\Sigma CO_2}$ with data expected for a closed system in which bio-fractionation is the only factor controlling $\delta^{13}C_{\Sigma CO_2}$, the North Atlantic values are on the negative side and the Southern Ocean values on the positive side of the Redfield fractionation line (Figure 4a). Neither value is in equilibrium with the atmosphere. The offset between measured and estimated equilibrium $\delta^{13}C_{\Sigma CO_2}$ is largest for the Southern Ocean which may be because of the short residence time of the surface water. For the North Atlantic, the offset between observed and estimated equilibrium $\delta^{13}C_{\Sigma CO_2}$ is smaller, partially because of the stronger influence of the poleward flow of surface waters which have been in contact with the atmosphere longer than surface waters in the

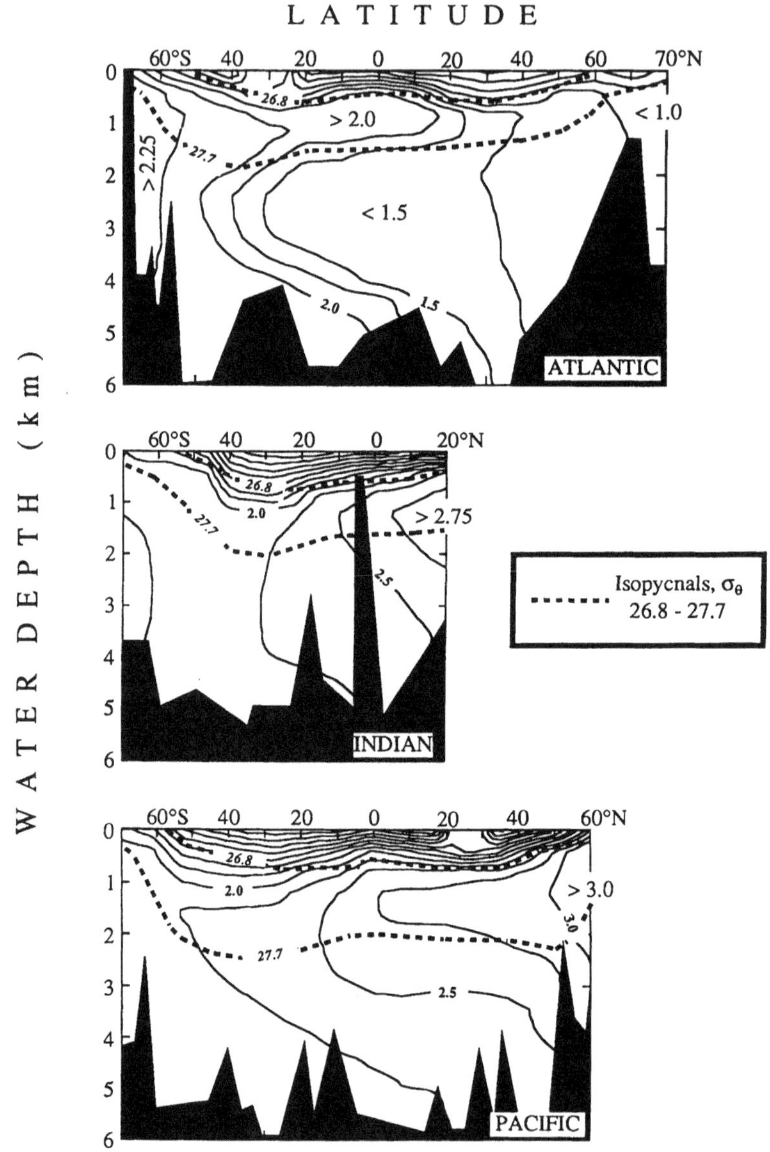

Figure. 2 North-south transects of phosphorus. Phosphorus isolines are in 0.25 μmol kg^{-1} intervals. Dotted lines envelope potential density range of 26.8-27.7. Data from within this density range were used to determine the $\delta^{13}C_{\Sigma CO2}$-$PO_4$ regressions listed in Table 2 and shown in Figure 3.

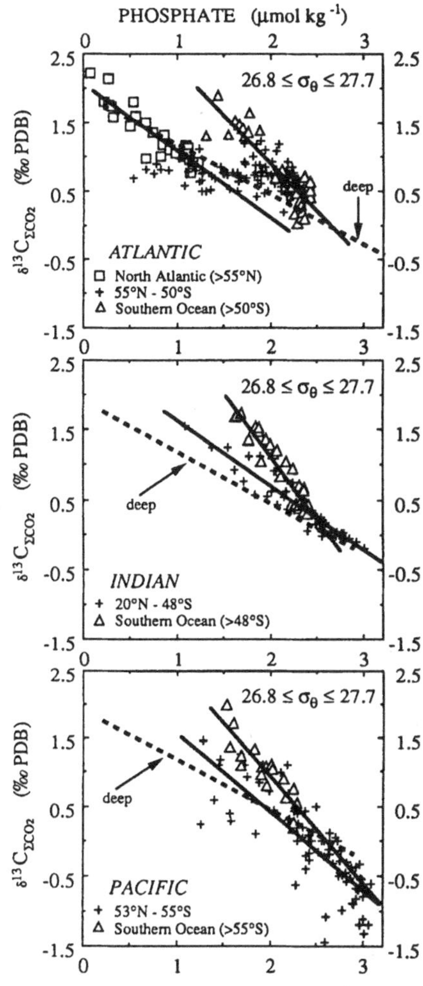

Figure. 3 $\delta^{13}C_{\Sigma CO_2}$-$PO_4$ regressions for the Atlantic, Indian and Pacific Oceans. Only data from the density range $26.8 < s_\Theta < 27.7$ have been used. Dotted line shows trend observed in deep ocean samples ($s_\Theta > 27.7$; Kroopnick, 1985). Southern Ocean samples (triangles) tend to have more positive $\delta^{13}C$ compared to values expected from the global trend. This points to the importance of gas exchange and the associated carbon isotope re-equilibration. There are no convection cells in the northern Indian and Pacific Oceans and thus, the discrepancy between the Southern Ocean data and the rest of the ocean basins is much less than in the Atlantic where nutrient-depleted surface waters convect in the north and nutrient enriched surface waters convect in the south. 'Preformed' $\delta^{13}C_{\Sigma CO_2}$ values are similar in both regions whereas PO_4 concentrations are higher in the south by more than a factor of three (see also Table 2).

Southern Ocean. Achievement of equilibrium with the atmosphere seems to be hampered, though, by lower summer surface PCO_2 which should produce a net invasion of isotopically light atmospheric CO_2 seasonally. The equilibrium value for warmer waters is more negative than that estimated for the cold Southern Ocean waters.

THE EQUATOR-TO-POLE $\delta^{13}C$ GRADIENT: TEMPERATURE VERSUS BIO-FRACTIONATION EFFECTS

The fractionation of carbon isotopes between sea and air is dependent on the water temperature, with cold waters exerting stronger fractionation towards more negative values (Mook et al., 1974). There has been some debate recently as to whether this has a significant influence on the distribution of $\delta^{13}C$ in the upper ocean (Broecker and Maier-Reimer, 1992; Lohmann et al., in press). As pointed out by Broecker and Peng (1986), by itself polar upwelling with high nutrients and low $^{13}C/^{12}C$ ratios would produce a negative equator-to-pole gradient in the surface water $\delta^{13}C$ that would be negatively correlated with nutrient concentration. In contrast, if the surface ocean came to isotopic equilibrium with the atmosphere, a positive equator-to-pole $\delta^{13}C$ gradient would be established due to the temperature effect on sea-air fractionation. The observed distribution is to a first approximation rather uniform (Kroopnick, 1985), and this may result from a dynamic balance of these two processes, where the evasion of ^{12}C enriched CO_2 from Antarctic surface waters into the atmosphere re-invades the warm surface ocean, raising the $\delta^{13}C$ of the Antarctic and decreasing $\delta^{13}C$ in the warm ocean. Alternatively, the relative uniform $\delta^{13}C$ distribution could be due to horizontal mixing (Lohmann et al., in press), but it is not clear why nutrients would not be equally dispersed, leaving the biological correlation with $\delta^{13}C$ intact.

In order to separate the influence of the temperature effect on surface water $\delta^{13}C$, we calculated the distribution of $\delta^{13}C$ by using both a temperature-dependent and a constant (i.e. temperature independent) sea-air fractionation for all surface boxes in an ocean-atmosphere box model (Keir, 1988). The resulting patterns of $\delta^{13}C$ versus PO_4 are compared in Figure 4b. Except for the Atlantic, all deep waters and most intermediate waters plot close to the biological $\delta^{13}C$-PO_4 correlation in both simulations. However, *even in the absence of the temperature effect*, air-sea gas exchange modifies the $\delta^{13}C$ distribution of the upper ocean and Atlantic deep water. The warm and northern Atlantic surface waters acquire a lower $\delta^{13}C$ than expected from biological fractionation alone, whereas the Antarctic surface $\delta^{13}C$ is somewhat more positive. This occurs because net biological production in the Antarctic is low relative to the upwelling fluxes of nutrients and dissolved carbon, allowing low $^{13}C/^{12}C$ ratios in the dissolved carbon to outcrop and exchange with the atmosphere. As a consequence, a dynamic balance is achieved where CO_2 evading from Antarctic waters to the atmosphere has a lower $^{13}C/^{12}C$ ratio than that invading, while the isotope ratio evading from the warm ocean is greater than that of the return flux of CO_2.

The results in Figure 4b show that in the model at least, the greater sea-air fractionation in the Antarctic amplifies the increase of Antarctic $\delta^{13}C$ and decrease of warm ocean $\delta^{13}C$. The temperature effect further decreases the warm surface by -0.6‰ and increases the Antarctic by an additional 0.3‰. Without the temperature effect in the model, the surface water $\delta^{13}C$ versus PO_4 appears only somewhat less than expected from biological fractionation, but the additional 0.9‰ reduction in the warm-to-Antarctic gradient due to the temperature effect produces a more uniform surface $\delta^{13}C$ distribution. As is shown in Figure 4a, the north-south $\delta^{13}C$ gradient were to be reversed if 'potential' (i.e. constrained by temperature only) equilibrium with the atmosphere were achieved. In reality, however, differential $\delta^{13}C$ fractionation in cold and warm surface waters combined with ^{12}C evasion from the cold Antarctic surface and ^{12}C invasion into the warm surface area tends to smoothen out the gradients.

CADMIUM-PHOSPHATE DISTRIBUTION

MODEL PROCEDURE

Phosphate and cadmium distributions are simulated with a 13-box ocean model (Keir, 1988) shown schematically in Figure 5. Both of these tracers are treated as conservative inventories which are redistributed internally within the ocean by the combined processes of biological cycling and ocean circulation. Factors which affect biological cycling are the degree of uptake of the dissolved constituent within the surface ocean and the depth distribution of the nutrient or trace metal remineralization of the vertical particulate flux beneath the surface. In the model, dissolved phosphate and cadmium are both completely removed from the warm surface ocean. A specified fraction of the consequent particulate flux remineralizes at intermediate depths and the remaining flux in the deep ocean. In the two high-latitude surface regions, the nutrient or trace metal uptake is set as a fraction of the upwelling dissolved flux and this fraction is removed as particles to the underlying box. Thus there are three variable parameters related to biological cycling which affect the distributions of phosphate and cadmium: (a) recycling fraction in intermediate water, f_{int}, (b) the fraction of upwelling dissolved flux taken up and removed as particulate matter from the circumpolar Antarctic surface, g_c, and (c) the fractional uptake in the northern Atlantic surface box, g_{na}. The phosphate distribution shown in Figure 5 is produced by the model transports given in Keir (1988, Figure 2) and the phosphate biological parameters, $f_{int}^P = 84\%$, $g_c^P = 24\%$ and $g_{na}^P = 13\%$. In general we examine the pattern of Cd versus PO_4 produced in the model when each of the biological parameters controlling cad

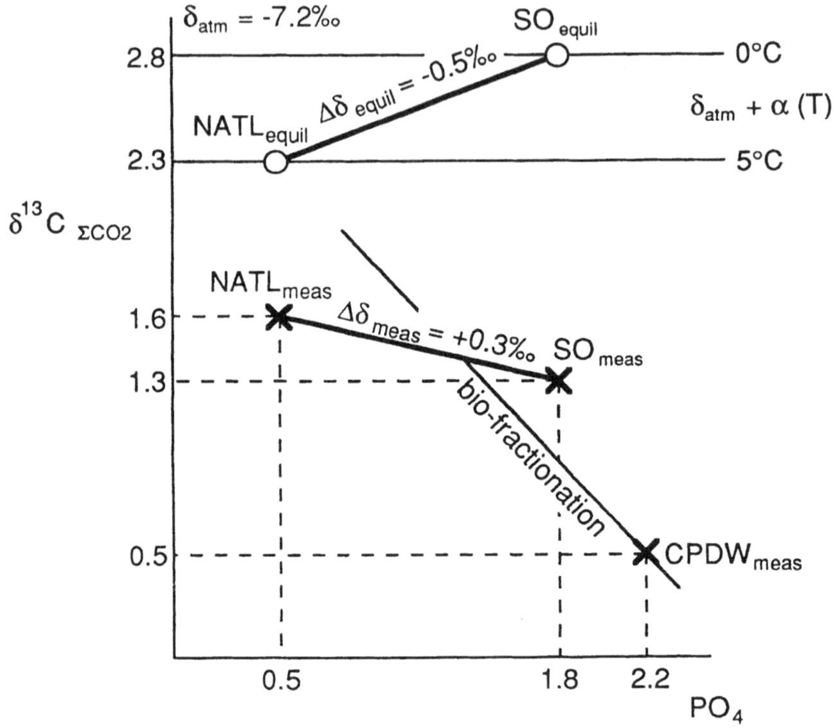

Figure 4a Schematic diagram showing the effect of (partial) carbon isotope equilibration in the Southern Ocean. Surface waters in the Southern Ocean are 5°C colder than in the North Atlantic (NATL). Thus, the Southern Ocean equilibrium $\delta^{13}C_{\Sigma CO2}$ value (open circle, SO_{equil}) should be more positive. Instead, in situ values (SO_{meas}) are 0.8‰ more negative than North Atlantic values. Compared to Redfield-driven bio-fractionation, Southern Ocean values are too positive and North Atlantic values too negative.

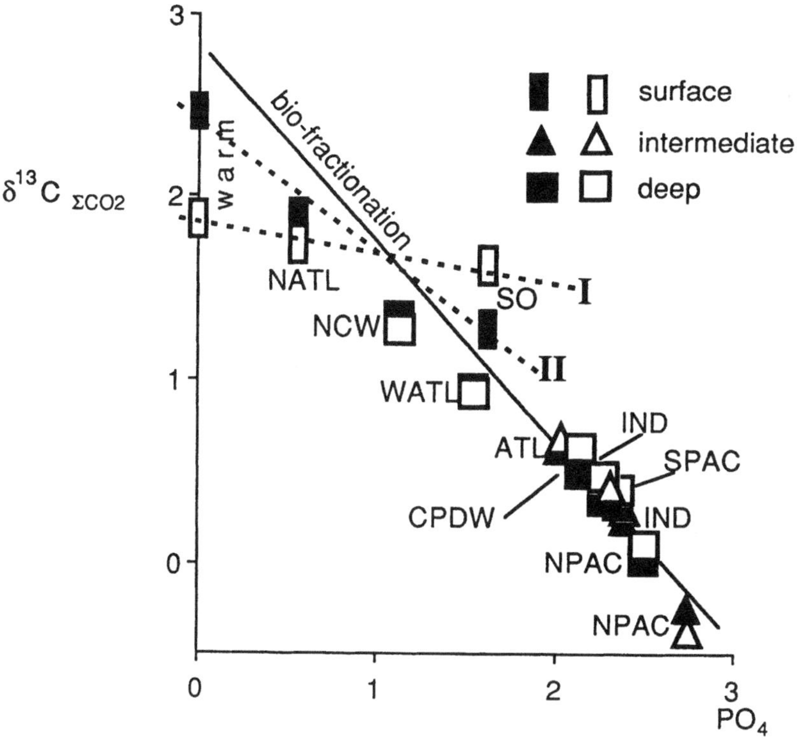

Figure 4b. Distribution of $\delta^{13}C$ from an ocean-atmosphere box model. Open symbols show results for temperature-dependent, closed symbols for constant (i.e. temperature independent) sea-air fractionation. Except for the Atlantic, all deep waters and most intermediate waters plot close to the biological $\delta^{13}C$-PO_4 correlation. The modeled anomalies in the North Atlantic and Antarctic are similar to the observed anomalies shown in Figure 4a. The anomalies are smaller if the temperature effect is eliminated (closed symbols), but they don't go away. We postulate that a dynamic balance exists between CO_2 evading from Antarctic waters to the atmosphere with a lower $^{13}C/^{12}C$ ratio than that invading and CO_2 invading the warm ocean with a lower $^{13}C/^{12}C$ ratio greater than that evading. Dotted lines 'I' and 'II' show cold-to-warm surface $\delta^{13}C$ gradients for the temperature-dependent and temperature-independent case.

mium varies from the corresponding value for phosphate. For example if cadmium uptake is more efficient in Antarctic surface waters, then $g_c^{Cd} > g_c^{P}$. In these simulations the mole ratio of Cd to PO_4 ocean inventory is about 0.3×10^{-3}, and the ocean circulation is not varied.

RESULTS

If cadmium behaved as a perfect analog to phosphate, the biological cycling parameters for the trace metal would be exactly the same as those for the nutrient. Figure 6 illustrates this "control" case, and as expected cadmium exhibits a constant ratio to phosphate corresponding the inventory ratio.

If Cd is completely stripped from the northern Atlantic surface, i.e. $g_{na}^{Cd} = 100\%$, the Cd concentration in this box is of course reduced to zero, but increases in Cd concentration elsewhere in the model are infinitesimal, on the order of 5×10^{-4} nmol kg^{-1} (Figure 5). This suggests that differential uptake of Cd in boreal Atlantic surface waters would only be detected locally. In the remainder of the ocean that is capped by other surface waters, a linear correlation of Cd and PO_4 would be observed throughout the water column. This result is due to the polar mediterranean nature of the circulation in the northern Atlantic, which consists of poleward flow at surface and intermediate depths, and the sinking and entrainment of these waters to form Northern Component Water (NCW). In such a system, the nutrient or trace metal concentration of this water mass depends only on the concentration in the Atlantic intermediate water and the amount of dilution by the surface water input, regardless of the biological uptake in the Atlantic polar surface water. Stripping cadmium from the northern Atlantic surface layer only adds its very small inventory to the rest of the world ocean.

Differential cadmium removal from Antarctic surface waters, in contrast to the northern Atlantic, produces a global rearrangement of the Cd:PO_4 ratio. Figure 7 shows the extreme case where Cd is completely taken up in the Antarctic surface box. In the model, the trace element is concentrated relative to PO_4 in the subsurface circumpolar box as well as in the deep Indo-Pacific because of mixing. Intermediate waters are depleted in Cd, especially in the Atlantic, which receives the strongest contribution of Antarctic surface water. Since Atlantic intermediate water feeds the NCW, the latter also is depleted in Cd relative to PO_4. The sequence connecting warm surface, northern component, deep Atlantic and deep Indo-Pacific waters exhibits a pronounced kink at the position of NCW on a Cd-PO_4 diagram of the model distributions. This occurs because the surface-NCW trend extrapolates to the low Cd:PO_4 ratio of Atlantic intermediate water, but the slope of the NCW-Pacific trend increases because of the enriched Cd recycling source in the circumpolar region.

The global effect of differential Cd uptake in Antarctic surface waters is a consequence of the estuarine type of circulation in this region. Part of the upwelling water subsequently ad-

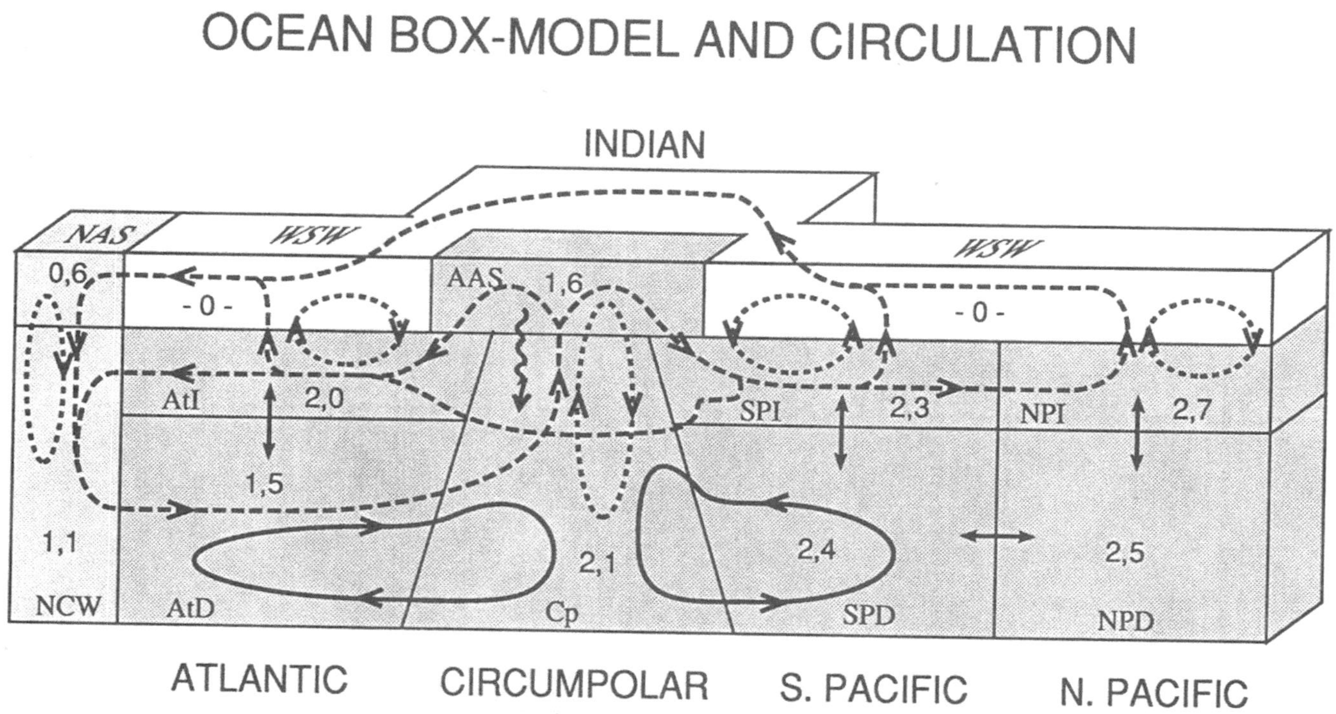

Figure 5. Schematic diagram of 13-box ocean model (Keir, 1988). Numbers indicate phosphate concentration in μmol kg^{-1} produced by the standard model.

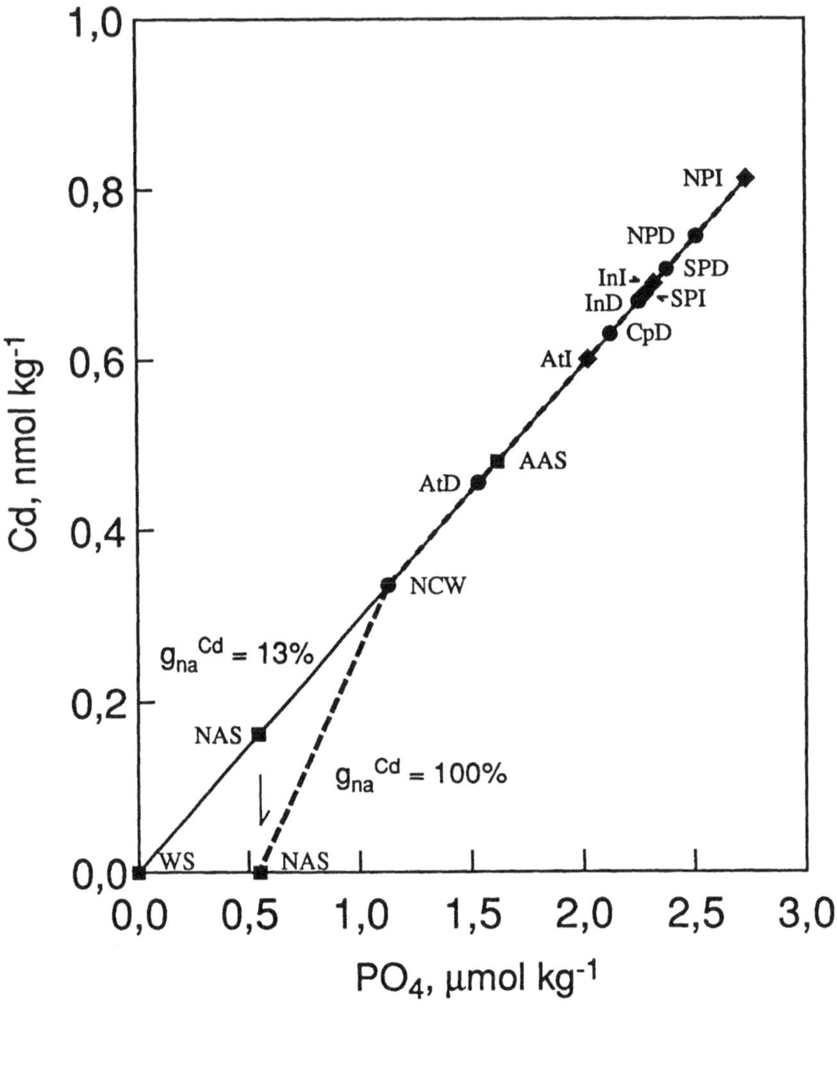

Figure 6. Model produced Cd versus PO_4. Solid line shows control case where Cd behaves analogously to PO_4. Dashed line indicates pattern when Cd is entirely depleted from the northern Atlantic surface. Symbols show discrete position of 'boxed' water masses in Figure 1. (InI and InD are Indian intermediate and deep boxes not shown).

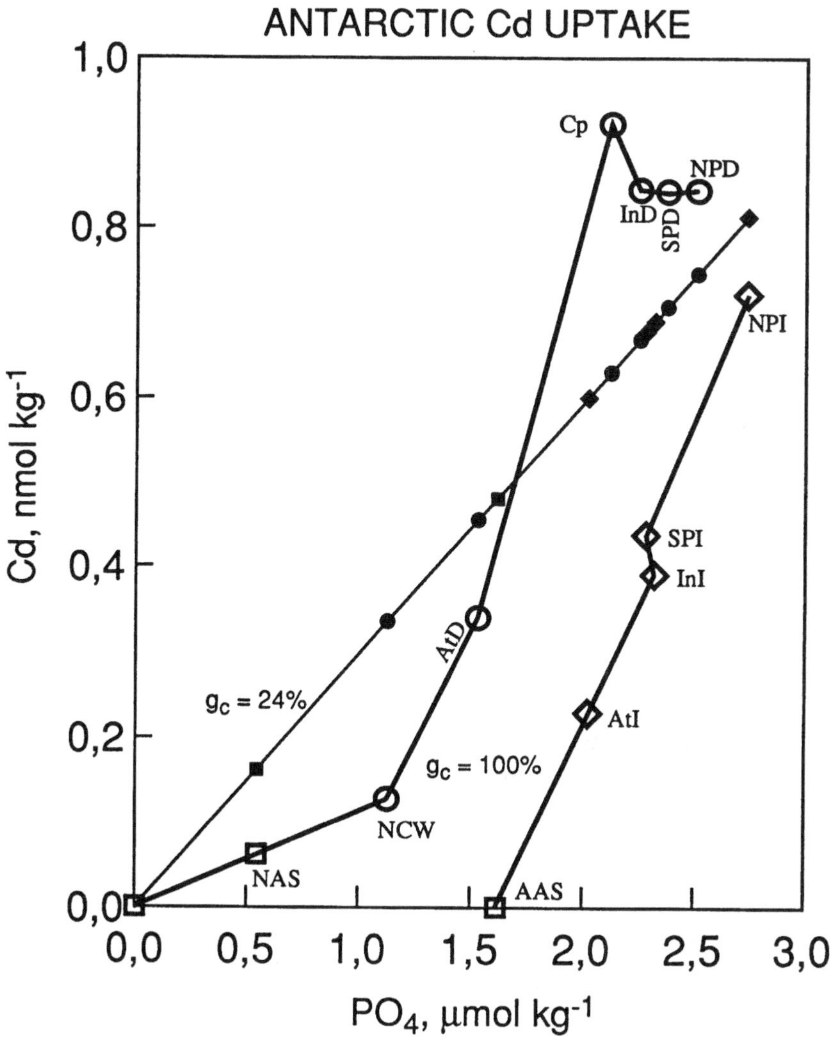

Figure 7. Model Cd-PO_4 produced by total depletion of Cd in Antarctic surface box (g_c^{Cd} = 100%; open points). Lines indicate separated trends for WSW-NCW-deep boxes and AAS-intermediate boxes. Solid points show perfect analog case (g_c^{Cd} = 24%) for comparison.

vects equatorward with low Cd:PO_4 ratio and becomes a source of intermediate waters, leaving a vertical particulate flux with high Cd:PO_4 ratio behind. In the extreme where no lateral mixing of subsurface Antarctic waters took place, complete biological stripping of the polar surface would concentrate all of the cadmium in the subsurface circumpolar water, leaving the rest of the ocean completely depleted of the trace metal.

MODEL-DATA COMPARISONS

In order to compare the box model simulations to observations, a subset of available data is plotted in Figure 8. This subset was selected for two reasons: (1) to reduce the scatter which tends to obscure comparison with the model, and (2) in the case of the Southern Ocean, to illustrate the sequence of various water masses in the overall pattern. The solid points in Figure 8 show data from the North Atlantic (Bruland and Franks, 1983, Boyle et al., 1985) and North Pacific (Bruland, 1980). The Pacific data sample the entire water column, and within the scatter of the measurements, no differentiation between the Cd-PO_4 trends of intermediate and deep North Pacific waters is evident. The solid points characterize the main trend shown in Boyle (1988, Figure 1), which includes other data from the Atlantic and Arctic oceans. The open points in Figure 8 illustrate the Cd-PO_4 pattern for certain water masses in the Southern Ocean, as follows: (a) surface waters south of the polar front with PO_4 concentrations of about 1.8 μmol kg^{-1}, (b) the shallow phosphate maximum south of the polar front (denoted P-max-60°S hereafter), (c) Antarctic Intermediate Water as characterized by the salinity minimum at 48°S in the southwest Pacific, and (d) Circumpolar Deep Water. Figure 9 illustrates the distribution of these water masses in the South Pacific, where the core of the PO_4 maximum indicates an outflow of older Pacific water. The Cd-PO_4 characteristics of these water masses fall on the Southern Ocean trend shown by Frew and Hunter (1992, Figure 3). In the sequence on the Cd-PO_4 diagram, the P-max-60°S water has the highest Cd:PO_4 ratio and lies in the field of the North Pacific observations (Figure 8), while CPDW has a somewhat lower ratio and lies inbetween the PO_4-max-60°S and AAIW characteristics, even though CPDW is not physically located between the PO_4 maximum and AAIW (Figure 9).

DIFFERENTIAL Cd UPTAKE IN ANTARCTIC WATERS

When 56% of the upwelling cadmium flux (as opposed to 24% for PO_4) is fixed in the Antarctic surface, the model ocean pattern of Cd versus PO_4 appears fairly similar to the observations (Figure 10). The kink in the main trend is reproduced very well and the model intermediate and Antarctic surface waters roughly duplicate the Cd-PO_4 trend observed in the Southern Ocean. However there are some interesting discrepancies. The greater Cd:PO_4 ratio in circumpolar deep waters predicted by the model does not appear in the measured CPDW val-

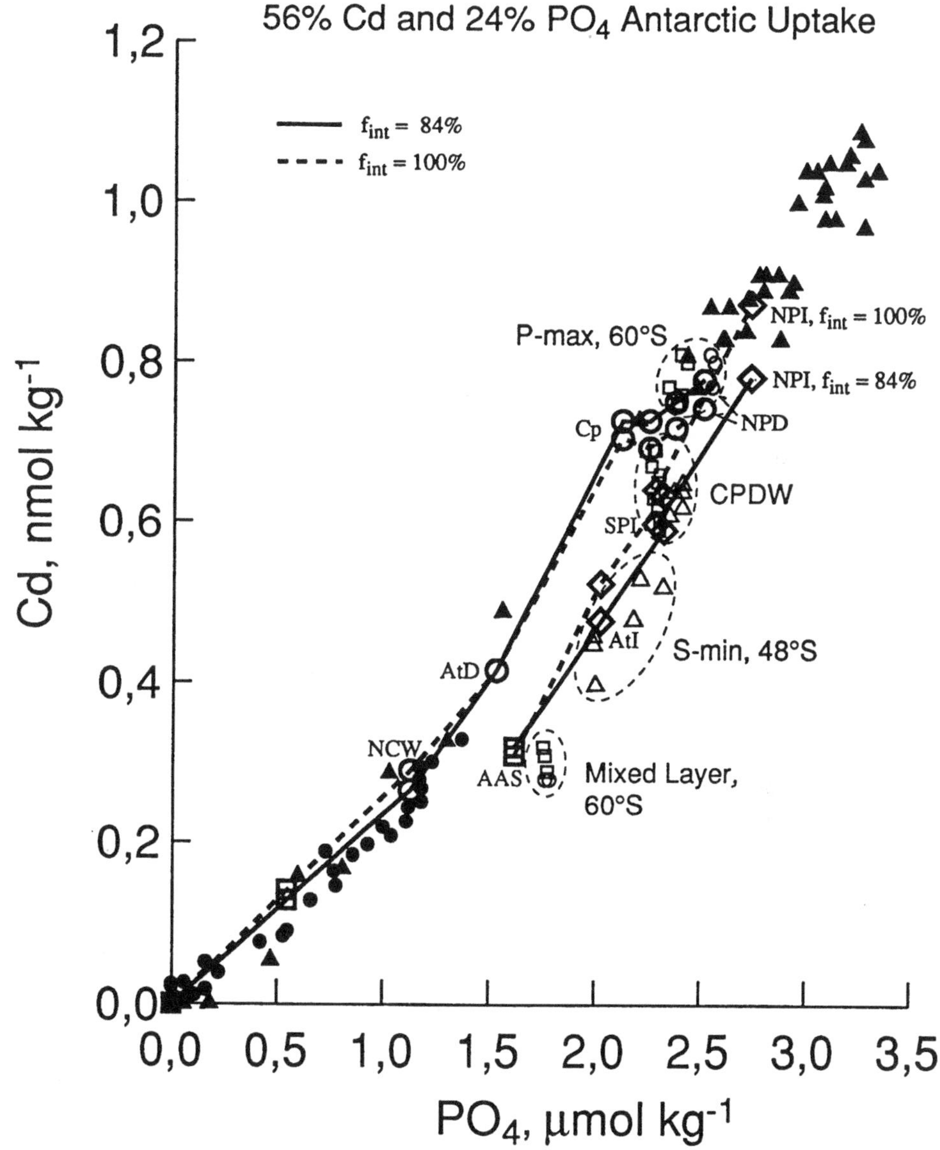

Figure 8. Measured Cd versus PO_4. Solid circles are from North Atlantic (Bruland and Franks, 1983; Boyle, 1985) and solid triangles from North Pacific (Bruland, 1980). Open symbols are from Southern Ocean stations: triangles - 48°S 164.5'E (Frew and Hunter, 1992), circles - 61°S 63'W (Martin et al., 1990), squares - 57°S 48'W (Nolting et al., 1991).

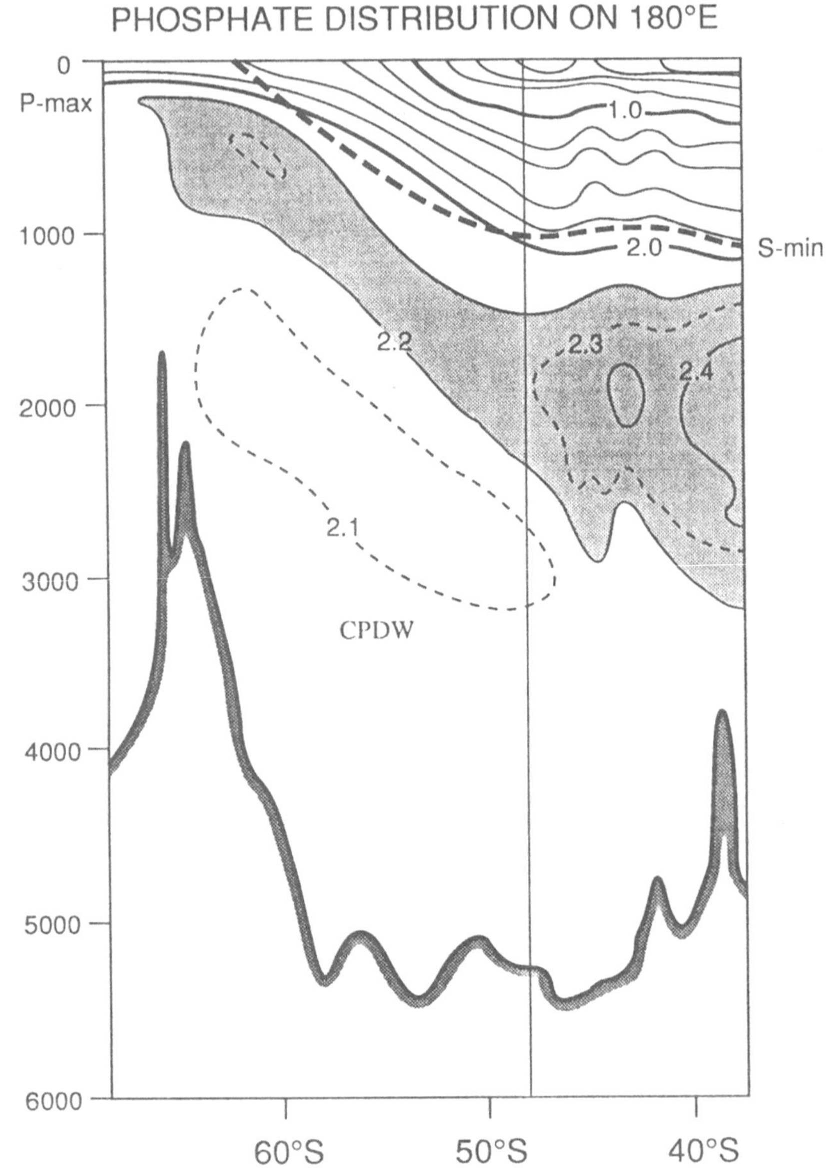

Figure 9. North-South phosphate (μmol kg^{-1}) section on approx. 170°W in the South Pacific (Craig et al., 1981). Dashed line indicates the salinity minimum which identifies AAIW.

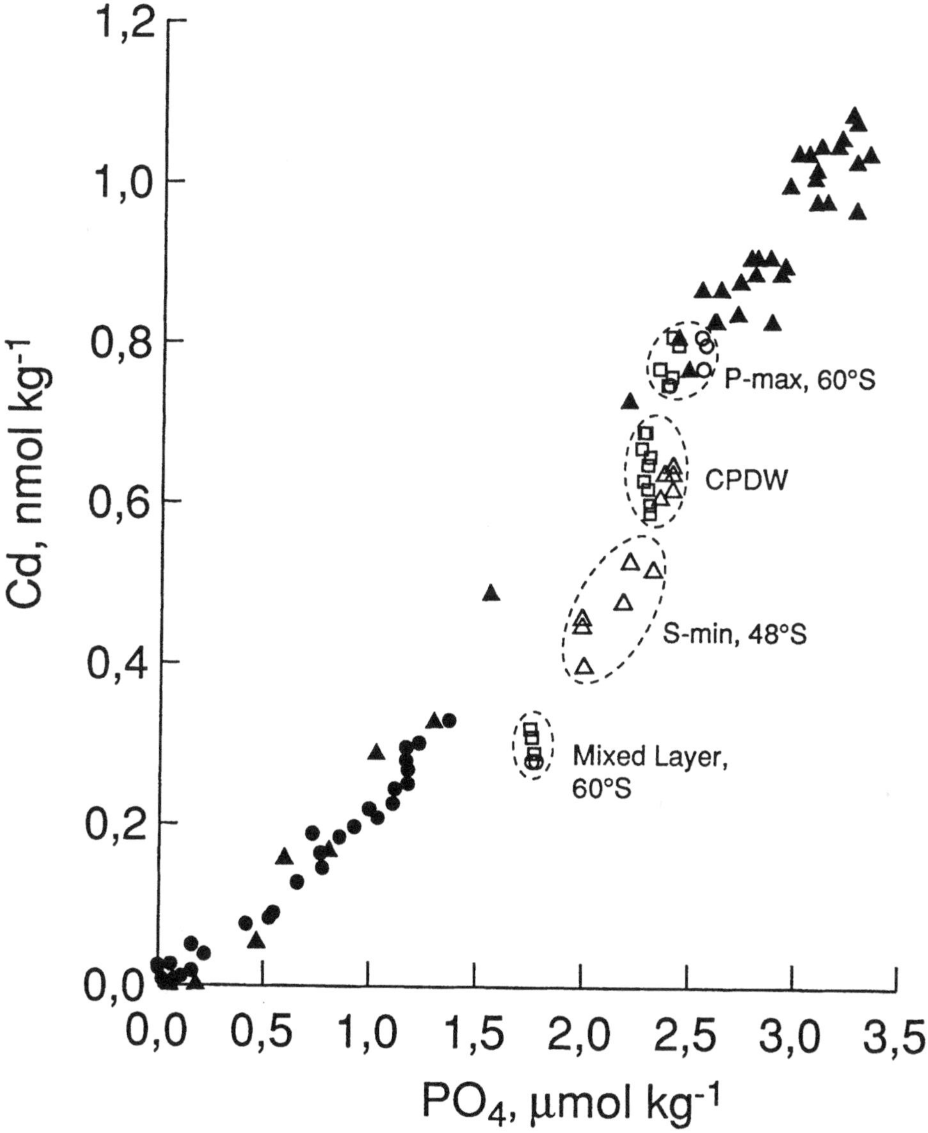

Figure 10. Model Cd-PO_4 patterns produced by 56% biological uptake of Cd flux and 24% uptake of PO_4 flux into Antarctic surface box (large symbols). Two variations are shown: (a) identical vertical recycling distribution for Cd and PO_4 (solid lines), and (b) 100% recycling of Cd in intermediate boxes (dashed lines). Small symbols show data in Figure 8.

ues. Instead the relatively shallow P-max waters south of the polar front appear to have the highest Cd:PO_4 ratio of the Southern Ocean. This suggests that cadmium is entirely remineralized within the upper few hundred meters in Antarctic waters. The lower Cd concentration observed in Pacific CPDW may be due to zonal advection of deep water from the Atlantic with a lower Cd:PO_4 ratio into the Pacific. The box model shown in Figure 5 does not take such a zonal transport into account. It should be noted that the higher Cd:PO_4 ratio found in the shallow phosphate maximum south of the polar front does not show up in the deep phosphate maximum lying on about the same isopycnal to the north at 48°S (Frew and Hunter, 1992). Thus the shallow phosphate maximum in Antarctic waters seems to be a cell of slightly greater cadmium, which may be a consequence of poleward flow and the greater Cd:PO_4 ratio of the particulate matter remineralizing at this depth.

If Cd is entirely remineralized in the upper water column of Antarctic waters, it may be this also occurs at low latitudes as well. The trends indicated by the dashed lines in Figure 10 indicates the model-produced Cd-PO_4 pattern when in addition to the differential Cd uptake in the Antarctic, 100% of the particulate Cd (as opposed to 84% for PO_4) is remineralized in intermediate waters. In this case Cd is somewhat more concentrated in North Pacific intermediate depths than it would be otherwise, which fits better with the field of observed North Pacific data. The fit to the observed data does not improve in other regions, and for example despite a small decrease in CPDW Cd, the discrepancy between observed and model Cd concentrations remains.

DEEPER RECYCLING OF Cd

Before data from the Southern Ocean became available, Boyle (1988) suggested the kink in the main Cd-PO_4 trend might be due to a slightly deeper recycling of Cd relative to PO_4. Figure 11 shows the model Cd-PO_4 diagram when this is taken to the extreme where all of the low-latitude Cd particulate flux is recycled in the deep boxes ($f_{int}^{Cd} = 0$). The deeper Cd recycling also reduces the intermediate water Cd relative to PO_4, and consequently a kink located at the position of NCW is produced in this scenario as well. However, whereas stripping Cd from Antarctic surface waters produces the greatest intermediate water depletion in the Atlantic (Figure 7), deeper recycling produces the greatest intermediate water Cd depletion in the North Pacific. The result shown in Figure 11 suggests that if deeper recycling of Cd were occurring, intermediate waters of the North Pacific would have noticeable lower Cd:PO_4 ratios than those of the deep waters in that ocean, ratios which are not observed in the measured concentrations (Figure 8).

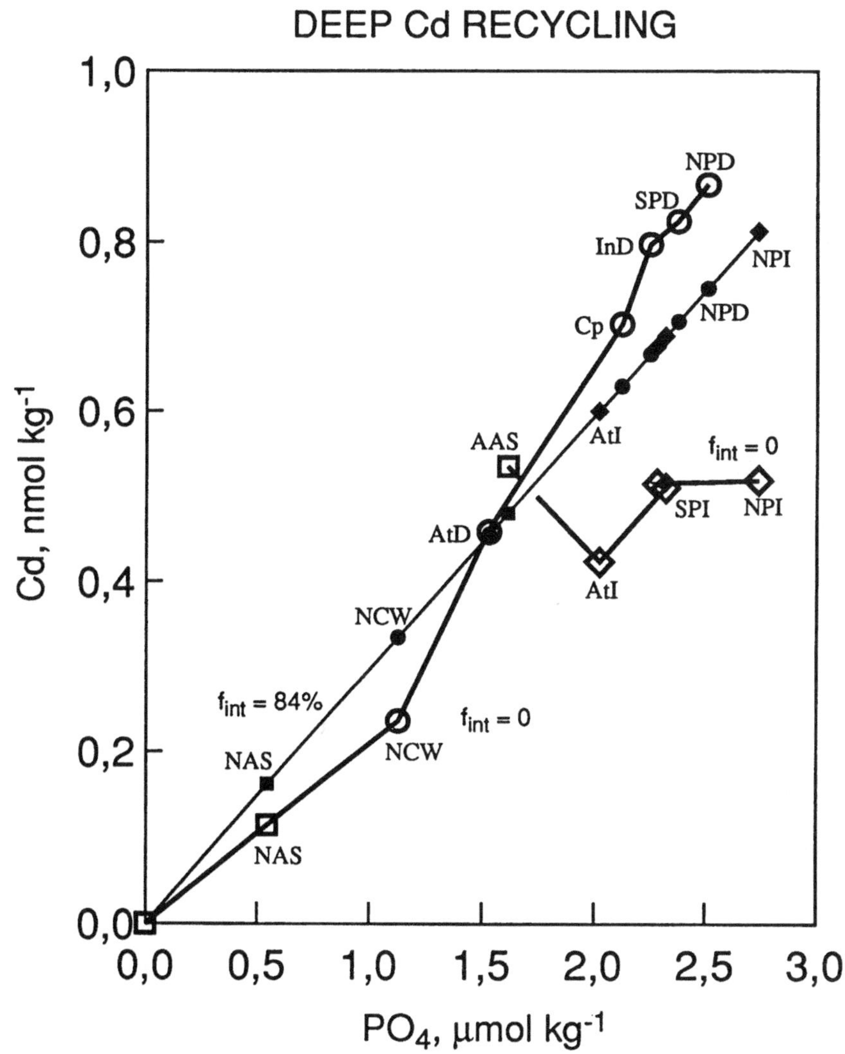

Figure 11. Model Cd-PO_4 produced by 100% recycling of particulate Cd in deep boxes ($f_{int}^{Cd} = 0$; open points). Solid points indicate perfect analog case.

CONCLUSIONS AND IMPLICATIONS

Model runs of ocean-atmosphere carbon exchange show that $\delta^{13}C_{\Sigma CO2}$ behaves less conservatively than nutrients and nutrient-related tracers (Keir, 1988, 1991; Broecker and Maier-Reimer, 1992). Tracers like PO_4 and Cd are recycled within the ocean so that their global distribution is defined by water mass mixing even if the recycling depths for PO_4 and Cd may not be exactly the same (see below). Carbon cycling on the other hand also includes the atmosphere. The associated carbon isotope fractionation during air-sea gas exchange is temperature dependent, sea-to-air fractionation being about 0.1‰ more negative for each 1°C temperature decrease. Since temperatures tend to be low in the ocean where nutrients are high, temperature effect and biological fractionation counter each other: temperatures in the Southern Ocean outcrop regions are low and tend to increase $\delta^{13}C_{\Sigma CO2}$ whereas PO_4 is high with the effect of lowering $\delta^{13}C_{\Sigma CO2}$. Re-equilibration of $\delta^{13}C_{\Sigma CO2}$ with the atmosphere basically works against the effects of bio-fractionation in a sense that surface $\delta^{13}C_{\Sigma CO2}$ is controlled by thermodynamic gas and isotope exchange regardless of what the nutrient concentrations in the surface waters are.

Water mass recycling seems to be a critical factor in determining the $\delta^{13}C_{\Sigma CO2}$:$PO_4$ distribution in the upper ocean. E.g., water masses which upwell in the Southern Ocean region are nutrient enriched and depleted in $\delta^{13}C_{\Sigma CO2}$. When they come close to the surface their $\delta^{13}C_{\Sigma CO2}$ partially re-equilibrates with the atmosphere while nutrient concentrations decrease by a relatively minor degree due to biological production. As a result, the ratio of $\delta^{13}C_{\Sigma CO2}$ increase over phosphorus decrease is larger than expected from the global Redfield trend. This means that the unit nutrient decrease per unit $\delta^{13}C_{\Sigma CO2}$ increase during upwelling may be much smaller than the unit nutrient increase per unit $\delta^{13}C_{\Sigma CO2}$ decrease during water mass "aging". This effect appears more important for the shallow ocean than for the deep ocean (Figure 3). The implication for using benthic foraminiferal $\delta^{13}C$ to monitor nutrient inventories of the shallow ocean (above, say, 2000 m water depth) is, that we may not simply slide forth and back along a $\delta^{13}C_{\Sigma CO2}$:$PO_4$ regression line in oder to estimate possible nutrient variations on the basis of benthic $\delta^{13}C$ increases and decreases. The effects of (partial) equilibration with the atmosphere will lead to deviations of $\delta^{13}C_{\Sigma CO2}$, and thus also benthic foraminiferal $\delta^{13}C$, from the closed-system $\Delta\delta$:P trend which is much better defined for the deep ocean.

From the observations as well as model output it is apparent that the reliability of foraminiferal $\delta^{13}C$ and Cd/Ca as paleoceanographic proxies is determined by the ocean's past physical and biological operation. Varying rates of sea-air gas exchange and possible variations of recycling depths for cadmium relative to phosphorus during the past may have scewed the proxy-nutrient relationship away from its present pattern. That is, $\delta^{13}C$ and Cd are not uniquely tied to phosphorus which makes a reliable interpretation of foraminiferal $\delta^{13}C$ and Cd/Ca records difficult, especiall if they come from shallow-ocean core sites at high latitudes or other areas of oceanic upwelling.

This has important consequences for the use of foraminiferal $\delta^{13}C$ and Cd/Ca as paleoceanographic proxies. For instance, a glacial foraminiferal $\delta^{13}C$ increase by some 0.5‰ has been observed at intermediate-depth core sites in the North Atlantic and has been used to infer better ventilation and lower nutrient contents for the Glacial North Atlantic Intermediate Water (GNAIW *sensu* Duplessy et al., 1988; Boyle and Keigwin, 1986; Cofer-Shabica and Peterson, 1986; Boyle and Keigwin, 1987; Oppo and Fairbanks, 1987; Zahn et al., 1987; de Menocal et al., 1992). This interpretation is supported by a glacial decrease of benthic Cd/Ca at the same water depth in the North Atlantic (Boyle and Keigwin, 1987; Boyle, 1992). The phosphorus decrease, however, which can be estimated from benthic Cd/Ca is less than that estimated from benthic $\delta^{13}C$. Even larger discrepancies exist for the Southern Ocean where benthic $\delta^{13}C$ records show large glacial-interglacial amplitudes ≥ 1‰ whereas benthic Cd/Ca shows only little or no change at all (Boyle, 1992).

It seems plausible to question the reliability of benthic $\delta^{13}C$ first since the effects of ocean-atmosphere gas exchange and the associated carbon isotope fraction on the $\delta^{13}C$-PO_4 distribution by far outnumbers the effects of varying Cd recycling depths on the Cd-PO_4 distribution. The large drop in glacial benthic $\delta^{13}C$ could well document limited gas exchange due to a more pronounced glacial sea-ice cover around Antarctica. But this could apply to the glacial North Atlantic as well, where sea ice should have been present at least during the winter seasons when convection was most likely to occur. Yet in the North Atlantic we observe an increase in benthic $\delta^{13}C$.

At present a satisfactory explanation for the observed discrepancies between the benthic $\delta^{13}C$ and Cd/Ca records does not seem to be in sight (see e.g. Broecker, 1993). However, even if we are to solve this mystery there is still the "non-ideal behavior" of the $\delta^{13}C$-PO_4 and Cd-PO_4 relations which apparently has to do with gas exchange and cadmium uptake/recycling in the Southern Ocean. When estimating nutrient concentrations from benthic $\delta^{13}C$ and/or Cd/Ca ratios we need to know if we are on the low-nutrient or high-nutrient side of the break in the global regression lines. The only chance we see right know to obtain this information is to measure a number of different proxies with different boundary conditions and compare the data to model results which assume a variety of plausible physical scenarios. The advantage here is that subtle differences between the different proxy records may tell us more about past ocean chemical cycling than coherent paleo-proxy records.

ACKNOWLEDGMENTS. We thank Delia Oppo and Dan McCorkle for reviews and critical comments.

REFERENCES

Boyle, E. A., Cadmium: chemical tracer of deepwater paleoceanography, *Paleoceanography*, *4*, 471-489, 1988.

Boyle, E.A., Cadmium and $\delta^{13}C$ paleochemical ocean distribution during the stage 2 glacial maximum, *Annu. Rev. Earth Planet. Sci.*, *20*, 245-287, 1992.

Boyle, E. A., S. D. Chapnick, X. X. Bai, and A Spivak, Trace metal enrichments in the Mediterranean Sea, *Earth Planet. Sci. Lett.*, *74*, 405-419, 1985.

Boyle, E.A., and L.D. Keigwin, Comparison of Atlantic and Pacific paleochemical records for the last 215,000 years: Changes in deep ocean circulation and chemical inventories, *Earth Planet. Sci. Lett*, *76*, 135-150, 1985/86.

Boyle, E. A. and L. D. Keigwin, North Atlantic thermohaline circulation during the past 20,000 years linked to high-latitude surface temperature, *Nature*, *330*, *35-40*, 1987.

Broecker, W.S., An oceanographic explanation for the apparent carbon isotope-cadmium discordancy in the glacial Antarctic?, *Paleoceanography*, *8*, 137-139, 1993.

Broecker, W. S., and T.-H. Peng, *Tracers in the Sea*, 690 pp., Lamont-Doherty Geological Observatory, Palisades, N. Y., 1982.

Broecker, W. S. and T.-H. Peng, Carbon cycle: 1985 Glacial to interglacial changes in the operation of the global carbon cycle, *Radiocarbon*, *28*, 309-327, 1986.

Broecker, W.S. and E. Maier-Reimer, The influence of air and sea exchange on the carbon isotope distribution in the sea, *Global Biogeochem. Cyc.*, *6*, 315-320, 1992.

Bruland, K. W. and R. P. Franks, Mn, Ni, Cu, Zn and Cd in the western North Atlantic, in *Trace Metals in Seawater*, edited by C.S. Wong, E. Boyle, K. W. Bruland, J. D. Burton and E. D. Goldberg, pp. 395-414, Plenum, New York, 1983.

Bruland, K. W., Oceanographic distributions of cadimium, zinc, nickel and copper in the North Pacific, *Earth Planet. Sci. Lett.*, *47*, 176-198, 1980.

Bryden, H.L., New polynomials for thermal expansion, adiabatic temperature gradient and potential temperature of sea water, *Deep-Sea Res.*, *20*, 401-408, 1973.

Charles, C.D. and R.G. Fairbanks, Glacial to interglacial changes in the isotopic gradients of Southern Ocean surface waters, in *Geological History of the Polar Oceans: Arctic Versus Antarctic*, edited by U. Bleil and J. Thiede, Kluwer Academic Publishers, 519-538, 1990.

Cofer-Shabica, N., and L. Peterson, Caribbean carbon isotope record for the last 300,000 years, *Geol. Soc. Am. Abstr. with programs*, *18*, 567, 1986.

Craig, H., W. S. Broecker, and D. W. Spencer, GEOSECS Pacific Expedition, vol. 4, U.S. Government Printing Office, Washington, D. C., 1981.

de Menocal, P.B., D.W. Oppo, R.G. Fairbanks, and W.L. Prell, Pleistocene $\delta^{13}C$ variability of North Atlantic Intermediate Water, *Paleoceanography*, *7*, 229-250, 1992.

Duplessy, J.-C., N. J. Shackleton, R. G. Fairbanks, L. D. Labeyrie, D. Oppo, and N. Kallel, Deepwater source variations during the last climatic cycle and their impact on the global deepwater circulation, *Paleoceanography*, *3*, 317-341, 1988.

Fofonoff, N.P., Computation of potential temperaturee of seawater for an arbitrary reference pressure, *Deep-Sea Res.*, *24*, 489-491, 1977.

Frew, R.D. and K.A. Hunter, Influence of Southern Ocean waters on the cadmium-phosphate properties of the global ocean, *Nature*, *360*, 144-146, 1992.

Hasunuma, K., Formation of the intermediate salinity minimum in the northwestern Pacific ocean, *Bull. Ocean Res. Inst. Univ. Tokyo*, *9*, 2-47, 1978.

Keir, R.S., On the late Pleistocene ocean geochemistry and circulation, *Paleoceanography*, *3*, 413-445, 1988.

Keir, R.S., The effect of vertical nutrient redistribution on surface ocean $\delta^{13}C$, *Global Biogeochem. Cycles*, *5*, 351-358, 1991.

Kroopnick, P.M., The distribution of ^{13}C of ΣCO_2 in the world oceans, *Deep-Sea Res.*, *32*, 57-84, 1985.

Lohmann, G.P., D.W. Oppo and W.B. Curry, Reassessmernt of the influence of temperature on the distribution of ^{13}C in the upper ocean, *Paleocaenography*, in press.

Martin, J.H., S.E. Fitzwater, and R.M. Gordon, We still say iron deficiency limits phytoplankton growth in the subarctic Pacific, *J. Geophys. Res.*, *96*, 20,699-20,700, 1991.

Nolting, R. F., H. J. W. De Baar, A. J. Van Bennekom and A Masson, Cadmium, copper and iron in the Scotia Sea, Weddell Sea and Weddell/Scotia Confluence (Antarctica), *Marine Chemistry, 35,* 219-243, 1991.

Oppo, D.W. and R.G. Fairbanks, Variability in the deep and intermediate water circulation of the Atlantic Ocean during the past 25,000 years: Northern Hemisphere modulation of the Southern Ocean, *Earth Plan. Sci. Lett., 86,* 1-15, 1987.

Östlund, H.G., C. Craig, W.S. Broecker, and D. Spencer (eds.), GEOSECS Atlantic, Pacific, and Indian Ocean Expeditions, Shorebased Data and Graphics, *GEOSECS Atlas Series, Vol. 7,* 200 pp., U.S. Government Printing Office, Washington, D.C. 20402, 1987.

Qasim, S. Z., Oceanography of the northern Arabian Sea. Deep-Sea Research, 29, 1041-1068, 1982.

Rau, G.H., P.N. Froelich, T. Takahashi, and D.J. DesMarais, Does sedimentary organic $\delta^{13}C$ record variations in Quaternary ocean [CO_2(aq)]?, *Paleoceanography, 6,* 335-347, 1991.

Talley, L.D., Distribution and formation of North Pacific Intermediate Water, *J. Phys. Oceanogr., 23,* 517-537, 1993.

UNESCO, Background papers and supporting data on the international equation of state of seawater 1980, *UNESCO Technical Papers in Marine Science, 38,* 192 pp., United Nations Educational Scientific and Cultural Organization, Paris, France, 1981.

Yuan, X. and L.D. Talley, On shallow salinity minima in the North Pacific, *J. Phys. Oceanogr., 50,* 1302-1316, 1992.

Zahn, R., M. Sarnthein, and H. Erlenkeuser, Benthic isotope evidence for changes of the Mediterranean outflow during the late Quaternary, *Paleoceanography, 2,* 543-559, 1987.

PART III:

GEOCHEMISTRY OF THE ORGANIC SEDIMENT FRACTION: CONSTRAINTS ON THE BIOLOGICAL CARBON PUMP

POSSIBLE EARLY DIAGENETIC ALTERATION OF PALAEO PROXIES

Gert J. De Lange, Bertil Van Os, Peter A. Pruysers, Jack J. Middelburg[1], Davide Castradori[2], Patrick Van Santvoort, Peter J. Müller[3], Hans Eggenkamp, and Fred G. Prahl[4].

Institute for Earth Sciences
Department of Geochemistry
Budapestlaan 4
3584 CD Utrecht
The Netherlands

ABSTRACT. The search for palaeo tracers in the last decade has resulted in a number of potentially useful tracers for past temperature/salinity, productivity, $pCO_2(aq)$, nutrients, etc. Recently, it has become evident that some limitations exist on the application of these tracers. Early diagenetic processes may considerably alter the primary signals. The diagenetic fates of palaeo proxies may vary for different environments, which can lead to preferential preservation, decomposition or relocation of particular tracers. Consequently, one must be very cautious when applying a palaeo-tracer that is sensitive to early diagenetic alteration. In this paper we present data from the Madeira Abyssal Plain, the eastern Mediterranean, and Kau Bay that clearly illustrate some of these alterations:

1. *Diagenetic changes in some organic proxies (% C, C/N,* $\delta^{13}C$, $\delta^{15}N$, U^K_{37}*).* Origin and diagenetic history of sedimentary organic matter may result in the preferential preservation of some compounds. As a consequence the organic matter content and its composition may be altered dramatically upon postdepositional decomposition.

2. *Variations in the sedimentary Ba content (un)related to palaeoproductivity.* Parameters other than productivity alone can determine the rain rate of Ba and consequently, its distribution in sediments. In addition, hydrothermal activity and anoxic mobilization coupled with reprecipitation are known to produce relatively high levels of Ba, unrelated to productivity.

3. *(Diagenetic) Mn-spikes at palaeo-productivity changes.* Although the occurrence of Mn spikes has been related to changes in palaeoproductivity, and to sudden changes of the ocean water from suboxic to oxic conditions, most of these spikes can as well be explained by the early diagenetic mobilization of Mn from within sediments, deposited under non-steady state conditions such as those fostered by the emplacement of turbidites, a sudden change in sediment or organic C accumulation rate, or a change in bottomwater oxygenation.

---

[1]Netherlands Inst. Ecology, Yerseke, The Netherlands; [2]Dipartimento di Scienze della Terra, Universita di Milano, 20133 Milano, Italy; [3]Fachbereich Geowissenschaften Universitat Bremen, Klagenfurter Strasse, D-28359 Bremen, Germany; [4]School of Oceanography, Univ. of Oregon, Corvallis, OR, USA

NATO ASI Series, Vol. I 17
Carbon Cycling in the Glacial Ocean:
Constraints on the Ocean's Role in Global Change
Edited by R. Zahn et al.

4. *Diagenetic dissolution/precipitation of carbonates.* Not only the preferential dissolution of some biogenic carbonate, but also the precipitation of carbonate (both by early diagenesis) may lead to variations in species- and isotopic composition that are unrelated to 'primary signals'. Despite these limitations, palaeogeochemical tracers can provide significant and highly useful insights into oceanographic processes, provided their use is properly tempered.

INTRODUCTION

The recognition that palaeoceanographic conditions can be established not only by traditional micropalaeontological or sedimentological indicators, but also by tracers from the relatively young field of geochemistry has resulted in an intensive search for palaeo geochemical proxies. As a result, a number of potentially useful tracers for determining past temperature / salinity, productivity, $pCO_2(aq)$, etc have been reported in the last decade (e.g. Table 1; Berger et al., 1983; Finney et al., 1988; Brassell et al., 1986; Prahl and Muehlhausen, 1989; Boyle, 1988; 1992; Boyle and Lea, 1989; Lea and Boyle 1989; 1990; Rau et al., 1991; Elderfield, 1991; Altabet, 1992).

One of the first successful methods used to estimate the palaeo ocean productivity was based on the organic carbon content of the sediment (e.g. Müller and Suess, 1979). On the basis of sediment trap data the 'Müller and Suess' equation was later modified for its depth dependency (Suess, 1980), and further refined by others (e.g. Sarnthein et al., 1988). However, the organic carbon content of sediments is not only dependent on the ocean productivity, export flux, water depth, and sedimentation rate, but also on the origin and diagenetic history of the organic matter. Origin and diagenetic fate may vary for different environments, which may lead to preferential preservation for some groups of compounds (e.g. Jumars et al., 1989; Tegelaar et al., 1989). As a consequence, sedimentary organic carbon content, and its isotopic characteristics ($\delta^{13}C$ and $\delta^{15}N$) may be altered dramatically during early diagenetic processes. In recent reports, a close correlation was shown between sedimentary $\delta^{13}C_{org}$ and the atmospheric CO_2 record from the Vostok ice core (Barnola et al., 1987; Jasper and Hayes, 1990; Rau et al., 1991), making the $\delta^{13}C_{org}$ a potential proxy for $PCO_2(aq)$ (Fontugne and Calvert, 1992; Rau, this volume). The $\delta^{15}N_{org}$ has recently been reported to be a possible recorder for palaeo nutrient utilization; diagenetic changes in the isotopic signal were not excluded but were thought to act uniformly (Altabet, 1992; Calvert et al., 1992; Altabet and Francois, this volume).

If $\delta^{13}C$ and $\delta^{15}N$ values are to be used as possible palaeo proxies for pCO_2 and surface water productivity, it is important to assess the effects of diagenetic alteration.

The occurrence of Mn spikes at climatic transitions has been reported for sediments

TABLE 1. Summary of palaeo proxies discussed in the text, and the relative effect of a diagenetic change on the interpretation of these palaeoproxies.

$\%C_{org}$	C/N	$\delta^{13}C_{org}$	$\delta^{15}N_{org}$	U^{K}_{37}	Ba	Mn	C.sp.	$\delta^{13}C_{carb}$	$\delta^{18}O$	^{14}C
pp	M/T	pp, M/T	pp, M/T	Temp	pp	δpp, ‰S	pp, Temp	pp, %S	‰S, Temp	age
-	+/-	+,T	+,T	0	displ	disapp (-?)	*	*	*	+/-

C.sp.: carbonate species; pp: palaeoproductivity; +/-: increases and/or decreases in signal; M/T: marine/terrestrial; Temp: temperature; ‰S: salinity; 0: no change; displ: displaced signal; disapp: disappeared signal; *: effect largely dependent on environment at time of diagenesis, may be +/-/disapp.

over large areas of the tropical Pacific (Berger et al., 1983). An interesting hypothesis for the origin of such a Mn-spike at the last glacial / interglacial transition in Pacific sediments is that of a sudden re-oxygenation of the oceanwater at the time of that very transition (Mangini et al., 1991; Mangini, 1992; Mangini et al., this volume). However, the more commonly reported view is that the formation of such Mn spikes is diagenetic, and may be influenced by the activity of benthic burrowing organisms, which in turn is dependent on ocean productivity. The formation of such a Mn spike in N.E. Atlantic and eastern Mediterranean sediments on the other hand, has been explained by geochemical mobilization alone (e.g. Thomson et al., 1984; Wallace et al., 1988; De Lange et al., 1989; Pruysers et al., 1991; 1993). In both areas, a clear change in the benthic flux of organic carbon has been invoked. Thus, the occurrence of a Mn spike could be a potential proxy for (changes in) ocean productivity, organic carbon accumulation rates, or bottomwater oxygenation. However, these very spikes may only be preserved under certain diagenetic conditions (e.g. Finney et al., 1988; Pruysers et al., 1993; and see Section 3).

In general, the distribution and relative concentration of excess Ba in marine sediments seem to be related to biogeochemical fluxes in the ocean (e.g. Dehairs et al., 1980; 1987; Bishop, 1988; Calvert, 1990). Therefore, Ba has been suggested as a possible palaeo-productivity proxy (e.g. Dymond, 1986; Jumars et al., 1989; Calvert, 1990; Von Breymann et al., 1992; Dymond et al., 1992; Suess et al., 1993; Shimmield et al., this volume). Variations in the Ba accumulation rate may not only be controlled by ocean productivity and water depth (Von Breymann et al., 1992), but may also arise from salinity excursions hydrothermal emanations (e.g. Schmitz, 1987), or variations in lithogenic Ba. These last three variations are likely to be of local importance only. Regardless of its origin, barite-Ba may become mobile under sulphate reducing conditions in the sediment (e.g. Church and Wolgemuth, 1972; Brumsack, 1989; Van Os et al., 1991). Diagenetic mobilization has been proposed to explain the enrichment of Ba in Cretaceous black shales (Vine and Tourtelot, 1970) although this hypothesis has more recently been questioned by Brumsack (1986). The various factors controlling the barite-Ba rain-rate and the postdepositional diagenetic mobilization of Ba suggest a need for the full understanding of its geochemistry before its ultimate utilization as a reliable palaeo-productivity proxy.

Diagenetic changes in the carbonate content and composition have been reported to occur in sediments of all ages (e.g. Emerson and Bender, 1981; Thierstein and Roth, 1991). In general, the distinction between primary (input) signals and diagenetic alterations becomes more difficult with increasing age of the sediments. Thorough study of examples in recent sediments and additional laboratory experiments may help us to better understand palaeo variations of carbonate-related proxies in the sedimentary record.

In view of the various diagenetic changes that may occur to possible palaeo proxies, it is evident that there are limitations to the inconsiderate application of these tracers. Despite these limitations some geochemical tracers have proven to be useful palaeo proxies. In addition to giving a brief overview, this paper not only discusses the possible diagenetic alterations of some palaeo proxies but also gives new and illustrative examples.

DIAGENETIC CHANGES TO PALAEO PROXIES

A description of some of the diagenetic processes and their effects on palaeo proxies will be made using pertinent examples mainly from three key areas that have recently been intensively studied:

a. Madeira Abyssal Plain (MAP): The late Quaternary sediments of the MAP consist of alternating thick (up to 6 m) distal turbidites and thin (cm to dm) pelagic deposits. Each turbidite is homogeneous in composition and has a distinct colour and lithology (Weaver and Kuijpers, 1983; De Lange et al., 1987). After emplacement, the tops of the organic-rich turbidites have been oxidized by oxygen diffusing into the sediment from the overlying seawater. Consequently, the C_{org} content of the upper portions of these turbidites has been reduced to low levels (e.g. Wilson et al., 1985; Jarvis and Higgs, 1987; De Lange et al., 1987; Prahl et al., 1989; McArthur et al., 1992). The homogeneous organic-rich turbidites of the MAP offer a unique opportunity to study the stability of redeposited organic matter, and of sedimentary constituents that are related to organic matter, under oxic versus suboxic / reducing conditions. In view of the time needed for these 'in-situ oxidation experiments' (ranging from 1- 50 ky), identical laboratory simulation experiments are not likely to be ever done.

b. Eastern Mediterranean (EM): Miocene to Holocene EM sediments consist of interbedded organic-rich (sapropel) and organic-poor (marl) intervals formed by pelagic deposition. The cyclic variation in organic carbon content appears to be climatically (monsoonally) controlled (Rossignol-Strick et al., 1983) and may originate from variations in the organic carbon supply rate (i.e. productivity) and changes in preservation i.e. bottom water oxygenation (e.g. Kullenberg, 1952; Olausson, 1961; Cita et al., 1977; Calvert, 1983; De Lange and Ten Haven, 1983; Hilgen, 1991; Calvert et al., 1992; Fontugne and Calvert, 1992). The sediment setting from the semi enclosed EM basin has often been mentioned as a potentially ideal recorder and magnifier of global climatic events.

c. Kau Bay, Indonesia (KB): KB (island of Halmahera, eastern Indonesia) is a 470 m deep basin separated from the Pacific Ocean by a sill that is at present only 40 m below sea-level. As a consequence, KB is at present a semi-euxinic marine basin, with sediments rich in organic carbon (3-4%) and sulphur (1-2%) (e.g. Middelburg et al., 1990; 1991). In general, anoxic sediments in such basins are considered to be ideal sites for micropalaeontological and climatic studies, because of their excellent preservation of carbonate and siliceous fossils.

1. DIAGENETIC CHANGES IN SOME ORGANIC PROXIES
(Organic Carbon, C/N, $\delta^{13}C$, $\delta^{15}N$, U^K_{37})

In most oceanic sediments, organic matter is decomposed by a sequential series of oxidants, in decreasing order of thermodynamic energy yield : oxygen, nitrate, Mn-hydroxides, Fe-hydroxides, and sulphate (e.g. Froelich et al., 1979; De Lange, 1986). Following the complete consumption of all of these oxidants, organic matter continues to be degraded by methanogenesis. Experiments on fresh planktonic material have shown that organic matter decomposition rates under oxidizing and under sulphate-reducing conditions are similar (e.g. Westrich and Berner, 1984). However, in comparing results obtained by various authors in the natural environment, different decomposition rates for oxic and for anoxic conditions have been reported (Emerson and Hedges, 1988; Henrichs and Reeburgh, 1987; Reimers, 1989; Henrichs, 1992). In addition, a number of areas are known where, despite the high amount of organic matter and of Mn-, Fe (hydr)oxides in these sediments, no appreciable decomposition of organic matter seems to take place once oxygen (and nitrate) has been exhausted (MAP: e.g. Wilson et al., 1985; Jumars et al., 1989; Prahl et al., 1989; De Lange, 1992; EM: De Lange et al., 1989; Pruysers et al., 1991;1993; and the Oman Margin: e.g. Pedersen and Shimmield, 1991).

It is often difficult to ascertain such differences in the natural environment. Being the first oxidant for organic matter decomposition, oxygen will be used by aerobic bacteria to decompose the most labile ('digestible') fraction of the organic pool, leaving the less labile fraction to be used by other micro-organisms such as sulphate-reducing bacteria. Consequently, the tendency in the natural environment is that decomposition rates decrease with time; hence, sulphate reducers are involved in the decomposition of less labile organic matter (e.g. Westrich and Berner, 1984; Middelburg, 1989; Middelburg et al., 1993). For the relatively fresh organic matter with a high energy yield upon decomposition, one expects minor if any differences between organic matter decomposition under oxic and under sulphate-reducing conditions. However, for the less labile organic fraction, the energy yield for the sulphate reducers may be small, whereas that for decomposition through oxygen is still substantial. It is within this fraction that one may expect to find large differences between the two lines of organic matter decomposition.

An alternative and more mechanistic view of this glaring contrast is the following: during the initial sulphate-reducing conditions, new molecular bonds/S-bearing compounds are formed ('vulcanization' or 'geopolymerization'; Sinninghe Damste et al., 1990 and refs herein; Berner, 1980). These new compounds may be stable under reducing, but not under oxic conditions. Upon renewed oxygen supply these sulphur-compounds may be attacked by aerobic micro-organisms which may lead to a resumed degradation of organic matter.

A few examples exist that seem to confirm this logic. The organic carbon content of sediments from the MAP is up to 2%, that of the EM up to 19%, and that of the Oman Margin around 3%. Although oxygen is depleted within the top 1 - 30 cm, no detectable sulphate reduction is taking place in most cores at least until depths of 36, 37, and 20-50 meters, respectively (Fig. 1; De Lange, 1992; Ten Haven et al., 1987; Van Santvoort and De Lange, in prep.; Pedersen and Shimmield, 1991).Furthermore, similar observations have been made in organic-rich sediments of the eastern Angola Basin (Schulz et al., 1994). It seems, therefore, that once it is beyond the reach of oxygen, a large fraction of the organic material in these areas is almost unaffected by subsequent diagenesis, i.e. its decomposition is largely 'oxygen-limited'. This is in line with a general observation that a large fraction of the organic matter in stored marine cores is degraded through oxygen relatively rapidly (within years), whereas under natural suboxic conditions there is virtually no change (within 1000's of years). Note that such post-sampling oxidation leads to considerable changes in organic matter composition too (e.g. ^{14}C age; Geyh et al., 1974), and may also result in changes to the carbonate fraction, and to the carbonate-related proxies (see under section 4 below).

To substantiate this further with evidence from natural systems, we need to compare the organic matter decomposition (rates) of this less-labile organic fraction under oxygenated and under suboxic / sulphate-reducing conditions. Two possible examples can be mentioned in this respect, both related to the well-known oxidation front phenomenon (see above and under Mn; e.g. MAP: Colley et al., 1984; Wilson et al., 1985; Thomson et al., 1987; De Lange et al., 1987; De Lange, 1992; Thomson et al., 1993; EM: De Lange et al., 1989; Pruysers et al., 1993). Briefly, if a distinct layer of organic-rich, sulphate-reducing sediments is buried by organic-poor sediments, an oxidation front may burn its way into the organic-rich sediments. Clearly, in the lower part of the organic-rich interval sub-oxic diagenesis may continue, whereas in the upper part the same sediments are decomposed through oxygen. Such a

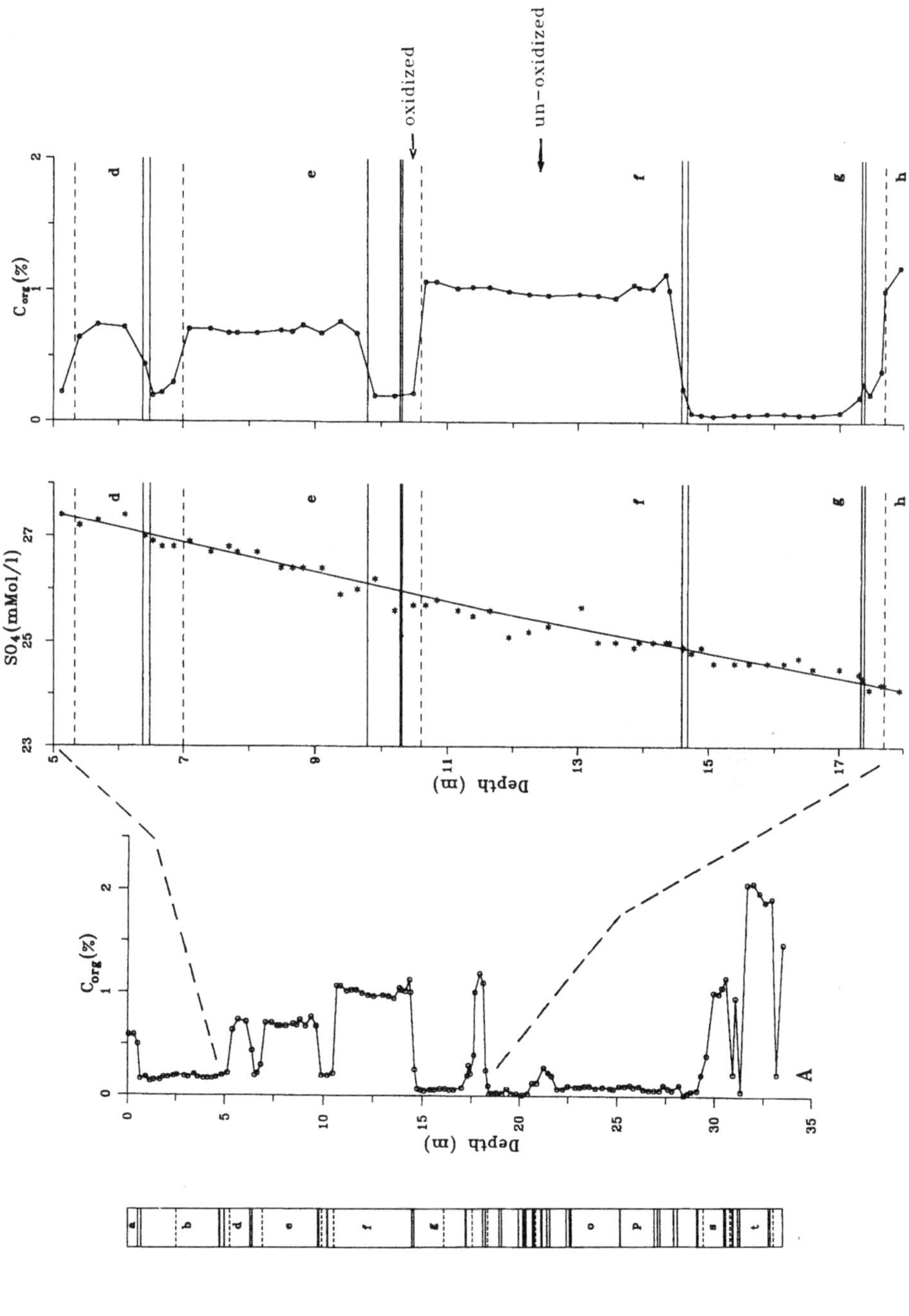

oxidized
un-oxidized
C$_{org}$(%)
SO$_4$(mMol/l)
Depth (m)
d
e
f
g
h
A

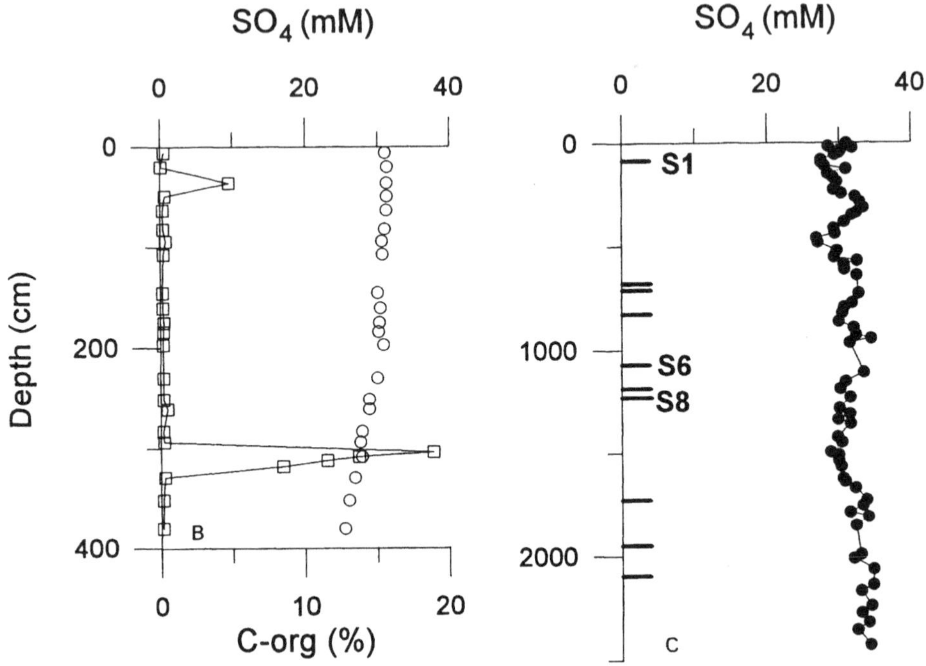

Figure 1.(a) (Previous page). Organic carbon solid phase, and porewater sulphate profiles versus depth for core MD10 from MAP; sulphate profile is shown for expanded section only, but profile is linear from 0-34 m, R^2=0.9 (after De Lange, 1992); (b). (Above). Sedimentary organic carbon and porewater sulphate profiles versus depth for core T83-30 from EM (after Ten Haven et al., 1987); (c) (Above). Porewater sulphate profile of core KC03 from EM; the slight increase with depth relates to a salinity increase; NB: cores KC01 and KC03 have similar sediments and sedimentation rates.

sedimentary setting allows for the in-situ comparison under natural conditions of decomposition rates under oxic and suboxic conditions of a naturally 'aged' (Middelburg, 1989) organic fraction. It must be noted that it is mainly this aged or pseudo-refractory fraction that is likely to be buried permanently.

The occurrence of the oxidation front phenomenon is clearly seen in EM and MAP sediments from Mn and Fe porewater and solid phase data (e.g. MAP: Colley et al., 1984; Wilson et al., 1985; Thomson et al., 1987; De Lange et al., 1987; De Lange, 1992; Thomson et al., 1993; EM: De Lange et al., 1989; Pruysers et al., 1993). In the case of MAP, turbidites each of homogeneous composition assure the direct comparison of initially identical sediment under oxic and under suboxic diagenetic conditions. However in the case of EM, one may argue that observed changes in sediment composition could be due to possible shifts in input fluxes too. To avoid such discussion, the importance of the 'oxygen limitation' will be illustrated with the well-controlled example from the MAP:

As stated before, we are certain that the composition was similar throughout each turbidite, at the start of the 'oxidation experiment' (De Lange et al., 1987). Therefore, any change that is observed must be due to postdepositional diagenetic changes. If the f-turbidite is taken as our example (age of deposition approx. 127 ky; burn-down time approx. 8 ky; Weaver and Rothwell, 1987; Buckley and Cranston, 1988; Colley et al., 1989), we observe a change in organic carbon content from 1% in the lower unoxidized part to 0.2% in the upper oxidized part of the same turbidite (Fig. 1a; De Lange, 1992). Although the total Hydrolysable Amino Acids (HAA) content also decreases to 20% of its initial (pre-'burn-down') value, conspicuous differences in the individual HAA proportions can be observed (Table 2). The unoxidized HAA composition closely resembles that observed in higher productivity sediments (e.g. Müller, 1975; Burdige and Martens, 1988) and marine plankton (Ittekkot et al., 1984; Müller et al., 1986) which are characterized by a predominance of acidic and neutral protein amino acids (e.g. AASP, GLU, GLY and ALA) and by relatively low proportions of the non-protein amino acids ß-alanine and γ-aminobutyric acid (Table 2; Appendix 1). The oxidized HAA composition on the other hand, is characterized by a relative enrichment of the non-protein amino acids which is a typical feature of pelagic sediments that are organic limiting (Schroeder, 1974; Whelan, 1977; Müller et al., 1988). Note also the similarity in degrees of decomposition of bulk organic carbon, total HAA, amino sugars, and alkenones in the f-turbidite (Tables 2, 3 and Appendix 1).

Coincident with the decrease in organic carbon are decreases in the nitrogen content, changes in the C/N ratio, $\delta^{13}C_{org}$, $\delta^{15}N_{org}$, but not in the total amino acid / amino sugar ratio (AA/AS; see Appendix 1), and the paleo temperature proxy U^{K}_{37} (Table 3). The latter proxy is the ratio of two chemically similar isomers. The stability of this proxy despite nearly 80% decomposition of its two isomers (Prahl et al., 1989) is not unexpected, and provides confidence for its use in ancient sediments.

Another example illustrating the oxygen-limited decomposition of organic matter is found in on-land sapropelic sections in Sicily (Fig. 2; Van Os et al., 1993a). Subaerial weathering of organic matter has caused a decrease in C_{org} and a concomitant shift of approximately -1.5‰ in the $\delta^{13}C_{org}$. Note that this shift is similar in direction and magnitude to that observed for the oxic decomposition of organic matter in the f-turbidite (Table 3).

The C/N ratio, $\delta^{15}N_{org}$, and the $\delta^{13}C_{org}$ are often used to establish the marine versus terrestrial contribution in marine sediments (e.g. Degens, 1969; Stevenson and Cheng, 1972; Peters et al., 1978; Hedges and Mann, 1979; Sweeney and Kaplan, 1980; Emerson et al., 1987; Calvert and Fontugne, 1988; Owens and Law, 1989; Fontugne and Calvert, 1992,

Table 2. Amino acid compostion of oxidized and unoxidized f-turbidite sediments, MAP (in nmol/g).

Amino Acid	Oxidized	Unoxidized
ASP	270	960
THR	70	310
SER	70	210
GLU	140	580
GLY	240	1100
ALA	150	750
VAL	100	580
MET+	11	70
aILE	4	33
ILE	28	310
LEU	45	450
TYR	10	100
PHE	25	260
βALA	210	290
γABA	255	270
HIS	7	50
ORN	23	180
LYS	66	620
ARG	100	510

For details and representation in mol %, see Appendix 1.

Calvert et al., 1992; Müller et al., this volume). If we use the relationship found between Σ n-alkanes and $\delta^{13}C_{org}$ for the sediments from the McKenzie river and the MAP f-turbidite (Prahl and Muehlhausen, 1989; Prahl et al., in prep.), it is possible to attribute endmember values to $\delta^{13}C_{org}$, and to calculate the relative marine and terrestrial contributions for sediments of the f-turbidite (Table 4). This shows that almost all of the marine organic matter is removed by oxic degradation, whereas less than half of the terrestrial organic matter is decomposed. This is in

TABLE 3. Carbonate content, and average organic composition of oxidized and un-oxidized samples from the f-turbidite, MAP. Average of f-turbidite from four different cores up to 200 km apart.

	$CaCO_3$ %	C_{org} %	N_{tot} %	C/N M/M	$\delta^{13}C_{org}$	$\delta^{15}N$*	alk.*	U^K_{37}*	SST (°C)*	n-A#	HAA *
Ox	46	0.20	0.037	6.3	-22.5	-1.7	1090	0.72	19.9	348	1830
Un-Ox	50	1.00	0.104	11.2	-20.3	+3.0	9180	0.70	19.5	96	7630

$\delta^{15}N$ data from Scholten (pers. comm.); *: average of 2 cores; #: Σ C_{25-31} in μg/gC, dominantly plantwax derived hydrocarbons; alk.: total alkenones in ng/g dry; HAA: total Hydrolyzable Amino Acids in nmol/g.

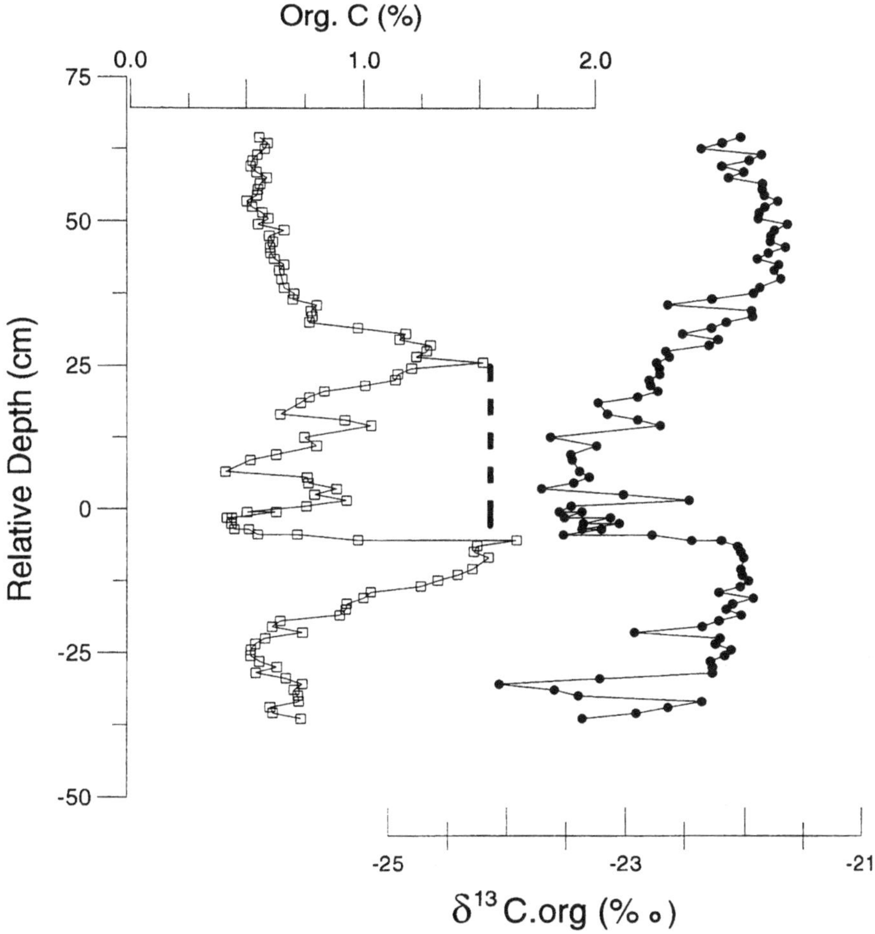

Figure 2. Organic carbon and $\delta^{13}C_{org}$ for Punta Piccola sapropel 12 (approx. 3 Ma BP; after Van Os et al., 1993c); dashed line indicates minimal initial C_{org} content.

line with earlier papers reporting that marine organic matter is easily digestible, whereas terrestrial organic matter is rather resistant (e.g. Goldhaber and Kaplan, 1974; Lyons and Gaudette, 1979). Such preferential decomposition of organic matter has great impact on the interpretation of the sedimentary record. Note that on the basis of Tables 3 and 4 it is possible to estimate the average marine and terrestrial $\delta^{15}N_{org}$ component (being +1.7 and -4.9‰ respectively), for the area and time frame considered.

In conclusion, it appears that the origin and diagenetic history of sedimentary organic matter may result in the preferential preservation of some compounds. As a consequence, palaeo proxies related to organic matter may be altered dramatically upon postdepositional diagenesis. Therefore, great care has to be taken in their interpretation. The use of two or more independent proxies is strongly recommended.

Table 4. Estimates of amounts of marine and terrestrial organic matter that have been oxidized in the f-turbidite, MAP.

	$\%C_{org}$ (total)	$\%C_{org}$ (terrigenous)	$\%C_{org}$ (marine)
Absolute concentration	0.80	0.05	0.75
Relative Percentage Oxidized	80	40	90

2. VARIATIONS IN THE SEDIMENTARY BA CONTENT (UN)RELATED TO PALAEOPRODUCTIVITY

Barite is a common mineral in marine sediments, where its occurrence is often associated with organic matter (Figs. 3,4; and see Goldberg and Arrhenius, 1958; Church, 1979; Brumsack, 1986; Dymond et al., 1992). Therefore, the distribution and relative concentration levels of Ba have often been linked to the intensity of biological productivity in the surface waters (e.g. Goldberg and Arrhenius, 1958; Dehairs et al., 1980; 1987; Schmitz, 1987; Bishop, 1988; Calvert, 1990). Consequently, Ba has been suggested as a possible palaeo productivity proxy (e.g. Dymond, 1986; Schmitz, 1987; Jumars et al., 1989; Calvert, 1990; Von Breymann et al., 1992; Dymond et al., 1992; Shimmield et al., this volume).

However, some studies have shown that great care must be taken in using Ba as a palaeo productivity indicator, as it is subject to diagenetic mobilization in anoxic environments (e.g. Church and Wolgemuth, 1972; Brumsack, 1989; Van Os et al., 1991). In addition, the accumulation rate of barite-Ba in sediments depends not only on the surface water productivity, but also on such factors as the water column dissolved Ba concentration, the type of substrate, and the depth of deposition. Apart from the barite-Ba, aluminium-silicate-Ba and carbonate-bound-Ba may also contribute significantly to the sedimentary Ba content, especially in nearshore environments. For all of these reasons, some authors have excluded the nearshore area for the use of Ba as a productivity proxy (Dymond et al., 1992; Von Breymann et al., 1992).

Factors controlling the barite-Ba rain rate

Settling fluxes of Ba and C_{org} are strongly related in settling particles (Fig. 3; Dymond et al., 1992). This, together with the Ba/C_{org} relation with depth (see below) clearly demonstrates that Ba fluxes are related to primary productivity in the upper ocean (cf. Dymond et al., 1992). These authors also suggest that the barite flux increases when particles descend through a watermass with enhanced Ba content. This is a logical consequence from the control of the barite solubility by its solubility product, i.e. by the concentrations of sulphate and Ba in the water. The sulphate concentration in the open ocean water is thought to have varied only

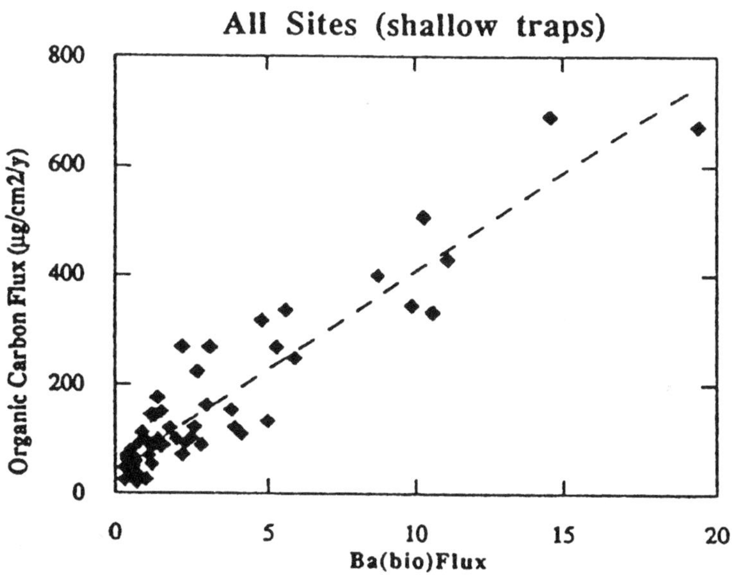

Figure 3. Ba flux versus Organic carbon flux from 10 sediment trap sites for depths between 500 - 1500 m level (for details see Dymond et al., 1992).

slightly due to small variations in salinity over the last million years. A much different situation may exist in semi-enclosed or silled basins where large fluctuations in salinity and sulphate concentration may have occurred due to climatic variations or due to past changes of sealevel. As a consequence, conditions for barite precipitation may have changed considerably independent of primary productivity. In addition, the transition from fresh to marine conditions may result in a Ba-enriched transitional interval in sediment (e.g. Suess, 1982).

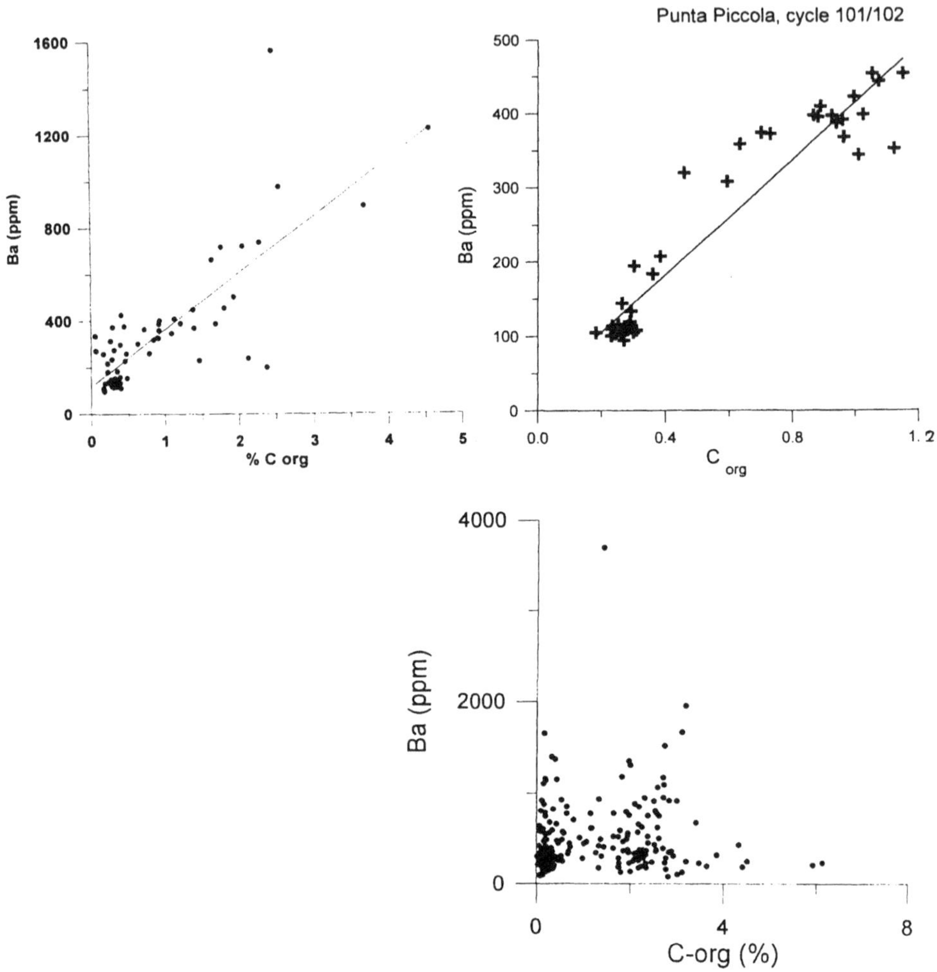

Figure 4. a) Ba versus organic carbon for sediments from a shallow site in the EM (600 m waterdepth; after Pruysers et al., 1991; $R^2 = 0.65$, n = 64); b) Ba versus organic carbon for sediments from a shallow site (approx. 600 m waterdepth; after Van Os et al., 1993c; $R^2 = 0.90$, n = 42); c) Ba versus organic carbon for sediments from a deep site in the EM (3000 m waterdepth; Van Santvoort and De Lange, in prep.). No plot was made for the Van Os et al., 1991 deep cores; however, the Ba/organic carbon plot for these data is similar to that in Fig. 4c).

The Ba concentration is known to increase with the age of a watermass (e.g. Broecker and Peng, 1982). A change in circulation pattern will, therefore, cause changes in the barite rain rate that are unrelated to changes in primary productivity. For example, a more sluggish circulation would result in higher concentrations of Ba in the seawater, and this would promote higher Ba-barite rain rates. A more sluggish circulation will also lead to lower oxygen concentrations in the bottom water, thereby enhancing the area of barite mobilization within the underlying sediments (see Section on diagenetic mobilization of Ba). In addition, the Ba rain rate, and consequently its distribution in sediments, appears to have a strong depth dependency (Von Breymann et al., 1992; Dymond et al., 1992). Reported values for the C_{org}/Ba ratio in present-day suspended matter range from around 200 for near surface waters to 20 at 4000 m (Broecker and Peng, 1982; Dymond et al., 1992). Such a shift follows directly from the proposed mechanism for the formation of barite, namely in microenvironments (within or related to siliceous micro-organisms) by decomposition of organic matter (Bishop, 1988). As a result, relatively low Ba rain rates occur at shallow sites, even at high productivities (e.g. Calvert and Price, 1983; Von Breymann et al., 1992).

A contrasting depth dependency seems to occur in EM sediments: in relatively shallow cores (600 m waterdepth, Pleistocene sediments; Pruysers et al., 1991; Pliocene sediments, Van Os et al., 1993c) a good correlation exists between the Ba and the organic carbon contents, whereas in deep cores (3000 m; Van Os et al., 1991; Van Santvoort and De Lange, in prep.) there is no correlation or even a negative correlation (Fig. 4). This deviation from open ocean sediments is thought to be due to differences in oxygenation between deep and shallow EM bottom waters and related differences in post-depositional diagenesis in the sediments (see below).

Diagenetic mobilization of Ba within sediments

Regardless of its origin, Ba may become mobile under certain diagenetic conditions in the sediment (e.g. Church and Wolgemuth, 1972; Brumsack, 1989; Van Os et al., 1991). Mobilization of Ba in marine sediments is governed by the porewater sulphate and Ba concentrations. In anoxic sediments, the porewater sulphate concentration may decrease to low levels, thereby enhancing the dissolution of barite (Brumsack and Gieskes, 1983; Trefry and Klinkhammer, 1988). The mobilized Ba will diffuse away and be reprecipitated at higher levels of sulphate as a 'diagenetic barite front'. The steeper the sulphate porewater profile versus depth, the sharper the diagenetic front will be.

Barium is often associated with Mn, so that upon the formation of a Mn-oxide diagenetic front (see Mn section) co-precipitation of Ba may also occur (Goldberg and Arrhenius, 1958; Brumsack, 1986; Van Os et al., 1991; Dymond et al., 1992). In the EM sites, these Ba/MnO_2 fronts are mostly much lower than the barite-Ba fronts.

The sediments of the EM form an excellent example to illustrate that present-day porewater conditions are not conclusive at all for the possible occurrence of sulphate reduction, hence diagenetic mobilization of barite, in the past. At present, EM sediments are suboxic with no detectable sulphate reduction until a depth of at least 37 m (De Lange et al., 1991; and see Fig. 1). In addition, the higher salinity of the eastern Mediterranean compared to average seawater (S = 38 vs 35) would enhance the preservation of a possible barite productivity signal. Furthermore, distinct periods of presumably enhanced productivity interbedded with periods of low productivity have been reported to occur as a result of monsoonally-driven climatic fluctuations (e.g. Hilgen, 1991). Finally, present-day EM productivity is reported to be low compared to open ocean systems (e.g. Murdoch and Onuf, 1974).

All of these factors suggest that the EM is an excellent environment for the preservation of Ba productivity signals. However, several organic-rich intervals do not have simultaneous Ba enrichments (e.g. Fig. 5; Van Os et al., 1991; 1993b; Van Santvoort and De Lange, -in prep.-). It is mainly in the deep sediments (water depth approx. 3000 m) that such deviations occur. In a similar high resolution study for relatively shallow sediments (water depth approx. 600 m), however, a good correlation was found between organic carbon and Ba contents in Pleistocene and Pliocene sediments (Figs. 4 and 6; Pruysers et al., 1991, R^2=0.65, n=64; and Van Os et al., 1993c, R^2=0.89, n=42). Sediment and organic C accumulation rates were similar in these studies for most sapropel intervals (Pruysers et al., 1991 and Van Os et al., 1991; Van Os et al., 1993c and Van Santvoort and De Lange, in prep.; for details see Appendix 2). Mobilization of Ba in the deep sediments is probably due to higher sulphate reduction rates leading to a low porewater sulphate content. This difference between shallow and deep sediments probably originates from low (or no) oxygen in the deep bottom waters, as sediment and organic carbon accumulation rates were similar. Briefly, during and shortly after deposition of the organic-rich sapropel intervals in the deep parts of the EM, sulphate reduction occurred in the sediments, resulting in pyrite formation and barite mobilization in these intervals. Subsequently, a barite precipitation front forms above and below the sapropel where the up- and downward diffusing Ba meets the suboxic high-sulphate porewater (see Fig. 5; and see Brumsack, 1986; Van Os et al., 1991).

After sapropel formation, when normal conditions with low organic carbon accumulation rates have been re-established, sulphate reduction ceases, and an oxidation front develops that slowly 'burns' its way down into the upper part of the sapropel (see section on Mn). The oxidation front not only decomposes labile organic matter, but also oxidizes pyrite in the sapropel. The oxidation of reduced sulphides will cause an increase in the porewater sulphate concentration, leading to supersaturation of the porewater with respect to barite (e.g. Goldhaber and Kaplan, 1974; Lord and Church, 1983; Van Os et al., 1991). As a consequence, barite precipitation will occur near the upper boundary of the un-oxidized sapropel sediments (Fig. 5).

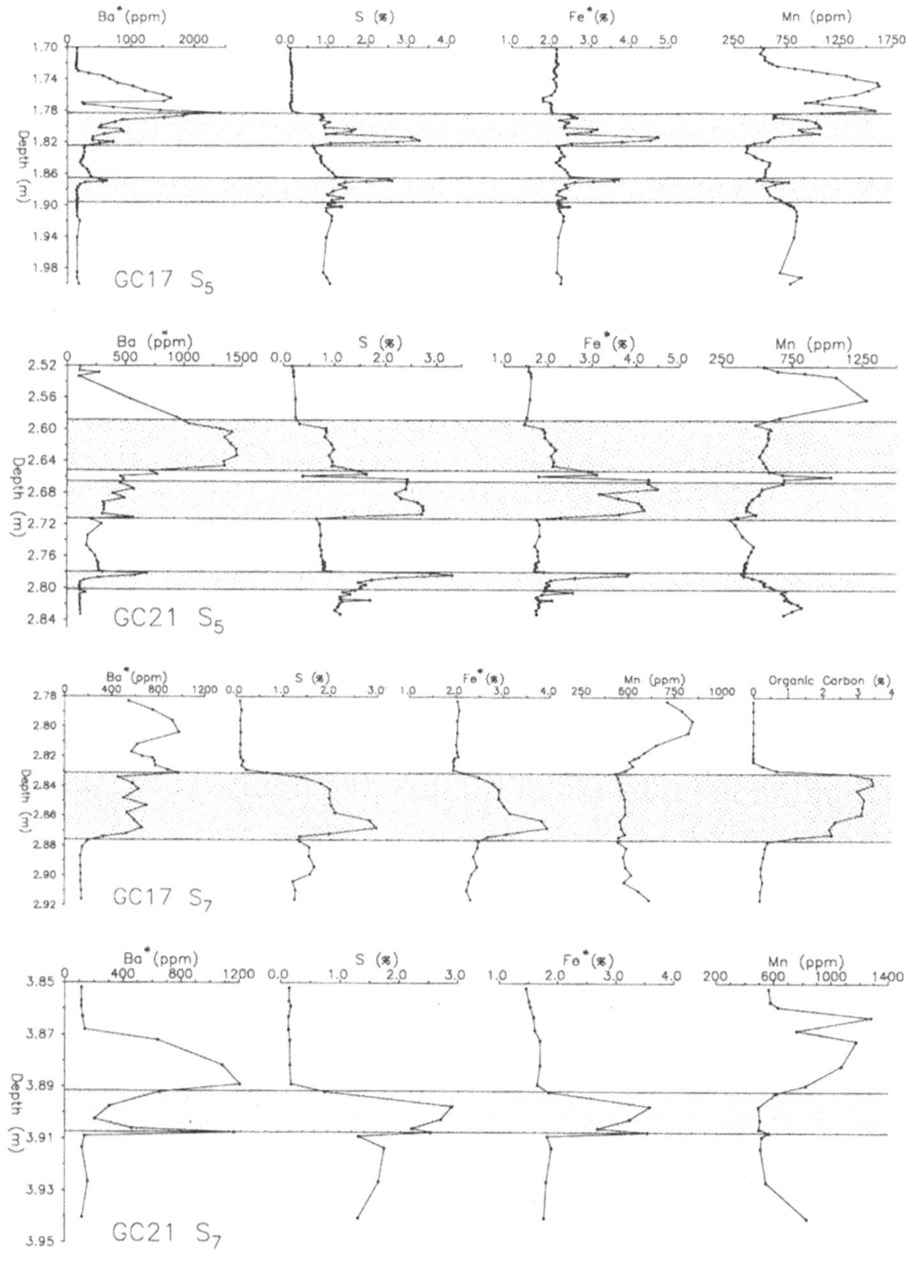
Ba* (ppm)
S (%)
Fe* (%)
Mn (ppm)
Depth (m)
GC17 S5
Ba (ppm*)
S (%)
Fe* (%)
Mn (ppm)
Depth (m)
GC21 S5
Ba* (ppm)
S (%)
Fe* (%)
Mn (ppm)
Organic Carbon (%)
Depth (m)
GC17 S7
Ba* (ppm)
S (%)
Fe* (%)
Mn (ppm)
Depth (m)
GC21 S7

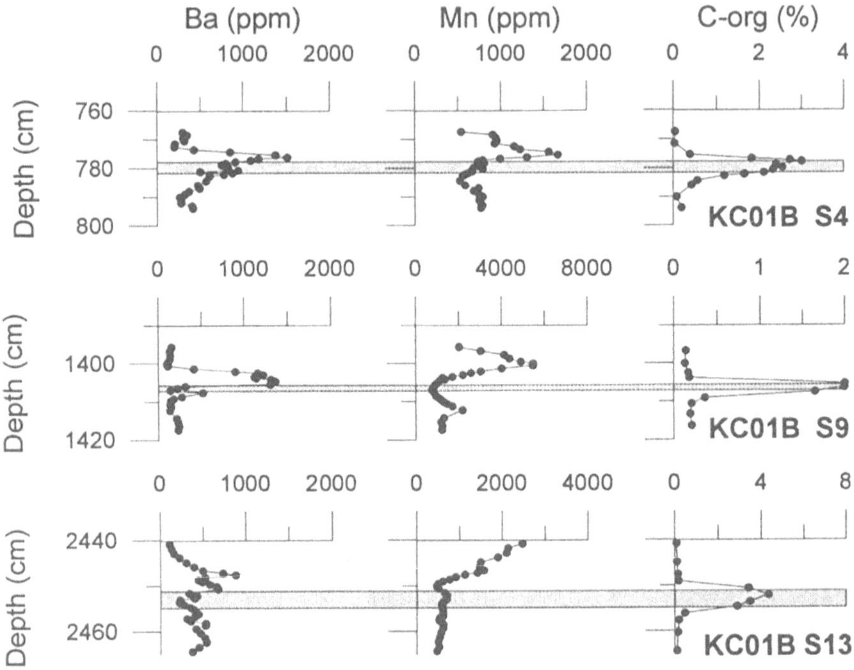

Figure 5 (preceding page and above). Selected Ba, C_{org}, S, Fe and Mn profiles versus depth for some sapropel intervals from deep sited cores in the EM; a) cores GC17, GC21; after Van Os et al. (1991); b) core KC01, Van Santvoort and De Lange (in prep.).

After burial of the sapropels, diagenetic conditions apparently do not change back to sulphate reducing, because barium enrichments are found just above and/or below sapropels S_1 to 'S_{14}' (approx. 1 Ma) (Fig. 5; Van Os et al., 1991; Van Santvoort and De Lange, in prep.). This could be due to the lack of sufficient organic matter that is labile under suboxic to reducing conditions.

It is possible that the integrated Ba peak in the sapropel and adjacent sediment intervals is still representative of the original productivity signal, but obviously more research has to be done before this statement can be confirmed. Moreover, such diagenetic Ba spikes are clear (palaeo) indicators of dramatically changed redox conditions in the sediments.

Finally, although only 30% of the Ba rain rate seems to be preserved in the sediments

(Dymond et al., 1992), this is still much more than the amount of organic carbon (< 1%; e.g. Emerson and Hedges, 1988; Jumars et al., 1989), or the amount of silica that is usually preserved. Clearly, although the degree of preservation of Ba is not constant, it is still one of the best geochemical proxies for productivity currently available. A continued search for more precise constraints for its utilization is, therefore, warranted.

3. (DIAGENETIC) MN-SPIKES AT PALAEO-PRODUCTIVITY CHANGES

Mn is known to be mobile under suboxic conditions. One of the first comprehensive overviews of the diagenetic sequence involved in the oxic/suboxic decomposition of organic matter was that of Froelich et al. (1979). The possible mechanism for the diagenetic formation of a solid-phase Mn-spike was further elaborated by others (e.g. Burdige and Gieskes, 1983; De Lange et al., 1989; Pruysers et al., 1993). It is mainly under non-steady-state conditions that pronounced Mn spikes that can be potentially preserved in the sedimentary record may form. Reported examples of non-steady state conditions are: 1) emplacement of an organic-rich turbidite in an organic-poor environment; 2) intervals with high organic carbon accumulation rate interbedded with intervals of low organic carbon accumulation rate; and 3) a suboxic Mn^{2+}-rich watercolumn suddenly being re-oxygenated. The latter mechanism was used to explain the Mn spike found at the glacial / interglacial transition in Pacific sediments (Mangini et al., 1990; 1991; this volume). We will focus here on the diagenetic changes and related redox changes within sediments. A fine example of 1) is the deposition of organic-rich turbidites interbedded with organic-poor pelagic sediments in the MAP (e.g. Wilson et al., 1986; Thomson et al., 1987; Thomson et al., 1993). An oxidation front progressing downwards into the turbidite also leads to the formation of an enhanced level of Mn-oxide just above the transition from oxic to suboxic sediment, i.e. from organic-poor to organic-rich sediment. The Mn spike is often associated with an enrichment of other trace elements, such as Fe, Co, Ni etc. (e.g. Jarvis and Higgs, 1987; Thomson et al., 1993; Van Os et al., 1993b, and refs therein). Although most of the recent information on such oxidation fronts comes from studies of the MAP, Mn spikes are also found in similar settings elsewhere (e.g. Nares Abyssal Plain, De Lange, 1986; N. Pacific, Dean et al., 1989).

2) An oxidation front may not only form by turbidite deposition, but also by other changes in environmental conditions, such as a sudden decrease in the influx rate or accumulation rate of organic C, or a sudden increase in the bottomwater oxygen content (e.g. Wilson et al., 1986). The formation of a Mn spike close to the Holocene/Pleistocene boundary in the N.E. Atlantic was reported to be due to a decrease in organic C flux and accumulation rate at the transition from the Pleistocene to the Holocene (Thomson et al., 1984; Wallace et al., 1988). In contrast, the occurrence of Mn spikes in N. Pacific sediments was not only attributed

to the deposition of turbidites, but also to a subsequent period of increased oxygen content of the bottomwater (Dean et al., 1989). Similar Mn spikes at or around the Pleistocene/Holocene transition have been reported for the eastern equatorial Pacific (Berger et al., 1983; Finney et al., 1988), and near the end of the last Glacial for the EM (De Lange et al., 1989; Pruysers et al., 1993). In all of these cases for mechanism 2), the Mn spike is found at the top of the transition from high to low organic carbon accumulation rates, and is reported to be diagenetic in nature.

Although the origin of the sudden transition in organic carbon content of the sediment is very different (namely turbidite deposition versus sudden environmental changes), the process by which the Mn spikes are formed is similar: under suboxic conditions Mn-oxides are reduced, and Mn^{2+} is transported by diffusion to an interval where redox conditions are more oxidizing. At this point Mn will reprecipitate as a Mn-oxide. A steep redox gradient will result in a sharp Mn spike (e.g. Burdige and Gieskes, 1983; Wilson et al., 1985; De Lange et al., 1989; Pruysers et al., 1993). On the basis of porewater/solid-phase results, Pruysers et al (1993) were able to explain size, shape and location of the Mn-oxide spike by diagenetic mobilization/reprecipitation of Mn as resulting from a sudden change in organic C accumulation rates at the end of the last glacial.

As long as some non-refractory organic matter remains, sedimentary conditions will at some time become suboxic again, and the Mn spike may be remobilized. Consequently, older Mn spikes may have disappeared from the record (e.g. Dean et al., 1989; Pruysers et al., 1991). Such remobilization is bound to take place, as at some depth oxygen and nitrate will be depleted, and decomposition of organic matter will be continued with Mn-oxides as oxidant. Favourable areas for the temporary preservation of Mn-oxide spikes are thought to be found in areas of low organic C accumulation rates such as large parts of the Pacific, but obviously not in those of relatively high accumulation rates such as those found in large parts of the N. Pacific, N.E. Atlantic, and possibly EM.

Finney et al. (1988) state that due to non-steady-state conditions it is possible that Mn spikes are initially preserved, probably as a Mn-carbonate. Upon further burial such Mn-carbonate spikes may remain in the sediments under suboxic conditions. It is not certain that the replacement of a Mn-oxide spike by a Mn-carbonate spike is a common feature for non-steady state situations. In several EM cores with obvious non-steady-state organic carbon accumulation rates, and with high resolution sampling, no Mn spikes are encountered below a depth of about 50 cm (e.g. Fig. 6; Pruysers et al., 1991; 1993). In contrast, in other cores from the same area Mn spikes seem to be preserved (see above; Fig. 5; Van Os et al., 1991). These latter cores were recovered from deep-water sites, whereas the former cores originated from shallow-water sites. In view of the depth dependency for organic C accumulation especially at low rain rates (see under section 1., above), the labile organic C content in sediments *between sapropels* is likely to be lower for the deep sites than

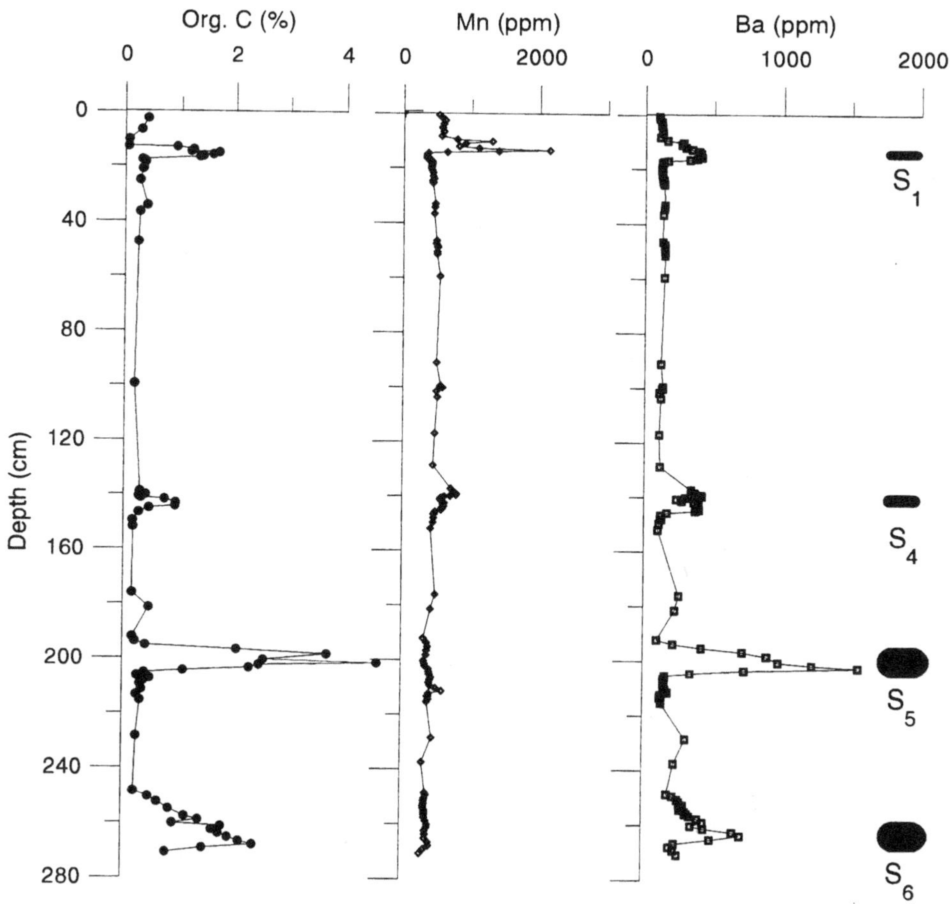

Figure 6. C_{org}, Mn and Ba profiles versus depth for a shallow core in the EM (core 9G1, after Pruysers et al., 1991). Note the disappearance of Mn-spikes initially present above all sapropels that are at a depth of 50 cm or more in this core.

for the shallow sites. Consequently, the preservation of MnO_2 spikes is better in the deep than in the shallow sites for sediments between sapropels having similar organic C rain rates, and sediment accumulation rates. Under conditions of continued diagenesis, it is possible that such MnO_2 spikes are later transformed into $MnCO_3$ or else are degraded. In addition, the site of the palaeo Mn spike may often be recognized by spikes of other trace elements (e.g. Pruysers et al., 1991; Van Os et al., 1993b).

Such lower labile organic carbon flux for sediments between sapropels seems to be in conflict with the higher labile organic carbon fluxes during sapropels (see under 2. 'Variations in sedimentary Ba content') both for the deep sites as compared to the shallow sites. However, hydrographic and biologic conditions (such as deep-water residence time, deep-water oxygen content, primary productivity) that were extremely different between sapropels and intersapropel periods may explain this paradox. It must be noted that alternative explanations are possible. However, an extensive discussion on this subject is beyond the scope of this paper.

In conclusion, it seems that a Mn-oxide spike resulting from non-steady state conditions will often disappear from the sedimentary record, but may sometimes be preserved as an oxide under low-labile-organic C conditions or as Mn-carbonate in suboxic conditions. In addition, the past presence of a Mn-oxide spike can sometimes be recognized by the presence of some initially coprecipitated trace metals that have subsequently been preserved in the sedimentary record.

4. DISSOLUTION / PRECIPITATION OF CARBONATES

The bulk carbonate content, its isotopic and trace element characteristics, and the relative concentration of some microfossils may be affected by early diagenetic processes. Oxic decomposition of organic carbon leads to dissolution of carbonates, whereas anoxic decomposition and oxidation of upward moving methane may lead to the precipitation of authigenic carbonate with a dramatically different isotope (e.g. Schneider et al., 1992) (and possibly trace element) signature. The selective dissolution of carbonate related to the oxic degradation of organic matter may change the relative distribution of microfossils.

Most of the organic matter that rains down to the sediment is decomposed at the interface or within the top few cms of the sediment (e.g. Emerson et al., 1980; Emerson and Bender, 1981). The CO_2 that is released dissolves some of the sedimentary carbonate via:

$138\ O_2 + C_{106}H_{263}O_{110}N_{16}P + 18\ HCO_3^- \rightarrow 124\ CO_2 + 16\ NO_3^- + HPO_4^{2-} + 140\ H_2O$ and $CO_2 + H_2O + CaCO_3 \rightarrow Ca^{2+} + 2\ HCO_3^-$. Emerson and Archer (1990) suggest that 40% or more of the particulate $CaCO_3$ rain at the calcite saturation horizon may potentially dissolve in response to oxic organic matter degradation. Similar diagenetic carbonate dissolution is observed during the 'burn down' of oxidation fronts (see Table 3). In the case of the f-turbidite, a 4% decrease in carbonate content is observed with a decrease of 0.8% organic carbon.

Diagenetic dissolution of carbonates may lead to the differential dissolution of microfossils. An often used 'dissolution index' is that of the ratio of benthic versus planktonic foraminifers, the former being more resistant to dissolution than the latter (e.g. Parker and Berger, 1971). However, the same B/P dissolution proxy is also used as a productivity proxy (e.g. see Mix, 1989). We will not discuss here all the other dissolution proxies that are more or less based on the same principles; instead we will focus on diagenetic aspects. Firstly, we will discuss the precipitation, recrystallization, and distribution of carbonate minerals in a recent semi-euxinic basin, Kau Bay. Secondly, we present evidence for the differential dissolution of carbonate microfossils, on the basis of laboratory experiments.

I. Kau Bay (KB)

Geochemical results from KB porewaters and sediments indicate the precipitation of dolomite in distinct horizons, namely at 3.5 and 7.5 m (see Middelburg et al., 1990). In addition, these results indicate significant precipitation, dissolution and re-equilibration of carbonates, which result in changes in the sedimentary record of % carbonate, $\delta^{13}C$-carb. (both for bulk and for pteropods), $\delta^{18}O$ -carb. and in the AMS- ^{14}C-carb. ages (Fig. 7). Although some of these individual changes have been reported previously, the KBay case demonstrates the full scale of such changes in a well-documented way, including the sediment and porewater data:

Below a depth of 5.5 m, there is a decrease in carbonate content and in the $\delta^{13}C$ of aragonite (pteropods) and bulk carbonate (Fig. 7a,b). The $\delta^{13}C$ values approach that observed for authigenic dolomite (-10‰) observed in this core (Middelburg et al., 1990). In addition, the ^{14}C-pteropod value remains constant with depth at 10 ky, despite an inferred constant sedimentation rate (Fig. 7c). The bulk carbonate $\delta^{18}O$ (-4.5‰) is well below the values for seawater equilibration (-3‰) above 3.5/4 m depth, whereas below this depth it is in equilibrium with the porewater at the ambient temperature of 28°C. This clearly indicates, that recrystallization begins at a depth of 3.5/4 m. This observation is confirmed by the Ion Activity Product for $CaCO_3$ which indicates undersaturation with respect to aragonite below 3.5 m. On the basis of these observations we may conclude that:

1. above a depth of 3.5 / 4 m the porewater is (super)saturated with respect to calcite and aragonite, and $CaCO_3$ phases as well as dolomite are being formed (e.g. see Middelburg et al., 1990); the bulk carbonate, which is a mixture of authigenic and biogenic components, still reproduces the correct ^{14}C ages, and the primary $\delta^{13}C$-pteropod signal is probably conserved.
2. at depths from 3.5/4 to 5.5 m, the porewater is undersaturated with respect to aragonite, and the $\delta^{18}O$ of bulk carbonate is in equilibrium with the porewater; i.e. recrystallization occurs.
3. below 5.5 m depth, there is carbonate dissolution and recrystallization leading to a resetting of the $\delta^{13}C$ of bulk carbonate and of pteropods, and of the ^{14}C age of pteropods.

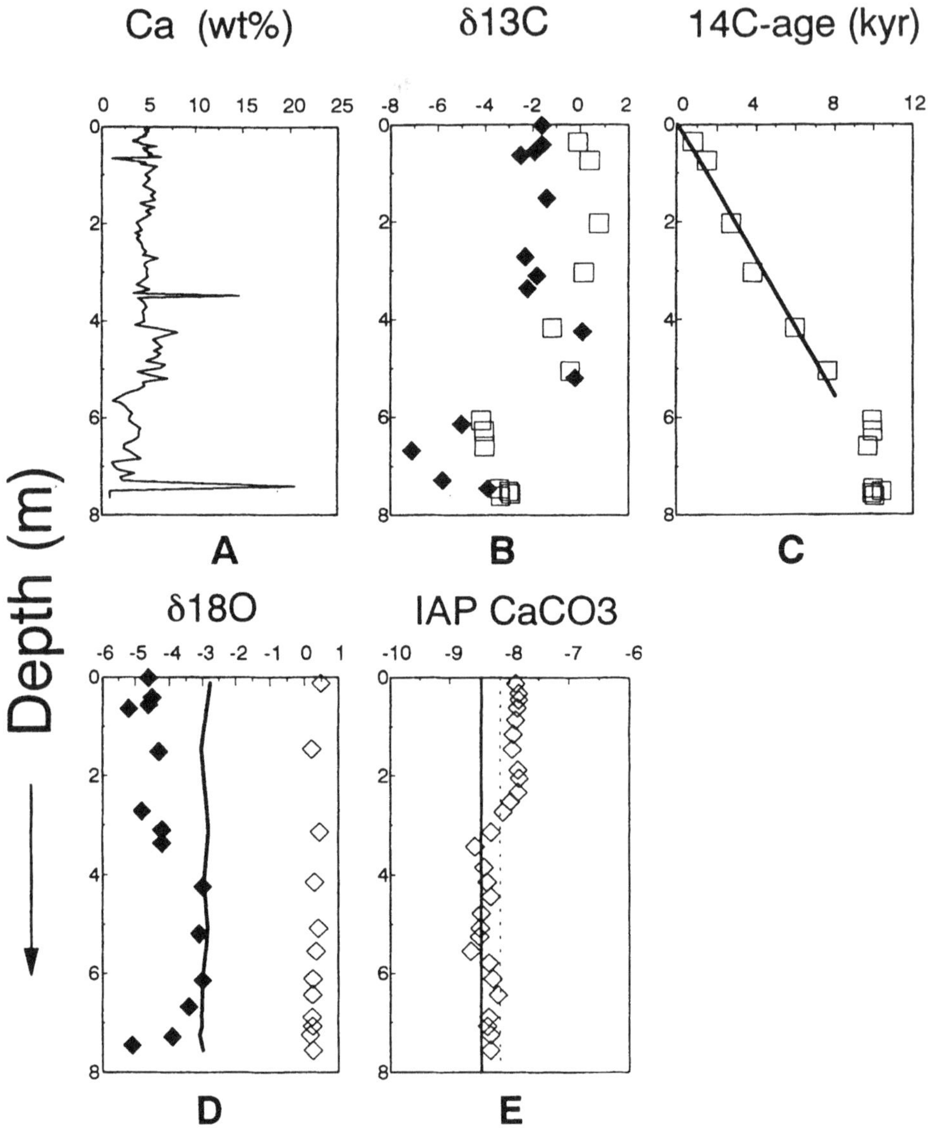

Figure 7. Concentration versus depth in core K4P3 from Kau Bay for: a) % Ca, b) $\delta^{13}C_{carbonate}$, c) ^{14}C age, d) $\delta^{18}O$, and e) the Ion Activity Product for $CaCO_3$ versus depth. Filled diamonds: bulk carbonate; open diamonds: porewater; open square: pteropods. The solid line in d) represents $\delta^{18}O$ in equilibrium with seawater at ambient temperature (28 °C). The solid and dashed lines in e) represent calcite and aragonite saturation, respectively.

II. Selective dissolution experiment

An EM pelagic, and sapropel S1 sample respectively were subjected to 12 consecutive dissolution steps using a 2 N NH_4Cl solution at pH 9.5 - 9.0. Each step was done for an 8-12 hour period in tightly-closed centrifuge tubes; after each step, a small subsample was taken for fossil examination. All nannofossils, studied by microscope, demonstrated some dissolution at the end of the experiment. One particular nanno, *Syracosphaera* sp.1, seemed much more severely dissolved than all others (Fig. 8). It is remarkable in this respect that this nannofossil is the only species that is typically missing from all EM sapropels, but that does occur in the sediments between sapropels (Violanti et al., 1991). We do not suggest at this point that its absence is due to diagenesis, but merely indicate that one has to be careful in interpreting the absence or presence of (nanno)fossils. It is clear that long-term in-situ differential dissolution may be much more selective than our simple experiment.

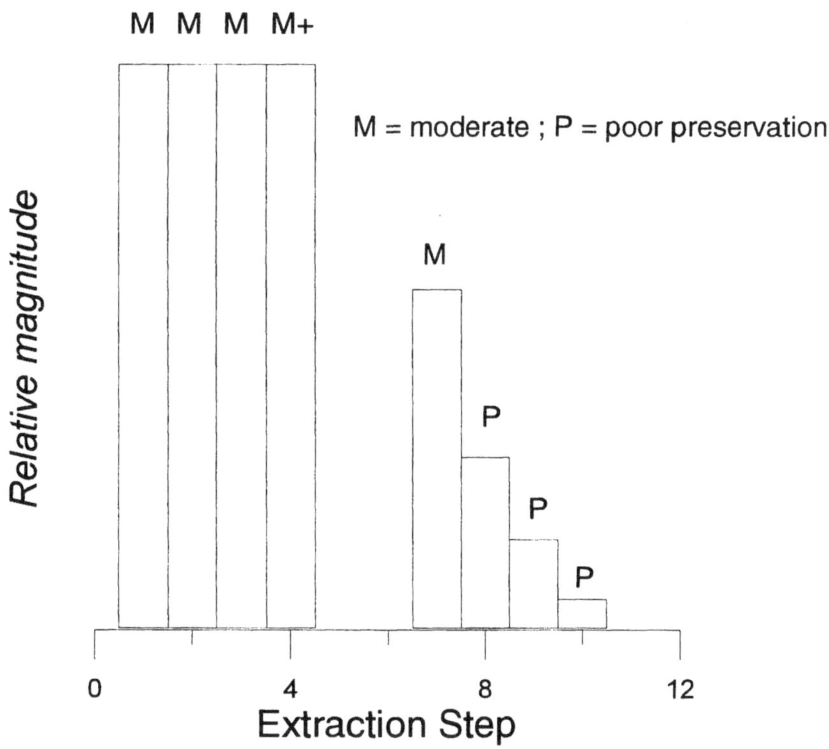

Figure 8. Visually observed preservation of the nannoplankton *Syracosphaera* sp.1 during sequential extration of inter-sapropelic Mediterrranean sediments; vertical axis: arbitrary units showing relative amounts 'surviving' dissolution, horizontal axis: consecutive extraction steps of 2N NH_4Cl; steps 1-5 at pH 9.5, and steps 6-10 at pH 9.0. Steps 5 and 6 were not analyzed.

CONCLUSIONS

Although a number of potentially useful palaeo tracers have been reported, it has become evident that some limitations exist regarding their application. Early diagenetic processes may considerably alter primary signals. Origin and diagenetic history may vary for different environments, and may therefore lead to preferential preservation, decomposition or relocation of particular tracers. Consequently, one must be very cautious when applying a palaeo-tracer that is sensitive to diagenetic alterations. Some of these alterations include:

1. *Diagenetic changes in some organic proxies (Organic C, C/N, $\delta^{13}C$, $\delta^{15}N$, U^{K}_{37})*

The origin and diagenetic history of sedimentary organic matter may result in the preferential preservation of some compounds. In addition, the postdepositional decomposition of organic matter may dramatically alter its composition, i.e. including related organic palaeo proxies. *In situ* comparison of decomposition rates for a naturally 'aged' or 'vulcanized' organic fraction of distal turbidites, demonstrates that this fraction can rapidly be decomposed under oxic sedimentary conditions, but not significantly under suboxic / reducing conditions. During the oxic decomposition, potential palaeo proxies such as C/N, $\delta^{13}C_{org}$, and $\delta^{15}N_{org}$ all change considerably, whereas the palaeotemperature indicator U^{K}_{37} remains unchanged.

2. *Variations in the sedimentary Ba content (un)related to palaeoproductivity*

Not only ocean productivity, but also factors such as seawater sulphate content, age of watermass, and waterdepth, seem to determine the Ba rain rates, and consequently its distribution in the underlying sediments. In addition, anoxic mobilization and hydrothermal activity are known to produce relatively high levels of Ba, unrelated to productivity. The influence of hydrothermal activity is probably limited to ocean ridge sediments. Anoxic mobilization results in the relocation of high levels of Ba associated with organic rich, sulphate-reducing sediments to intervals with low labile organic matter content, and higher sulphate porewater concentrations.

3. *(Diagenetic) Mn-spikes at palaeo-productivity changes*

Although the occurrence of Mn spikes has been related to changes in palaeoproductivity, and to sudden changes of the ocean water from suboxic to oxic conditions, these observations can as well be explained by the early diagenetic mobilization of Mn from within sediments, deposited under non-steady state conditions.

A Mn spike resulting from non-steady-state conditions often disappears from the sedimentary record, but may be preserved as MnO_2 under low-labile-organic carbon conditions or as Mn-carbonate under suboxic conditions. Furthermore, (b)leached Mn spikes may sometimes be recognized by the preserved presence of initially coprecipitated trace metals.

Pertinent examples from non-steady-state areas show, that such Mn spikes can be preserved only under well defined conditions.

4. Diagenetic dissolution/precipitation of carbonates

Not only the preferential dissolution of some biogenic carbonate, but also the precipitation and recrystallization of carbonate (both by early diagenesis) may lead to variations in species-and isotopic composition that are unrelated to 'primary signals'.

ACKNOWLEDGEMENTS

G. Nobbe, M. Klein-Tank, H. de Waard, M. and E. Koning are acknowledged for doing most of the analyses. We thank our colleagues on the Marion Dufresne MARFLUX-1 cruise to the eastern Mediterranean for their cordial cooperation. This project was partially supported by the Netherlands Foundation for Sea Research (SOZ), and by the MARFLUX project funded by the EC MAST programme 0022-C. We acknowledge two anonymous reviewers and T. Pedersen for their thorough and constructive reviews that considerably improved the manuscript. P.J. Müller acknowledges support from the Bundesminister für Forschung and Technologie (BMFT), Bonn.

REFERENCES

ALTABET MA (1992) The use of nitrogen isotopic ratio for reconstruction of past changes in the marine nitrogen cycle and its influence on palaeoproductivity. In: Fourth International Conference on Palaeoceanography ICP IV, p. 321.

BARNOLA JM, RAYMOND D, KOROTKEVICH KS, LORIUS C (1987) Vostok ice core provides 160,000 year record of atmospheric CO_2. *Nature* **329**, 408-413.

BATHMANN U, FISCHER G, MÜLLER PJ, GERDES D (1991) Short-term variations in particulate matter sedimentation off Kapp Norvegia, Weddell Sea, Antarctica: relation to water mass advection, ice cover, plankton biomass and feeding activity. *Polar Biol* **11**, 185-195.

BERGER WH, FINKEL RC, KILLINGLEY JS, MARCHIG V (1983) Glacial-Holocene transition in deep-sea sediments: manganese-spike in the east-equatorial Pacific. *Nature* **303**, 231-233.

BERNER RA (1980). Early Diagenesis, Princeton University Press, 241 pp.

BISHOP JKB (1988) The barite-opal-organic carbon association in oceanic particulate matter. *Nature* **332**, 341-343.

BOYLE EA (1988) Cadmium: Chemical tracer of deepwater palaeoceanography. *Palaeoceanography* **3**, 471-489.

BOYLE EA (1992) Cadmium and $\delta^{13}C$ palaeochemical ocean distributions during the stage 2 glacial maximum. *Ann. Rev. Earth Planet. Sci.* **20**, 245-287.

BOYLE EA, LEA DW (1989) Cd and Ba in planktonic foraminifera from the Eastern Mediterranean: evidence for river outflow and enriched nutrients during sapropel formation. *EOS* **70**, 1134.

BRASSELL SC, EGLINTON G, MARLOWE LT, PFLAUMANN U, SARNTHEIM M. (1986) Molecular stratigraphy: a new tool for climatic assessment Nature **320**, 129-133.
BROECKER WS, PENG T-H (1982) Tracers in the sea. LDGO, Columbia Univ., NY; pp. 690.
BRUMSACK HJ, GIESKES JM(1983) Interstitial water trace element chemistry of laminated sediments from the Gulf of California, Mexico. *Mar. Chem.* **14,** 89-106.
BRUMSACK HJ (1986) The inorganic geochemistry of Cretaceous black shales (DSDP Leg 41) in comparison to modern upwelling sediments from the Gulf of California. In: *North Atlantic Palaeoceanography. Geol Soc. Spec. Pub.* **21,** 447-462.
BRUMSACK HJ (1989) Geochemistry of recent TOC-rich sediments from the Gulf of California and the Black Sea. *Geol. Rundschau* **78/3**, 851-882.
BUCKLEY DE, CRANSTON RE (1988) Early diagenesis in deep sea turbidites: The imprint of palaeo-oxidation zones. *Geochim. Cosmochim. Acta* **52**, 2925-2939.
BURDIGE DJ, GIESKES JM (1983) A pore water/solid phase diagenetic model for manganese in marine sediments. *Am. J. Sci.* **283**, 29-47.
BURDIGE DJ, MARTENS CS (1988) Biogeochemical cycling in an organic-rich coastal marine basin : 10. Role of amino acids in sedimentary carbon and nitrogen cycling. *Geochim. Cosmochim. Acta* **52**, 1571-1584.
CALVERT SE (1983) Geochemistry of Pleistocene sapropels and associated sediments from the Eastern Mediterranean. *Oceanol. Acta* **6,** 255-267.
CALVERT SE, PRICE NB (1983) Geochemistry of the recent sediments of the Namibian shelf. In : *Coastal Upwelling: its sedimentary record*, Suess E, Thiede J (eds) Plenum Press v. **1**, 337-375.
CALVERT SE, FONTUGNE MR (1988) Geochemistry and origin of Mediterranean sapropels. *EOS* **69,** 1234.
CALVERT, SE (1990) Geochemistry and origin of the Holocene sapropel in the Black Sea. In: *Facets of Modern Biogeochemistry*, Ittekkot V, Kempe S, Michaelis W, Spitzy A (eds), Springer Verlag; p. 328-353.
CALVERT SE, NIELSEN B, FONTUGNE MR (1992) Evidence from nitrogen isotope ratios for enhanced productivity during formation of eastern Mediterranean sapropels. *Nature* **359**, 223-225.
CHURCH TM (1979) Marine Barite. In *Marine Minerals* (R.G. Burns ed.). Mineralogical Society of America. Short Course Notes. **6,** 175-209.
CHURCH TM, WOLGEMUTH, K. (1972) Marine barite saturation. *Earth Planet. Sci. Lett.* **15,** 35-44.
CITA MB, VERGNAUD-GRAZZINI C, ROBERT C, CHAMLEY H, CIARANFI N, D'ONOFRIO S (1977) Palaeoclimatic record of a long deep sea core from the eastern Mediterranean. *Quat. Res.* **8,** 205-235.
COLLEY S, THOMSON J, WILSON TRS, HIGGS NC (1984) Post-depositional migration of elements during diagenesis in brown clay and turbidite sequences in the North East Atlantic. *Geochim. Cosmochim. Acta* **48,** 1223-1235.
COLLEY S, THOMSON J, TOOLE J (1989) Uranium relocations and derivation of the quasi-isochrons for a turbidite/pelagic sequence in the Northeast Atlantic. *Geochim. Cosmochim. Acta* **53,** 1223-1234.
DEAN WE, GARDNER JV, HEMPHILL-HALEY E (1989) Changes in redox conditions in deep-sea sediments of the subarctic north Pacific Ocean: Possible evidence for the presence of north Pacific deep water. *Palaeoceanography* **4**, 639-653.
DEGENS ET (1969) Biogeochemistry of stable carbon isotopes. In: *Organic Geochemistry,* Eglinton G, Murphy MTJ (eds), Springer Verlag; pp. 304-329.
DEHAIRS F, CHESSELET R, JEDWAB J (1980) Discrete suspended particles of barite and the barium cycle in the open ocean. *Earth Planet. Sci. Lett.* **49**, 529-550.
DEHAIRS F, LAMBERT CE, CHESSELET R, RISLER N (1987) The biological production of marine suspended barite and the barium cycle in the Western Mediterranean Sea. *Biogeochem.* **4,** 119-139.
DE LANGE GJ (1986) Early diagenetic reactions in interbedded pelagic and turbiditic sediments in the Nares Abyssal Plain (western North Atlantic): Consequences for the

composition of sediment and interstitial water. *Geochim. Cosmochim. Acta* **50**, 2543-2561.
DE LANGE GJ, TEN HAVEN HL (1983) Recent sapropel formation in the eastern Mediterranean. *Nature* **305**, 797-798.
DE LANGE GJ, JARVIS I, KUIJPERS A (1987) Geochemical characteristics and provenance of late Quaternary sediments from the Madeira Abyssal Plain, eastern North Atlantic. In : *Geology and Geochemistry of Abyssal Plains* , Weaver PPE, Thomson J,(eds) *Spec. Pub. Geol. Soc. Lond* **31**, 147-165.
DE LANGE GJ, MIDDELBURG JJ, PRUYSERS PA (1989) Middle and Late Quaternary depositional sequences and cycles in the eastern Mediterranean. *Sedimentology* **36**, 151-158.
DE LANGE GJ, Shipboard Party (1991) Shipboard Report MD69-MARFLUX Expedition in the eastern Mediterranean. pp 101.
DE LANGE GJ (1992) Distribution of exchangeable, fixed, organic and total nitrogen in interbedded turbiditic/pelagic sediments of the Madeira Abyssal Plain, eastern North Atlantic. *Mar. Geol.* **109**, 95-114.
DYMOND J, (1986) Particulate barium fluxes in the oceans:an indicator of new productivity. *EOS* Am. Geophys Un. **56,** 1275
DYMOND J, SUESS E, LYLE M (1992) Barium in deep-sea sediment: geochemical proxy for palaeoproductivity. *Palaeoceanography* **7**, 163-181.
ELDERFIELD H (1991) tracers of ocean palaeo-productivity and palaeochemistry: An introduction. *Palaeoceanography* **5**, 711-717.
EMERSON S R, BENDER ML (1981) Carbon fluxes at the sediment-water interface of the deep sea: calcium carbonate preservation. *J. Mar. Res.* **39**, 139-162.
EMERSON S, JAHNKE R, BENDER M, FROELICH P, KLINKHAMMER G, BOWSER C., SETLOCK G. (1980) Early diagenesis in sediments from the Eastern Equatorial Pacific, I. Pore water nutrient and carbonate results. *Earth Planet. Sci. Lett.* **49**, 57-80.
EMERSON S, STUMP C, GROOTES PM, STUIVER M, FARWELL GW, SCHMIDT FH (1987) Estimates of degradable organic carbon in deep-sea surface sediments from ^{14}C concentrations. *Nature* **329**, 51-53.
EMERSON S, HEDGES JI (1988) Processes controlling the organic carbon content of open ocean sediments. *Palaeoceanography* **3**, 621-634.
EMERSON S, ARCHER D (1990) Calcium carbonate preservation in the ocean. *Phil. Trans. Roy. Soc. London*, Ser. A **331**, 29-40.
FINNEY BP, LYLE MW, HEATH GR (1988) Sedimentation at MANOP site H (eastern equatorial Pacific) over the past 400,000 years: Climatically induced redox variations and their effects on transition metal cycling. *Palaeoceanography* **3**, 169-189.
FONTUGNE MR, CALVERT SE (1992) Late Pleistocene variability of the carbon isotopic composition of organic matter in the eastern Mediterranean: Monitor of changes in carbon sources and atmospheric CO_2 concentrations. *Palaeoceanography* **7**, 1-20.
FROELICH PN, KLINKHAMMER GP, BENDER ML, LUEDTKE NA, HEATH GR, CULLEN D, DAUPHIN P, HAMMOND D, HARTMAN B, MAYNARD V (1979) Early oxidation of organic matter in pelagic sediments of the Eastern Equatorial Atlantic: Suboxic diagenesis. *Geochim. Cosmochim. Acta* **43,** 1075-1090
GEYH MA, KRUMBEIN WE, KUDRASS H-R (1974) Unreliable ^{14}C dating of long-stored deep-sea sediments due to bacterial activity. *Mar. Geol.* **17**, M45-M50.
GOLDBERG ED, ARRHENIUS GOS (1958) Chemistry of Pacific pelagic sediments. *Geochim. Cosmochim. Acta* **13,** 153-212.
GOLDHABER MB, KAPLAN IR (1974) The sulfur cycle. In: Goldberg, E.D. (Editor),*The Sea*, vol.**5**, Wiley, 595-655.
HEDGES J I, MANN DC (1979) The lignin geochemistry of marine sediments from the southern Washington coast. *Geochim. Cosmochim. Acta* **43**, 1809-1818.
HENRICHS SA, REEBURGH WS (1987) Anaerobic mineralization of marine sediment organic matter: rates and the role of anaerobic processes in the ocean carbon economy. *Geomicrobiol. J.* **5**, 191-237.
HENRICHS SA (1992) early diagenesis of organic matter in marine sediments: progress and

perplexity. *Mar. Chem.* **39**, 119-149.
HILGEN F J (1991) Extension of the astronomically calibrated (polarity) time scale to the Miocene / Pliocene boundary. *Earth Planet. Sci. Lett.* **104**, 226-244.
ITTEKOT V, DEUSER WG, DEGENS ET (1984) Seasonality in the fluxes of sugars, amino acids and amino sugars to Deep Ocean. Part I. Sargasso Sea. *Deep Sea Res.* **31**, 1057-1069.
JARVIS I, HIGGS NC (1987) Trace-element mobility during early diagenesis in distal turbidites: Late Quaternary of the Madeira Abyssal Plain, N. Atlantic In: *Geology and Geochemistry of Abyssal Plains* , Weaver PPE, Thomson J (eds) *Spec. Pub. Geol. Soc. Lond* **31**, 179-213.
JASPER JP, HAYES JM (1990) A carbon-isotopic record of CO_2 levels during the late Quaternary. *Nature* **347**, 462-464.
JUMARS PA, ALTENBACH AV, DE LANGE GJ, EMERSON SR, HARGRAVE BT, MULLER PJ, PRAHL FG, REIMERS CE, STEIGER T, SUESS E (1989) Transformation of seafloor arriving fluxes into the sedimentary record; *In: Productivity of the Oceans: Present and Past* , Berger WH, Smetacek VS, Wefer G, (eds); J. Wiley, NY; p. 291-311.
KULLENBERG B (1952) On the salinity of the water contained in marine sediments. *Kungl. Vetnsk. Vitter. Handl, Goteborg* **21**, 3-38.
LEA DW, BOYLE EA (1989) Barium content of benthic foraminifera controlled by bottom-water composition. *Nature* **338**, 751-753.
LEA DW, BOYLE EA (1990) Foraminiferal reconstruction of barium distributions in water masses of the glacial oceans. *Palaeoceanography* **5**, 719-742.
LORD CJ, CHURCH TM (1983) The geochemistry of salt marshes: sedimentary ion diffusion, sulfate reduction and pyritization. *Geochim. Cosmochim. Acta* **47**, 1381-1391.
LYONS WB, GAUDETTE HE (1979) Sulfate reduction and the nature of organic matter in estuarine sediments. *Org. Geoch.* **1**, 151-155.
MANGINI A (1992) ^{230}Th, ^{231}Pa and manganese: indicators of changes of oceanchemistry. In : *ICP IV, Fourth International Conference on Palaeoceanography*, p. 334.
MANGINI A, EISENHAUER A, WALTER P (1990) Response of manganese in the ocean to the climatic cycles in the Quaternary. *Paleoceanography* **5**, 811-821.
MANGINI A, EISENHAUER A, WALTER P (1991) A spike of CO_2 in the atmosphere at glacial-interglacial boundaries induced by rapid deposition of manganese in the oceans. *Tellus* **43B**, 97-105.
MCARTHUR JM, TYSON RV, THOMSON J, MATTEY D (1992) Early diagenesis of marine organic matter: Alteration of the carbon isotopic composition. *Mar. Geol.* **105**, 51-61.
MIDDELBURG JJ (1989) A simple rate model for organic matter decomposition. *Geochim. Cosmochim. Acta* **53**,1566-1571.
MIDDELBURG JJ, DE LANGE GJ, KREULEN R. (1990) Dolomite formation in anoxic sediments of Kau Bay, Indonesia. *Geology* **18**, 399-402.
MIDDELBURG JJ, CALVERT SE, KARLIN R (1991) Organic-rich transitional facies in silled basins: Response to sea-level change. *Geology* **19**, 679-682.
MIDDELBURG JJ, VLUG T, VAN DER NAT FJWA (1993) Organic matter decomposition in the marine environment. *Global Planet. Change* **8**, 47-58.
MIX A (1989) Pleistocene Palaeoproductivity: Evidence from organic carbon and foraminiferal species. *In: Productivity of the Oceans: Present and Past* , Berger WH, Smetacek VS, and Wefer G, (eds); J. Wiley, NY; p. 313-340.
MÜLLER PJ (1975) Diagenese stickstoffhaltiger organischer Substanzen in oxischen and anoxischen marinen Sedimenten. *Meteor Forsch. Ergebn.* **22**, 1-60.
MÜLLER PJ, SUESS E (1979) Productivity, sedimentation rate, and sedimentary organic carbon in the ocean - I. Organic carbon preservation. *Deep-Sea Res.* **26A**, 1347-1362.
MÜLLER PJ, SUESS E, UNGERER CA (1986) Amino acids and amino sugars of surface particulate and sediment trap material from waters of the Scotia Sea. *Deep-Sea Res.* **33**, 819-838.

MÜLLER PJ, HARTMANN M, SUESS E (1988) The chemical environment of pelagic sediments. In: *The Manganese Nodule Belt of the Pacific Ocean.* (Halbach P., Friedrich G., von Stacklenberg U. (Eds.) Enke, Stuttgart, pp. 70-90.

MURDOCH WW, ONUF CP (1974) The Mediterranean as a system, part I - large ecosystems. *Intern. J. Environmental Stud.* **5**, 275-284.

OLAUSSON E (1961) Studies of deep-sea cores. Rep. Swedish Deep-Sea Exp. 8/6, 337-391.

OWENS NJP, LAW CS (1989) Natural variations in ^{15}N content of riverine and estuarine sediments. *Estuar. Coast. Shelf Sci.* **28**, 407-416.

PARKER FL, BERGER WH (1971) Faunal and solution patterns of planktonic foraminifera in surface sediments of the South Pacific. *Deep-Sea Res.* **18**, 73-88.

PEDERSEN TF, SHIMMIELD GB (1991) Interstitial water chemistry, Leg 117: contrasts with the Peru Margin. *In: Proc. Ocean Drill. Prg., Sci. Results*, Prell WL, Niitsuma N, et al.(eds), p. 499-513.

PETERS KE, SWEENEY RE, KAPLAN IR (1978) Correlation of carbon and nitrogen stable isotope ratios in sedimentary organic matter. *Limnol. Oceanogr.* **23**, 598-604.

PRAHL FG, MUEHLHAUSEN LA (1989) Lipid biomarkers as geochemical tools for palaeoceanographic study. *In: Productivity of the Oceans: Present and Past*, BergerWH, Smetacek VS, Wefer G (eds); J. Wiley, NY; p. 271-289.

PRAHL FG, DE LANGE GJ, LYLE M, SPARROW MA (1989) Post-depositional stability of long-chain alkenones under contrasting redox conditions. *Nature* **341**, 434-437.

PRUYSERS PA, DE LANGE GJ, MIDDELBURG JJ (1991) Geochemistry of eastern Mediterranean sediments: Primary sediment composition and diagenetic alterations. *Mar. Geol.* **100**, 137-154.

PRUYSERS PA, DE LANGE GJ, MIDDELBURG JJ, HYDES DJ (1993) The diagenetic formation of metal-rich layers in sapropel-containing sediments in the eastern Mediterranean. *Geochim. Cosmochim. Acta* **57**, 527-536.

RAU GH, FROELICH PN, TAKAHASHI T, DESMARAIS DJ (1991) Does sedimentary organic $\delta^{13}C$ record variations in Quaternary ocean [CO_2(aq)] ? *Palaeoceanography* **6**, 335-347.

REIMERS CE (1989) Control of benthic fluxes by particulate supply. *In: Productivity of the Oceans: Present and Past* (W.H. Berger, V.S. Smetacek, and G. Wefer, Eds); J. Wiley, NY; p. 217-233.

ROSSIGNOL-STRICK M, NESTEROFF WD, OLIVE P, VERGNAUD-GRAZZINI C (1982) After the deluge: Mediterranean stagnation and sapropel formation. *Nature* **295**, 105-110.

SARNTHEIN M, WINN K, DUPLESSY J-C, FONTUGNE MR (1988) Global variations of surface ocean productivity in low and mid-latitudes: influence on CO_2 reservoirs of the deep ocean and atmosphere during the last 21,000 years *Palaeoceanography* **3**, 361-399.

SCHMITZ B (1987) Barium, equatorial high productivity, and the northward wandering of the Indian continent. *Palaeoceanography* **2**, 63-77.

SCHNEIDER R, DAHMKE A, KOLLING A, MULLER PJ, SCHULZ HD, WEFER G. (1992) Strong deglacial minimum in the $\delta^{13}C$ record from planktonic foraminifera in the Benguela upwelling region: palaeoceanographic signal or early diagenetic imprint. In: *Upwelling systems: Evolution Since the Early Miocene*, Summerhayes CP, Prell WL, Emeis KC (eds) *Geol. Soc. Spec. Pubs.* **64**, 285-297.

SCHROEDER RA (1974) Kinetics, mechanism and geochemical applications of amino acid racemization in various fossils. *Thesis Univ. California, San Diego*, pp 275. (Univ. Microfilm 75-4656).

SCHULZ HD, DAHMKE A, SCHINZEL U, WALLMANN K, ZABEL M (1994) Early diagenetic processes, fluxes and reaction rates in sediments of the South Atlantic. *Geochim. Cosmochim. Acta* (in press).

SINNINGHE DAMSTE JS, EGLINTON TI, RIJPSTRA WIC, DE LEEUW JW (1990) Molecular characterization of organically-bound sulphur in sedimentary high-molecular-weight organic matter using flash pyrolysis and Raney Ni desulphurization. *In:*

Geochemistry of Sulfur in Fossil Fuels , Orr W, White CM (eds.), ACS Symposium Series **249**, p. 486-528 (ACS, Washington DC).
STEVENSON FJ, CHENG C-N (1972) Organic geochemistry of the Argentine basin sediments: carbon-nitrogen relationships and Quaternary correlations. *Geochim. Cosmochim. Acta* **36**, 653-671.
SUESS E (1980) Particulate organic carbon flux in the oceans - Surface productivity and oxygen utilization. *Nature* **288**, 260-263.
SUESS E.(1982) Authigenic barite in varved clays: Result of marine transgression over freshwater deposits and associated changes in interstitial water chemistry. *In: The dynamic environment of the ocean floor,* Fanning K, Manheim, FT (eds), Lexington Books, p. 339-355.
SWEENEY RE, KAPLAN IR (1980) Natural abundances of ^{15}N as a source indicator for near-shore marine sedimentary and dissolved nitrogen. *Mar. Chem.* **9**, 81-94.
TEGELAAR EW, DE LEEUW JW, DERENNE S, LARGEAU C (1989) A reappraisal of kerogen formation. *Geochim. Cosmochim. Acta* **53**, 3103-3106.
TEN HAVEN HL, DE LANGE GJ, McDUFF RE (1987) Interstitial water studies of Late Quaternary eastern Mediterranean sediments with emphasis on early diagenetic reactions and evaporitic salt influences. *Mar. Geol.* **75**, 119-136.
THIERSTEIN HR, ROTH PH (1991) Stable isotopic and carbonate cyclicity in Lower Cretaceous deep-sea sediments: Dominance of diagenetic effects. *Mar. Geol.* **97**, 1-34.
THOMSON J, CARPENTER MSN, COLLEY S, WILSON TRS, ELDERFIELD H, KENNEDY H. (1984) Metal accumulation rates in northwest Atlantic pelagic sediments. *Geochim. Cosmochim. Acta* **48**, 1935-1948.
THOMSON J, COLLEY S, HIGGS NC, HYDES D, WILSON TRS, SORENSEN J (1987) Geochemical oxidation fronts in NE Atlantic distal turbidites and their effects on the sedimentary record. In: *Geology and Geochemistry of Abyssal Plains* , Weaver PPE, Thomson J (eds) *Spec.Pub. Geol. Soc. Lond* **31**, 167-177.
THOMSON J, HIGGS NC,CROUDACE IW, COLLEY S, HYDES D (1993) Redox zonation of elements at an oxic/post-oxic boundary in deep-sea sediments *Geochim. Cosmochim. Acta* **57**, 579-595.
TREFRY JH and KLINKHAMMER GP (1988) Remobilization of barium in sulfate reducing interstitial waters. *EOS* **69**,1234.
VAN OS BJH, MIDDELBURG JJ, DE LANGE GJ (1991) Possible diagenetic mobilisation of barium in sapropelic sediment from the eastern Mediterranean. *Mar. Geol.* **100**, 125-136.
VAN OS BJH, MIDDELBURG JJ, DE LANGE G.J. (1993a) Extensive remineralization and fractionation of organic matter during weathering. *Geochim. Cosmochim. Acta* (revised)
VAN OS BJH, VISSER HJ, MIDDELBURG JJ, DE LANGE GJ (1993b) occurrence of thin, metal-rich layers in deep-sea sediments: A geochemical characterization of copper remobilization. *Deep-Sea Res.* (in press).
VAN OS BJH, LOUREN SL, DE LANGE GJ (1993c) The formation of Pliocene sapropels and carbonate cycles in the Mediterranean: Dilution, Diagenesis and productivity. (in prep)
VINE JD, TOURTELOT EG (1970) Geochemistry of black shale deposits: a summary report. *Econ. Geol.* **65**, 253-272.
VIOLANTI D, GRECCHI G, CASTRADORI D (1991) Paleoenvironmental interpretration of core Ban88-11GC (Eastern Mediterranean, Pleistocene-Holocene) on the grounds of Foraminifera, Thecosomata and calcareous nannofossils. *Il Quaternario* **4 (1a)**, 13-39.
VON BREYMANN MT, EMEIS K-C, SUESS E (1992) Water depth and diagenetic constraints on the use of barium as a palaeoproductivity indicator. *In: Upwelling Systems: Evolution since the Early Miocene;* Summerhayes CP, Prell WL, Emeis K-C (eds); *Spec. Pub. Geol. Soc. Lond.* **64**, 273-284.
WALLACE HE, THOMSON J, WILSON TRS, WEAVER PPE, HIGGS NC, HYDES DJ (1988) Active diagenetic formation of metal-rich layers in NE Atlantic sediments. *Geochim. Cosmochim. Acta* **52**, 1557-1569.
WEAVER PPE, KUIJPERS A (1983) Climatic control on turbidite deposition on the Madeira

Abyssal Plain. *Nature* **306**, 360-363.
WEAVER PPE, ROTHWELL RG (1987) Sedimentation on the Madeira Abyssal Plain over the last 300,000 years. *In: Geology and Geochemistry of Abyssal Plains*, Weaver PPE, Thomson J (eds) *Spec. Pub. Geol. Soc. Lond* **31**, 71-86.
WESTRICH JT, BERNER RA (1984) The role of sedimentary organic matter in bacterial sulfate reduction: The G-model tested. *Limnol. Oceanogr.* **29**, 236-249.
WHELAN JK (1977) Amino acids in a surface sediment core of the Atlantic abyssal plain. *Geochim. Cosmochim. Acta* **41**, 803-810.
WILSON TRS, THOMSON J, COLLEY S, HYDES DJ, HIGGS NC, SøRENSEN J (1985) Early organic diagenesis: The significance of progressive subsurface oxidation fronts in pelagic sediments. *Geochim. Cosmochim. Acta* **49**, 811-822
WILSON TRS, THOMSON J, HYDES DJ. COLLEY S, CULKIN F, SøRENSEN J. (1986) Oxidation fronts in pelagic sediments: Diagenetic formation of metal-rich layers. *Science* **232**, 972-975

APPENDIX 1. Total hydrolyzable amino acids and amino sugars (glucosamine, galactosamine) in oxidized and unoxidized f-turbidite samples of MAP cores 86P5 and 86P25.

Core		86P5			86P25	
Depth interval (cm)	704-714	751-761	772-782	780-789	855-868	905-916
Turbidite	f (ox)	f (un-ox)	f (un-ox)	f (ox)	f (un-ox)	f (un-ox)
Aspartic acid	14.2	12.5	12.4	15.3	12.9	12.6
Threonine	3.7	3.9	4.1	3.8	4.0	4.1
Serine	4.0	3.2	4.2	3.9	3.1	3.5
Glutamic acid	7.6	7.7	7.3	8.0	7.8	7.6
Glycine	13.1	13.5	13.6	13.4	13.9	14.0
Alanine	7.8	9.7	9.8	8.3	9.8	9.9
Valine	5.0	7.6	7.5	5.6	7.7	7.7
Methionine+	0.6	0.9	0.9	0.5	0.9	0.9
allo-Isoleucine	0.2	0.4	0.4	0.2	0.4	0.4
Isoleucine	1.5	4.0	3.9	1.6	4.1	4.1
Leucine	2.4	5.8	5.6	2.5	5.9	5.8
Tyrosine	0.5	1.3	1.4	0.6	1.3	1.4
Phenylalanine	1.3	3.4	3.3	1.4	3.4	3.5
β-Alanine	12.3	4.1	3.8	11.1	3.7	3.6
γ-Aminobutyric acid	15.4	3.9	3.7	12.6	3.3	3.2
Histidine	0.4	0.7	0.8	0.4	0.7	0.7
Ornithine	1.1	2.4	2.8	1.4	2.2	2.3
Lysine	3.5	8.1	7.9	3.8	8.1	8.0
Arginine	5.5	7.0	6.6	5.5	6.9	6.6
Total amino acids (nM/g)	1598	6986	7533	2065	8011	7985
Glucosamine (nM/g)	52.4	178.2	203.4	63.9	211.6	213.5
Galactosamine (nM/g)	21.7	132.3	149.9	30.9	159.5	160.6
AA/AS ratio	21.6	22.5	21.3	21.8	21.6	21.4

Individual acids given in Mol %. Analytical conditions as described by Bathmann et al. (1991).

APPENDIX 2. Comparison of sedimentation rates discussed in the text.

Deposition of EM sapropels is assumed to be isochronous; only positively (microfossil assemblages) identified,un-redeposited,and numbered sapropels are considered. Examples come from two different regions in the EM.

1. Deep- and shallow-sited Pleistocene sediments (Van Os et al., 1991, and Pruysers et al., 1991 respectively):

After correcting for the enhanced sedimentation rate during approximately the last 20 ky,on average 200 cm of sediment has been deposited to the S_5 sapropel (approx. 125 ky) for the deep-sited cores (Fig. 5; Van Os et al.,1991). For the shallow-site cores not only a similar sedimentation rate is found for this interval,but also similar organic carbon accumulation rates (Fig. 6; Pruysers et al., 1991).

2. *Shallow-sited Pliocene and deep-sited Pleistocene sediments* (Van Os et al., 1993c, and Van Santvoort and De Lange, in prep.; see De Lange et al., 1991)

Pleistocene: inferred sedimentation rate of 5.5 cm/ky for 0-1 My BP
Pliocene: inferred sedimentation rate approx. 5-8 cm/ky

NITROGEN ISOTOPE FRACTIONATION IN THE MODERN OCEAN: IMPLICATIONS FOR THE SEDIMENTARY RECORD

Joseph P. Montoya
The Biological Laboratories
Harvard University
16 Divinity Avenue
Cambridge, MA 02138 USA

ABSTRACT. A variety of biological processes associated with upper ocean phyto- and zooplankton may be important in determining the isotopic composition of organic matter sedimenting to the sea floor. In the modern ocean, the natural abundance of ^{15}N ($\delta^{15}N$) varies in a regular way within planktonic ecosystems as a result of biological isotopic fractionation: the $\delta^{15}N$ of the primary producers depends on the isotopic composition of the inorganic forms of nitrogen used as a substrate for growth, while the natural abundance of ^{15}N in zooplankton increases as a function of trophic position, with an average increment of about 3.5‰ per trophic step. Among the processes that are likely to be of critical importance in determining the $\delta^{15}N$ of sedimenting organic matter are the uptake of inorganic nitrogen by phytoplankton, the remineralization of nutrients within the euphotic zone and the production of rapidly-sinking fecal pellets and detrital particles by zooplankton. In addition, a number of recent studies provide evidence that the isotopic signature of different components of planktonic ecosystems may vary dynamically on time scales of days to months in response to perturbations in the nitrogen cycle such as wind-induced vertical mixing or seasonal variations in production. As a result of such temporal changes and the filtering of the isotopic signal that occurs as nitrogen moves through the planktonic food web, the $\delta^{15}N$ of sedimentary organic material is likely to be most useful in studying processes that affect planktonic ecosystems as a whole. The utility of $\delta^{15}N$ measurements in paleoceanographic studies will be greatly enhanced by additional study of the ways in which the physiological processes of phyto- and zooplankton discriminate between the nitrogen isotopes. In this paper, I review the areas of plankton biology that are most relevant to paleoceanographic applications of ^{15}N natural abundance measurements, and provide a brief overview of areas of current research interest.

NATO ASI Series, Vol. I 17
Carbon Cycling in the Glacial Ocean:
Constraints on the Ocean's Role in Global Change
Edited by R. Zahn et al.

INTRODUCTION

Nitrogen is an essential constituent in living systems, and is in limiting supply in many environments. Our understanding of the role of nitrogen as a limiting nutrient in the ocean has been expanded by studies of the concentration of dissolved inorganic nutrients in natural waters (Smith et al. 1986), enrichment experiments in coastal waters (Ryther and Dunstan 1971), and box-model calculations (Smith 1984), among others. All these approaches have provided evidence that the availability of nitrogen often limits primary production in marine systems, creating a tight linkage between the marine carbon and nitrogen cycles.

The major biological transformations of nitrogen in marine systems include utilization of the dissolved forms of inorganic nitrogen (NO_2^-, NO_3^-, and NH_4^+) by phytoplankton, consumption of phytoplankton by grazers, and the remineralization of organic nitrogen by animals and bacteria (Figure 1). In addition, various marine prokaryotes are able to carry out a suite of reactions that move nitrogen in and out of the major inorganic pools; these include dinitrogen fixation and dissimilatory transformations of nitrogen such as nitrification and denitrification. Thus, the marine nitrogen cycle involves multiple inorganic and organic pools which are coupled by rapid biological transformations of nitrogen. For example, the ability of phytoplankton to utilize ammonium efficiently has been well established in laboratory and field experiments (Goldman and Glibert 1983; McCarthy 1980; 1981). Since most marine animals excrete the bulk of their nitrogenous wastes as ammonium (Bidigare 1982), the utilization of ammonium by phytoplankton creates a tight link between the dissolved inorganic pools, the primary producers, and the higher trophic levels of the food web. As a result, perturbations in one part of the nitrogen cycle can propagate to other pools relatively quickly. In addition, it is worth noting that the marine nitrogen cycle is inherently more complex than the carbon cycle; the multiple oxidation states of inorganic nitrogen and the multiplicity of biological pathways interconnecting different pools of nitrogen (Figure 1) often make it difficult to identify the key processes that control the movement of nitrogen through ecosystems.

In recent years, measurements of the isotopic composition of nitrogen in natural systems have been used to complement traditional experimental studies of the nitrogen cycle using ^{15}N-labelled substrates (Horrigan et al. 1990a; Horrigan et al. 1990b; Montoya et al. 1990; Montoya et al. 1991). For example, measurements of the $\delta^{15}N$[1] of different components of an ecosystem can contribute to an understanding of the bulk flow of nitrogen in the system as a whole (Heaton 1986; Owens 1987; Peterson and Fry 1987 provide good general reviews of this topic). This approach integrates over space and time and is a powerful complement to experimental studies of elemental cycles, which typically involve the measurement of short–term changes in small, isolated volumes. To date, however, the application of this approach has been limited by an incomplete understanding of the processes that affect the distribution of

---

[1]The $\delta^{15}N$ of a sample represents the per mil (‰) difference between the isotope ratio (^{15}N:^{14}N) of the sample and the isotope ratio (^{15}N:^{14}N) of tropospheric dinitrogen.

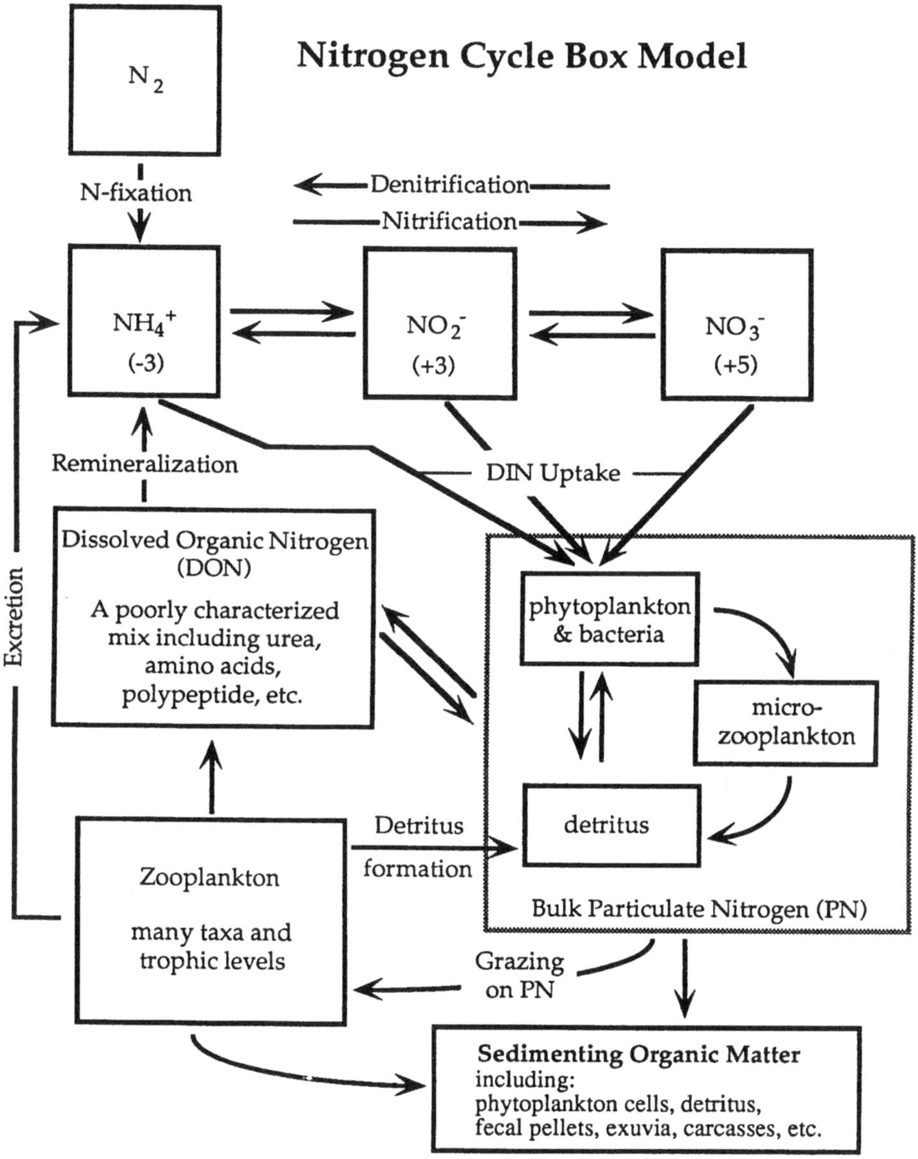

Figure 1: Box model of the marine nitrogen cycle. Each box represents one major functional pool of nitrogen in the ocean. Some of the important components of heterogeneous pools are listed beneath the name of the pool, as is the oxidation state of the dissolved inorganic forms. Arrows depict the major biological transformations of nitrogen. Note that a variety of organisms at different trophic positions contribute to the vertical flux of organic N.

nitrogen isotopes in aquatic ecosystems. Such concerns are equally relevant to paleoceanographic studies that attempt to relate variations in the isotopic composition of sedimentary nitrogen to the biological processes operating in the ocean at the time of deposition. Our current challenge is to develop the level of understanding necessary to use ^{15}N stable isotope data in quantitative studies of the marine nitrogen cycle through time. In this paper, I review our current understanding of the factors that regulate the distribution of ^{15}N within planktonic ecosystems, and outline several promising avenues for future applications of stable isotope techniques to studies of the marine nitrogen cycle.

NATURAL ABUNDANCE OF ^{15}N IN MODERN MARINE ECOSYSTEMS

General Patterns

During the last decade, most of the research on the stable isotopes of nitrogen in marine systems has been oriented toward elucidating spatial and ecological variations in the ^{15}N content of the major inorganic and organic forms of nitrogen (Owens 1987, and Peterson and Fry 1987 have recently reviewed this topic). This body of research has demonstrated that the natural abundance of ^{15}N varies in a regular way within marine ecosystems as a result of biological processes that discriminate between the two isotopes (Ehleringer et al. 1986; Owens 1987; Peterson and Fry 1987). Most biologically-mediated transformations of nitrogen are accompanied by measurable isotopic fractionation (Table 1), but in many cases, the mechanisms that lead to this isotopic fractionation are not understood in detail. As noted above, our knowledge of the stable isotope biogeochemistry of nitrogen is much less detailed than our understanding of the factors that regulate the distribution of the carbon isotopes in natural systems.

Since most marine autotrophs require combined nitrogen as a substrate for growth, the isotopic composition of dissolved inorganic nitrogen (DIN) acts as a sort of master variable in setting the baseline isotopic composition of marine plankton. A number of biological processes may alter the isotopic composition of the marine pool of DIN: these include nitrogen fixation, denitrification, and nitrification, all of which may lead to a net transfer between the oceanic pool of combined nitrogen and the atmosphere. As a result, the fractionations associated with these processes may, over time, lead to alterations in the overall isotopic inventory of marine nitrogen. Relatively few measurements of the isotopic composition of DIN are currently available in the literature: NO_3^- dominates the oceanic pool of combined nitrogen, and the available data indicate that the $\delta^{15}N$ of NO_3^- from oxygenated deep waters ranges between about 3‰ and 7‰ (Altabet 1988; Altabet and McCarthy 1985; Liu and Kaplan 1989; Miyake and Wada 1967; Wada et al. 1975), with a mean value around 6‰ (Liu and Kaplan 1989). The $\delta^{15}N$ of NO_3^- is significantly higher than this mean in and around anoxic water masses where denitrification can occur (Cline and Kaplan 1975; Liu and Kaplan 1989). It is worth noting that on the time scales relevant to paleoceanographic studies, changes in the global rate of denitrification could readily lead to significant increases in the $\delta^{15}N$ of marine NO_3^- (Altabet and Curry 1989). This in turn would affect the $\delta^{15}N$ of marine plankton, since the isotopic composition of the phytoplankton at the base of the food web depends at least in part on the $\delta^{15}N$ of the available DIN.

The isotopic fractionation associated with the uptake and metabolism of DIN by phytoplankton (Table 1) typically produces phytoplankton biomass depleted in ^{15}N relative to the DIN used as a substrate for growth (Cifuentes et al. 1989; Cifuentes et al. 1988; Montoya and McCarthy in prep.; Pennock et al. 1988; Wada 1980; Wada and Hattori 1978). The degree of isotopic discrimination associated with primary production varies between taxa and with growth conditions, and work with phytoplankton cultures suggests that diatoms may discriminate more strongly than flagellates during growth on NO_3^- (Montoya and McCarthy in prep.; Pennock et al. 1988; Wada 1980; Wada and Hattori 1978).

Nitrogen-fixing cyanobacteria belonging to the genus *Oscillatoria* are abundant in many oligotrophic regions of the ocean. Consequently, nitrogen fixation could potentially make a significant contribution to the overall nitrogen budget in the upper water column (Carpenter 1983). This would also affect the isotopic balance of the ecosystem since dinitrogen dissolved in seawater has an isotopic composition very near 0‰ (Emerson et al. 1991; Klots and Benson 1963). Nitrogen fixation is accompanied by little or no isotopic fractionation (Table 1), so any inputs of combined nitrogen to the ecosystem through this pathway will have an isotopic signature very near 0‰, which is significantly lower than the $\delta^{15}N$ of roughly 6‰ that appears to be characteristic of deep water NO_3^- (Liu and Kaplan 1989). The generally low $\delta^{15}N$ values measured for plankton in regions of the North Pacific where *Oscillatoria* sp. is abundant may reflect just such an input of atmospheric nitrogen through nitrogen fixation (Minagawa and Wada 1984; Wada and Hattori 1976). Despite their abundance, marine planktonic cyanobacteria appear to make a relatively small contribution to the planktonic nitrogen budget in most regions of the ocean (Carpenter 1983), and a variety of mechanisms other than nitrogen fixation may contribute to low $\delta^{15}N$ values in upper ocean plankton (see below).

Finally, a complex example of isotopic fractionation occurs during the trophic transfer of nitrogen within food webs, where a regular increase in $\delta^{15}N$ with trophic position characterizes a wide variety of animals taken from marine, freshwater, and terrestrial ecosystems (Figure 2). In general, the nitrogen in an animal's tissues is roughly 3.5‰ more enriched in ^{15}N than the nitrogen in its food (DeNiro and Epstein 1981; Fry 1988; Minagawa and Wada 1984; Montoya et al. 1990; Montoya et al. 1992; Mullin et al. 1984; Pang and Nriagu 1977; Wada et al. 1987). Although the mechanisms responsible for this trophic effect on $\delta^{15}N$ are not currently known in detail, a variety of evidence derived from both laboratory and field studies suggests that the difference in $\delta^{15}N$ between an animal's tissues and its food is the net result of isotopic fractionation during the assimilation of ingested food and the excretion of nitrogenous wastes (Altabet and Small 1990; Checkley and Entzeroth 1985; Checkley and Miller 1989;

Table 1: Selected estimates of isotopic fractionation factors for biologically-catalyzed transformations of nitrogen. Fractionation factors are expressed as a ratio of rate constants, $\alpha = ({}^{14}k{:}{}^{15}k)$, such that $\alpha = 0$ implies no isotopic discrimination and $\alpha > 0$ implies discrimination against ^{15}N. The organisms used include bacteria (*Azotobacter*, *Nitrosomonas*, *Pseudomonas*, *Delsulfovibrio*), cyanobacteria (*Anabaena*, *Nostoc*, *Trichodesmium*), a variety of phytoplankton (diatoms, haptophytes, cryptophytes, and chlorophytes), and zooplankton (copepods).

Process	Reaction	$\bar{\alpha}$	Range of α	Organisms Used	Reference
N_2 fixation	$N_2 \rightarrow$ organic N	1.0000	0.9910 - 1.0035	*Azotobacter* spp. (4 clones)	(Hoering and Ford 1960)
		1.0039	1.0031 – 1.0054	*Azotobacter vinelandii*	(Delwiche and Steyn 1970)
		•	1.001 – 1.002	*Anabaena cylindrica*, *Nostoc* sp.	(Mitsubishi-Kasei 1978)
		1.0024	1.0022 – 1.0029	*Anabaena* sp.	(Macko et al. 1987)
Denitrification	$NO_3^- \rightarrow N_2$	•	1.024 – 1.044	Field Data	(Richards and Benson 1961)
		1.0173	1.0134 – 1.0208	*Pseudomonas denitrificans*	(Delwiche and Steyn 1970)
		1.020	•	Marine Denitrifier	(Miyake and Wada 1971)
		•	1.030 – 1.040	Field Data	(Cline and Kaplan 1975)
		•	1.002 – 1.012	Marine Denitrifier	(Wada 1980)
		1.0294	•	Soil Bacteria (T = 20°C)	(Mariotti et al. 1981)
		1.0246	•	Soil Bacteria (T = 30°C)	(Mariotti et al. 1981)
	$NO_3^- \rightarrow NH_4^+$	•	1.021 – 1.032	*Desulfovibrio* sp.	(McCready et al. 1983)
	$NO_2^- \rightarrow N_2O$	•	1.011 – 1.033	Soil Bacteria	(Mariotti et al. 1982)
		1.0355	1.0349 – 1.0362	*Nitrosomonas europaea*	(Yoshida 1988)
Nitrification	$NH_4^+ \rightarrow NO_2^-$	1.0260	1.0176 – 1.0360	*Nitrosomonas europaea*	(Delwiche and Steyn 1970)
		1.019	1.016 – 1.026	Unidentified marine nitrifier	(Miyake and Wada 1971)
		1.0362	1.0082 – 1.0943	*Nitrosomonas europaea*	(Mariotti et al. 1981)
		1.0288	1.0246 – 1.0320	*Nitrosomonas europaea*	(Yoshida 1988)
	$NH_4^+ \rightarrow N_2O$	1.0653	1.0604 – 1.0684	*Nitrosomonas europaea*	(Yoshida 1988)

(continued)

Table 1 (continued)

Process	Reaction	$\bar{\alpha}$	Range of α	Organisms Used	Reference
DIN Uptake	$NO_3^- \rightarrow$ organic N	•	1.0007 – 1.0230	*Pheodactylum tricornutum* (diatom)	(Wada and Hattori 1978)
		•	1.0009 – 1.0045	*Chaetoceros* sp. (diatom)	(Wada and Hattori 1978)
		1.0098	•	*Skeletonema costatum* (diatom)	(Pennock et al. 1988)
		1.0090	1.0064 - 1.0119	*Skeletonema costatum* (diatom)	(Montoya and McCarthy in prep.)
		1.0121	1.0097 – 1.0154	*Thalassiosira weissflogii* (diatom)	(Montoya and McCarthy in prep.)
		1.0032	1.0022 - 1.0044	*Isochrysis galbana* (haptophyte)	(Montoya and McCarthy in prep.)
		1.0009	1.0003 - 1.0039	*Pavlova lutheri* (haptophyte)	(Montoya and McCarthy in prep.)
		1.0022	1.0012 - 1.0029	*Chroomonas salina* (cryptophyte)	(Montoya and McCarthy in prep.)
		1.0034	1.0007 - 1.0061	*Dunaliella tertiolecta* (chlorophyte)	(Montoya and McCarthy in prep.)
		1.0040	•	Auke Bay plankton	(Goering et al. 1990)
		1.0070	•	Chesapeake Bay plankton	(Horrigan et al. 1990)
	$NO_2^- \rightarrow$ organic N	1.0007	•	*Pheodactylum tricornutum* (diatom)	(Wada and Hattori 1978)
		•	1.0000 – 1.0025	*Chaetoceros* sp. (diatom)	(Wada 1980)
	$NH_4^+ \rightarrow$ organic N	•	0.9903 – 0.9947	*Chaetoceros* sp. (diatom)	(Wada and Hattori 1978)
		•	1.0000 – 1.0010	*Dunaliella tertiolecta* (chlorophyte) & *Criscosphaera carterae* (haptophyte)	(Wada 1980)
		1.0194	•	*Skeletonema costatum* (diatom)	(Pennock et al. 1988)
		1.0091	•	Delaware Bay plankton	(Cifuentes et al. 1989)
		•	1.0065 - 1.0080	Chesapeake Bay plankton	(Montoya et al. 1991)
Decomposition	PN $\rightarrow$ DIN + DON	•	0.999 – 1.002	*Scenedesmus* sp. (chlorophyte)	(Miyake and Wada 1971)

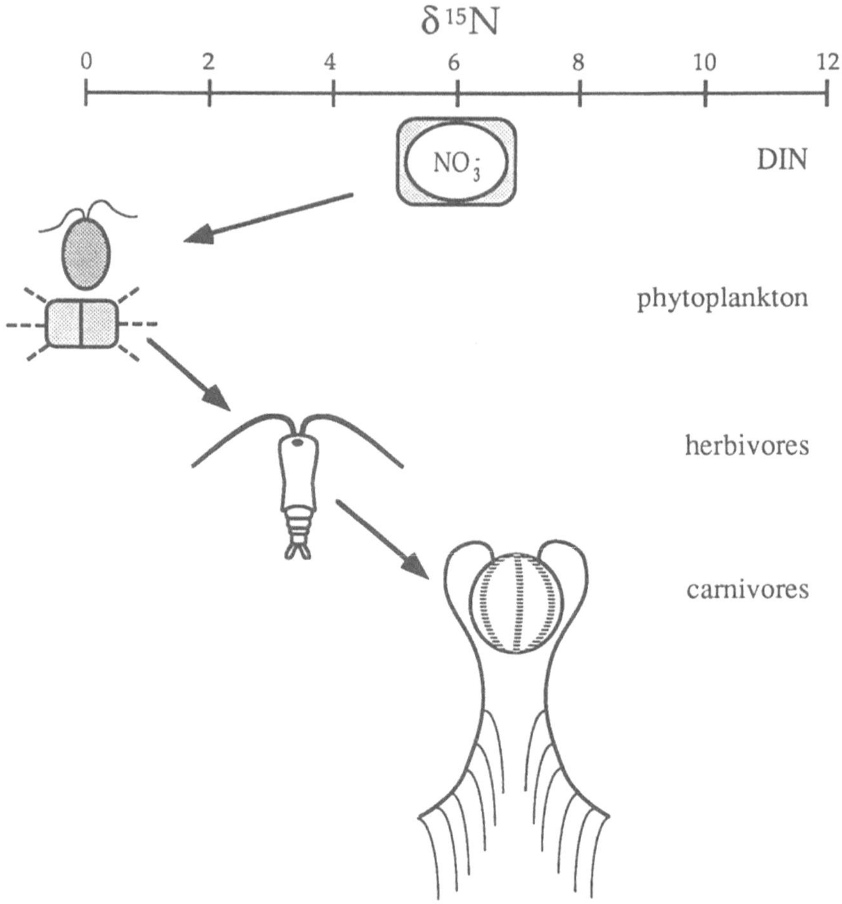

Figure 2: Food chain schematic showing the general increase in $\delta^{15}N$ with trophic level typically observed in marine ecosystems. Isotopic fractionation during the uptake of DIN produces phytoplankton with a $\delta^{15}N$ lower than that of the source pool. Arrows depict the major biological transformations of nitrogen. The effects of an average trophic level effect of ca. 3.5‰ are shown for an herbivorous copepod and a carnivorous ctenophore.

DeNiro and Epstein 1981; Steele and Daniel 1978). Because of the interconnections between dissolved inorganic nitrogen, autotrophs, and heterotrophs that characterize the marine nitrogen cycle (Figure 1), isotopic fractionation by zooplankton has the potential to affect the $\delta^{15}N$ of the phytoplankton and the dissolved pools of nitrogen as well as the $\delta^{15}N$ of the organisms themselves.

Seasonal Changes in Isotopic Composition of Planktonic Ecosystems

In addition to creating a trophic scaling of $\delta^{15}N$, biological isotopic fractionation can lead to significant temporal changes in the $\delta^{15}N$ of various components of an ecosystem. For example, primary production during a bloom will produce phytoplankton biomass with a $\delta^{15}N$ lower than that of the DIN used as a substrate for growth. Over time, this selective removal of ^{14}N will lead to a progressive increase in the $\delta^{15}N$ of the residual pool of DIN, which will lead in turn to a steady increase in the $\delta^{15}N$ of the phytoplankton produced. In the end, when all the substrate has been consumed, the integrated $\delta^{15}N$ of all the phytoplankton produced will be equal to the $\delta^{15}N$ of the original pool of substrate, reflecting the conservation of isotopes. This process is equivalent to a fractional distillation, and the time-course of change in the $\delta^{15}N$ of the residual substrate, the instantaneous product, and the integrated product (Figure 3) can all be described as a function of the fractionation factor for the reaction and the progress of the reaction (Mariotti et al. 1981). Several recent studies have demonstrated the potential for changes in the $\delta^{15}N$ of phytoplankton on time scales ranging from days to months. In an extensive field study of the $\delta^{15}N$ of dissolved and particulate nitrogen, Cifuentes and his colleagues (Cifuentes et al. 1989; Cifuentes et al. 1988) showed that $\delta^{15}N$ of NH_4^+ and particulate nitrogen (PN) in the Delaware River and Bay change dramatically through the seasonal cycle as a result of phytoplankton uptake of NH_4^+. More recently, Goering et al. (1990) have documented significant changes in the $\delta^{15}N$ of PN in the course of the spring bloom in an Alaskan embayment (Auke Bay), and Montoya et al. (1991) have described rapid changes in the $\delta^{15}N$ of PN in the Chesapeake Bay following an autumn storm event that injected NH_4^+ into the surface layer of the Bay. Finally, Altabet et al. (1991) have documented clear changes in the $\delta^{15}N$ of both suspended and sinking particles in the course of the spring bloom in the North Atlantic. In all of these studies, the temporal changes in the $\delta^{15}N$ of PN were consistent with the effects of isotopic fractionation during the consumption of the inorganic substrate (either NH_4^+ or NO_3^-) by phytoplankton (shown schematically in Figure 3 and Figure 4).

Temporal changes in the $\delta^{15}N$ of PN can propagate through ecosystems as nitrogen moves through the food web (Figure 4). Two recent studies provide clear evidence for changes in the $\delta^{15}N$ of zooplankton that reflect shifts in the $\delta^{15}N$ of the PN at the base of the food web (Goering et al. 1990; Montoya et al. 1991). Similarly, Montoya et al. (1992) provide circumstantial evidence that a similar shift occurs in the $\delta^{15}N$ of a wide variety of crustacean and gelatinous zooplankton during and after the spring bloom in the Gulf Stream Region. Since an animal's nitrogen content represents the integrated effects of feeding over a finite time interval, an animal's isotopic composition will always change more slowly than that of its food. That is, animals act as low pass filters with a time constant that reflects the turnover time for nitrogen within their tissues. Animals that are several trophic transfers removed from the primary producers and large animals with a relatively long turnover time for body nitrogen should show relatively little change in $\delta^{15}N$ as a result of seasonal changes in the $\delta^{15}N$ of

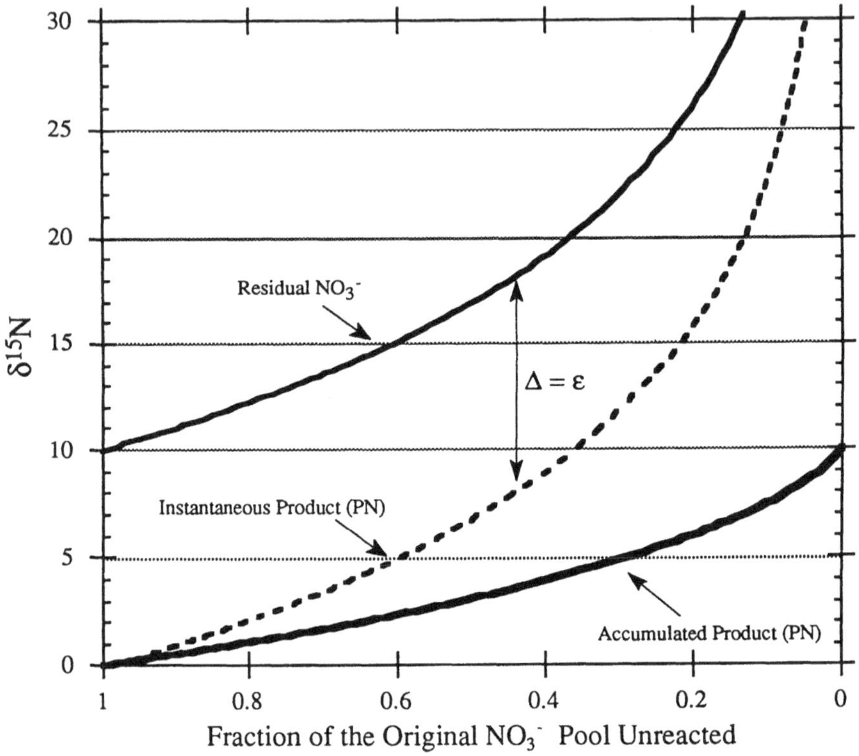

Figure 3: Effects of a 10‰ discrimination against ^{15}N (i.e. $\alpha = 1.010$, $\varepsilon = 10‰$) on the $\delta^{15}N$ of the residual substrate (solid line), the instantaneous product (broken line), and the accumulated product (heavy line) of a reaction. The fraction of the original pool of substrate remaining unreacted is plotted on the horizontal axis so that the time course of the reaction proceeds from left to right. The time course for each pool was calculated using standard equations developed by Mariotti et al. (1981). Note that the instantaneous difference between the substrate and the product of the reaction is equal to the enrichment factor, ε, and that the accumulated product converges on the $\delta^{15}N$ of the original pool of substrate as the reaction nears completion.

phytoplankton (Figure 4). Conversely, short-lived and small animals are likely to show pronounced seasonal changes in $\delta^{15}N$ in regions with a distinct seasonal production cycle. Thus, different components of the zooplankton community in effect sample isotopic variation on different time scales and the isotopic composition of those animals may be a useful index to isotopic variation on these time scales.

Recycling and Export of Nitrogen From the Upper Ocean by Zooplankton

Zooplankton processes are important in determining the net balance of nitrogen isotopes within the euphotic zone. For example, NH_4^+ excreted by zooplankton may contribute a

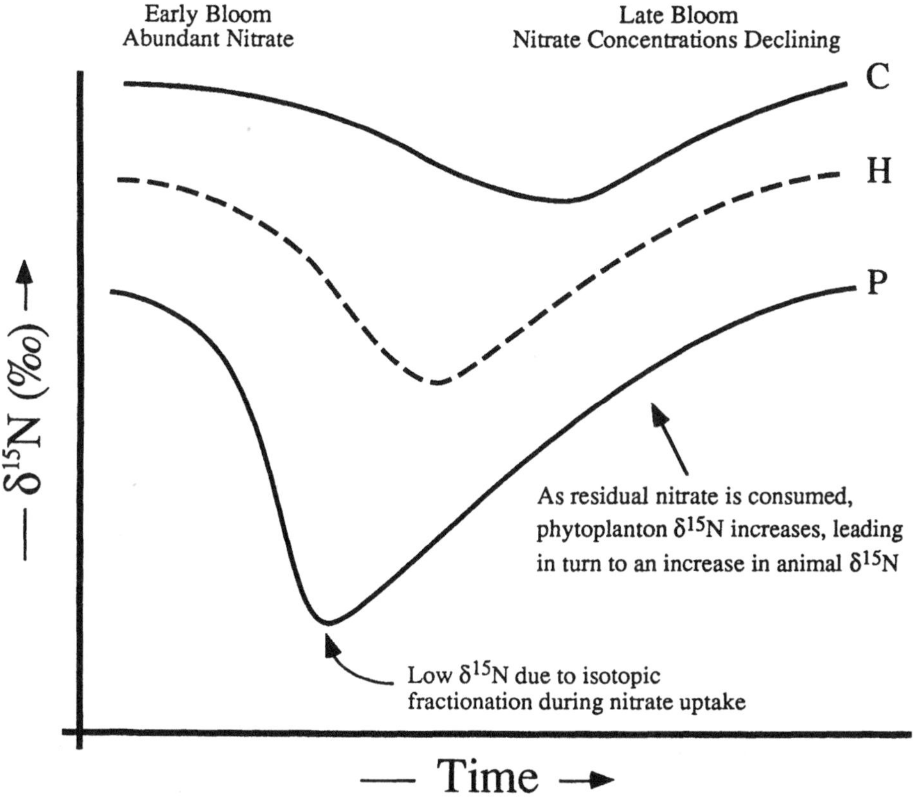

Figure 4: Schematic representation of changes in the $\delta^{15}N$ of phytoplankton (P), herbivorous zooplankton (H) and carnivorous zooplankton (C) in the course of a bloom. Note that an analogous spatial pattern of variation in $\delta^{15}N$ will result as as result of the consumption of NO_3^- in upwelled water as it is advected away from its source.

significant fraction of the total nitrogen ration of phytoplankton, especially in stratified water columns where the surface layer is depleted of NO_3^-. Conversely, sinking fecal pellets and exuvia represent an important route for exporting nitrogen from the upper ocean. As a result, differences between the $\delta^{15}N$ of excreted NH_4^+ and the $\delta^{15}N$ of fecal pellets and exuvia leaving the mixed layer may lead to significant changes in the average isotopic composition of nitrogen in the upper ocean (Altabet 1988; Altabet and Small 1990; Checkley and Entzeroth 1985; Checkley and Miller 1989; Montoya et al. 1992; Mullin et al. 1984). An additional consideration in paleoceanographic studies is the way in which the zooplankton processes mediating the vertical flux of organic matter to the deep sea act as a sort of biological filter affecting the isotopic composition of the organic matter contributing to the sedimentary record.

In view of the role of animals in mediating the vertical flux of nitrogen out of the mixed layer, a zooplankton isotope budget would be a significant contribution to the use of ^{15}N as a natural tracer in studies of the upper ocean nitrogen cycle as well as in paleoceanographic studies based on sedimentary nitrogen isotope ratios. Unfortunately, I am aware of no such budget in the literature, though several researchers have measured one or more of the necessary terms: the $\delta^{15}N$ of the tissues, the $\delta^{15}N$ of excreted NH_4^+, the $\delta^{15}N$ of fecal pellets, and the $\delta^{15}N$ of the food (Table 2). For example, most of the available data show that the fecal pellets of crustacean zooplankton have a higher $\delta^{15}N$ than the animal's tissues (Checkley and Entzeroth 1982; Checkley and Entzeroth 1985; Montoya et al. 1992), though some euphausiids (Montoya et al. 1992) and mixed samples of zooplankton (Altabet and Small 1990) produce fecal pellets with a $\delta^{15}N$ lower than that of the whole animal. In contrast, the NH_4^+ excreted by zooplankton is consistently depleted in ^{15}N relative to the animal's tissues by roughly 2.7‰ (Checkley and Miller 1989). Finally, the $\delta^{15}N$ of crustacean exuvia is poorly characterized, though two species of amphipod collected in the Gulf Stream region produced molts with $\delta^{15}N$ values 3.8 and 4.6‰ lower than the $\delta^{15}N$ of the whole animals (Montoya et al. 1992). Interestingly, DeNiro and Epstein (1981) have found that chitin isolated from insects is also highly depleted in ^{15}N, suggesting that this may be a general feature of arthropodan chitin.

Taken together, the available data can be combined into a composite isotope budget in which excreted NH_4^+ and molted exoskeletons have a lower $\delta^{15}N$ than an animal's tissues, while fecal pellets are usually enriched in ^{15}N relative to tissues, and are always enriched in ^{15}N relative to the animal's food (Table 2, Figure 5). These observations indicate that significant isotopic fractionation occurs in the assimilation of ingested nitrogen (leading to feces enriched in ^{15}N relative to the food) as well as in the excretory process (leading to dissolved excreta depleted in ^{15}N relative to the tissues). Note that although these processes have opposite effects on the isotopic enrichment of the animal, together they both tend to retain ^{14}N within the euphotic zone while exporting ^{15}N to the deep ocean. At times when crustacean zooplankton are abundant and growing actively, sinking exoskeletons may act as a counterbalance to this trend by preferentially exporting ^{14}N from the upper ocean, though additional data are clearly needed before the role of exuvia in the movement of the nitrogen isotopes can be evaluated quantitatively.

Although all of these data indicate that zooplankton processes play an important role in determining the isotopic composition of organic nitrogen in the upper ocean, very few studies have sought to address this role directly. Perhaps the most direct evidence bearing on this question comes from a comparison of the isotopic composition of oceanic and coastal zooplankton by Mullin et al. (1984), who showed that copepods and chaetognaths from the North Pacific Central Gyre have a lower $\delta^{15}N$ than their analogues from the Southern California Bight. In addition, the $\delta^{15}N$ of zooplankton from the Southern California Bight showed a significant positive correlation with the rate of primary production measured at the time of sample collection. This correlation, and a consideration of the assimilation rates of

Table 2: Isotopic discrimination associated with animal processes, expressed as the difference in isotopic composition between a product and either the animal's bulk tissues or the animal's food if known. Note that $\Delta\delta^{15}N$ is not a true isotopic discrimination factor since it does not represent the difference in $\delta^{15}N$ between the reactants and product of a specific reaction. Instead, these values provide an index to the gross partitioning of nitrogen isotopes by zooplankton processes.

Process	Difference Calculated	$\Delta\delta^{15}N$	Animals Used	Reference
Feeding	$\delta^{15}N$(fecal pellets) - $\delta^{15}N$(animal)	-1.3‰	mixed zooplankton (various size fractions)	(Altabet and Small 1990)
		+2.2‰	*Temora longicornis* (copepod)	(Checkley and Entzeroth 1985)
		-0.6‰ – 4.8‰	*Euphausia krohnii* (euphausiid)	(Montoya et al. 1992)
		2.1‰	*Euphausia mutica* (euphausiid)	(Montoya et al. 1992)
		-2.7‰ , -3.7‰	*Meganyctiphanes norvegica* (euphausiid)	(Montoya et al. 1992)
		1.9‰ – 3.5‰	*Vibilia* sp. (amphipod)	(Montoya et al. 1992)
		1.4‰	*Parathemisto gaudichaudii* (amphipod)	(Montoya et al. 1992)
	$\delta^{15}N$(fecal pellets) - $\delta^{15}N$(food)	+ 8.0‰	*Temora longicornis* (copepod)	(Checkley and Entzeroth 1985)
Excretion	$\delta^{15}N(NH_4^+)$ - $\delta^{15}N$(animal)	-2.7‰	various zooplankton	(Checkley and Miller 1989)
Ecdysis	$\delta^{15}N$(exoskeleton) - $\delta^{15}N$(animal)	-3.8‰	*Vibilia* sp. (amphipod)	(Montoya et al. 1992)
	$\delta^{15}N$(exoskeleton) - $\delta^{15}N$(animal)	-4.6‰	*Parathemisto gaudichaudii* (amphipod)	(Montoya et al. 1992)

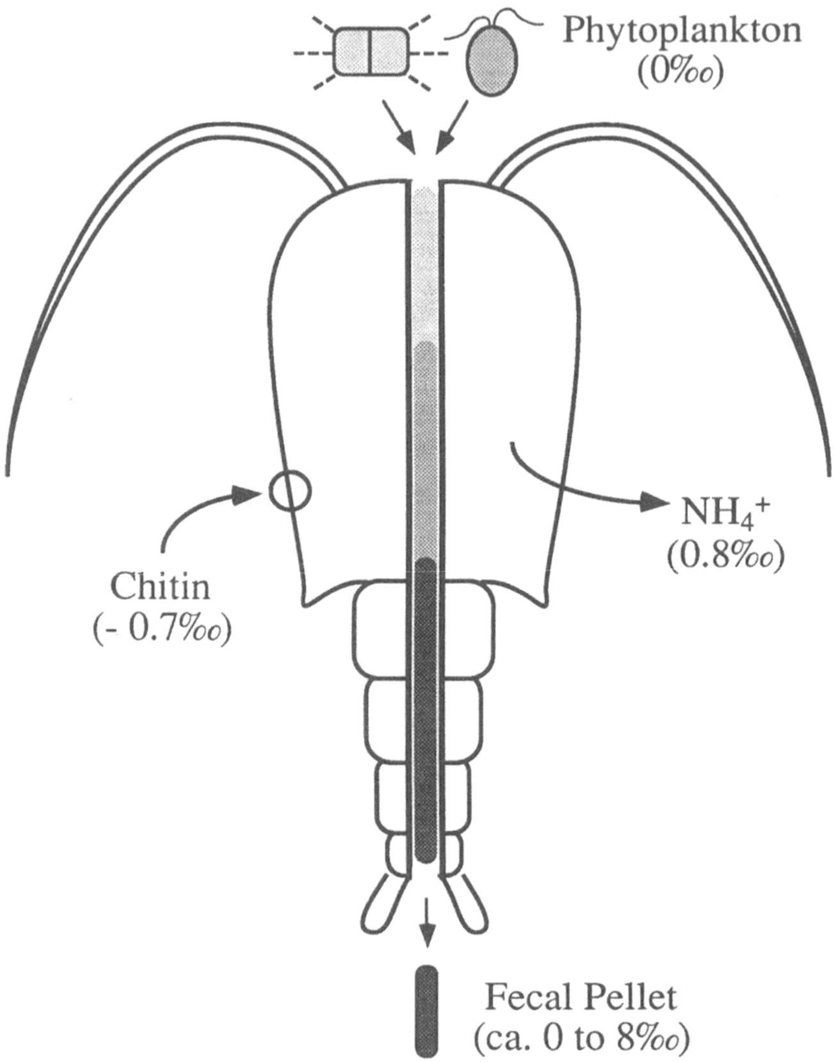

Figure 5:
Schematic of fractionation associated with zooplankton processes. Given a phytoplankton $\delta^{15}N$ of 0‰, herbivorous zooplankton will have a $\delta^{15}N$ of ca. 3.5‰, and will produce NH_4^+ that is about 2.7‰ depleted relative to their tissues (Checkley and Miller 1989). Fecal pellets show greater isotopic variation, but are enriched relative to the food (Altabet and Small 1990; Checkley and Entzeroth 1985; Montoya et al. 1992). Exoskeletons are significantly depleted relative to bulk tissues (Montoya et al. 1992).

NO_3^- and NH_4^+ by the phytoplankton, led Mullin et al. (1984) to suggest that the $\delta^{15}N$ of zooplankton reflects the relative importance of NO_3^- and NH_4^+ in supplying the phytoplankton demand for inorganic nitrogen. In other words, zooplankton production of isotopically light NH_4^+, and its subsequent uptake by phytoplankton, leads to a selective retention of ^{14}N in the euphotic zone, and an inverse correlation between the $\delta^{15}N$ of phytoplankton and the degree to which primary production is fueled by NH_4^+ rather than NO_3^-. Although Mullin et al. (1984) did not actually sample the suspended particles in their study areas, the $\delta^{15}N$ of PN in the oligotrophic central gyre of the North Pacific does appear to be low in comparison to PN in the more eutrophic regions of that ocean (Saino and Hattori 1987; Wada and Hattori 1976). This relationship clearly suggests that $\delta^{15}N$ measurements may be useful in evaluating the relative importance of new and regenerated production in planktonic ecosystems, though further study of the mechanisms by which zooplankton influence the isotopic balance of the upper ocean is clearly warranted.

Transport of Organic Nitrogen to the Deep Sea

The most extensive studies to date of the nitrogen isotopic composition of sinking organic matter have been carried out in the Sargasso Sea (Altabet 1988; Altabet 1989; Altabet and Deuser 1985; Altabet et al. 1991) and in the Greenland Sea (Voss 1991). These studies have provided clear evidence linking the nitrogen isotopic composition of sinking organic matter to biological processes occurring in the upper water column, though the impact of remineralization processes acting on sinking particles remains difficult to evaluate.

In his extended studies at a deep-water station near Bermuda, Altabet (1988) has shown that the mean $\delta^{15}N$ of organic matter leaving the mixed layer is equal to the mean $\delta^{15}N$ of deep water NO_3^-, suggesting that the upper ocean is in isotopic balance in oligotrophic systems like the Sargasso Sea. Although isotopic balance is maintained on an annual time scale, significant variations in the $\delta^{15}N$ of sedimenting organic matter occur on shorter time scales reflecting the changes in upper-ocean $\delta^{15}N$ associated with the seasonal production cycle. The large changes in $\delta^{15}N$ that occur in the course of a phytoplankton bloom (e.g., Figure 3) may be reflected almost immediately in the isotopic composition of sinking particles (Altabet et al. 1991; Voss 1991). Despite the rapid transmission of changes in mixed layer isotopic composition to the particles sinking at depth, a variety of diagenetic processes may lead to significant differences between the $\delta^{15}N$ of near-surface PN and the particles entering the deep sea. For example, Altabet (1988) has used the difference between the $\delta^{15}N$ of rapidly sinking particles and the $\delta^{15}N$ of upper ocean bulk particulate nitrogen as an index to the number of trophic transfers separating the phytoplankton and the zooplankton making the major contribution to the vertical flux of organic matter. While en route to the deep sea, remineralization of organic nitrogen may lead to further increases in the $\delta^{15}N$ of sinking particles, producing a subsurface maximum in the $\delta^{15}N$ of PN (Altabet and McCarthy 1985; Montoya et al. 1992; Saino and Hattori 1980; Saino and Hattori 1985; Saino and Hattori 1987), though such diagenetic alterations are not likely to have a major effect on the isotopic composition of large, rapidly sinking particles during their transit through the water column.

These observations clearly suggest that the form in which organic nitrogen leaves the mixed layer is an important determinant of the likelihood that it will reach and be incorporated into deep sea sediments. Since large, rapidly sinking particles dominate the vertical flux of organic matter to the deep sea (McCave 1975), fecal pellets and exuvia are likely to exert a strong influence on the isotopic composition of nitrogen reaching the sea floor. As a result, the isotopic fractionation associated with zooplankton feeding and growth may be incorporated into the isotopic composition of sedimenting organic matter, perhaps as a sort of overprint on the basic isotopic signature derived from phytoplankton processes in the upper ocean. In the absence of techniques for assessing the effects of zooplankton processes on the sedimentary record, the isotopic composition of bulk sedimentary nitrogen is perhaps best viewed as a record of the flux-weighted average $\delta^{15}N$ of upper ocean phyto- and zooplankton.

IMPLICATIONS FOR PALEOCEANOGRAPHY

Because the nitrogen and carbon cycles are interconnected by biological production, the nitrogen isotopic signature of deep sea sediments may provide a useful record of past changes in marine production. As noted above, the complexity of the marine nitrogen cycle and the effects of food web processes in filtering the isotopic signature of organic matter that reaches the deep sea both suggest that marine sediments are most likely to preserve a record of upper ocean processes that affect the entire euphotic zone ecosystem. For example, sedimentary nitrogen is unlikely to provide a simple record of the mean $\delta^{15}N$ of organic matter in the upper water column since the annual vertical flux of organic matter may be dominated by the spring bloom, biasing the inputs to the sediments toward values that are significantly different from the annual mean $\delta^{15}N$ of the upper ocean (Figure 4). Similarly, the contribution of zooplankton fecal pellets to the vertical flux of organic matter will tend to elevate the mean $\delta^{15}N$ of sedimenting organic matter relative to the mean value for euphotic zone phytoplankton. Thus, the mean $\delta^{15}N$ of sediments is likely to be a weighted integration of the various fluxes out of the upper ocean, with a sampling bias toward the spring bloom and toward the fecal pellets produced by the abundant zooplankton.

Although this sampling bias complicates the interpretation of the sedimentary record, it is interesting to speculate on the sorts of processes that are likely to affect the entire ecosystem in ways that would alter the $\delta^{15}N$ of organic nitrogen in sediments. For example, changes in the global rate of denitrification could significantly alter the isotopic composition of marine NO_3^-, causing global shifts in the average $\delta^{15}N$ of marine ecosystems (Altabet and Curry 1989). A second, smaller-scale example in which the $\delta^{15}N$ of sedimentary nitrogen may be used profitably is in studying the relative extent of utilization of available nutrients in the water column. This approach relies on the isotopic fractionation that accompanies the uptake of NO_3^- by phytoplankton (Figure 3) to provide a measure of the degree to which phytoplankton deplete the available pool of NO_3^-. In this scenario, low $\delta^{15}N$ values result from incomplete consumption of the available nutrients, resulting in significant differences in isotopic composition between the residual nutrients and the organic matter formed. For example, Calvert et al. (1992) have argued that the low $\delta^{15}N$ of sapropels in the eastern Mediterranean reflects primary production of organic matter in the presence of high concentrations of nutrients, suggesting that the sapropels were formed during periods of elevated primary production in the eastern Mediterranean.

In systems where upwelled water containing elevated concentrations of NO_3^- is advected laterally, the isotopic fractionation that accompanies the consumption of NO_3^- by primary producers (Figure 3) will lead to an increase in the $\delta^{15}N$ of the residual NO_3^- along the path of flow. In effect, this creates a spatial analog to the temporal changes in $\delta^{15}N$ of NO_3^-, primary producers, and zooplankton that arise in the course of a bloom. For example, Altabet and François (François et al. 1992; Altabet and François, this volume; Altabet and François submitted) have found an inverse relationship between the $\delta^{15}N$ of core-top sediments and concentration of NO_3^- in the overlying waters in both the equatorial Pacific and in the Indian sector of the Southern Ocean. In both cases, the spatial variations in core-top $\delta^{15}N$ appear to provide an accurate record of the effects of isotopic fractionation during phytoplankton consumption of NO_3^- as the water moves away from the site of upwelling, albeit with some diagenetic alteration of the isotopic signature preserved in the sediments. The origin of this diagenetic offset is not currently known, but diagenetic processes appear to have had similar effects in different cores, leading to $\delta^{15}N$ values generally elevated relative to expectation, but with no significant effect on the difference in $\delta^{15}N$ between cores. In the Southern Ocean, François et al. (François et al. 1992) extended this approach to samples representing the last glacial maximum (LGM); these show a pattern of variation similar to the modern ocean, though the strongest latitudinal gradient in sediment $\delta^{15}N$ occurred somewhat farther south at the LGM. This observation suggests that the boundary between nitrate-rich and nitrate-poor waters was south of the LGM Polar Front, a scenario consistent with increased utilization of nutrients by phytoplankton in Subantarctic waters.

Finally, a variety of evidence suggests that the degree of recycling of nutrients within the upper ocean exerts a strong control on the net isotopic balance of nitrogen in the mixed layer (Altabet 1988; Checkley and Entzeroth 1985; Checkley and Miller 1989; Montoya et al. 1992; Mullin et al. 1984). As noted earlier, this leads to an inverse correlation between the degree of recycling of nitrogen within the mixed layer and the average $\delta^{15}N$ of modern planktonic systems. This suggests that sedimentary nitrogen isotope measurements may provide an index to temporal variations in the relative importance of vertical mixing, and therefore new and regenerated production in a particular locale. That is, episodes of relatively intense mixing would be characterized by elevated $\delta^{15}N$ values in the euphotic zone leading to a corresponding elevation in the $\delta^{15}N$ of organic matter in sediments. Conversely, periods characterized by strong vertical stratification and high recycling efficiencies would produce low $\delta^{15}N$ values both in upper ocean plankton and subsequently in deep sea sediments.

These examples clearly show the potential for using nitrogen isotope measurements in paleoceanographic studies. Our ability to interpret the $\delta^{15}N$ of sedimentary nitrogen is currently limited by uncertainty regarding the effects of diagenesis on the $\delta^{15}N$ of sediments, and our relatively poor knowledge of the mechanisms of biological isotopic fractionation in the upper water column. Laboratory studies of the isotopic fractionation that accompanies primary production and the subsequent movement of nitrogen through planktonic ecosystems will be instrumental in clarifying the nature of the biological filtering that occurs in the transfer of nitrogen from the upper ocean to the sea floor. In the longer term, single-compound isotopic analyses (Bidigare et al. 1991; Boreham et al. 1989; Coffin et al. 1990; Macko et al. 1987) may prove to be the most sensitive tools for extracting information from marine sediments. For example, the isotopic composition of molecules specific to particular components of marine communities (e.g., photosynthetic pigments from phytoplankton and chitin from crustacean zooplankton) could potentially provide information on the structure and function of ancient planktonic communities.

CONCLUSIONS

Despite the complexity of the nitrogen cycle, several recent studies have clearly demonstrated the potential utility of nitrogen isotope measurements in paleoceanographic studies. A consideration of the food web processes that appear to mediate the vertical flux of nitrogen to the deep sea suggests that the isotopic composition of sedimentary nitrogen may not be related to the $\delta^{15}N$ of upper ocean particles in a straightforward way. Instead, the $\delta^{15}N$ of sedimentary nitrogen is most likely to provide a record of factors that affect the isotopic composition of the plankton community in the mixed layer as a whole. The degree of recycling of nitrogen within the euphotic zone, and the degree of NO_3^- utilization during the lateral advection of water masses initially containing high concentrations of NO_3^- are two examples of ecosystem-level factors that are likely to be recorded in marine sediments. Our ability to utilize nitrogen isotope measurements in paleoceanographic studies will be greatly enhanced by controlled studies of biological isotopic fractionation, and by the continuing develoment of techniques for isolating specific compounds of known origin for isotopic analysis.

REFERENCES

Altabet MA (1988) Variations in nitrogen isotopic composition between sinking and suspended particles: implications for nitrogen cycling and particle transformation in the open ocean. Deep-Sea Res. 35: 535-554

Altabet MA (1989) A time-series study of the vertical structure of nitrogen and particle dynamics in the Sargasso Sea. Limnol. Oceanogr. 34: 1185-1201
Altabet MA, Curry WB (1989) Testing models of past ocean chemistry using foraminifera $^{15}N/^{14}N$. Glob. Biogeochem. Cycles 3: 107-119
Altabet MA, Deuser WG (1985) Seasonal variations in natural abundance of ^{15}N in particles falling to the deep Sargasso Sea. Nature 315: 218-219
Altabet MA, Deuser WG, Honjo S, Stienen C (1991) Seasonal and depth-related changes in the source of sinking particles in the North Atlantic. Nature 354: 136-139
Altabet MA, François R (submitted) Sedimentary nitrogen isotopic ratio records surface ocean nitrate utilization. Glob Biogeochem Cycl
Altabet MA, McCarthy JJ (1985) Temporal and spatial variation in the natural abundance of ^{15}N in PON from a warm-core ring. Deep-Sea Res. 32: 755-772
Altabet MA, Small LF (1990) Nitrogen isotopic ratios in fecal pellets produced by marine zooplankton. Geochim. Cosmochim. Acta 54: 155-163
Bidigare R (1982) Nitrogen excretion by marine zooplankton, pp. 385-409. *In* Carpenter, EJ, Capone, DG (eds), *Nitrogen in the Marine Environment.* Academic Press, London and New York.
Bidigare RR, Kennicutt MC II, Keeney-Kennicutt WL, Macko SA (1991) Isolation and purification of chlorophylls *a* and *b* for the determination of stable carbon and nitrogen isotope compositions. Anal. Chem. 63: 130-133
Boreham CJ, Fookes CJR, Popp BN, Hayes JM (1989) Origins of etioporphyrins in sediments: Evidence from stable carbon isotopes. Geochim. Cosmochim. Acta 53: 2451-2455
Calvert S, Nielsen B, Fontugne MR (1992) Evidence from nitrogen isotope ratios for enhanced productivity during formation of eastern Mediterranean sapropels. Nature 359: 223-225
Carpenter EJ (1983) Nitrogen fixation by marine *Oscillatoria* (*Trichodesmium*) in the world's oceans, pp. 65-103. *In* Carpenter, EJ (ed), *Nitrogen in the Marine Environment.* Academic Press, London and New York
Checkley DM, Entzeroth LC (1982) Food type selection by particle-grazing copepods (abstract). EOS 63: 53-54
Checkley DM, Entzeroth LC (1985) Elemental and isotopic fractionation of carbon and nitrogen by marine, planktonic copepods and implications to the marine nitrogen cycle. J. Plankton Res. 7: 553-568
Checkley DM, Miller CA (1989) Nitrogen isotope fractionation by oceanic zooplankton. Deep-Sea Res. 36: 1449-1456
Cifuentes LA, Fogel ML, Pennock JR, Sharp JH (1989) Biogeochemical factors that influence the stable nitrogen isotope ratio of dissolved ammonium in the Delaware Estuary. Geochim. Cosmochim. Acta 53: 2713-2721
Cifuentes LA, Sharp JH, Fogel ML (1988) Stable carbon and nitrogen isotope biogeochemistry in the Delaware estuary. Limnol. Oceanogr. 33: 1102-1115
Cline JD, Kaplan IR (1975) Isotopic fractionation of dissolved nitrate during denitrification in the eastern tropical North Pacific Ocean. Mar. Chem. 3: 271-299
Coffin RB, Velinsky DJ, Devereux R Price, W A, Cifuentes LA (1990) Stable carbon isotope analysis of nucleic acids to trace sources of dissolved substrates used by estuarine bacteria. Appl. Environ. Microbiol. July 1990: 2012-2020
Delwiche CC, Steyn PL (1970) Nitrogen isotope fractionation in soils and microbial reactions. Env. Sci. Technol. 4: 929-935
DeNiro MJ, Epstein S (1981) Influence of diet on the distribution of nitrogen isotopes in animals. Geochim. Cosmochim. Acta 45: 341-351
Ehleringer JR, Rundel PW, Nagy KA (1986) Stable isotopes in physiological ecology and food web research. Trends Ecol. Evol. 1: 42-45
Emerson S, Quay P, Stump C, Wilbur D, Knox M (1991) O_2, Ar, N_2, and ^{222}Rn in surface waters of the subarctic ocean: net biological O_2 production. Glob. Biogeochem. Cycles 5: 49-69

François R, Altabet MA, Burckle LH (1992) Glacial to interglacial changes in surface nitrate utilization in the Indian sector of the Southern Ocean as recorded by sediment $\delta^{15}N$. Paleoceanography 7: 589-606

Fry B (1988) Food web structure on Georges Bank from stable C, N, and S isotopic compositions. Limnol. Oceanogr. 33: 1182-1190

Goering J, Alexander V, Haubenstock N (1990) Seasonal variability of stable carbon and nitrogen isotope ratios of organisms in a North Pacific Bay. Est. Coast. Shelf Sci. 30: 239-260

Goldman JC, Glibert PM (1983) Kinetics of inorganic nitrogen uptake by phytoplankton, pp. 233-274. *In* Carpenter EJ, Capone DG (eds), *Nitrogen in the Marine Environment.* Academic Press, London and New York

Heaton THE (1986) Isotopic studies of nitrogen pollution in the hydrosphere and atmosphere: a review. Chem. Geol. (Isot. Geosci. Sect.) 59: 87-102

Hoering TC, Ford HT (1960) The isotope effect in the fixation of nitrogen by *Azotobacter.* J. Amer. Chem. Soc. 82: 376-378

Horrigan SG, Montoya JP, Nevins JL, McCarthy JJ (1990a) Natural isotopic composition of dissolved inorganic nitrogen in the Chesapeake Bay. Est. Coast. Shelf Sci. 30: 393-410

Horrigan SG, Montoya JP, Nevins JL, McCarthy JJ, Ducklow HW, Goericke R, Malone T (1990b) Nitrogenous nutrient transformations in the spring and fall in the Chesapeake Bay. Est. Coast. Shelf Sci. 30: 369-391

Klots CE, Benson BB (1963) Isotope effect in the solution of oxygen and nitrogen in distilled water. J. Chem. Phys. 38: 890-892

Liu K-K, Kaplan IR (1989) The eastern tropical Pacific as a source of ^{15}N-enriched nitrate in seawater off southern California. Limnol. Oceanogr. 34: 820-830

Macko SA, Fogel ML, Hare PE, Hoering TC (1987) Isotopic fractionation of nitrogen and carbon in the synthesis of amino acids by microorganisms. Chem. Geol. (Isot. Geosci. Sect.) 65: 79-92

Mariotti A, Germon JC, Hubert P, Kaiser P, Letolle R, Tardieux A, Tardieux P (1981) Experimental determination of nitrogen kinetic isotope fractionation: some principles; illustration for the denitrification and nitrification processes. Plant and Soil 62: 413-430

Mariotti A, Germon JC, LeClerc A, Catroux G, Letolle R (1982) Experimental determination of kinetic isotopic fractionation of nitrogen isotopes during denitrification, pp. 459-464. *In* Schmidt HL, Forstel H, Heinzinger K (eds), *Stable Isotopes.* Elsevier, Amsterdam

McCarthy JJ (1980) Nitrogen, pp. 191-234. *In* I. Morris [ed.], *The Physiological Ecology of Phytoplankton.* Blackwell, Oxford

McCarthy JJ (1981) The kinetics of nutrient utilization., pp. 211-233. *In* Platt T (ed) *Physiological Bases of Phytoplankton Ecology.* Bulletin 210 Can. Bull. Fish. Aquat. Sci., Bulletin 210, Ottawa

McCave IN (1975) Vertical flux of particles in the ocean. Deep-Sea Res. 22: 491-502

McCready RGL, Gould WD, Barendregt RW (1983) Nitrogen isotope fractionation during the reduction of NO_3^- to NH_4^+ by *Desulfovibrio* sp. Can. J. Microbiol. 29: 231-234

Minagawa M, Wada E (1984) Stepwise enrichment of ^{15}N along food chains: further evidence and the relation between $\delta^{15}N$ and animal age. Geochim. Cosmochim. Acta 48: 1135-1140

Mitsubishi-Kasei (1978) Nitrogen isotope ratios in biogeochemical systems. Mitsubishi-Kasei Inst. Life Sci. Ann. Rept. 7: 98-100

Miyake Y, Wada E (1967) The abundance ratio of $^{15}N/^{14}N$ in marine environments. Rec. Oceanogr. Works Japan 9: 37-53

Miyake Y, Wada E (1971) The isotope effect on the nitrogen in biochemical oxidation-reduction reactions. Rec. Oceanogr. Works Japan 11: 1-6

Montoya JP, Horrigan SG, McCarthy JJ (1990) Natural abundance of ^{15}N in particulate nitrogen and zooplankton in the Chesapeake Bay. Mar. Ecol. Prog. Ser. 65: 35-61

Montoya JP, Horrigan SG, McCarthy JJ (1991) Rapid, storm-induced changes in the natural abundance of ^{15}N in a planktonic ecosystem. Geochim. Cosmochim. Acta 55: 3627-3638

Montoya JP, McCarthy JJ (in prep.) Nitrogen isotope fractionation during nitrate uptake by marine phytoplankton in continuous culture

Montoya JP, Wiebe PH, McCarthy JJ (1992) Natural abundance of ^{15}N in particulate nitrogen and zooplankton in the Gulf Stream region and Warm-Core Ring 86A. Deep-Sea Res. 39, Suppl. 1: S363-S392

Mullin MM, Rau GH, Eppley RW (1984) Stable nitrogen isotopes in zooplankton: Some geographic and temporal variations in the North Pacific. Limnol. Oceanogr. 29: 1267-1273

Owens NJP (1987) Natural variations in ^{15}N in the marine environment. Adv. Mar. Biol. 24: 390-451

Pang PC, Nriagu JO (1977) Isotopic variations of the nitrogen in Lake Superior. Geochim. Cosmochim. Acta 41: 811-814

Pennock JR, Sharp JH, Ludlam J, Velinsky DJ, Fogel ML (1988) Isotopic fractionation of nitrogen during the uptake of NH_4^+ and NO_3^- by *Skeletonema costatum*. EOS 69: 1098

Peterson BJ, Fry B (1987) Stable isotopes in ecosystem studies. Ann. Rev. Ecol. Syst. 18: 293-320

Richards FA, Benson BB (1961) Nitrogen/argon and nitrogen isotope ratios in two anaerobic environments, the Cariaco Trench in the Caribbean Sea and Dramsfjord, Norway. Deep-Sea Res. 7: 254-264

Ryther JH, Dunstan WM (1971) Nitrogen, phosphorus, and eutrophication in the coastal marine environment. Science 171: 1008-1013

Saino T, Hattori A (1980) ^{15}N natural abundance in oceanic suspended particulate matter. Nature 283: 752-754

Saino T, Hattori A (1985) Variation of ^{15}N natural abundance of suspended organic matter in shallow oceanic waters, pp. 1-13. *In* Sigleo AC, Hattori A (eds), *Marine and Estuarine Geochemistry*. Lewis Publishers, Chelsea, MI

Saino T, Hattori A (1987) Geographical variation of the water column distribution of suspended particulate organic nitrogen and its ^{15}N natural abundance in the Pacific and its marginal seas. Deep-Sea Res. 34: 807-827

Smith SV (1984) Phosphorus versus nitrogen limitation in the marine environment. Limnol. Oceanogr. 29: 1149-1160

Smith SV, Kimmerer WJ, Walsh TW (1986) Vertical flux and biogeochemical turnover regulate nutrient limitation of net organic production in the North Pacific Gyre. Limnol. Oceanogr. 31: 161-167

Steele KW, Daniel RM (1978) Fractionation of nitrogen isotopes by animals: a further complication to the use of variations in the natural abundance of ^{15}N for tracer studies. J. Agric. Sci., Camb. 90: 7-9

Voss M (1991) Räumliche und zeitliche Verteilung stabiler Isotope ($\delta^{15}N$, $\delta^{13}C$) In suspendierten und sedimentierten Partikeln im Nördlichen Nordatlantik. PhD Disertation, Christian-Albrechts Universität zu Kiel. 102 pp

Wada E (1980) Nitrogen isotope fractionation and its significance in biogeochemical processes occurring in marine environments, pp. 375-398. *In* Goldberg E, Horibe Y, Saruhashi K (eds), *Isotope Marine Chemistry*. Uchida Rokakuho, Tokyo

Wada E, Hattori A (1976) Natural abundance of ^{15}N in particulate organic matter in the North Pacific Ocean. Geochim. Cosmochim. Acta 40: 249-251

Wada E, Hattori A (1978) Nitrogen isotope effects in the assimilation of inorganic nitrogenous compounds. Geomicrobiol. J. 1: 85-101

Wada E, Kadonaga T, Matsuo S (1975) ^{15}N abundance in nitrogen of naturally occurring substances and global assessment of denitrification from isotopic viewpoint. Geochem. J. 9: 139-148

Wada E, Terazaki M, Kobaya Y, Nemoto T (1987) ^{15}N and ^{13}C abundances in the Antarctic Ocean with emphasis on biogeochemical structure of the food web. Deep-Sea Res. 34: 829-841

Yoshida N (1988) ^{15}N-depleted N_2O as a product of nitrification. Nature 335: 528-529

THE USE OF NITROGEN ISOTOPIC RATIO FOR RECONSTRUCTION OF PAST CHANGES IN SURFACE OCEAN NUTRIENT UTILIZATION

M. A. Altabet and R. Francois
Dept. of Marine Chemistry and Geochemistry
Woods Hole Oceanographic Institution
Woods Hole, MA 02543

ABSTRACT. Sedimentary $^{15}N/^{14}N$ ratios can be used as a unique measure of past changes in surface ocean nutrient utilization which in turn is a function of past changes in productivity and nutrient input from deeper waters. Due to isotopic fractionation during nitrate uptake by phytoplankton, partial nutrient utilization produces substantial ^{15}N enrichment in particulate matter reaching the seafloor. Examples are given from the N. Atlantic, Equatorial Pacific, and Southern Ocean. In the latter two regions large variations in nutrient utilization with latitude are observed which go along with gradients in both near-surface ocean and core top $^{15}N/^{14}N$. Thus isotopic signals generated in surface waters are transferred and preserved in sediments. First down core results from the Southern Ocean indicate that only modest increases in nutrient utilization and perhaps productivity occured in Subantarctic waters uring the last glacial maximum.

INTRODUCTION

The availability of nutrients in the surface ocean is an important control on primary production, whose variability over geologic time may have altered major biogeochemical cycles and influenced climate (e.g. Sarmiento and Toggweiler, 1984; Siegenthaler and Wenk, 1984; Knox and McElroy, 1984; Toggweiler and Sarmiento, 1985). Cd/Ca and $\delta^{13}C$ in foraminiferal calcite have separately been used as indicators for surface nutrient concentrations (Boyle, 1988; Keigwin and Boyle, 1989; Labeyrie and Duplessy, 1985; Charles and Fairbanks, 1990). Ge/Si ratios in preserved diatom frustrules have been used to reconstruct past changes in surface silicate utilization (Froelich et al., 1989) but now appear to actually be an indicator for changes in oceanic Ge/Si (Froelich, pers. comm.). Despite the importance of fixed nitrogen as a limiting nutrient, means have not been previously available for specific reconstruction of past surface concentrations or utilization.

NATO ASI Series, Vol. I 17
Carbon Cycling in the Glacial Ocean:
Constraints on the Ocean's Role in Global Change
Edited by R. Zahn et al.

Natural variations in nitrogen isotopic ratio have been widely used as a contemporary tracer of nitrogen biogeochemistry in the marine environment (see Montoya, this volume for a general review), but paleoceanographic applications have been rather limited (Rau et al., 1987; Altabet and Curry, 1989). It is only recently that sedimentary $^{15}N/^{14}N$[1] have been applied as the only recorders available of surface nutrient utilization (Francois et al., 1992; Francois et al., 1993; Calvert et al., 1992; Altabet and Francois, 1993). Findings to date indicate substantial potential for sedimentary $\delta^{15}N$, by detecting the degree of nutrient depletion of overlying surface waters, to make important contributions to our understanding of past changes in surface productivity and its influence on ocean biogeochemistry.

This application of sedimentary $\delta^{15}N$ is based on the observation that, in contemporary surface waters, nutrient depletion is strongly related to large changes in the $\delta^{15}N$ of particulate organic matter (POM) as a result of isotopic fractionation during phytoplankton utilization of nitrate (e.g. Altabet et al., 1991). Paleoceanographic use, however, hinges in part on the fidelity with which this isotopic signal generated in surface waters is preserved in sediments. Unlike Cd/Ca and calcite $\delta^{13}C$ which are carried in mineral phases of marine microfossils, the $\delta^{15}N$ signal is carried in organic matter raining down from surface waters. Therefore, $\delta^{15}N$ values can potentially be substantially altered by particle decomposition (Saino and Hattori, 1980; Altabet and McCarthy, 1986; Saino and Hattori, 1987; Altabet, 1988) and trophic exchange of nitrogen (DeNiro and Epstein, 1981; Minagawa and Wada, 1984; Fry, 1988). Despite this perspective, recent work summarized below does demonstrate that the surface signal is coherently transferred to and preserved in the sediments of the Southern Ocean and Equatorial Pacific. The following sections also include a brief methodological description, and discussion of the factors influencing generation and magnitude of the surface-ocean signal and its transfer to the seafloor.

MATERIALS AND METHODS

In all the examples given below (Table 1), suspended PN (particulate nitrogen) was collected from near-surface waters by filtering water samples onto precombusted (450° C), Whatman GF/F filters. These filters have been found to have very low blank levels when precombusted, retain particles at small enough sizes to collect the majority of open ocean PN, and are sufficiently robust for convenient handling (Altabet, 1990). Protocols have been designed to minimize contamination from air-born particles (e.g. in-line filter holders were used

---

[1]measured as:

$$\delta^{15}N\,(‰) = \frac{^{15}N/^{14}N_{sample} - {}^{15}N/^{14}N_{atm.\ N_2}}{^{15}N/^{14}N_{atm.\ N_2}} \times 1000$$

TABLE 1. Sample types and locations for nitrogen isotopic data discussed.

Region	Sample Type	Latitude	Longitude
Subarctic Pacific	NO_3^-	50°N	145°W
Temperate N. Atlantic	Suspended particles	47°N	20°W
	Sinking particles	47°N	20°W
Equatorial Pacific	Suspended particles	12°N to 12°S	135° to 140°W
	Sinking particles	12°N to 12°S	135° to 140°W
	Sediment core tops	11°N to 12°S	140°W
Southern Ocean	Suspended particles	29° to 59°S	51° to 80°E
	Sediment core tops	30° to 61°S	101° to 115°E

instead of uncovered filter funnels). When dried at 60° C in the laboratory, sample material can be kept indefinitely prior to preparation for isotopic analysis. In contrast to small, suspended particles, large, fast-sinking particles are the actual carriers of organic matter to the seafloor but are too rare to be adequately sampled by filtration of small to moderate volumes. For this study, samples of sinking PN were collected using either floating (PIT - style; Knauer et al., 1979) or time-series, moored sediment trap (Honjo and Doherty, 1987) After retrieval, zooplankton that actively entered the traps ('swimmers') were removed from samples.

Equatorial Pacific sediments were collected along 135°W between 11° N and 12° S in 1988 using a Tetrahedron piston corer designed to preserve the sediment/water interface (W. Martin, pers. comm.). Southern Ocean sediments were collected using piston cores during *Eltanin* cruises #45 and #49 to the southeast sector of the Indian ocean in the early 1970's and subsequently stored at the Antarctic Core Facility, Department of Geology at Florida State University. To ensure that samples from the Southern Ocean were at or near the core top, trigger cores were primarily used. Sediment samples were dried at 60°C and homogenized with a mortar and pestle prior to subsampling for isotopic analysis.

Most samples were prepared using combustion in evacuated quartz tubes (Minagawa et al., 1985). They were placed in quartz ampules with CuO, Cu, and Ag metal as reagents. Following evacuation on a vacuum line, the ampules were flame sealed, combusted at 850° C, and cooled overnight. The N_2 gas evolved was separated and purified on a manifold attached to the mass spectrometer inlet. Isotopic analyses were made using a Finnigan MAT 251 isotope ratio mass spectrometer with atmospheric N_2 as the standard. About 100 mg of sediment was used per analysis depending on nitrogen content. Typical reproducibility for the $\delta^{15}N$ of sediments samples and suspended particles collected from surface waters is ± 0.35 ‰ and ± 0.15 ‰ respectively. Sinking and suspended particles from the Equatorial Pacific (JGOFS EqPac study, Fall Survey cruise) were analyzed using an elemental analyzer (Europa Euroboprep) attached to the Finnigan MAT 251. Samples were automatically combusted and purified to remove CO_2 and H_2O. N_2 in the continuous flow of He from the elemental analyzer is introduced directly into the mass spectrometer via a splitter valve and crimp which regulated the flow. Reproducibility and sample size requirements have been found to be superior to the sealed ampule technique. Another principal advantage of this technique is a ten-fold reduction in preparation and analysis time. The increased sample throughput will be critical to future paleoceanographic studies requiring high-resolution data sets.

GENERATION OF THE SURFACE NITROGEN ISOTOPIC SIGNAL

Laboratory studies (Wada and Hattori, 1978; Montoya, 1990; Montoya, this volume) have shown that phytoplankton utilization discriminates against $^{15}NO_3^-$ relative to $^{14}NO_3^-$ such that phytoplankton PN has low $\delta^{15}N$ values relative to the NO_3^- source. As NO_3^- is depleted, its $\delta^{15}N$ rises as dictated by 1st-order Raleigh fractionation kinetics producing a subsequent increase in the $\delta^{15}N$ of newly produced PN:

1) $\delta^{15}NO_3^-{}_{(f)} = \delta^{15}NO_3^-{}_{(f=1)} - \varepsilon \times \ln(f)$

2) $\delta^{15}N\text{-}PN_{(f)} = \delta^{15}NO_3^-{}_{(f)} - \varepsilon$

where f is the fraction of unutilized NO_3^- remaining ($[NO_3^-]/[NO_3^-]_{initial}$), ε_u[2] is the fractionation factor in ‰ units, and $\delta^{15}N$-PN refers to the $\delta^{15}N$ of PN produced at any single point in the 'reaction' and not an accumulating pool. These equations are close approximations of those given by Miyake and Wada (1967) to permit convenient use of the 'δ' notation. They also underscore that changes in $\delta^{15}N$ are in reality a function of relative NO_3^- utilization (depletion) and not concentration, *per se*. If there is no removal of PN from the system (almost never the case), the change in $\delta^{15}N$ of this accumulating pool as a function of f is given by the integral of eqs. 1 and 2 (Miyake and Wada, 1967):

[2] $\varepsilon = (k/k^* - 1) \times 1000$, where k is first-order rate constant and '*' denotes the heavier isotope.

TABLE 2. Estimates for the isotope fractionation effect (ε_u) assocated with NO_3^- utilization by phytoplankton.

Citation	Method for Estimate	ε_u	Notes
Culture Experiments			
1) Wada and Hattori (1978)	Changing $\delta^{15}N$ of PN with NO_3^- depletion	0 to 19‰	Varied with growth rate
2) Montoya (1990)	Changing $\delta^{15}N$ of PN with NO_3^- depletion	0.7 to 6.3‰	Varied with species but not with growth rate
Field Studies			
3) Wada (1980) N.W. Pacific	a. Difference in $\delta^{15}N$ between suspended PN and NO_3^-	5‰	Assumed $\delta^{15}N$ of sinking PN equaled $\delta^{15}N$ of suspended PN
	b. Vertical change in $\delta^{15}NO_3^-$- with changing $[NO_3^-]$ (our estimate)	8‰	
4) Biggs et al. (1988) Weddell Sea	Difference in $\delta^{15}N$ between sinking PN and NO_3^-	8‰	
5) This report Subarctic Pacific	Vertical change in $\delta^{15}NO_3^-$ with changing $[NO_3^-]$	9‰	(See Fig. 1)
6) Altabet et al. (1991) Temp. N. Atlantic	Difference in $\delta^{15}N$ between suspended PN and NO_3^-	8 to 9‰	$\delta^{15}N$ of suspended PN shown to equal sinking PN

Perhaps more relevant, though, for our study are field estimates for ε_u which fall within the more narrow range of 8 to 9 ‰ (Table 2) from the temperate N. Atlantic (Altabet et al., 1991; Fig. 2), the Weddell sea (Biggs et al., 1988) and the Subarctic Pacific (Fig. 1). Variability in ε_u for conditions of moderate to high surface $[NO_3^-]$ may be rather small. As seen below, the Equatorial Pacific may be an exception. In the northwest Pacific, Wada (1980) estimated ε_u to be 5‰. However applying eq. 2 to their vertical profile for $[NO_3^-]$ and $\delta^{15}NO_3^-$ in the euphotic zone as was done for the Subarctic Pacific (Fig. 1), an estimate of ε_u near 8 ‰ is derived. Despite the apparent concordance of field estimates for ε_u, the available data set is still rather sparse and additional studies are required in specific regions of interest.

POSSIBLE DIAGENETIC ALTERATIONS OF NITROGEN ISOTOPIC RATIOS

In the transfer of PN from the surface ocean to the sediments, organic matter transformation and decay has the potential for producing large alterations of $\delta^{15}N$ values which would severely compromise their use as a paleoceanographic tool. Breakdown of organic matter may preferentially release isotopically depleted dissolved nitrogen with subsequent enrichment of the remainder (Saino and Hattori, 1980). Isotopic fractionation can occur during bond cleavage or there may be selective removal of molecular classes which vary in isotopic ratio. The first step in the downward transfer of PN is the conversion of small, suspended particles into large fast sinking ones which can sink at speeds of 50 to 200 m/day. In the case of the JGOFS N. Atlantic Bloom Experiment, sinking and suspended particles had the same $\delta^{15}N$ values (Fig. 2). Apparently, formation of large particles was primarily by physical aggregation with little alteration of suspended PN (Altabet et al., 1991). However, macrozooplankton are generally known to be important facilitators of large particle formation by production of mucus webs and/or fecal pellets (Fowler and Knauer, 1986). The latter were found to be enriched in ^{15}N by 2‰ relative to zooplankton food sources (Altabet and Small, 1990). Trophic transfer (ε_u.g. phytoplankton to zooplankton) is well known to increase $\delta^{15}N$ by 3 to 4‰ per trophic step (DeNiro and Epstein, 1981; Minagawa and Wada, 1984; Fry, 1988). Since suspended PN in surface waters is usually a mixture of phytoplankton, bacteria, microzooplankton, and detritus, selection by macrozooplankton of components with distinct $\delta^{15}N$ in addition to fecal pellet production would alter $\delta^{15}N$ values between primarily produced organic matter and sinking particles. In fact, observations in the Sargasso Sea show an enrichment in ^{15}N in sinking particles of up to 5‰ (Altabet, 1988). Even larger differences were observed in the Equatorial Pacific (see below).

Nevertheless, this 'particle removal' effect does not perturb the relationship between the $\delta^{15}N$ of sinking particles and degree of NO_3^- utilization. Though counter-intuitive, this conclusion is derived from consideration of a simple mass balance for nitrogen in the euphotic zone:

Recycled N
↑ | ↓
ε_u ε_{pr}
NO_3^- ⇒ susp. PN ⇒ sinking PN

Since there needs to be, on average, mass balance between euphotic zone NO_3^- utilization (the source of 'new' nitrogen) and the downward export of organic N (Eppley and Peterson, 1979),

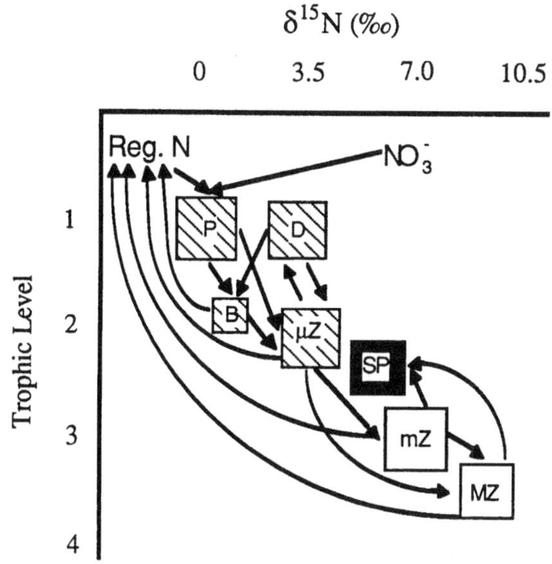

FIGURE 3. Simplified schematic of nitrogen cycling in the upper ocean illustrating the isotopic effects of particle removal from the euphotic zone. Phytoplankton (P), bacteria (B), microzooplankton (μZ), and detritus (D) are operationally within the pool of suspended particles (cross-hatched boxes). Mesozooplankton (mZ) and macrozooplankton (MZ) have key roles as contributors to the formation of sinking particles (SP). In a relatively complex food web, there are multiple trophic steps between phytoplankton and the production of sinking particles when zooplankton select higher trophic level components of the suspended particle pool. Since $\delta^{15}N$ increases by 3.5‰ per step , sinking particles are enriched in ^{15}N relative to the average for suspended particles. Mass balance dictates that $\delta^{15}N$ values for assimilated NO_3^- and sinking particles (the source and sink for N from the system respectively) need to be equivalent. As a result, particle removal produces relative depletion of ^{15}N in suspended particles. A necessary mechanism for this phenomenon is the assimilation by phytoplankton of regenerated N low in $\delta^{15}N$ (Checkley and Miller, 1989).

isotopic balance between the two is also required (Fig. 3). The $\delta^{15}N$ of sinking particles exiting the euphotic zone should match the $\delta^{15}N$ of NO_3^- utilized by phytoplankton; $\delta^{15}NO_3^-$ minus ε_u when there is little utilization, $\delta^{15}NO_3^-$ when there is complete utilization. An example of the latter is the Sargasso Sea, where $\delta^{15}N$ values for sinking PN and NO_3^- below the euphotic zone were equivalent despite up to 5‰ lower values for suspended PN (Altabet , 1988; Altabet, 1989). If nitrogen fixation or atmospheric sources of new nitrogen were significant this view of the isotopic budget would not hold. However, these sources have been shown to be generally trivial compared to marine NO_3^- even under oligotrophic conditions (McCarthy and Carpenter, 1983) and even more so in the settings discussed below in which

NO_3^- is found at elevated concentrations. In contrast to sinking PN, the isotopic composition of suspended PN at steady state (more precisely suspended PN + recycled N pools, though the latter is usually found at very low concentrations in surface waters) is in addition a function of particle removal effect (ε_{pr}); $\delta^{15}N_{PN} = \delta^{15}NO_3^- - \varepsilon_u + \varepsilon_{pr}$. Because sinking particles are enriched in ^{15}N relative to suspended PN, ε_{pr} is negative. Production and consumption of recycled N may also have significant fractionation effects associated with these processes (Checkley and Miller, 1989) but their principal influence will be on the difference in $\delta^{15}N$ between suspended PN and recycled N and not on the $\delta^{15}N$ for sinking PN. Previously, low $\delta^{15}N$ values for euphotic zone PN had been mistakenly cited as evidence for dependence on nitrogen fixation (e.g. Minagawa and Wada, 1986). As shown in the above discussion, valid interpretation of isotopic data for any system must take into account both its sources and sinks.

During the downward descent of sinking particles, there is further opportunity for alteration of isotopic composition; only a small fraction of the PN flux from near-surface waters reaches the seafloor. In fact, suspended PN exhibits almost universally large increases in $\delta^{15}N$ (5 to

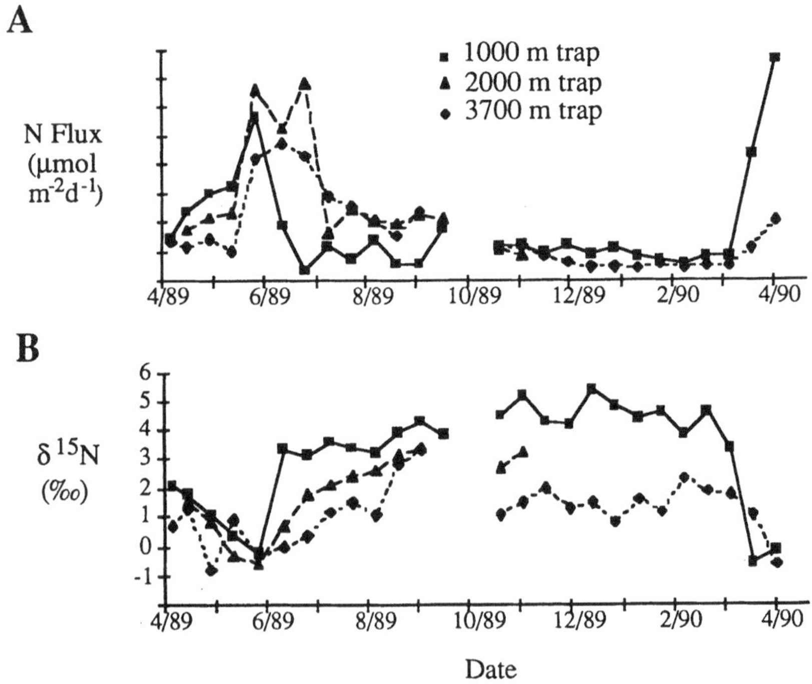

FIGURE 4. Deep moored sediment trap time-series taken between April 1989 and April 1990 during the JGOFS N. Atlantic Bloom Experiment (NABE). **A.** PN flux at 1000, 2000 and 3700 m. **B.** Corresponding $\delta^{15}N$ values.

10‰) with depth below the euphotic zone that have been ascribed to the effects of decomposition (Saino and Hattori, 1980; Altabet and McCarthy, 1986; Saino and Hattori, 1987). Altabet (1988), however, concluded that contributions to the deep suspended PN pool from the breakup of sinking particles enriched in ^{15}N could also account for these observations. In contrast, deeply sinking PN during the JGOFS N. Atlantic Bloom experiment exhibited no increase in $\delta^{15}N$ with depth (Altabet et al., 1991; Fig. 4). During bloom periods of high flux (which would dominate the sedimentary record) there is no change in $\delta^{15}N$ with depth. During periods of low flux there is a modest (and curious) decrease in $\delta^{15}N$ which may reflect low-level sources to sinking particles in the ocean's interior. These observations have been shown to be general for the ocean with similar results in the Norwegian and Greenland Seas (Voss, 1991), the Sargasso Sea (Altabet et al., 1991), the Southern Ocean (Wada et al., 1987) and the S. California Bight (Williams et al., 1992).

Further removal of organic matter occurs at the sediment/water interface and in the upper sections of the sediment. Alterations of isotopic composition that may accompany incorporation into the sediments was partially assessed using 1 mm sections of a core from the Equatorial Pacific In the upper 2 cm, % organic C decreases from 0.55 to 0.26 % (W. Martin, unpublished) and % nitrogen decreases from 0.065 to 0.045 (Fig. 5). Despite the clear

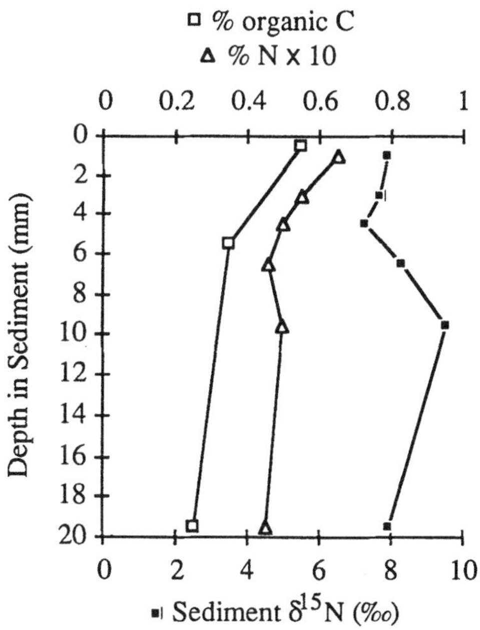

FIGURE 5. Vertical variations in % organic Carbon (W. Martin, unpublished data), % Nitrogen, and $\delta^{15}N$ within the upper 20 mm of the sediment in core T42 (collected in the Equatorial Pacific at 1°S 135°W).

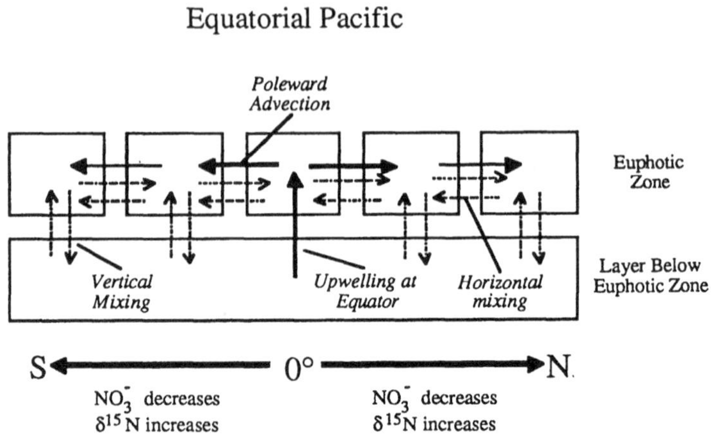

FIGURE 6. Schematic representation of the physical processes that may influence the latitudinal distribution of $\delta^{15}N$ in the Equatorial Pacific. New nitrogen sources for this region are dominated by NO_3^- upwelled at the equator. Progressive utilization of this NO_3^- by phytoplankton that occurs with poleward advection of the upwelled water mass produces subsequent enrichment in ^{15}N. It is expected that near the boundaries of the system local vertical transport becomes more important as a source of NO_3^-. However, there may be some influence from this process throughout the region (see text).

removal of organic matter with depth (bioturbation was not rapid enough to homogenize the upper sediment column), there is no corresponding shift in $\delta^{15}N$ over this depth interval. As with removal of PN from the euphotic zone and subsequent downward transport , early diagenesis in sediments does not appear to cause significant alteration of nitrogen isotopic ratios as compared to the surface-generated signals discussed below. The most conclusive evidence that sediments accurately record the surface-generated ^{15}N signal though would be a direct comparison between the two as discussed in the next section.

THE EQUATORIAL PACIFIC STUDY

In the Equatorial Pacific, continuous upwelling at 0° latitude (except strong El Niño years) is the single primary source for NO_3^- for this region (Fig. 6). At the equator, surface [NO_3^-] (~5 to 8 μmol/l at 135 to 140° W) is relatively high for tropical waters. As waters upwelled at the equator are advected to the north and south, utilization of NO_3^- by phytoplankton reduces concentrations to nearly undetectable levels (Fig. 7). At 135° W, this effect is well known to occur more rapidly to the north than to the south whereas further to the west at 150° W, NO_3^- distribution is rather symmetrical about the equator. Variability in surface nutrient distributions occurs in response to El Niño events and the passage of equatorial Kelvin waves. Surface

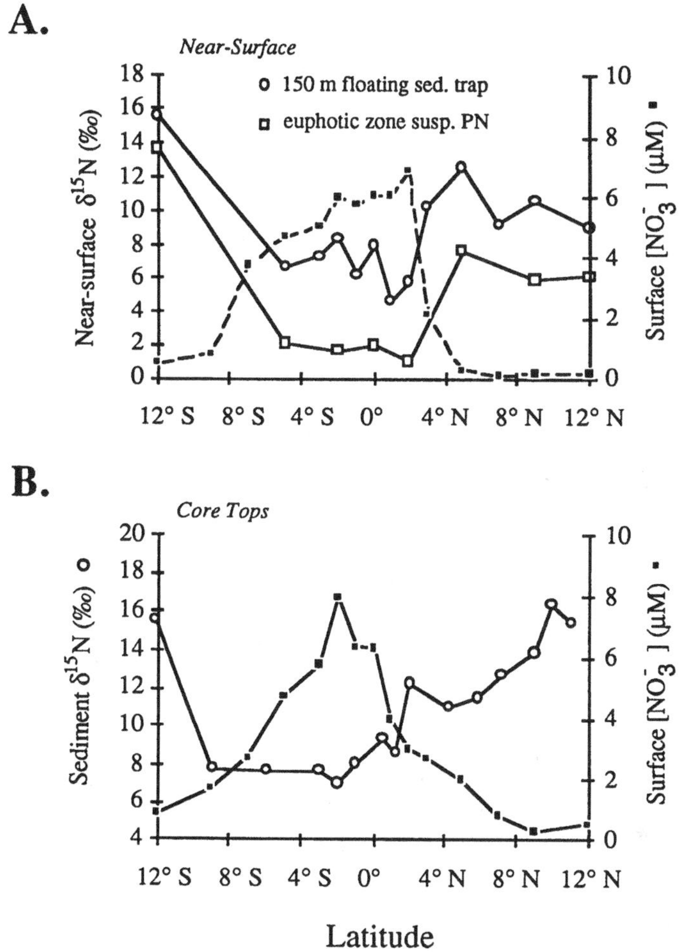

FIGURE 7. Results for north-south transects of the Equatorial Pacific between 135 and 140° W. **A.** Comparison of latitudinal variations in surface [NO_3^-] and $\delta^{15}N$ for sinking particles collected at 150 m by floating sediment trap during the U.S. JGOFS EqPac Fall Survey cruise. **B.** Comparison of latitudinal variations in surface [NO_3^-] and $\delta^{15}N$ for core-top sediments. NO_3^- data are the climatological data of Barber and Chavez (1991).

sediments in this region, though, would record signals averaged over the last 100's to 1000 years based on sediment accumulation rates (W. Martin pers. comm.). Barber and Chavez (1991) have made an 'historical' average of NO_3^- data for the region which we will use as the best data set for comparison to sedimentary $\delta^{15}N$. We need to take into account in our analysis, however, that the historical data of Barber and Chavez averages all data between 110° and 150° W into latitudinal bins and the asymmetry at 135 to 140°W is not apparent (Fig. 7b).

The $\delta^{15}N$ of bulk sediment from core tops along 135° W varies dramatically with latitude and mirrors the surface distribution of NO_3^- (Fig. 7b). The lowest values are found near the equator (~7‰) with highest ones (~16‰) at the northern and southern boundaries of the system. Also apparent in the data is the southwardly skewed distribution in $\delta^{15}N$. Despite large differences in time scale (days as compared to 10^2 to 10^3 years), $\delta^{15}N$ for sinking particles collected at 150 m agrees well with the sediment data both in general magnitude and covariation with surface NO_3^- values (Fig. 7a). In the Equatorial Pacific, sediments do indeed record the surface nitrogen isotopic signal generated by NO_3^- utilization. Differences north of the equator are associated with a sharper nutrient 'front' found during the EqPac Fall Survey Cruise than in the historical average. At the time the near-surface samples were collected, the equatorial system had nearly recovered from El Niño conditions (reduced upwelling of cold, nutrient-rich water), but normal conditions apparently had not fully returned to the region between 3° and 7°N. Suspended particles collected at 40 m depth show latitudinal variations in $\delta^{15}N$ similar in shape and amplitude as found in sinking particles. $\delta^{15}N$ values, though, are substantially lower probably due to particle removal effects discussed above.

The Equatorial Pacific presents are good example of how quantitative reconstruction of past variations in nutrient utilization will be strongly model dependent. Choice of equation (eqs. 1 and 2 vs. 3) to model the relationship between NO_3^- utilization and $\delta^{15}N$ depends in part on the residence time for suspended PN in surface waters compared to NO_3^- depletion time. Short particle residence times would approximate the conditions of the 'instantaneous' product eqs. 1 and 2. Long particle residence times relative to NO_3^- depletion would be best fit by the 'accumulated product'; eq. 3. Eppley et al. (1992) found particle residence times in the surface layer of the Equatorial Pacific to be only 4 to 15 days whereas NO_3^- consumption times are on the order of hundreds of days (approximated from poleward current velocities and horizontal extent of NO_3^- rich waters). As a result, it is expected that variations in $\delta^{15}N$ for sinking particles with latitude should follow eqs. 1 and 2. Assuming maximum surface $[NO_3^-]$ observed during the Fall EqPac Survey cruise equivalent to initial $[NO_3^-]$, most of the sinking particle data fit well the change in $\delta^{15}N$ expected using eq. 1 & 2 for $\varepsilon_u = 2.5‰$ (Fig. 8). The principal exceptions are points associated with the lowest NO_3^- concentrations. These are for stations at the northern boundary of the system where otherwise trivial terms in our implicitly assumed nitrogen budget become important (ε_u.g. local vertical flux of NO_3^-). Using the maximum surface $[NO_3^-]$ for the historical data as initial $[NO_3^-]$, the core top data also fit eq. 1 & 2 well with $\varepsilon_u = 2.5‰$ (Fig. 8). The one point that significantly falls off the relationship is for 5° S where we believe the climatological data does not accurately portray the southward asymmetry in NO_3^- distribution near 140° W observed during the EqPac cruises and in other data sets (e.g. Peña et al., 1991).

In the Equatorial Pacific, data for sinking particles and core tops are not only generally similar in $\delta^{15}N$, but also provide good fits to eq. 1 & 2 for the same value of ε_u when the most

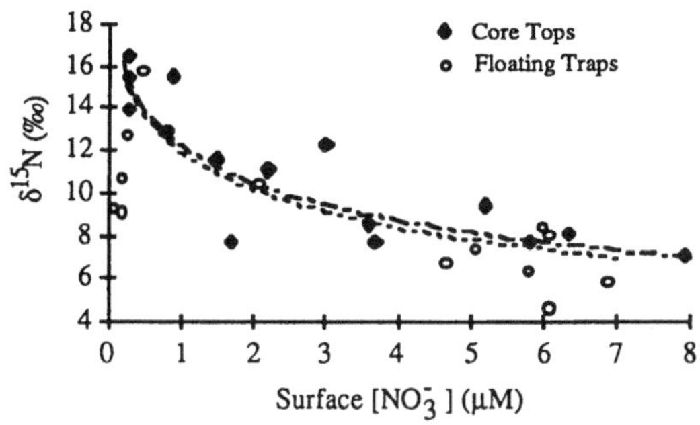

FIGURE. 8 $\delta^{15}N$ of sinking particles and core top sediments vs. surface [NO_3^-] in the equatorial Pacific. Lines show fit of Raleigh fractionation equations for the instantaneous product (eqs. 1 and 2 in text). Upper curve is the fit to sediment data and the lower curve is the fit for the sediment trap data. In each case, [NO_3^-]$_{initial}$ was assumed to be the maximum [NO_3^-] observed in each of the 2 data sets. *In lieu* of $\delta^{15}N$ data for NO_3^-, $\delta^{15}NO_3^-{}_{initial}$ was set, taking the station of maximum [NO_3^-], as $\delta^{15}N$(particulate) + ε_u. ε_u was set at 2.5‰ for both curves.

appropriate data sets for NO_3^- are used (synoptic data for comparison to sinking particles and climatological data for comparison to the sediments). The isotopic signal generated in surface waters by the progressive depletion of NO_3^- away from the equator is seen to be reliably transferred to and preserved in the sediments. The low value for the fractionation factor relative to other field estimates for the open ocean (Table 2) may be explained by the predominance of forms of phytoplankton other than diatoms in these tropical waters (Chavez, 1989) which have been shown in culture studies to have more modest values for ε_u (Montoya, 1990). Alternatively, there may be, contrary to assumptions (Fig. 6), significant contributions from NO_3^- that is locally, vertically mixed into the euphotic zone along the poleward advection paths. Since there has been no prior utilization, this source would be relatively depleted in ^{15}N resulting in dilution of the latitudinal signal. Another uncertainty involves the actual $\delta^{15}N$ for NO_3^- upwelled at the equator. Our lowest sinking particle $\delta^{15}N$ value (5‰) is similar to values for deep water NO_3^- (5 to 6 ‰) instead of being depleted by 2.5‰. Upwelled waters at the equator have the equatorial undercurrent as their source (Bryden and Brady, 1985), and this source of NO_3^- may be relatively enriched in ^{15}N as compared to deep water. Both these questions will be resolved with planned isotopic analyses of NO_3^- itself.

THE SOUTHERN OCEAN STUDY

Our suite of measurements in this region span the north to south transition from NO_3^- depleted, oligotrophic waters to nutrient-rich Subantarctic and Polar surface waters (NO_3^- reaches >25 μM). These changes in [NO_3^-] are brought about by a combination of the north to south increase in the nutrient content of winter-time source waters for the surface layer (denser isopycnals outcrop to the south as seen by temperature changes) and concomitant north to south reduction in the seasonal utilization of nutrients (Fig. 9). In the S.W. Indian Ocean, where our surface $\delta^{15}N$ samples were taken, topographic narrowing of the Antarctic Circumpolar Current (ACC) causes the transition from low to high surface [NO_3^-] to occur within a few degrees latitude (Figs. 10 and 11) at the Subantarctic front. Despite high NO_3^- in the colder polar water, austral winter to summer differences in near surface NO_3^- indicate greater nutrient utilization north of the front. Unlike the Equatorial Pacific, latitudinal changes in near-surface [NO_3^-] are not a direct indicator of utilization. Instead, seasonal amplitudes of variation best reflect new nutrient utilization. Unfortunately austral winter data sets such as the one shown in Fig. 11 are rare, making calculations of seasonal utilization (f) difficult. As expected from the Equatorial Pacific study, latitudinal changes in $\delta^{15}N$ for near-surface, suspended PN are inversely related to these variations in NO_3^- particularly in the frontal region (Fig. 12). An 11‰ decrease in $\delta^{15}N$ (north to south) is observed within the same narrow latitude band in which [NO_3^-] dramatically increases and the correlation between the two is highly statistically significant in this region (Fig. 12b).

Isotopic analysis of core tops from collections taken in the southeast Indian Ocean provided the best available comparison with our surface observations (Fig. 10). Regardless of longitude, the northern half of the Southern Ocean is dominated by the Subtropical, Subantarctic, and Polar fronts between which lies the strong east to west flow of the Antarctic Circumpolar Current (ACC). It is expected that variation in $\delta^{15}N$ along lines of latitude would only result

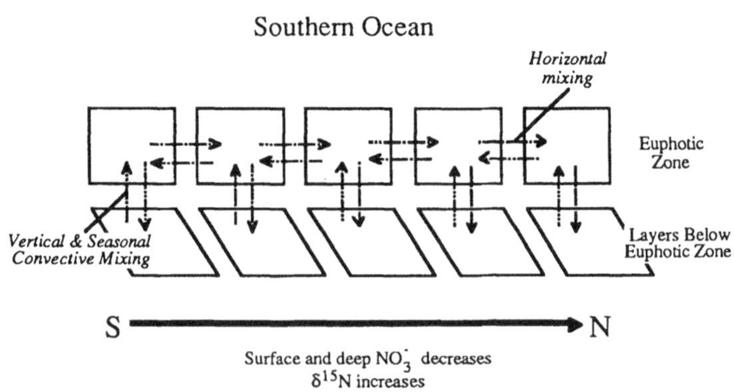

FIGURE 9. Schematic representation of the physical processes that may influence the latitudinal distribution of $\delta^{15}N$ in the Southern Ocean. This region is dominated by seasonal local replenishment of surface NO_3^- by deep convective mixing in Austral winter followed by partial depletion of NO_3^- in Austral spring and summer. Due to descending isopycnal surfaces, there is an equatorward decrease in [NO_3^-] in the region below the euphotic zone. There is a corresponding equatorward decrease in wintertime surface [NO_3^-] (see Fig. 11).

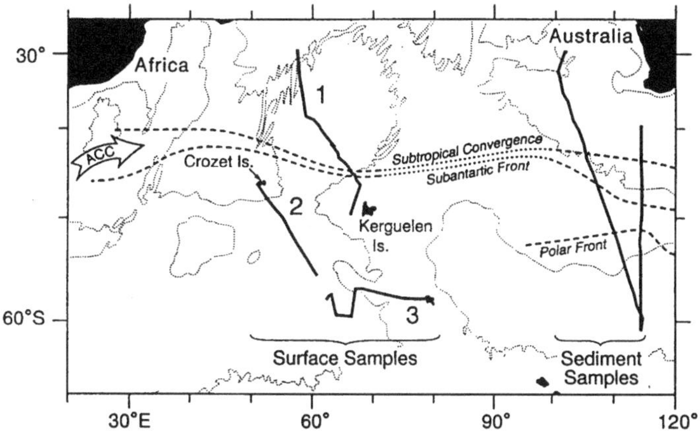

FIGURE 10. Map of transects made in the Southern Ocean, Indian Ocean sector. The frontal system divides nutrient depleted subtropical waters to the north from nutrient rich waters to the south. Dotted portions indicate the longitudinal continuity of the fronts and do not represent their actual latitudinal position in the south central Indian Ocean. Samples of near-surface suspended particles were collected along track lines in the S.W. Indian Ocean. Here, topographic steering of the Antarctic Circumpolar current by the ocean ridge system horizontally squeezes the frontal zone. In the S.E. Indian ocean, core-top sediments were analyzed for $\delta^{15}N$ along the track lines shown. Going west to east in this region, topographic steering of the Antarctic Circumpolar horizontally displaces the frontal zone to the south.

from changes in the position and extent of the frontal system. Similar to the southwest sector, data for the southeast Indian Ocean from the NODC archive (data adjustments made as suggested by Gordon and Molinelli, 1986) and GeoSecs program (Weiss et al., 1983) also show large north to south increases in surface nutrient concentration (characteristic of the entire Southern Ocean). Here, though, the transition from low to high surface [NO_3^-] occurs over a substantially wider latitude band of Subantarctic water between the subtropical and polar fronts (37 to 54° S). Similar to the southwest Indian Ocean, available data, though variable and uneven in distribution, also suggests substantial seasonal variability in surface [NO_3^-] in the frontal region.

$\delta^{15}N$ values for surficial sediments were measured along two converging transects across these frontal systems (Fig. 10, for detailed description see Francois et al., 1992). As found for surface PN, sedimentary $\delta^{15}N$ is relatively high in sediments underlying oligotrophic waters and decreases southward by 7‰ (Fig. 13). For our most southern samples, our results are not significantly different from those of Wada et al. (1987) for nearby sites in the Australian sector of the Southern Ocean (5 to 6‰). These sedimentary values, though, are overall significantly higher than our near-surface results. The southward shift in the latitudinal distribution in $\delta^{15}N$ for the eastern transect is associated with a documented west to east bend to the south in the Subantarctic front at this longitude (Francois et al., 1992) also produced by topographic

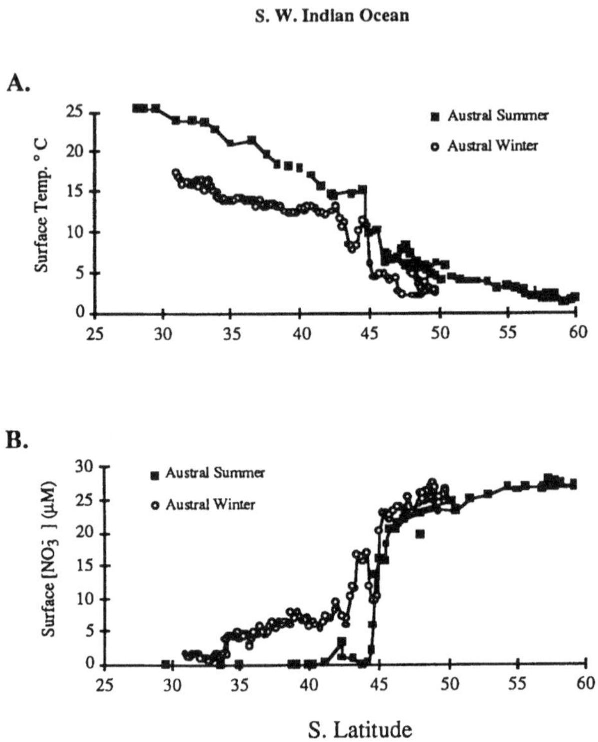

FIGURE 11. Results for the S.W. Indian Ocean. Austral summer data were taken during a cruise of the *Marion Dufresne*. Winter data are taken from Chen et al. (1986) for a nearby cruise track. **A.** Latitudinal variation in surface temperature. **B.** Latitudinal variation in near-surface $[NO_3^-]$.

steering of the ACC (M. McCartney, pers. comm.). Sedimentary $\delta^{15}N$ not only records the primary signal of southward decrease in NO_3^- utilization but a secondary feature; the east to west change in latitudinal position of the frontal system in the southeast Indian Ocean.

The particle removal effect observed in the Equatorial Pacific and discussed above likely accounts for at least part of the difference in $\delta^{15}N$ values observed for near-surface ocean suspended particles and core-top sediments in the Southern Ocean. Wada et al. (1987) observed sinking particles in the Australian sector to have $\delta^{15}N$ values of -3 to 1‰, at the high end of our suspended particle data for polar waters. Similarly, Biggs et al. (1989) showed in Prydz Bay, Antarctica that sinking particles were up to 4‰ enriched over suspended ones. Our S.W. Indian Ocean data is therefore, at best, good qualitative evidence for the isotopic signal generated by latitudinal variations in nutrient depletion across the Southern Ocean that is transferred to the sediments. Unfortunately, a transect of sinking PN has not been made. A range for ε_u in Subantarctic waters can be estimated if assumptions regarding the particle

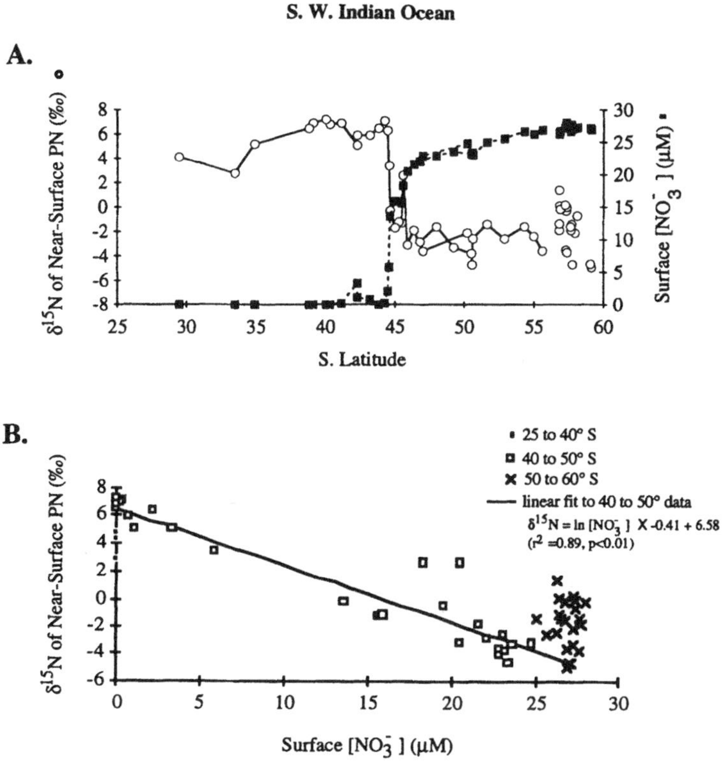

FIGURE 12. δ^{15}N results for the S.W. Indian Ocean (see Fig. 10; data synoptic with summertime observations in Fig. 11) **A.** Latitudinal variation in δ^{15}N for near-surface suspended PN and [NO_3^-]. **B.** δ^{15}N for near-surface suspended PN vs. near-surface [NO_3^-].

the removal effect are made. Biggs et al. (1988) found δ^{15}N values for surface NO_3^- in the Weddell Sea to average 8.2‰ during austral summer. Assuming no particle removal effect, values of ε_u of up to 10 to 12‰ would be derived using eq. 2 (f is sufficiently high that particle residence time is not an issue). With the near maximum particle removal effect previously observed, ε_u would range from 5 to 7. While large, this range in estimates is consistent with previous field observations (Table 2).

The across-front amplitude in δ^{15}N for core top sediments in the S.E. Indian Ocean is somewhat less than for suspended particles (7 vs. 11‰). The difference may, in part, be due to the integration by sediments of seasonal variations in $\delta^{15}N$ of sinking particles. In contrast to the Equatorial Pacific, east/west flow is much larger than north/south flow (McCartney, pers.

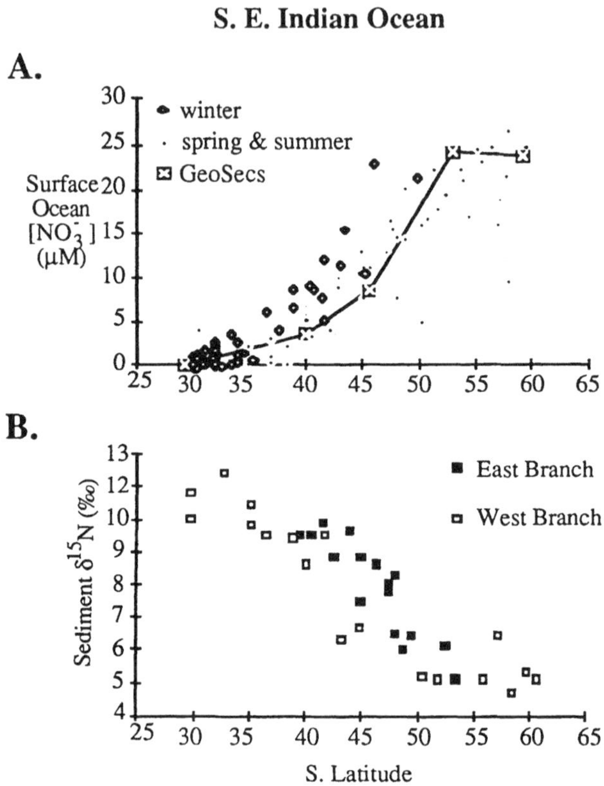

FIGURE 13. Results for the S.E. Indian Ocean (see Fig. 10). **A.** Latitudinal variations in near-surface [NO_3^-]. Because of the relative paucity of NO_3^- data from this sector of the Southern Ocean, data taken from the NODC data are actually calculated from [PO_4^{-3}] data using relationships observed in the GeoSecs data set. **B.** Core top sediment $\delta^{15}N$ vs. latitude along two meridional transects converging at 60°S, 115°E. The latitudinal offset in sediment $\delta^{15}N$ between the two transects can be ascribed to a west to east southward bend in the Subantarctic front (Francois et al., 1992).

comm.) in the Southern Ocean. At a given site, the sediments would be expected to receive material produced over an entire seasonal cycle of NO_3^- depletion during austral summer. In effect, the latitudinal variation in $\delta^{15}N$ for sediments should be best described by eq. 3 since the sediments have effectively sampled the accumulated seasonal production in surface waters. With annual NO_3^- depletion in polar waters at 10 to 20% (f= 0.9 to 0.8, from Fig. 11 and analysis of the NODC data[3]) and depletion in subtropical waters at 100% (f=0), the 6‰ south to north variation observed in the southeast Indian Ocean corresponds to a value for ε_u of ~ 7‰ using eq. 3. This estimate falls within the range in ε_u estimated from the near-surface suspended particle data. However, an apparent diagenetic enrichment in ^{15}N for polar waters (discussed below) may have led to an underestimate in ε_u based on sedimentary $\delta^{15}N$.

Because the particle removal effect leads to ^{15}N depletion in near-surface suspended PN, our suspended particle data cannot be used to ascertain the fidelity with which the surface-generated isotopic signal is preserved in Southern Ocean sediments. Nevertheless, there are grounds to suspect that, in contrast to the Equatorial Pacific, sediments south of the polar front have experienced an apparent diagenetic enrichment in ^{15}N. Wada et al's (1987) data for deeply sinking particles collected near our site (60° S, 150° ε_u) are significantly lower than for underlying sediments (-3 to 1‰ vs. 5 ‰ respectively). A separate data set from this region for sinking particles collected nearer to the surface (50 to 200 m), show values to average 1.4‰ (Biggs et al., 1988, excluding one anomalous site). In addition, polar sediments are also substantially enriched in ^{15}N compared to expected values calculated using eq. 3. Assuming the $\delta^{15}N$ for NO_3^- in surface waters prior to depletion to be the same as found for deep ocean water (5 to 6‰, Liu and Kaplan, 1989), ε_u to be 7 to 11‰ (range of estimates discussed above, Biggs et al., 1988 estimated 8‰ for the Weddell sea), and depletion each year to be 10 to 20%, the calculated range in maximum $\delta^{15}N$ for material accumulating at the seafloor is -5 to 0‰. This range is consistent with Wada et al.'s sinking particle data for early austral summer (late Dec. to Jan.). From these analyses, polar sediments appear to be 5 to 10 ‰ enriched in $\delta^{15}N$ compared to the surface-generated signal.

It is not clear what factors may distinguish the polar Southern Ocean from the Equatorial Pacific to produce apparent diagenetic ^{15}N enrichment in sediments. It is not likely that processes that attenuate the downward flux of sinking particles in the polar Southern Ocean operate very differently from other regions. In fact, diverse regions of the ocean (summarized by Altabet et al. 1991) including the Southern Ocean (Wada et al., 1987) exhibit no enrichment in the $\delta^{15}N$ of sinking particles with depth (1000 to 4200 m). However, polar Southern Ocean sediments are distinguished from the Equatorial Pacific in being primarily composed of siliceous (mostly diatoms and radiolaria) as compared to carbonate microfossils (coccoliths and foraminifera) for most of our Equatorial Pacific samples. A possible explanation may be that most of the nitrogen in our polar sediment samples is, in fact, associated with the diatom microfossils as part of an organic template set down by biomineralization in surface waters. This material may be shifted in isotopic composition relative to the whole organism. Purified and cleaned fossils diatom frustrules (provided by A. Shemesh) from southern polar waters were found to have relatively high $\delta^{15}N$ values of 5‰ similar to our observations for whole

[3] f estimated as the ratio of summertime surface $[NO_3^-]$ to $[NO_3^-]$ in the subsurface temperature minimum. The latter is considered a relic of wintertime surface water.

sediment. If the significance of this mechanism for apparent diagenetic enrichment in ^{15}N is born out by further work, bulk sedimentary $\delta^{15}N$ could be adjusted by simply knowing % opal. Alternatively, sediment fractions could be separated and cleaned for separate isotopic

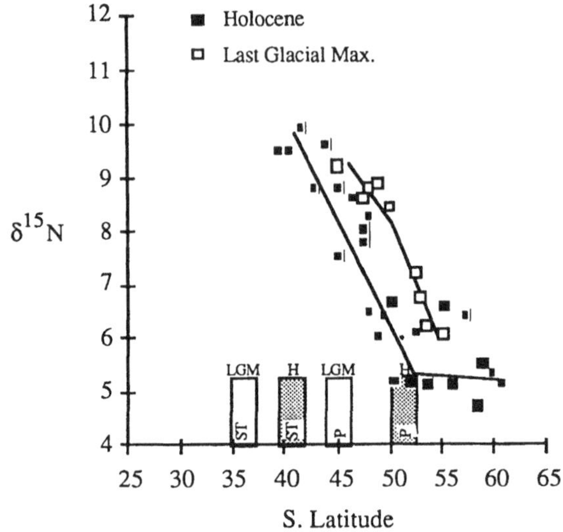

FIGURE 14. Contrast in sediment $\delta^{15}N$ for core top and LGM samples in the S.E. Indian Ocean. For easier comparison, adjustments were made to account for west to east changes in latitudinal position of the frontal system between the two transects (see Fig. 13). The change in latitudinal position of the subtropical and polar fronts between the Holocene and LGM are shown. While the fronts appeared to shift to the north, the 'nutrient utilization' front shifted to the south.

analysis. The Equatorial Pacific results indicate that the carbonate-rich sediments north of the polar front (in the sector we studied) more reliably record the surface-generated ^{15}N signal. If polar sediments are adjusted downward in $\delta^{15}N$ by a minimum of 5‰, the latitudinal amplitude for Southern Ocean sediments would be 11‰, similar to near-surface observations for this region and the Equatorial Pacific.

We have also analyzed sequences from the last glacial maximum (LGM ~ 18 kyr before present) in number of cores from the S.E. Indian Ocean (Fig. 14; Francois et al., 1992). $\delta^{15}N$ values are 1 to 3‰ higher than during the Holocene for cores underlying Subantarctic waters. These results suggest some increase in relative nutrient utilization during the LGM that could have been brought about by either an increase in productivity or a decrease in the supply of NO_3^- (with productivity constant). However, the data does not support a massive depletion of nutrients in Southern Ocean waters required by some models attempting to explain the decrease in atmospheric pCO_2 during the last glaciation. Such an effect would have increased the $\delta^{15}N$ of Subantarctic and Polar sediments to values similar to those for subtropical waters, in effect eliminating the latitudinal gradient in $\delta^{15}N$ observed.

SUMMARY AND CONCLUSIONS

Isotope fractionation during NO_3^- utilization by phytoplankton produces a nitrogen isotopic signal related to the degree of nutrient depletion. The amplitude of this effect is dependent on the fractionation factor associated with phytoplankton uptake (ε_u). While processes which remove PN from surface waters and transport it to deep-ocean sediments could potentially produce large alterations of this signal, observations show these effects not to be significant relative to the signals observed. Recent studies in the Equatorial Pacific, in fact, show a good correspondence in $\delta^{15}N$ between sinking PN leaving the surface layer of the ocean and core top values. Such a comparison is hindered for the Southern Ocean by the lack of data for sinking PN. As demonstrated above, $\delta^{15}N$ values for suspended PN cannot be used to directly compare with sediments because of isotopic fractionation effects associated with removal of PN from surface waters. However, indirect evidence suggests polar sediment are enriched by a least 5‰ relative to inputs from surface waters in contrast to our results for the Equatorial Pacific but further verification is needed.

Modelling the dependence of the large latitudinal variations observed in near-surface and sedimentary $\delta^{15}N$ on nutrient utilization (f) is strongly dependent on hydrodynamic regime. The Equatorial Pacific data is best modeled using the instantaneous product equation (eq. 1 & 2) as a result of the dominance of upwelled NO_3^- at the equator as the source of new nitrogen for the region. Upwelled NO_3^- is progressively depleted with poleward advection. The Southern Ocean data is best modeled by the accumulated product equation (eq. 3) since sediments there integrate the $\delta^{15}N$ signal over each annual cycle of nutrient depletion and replenishment. Local wintertime convective mixing resupplies nutrients to the euphotic zone that are depleted during Austral spring and summer. Inspection of eqs. 1 and 3 suggest that this difference may, in part, explain why the Equatorial Pacific has a greater latitudinal amplitude in $\delta^{15}N$ despite a possible lower value for ε_u.

Our findings indicate great potential for using sedimentary $\delta^{15}N$ as a paleo-recorder for surface nutrient utilization. Work to date has already provided important evidence that the Southern Ocean did not become completely nutrient depleted during the LGM. Nevertheless, future work needs to better constrain possible variations in ε_u between and within regions of interest to permit quantitative use of $\delta^{15}N$ data. For example, our data suggests lower values for ε_u in the Equatorial Pacific than in other regions studied. In addition, the influence on $\delta^{15}N$ values of different sedimentary nitrogen reservoirs (for example, N bound into diatom frustrules) as an explanation for (and possible solution for) the apparent diagenetic effect in polar Southern Ocean waters needs to be rigorously evaluated.

ACKNOWLEDGMENTS

This work was principally funded by NSF grants OCE 9115641 and DPP 9118031. The comments of two anonymous reviewers and editing by Lisa Michael improved the manuscript. This work is WHOI contribution# 8473.

REFERENCES

Altabet MA (1988) Variations in nitrogen isotopic composition between sinking and suspended particles: implications for nitrogen cycling and particle transformation in the open ocean. Deep-Sea Res 35: 535-554

Altabet MA (1989). A time-series study of the vertical structure of nitrogen and particle dynamics in the Sargasso Sea. Limnol Oceanogr 24: 1185-1201

Altabet MA (1990). Organic C, N, and stable isotopic composition of particulate matter collected on glass-fiber and aluminum oxide filters. Limnol Oceanogr 34: 902-909.

Altabet MA, Curry WB (1989). Testing models of past ocean chemistry using foraminiferal $^{15}N/^{14}N$. Glob Biogeochem Cycles 3: 107-119

Altabet MA, Deuser WG (1985) Seasonal variations in natural abundance of ^{15}N in particles sinking to the deep Sargasso Sea. Nature 315: 218-219

Altabet MA, Deuser WG, Honjo S, Stienen S (1991) Seasonal and depth-related changes in the source of sinking particles in the N. Atlantic. Nature 354: 136-139.

Altabet MA, Francois R (1993) Sedimentary nitrogen isotopic ratio records surface ocean nitrate utilization. Glob Biogeochem Cycl: submitted

Altabet MA, McCarthy JJ (1985) Temporal and spatial variations in the natural abundance of ^{15}N in PON from a warm-core ring. Deep-Sea Res 32: 755-772

Altabet MA, McCarthy JJ (1986) Vertical patterns in ^{15}N natural abundance in PON from the surface waters of warm-core rings. J Mar Res 44: 185-201

Altabet MA, Small LF (1990) Nitrogen isotopic ratios in fecal pellets produced by marine zooplankton. Geochim Cosmochim Acta 54: 155-163

Barber RT, Chavez FP (1991) Regulation of primary productivity rate in the Equatorial Pacific Ocean. Limnol Oceanog 36, 1803-1815

Biggs DC, Berkowitz SP, Altabet MA, Bidigare RR, DeMaster DJ, Dunbar RB, Leventer A, Macko SA, Nittrouer CA, Ondrusek ME (1988) A cooperative study of upper-ocean particulate fluxes in the Weddell Sea. In: Proceedings of the Ocean Drilling Program - Initial Reports, Vol 113 Part A. N S F Joint Oceanographic Institutions Inc. pp 77-86

Biggs DC, Berkowitz SP, Altabet MA, Bidigare RR, DeMaster DJ, Macko SA, Ondrusek ME, Noh IL (1989) A cooperative study of upper ocean particulate fluxes. In: Proceedings of the Ocean Drilling Program - Initial Reports, Vol 119 Part A. N S F Joint Oceanographic Institutions Inc, pp 109-120

Boyle EA (1988) Cadmium: Chemical tracer of deepwater paleoceanography. Paleoceanogr 3: 471-489

Bryden HL, Brady EC (1985) Diagnostic model of the three-dimensional circulation in the upper Equatorial Pacific Ocean. J Phys Oceanogr 15: 1255-1273

Calvert SE, Nielson B, Fontugne MR (1993) Evidence from nitrogen isotopic ratios for enhanced productivity during formation of eastern Mediterranean sapropels. Nature 359: 223-225

Charles CD, Fairbanks RF (1990) Glacial-interglacial changes in the isotopic gradients of southern ocean surface water. In: Bleil U, Thiede J (eds)The Geologic History of Polar Oceans: Arctic vs Antarctic, NATO ASI Series 308, Kluwer Academic Pub, pp 519-538

Chavez FP (1989) Size distributions of phytoplankton in the Central and Eastern Tropical Pacific. Glob Biogeochem Cycles 3: 27-36

Checkley DM Jr., Miller CA (1989) Nitrogen isotope fractionation by oceanic zooplankton. Deep-Sea Res 36: 1449-1456
Chen CA, Poisson A, Goyet C (1986) Preliminary Data Report for the INDIVAT I and INDIGO/INDIVAT 3 Cruises in the Indian Ocean. US DOE data report# DOE/NBB-0074
DeNiro MJ, Epstein S (1981) Influence of diet on the distribution of nitrogen isotopes in animals. Geochim Cosmochim Acta 45: 341-351
Eppley RW, Chavez FP, Barber RT (1992) Standing stocks of particulate carbon and nitrogen. J Geophy Res 97: 655-661
Eppley RW, Peterson BJ (1979) Particulate organic matter flux and planktonic new production in the deep ocean. Nature 282: 677-680
Fowler SW, Knauer GA (1986) Role of large particles in the transport of elements and organic compounds through the oceanic water column. Prog Oceanogr 16: 147-194
Francois R, Altabet MA, Burkle LH (1992) Glacial to interglacial changes in surface nitrate utilization in the Indian sector of the Southern Ocean as recorded by sediment $\delta^{15}N$. Paleoceanogr 7: 589-606
Francois R, Bacon MP, Altabet MA, Labeyrie LD (1993) Glacial/intergalcial changes in sediment rain rate in the S W Indian sector of Subantarctic waters as recorded by ^{230}Th, ^{231}Pa, U, and $\delta^{15}N$. Paleoceanogr, in press
Froelich PN, Mortlock RA, Shemesh A (1989) Inorganic germanium and silica in the Indian Ocean: biological fractionation during (Ge/Si) opal formation. Global Biogeochem Cycles 3: 79-88
Fry B (1988) Food web structure on Georges Bank from stable C, N, and S isotopic compositions. Limnol Oceanogr 33: 1182-1190
Goering J, Alexander V, Haubenstock N (1990) Seasonal variability of stable carbon and nitrogen isotopic ratios of organisms in a North Pacific bay. Est Coast Shelf Sci 30: 239-260
Gordon AL, Molinelli EJ (1986) Southern Ocean Atlas: Thermohaline and Chemical Distribution Data Set. Amerind Publishing Co.
Honjo S, Doherty KW (1987) Large aperture time-series sediment traps: design objectives, construction and application. Deep-Sea Res 29: 133-149
Keigwin LD, Boyle EA (1989) Late quaternary paleochemistry of high-latitude surface waters. Paleogeogr Palaeoclim Paleoecol 73: 85-106
Knauer GA, Martin JH, Bruland K (1979) Fluxes of particulate carbon, nitrogen, and phosphorous in the upper water column of the northeast Pacific. Deep-Sea Res 26: 97-108
Knox F, McElroy M (1984) Changes in atmospheric CO_2 influence of the marine biota at high latitude. J Geophys Res 89: 4629-4637
Labeyrie LD, Duplessy J-C (1985) Changes in the oceanic $^{13}C/^{12}C$ ratio during the last 140,000 years: high latitude surface water records. Palaeogeogr Palaeoclimat Palaeoecol 50: 217-240
Liu K-K, Kaplan IR (1989) The eastern tropical Pacific as a source of ^{15}N-enriched nitrate in seawater off southern California. Limnol Oceanogr 34: 820-830
McCarthy JJ, Carpenter EJ (1983) Nitrogen cycling in near-surface waters of the open ocean. In: Carpenter EJ, Capone DG (eds) Nitrogen in the Marine Envirionment. Academic Press, pp 487-512
Minagawa M, Wada E (1984) Stepwise enrichment of ^{15}N along food chains: Further evidence and the relation between $\delta^{15}N$ and animal age. Geochim Cosmochim Acta 48: 1135-1140
Minagawa M, Wada E (1986) Nitrogen isotope ratios of red tide organisms in the East China Sea: A characterization of biological nitrogen fixation. Mar Chem 19: 245-249
Minagawa M, Winter DA, Kaplan IR (1985) Comparison of Kjeldahl and combustion methods for measurement of nitrogen isotope ratios in organic matter. Anal Chem 56: 1859-1861
Miyake Y, Wada E (1967) The abundance of $^{15}N/^{14}N$ in marine environments. Rec Oceanog Works Japan 9: 37-53
Montoya JP (1990) Natural abundance of ^{15}N in marine and estuarine plankton: Studies of biological isotopic fractionation and plankton processes. PhD, Harvard University
Nakatsuka T, Handa N, Wong CS (1992) The dynamic changes of stable isotopic ratios of carbon and nitrogen in suspended and sedimented particulate organic matter during a phytoplankton bloom. J Mar Res 50: 267-296

Peña MA, Lewis MR, Harrison WG (1990) Primary productivity and size structure of phytoplankton biomass on a transect of the equator at 135°W in the Pacific Ocean. Deep-Sea Res 37: 295-315
Rau GH, Arthur MA, Dean WE (1987) $^{15}N/^{14}N$ variations in cretaceous Atlantic sedimentary sequences: Implication for past changes in marine nitrogen biogeochemistry. Earth Plan Sci Lett 82: 269-279
Saino T, Hattori A (1980) ^{15}N natural abundance in oceanic suspended particulate matter. Nature 283: 752-754
Saino T, Hattori A (1987) Geographical variation in the water column distribution of suspended particulate organic nitrogen and its ^{15}N natural abundance in the Pacific and its marginal seas. Deep-Sea Res 34: 807-827
Sarmiento JL, Toggweiler R (1984) A new model for the role of the oceans in determining atmospheric pCO_2. Nature 308: 621- 624
Siegenthaler U, Wenk T (1984) Rapid atmospheric CO_2 variations and ocean circulation. Nature 308: 624-626
Toggweiler R, Sarmiento JL (1985) Glacial to interglacial changes in atmospheric carbon dioxide: the critical role of ocean surface water in high latitudes. In: Sundquist ET and Broecker WS (eds) The Carbon Cycle and Atmospheric CO_2: Natural Variations Archean to Present, AGU Monograph 32, pp 163-184
Voss M (1991) Räumliche und verteilung stabiler isotope ($\delta^{15}N$, $\delta^{13}C$) in suspendierten and sedimentierten partikeln im Nördlichen Nordatlantik. Ph D, Christian -Albrechts Universität zu Kiel
Wada E (1980) Nitrogen isotope fractionation and its significance in biogeochemical processes occurring in marine environments. In: Goldberg ED, Horibe Y (eds) Isotope Marine Chemistry, pp 375-398
Wada E, Hattori A (1976) Natural abundance of ^{15}N in particulate organic matter in the North Pacific Ocean. Geochim Cosmochim Acta 40: 249-251
Wada E, Hattori A (1978) Nitrogen isotope effects in the assimilation of inorganic nitrogenous compounds by marine diatoms. Geomicrobio J 1: 85-101
Wada E, Terazaki M, Kabaya Y, Nemoto T (1987) ^{15}N and ^{13}C abundances in the Antarctic Ocean with emphasis on the biogeochemical structure of the food web. Deep-Sea Res 34: 829-841
Weiss RF, Boecker WS, Craig H, and Spencer D (1983) GeoSecs Indian Ocean Expedition, Vol 5, Hydrographic Data. U S Gov Printing Office, Washington DC
Williams PM, Robertson KJ, Soutar A, Griffin SM, Druffel ERM (1992) Isotopic signatures (^{14}C, ^{13}C, ^{15}N) of tracers of sources and cycling of soluble and particulate organic matter in the Santa Monica Basin, California. Prog Oceanogr 30: 253-290

VARIATIONS IN SEDIMENTARY ORGANIC $\delta^{13}C$ AS A PROXY FOR PAST CHANGES IN OCEAN AND ATMOSPHERIC CO_2 CONCENTRATIONS

Greg H. Rau
Institute of Marine Sciences
University of California, Santa Cruz
Santa Cruz, California 95064

ABSTRACT. Theory, laboratory studies, and ocean observations indicate that the $\delta^{13}C$ of marine plankton biomass and sedimentary remains thereof can be used as a proxy for ambient molecular CO_2 concentration, $[CO_{2(aq)}]$, in ocean surface waters. A compilation of *in situ* ocean data suggests that about 89 % of the global $\delta^{13}C_{org}$ variation within bulk plankton or seston can be explained by a simple negative linear response to ambient $[CO_{2(aq)}]$, with the slope of the best-fit line = -0.6 ‰ μM^{-1}. With this model the standard error of the estimate of surface ocean $[CO_{2(aq)}]$ is ±2.0 μM when $\delta^{13}C_{org}$ is specified. This residual variability may be largely due to effects on plankton $\delta^{13}C_{org}$ imparted by changes in phytoplankton CO_2 demand that are independent of $[CO_{2(aq)}]$. Within this variability and within the current range of ocean $[CO_{2(aq)}]$ there are slight differences between this model and various proposed nonlinear fits to observed global data. While an inverse relationship that can be influenced by both CO_2 demand as well as concentration is theoretically expected, it does not provide an improved fit to observations over the negative linear model. When applied to the sedimentary $\delta^{13}C_{org}$ record, the latter model predicts that the approximate 80 μatms increase in atmospheric pCO_2 during the last glacial-interglacial transition (as documented by ice core analyses) should have resulted in a 1-2 ‰ decrease in plankton $\delta^{13}C$. Indeed, changes of this direction and magnitude are evident in most low-latitude Pleistocene/Holocene sediment core profiles of $\delta^{13}C_{org}$ thus far reported. However, some geographic and temporal differences in past plankton isotopic response are present and expected due to i) regional non-equilibrium between ocean and atmospheric $[CO_2]$, and ii) changes in phytoplankton CO_2 demand.

INTRODUCTION

Because of the role of carbon dioxide in global heat balance and climate, it is of interest to learn how and by what mechanisms atmospheric CO_2 levels have changed in the past and how they will change in the future. An important piece of evidence that CO_2 concentrations have indeed varied significantly over geologic time is provided by analyses of air trapped in glacial

NATO ASI Series, Vol. I 17
Carbon Cycling in the Glacial Ocean:
Constraints on the Ocean's Role in Global Change
Edited by R. Zahn et al.

ice, where Holocene atmospheric pCO_2 has been found to be some 40 % higher than during the preceding glacial maximum (e.g. Barnola et al., 1987 and this volume). Further work using this approach will provide additional details on the nature and timing of such events, but the ice record of atmospheric CO_2 is clearly limited to the relatively recent geologic past. Other measurements or proxies are needed both to confirm the ice core results and to extend CO_2 reconstructions back in time so as to address hypotheses about CO_2 variations earlier in Earth's history (e.g., Owen et al., 1979; Berner, 1990; Crowley, 1991).

A variety of evidence indicates that variations in ocean/atmosphere [CO_2] may be recorded as changes in marine organic matter $^{13}C/^{12}C$ within the sedimentary record (e.g., reviews by Rau et al., 1991a; Freeman and Hayes, 1992). Laboratory and field studies have shown that the $\delta^{13}C$ of the bulk organic fraction of plankton or seston ($\delta^{13}C_{org}$, see Figure 1 legend) generally decreases as the dissolved molecular CO_2 concentration, [$CO_{2(aq)}$], increases. While a CO_2 concentration effect on plant biomass $\delta^{13}C$ was reported earlier (e.g., Park and Epstein, 1960; Abelson and Hoering, 1961), Degens and co-workers (1968) were the first to observe that the $\delta^{13}C_{org}$ of marine phytoplankton (relative to bicarbonate) significantly decreased when higher concentrations of inorganic carbon were supplied. Degens (1969) later described the potential paleoceanographic significance of such a relationship in interpreting carbon isotopic variations within the marine sedimentary organic matter record. Subsequently, Seckbach and Kaplan (1973), Calder and Parker (1973), Pardue et al. (1976), Mizutani and Wada (1982), and McCabe (1985), among others produced additional evidence that a negative relationship exists between plankton $\delta^{13}C_{org}$ and [$CO_{2(aq)}$]. All of these experimental studies implied that by one of several plausible mechanisms, high ocean/atmosphere [CO_2] in the past could have resulted in greater apparent photosynthetic carbon isotope fractionation and therefore lower $\delta^{13}C_{org}$ as frequently found earlier in the sedimentary record (e.g., Degens, 1969; Mizutani and Wada, 1982; Arthur et al., 1985; Popp et al., 1989; Des Marais et al, 1992).

Nevertheless, other researchers have noted that phytoplankton productivity, carbon fixation pathways, and species composition among other factors can significantly influence plankton $\delta^{13}C_{org}$ (e.g., Deuser et al., 1968; Wong and Sackett, 1978; Descolas-Gros and Fontugne, 1990; Falkowski, 1991; Fry and Wainwright, 1991; Rau et al., 1992; Goericke et al., in press). These effects, if operating independently of [$CO_{2(aq)}$], could serve to weaken plankton isotopic response to [$CO_{2(aq)}$] and thus attenuate or eliminate such isotopic signals both in plankton and in their sedimentary remains. An important test of the net effect of [$CO_{2(aq)}$] on marine $\delta^{13}C_{org}$ is therefore the degree to which these two parameters (or arithmetic transformations) are correlated both in the modern ocean and in the sedimentary record.

MODERN-OCEAN OBSERVATIONS

Based on unpublished ocean PCO_2 (partial pressure of molecular CO_2 in seawater), temperature, and salinity measurements by T. Takahashi and coworkers, Rau et al. (1989) calculated that as much as 2.5 times more $CO_{2(aq)}$ is present at high- versus low-latitude South Atlantic surface waters. This latitudinal $[CO_{2(aq)}]$ increase was found to coincide closely with a previously observed (e.g., Rau et al., 1982) 12 ‰ decrease in plankton $\delta^{13}C_{org}$. The relationship between these two parameters appeared to be linear, with a slope estimated to be -0.8 ‰ μM^{-1} (Rau et al., 1989). A negative linear relationship is further supported by a more recent compilation of data from various ocean regions (Fig. 1a), where the best-fit line has a slope of -0.6 (± 0.01) ‰ μM^{-1} with an r^2 of 0.892. Inverting this relationship yields a linear predictor of $[CO_{2(aq)}]$ with an estimation error of ±2.0 μM when $\delta^{13}C_{org}$ is specified (Fig. 2, Line A). Note, however, that the greater number of data points at either end of the $[CO_{2(aq)}]$ range and the relative lack of observations at mid-range (Figs. 1a and 2) may generate correlation statistics that would not be typical of a more random global sampling. Also, most of the $[CO_{2(aq)}]$ values in these figures were not measured directly and in concert with $\delta^{13}C_{org}$ determinations, nor was there standardized POM sampling among the various data sets. These circumstances further question the representativeness of these data and trends, and call for additional field sampling and analysis to confirm these preliminary observations.

Given the existing data and their shortcomings, the linear relationship between $\delta^{13}C_{org}$ and $[CO_{2(aq)}]$ described above is not much different from a logarithmic fit to observed data proposed by Freeman and Hayes (1992) (Fig. 2, Line C). Significant deviations from the observed ocean trend, however, are present among some parameterizations of the logarithmic model (e.g., Lines E and F, Fig. 2). These deviations may arise because non-marine or non-modern ocean data were used for model calibration in these instances. Model differences may also arise because it is not straightforward to interrelate trends based on estimates of phytoplankton isotope fractionation (used by the advocates of the logarithmic modeling approach) to trends based on direct, bulk plankton $\delta^{13}C_{org}$ measurements (Rau et al., 1991a; Freeman and Hayes, 1992).

There is also reason to suspect that neither a linear nor a logarithmic model is appropriate for describing the marine $\delta^{13}C_{org}$ vs $[CO_{2(aq)}]$ relationship. As discussed by Rau et al. (1992) and Rau and Des Marais (1992), the plant carbon-isotope fractionation scheme of Farquhar et al. (1982) requires that plant $\delta^{13}C_{org}$ be inversely and therefore nonlinearly related to ambient $[CO_2]$, and furthermore that changes in CO_2 demand within plants can significantly alter this relationship. Such a model is appealing because unlike linear and logarithmic trends, plankton $\delta^{13}C_{org}$ must asymptotically approach a minimum value rather than decrease infinitely as $[CO_{2(aq)}]$ increases. This makes sense given that an upper limit of possible isotope fractionation by plant (and therefore a lower limit of plant $\delta^{13}C_{org}$ under relatively invariant inorganic

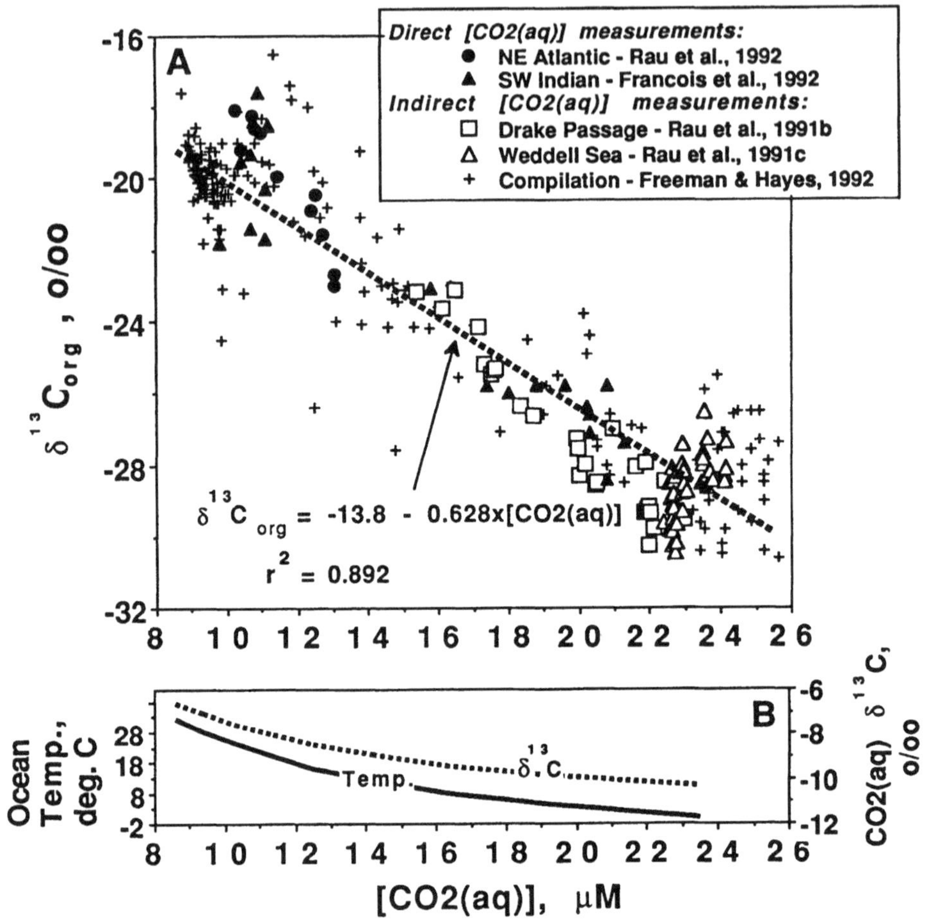

Figure 1. A) Compilation of surface ocean $\delta^{13}C_{org}$ and corresponding $[CO_{2(aq)}]$. $\delta^{13}C_{org}$ is defined here as the $\delta^{13}C$ of bulk plankton organic biomass or organic seston sampled by various means from surface or near-surface waters such that $\delta^{13}C = [\{(^{13}C/^{12}C_{sample}) - (^{13}C/^{12}C_{PDB})\}/(^{13}C/^{12}C_{PDB})] \times 1{,}000$ in units of parts per thousand, ‰, and where "PDB" refers to the Peedee Belemnite standard. Without excluding the potential variability introduced into this compilation by i) nonuniform sampling of organic matter (various particle sizes, seasons, depths, etc.; e.g., Rau et al. 1990) and ii) non-coincident and/or indirect $[CO_{2(aq)}]$ determinations, a negative linear relationship (dotted line) can explain about 89 % of the observed variation of $\delta^{13}C_{org}$, with a standard error of the estimate, $s_{y|x}$, of ± 1.3 ‰ when $[CO_{2(aq)}]$ is specified, where $s_{y|x} = [(n-1)(s_y)^2(1-r^2)/(n-2)]^{0.5}$. B) Estimation of surface ocean temperature and $\delta^{13}C_{CO2(aq)}$ trends associated with variations in $[CO_{2(aq)}]$ assuming a constant surface ocean PCO_2 of 350 μatms, and assuming a temperature-sensitive isotopic offset between ocean HCO_3^- and $CO_{2(aq)}$ (Mook et al. 1974) under a constant $\delta^{13}C_{HCO3-}$ of +1.5 ‰ (e.g., Rau et al., 1992).

carbon $\delta^{13}C$) is set by the kinetic isotope fractionation at the CO_2 fixation step (e.g., $\delta^{13}C_{CO2(aq)}$ - plant $\delta^{13}C_{org}$ = 29 ‰ for C_3 plants; e.g., Roeske and O'Leary, 1984).

The inverse model is also attractive because it takes into account isotopic effects imparted by plant carbon demand (independent of $[CO_{2(aq)}]$) which have been observed experimentally (e.g., Deuser et al., 1968; Fry and Wainwright, 1991) as well as in the field (e.g., Deuser, 1970; Fry and Wainwright, 1991; Rau et al., 1992). However, the correlation between plankton $\delta^{13}C_{org}$ and $1/[CO_{2(aq)}]$ ($r^2 = 0.858$) is not superior to the simple linear model (Fig 2, Line B vs A). This circumstance can be reconciled if phytoplankton carbon demand varies significantly across the current global range of ocean $[CO_{2(aq)}]$ (Rau et al., 1992), and thus no single parameterization of the inverse model can be representative of the global trend. Also, an inverse relationship may not be strongly evident in the modern ocean because surface water $[CO_{2(aq)}]$ maxima are presently not high enough to allow full expression of ^{13}C discrimination during organic matter formation, thus marine $\delta^{13}C_{org}$ values are not observed to approach asymptotically some minimum value (Figs. 1a and 2). Such may not have been the case earlier in Earth's history when higher ocean/atmospheric $[CO_2]$ and lower $\delta^{13}C_{org}$ values were prevalent (e.g., Degens, 1969; Arthur et al., 1985; Des Marais et al., 1992).

RECONSTRUCTING PAST VARIATIONS IN PLANKTON $\delta^{13}C$ AND OCEAN $[CO_{2(aq)}]$

If organic matter preserved in the marine sedimentary record is isotopically representative of the organic matter produced by past plankton communities, one could determine its $\delta^{13}C$ within chronologically-defined strata and reconstruct variations in plankton $\delta^{13}C_{org}$. However, a number of factors potentially complicate such an approach. One concern is whether or not the organic matter in marine sediments is in fact of marine origin. A significant admixture of terrestrial organic matter can be present in marine sediments and may be pervasive, even in open-ocean settings (e.g., Dean et al., 1986; Prahl et al., 1989; Jasper and Gagosian, 1990; 1993). Also, the isotopic composition of bulk plankton organic matter may be altered metabolically and diagenetically at the site of its origin (surface waters), during its transit to the ocean bottom, and during its residence in sediments (e.g., Dean et al., 1986; Westerhausen et al., 1993 and this volume).

The magnitude of these effects at any given location may, however, be relatively small and/or can be corrected for in interpreting bulk sediment $\delta^{13}C_{org}$. For example, by assuming that the $\delta^{13}C$ of sedimentary porphyrins has over time been uniformly very similar to the $\delta^{13}C$ of contemporaneous algal biomass, Hayes et al. (1989) found that the algal $\delta^{13}C$ variations through a Cretaceous sedimentary sequence were 0.6 to 2.8 ‰ lower than the $\delta^{13}C$ of the corresponding bulk sedimentary organic material. That is, this approximate 1 to 3 ‰ offset

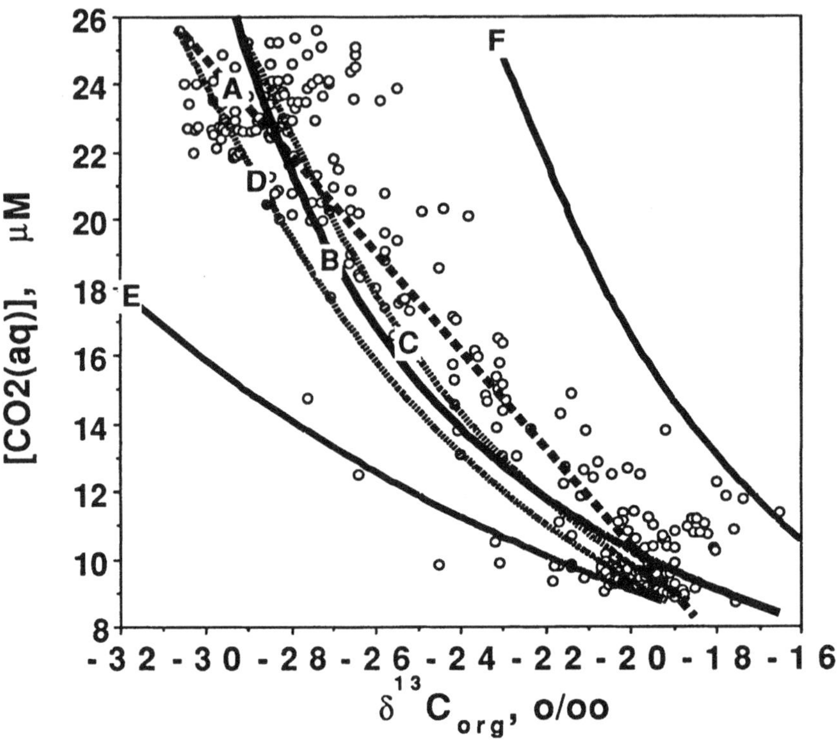

Figure 2. Various models that predict ocean $[CO_{2(aq)}]$ given $\delta^{13}C_{org}$ as follows: Line A) $[CO_{2(aq)}] = -1.42 \times \delta^{13}C_{org} - 17.9$ where $s_{y|x} = \pm 2.0$ µM (linear fit to observed data, this report); Line B) $[CO_{2(aq)}] = 134.84/(\delta^{13}C_{org} + 33.648)$ (inverse model of Rau et al. (1992) fitted to observed data); Lines C-F) $[CO_{2(aq)}] = \exp[(Ep - b)/a]$, where $Ep = \{[(\delta^{13}C_{CO2(aq)} + 1000)/(\delta^{13}C_{org} + 1000)] - 1\}1000$, where $\delta^{13}C_{CO2(aq)}$ is specified in Fig. 1b, and where: C) a = 14.78, b = -1.83 (Freeman and Hayes, 1992 - best fit to observed marine data); D) a = 17.0, b = -3.4 (lake-based calibration - Popp et al., 1989); E) a = 32.9, b = -10.5 (paleo-calibration of alkenones + 3.8 ‰ - Jasper and Hayes, 1990); F) a = 11.64, b = -3.56 (lake-based calibration - Hollander and McKenzie, 1991). Observed present-ocean data (circles) from Fig. 1a.

presumably represents the total isotopic deviation imparted on bulk organic matter by post-production metabolism and diagenesis together with the effects (if any) of non-marine inputs. Furthermore, both the algal and the bulk $\delta^{13}C_{org}$ variations through this sequence were highly synchronous, indicating that both signals were responding in near-identical fashion to a common environmental influence.

In contrast, Jasper and Hayes (1990) found large and inconsistent down-core differences between phytoplankton biomarker- and bulk organic- $\delta^{13}C$ as a consequence of significant terrestrial contamination at a Gulf of Mexico site. Such contamination, however, was evident by elevated C/N and terrestrial organic compound concentrations in affected strata (Jasper and Gagosian, 1990; 1993). Using ancillary sediment measurements such as these, it may be possible (with some level of accompanying uncertainty) to correct or reject the bulk sediment $\delta^{13}C_{org}$ for the reconstruction of the marine $\delta^{13}C_{org}$ signal (e.g., Jasper and Gagosian, 1993; Westerhausen et al., 1993 and this volume). This may be important in cases where limited sample size, inappropriate core storage, or lack of resources may not permit a more direct reconstruction of marine $\delta^{13}C_{org}$ via extraction and analysis of phytoplankton-specific compounds (e.g., Popp et al., 1989; Jasper and Hayes, 1990).

ATMOSPHERIC pCO_2 RECONSTRUCTION

While the preceding discussions address a means of reconstructing surface ocean $[CO_{2(aq)}]$ from bulk sediment $\delta^{13}C_{org}$ measurements, a calculation of past ocean PCO_2 and hence atmospheric pCO_2 variations is desired. [Note that the partial pressure of CO_2 in seawater or in air is indicated by "P" or "p", respectively (e.g., Skirrow, 1975)] The conversion of seawater $[CO_{2(aq)}]$ to PCO_2 is readily done using Henry's Law when seawater temperature and salinity are provided (Fig. 3). These parameters are routinely measured in the modern ocean, but temperature and salinity estimates for paleo-marine settings are problematic. However, changes in ocean salinity of only several ‰ (of the magnitude relevant to the recent geologic past; e.g., Broecker, 1982) result in relatively minor changes in CO_2 solubility and hence ocean PCO_2 calculations. Of more significance is temperature. Several means have been offered for reconstructing paleo-ocean surface temperature (e.g., CLIMAP, 1981; Sikes et al., 1991; Shemesh et al., 1992), and the inaccuracies of these estimates ($\pm$ 1 to 3°C) when used in conjunction with $[CO_{2(aq)}]$ reconstructions will degrade the precision with which relevant, temperature-sensitive CO_2 solubilities and thus past surface ocean PCO_2 can be determined.

Lastly, after paleo-ocean PCO_2 variations have been estimated for a given location, paleo-atmospheric pCO_2 variations can be inferred only under the assumption that the partial pressures of CO_2 in the ocean and in the atmosphere have been equal at that location. Yet as has been demonstrated in the present ocean-atmosphere system, such CO_2 equilibrium is rarely found at

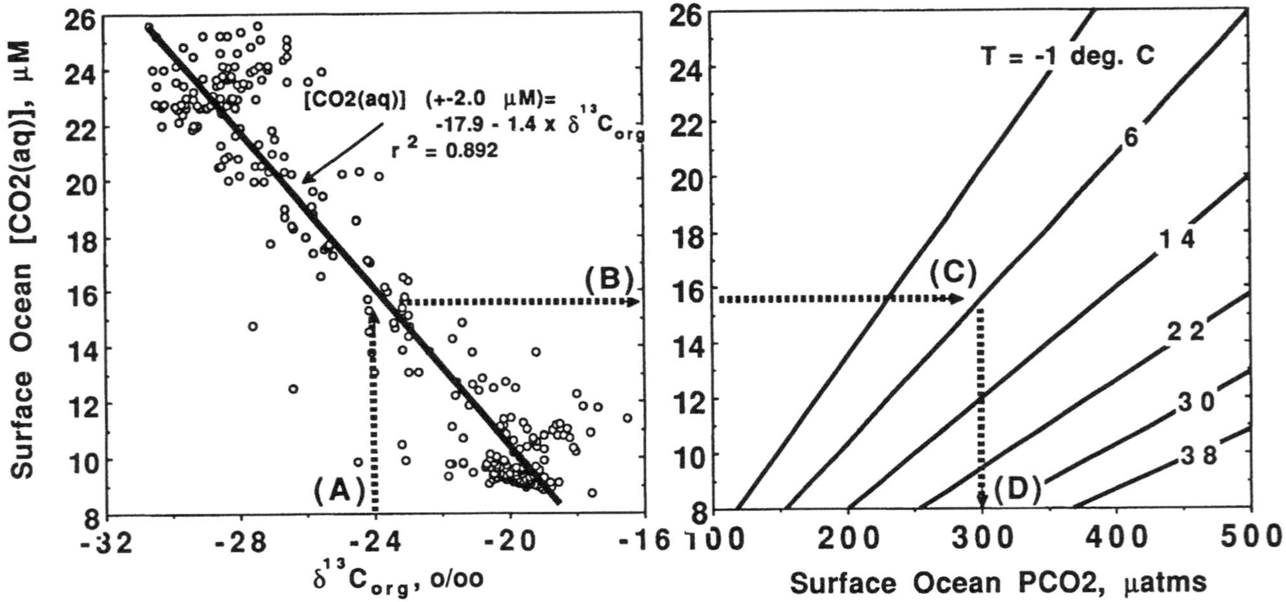

Figure 3. Procedure for estimating surface ocean $[CO_{2(aq)}]$ and PCO_2 when $\delta^{13}C_{org}$ and surface ocean temperature are provided and assuming an ocean salinity of 35 ‰. For example, A) $\delta^{13}C_{org}$ is specified and the linear model (Fig. 2, line A) is used to estimate $[CO_{2(aq)}]$, B; C) by specifying ocean temperature and thus CO_2 solubility, the estimated $[CO_{2(aq)}]$ can be related to surface ocean PCO_2 (D).

any one place; ocean-atmosphere pCO_2 differences can be >100 μatms in some regions (e.g., Tans et al., 1990). This adds an additional uncertainty in the reconstruction of paleo-atmospheric pCO_2 from an individual marine sedimentary $\delta^{13}C_{org}$ profile. As illustrated below, in the absence of relevant paleoceanographic information, the potential for significant regional variations in surface ocean PCO_2 appears to require that isotopic data from several ocean regions be provided so as to discern regional from global isotopic signals and atmospheric pCO_2 inferences.

COMPARISONS BETWEEN ICE CORE pCO_2 AND MARINE SEDIMENT $\delta^{13}C_{org}$

With the preceding models and concerns in mind, several published sets of isotopic records can be used to see if bulk sedimentary $\delta^{13}C_{org}$ actually has responded to past global changes in ocean/atmosphere $[CO_2]$. It is relevant to focus on the period encompassing the Pleistocene/Holocene transition because a significant increase in atmospheric pCO_2 (and by inference surface ocean PCO_2) as been documented by ice core analyses (e.g., Barnola et al. 1987 and this volume). Figure 4 presents six low-latitude bulk sediment $\delta^{13}C_{org}$ profiles spanning the past 30,000 yr BP. In all but one profile, Holocene sediment $\delta^{13}C_{org}$ was found to be uniformly lower by some 1 to 3 ‰ than during the last glacial maximum. Similar sediment $\delta^{13}C_{org}$ responses over this period have also recently been reported by Müller et al. (1994) and Westerhausen et al. (this volume).

Based on the modern-ocean relationship among surface ocean $\delta^{13}C_{org}$, $[CO_{2(aq)}]$, temperature, and PCO_2 (Fig. 3), the preceding sediment $\delta^{13}C_{org}$ trends are generally consistent with the observation that atmospheric pCO_2 (and presumably mean surface ocean PCO_2) increased approximately 80 μatms across the glacial-interglacial transition, resulting in some 2 - 3 μM lower surface ocean $[CO_{2(aq)}]$ (Rau et al. 1991a). Note, however, that the size of such a glacial-interglacial $[CO_{2(aq)}]$ change is within the "noise" (±1 standard estimation error = ±2.0 μM) of the present-day ocean $\delta^{13}C_{org}$ versus $[CO_{2(aq)}]$ relationship (Figs. 3). The isotopic responses evident in Figure 4 therefore suggests that plankton/sediment $\delta^{13}C_{org}$ at the locales represented by each of these cores may in fact be more closely linked to changes in surface ocean $[CO_{2(aq)}]$ (i.e., less "noisy") than is evident in the global-scale relationship compiled in Figure 1a.

Additional evidence that the observed glacial-age sediment $\delta^{13}C_{org}$ elevation is indeed a true reflection of ^{13}C enrichment in marine plankton can be found in the alkenone-derived $\delta^{13}C_{org}$ profile (Fig. 4) reported by Jasper and Hayes (1990). Here, the higher glacial-age $\delta^{13}C$ of phytoplankton-specific alkenones argues that the generally elevated $\delta^{13}C_{org}$ in glacial-age bulk sediments has also documented a widespread plankton ^{13}C enrichment present during the last glacial maximum. It is noted, however, that the glacial-interglacial change in alkenone $\delta^{13}C$ is several per mil larger than observed in the bulk $\delta^{13}C_{org}$ profiles (Fig. 4), and is larger than anticipated by the modern-ocean $\delta^{13}C_{org}$ vs $[CO_{2(aq)}]$ linear relationship (Fig. 3) assuming a 2-3 μM glacial-interglacial increase in $[CO_{2(aq)}]$.

At any given time horizon there are sometimes large differences among the various bulk sediment $\delta^{13}C_{org}$ records (Fig. 4). These may be due to i) local contamination of the marine sediment $\delta^{13}C_{org}$ by nonmarine inputs, ii) significant but temporally non-uniform diagenetic modifications of sediment $\delta^{13}C_{org}$, iii) inaccuracies in core chronologies (Petit et al., 1990; Bard et al., 1990), iv) temporally non-uniform plankton $\delta^{13}C_{org}$ response to ocean $[CO_{2(aq)}]$ (e.g., variable phytoplankton carbon demand), and v) true differences in ambient ocean $[CO_{2(aq)}]$ and PCO_2 (estimated from sediment $\delta^{13}C_{org}$) versus those values expected at atmospheric equilibrium as determined from ice core analyses. Speaking to this last point, a sediment $\delta^{13}C_{org}$ profile from the E. equatorial Pacific Ocean does not contain elevated glacial-period values as observed at the other sites (Fig. 4). Pedersen et al. (1991) suggest that the relative lack of isotopic response in this core likely reflects a persistent pCO_2 disequilibrium between CO_2-rich upwelled water and the atmosphere at this site. Such past disequilibrium is also implied by an alkenone $\delta^{13}C$ profile from this region (Jasper and Hayes, this volume). These important observations underline the potential difficulty in reconstructing paleo-atmospheric pCO_2 from estimates of ocean PCO_2 at any given ocean location.

The cause of the widespread elevation in plankton $\delta^{13}C_{org}$ during the last glacial maximum was not due to an elevation in $\delta^{13}C$ of the surface ocean ΣCO_2 available for organic matter formation. On average, the $\delta^{13}C$ of the carbonate of single-species planktonic foraminifera tests shows relatively little variation (<0.5 ‰) between the recent glacial and interglacial periods (Broecker and Peng, 1982; Shackleton and Pisias, 1985; Curry and Crowley, 1987). Curry et al. (1988) and Duplessy et al. (1988) report a glacial to interglacial increase in deep-ocean $\delta^{13}C_{\Sigma CO2}$ of 0.3 to 0.5 ‰, a trend opposite to the general decrease seen in most sediment $\delta^{13}C_{org}$ profiles (Figure 4). If $CO_{2(aq)}$ is assumed to be the sole inorganic carbon substrate for organic matter formation during this period, the isotopic variation in this compound (as a contributor to $\delta^{13}C_{org}$ variations) can be approximated by assuming that the measured planktonic foram $\delta^{13}C$ is representative of surface water $\delta^{13}C_{HCO3-}$ and that $\delta^{13}C_{CO2(aq)}$ is offset from these values by an amount determined by temperature-sensitive equilibrium isotope fractionation (Mook et al., 1974). The general glacial-interglacial increase in surface ocean $\delta^{13}C_{CO2(aq)}$ values thus calculated (e.g., Jasper and Hayes, 1990) are consistent with a global trend to higher $\delta^{13}C$ values of atmospheric (and, by inference, ocean) CO_2 during this period (Marino et al., 1992; Leuenberger et al., 1992). Such trends are again opposite to the general shift from higher to lower marine sedimentary $\delta^{13}C_{org}$ across the Pleistocene/Holocene transition (Fig. 4). The decrease in marine $\delta^{13}C_{org}$ during this period therefore was not due to changes in inorganic carbon $\delta^{13}C$, and may have resulted from an increase in ^{13}C discrimination by Holocene phytoplankton (Jasper and Hayes, 1990) associated with a widespread increase in

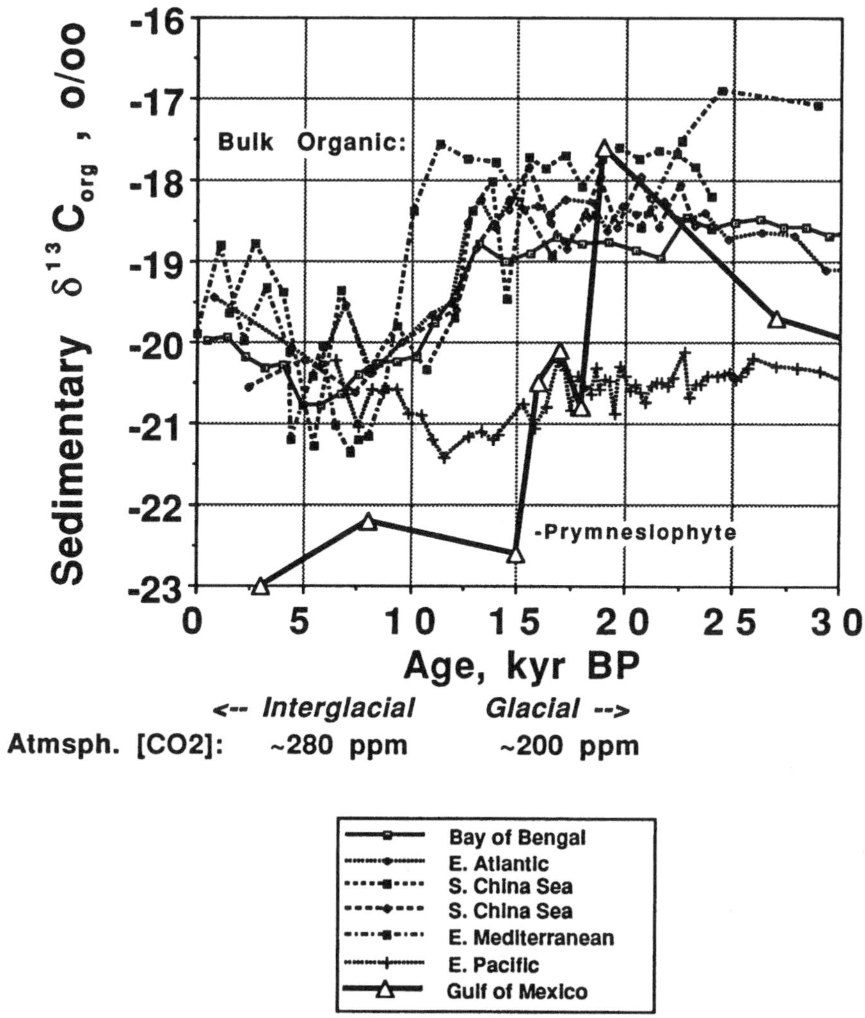

Figure 4. Low-latitude bulk sedimentary $\delta^{13}C_{org}$ versus sediment age from cores in the Bay of Bengal (Fontugne and Duplessy, 1986), E. Atlantic (Muller et al., 1983), S. China Sea (Thunell et al., 1992), E. Mediterranean (Fontugne and Calvert, 1992), and E. Pacific (Pedersen et al., 1991). Phytoplankton (Prymnesiophyte) $\delta^{13}C_{org}$ reconstruction based on alkenone $\delta^{13}C$ (+ 3.8 ‰) from the Gulf of Mexico (Jasper and Hayes, 1990). Mean atmospheric $[CO_2]$ data from Barnola et al. (1987).

$CO_{2(aq)}$ availability and/or decrease in phytoplankton carbon demand (e.g., Rau et al., 1992). Alternatively, the elevated glacial $\delta^{13}C_{org}$ values could reflect increased phytoplankton utilization of ^{13}C-rich HCO_3^- (relative to $CO_{2(aq)}$) during that time (Descolas-Gros and Fontugne, 1985).

CONCLUSIONS

By whatever mechanism, it is encouraging that a bulk sedimentary $\delta^{13}C_{org}$ response of appropriate direction and magnitude has been generally observed in association with an approximated 80 μatms glacial-interglacial change in ocean/atmospheric pCO_2. Such an observation is significant given current uncertainties in: i) the modern $[CO_{2(aq)}]$ vs plankton $\delta^{13}C_{org}$ relationship, ii) the potential effects of terrestrial contamination and diagenesis on downcore isotopic variations in bulk sedimentary organic material, iii) reconstructing relevant surface ocean temperature, CO_2 solubility, and PCO_2, and iv) relating local ocean PCO_2 variations to global atmospheric pCO_2 as monitored by ice core analyses. These initial observations suggest that atmospheric pCO_2 variations of equal or greater magnitude to that above (and hypothesized to have occurred earlier in the geologic record) may be recorded as down-core isotopic changes within the marine organic sedimentary record. However, further research is clearly needed to characterize more accurately the relationship between marine $\delta^{13}C_{org}$ and $[CO_{2(aq)}]$, to document better past variations in plankton $\delta^{13}C_{org}$ and surface ocean temperature, and to delineate local from global signals within the sedimentary record.

ACKNOWLEDGMENTS. I thank Drs. D. Des Marais, T. Takahashi, P. Froelich, M. Raymo, K. Freeman, and T. Pedersen for their contributions to this work. The technical assistance of A. Tharpe, C. Low, and D. Chipman is gratefully acknowledged. R. François, J. Jasper, and P. Müller provided very helpful comments on an earlier draft of this paper. Support has been provided by U.S. National Science Foundation grants DPP-8613981, and OCE-9012172 and -9017773.

REFERENCES

Abelson PH, Hoering TC (1961) Carbon isotope fractionation in formation of amino acids by photosynthetic organisms. Proc Nat Acad Sci 47:623-632

Arthur MA, Dean WE, Claypool GE (1985) Anomalous ^{13}C enrichment in modern marine organic carbon. Nature 315:216-218

Bard E, Hamelin B, Fairbanks RG, Zindler A (1990) Calibration of the ^{14}C timescale over the past 30,000 years using mass spectrometric U-Th ages from Barbados corals. Nature 345:405- 410

Barnola JM, Raynaud D, Korotkevich DYS, Lorius C (1987) Vostok ice core provides 160,000-year record of atmospheric CO_2. Nature 329:408-414

Berner RA (1990) Atmospheric CO_2 over Phanerozoic time. Science 249:1382-1386

Broecker WS (1982) Ocean chemistry during glacial time. Geochim. Cosmochim. Acta 46:1689-1705

Broecker WS, Peng T- H (1982) Tracers in the Sea. Eldigio, Palisades

Calder JA, Parker PL (1973) Geochemical implications of induced changes in ^{13}C fractionation by blue-green algae. Geochim Cosmochim Acta 37:133-140

CLIMAP (1981) Seasonal reconstructions of the Earth's surface at the last glacial maximum. Geol Soc Am Chart Ser MC36

Crowley T (1991) Past CO_2 changes and tropical sea surface temperatures. Paleoceanog 6:387-395

Curry WB, Crowley TJ (1987) The $\delta^{13}C$ of equatorial Atlantic surface waters: Implications for ice age pCO_2 levels. Paleoceanog 2:485-517

Curry WB, Duplessy JC, Labeyrie LD Shackleton NJ (1988) Changes in the distribution of $\delta^{13}C$ of deep water TCO_2 between the last glaciation and the Holocene. Paleoceanog 3:317-341

Dean WE, Arthur MA, Claypool GE (1986) Depletion of ^{13}C in Cretaceous marine organic matter: Source, diagenetic, or environmental signal? Mar Geol 70:119-157

Degens ET (1969) Biogeochemistry of stable carbon isotopes. In: Eglinton G, Murphy MTJ (eds) Organic geochemistry. Springer, New York

Degens ET, Guillard RRL, Sackett WM, Hellebust JA (1968) Metabolic fractionation of carbon isotopes in marine plankton - I: Temperature and respiration experiments Deep-Sea Res 15:1-9

Descolas-Gros C, Fontugne MR (1985) Carbon fixation in marine phytoplankton: carboxylase activities and stable carbon isotope ratios; physiological and paleoclimatological aspects. Mar Biol 87:1-6

Descolas-Gros C, Fontugne MR (1990) Stable carbon isotope fractionation by marine phytoplankton during photosynthesis. Plant Cell Environ 13:207-218

Des Marais DJ, Strauss H, Summons RE, Hayes JM (1992) Carbon isotopic evidence for the stepwise oxidation of the Proterozoic environment. Nature 359: 605-609

Deuser WG (1970) Isotopic evidence for diminishing supply of available carbon during diatom bloom in the Black Sea. Nature 225:1069-1071

Deuser WG, Degens ET, Guillard RRL (1968) Carbon isotope relationships between plankton and sea water. Geochim Cosmochim Acta 32:657-660

Duplessy JC, Shackleton NJ, Fairbanks RG, Labeyrie L, Oppo D, Kallel N (1988) Deepwater source variations during the last climatic cycle and their impact on the global deepwater circulation. Paleoceanog 3:343-360

Falkowski PG (1991) Species variability in the fractionation of ^{13}C and ^{12}C by marine phytoplankton. J Plank Res 13: 21-28

Farquhar GD, O'Leary MH, Berry JA (1982) On the relationship between carbon isotope discrimination and the intercellular carbon dioxide concentration in leaves. Aust J Plant Physiol 9:121-137

Fontugne MR, Calvert SE (1992) Late Pleistocene variability of the carbon isotopic composition of organic matter in the Eastern Mediterranean: Monitor of changes in the carbon sources and atmospheric CO_2 concentrations. Paleoceanog 7:1-20

Fontugne MR, Duplessy J-C (1986) Variations of the monsoon regime during the upper Quaternary: Evidence from carbon isotopic record of organic matter in North Indian Ocean sediment cores. Palaeogeogr Palaeoclimatol Palaeoecol, 56:69-88

François R, Altabet MA, McCorkle D, Brunet C, Poisson A (1992) Changes in the $\delta^{13}C$ of particulate organic matter suspended in the surface waters across the Subtropical Convergence (STC) in the S.W. Indian Ocean. EOS 73(43): 285

Freeman KH, Hayes JM (1992) Fractionation of carbon isotopes by phytoplankton and estimates of ancient CO_2 levels. Glob Biogeochem Cycles 6:185-198

Fry B, Wainwright SC (1991) Diatom sources of ^{13}C-rich carbon in marine food webs. Mar Ecol - Prog Ser 76:149-157

Goericke R, Montoya JP, Fry B (in press) Physiology of isotope fractionation in algae and cyanobacteria. In: Lajtha K Michener B (eds), Stable isotopes in ecology. Blackwell, Oxford

Hayes JM, Popp BN, Takigiku R, Johnson MW (1989) An isotopic study of biogeochemical relationships between carbonates and organic carbon in the Greenhorn Formation. Geochim Cosmochim Acta 53:2961-2972

Hollander DJ, McKenzie JA (1991) CO_2 control on carbon isotope fractionation during aqueous photosynthesis: A paleo-pCO_2 barometer. Geology 19:929-932

Jasper JP, Gagosian RB (1990) The sources and deposition of organic matter in the Late Quaternary Pigmy Basin, Gulf of Mexico. Geochim Cosmochim Acta 54:1117-1132

Jasper JP, Gagosian RB (1993) The relationship between sedimentary organic carbon isotopic composition and organic biomarker compound concentration. Geochim Cosmochim Acta 57:167-186

Jasper JP, Hayes JM (1990) A carbon isotope record of CO_2 levels during the late Quaternary. Nature 347:462-464

Leuenberger M, Siegenthaler U, Langway CC (1992) Carbon isotopic composition of atmospheric CO_2 during the last ice age from an Antarctic ice core. Nature 357:488-490

Marino BD, McElroy MB, Salawitch RJ, Spaulding WG (1992) Glacial to interglacial variations in the carbon isotopic composition of atmospheric CO_2. Nature 357:461-466

McCabe B (1985) The dynamics of ^{13}C in several New Zealand lakes. Ph.D. thesis, Univ. Waikato, Hamilton, New Zealand

Mizutani H, Wada E (1982) Effect of high atmospheric CO_2 concentration on $\delta^{13}C$ of algae Orig Life 12:377-390

Mook WG, Bommerson JC, Staverman WH (1974) Carbon isotope fractionation between dissolved bicarbonate and gaseous carbon dioxide. Earth Planet Sci Lett 22:169-176

Müller PJ, Erlenkeuser H, von Grafenstein R (1983) Glacial-interglacial cycles in oceanic productivity inferred from organic carbon contents in eastern North Atlantic sediment cores. In: Thiede J, Suess E (eds) Coastal upwelling - its sediment record, Part B. Plenum, New York

Owen T, Cess RD, Ramanathan V (1979) Early earth: An enhanced carbon dioxide greenhouse to compensate for reduced solar luminosity. Nature 277:640-642

Pardue JW, Scanlan RS, van Baalen CB, Parker PL (1976) Maximum carbon isotope fractionation in photosynthesis by blue-green algae and a green alga. Geochim Cosmochim Acta 40:309-312

Park R, Epstein S (1960) Carbon isotope fractionation during photosynthesis. Geochim Cosmochim Acta 21:10-126

Pedersen TF, Nielsen B, Pickering M (1991) Timing of late Quaternary productivity pulses in the Panama Basin and implications for atmospheric CO_2. Paleoceanog 6:657-677

Petit JR, Mounier L, Jouzel J, Korotkevich YS, Kotlyakov VI, Lorius C (1990) Palaeoclimatological and chronological implications of the Vostok core dust record. Nature 343;56-58

Popp BN, Takigiku R, Hayes JM, Louda JW, Baker EW (1989) The post-Paleozoic chronology and mechanisms of ^{13}C depletion in primary marine organic matter. Am J Sci, 289:436-454

Prahl FG, Muehlhausen LA, Lyle M (1989) An organic geochemical assessment of oceanographic conditions at MANOP Site C over the past 26,000 years. Paleoceanog 4:495-510

Rau, GH, Des Marais DJ (1992) A model relating plankton $\delta^{13}C$ to CO_2 supply and demand. Trans Am Geophys Union 72: 65

Rau, GH, Froelich PN, Takahashi T, Des Marais DJ (1991a) Does sedimentary organic $\delta^{13}C$ record variations in Quaternary ocean [$CO_{2(aq)}$]? Paleoceanog 6:335-347

Rau, GH, Sullivan CW, Gordon LI (1991b) $\delta^{13}C$ and $\delta^{15}N$ variations in Weddell Sea particulate organic matter. Mar Chem 35:355-369

Rau, GH, Sweeney RE, Kaplan IR (1982) Plankton $^{13}C/^{12}C$ ratio changes with latitude: Differences between northern and southern oceans. Deep-Sea Res 29:1035-1039

Rau, GH, Takahashi T, Des Marais DJ (1989) Latitudinal variations in plankton $\delta^{13}C$: implications for CO_2 and productivity in past oceans. Nature 341:516-518
Rau, GH, Takahashi T, Des Marais DJ, Repeta DJ, Martin JH (1992) The relationship between $\delta^{13}C$ of organic matter and $[CO_{2(aq)}]$ in ocean surface water: Data from a JGOFS site in the northeast Atlantic Ocean and a model. Geochim Cosmochim Acta 56:1413-1419
Rau, GH, Takahashi T, Des Marais DJ, Sullivan CW (1991c) Particulate organic matter $\delta^{13}C$ variations across the Drake Passage. J Geophys Res 96C:15131-15135
Rau GH, Teyssie J-L, Rassoulzadegan R, Fowler SW (1990) $^{13}C/^{12}C$ and $^{15}N/^{14}N$ variations among size-fractionated marine particles: Implications for their origin and trophic relationships. Mar Ecol - Prog Ser 59:33-38
Roeske CA O'Leary MH (1984) Carbon isotope effects on the enzyme-catalyzed carboxylation of ribulose bisphosphate. Biochem 23:6275-6285
Seckbach J, Kaplan IR (1973) Growth pattern and $^{13}C/^{12}C$ isotope fractionation of *Cyanidium caldarium* and hot spring algal mats. Chem Geol 12:161-169
Shackleton NJ, Pisias NG (1973) Atmospheric carbon dioxide, orbital forcing, and climate. In: Sundquist ET, Broecker WS (eds) The carbon cycle and atmospheric CO_2: Natural variations Archean to present. Geophys Monogr Ser, vol. 32, AGU, Washington, D.C.
Shemesh A, Charles CD, Fairbanks RG (1992) Oxygen isotopes in biogenic silica: Global changes in ocean temperature and isotopic composition. Science 256:1434-1436
Sikes EL, Farrington JW, Keigwin LD (1991) Use of the alkenone unsaturation ratio U^{k}_{37} to determine past sea surface temperatures: Core top SST calibrations and methodology considerations. Earth Planet Sci Lett 104:36-47
Skirrow G (1975) The dissolved gases - carbon dioxide. In: Riley JP, Skirrow G (eds) Chemical oceanography. Academic, London
Tans PP, Fung IY, Takahashi T (1990) Observational constraints on the global atmospheric carbon dioxide budget. Science 247:1431-1438
Thunell RC, Qingmin M, Calvert SE, Pedersen TF (1991) Glacial-Holocene biogenic sedimentation patterns in the South China Sea: Productivity variations and surface water pCO_2. Paleoceanog 7:143-162
Westerhausen L, Poynter J, Eglinton G, Erlenkeuser H, Sarnthein M (1993) Marine and terrigenous origin of organic matter in modern sediments of the equatorial east Atlantic: the $\delta^{13}C$ and molecular record. Deep-Sea Res (in press)
Wong WW, Sackett WM (1978) Fractionation of stable carbon isotopes by marine phytoplankton. Geochim Cosmochim Acta 42:1809-1815

RECONSTRUCTION OF PALEOCEANIC PCO_2 LEVELS FROM CARBON ISOTOPIC COMPOSITIONS OF SEDIMENTARY BIOGENIC COMPONENTS

John P. Jasper and J. M. Hayes
Biogeochemical Laboratories
Departments of Chemistry and of Geological Sciences
Indiana University
Bloomington, Indiana 47405-1403 U.S.A.

ABSTRACT. The principles and practice of reconstruction of concentrations of dissolved CO_2 from carbon isotopic compositions of contemporaneous, coeval marine organic matter and carbonates are summarized here. Techniques for determination of the isotopic fractionation accompanying photosynthetic fixation of carbon are exemplified by three oceanic case histories. The northern Gulf of Mexico during the late Quaternary represents an environment with near-equilibrium distribution of CO_2 between the atmosphere and ocean surface, whereas non-equilibrium conditions prevail in the central equatorial Pacific. The Black Sea, with surface waters buffered strongly by high concentrations of dissolved CO_2 at depth, represents an exotic PCO_2 environment. Consideration of a preliminary water-column model focussed on the loci of production of biogenic phases used for reconstruction of paleoceanic PCO_2 illuminates some of the major issues to be considered as this technique is refined.

INTRODUCTION

To reconstruct the oceanic history of the sources and sinks of carbon dioxide, the mechanisms by which oceanic and atmospheric levels of CO_2 are maintained must be better defined. A regionally useful indicator of oceanic concentrations of CO_2(aq) is thus required. The record of the isotopic fractionation associated with fixation of CO_2 by marine phytoplankton may be useful because that fractionation is largely a function of ambient CO_2(aq) concentration (Popp *et al.*, 1989; Rau *et al.*, 1989,1991; Jasper and Hayes, 1990; Freeman and Hayes, 1992; Rau, this volume). The geological history of ε_P (the photosynthetic isotope effect) is recorded in the carbon isotopic compositions ($\delta^{13}C$) of phytoplanktonic products and of contemporaneously formed biogenic carbonates (Popp *et al.*,

NATO ASI Series, Vol. I 17
Carbon Cycling in the Glacial Ocean:
Constraints on the Ocean's Role in Global Change
Edited by R. Zahn et al.

1989). In 1989, Popp *et al.* published an ~ 250-Ma record of ε_P based on ^{13}C contents of geoporphyrins and carbonates. Its details corresponded with those expected from the Fischer "Icehouse-Greenhouse" Cycles (Fischer, 1981; Popp *et al.*, 1989). More recently, Jasper and Hayes (1990) studied the carbon isotopic expression of oceanic and atmospheric CO_2 variations during the last Quaternary glacial-interglacial cycle. By assuming equilibration of CO_2 between the atmosphere and the surface waters of the northern Gulf of Mexico throughout the cycle, they were able to use stratigraphically robust points from an ~100 ka record of ε_P, together with the Vostok pCO_2 record, to calibrate the relationship between ε_P and oceanic PCO_2. That isotopic record was derived from the sedimentary remains of Prymnesiophyte microalgae (specifically a C_{37} alkadienone, heptatriconta-15,22-dien-2-one) and the carbonate tests of a coeval planktonic foraminifera (*Globigerinoides ruber*). The calibration permitted the estimation of ~100 ka records of CO_2(aq) concentration and "paleoceanic PCO_2." More recently, a contemporary oceanic calibration of ε_P (based on paired $\delta^{13}C$ measurements of marine particulate organic carbon and dissolved inorganic carbon, ΣCO_2) versus [CO_2(aq)] allowed the estimation of paleoceanic PCO_2 spanning the last ~ 250 Ma (Freeman and Hayes, 1992).

In this paper, three case histories of compound-specific isotopic reconstructions of CO_2(aq) and equilibrium atmospheric pCO_2 levels are presented in the context of the underlying principles of this recently developed method. Paleoceanic PCO_2 records from the late Quaternary exemplifying near equilibrium (the northern Gulf of Mexico), non-equilibrium (the central equatorial Pacific), and exotic PCO_2 environments (the Black Sea) will be examined in turn.

METHODOLOGY

Analyses

Foraminiferal $\delta^{13}C$. Foraminiferal carbonates are analyzed by standard methods, using a dual inlet mass spectrometer. In these analyses, CO_2 is prepared by reaction of carbonates with anhydrous phosphoric acid at 90°C and is cryogenically distilled for mass spectrometric analysis.

Alkenone $\delta^{13}C$. Liquid chromatography is used to isolate fractions for gas chromatographic analysis (Prahl *et al.*, 1989). Selected fractions are analyzed by electron-impact (70 eV) gas chromatography/mass spectrometry to confirm the presence of the alkenones and the absence of interfering compounds.

The $\delta^{13}C$ of heptatriconta-15,22-dien-2-one, the major C_{37} alkadienone (hereafter, $C_{37:2}$; de Leeuw *et al.*, 1979) is determined by isotope-ratio-monitoring gas chromatography/mass spectrometry (irmGCMS: Hayes *et al.*, 1990; Freeman *et al.*, 1990). Coinjected *n*-alkanes (C_{38} and C_{41}) are employed as isotopic standards. The precision and accuracy obtained in each run are internally checked by analysis of a two other coinjected *n*-alkanes (C_{36} and C_{37}). Instrument performance is monitored by comparison of d values obtained by combustion and by injection of CO_2 gas calibrated against NBS standards. Recently, typical standard deviations determined by pooling of replicate measurements have been ~0.30‰. Carried through the equilibrium PCO_2 calculation outlined below, this corresponds to an uncertainty of ~28 μatm at $PCO_2 = 250$ μatm.

Prior to carbon isotopic analysis, each alkenone fraction is gas chromatographically analyzed to determine whether significant quantities of a coeluting alkenoate are present. Since the alkenoate coelutes with $C_{37:2}$, its presence can cause systematic errors. If concentrations of alkenoates are ≥3% of $C_{37:2}$, the alkenone-containing fraction is saponified in 0.5 N KOH in CH_3OH/H_2O (95/5, v/v) to allow removal of the coeluting compound.

Calculations

The calculation of ε_P-based equilibrium levels of PCO_2 is outlined in Figure 1. The δ values of biogenic organic (alkenones, $\delta_{37:2}$) and carbonate phases (planktonic forams, δ_f) allow reconstruction of the ^{13}C contents of primary photosynthate (δ_P) and of dissolved CO_2 (δ_d). Values of ε_P are calculated from δ_P and δ_d. Using the ε_P *versus* [CO_2(aq)] calibration, concentrations of dissolved CO_2 are calculated from measured values of ε_P. Equilibrium PCO_2 can then be estimated using Henry's Law. For ε_P-based PCO_2 records of regions which approach air-sea equilibrium, paleoatmospheric pCO_2 records can be estimated.

The following definitions are pertinent.

$\varepsilon_{m/d}$(T): the temperature-dependent, equilibrium fractionation factor relating ^{13}C contents of mineral calcite (δ_m) and CO_2(aq) (δ_d)

$$\varepsilon_{m/d} \int (\alpha_{m/d} - 1)10^3 = [(\delta_m + 1000)/(\delta_d + 1000) - 1]10^3 \qquad (1)$$

$\Delta_{m/f}$: the isotopic difference between equilibrated calcite and foraminiferal calcite

$$\Delta_{m/f} \int \delta_m - \delta_f \qquad (2)$$

where δ_f is the measured $\delta^{13}C$ of foraminiferal calcite.

δ_P: the estimated $\delta^{13}C$ of Prymnesiophyte biomass, which is distinct from the measured isotopic composition of $C_{37:2}$

$$\delta_P = \delta_{37:2} + \Delta_{P/l} \qquad (3)$$

where $\delta_{37:2}$ is the $\delta^{13}C$ of $C_{37:2}$ and $\Delta_{P/l}$ is the isotopic difference between total biomass and $C_{37:2}$ (the subscipt, P/l, refers to net **P**roduction and to **l**ipid carbon).
ε_P: the isotopic fractionation accompanying photosynthetic fixation of carbon

$$\varepsilon_P \int (\alpha_P - 1)10^3 = [(\delta_d + 1000)/(\delta_P + 1000) - 1]10^3 \qquad (4)$$

Where observations of δ_d and δ_P pertained to different core depths, values of δ_d are interpolated linearly with respect to time. This definition of ε_P reverses the sign (from negative to positive values) from earlier work (Jasper and Hayes, 1990). It is now consistent with the plant-physiological literature and provides a more intuitive measure in which larger values correspond to larger fractionations. Absolute values of ε_P are not significantly affected.

Determination of ε_P From δ_f and $\delta_{37:2}$

The equilibrium fractionation factor relating $\delta^{13}C$ values of $HCO_3^-(aq)$ and $CO_2(aq)$ has been measured by Mook *et al.* (1974)

$$\varepsilon_{d/b} \int [(\delta_d + 1000)/(\delta_b + 1000) - 1]10^3 = 24.12 - 9866/T \qquad (5)$$

where δ_b is the $\delta^{13}C$ of dissolved bicarbonate and T is in °K. Similarly, Morse and McKenzie (1990) have reviewed data relevant to equilibrium fractionation of carbon isotopes between dissolved bicarbonate and calcite and shown that

$$\varepsilon_{m/b} \int [(\delta_m + 1000)/(\delta_b + 1000) - 1]10^3 = 10.51 - 2980/T \qquad (6)$$

Combination of equations 5 and 6 yields quantitative values for $e_{m/b}$ (*cf.* equation 1)

$$\varepsilon_{m/d} = -14.07 + 7050/T \qquad (7)$$

Evaluation of data presented by Fairbanks *et al.* (1982) indicates that, for *N. dutertrei*, $\Delta_{m/f}$ = -0.70‰. Application of equations 1 and 2 then allows calculation of δ_d from measured values of δ_f. Evaluation of $\varepsilon_{m/d}$ requires knowledge of the temperature at which carbonate species have equilibrated. For the equatorial Pacific PCO_2 calculations, we used temperatures derived from measurements of the alkenone-unsaturation parameter, $U^K{}_{37}$, in the same samples (Lyle *et al.*, 1992).

Provisionally, $\Delta_{P/l}$ = +3.8‰. Although this value derives from results of a single culture experiment (Jasper and Hayes, 1990), it is typical of the isotopic difference observed between the biomass and extractable lipids of marine plankton (~4‰; Degens *et al.*, 1968) and the likelihood that it will be adjusted significantly is small. Moreover, since any change would affect equally both the estimation of ε_P and the calibration of ε_P-$[CO_2(aq)]$ relationship, there would be no effect on PCO_2 estimates. The $\Delta_{P/l}$ concept is valuable because it facilitates discussion of all photosynthetic fractionations in terms of biomass, thus referring to total fixed carbon rather than to specific compounds. As paleobarometric techniques are extended to diverse species having different biomarkers, continued use of the compound-specific-Δ concept will provide a standardized frame of reference for all observations of photosynthetic fractionation. In nature, $\Delta_{P/l}$ will depend on isotope effects accompanying lipid biosynthesis and on the division of cellular carbon flows among biosynthetic pathways (Monson and Hayes, 1980). The first of these factors will be constant for all cells of given species, and probably for a wide variety of algae. Variations in the second factor would cause variations in cellular composition (*i. e.*, lipids *vs.* carbohydrates *vs.* proteins). Consequential variations are not expected, but the point deserves further study.

Determination of [$CO_2(aq)$] and PCO_2 From δ_P

A set of eight calibration points for sedimentary stratigraphic intervals corresponding to Vostok pCO_2 ranges < 20 matm was used to calibrate the Gulf of Mexico (DSDP 619) record against equilibrium [CO_2(aq)] values calculated from the Vostok ice core CO_2 record (Jasper and Hayes, 1990). Here (Figure 1), the original calibration has been updated by incorporation of the effects of the glacial-interglacial temperature change (from 23° to 25°C, ^{18}O-derived temperatures discussed by Jasper and Gagosian, 1989), which has a small effect on both $\varepsilon_{m/d}$ and [CO_2(aq)]. The relationship found is

$$\varepsilon_P = 29.2 - 109/d \qquad (8)$$

where d = [CO_2(aq)], μmol/L (Jasper *et al.*, 1993).

Equilibrium pCO_2 values were calculated from values of d, ambient water temperatures, and Henry's Law

$$PCO_2 = d/K_H(T) \qquad (9)$$

where $K_H(T)$ is the temperature-dependent Henry's Law constant (Weiss, 1974).

CASE HISTORIES

A Near-Equilibrium Environment: The Pigmy Basin, Northern Gulf of Mexico

The availability of a previously identified suite of marine compounds (Jasper and Gagosian, 1989, 1993) from a region whose surface waters were relatively close to equilibrium (Tans *et al.*, 1990) allowed examination of the concept of an ε_P-based CO_2 paleo-

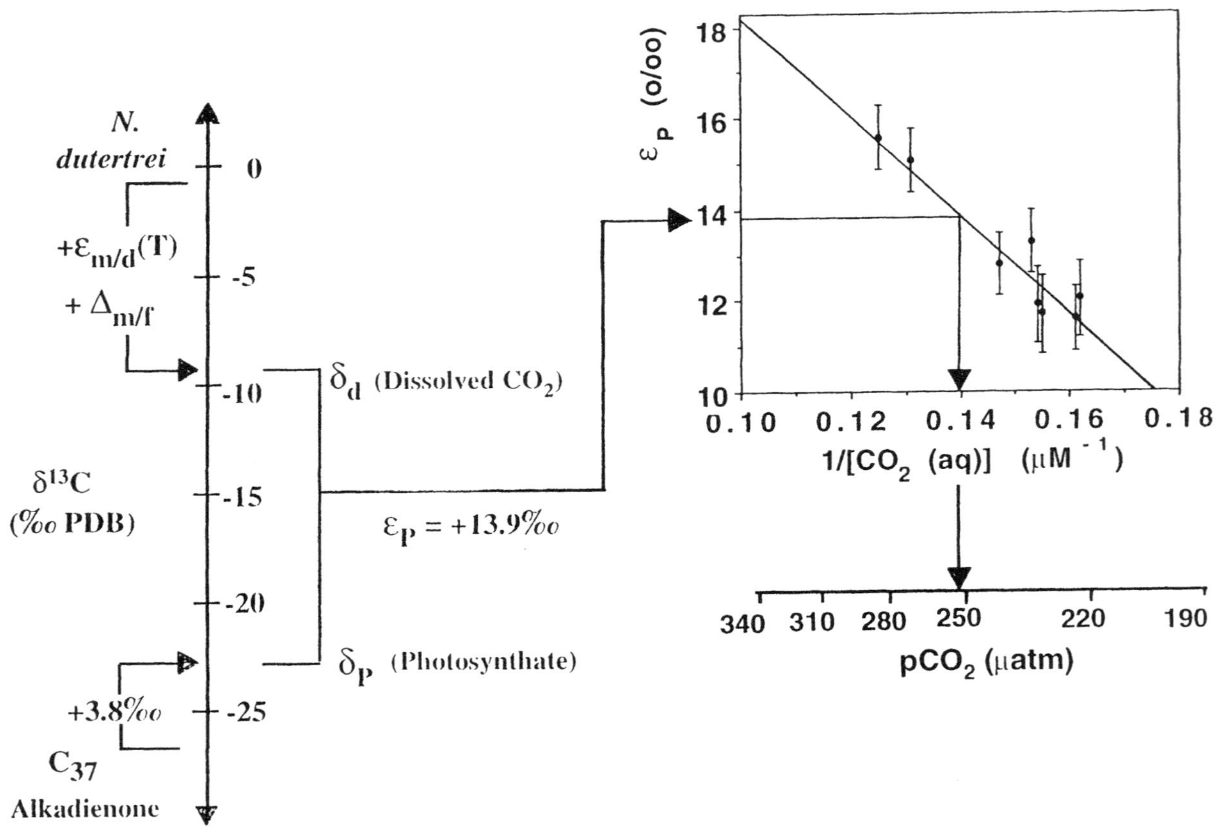

Figure 1. Diagram outlining the calculation of equilibrium, or paleoceanic, PCO_2 values from the photosynthetic fractionation of CO_2 (ε_P) (from Jasper *et al.*, 1993). Calibration of ε_P points against CO_2 (aq) concentrations allows reconstruction of equilibrium pCO_2 values from sedimentary components. The diagram and calculations are explained in the text.

barometer (Popp *et al.*, 1989). In 1990, Jasper and Hayes produced an ~100-ka record of the $\delta_{37:2}$ with the intention of testing whether evidence of the [CO_2(aq)]-dependent photosynthetic fractionation of carbon isotopes had been recorded. A glacial-to-interglacial difference of 5.3‰ was found in the $\delta_{37:2}$ record. A previous record of δ_f for *Globigerinoides ruber* from the same core varied by ~1.4‰ (Williams and Kohl, 1986). Together, these records allowed calculation of the history of ε_P for the surface waters of the northern Gulf of Mexico. Related concentrations of CO_2(aq) were estimated from Vostok-ice-core values of atmospheric pCO_2 (Barnola *et al.*, 1987) and paleo-sea-surface temperatures (Jasper and Gagosian, 1989). Selection of intervals for which uncertainities in the stratigraphic correlation between the ice core and the sediment core did not significantly affect relationships between [CO_2(aq)] and ε_P yielded the calibration line shown in Figure 1.

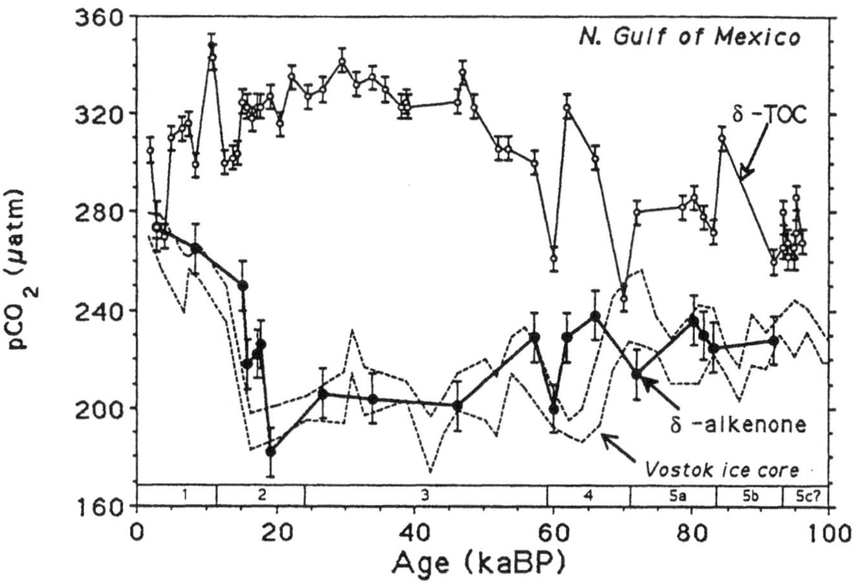

Figure 2. Time-series of PCO_2 values reconstructed from compound-specific $\delta^{13}C_{37:2}$ alkenone values and total organic carbon $\delta^{13}C$ values from the northern Gulf of Mexico (Pigmy Basin, DSDP Site 619; Jasper and Hayes, 1990). The Vostok ice core pCO_2 records is shown for comparison. In this case, the large contribution of terrigenous organic carbon significantly interferes with air-sea equilibrium pCO_2 reconstruction because the isotopic paleobarometer is based on the reconstruction of primary marin organic $\delta^{13}C$ values

Paleoceanic PCO_2 values were calculated for all the alkenone-based ε_P points from the Pigmy Basin record using that calibration. In Figure 2, they are compared with the Vostok ice core pCO_2 record and with values of PCO_2 that would be obtained if the $\delta^{13}C$ of total organic carbon (TOC) had been used as the basis for the calculation. The correspondence between the alkenone-based and ice-core records is good. However, since (i) the TOC in the Pigmy Basin is climatically-forced mixture of terrigenous and marine organic carbon (Jasper and Gagosian, 1990) and (ii) CO_2-related variations in ε_P are much greater for marine than for terrestrial plants (Popp *et al.*, 1989), the pCO_2 record based on TOC approaches the ice core record only when the TOC is dominated by marine organic carbon

The "test-of-concept" study in the Pigmy Basin encouraged broader application (Rau *et al.*, 1991; Freeman and Hayes, 1992; Jasper *et al.*, 1991, 1993) of the ε_P-based pCO_2 paleobarometer. Refinement of the method will come from water-column calibrations of compound-specific ε_P values against CO_2(aq) concentrations, as well as from controlled-culture experiments. The most useful oceanic calibrations will be those in which paired values of ε_P and [CO_2(aq)] are measured through the chemical gradients of the euphotic zone.

A Non-Equilibrium Environment: MANOP SITE C, Central Equatorial Pacific

The surface waters of the central equatorial Pacific are presently supersaturated with CO_2(aq) and are a significant source of CO_2 to the atmosphere (Volk and Bacastow, 1989; Tans *et al.*, 1990). Ideally, an interglacial record of concentrations of CO_2(aq) would record this disequilibrium. To examine ocean-atmosphere relationships during Quaternary glacial-interglacial cycles, an ~255-ka record of CO_2 levels was generated from a central equatorial Pacific sediment core raised from MANOP Site C at 0°57.2'N, 138°57.3'W. A record of δ_P values was reconstructed *via* isotopic analyses of $C_{37:2}$ alkadienones (eqn. 3). The prymnesiophyte algae that are the exclusive source of this compound grow most abundantly at depths of ~60-80 m in the subsurface eutrophic zone in the central equatorial Pacific (Honjo and Okada, 1974; Chavez *et al.*, 1990; Pena *et al.*, 1990; Marlowe *et al.*, 1990). Using the mathematical methods described above, the record of isotopic compositions of CO_2(aq) was reconstructed from observed δ values of *N. dutertrei*, a planktonic foram which apparently calcifies at the depth of the chlorophyll maximum (Ravelo *et al.*, 1990), *i. e.*, near the depth at which the prmnesiophytes grow. From the δ_P and δ_d data, the history of ε_P in the central Equatorial Pacific was calculated. Concentrations of CO_2(aq) were then estimated from the ε_P record, using the calibration from the Gulf of Mexico sediment core. Finally, paleoceanic pCO_2 levels were calculated using Henry's Law and assuming that the waters in which the forams and prymnesiophytes grew had been in equilibrium with the atmosphere at the temperature given by the alkenone $U^K{}_{37}$ value (Lyle *et al.*, 1992).

As shown in Figure 3, the resulting range of pCO_2 values (208-260 matm) is markedly smaller than that in the ice core (190-280 matm) and, for the late Holocene, concentrations of CO_2 are at sharp variance with modern observations. Since the alkenones and *N. dutertrei* tests are biosynthesized in the subsurface eutrophic layer, not in the upper oligotrophic layer

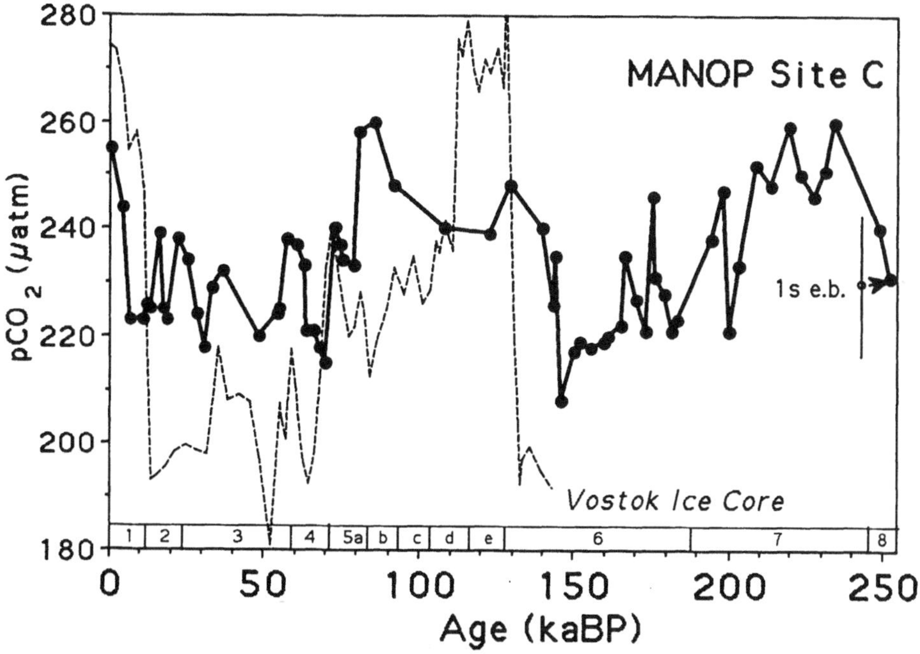

Figure 3. A record of paleoceanic PCO_2 calculated from ε_P values derived from C_{37} alkadienines and *N.dutertrei* calcite from a central equatorial Pacific (MANOP Site C; 0°57.2'N, 138°57.3'W) sediment core superposed on the Vostok ice core record of Barnola *et al.* (1987) (from Jasper *et al.*, 1933a). A typical 1σ error bar is given.

(as described by Coale and Bruland, 1987), and since those waters are not in direct communication with the atmosphere, we must consider whether the attenuated range and unexpectedly low levels might reflect the presence of a subsurface CO_2 minimum and glacial-to-interglacial variations in the phenomena shaping it.

Understanding the CO_2(aq) balance in the photic zone requires insight into the structure of the upper ocean. Feely *et al.*, (1987) observed subsurface PCO_2 minima in the upper ~50-125 m of the central equatorial Pacific within 3° of the equator at 150°W. Although the sampling resolution was relatively coarse (~40- to 60-m intervals), the results are relevant here. Research in progress at 140°W in the eastern equatorial Pacific reveals subsurface minima in *total* CO_2 concentration approximately coincident with the chlorophyll maximum (Feely *et al.*, 1992), although oceanic PCO_2 profiles were not available in that preliminary report. Earlier work had indicated that, within 15° of the equator in the Pacific and Atlantic oceans, subsurface ^{13}C maxima in ΣCO_2 were probably due to the preferential fixation of ^{12}C during photosynthesis (Kroopnick, 1985). These subsurface PCO_2 minima are likely due to a com-

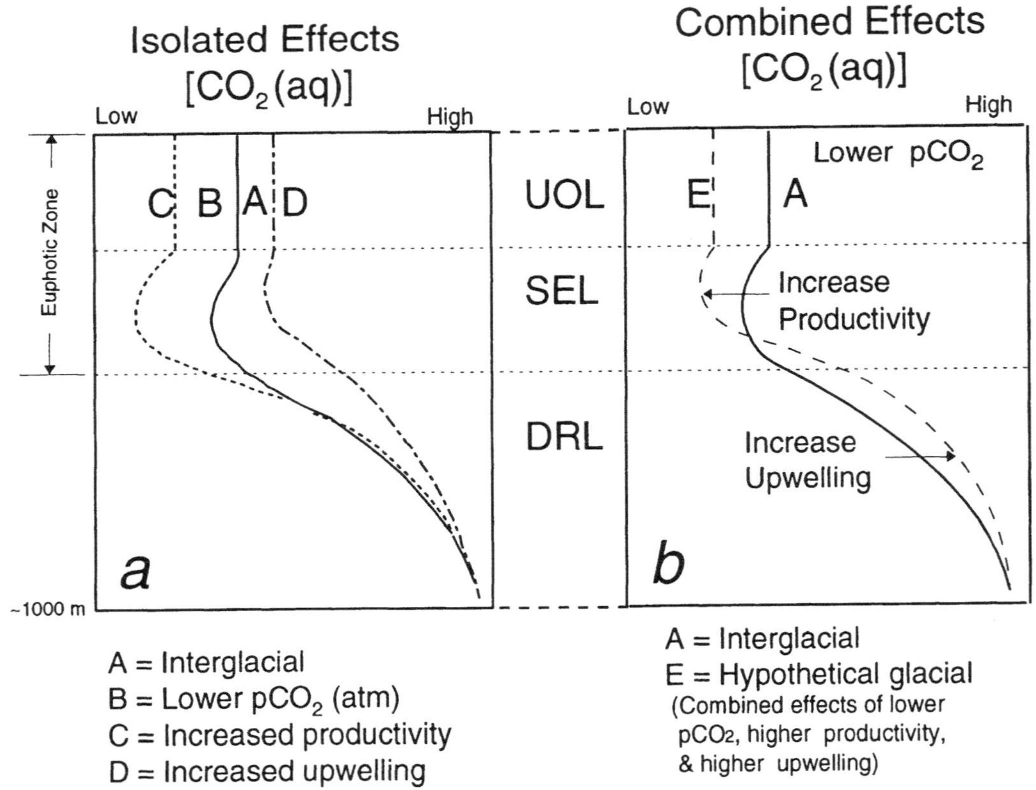

Figure 4. Diagram summarizing the contemporary CO_2(aq) concentration profile (curve A) of equatorial surface water in the upper oligotrophic layer (UOL), the subsurface eutrophic layer (SEL), and the deep recycling layer (DRL) as it affected by the (a) isolated effects of variations in atmospheric pCO_2 (curve B), primary productivity (curve C), and upwelling (curve D), and (b) the combined effects of those processes (curve E). These isolated and combined scenarios are discussed in the text.

bination of photosynthetic drawdown of CO_2(aq) and advection (geostrophic inflow) of waters with low concentrations of CO_2(aq) (Wyrtki, 1981; Feely *et al.*, 1987).

Combining these concepts of water-column chemistry with recent ideas about water-column stratification, particularly the "stratified euphotic zone" (*e.g.*, Coale and Bruland, 1987; Longhurst and Harrison, 1989), Jasper *et al.* (1993) developed a conceptual model for budgets of CO_2 in the upper water column (Figure 4). In Figure 4 the contemporary (interglacial) [CO_2(aq)] profile is represented by curve **A**. A profile that might prevail if atmospheric pCO_2 decreased, all factors remaining constant, is represented by curve **B**. Isolated effects of increased primary productivity and increased upwelling are represented by curves **C** and **D** (Figure 4a). Curve **A** shows high concentrations of CO_2(aq) in the deep recycling layer (DRL). These decrease upward to the euphotic zone, where a weak minimum is found in the subsurface eutrophic layer (SEL), the zone of maximum net export productivity (Eppley and Peterson, 1979). In Curve **B**, the effect of lowering atmospheric pCO_2 (*e. g.*, to glacial values) is reflected in lower concentrations of CO_2(aq) in the upper oligotrophic layer (UOL), a CO_2(aq) minimum in the SEL, and a return to interglacial CO_2(aq) concentrations at depth. Thus, alkenone producers living in the SEL could record the interglacial-to-glacial decrease of [CO_2(aq)], but the amplitude of the fluctuation would be reduced.

The effect of increased primary production is illustrated by curve **C**, which depicts drawdown of CO_2(aq) in both layers of the euphotic zone. In fact, multiple lines of evidence suggest that export productivity in glacial periods was higher than that in interglacials (Lyle *et al.*, 1988; Prahl *et al.*, 1989; Pedersen *et al.*, 1991). The subsurface CO_2(aq) minimum would be enhanced as the increase in export production drew down CO_2(aq) in the SEL, perhaps causing a net influx of CO_2 from the atmosphere. In this case, alkenone-producers living in the SEL would record a greater amplitude of [CO_2(aq)] lowering.

The effect of increased upwelling on the CO_2(aq) profile is indicated by curve **D**, which indicates that concentrations of CO_2(aq) would be higher throughout the upper water column. In fact, application of a simple physical oceanographic model to the $U^K{}_{37}$ temperature record from MANOP Site C (Lyle *et al.*, 1992) indicates that upwelling rates during glacial time were up to ~2.5 times (but typically <1.5 times) the interglacial rate. This general increase in concentrations of CO_2(aq) could lead to evasion of CO_2 from the ocean to the atmosphere. With constant primary productivity, alkenone-producers in the SEL would record a [CO_2(aq)] increase relative to the contemporary case (curve **A**).

A hypothetical combination of these effects, represented here as curve **E** in Figure 4b, provides a possible explanation the variations in paleoceanic PCO_2 indicated in Figure 3. Because of the combined effects of lower atmospheric pCO_2 and increased export productivity, glacial concentrations of CO_2(aq) in the Upper Oligotrophic Layer and part of the Subsurface Eutrophic Layer could have been decreased relative to the interglacial case (**A**), whereas those in the lower SEL and the DRL might have been increased due to upwelling of CO_2(aq) and remineralization of higher fluxes of organic matter. The isotopic signal recorded by the alkenone producers is thus dependent on their habitat. If they live in the UOL, they will more fully record the [CO_2(aq)] signal related to oceanic-atmospheric exchange (Figure 4b). If they live in the SEL, variations in the [CO_2(aq)] signal could be significantly attenuated or even (below the crossover of curves **E** and **A**) reversed.

The most recent observations, however, indicate that Prymnesiophytes tend to grow in these waters at depths above the chlorophyll maximum (Fluegge *et al.*, 1993). An alternative to the CO_2-minimum hypothesis is thus required, and is provided by suggestions that ε_P-[CO_2(aq)] relationships can be dependent on algal growth rates and thus on nutrient levels (Francois *et al.*, 1993). Changes in such relationships amount to redefinitions of the calibration line; notably, changes in the coefficient for the 1/d term in equation 8 (Jasper *et al.*, 1993; Hayes and Jasper, 1994). On balance, such an adjustment, taking into account the much higher nutrient levels and growth rates in the equatorial Pacific (*vs.* the northern Gulf of Mexico, the site of the previous calibration), appears now to provide the best interpretation of the ε_P record at the MANOP site. On this basis, it appears that the equatorial Pacific has continuously been a source of CO_2 and, if anything, that fluxes from the ocean to the atmosphere in this region were slightly higher during glacial intervals (Jasper *et al.*, 1993).

The subsurface-CO_2-minimum hypothesis, however, was not without support. If the "oligotrophic" (*i. e.*, Gulf of Mexico) calibration were applied, low-amplitude variations in [CO_2(aq)] would be indicated at MANOP Site C (Figure 3). This could indicate that the alkenone producers lived in the lower portion of the subsurface eutrophic layer, recording the [CO_2 (aq)] at that depth range, which is plausibly lower than that of the UOL (Figure 4b). In support of this inference, it could be noted that the nannoplankton-related concentration of chlorophyll-*a* (associated mainly with *E. huxleyi* in the equatorial Pacific) reaches its maximum below 80 m (Honjo and Okada, 1974; Chavez *et al.*, 1990; Pena *et al.*, 1990), and that a relatively large portion of the alkenones biosynthesized in the SEL might sink as new production.

In the northeast Pacific, alkenone-based temperatures partially support this inference since they correspond to the temperatures at the subsurface chlorophyll maximum during summer stratification (Prahl *et al.*, 1992). However, the locus of alkenone production apparently is seasonally variable, moving from the UOL in the winter to the SEL in the summer (Prahl *et al.*, 1992). The seasonal changes in hydrography in the northeast Pacific probably lead to a more complex series of events in the water-column flux (and biosynthesis) of alkenones (Prahl *et al.*, 1992) than that inferred for the relatively hydrographically stable equatorial Pacific. To the extent that the summer stratification in the northeast Pacific typifies the average water-column structure of the central equatorial Pacific, the average composition of sedimentary alkenones could record the PCO_2 of the subsurface eutrophic layer, not that of the upper oligotrophic layer. The subsurface-CO_2-minimum model suggests that concentrations of CO_2(aq) in the upper SEL covary with atmospheric pCO_2, but that proportional variations are attenuated.

This model illustrates the importance of understanding the depth habitat of the biological monitors of paleoceanic PCO_2 (selected phytoplankton and planktonic foraminifera) when seeking to reconstruct paleoceanic sources and sinks of CO_2 through the $\delta^{13}C$ of biogenic carbon phases. Accordingly, the loci of production of alkenones and foraminferal carbonates will be important questions in forthcoming research into the recalibration of the ε_P-based CO_2 paleobarometer.

If primary productivity during glacial periods responded to other variables such as concentrations of micronutrients (*e. g.*, iron: Martin, 1990), then a net reduction of paleoceanic PCO_2 could have occurred at such times. Pedersen *et al.* (1991) argued for iron inputs associated with eolian dust as a likely cause of enhanced fertility of the equatorial Pacific region during glacial intervals.

AN EXOTIC CO_2 ENVIRONMENT

The Black Sea

The availability of a suite of marine and terrigenous organic biomarker compounds from a finely laminated, ^{210}Pb-dated sequence of sediments from the western basin of the Black Sea (Box Core 9, 42° 55'N, 31°22' E, 2094 m; Crusius and Anderson, 1991) has allowed further examination of CO_2 budgets in surface waters. In particular, it has been possible to prepare and analyze isotopically both alkenones and related coccolith carbonates as well as additional planktonic biomarkers. The resulting isotopic records can be considered in the light of new hydrographic data and related studies of the inorganic-carbon system in the Black Sea, known changes in atmospheric levels of CO_2, and possible changes in terrigenous inputs (Jasper et al. 1993b).

The $\delta^{13}C$ of total carbonate in sediment fractions enriched in coccolith carbonate was measured using materials separated from 1-2 g sediment samples (Figure 5a). A time-series record shows a general decrease of 0.5‰ over the last ~150 years. A similar trend is also recorded in Bermuda coral (Nozaki *et al.*, 1978) and in the CO_2(g) from a Siple Station ice core (Friedli *et al.*, 1986). The correspondence between these $\delta^{13}C$ records indicates that the surface waters of the Black Sea isotopically exchange with the atmosphere on relatively short (decadal) timescales. Both the coral and ice-core records have been interpreted as manifestations of the Suess Effect, which is a historical record of depletion of ^{13}C in atmospheric CO_2

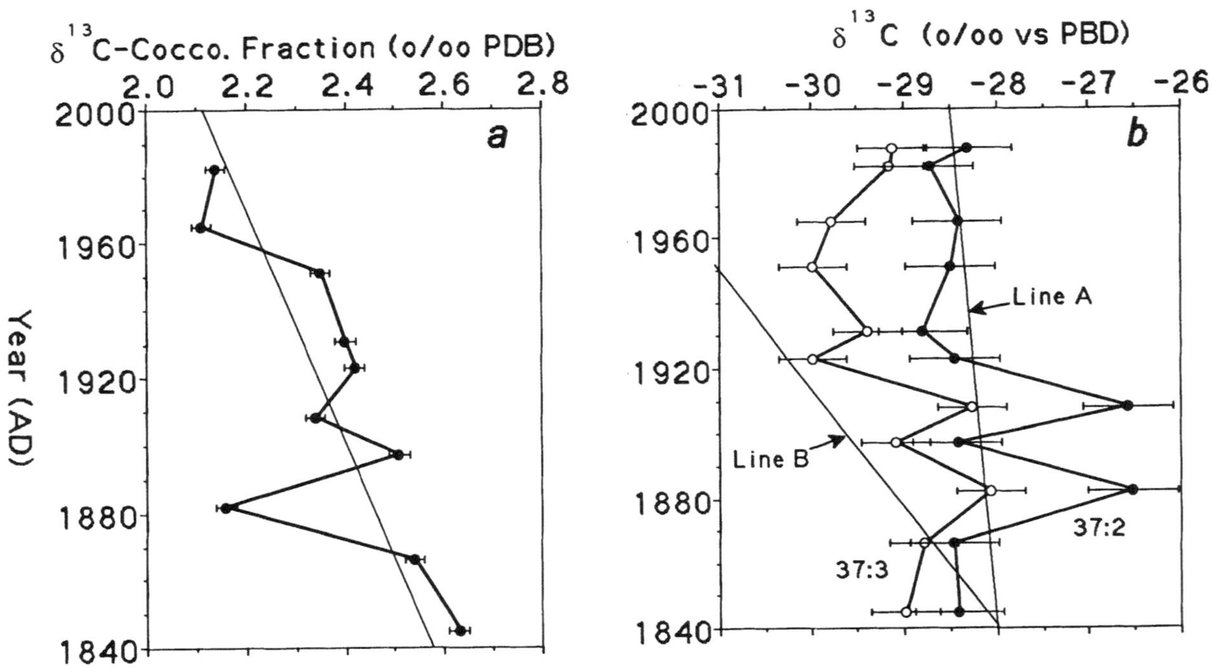

Figure 5. Historical records of (a) the $\delta^{13}C$ of a sediment fraction enriched in coccolith-carbonate, and (b) the $\delta^{13}C$ of $C_{37:2}$ and $C_{37:3}$ alkenones ($C_{37:2}$: filled symbols; $C_{37:3}$: open symbols). Line A in Fig. 5b is a superposition of the slope of linear regression line ($t = 7.89 + 0.0029 \bullet \delta^{13}C$-coccolith) from Fig. 5a. It indicates that the photosynthetic fractionation of carbon isotopes (ε_P) was constant (assuming constant T, alkalinity, and ΣCO_2) while the $\delta^{13}C$-coccolith values were generally decreasing. Line B shows the calculated effect of paleoceanic PCO_2 increasing from 280 to 360 µatm over the last ~150 years if Black Sea surface waters were continuously in atmospheric equilibrium, using the ε_P-vs.-PCO_2 slope of Jasper *et al.* (1993a), 35.9‰/$\log_{10}[CO_2(aq)]$. The divergence of Line B from the observed results shows the implausibility of atmospheric equilibration in determining the $[CO_2(aq)]$ in the surface waters of the Black Sea.

and surface oceanic CO_2 caused by the combustion of ^{13}C-depleted fossil fuels during the Industrial Age.

The carbon isotopic record in alkenones is shown in Figure 5b. The signal is more variable than that derived from the carbonates and, as a result, a coarser scale has been employed. In spite of appearances, therefore, Line A in Figure 5b is parallel to the Suess-effect trend in Figure 5a. The fit of this line to the alkenone data is good enough that it is reasonable to conclude that (i) the isotopic fluctuations observed in sediment samples from approximately 1885 and 1908 likely represent biological or palaeoecological events in which Prymnesiophytes either changed their habitat or a distinct species (with different isotopic characteristics) temporarily became dominant, and (ii) apart from this, there has been no major change in the carbon isotopic difference between coccoliths and alkenones over the time span examined. The latter point is notable because atmospheric levels of CO_2 have increased significantly during the same time interval. Had this change been paralleled by an increase in concentrations of dissolved CO_2 at depths of Prymnesiophyte growth, an increase in isotopic fractionation would be expected. Specifically, since atmospheric pCO_2 has increased by a factor of 1.25, ε_P — and the coccolith-alkenone difference — would be expected to increase by 3.6‰ if the ε_P-*vs.*-PCO_2 slope observed previously (Jasper and Hayes, 1990) also prevailed in this system. The corresponding "changing [CO_2(aq)] trend" is indicated by Line B. The divergence of the observed record from line B is apparent. The results strongly favor either (i) independence of isotopic compositions and concentrations of dissolved CO_2 or (ii) constancy of concentrations of dissolved CO_2.

Isotopic compositions of other planktonic products, including, thus far, dinosterol, a C_{30} alkanediol, a C_{30} keto-ol, and *n*-C_{26} alcohol, have also been determined. Isotopic records for all of these are, like those of the alkenones, parallel to line A, not line B. Since Hollander and coworkers (private communication) have found that ^{13}C contents of all primary producers respond to variations in [CO_2(aq)], this observation provides further evidence that concentrations of dissolved CO_2 have varied much less strongly than atmospheric pCO_2; *i. e.*, that the second alternative is correct.

The molecular-isotopic conclusions are consistent with results of recent hydrographic studies of the inorganic carbon system in the Black Sea (Goyet *et al.*, 1991). Specifically, it has been found that surface waters of the Black Sea are supersaturated with CO_2. The source of this CO_2 is the deep, anoxic waters of the Black Sea. It is inconceivable that the very high levels of CO_2 in those waters have accumulated in only the last few years, so it is very likely that the present flux from the deep waters has a long history and that, as indicated by the present results, this deep-water control of surface [CO_2(aq)] has persisted for at least 150 years.

CONCLUSION

Compound-specific isotopic analyses, which allow definitive recovery of marine, algal ^{13}C contents, independent of terrigenous inputs and effects of consumers, allow reconstruction of the carbon-isotopic fractionations accompanying fixation of carbon by ancient producers. As a result, paleoceanic CO_2(aq) concentrations and paleoatmospheric pCO_2 levels can be estimated. The first late Quaternary reconstruction of an ~100-ka ε_P record through the last glacial-interglacial cycle permitted a preliminary calibration of equlibrium CO_2(aq) concentrations calculated from the Vostok ice pCO_2 record. The long chain alkenones were an opportune, informed choice of compounds for paleoceanic PCO_2 research for a number of reasons: a nearly global distribution, an internal paleothermometer, good preservational characteristics, and relatively facile chromatographic separation. There remain many other biomarkers that may prove to be useful CO_2 paleobarometers, but which have yet to be explored. For all those compounds, calibrations against ambient CO_2(aq) concentrations, whether in the water column or in culture, will be essential to the refinement of CO_2 paleobarometry. Similarly, improved understanding of the isotopic "life histories" of planktonic foramiinfera will be important in the refinement of the technique. The control of Quaternary atmospheric pCO_2 via paleoceanic PCO_2 is a most significant problem for which ε_P-based CO_2 paleobarometry will find useful application.

ACKNOWLEDGEMENTS

A. C. Mix, F. G. Prahl, and S. G. Wakeham provided refined samples and various expertise related to the case histories reviewed here. A. C. Mix was particularly helpful in the paleoceanographic interpretation of the MANOP Site C results, including the development of the water-column model of [CO_2(aq)]. N.S.F. Ocean Sciences and N.A.S.A. supported this research.

REFERENCES

Barnola JM, D Raynaud, YS Korotkevich, and L Lorius (1987) Vostok ice core provides 160,000-year record of atmospheric CO_2. *Nature* **329**, 408-414.

Chavez FP, KR Buck, and RT Barber (1990) Phytoplankton taxa in relation to primary production in the equatorial Pacific. *Deep-sea Res.* **37**:1733-1752.

Coale KH, and KW Bruland (1987) Oceanic stratified euphotic zone as elucidated by ^{234}Th:^{238}Th disequilibria. *Limnol. Oceanogr.* **32**:189-200.

Crusius J, and RF Anderson (1992). Inconsistencies in accumulation rates of Black Sea sediments inferred from records of laminae and ^{210}Pb. *Paleoceanography* **7**:215-228.

Degens ET, M Behrendt, B Gotthardt, and E Reppmann (1968) Metabolic fractionation of carbon isotopes in marine plankton. *Deep-Sea Res.* **15**, 11-20.

Eppley RW and BJ Peterson (1979) Particulate organic carbon flux and planktonic new production in the deep ocean. *Nature* **282**:677-680.

Fairbanks RG, M Sverdlove, R Free, PH Wiebe, and AWH Be (1982) Vertical distribution and isotopic fractionation of living planktonic foraminifera from the Panama Basin. *Nature* **298**, 841-844.

Feely RA, RH Gammon, BA Taft, PE Pullen, LS Waterman, TJ Conway, JF Gendron, and JF Wisegarver (1987) Distribution of chemical tracers in tracers during and after the 1982-1983 El Nino/Southern Oscillation Event. *J. Geophys. Res.* **92**, 6545-6558.

Feely RA, RH Wanninkhof, and DK Atwood (1992) NOAA completes successful Spring EqPac cruise. *U. S. JGOFS News* 3(4), 5.

Fischer AG (1981)*Climatic oscillations in the biosphere*, in Nitecki, M. ed., Biotic Crises in Ecological and Evolutionary Time, New York, Academic, pp.103-131.

Fluegge A, AC Mix, JM Hayes, and JP Jasper (submitted 1993) Photosynthetic fractionation of carbon isotopes in surface waters from the equatorial Pacific and its record in surface sediments. *Trans. Am. Geophys. Un. Eos*

Francois R, MA Altabet, R Goericke, DC McCorkle, C Brunet, and A Poisson (1993) Changes in the $\delta^{13}C$ of surface water particulate organic matter across the subtropical convergence in the S. W. Indian Ocean. *Global Biogeochemical Cycles* **7**, 000-000.

Freeman KH and JM Hayes (1992) Fractionation of carbon isotopes by phytoplankton and estimates of ancient CO_2 levels. *Glob. Biogeochem. Cycle* **6**, 185-198.

Freeman KH, JM Hayes, J-M Trendel, and P Albrecht (1990) Evidence from carbon isotope measurements for diverse origins of sedimentary biomarkers. *Nature* **353**, 254-256.

Friedli H, H Lotscher, H Oeschger, U Siegenthaler, and B Stauffer (1986) Ice core record of the $^{13}C/^{12}C$ ratio of atmospheric CO_2 in the past two centuries. *Nature* **324**:237-238.

Goyet C, AL Bradshaw, and PG Brewer (1991) The carbonate system in the Black Sea. *Deep-Sea Res.* **38S**:1049-1068.

Hayes JM, KH Freeman, BN Popp, and CH Hoham (1990) Compound-specific isotopic analyses, a novel tool for reconstruction of ancient biogeochemical processes. *Org. Geochem.* **16**, 1115-1128.

Hayes JM, and JP Jasper (submitted) Relationships between CO_2 levels and ^{13}C in organic matter, An overview and prospects. In *Advances in Organic Geochemistry 1993*. Pergamon Press, 1994.

Honjo S and H Okada (1974) Community structure of coccolithophores in the photic layer of the mid-Pacific. *Micropaleontology* **20**:209-230.

Jasper JP, and RB Gagosian (1989) Alkenone molecular stratigraphy in an oceanic environment affected by glacial meltwater events. *Paleoceanography* **4**:603-614.

Jasper JP, and RB Gagosian (1990) The sources and deposition of organic matter in the late Quaternary Pigmy Basin. *Geochim. Cosmochim. Acta* **54**:1117-1132.

Jasper JP, and RB Gagosian (1993) The relationship between sedimentary organic carbon isotopic composition and lipid biomarker concentrations in the late Quaternary Pigmy Basin. *Geochim. Cosmochim. Acta* **57**, 167-186.

Jasper JP and JM Hayes (1990) A carbon isotope record of CO_2 levels during the late Quaternary. *Nature* **347**, 462-464.

Jasper JP, FG Prahl, AC Mix, and JM Hayes (submitted) Photosynthetic ^{13}C fractionation and estimated CO_2 levels in the Central Equatorial Pacific over the last 255,000 years. *Paleoceanography*, 1993b.

Jasper JP, FG Prahl, AC Mix, SG Wakeham, RF Anderson, J Crusius, and JM Hayes (1991) Isotopically-derived estimates of dissolved CO_2 and equilibrium pCO_2 from late Quaternary sedimentary records in the equatorial Pacific and the Black Sea. *Trans. Am. Geophys. Un. Eos* **72**(44):272.

Jasper JP, SG Wakeham, and JM Hayes (1993) The state of dissolved CO_2 in the Black Sea photic zone recorded by the $\delta^{13}C$ of phytoplanktonic biomarker compounds and coccolith-carbonate during the Industrial Age. *Abstracts of the European Assoc. Org. Geochem.*, Stavanger, Norway.

Kroopnick, PM (1985) The distribution of ^{13}C of ΣCO_2 in the world oceans. *Deep-Sea Res.* **32**, 57-84.
de Leeuw JW, FW van de Meer, WIC Rijpstra, and PA Schenck (1979) On the occurrence and structural identification of long chain unsaturated ketones and hydrocarbons in sediments. In *Advances in Organic Geochemistry 1979* (eds. A. G. Douglas and J. R. Maxwell), pp. 2ll-217. Pergamon Press.
Longhurst AR, and WG Harrison (1989) The biological pump: Profiles of plankton production and consumption in the upper ocean. *Prog. Oceanogr.* **22**:41-123.
Lyle M, DM Murray, BP Finney, J Dymond, JM Robbins, and K Brooksforce (1988) The record of late Pleistocene biogenic sedimentation in the eastern tropical Pacific Ocean. *Paleoceanography* **3**, 39-59.
Lyle M, FG Prahl, and MA Sparrow (1992) Upwelling and productivity changes inferred from a temperature record in the central equatorial Pacific. *Nature* **355**, 812-815.
Marlowe IT, Brassell SC, Eglinton G and Green J C (1990) Long-chain alkenones and alkyl alkenoates and the fossil record of marine sediments. *Chem. Geol.* **88**, 349-375.
Martin JH (1990) Glacial-interglacial CO_2 change: The Iron Hypothesis. *Paleoceanography* **5**:1-14.
Monson KD, and JM Hayes (1982) Biosynthetic control of the natural abundance of carbon-13 at specific positions within fatty acids in *Escherechia coli*, evidence regarding the coupling of fatty acid and phospholipid synthesis. *J. Biol. Chem.* **255**, 11435-11441.
Mook WG, JC Bommerson, and WH Staberman (1974) Carbon isotope fractionation between dissolved bicarbonate and gaseous carbon dioxide. *Earth Plan. Sci. Lett.* **22**:169-176.
Morse JW and FT Mackenzie (1990) Geochemistry of Sedimentary Carbonates, Dev. Sedimentol., vol. 48, 417 pp., Elsevier, New York.
Nozaki Y, DM Rye, KK Turekian and RE Dodge (1978) A 200 year record of carbon-13 and carbon-14 variations in a Bermuda coral. *Geophys. Res. Lett.* **5**:825-828.
Pedersen TF, B Nielsen, and M Pickering (1991) Timing of late Quaternary productivity pulses in the Panama Basin and implications for atmospheric CO_2. *Paleoceanography* **6**, 657-677.
Pena MA, MR Lewis, and WG Harrison (1990) Primary productivity and size structure of phytoplankton biomass on a transect of the equator at 135°W in the Pacific Ocean. *Deep-sea Res.* **37**:295-315.
Popp BN, R Takigiku, JM Hayes, JW Louda, and EW Baker (1989) The post-Paleozoic chronology and mechanism of ^{13}C depletion in primary marine organic matter. *Am. J. Sci.* **289**, 436-454.
Prahl FG, LA Muehlhausen, and DL Zahnle (1992) Further evaluation of long-chain alkenones as indicators of paleoceanographic conditions. *Geochim. Cosmochim. Acta* **52**:2303-2310.
Prahl FG, LA Muehlhausen, and M Lyle (1989) An organic geochemical assessment of oceano-graphic conditions at MANOP Site C over the past 26,000 years. *Paleoceanography* **4**, 495-510.
Rau GH (1994) Variations in sedimentary organic $\delta^{13}C$ as a proxy for past changes in ocean and atmospheric CO_2 concentrations. *In*: Carbon Cycling in the Glacial Ocean: Constraints on the Ocean's Role in Global Change, eds. M. Kaminiski, L. D. Laberyrie, T. F. Pedersen, and R.Zahn, Berlin: Springer-Verlag.
Rau GH, RE Sweeney, and DJ DesMarais (1989) Latitudinal variations in plankton $\delta^{13}C$: Implications for CO_2 and productivity in past oceans. *Deep Sea Res.* **341**, 516-518.
Rau GH, PN Froelich, T Takahashi, and DJ DesMarais (1991) Does sedimentary organic $\delta^{13}C$ record variation in Quaternary ocean [CO_2(aq)]? *Paleoceanography* **6**, 335-347.
Ravelo C, RG Fairbanks, and SGH Philander (1990) Reconstructing tropical Atlantic hydrography using planktonic foraminifera and an ocean model. *Paleoceanography* **5**, 409-431.
Tans PP, IY Fung, and T Takahashi (1990) Observational constraints on the global atmospheric CO_2 budget. *Science* **247**, 1431-1438.
Volk T and R Bacastow (1989) The changing patterns of $DpCO_2$ between ocean and atmosphere. *Glob. Biogeo. Cycles* **3**, 179-189.

Weiss RF (1974) Carbon dioxide in water and seawater: the solubility of a non ideal gas. *Mar. Chem.* **2**, 203-215.
Williams DF and B Kohl (1986) Isotope chronostratigraphy and carbonate record for Quaternary site 619, Pigmy Basin, Louisiana Continental Slope, *Initial Rep. Deep Sea Drill. Proj.*, ***96***, pp. 671-676.
Wyrtki K (1981) An estimate of equatorial upwelling in the Pacific. *J. Phys. Oceanogr.* **11**:1205-1214.

LATE QUATERNARY PCO_2 VARIATIONS IN THE ANGOLA CURRENT: EVIDENCE FROM ORGANIC CARBON $\delta^{13}C$ AND ALKENONE TEMPERATURES

Peter J. Müller, Ralph Schneider and Götz Ruhland
FB 5 - Geowissenschaften
Universität Bremen
Klagenfurter Strasse
D-28359 Bremen, Germany

ABSTRACT. The $\delta^{13}C$ record for organic carbon in Late Quaternary sediments of the Eastern Angola Basin shows a pronounced cyclicity with high values (-18.3 to -20‰) in glacial isotopic stages 2 to 4 and stage 6, and lower values (-20 to -21.2‰) in interglacial stages 1 and 5. Sea-surface temperatures, derived from the alkenone U^{k}_{37} index, vary in the opposite sense, ranging from 20-22°C in glacial stages up to 24-26°C in interglacial stages. The inverse relationship between $\delta^{13}C_{org}$ and SST values suggests that the isotopic variations are not due to temperature-dependent isotopic fractionation during photosynthesis, in which case $\delta^{13}C_{org}$ values would be expected to increase with increasing temperature. Relationships between sedimentary $\delta^{13}C_{org}$ and C/N ratios indicate that differing proportions of marine and terrigenous organic matter can also be ruled out as a cause for the $\delta^{13}C_{org}$ variability. Instead we conclude that changes in surface water CO_2 concentrations were responsible for the observed glacial to interglacial isotopic variations.

Employing the relationships between planktonic $\delta^{13}C_{org}$ and dissolved $CO_2(aq)$ derived by POPP et al. (1989) and RAU et al. (1989) we estimate that the partial pressure of carbon dioxide (PCO_2) in the surface water of the Angola Current (AC) has been consistently higher than values given by the Vostok atmospheric CO_2 record for the last 160,000 years. This indicates that the AC has generally acted as a source region for atmospheric CO_2 throughout this period, which is consistent with the modern situation. However, the difference between surface-water and atmospheric pCO_2 was larger in interglacial ($pCO_2 \approx$ 100-120 µatm) than in glacial times (mean $pCO_2 \approx$ 70 µatm). This suggests that the AC was a weaker CO_2 source during glacial periods, presumably due to lower sea-surface temperatures and higher biological productivity.

NATO ASI Series, Vol. I 17
Carbon Cycling in the Glacial Ocean:
Constraints on the Ocean's Role in Global Change
Edited by R. Zahn et al.

INTRODUCTION

Increased biological productivity in the surface ocean has been proposed as one major mechanism responsible for the glacial reduction in the atmospheric concentration of carbon dioxide as deduced from ice-core measurements (BERGER et al. 1989). Models invoking enhanced "biological pumping" of carbon to the deep sea have concentrated on the Southern Ocean, which today has high nutrient concentrations in its surface waters (KNOX & McELROY 1984, VOLK & HOFFERT 1985, MARTIN 1990, KEIR 1990). However, recent tracer records of paleoproductivity do not support the hypothesis of a generally increased glacial productivity in the Antarctic Ocean (CHARLES et al. 1991, MORTLOCK et al. 1991, KUMAR et al. 1993).

In temperate to tropical regions, however, there is abundant evidence for increased biological productivity during glacial times, which may have contributed to the glacial drop in atmospheric CO_2 (MÜLLER et al. 1983, LYLE et al. 1988, SARNTHEIN et al. 1988, MIX 1989, SHIMMIELD et al. 1990, SCHNEIDER 1991, BERGER & HERGUERA 1992). On the other hand, low-latitude upwelling regions have been disregarded as playing a significant role in the glacial CO_2 decrease because they are, in principle, sources for atmospheric CO_2. Examples are the equatorial divergence zones where upwelling and warming of CO_2-rich subsurface waters result in positive PCO_2 anomalies and net CO_2 fluxes from the sea to the atmosphere (ANDRIÉ et al. 1986, TANS et al. 1990, MURPHY et al. 1991).

There is also evidence, however, that the production of phytoplankton significantly influences the exchange of CO_2 between ocean and atmosphere. Direct measurements made in the North Atlantic (WATSON et al. 1991) and in the Indian sector of the Antarctic Ocean (METZL et al. 1991) have revealed that the surface-water PCO_2 is strongly modulated by biological activity and is variable on short spatial and temporal scales. In spring 1989, for example, the surface PCO_2 dropped from 350 to 290 µatm within 5 days at 60°N, 20°W in the North Atlantic (ROBERTSON et al. 1993). Hence, one question which must be addressed is whether the pCO_2 difference between surface water and atmosphere (pCO_2) became lower in temperate to low latitudes during glacial times as a result of increased biological productivity.

To establish a record of paleo-PCO_2 in the eastern subtropical Atlantic, we used $\delta^{13}C$ measurements of organic carbon in conjunction with sea-surface temperatures derived from alkenone analyses, adopting the models of POPP et al. (1989) and RAU et al. (1989, 1991) to convert the measurements to CO_2. Our study is based on core GeoB 1016-3 from the high-productivity area in the eastern Angola Basin (Fig. 1). According to BERGER (1989), this region is characterized by primary production rates of 125-180 gC m^{-2} y^{-1}, the highest rates in the eastern South Atlantic outside the coastal upwelling centers.

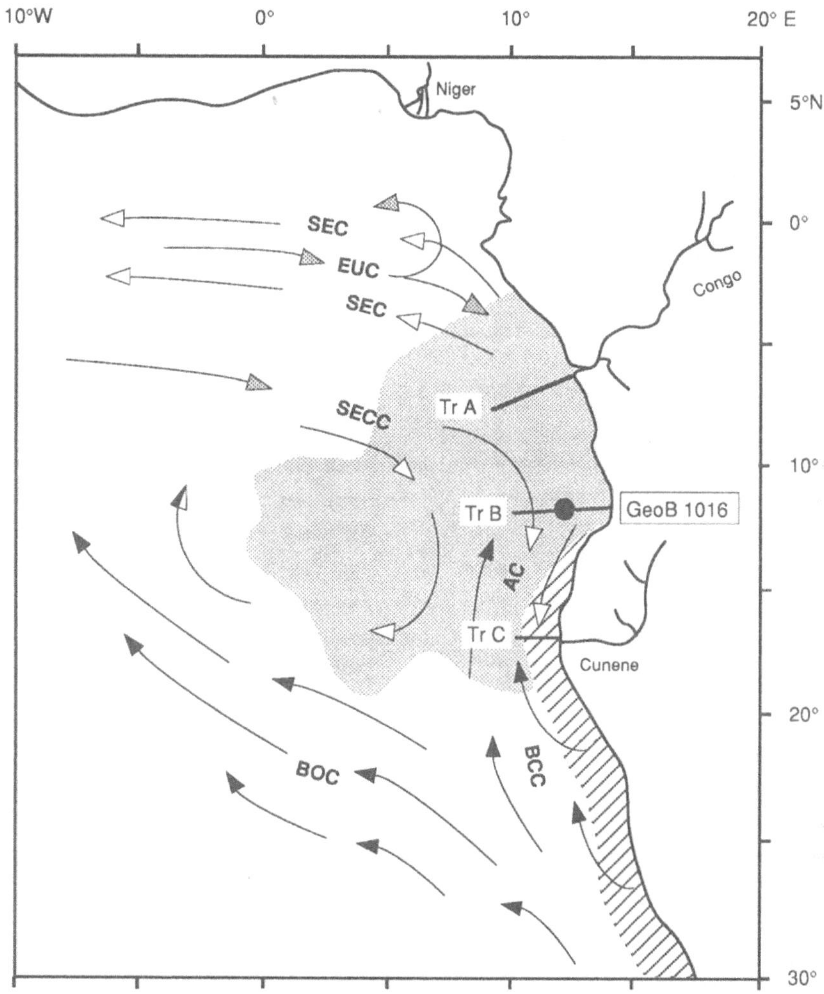

Figure 1 Study area in the eastern Angola Basin (Transects A, B, and C of "Meteor" cruise 6/6; WEFER et al. 1988). Large dot denotes position of core GeoB 1016-3 which was used for PCO_2 reconstructions. Arrows indicate surface current system: white - warm surface current, black - cold surface current, gray - subsurface current.; SEC - South Equatorial Current, EUC - Equatorial Undercurrent; SECC - South Equatorial Counter Current; AC - Angola Current; BCC - Benguela Coastal Current; BOC - Benguela Oceanic Current. Dotted and striped areas indicate oceanic (shallow thermocline) and coastal (upwelling) high-productivity regions, respectively (modified from SCHNEIDER 1991).

METHODS

Samples. The sediment cores of this study were recovered during RV "Meteor" cruise 6/6 (1988) in the eastern Angola Basin (WEFER et al. 1988). The study area and core locations are presented in Fig. 1 and Table 1. Holocene sediments were obtained from short box cores along three transects which cross the continental margin off Zaire and Angola at about 6°S, 12°S and 17°S (Fig. 1). Water depths at these stations ranged from 45 to 4600 m. The dot in Fig. 1 denotes the location of gravity core GeoB 1016-3 which is used here to reconstruct surface water PCO_2 variations for the past 200,000 years.

Shortly after core recovery on board the ship, subsamples of known volume were taken at 3 cm (box cores) and 5 cm (gravity core) intervals using 10 cm^3 plastic syringes. The syringes were sealed using electrical tape and cooled to a temperature of about 4°C, which was maintained until further processing. In the shore-based laboratory, the syringe samples were weighed, freeze-dried and re-weighed to determine the water content, and the wet and dry bulk densities. The dry sediment was then ground in an agate mortar and used for elemental, $\delta^{13}C_{org}$ and alkenone analyses. A second series of syringe samples, taken parallel to the first, was wet sieved to obtain the coarse fraction (>63 µm) from which planktonic foraminifera were obtained for stable oxygen and carbon isotope measurements.

Foraminiferal oxygen and carbon isotopes. Oxygen and carbon isotopic measurements were performed on the planktonic foraminiferal species *Globigerinoides ruber* (pink variety) using a Finnigan MAT 251 mass spectrometer and a manual preparation line (SCHNEIDER 1991). Calibration of the lab-internal standard gas to the PDB standard scale was achieved using NBS 18, 19 and 20 $CaCO_3$ standards. The analytical precision (±1), based on replicate analyses of an internal laboratory $CaCO_3$ standard, was 0.07‰ for ^{18}O and 0.05‰ for $\delta^{13}C$.

Organic carbon, total nitrogen and carbonate. Total carbon (C_{tot}) and nitrogen (N_{tot}), and total organic carbon (C_{org}) values were obtained by combustion at 1,050°C using a Heraeus CHN-O-Rapid elemental analyser. For the determination of organic carbon, 25 mg aliquots of the dried sediment were weighed in silver foil crucibles, wetted with a few drops of ethanol to suppress foaming during acidification, decalcified by drop-wise addition of 6 M HCl, and dried on a hot plate at about 80°C. Total carbon and nitrogen were determined on untreated sample aliquots (25 mg) using tin crucibles for weighing. The carbonate content of the samples was calculated from the difference between total and organic carbon and expressed as calcite [$CaCO_3 = 8.33 * (C_{tot} - C_{org})$]. All analyses were performed in duplicate and expressed in weight percent dry, salt-free sediment assuming a pore-water salinity of 35‰. The relative precision of the measurements, based on duplicates and control analyses of a lab-internal reference sediment sample (WS1), was better than 3 %.

Stable isotopes of organic carbon. Isotopic measurements of sedimentary organic carbon ($\delta^{13}C_{org}$) were performed using a Finnigan MAT delta E mass spectrometer and a Heraeus

CHN-Rapid elemental analyser interfaced by an automated trapping box. The samples were prepared as described above for total organic carbon measurements. The overall analytical precision (±1), based on duplicates and repeated analyses of a lab-internal reference sediment sample (WS1), was better than 0.1‰. The isotopic data are reported in notation relative to the PDB standard.

Alkenones and U^k_{37}. Long-chain, unsaturated ketones (alkenones) were extracted from 0.5 or 1 g aliquots of the freeze-dried sediment by a modified flow-blending technique (RADKE et al. 1978) using an Ultra-Turrax T25 at 24,000 rpm and successively less polar solvent mixtures (MeOH, MeOH/$CHCl_3$ 1:1, $CHCl_3$), each for five minutes. The three extracts

Table 1 Positions and water depths for sediment cores used in this study and geochemical characteristics of surface sediments (0-1.4 cm depth). Results are given in wt.% on a dry salt-free basis.

Core	Position		Water depth (m)	$CaCO_3$ (%)	C_{org} (%)	N_{tot} (%)	C/N wt.ratio	$\delta^{13}C_{org}$ (‰)
			Transect A, Congo Fan					
GeoB 1001-2	05°51.8'S	11°58.3'E	45	3.0	3.34	0.268	12.5	-25.51
GeoB 1000-1	05°41.9'S	11°42.6'E	106	3.8	3.43	0.292	11.7	-24.58
GeoB 1005-2	06°14.1'S	11°30.1'E	237	7.8	2.84	0.276	10.3	-23.03
GeoB 1006-2	06°17.2'S	11°18.1'E	682	1.9	4.51	0.476	9.5	-22.73
GeoB 1007-2	06°23.8'S	10°56.3'E	1533	0.5	3.22	0.346	9.3	-21.83
GeoB 1008-6	06°35.6'S	10°19.1'E	3100	0.3	2.31	0.280	8.2	-21.80
GeoB 1009-3	06°55.1'S	08°59.8'E	4074	0.4	1.25	0.161	7.8	-20.55
GeoB 1010-3	07°25.7'S	07°25.2'E	4474	16.9	0.96	0.151	6.4	-20.46
			Transect B, Middle Angola					
GeoB 1011-2	11°48.2'S	13°39.7'E	73	20.2	0.95	0.100	9.5	-21.19
GeoB 1012-1	11°48.5'S	13°35.4'E	99	17.9	0.86	0.089	9.7	-20.95
GeoB 1013-2	11°47.9'S	13°26.8'E	250	18.6	1.83	0.196	9.3	-21.04
GeoB 1014-2	11°47.2'S	13°18.3'E	701	14.0	5.16	0.552	9.4	-20.78
GeoB 1015-2	11°46.7'S	12°50.8'E	1638	6.4	2.49	0.333	7.5	-20.24
GeoB 1016-2	11°35.2'S	11°41.8'E	3410	6.2	1.96	0.254	7.7	-20.37
GeoB 1016-3	11°46.2'S	11°40.9'E	3410	-	-	-	-	-
GeoB 1017-3	11°44.4'S	10°33.0'E	3809	23.8	1.06	0.141	7.5	-19.56
			Transect C, South Angola					
GeoB 1020-1	17°10.1'S	11°32.9'E	110	7.3	1.99	0.236	8.4	-21.12
GeoB 1023-2	17°09.7'S	11°00.6'E	1965	7.9	4.27	0.535	8.0	-20.11
GeoB 1024-3	17°09.7'S	10°40.9'E	2802	11.1	4.32	0.561	7.7	-19.99
GeoB 1025-2	17°09.9'S	10°14.5'E	3565	21.8	3.18	0.451	7.1	-19.98
GeoB 1026-3	17°10.2'S	08°54.4'E	4601	61.1	0.94	0.146	6.4	-20.03

were combined, desalted, dried over Na_2SO_4, concentrated and finally taken up in 25 µl of the 1:1 solvent mixture. A 2µl aliquot of this extract was analysed by capillary gas chromatography using an HP 5890A gas chromatograph and a 50 m x 0.2 mm i.d. HP Ultra 1 fused silica column, split/splitless injection, temperature programming, helium as carrier gas and flame ionization detection. Quantification of di- and tri-unsaturated C_{37}-alkenones was achieved by an internal standard method using the relative response factor of the C_{38} n-alkane for both alkenones. A more detailed description of the analytical procedure will be given elsewhere (RUHLAND, in prep.).

The ketone unsaturation index U^k_{37} (BRASSELL et al. 1986) was calculated from the concentrations of di- and tri-unsaturated C_{37}-alkenones using the expression U^k_{37} = [37:2]/[37:2+37:3] and translated into sea-surface temperature (SST) using the calibration derived by PRAHL et al. (1988): SST (°C) = (U^k_{37}-0.039)/0.034. The precision (±1) of the measurements was better than 0.01 U^k_{37} units (or 0.3 °C) based on duplicates and multiple extractions of a sediment sample used as lab-internal reference sample.

RESULTS AND INTERPRETATION

Holocene sediments

To determine the modern and Late Holocene distribution patterns of terrigenous organic matter in the eastern Angola Basin, we analysed a series of short box cores, taken along the Transects A , B and C (Fig. 1), for total organic carbon and nitrogen, and the stable isotope ratio of organic carbon ($\delta^{13}C_{org}$). The results for the surface sediments of these cores are included in Table 1.

As depicted in Fig. 2, shelf sediments near the mouth of the Congo River (Transect A) show the most negative $\delta^{13}C_{org}$ values ranging from -24.6 to -27.0‰. The latter value was reported by JANSEN et al. (1984) for a sample of core T78-23 from the shelf at 75 m water depth. These low isotopic values suggest a terrestrial C_3 plant source for the organic material supplied by the Congo River. SCHWARTZ et al. (1986), for example, reported $\delta^{13}C_{org}$ values of -26.6 and -27.8‰ for the A1 horizons of soils under forests with C_3-type vegetation in southern Congo and MARIOTTI et al. (1991) showed that the isotopic composition of particulate organic carbon in the Congo River remains remarkably constant at -26.7 ± 0.4‰ (mean ±) during all seasons. Fig. 2 also shows an increase of the $\delta^{13}C_{org}$ values with increasing water depth along Transect A, indicating that the influence of terrigenous organic matter decreases with increasing distance from the river mouth.This interpretation is supported by the negative correlation between C/N and $\delta^{13}C_{org}$ ratios shown in Fig. 3 (squares) for the Holocene Congo Fan sediments. The inverse relationship clearly reflects mixing of nitrogen-poor, isotopically

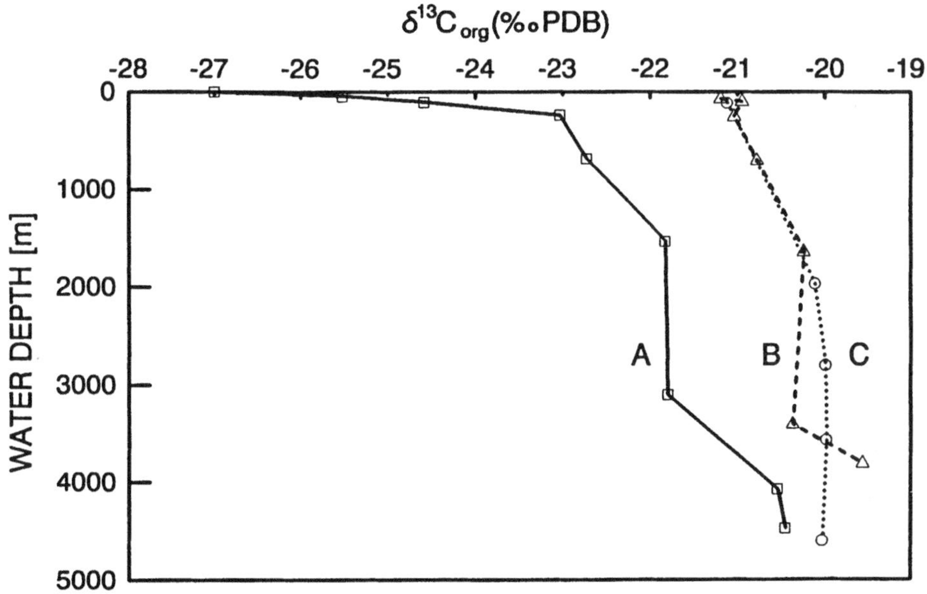

Figure 2 Bathymetric profiles of organic carbon isotopic composition ($\delta^{13}C_{org}$) for surface sediments (1cm depth) of the Congo Fan (Transect A) and the continental margin off Angola at 12°S (Transect B) and 17°S (Transect C).

light terrestrial organic matter (C/N > 15, $\delta^{13}C_{org} \approx$ -27 to -28‰) with nitrogen-rich, isotopically heavy marine organic matter (C/N ≈ 7-9, $\delta^{13}C_{org}$ > -21‰). A similar relationship was reported by JASPER & GAGOSIAN (1990) for Late Quaternary sediments of the Pigmy Basin, Gulf of Mexico. However, this connection may not hold true in regions away from major rivers (WESTERHAUSEN et al. 1993).

In contrast to the Congo Fan, the sediments of Transects B and C show heavy and constant $\delta^{13}C_{org}$ values of -20.0 ± 0.3‰. Only the shelf sediments had slightly lower values (ca. -21‰). The same range of values was found in plankton net samples (0-50 m water depth) from these stations (FISCHER, unpubl. data). These values are within the range generally associated with marine planktonic organic matter at water temperatures >20°C, i.e. -18 to -21‰ (FONTUGNE & DUPLESSY 1981, FREEMAN & HAYES 1992).

Core GeoB 1016-3

This core was recovered from the continental slope off Angola at a water depth of 3411 m (Fig. 1, Table 1). It is composed predominantly of fine-grained, hemipelagic mud with carbonate and biogenic opal contents ranging from 1-30 % and 0.6-4.7 %, respectively. The chronostratigraphy of the core is based on the ^{18}O record of the planktonic foraminifer *Globigerinoides ruber* (pink), applying the SPECMAP time scale (IMBRIE et al. 1984).

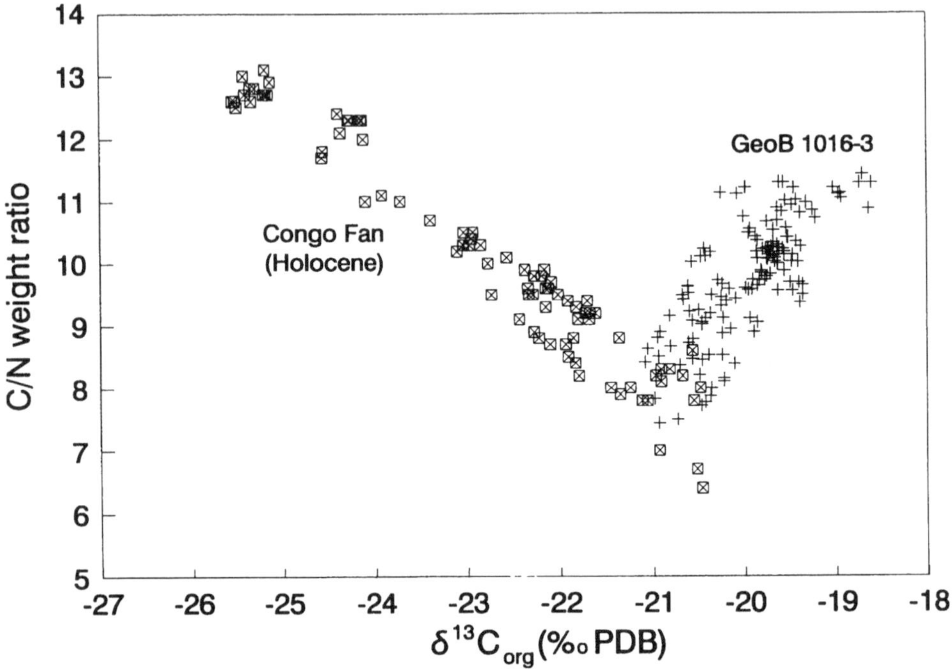

Figure 3 Plots of $\delta^{13}C_{org}$ against C/N weight ratio for Holocene sediments of the Congo Fan (squares) and for Late Quaternary sediments of core GeoB 1016-3 (crosses; <200,000 years) at the continental margin off Angola (see Fig. 1).

Accordingly, core GeoB 1016-3 records the last 8 oxygen isotope stages and bottoms in stage 9, resolving the past 300,000 years (SCHNEIDER 1991). The average sedimentation rate is 4 cm/1000 yrs, ranging from 2 to 8 cm/1000 yrs. The time resolution of samples taken at intervals of 5 cm is thus 600 to 2500 years (Table 2). In this study, we concentrate on the record of the past 200,000 years, since alkenones have not been analysed in the older core sections.

Fig. 4 shows the ^{18}O record of *G. ruber* for the past 200,000 years in comparison to time series of other basic measurements from core GeoB 1016-3. The organic carbon record (Fig. 4B) reveals that sediments of interglacial stages 1, 5.1, 5.3, 5.5, and 7.1 have significantly lower organic matter concentrations (typically 0.7-1.4 % C_{org}) than sediments of glacial stages 2-4 and 6 (2-4 % C_{org}). Most noteworthy are the pronounced C_{org} maxima in interglacial sub-stages 5.2 and 5.4. Similar profiles have been found in other sediment cores from the eastern Angola Basin (SCHNEIDER et al. 1993, 1994, MÜLLER, unpubl. data). Time series analyses indicate that the organic carbon cycles reflect variations in paleoproductivity with a significant

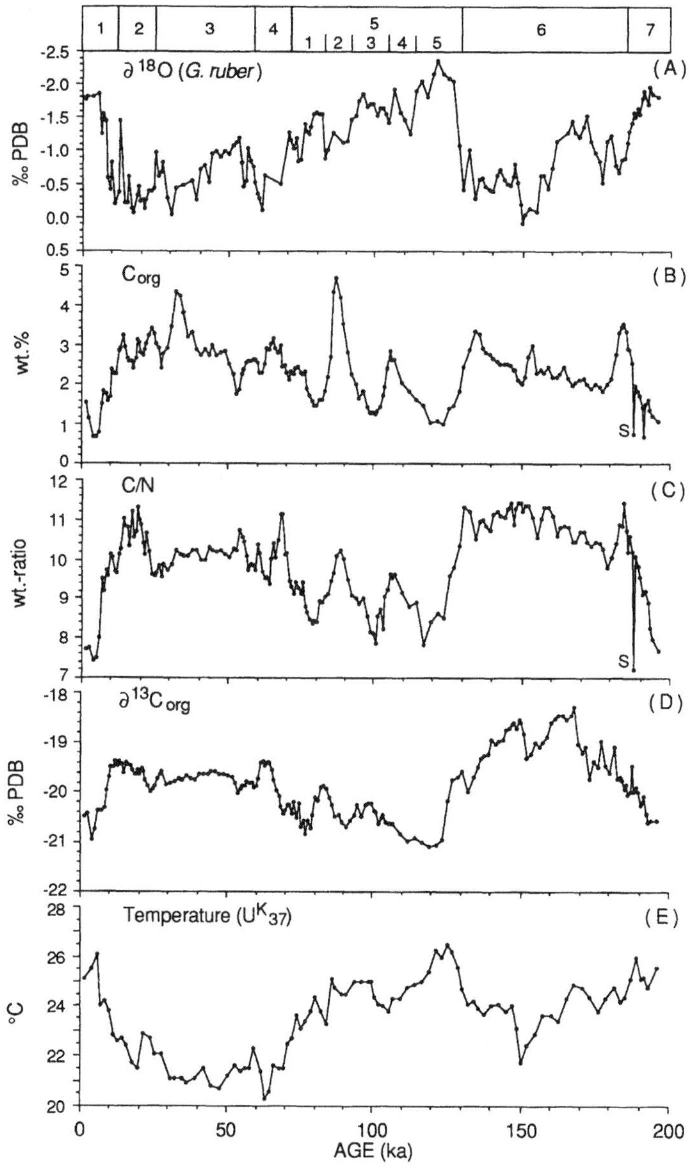

Figure 4 Time series (ka = 1,000 years) of basic isotopic and geochemical measurements on core GeoB 1016-3: (A) Oxygen isotope composition of the planktonic foraminiferal species *Globigerinoides ruber* (pink variety), (B) total organic carbon (C_{org}) content, 'S' denotes sand layer , (C) organic carbon/total nitrogen weight ratio, (D) stable isotopic composition of total organic carbon ($\delta^{13}C_{org}$), and (E) sea surface temperatures (SSTs) derived from the alkenone U^{k}_{37} index.

23,000 year periodicity which is coherent with the precessional cycle in the earth-sun distance (SCHNEIDER 1991).

The $\delta^{13}C_{org}$ record (Fig. 4D) also shows significant glacial-interglacial differences. In general, the sedimentary organic matter is depleted in $\delta^{13}C$ in the interglacial stages and enriched in the glacial stages, the total range being -18.3 to -21.1‰. The lightest isotopic values occur in warm sub-stages 1.1 and 5.5 and the heaviest in glacial stage 6. Glacial stages 2-4 have intermediate and rather constant $\delta^{13}C_{org}$ values of -19.7 ± 0.3‰.

The sea-surface temperatures (SSTs) derived from the alkenone U^{k}_{37} index are displayed in Fig. 4E. These temperatures probably represent annual mean values rather than seasonal extremes, based on a comparison of alkenone-derived temperatures from surface sediments with actual SST measurements. For example, the alkenone SST values obtained from the surface sediments along Transects A to C closely agree with the annual mean temperatures in the overlying surface water (RUHLAND et al. 1992). Moreover, sea surface temperatures reconstructed from the ^{18}O record (Fig. 4A) of the "warm-season" planktonic foraminiferal species *G. ruber* (pink) (e.g. GANSSEN 1983, DEUSER & ROSS 1989, VAN LEUUWEN 1989) were found to be systematically higher than the alkenone-derived temperatures by 2-4 °C, although the two SST records show a high degree of co-variation (SCHNEIDER, unpubl. data). Hence there is no reason to suspect significant deterioration of the temperature signal recorded by alkenones over the time span discussed here. In fact, there is good evidence that the U^{k}_{37} ratio remains unaffected during surficial (CONTE et al. 1992) and post-depositional diagenesis (BRASSELL et al. 1986), even if >85% of the sedimentary alkenone content is reduced by oxic decomposition (PRAHL et al. 1989).

As shown in Fig. 4E, SST was generally lower in glacial than in interglacial stages, the total amplitude being 6.2°C. The highest SST values occurred in stages 1.1 (26.1°C) and 5.5 (26.5°C) and the lowest in stage 4 (20.3°C). In general, the SST estimates vary from 20 to 23°C in the cold stages 2-4 and from 23 to 26°C in the warm stages 1 and 5. Noteworthy is that glacial stage 6 shows considerably higher temperatures (24-25°C) than stage 2, except for a cooling period during sub-stage 6.4. The glacial to interglacial temperature increase was thus 5°C at Termination I but only about 3°C at Termination II. Essentially the same temperature patterns have been observed in a sediment core from Transect A (RUHLAND et al. 1992), located some 650 km to the north of site GeoB 1016-3, indicating that the variations reflect general temperature changes in the Angola Current rather than local phenomena.

DISCUSSION

The organic carbon isotopic results from core GeoB 1016-3 (Table 2, Fig. 4D) clearly demonstrate that the organic carbon is depleted in $\delta^{13}C$ in interglacial compared to glacial core sections. This could reflect a higher proportion of isotopically light terrigenous organic carbon in Holocene and stage 5 sediments. However, the isotopic and elemental characteristics of the Holocene sediments (Table 1, Figs. 2 and 3) show that the organic material supplied by the Congo River is deposited in the Congo Fan region, while Holocene sediments off middle and southern Angola (Transects B and C) do not contain significant amounts of terrigenous organic matter. This is consistent with results from other riverine-offshore transects which show that the transition to marine $\delta^{13}C_{org}$ values (<-21‰) occurs fairly close to the river mouths (see e.g. the review of FRY & SHERR 1984). Moreover, if different mixtures of marine and C_3-terrigenous organic matter were responsible for the isotopic variations in core GeoB 1016-3, C/N and $\delta^{13}C_{org}$ ratios should show an inverse relationship, similar to that of the Holocene Congo Fan sediments (squares in Fig. 3). The positive correlation between these ratios (crosses in Fig. 3) argues against this possibility.

Another factor that could cause shifts in the $\delta^{13}C$ record of sedimentary organic carbon is pre- and post-burial diagenesis (e.g. McARTHUR et al. 1992). However, as shown by the reviews of FRY & SHERR (1984) and FONTUGNE & CALVERT (1992), it is difficult to document consistent unambiguous diagenetic trends for different depositional environments. Recent time-series sediment trap experiments in the Guinea Basin and on the Walvis Ridge indicate an average $\delta^{13}C$ enrichment of 0.5 to 0.8‰ in the material settling to the sea floor (FISCHER & WEFER, unpubl. data, MÜLLER et al. 1994). Surface sediments of Transect B show a similar $\delta^{13}C$ enrichment (by about 0.5‰) when compared to plankton samples from the same stations (FISCHER, unpubl. data). These observations indicate a slight preferential loss of ^{12}C during settling and surficial diagenesis, on the order of 1‰ or less. However, there is no evidence for significant post-depositional diagenetic isotopic changes in sediments of the eastern Angola Basin. For example, the average $\delta^{13}C_{org}$ values (±) of the sediments from interglacial stages 1, 5, 7, and 9 in core GeoB 1016-3 (Table 2) are -20.4 ± 0.2‰, -20.5 ± 0.3‰, -20.2 ± 0.3‰, and -20.3 ± 0.3‰, respectively, compared to a surface sediment value of -20.37‰ (core GeoB 1016-2, Table 1).

Comparison between the $\delta^{13}C_{org}$ (Fig. 4D) and SST (Fig. 4E) records reveals that they vary in the opposite sense. Heavy isotopic values in the glacial stages are associated with low SST values, while interglacial stages show lighter isotopic and higher SST values. This inverse relationship between $\delta^{13}C_{org}$ and SST values indicates that the isotopic variations are not due to temperature-dependent isotopic fractionation during photosynthesis, which would lead to the opposite relationship, e.g. an increase of $\delta^{13}C_{org}$ values with increasing temperature (FONTUGNE & DUPLESSY 1981).

On this basis we conclude that there must be another reason, most likely CO_2 variations, responsible for the observed glacial to interglacial isotopic variations in core GeoB 1016-3. Experimental and field studies have shown that the fractionation of stable carbon isotopes during photosynthesis by plankton depends on the concentration of ambient dissolved molecular carbon dioxide, [$CO_2(aq)$] (DEGENS et al. 1968, McCABE 1985, RAU et al. 1989, RAU, this volume). This has led to suggestions that the sediment $\delta^{13}C_{org}$ may serve as a proxy for surface water $CO_2(aq)$ (POPP et al. 1989, RAU et al. 1991) and recent studies using this approach with late Quaternary sedimentary records, for example in the Gulf of Mexico (JASPER & HAYES 1990), the Panama Basin (PEDERSEN et al. 1991), the Mediterranean (FONTUGNE & CALVERT 1992), the South China Sea (THUNELL et al. 1992) and the Equatorial Atlantic (WESTERHAUSEN et al., this volume) have been quite successful in establishing $\delta^{13}C_{org}$ as a PCO_2 proxy.

Accordingly, the $\delta^{13}C_{org}$ record of core GeoB 1016-3 (Fig. 4D) was converted to dissolved CO_2 concentrations using two different approaches in order to reconstruct late Quaternary CO_2 changes in the Angola Current:

A) The model of POPP et al. (1989) is based on the relationship determined by McCABE (1985):

$$\varepsilon_p = (-17.0 \pm 2.2) \log [CO_2(aq)] + 3.4 \qquad (1)$$

where $_p$ (‰) is the isotope effect associated with photosynthetic fixation of carbon. Values of the $_p$ term were determined from the equation:

$$\varepsilon_p = 1{,}000 \, [(\delta_p + 1{,}000) \, / \, (\delta_d + 1{,}000) - 1] \qquad (2)$$

where δ_p (‰) and δ_d (‰) are the carbon isotopic compositions of primary photosynthate and of $CO_2(aq)$, respectively (POPP et al. 1989). Assuming that the organic material in core GeoB 1016-3 was predominantly derived from marine primary producers, and that diagenetic isotopic fractionations can be neglected, we substituted our measured $\delta^{13}C_{org}$ values (Table 2) directly for δ_p. Values of δ_d were estimated from the $\delta^{13}C$ record of the surface-dwelling planktonic foraminifer *Globigerinoides ruber* (pink) (Table 2) assuming that the tests of this species are depleted in ^{13}C relative to total dissolved CO_2 (ΣCO_2) by a constant of about 0.5 ‰ (FAIRBANKS et al. 1982, CURRY & CROWLEY 1987) and the temperature-dependent carbon isotope fractionation between $CO_2(aq)$ and dissolved bicarbonate ($\varepsilon_{b(a)}$ in‰) determined by MOOK et al. (1974):

$$\delta_{\Sigma CO2} = \delta_{G.ruber} + 0.5 \qquad (3)$$

$$\delta_d = \delta_{\Sigma CO2} - {}_{b(a)} \qquad (4)$$

Values of $\varepsilon_{b(a)}$ were calculated according to MOOK et al. (1974) using the alkenone-derived SST values from Table 2. For the temperature range of interest here (20-26°C, Fig. 4E), $\varepsilon_{b(a)}$ values range from -9.54 to -8.86‰. We did not consider the temperature dependent carbon isotope fractionation between the foraminiferal calcite and dissolved bicarbonate (0.035‰ / °C; EMRICH et al. 1970) which would be insignificant at the temperature range of only 6°C. Values of $CO_2(aq)$ concentrations (in µmol/l) were then estimated by substituting equation 2 into equation 1 and solving for $CO_2(aq)$:

$$CO_2(aq) = 10 \exp [(\varepsilon_p - 3.4) / - 17] \qquad (5)$$

B) The model by RAU et al. (1991) is based on an empirical relationship between plankton $\delta^{13}C_{org}$ (‰) and $CO_2(aq)$ (µmol/l) as observed in the South Atlantic and Southern Oceans (RAU et al. 1989):

$$\delta^{13}C_{org} = -0.8 [CO_2(aq)] - 12.6 \qquad (6)$$

or

$$[CO_2(aq)] = (\delta^{13}C_{org} + 12.6) / -0.8 \qquad (7)$$

To correct for changes through time in the isotopic composition of total surface water CO_2 (ΣCO_2), which would affect the isotopic compositions of $CO_2(aq)$ and hence primary produced organic carbon (e.g. equations 4 and 6), we have subtracted the deviations in the *G. ruber* $\delta^{13}C$ record relative to the core top value from the $\delta^{13}C_{org}$ record. The *G.ruber* $\delta^{13}C$ values range from 0.5 to 2.0‰, and the core-top value is 1.59‰ (Table 2). SCHNEIDER (1991) compared the *G.ruber* $\delta^{13}C$ record of core GeoB 1016-3 to a record from a less productive region in the tropical Atlantic (CURRY & CROWLEY 1987), which is believed for the most part to represent the global signal, and showed that negative $\delta^{13}C$ deviations in core GeoB 1016-3 were related to past periods of increased productivity.

The $CO_2(aq)$ concentrations derived by both approaches (eqs. 5 and 7) were then converted to CO_2 partial pressure values (PCO_2, µatm) using Henry's Law:

$$PCO_2 = CO_2(aq)/\alpha \qquad (8)$$

The solubility constant α was calculated according to WEISS (1974) using the alkenone-derived SST values (Table 2, Fig. 4E), assuming a constant salinity of 35‰.

According to these calculations, the $CO_2(aq)$ concentrations in the surface water at station GeoB 1016-3 have ranged from about 6.8 to 11.2 µmol/l over the past 200,000 years, and were generally higher in interglacial than in glacial periods. Both approaches yielded nearly the same CO_2 concentrations, the maximal deviation being 0.6 µmol/l, or 20 µatm in terms of PCO_2 at 25°C. Fig. 5 compares the surface-water PCO_2 records of the sediment core to that of the ice core. The original Vostok CO_2 data (BARNOLA et al. 1987) are indicated by a solid line without symbols, while open squares denote new measurements of BARNOLA et al.

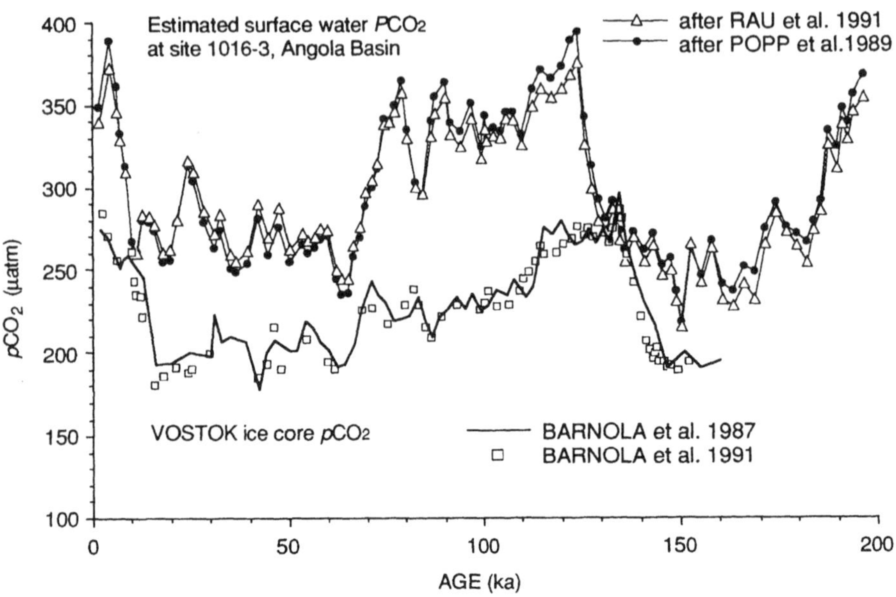

Figure 5 Comparison of the Vostok ice-core atmospheric pCO_2 record (solid line: BARNOLA et al. 1987, open squares: BARNOLA et al. 1991) with surface water PCO_2 estimates for core GeoB 1016-3 in the eastern Angola Basin (note that the partial pressure of CO_2 in seawater or in air is indicated by "P" or "p", respectively, e.g. SKIRROW 1975).

(1991). The time scale used for the ice core is that of BARNOLA et al. (1987). The phase shifts between the two records probably reflect stratigraphic problems rather than actual leads or lags, particularly in ice-core sections older than 110,000 years (JOUZEL et al. 1987, SHACKLETON et al. 1992). Ignoring the stratigraphic uncertainties, the most significant implication of Fig. 5 is that the surface water PCO_2 of the Angola Current (AC) was consistently higher than the atmospheric level during the past 200,000 years. This indicates that the AC generally acted as source region for atmospheric CO_2 during this period, which is qualitatively in agreement with the modern situation (SKIRROW 1975). According to the PCO_2 measurements of ANDRIÉ et al. (1986) the surface waters of the South Equatorial Counter Current, which feeds the Angola Current, are oversaturated with respect to the atmosphere by 20 to 60 µatm, on average.

However, as displayed in Fig. 6, the pCO_2 difference between the surface water and atmospheric records (ΔpCO_2), which determines, among other parameters, the CO_2 flux across the air/sea interface (e.g. GARÇON et al. 1989), was generally lower in the glacial stages. To better illustrate this effect, a constant value of 70 µatm was added to the Vostok CO_2 record in Fig. 6. This value represents the average difference between the two records for glacial stages 2-4 and 6. Since the time scales of the Vostok ice core and of sediment core GeoB 1016-3 are not in phase, we first correlated the ice core to the time scale of our core (using the LINEAGE software developed by D. PAILLARD & L.D. LABEYRIE, Centre des Faibles Radioactivités, Gif-sur-Yvette), assuming that the phase shifts are mainly due to dating problems. It is evident from Fig. 6, that the glacial to interglacial PCO_2 increase at Terminations I and II was larger by 20-30 µatm at site GeoB 1016-3 than in the Vostok record. Moreover, during most of the interglacial periods, pCO_2 was significantly higher, ranging from 100-120 µatm, than during glacials ($\Delta pCO_2 \approx 70$ µatm).

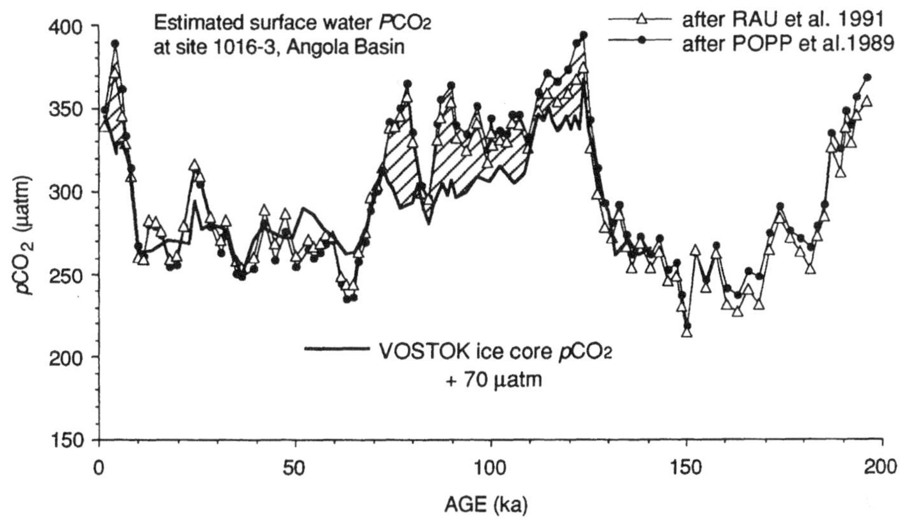

Figure 6 Comparison of surface water PCO_2 estimates for core GeoB 1016-3 with the original Vostok ice-core CO_2 record (BARNOLA et al. 1987) using a normalized time scale (see text). The average ice to sediment-core difference in glacial pCO_2 (70 µatm) was added to the ice-core data to emphasize that ΔpCO_2 was higher in interglacial periods.

These results suggest that the surface waters in the eastern Angola Basin generally acted as a CO_2 source during the past 200,000 years, but that the CO_2 flux to the atmosphere was reduced during glacial periods. At the present time, the northeastern Angola Basin is a region of high primary productivity (e.g. BERGER et al. 1989) due to a shallow thermocline (20-25 m) and the occurrence of oceanic upwelling which causes "doming" of the thermocline and brings nutrients into the photic zone (VAN BENNEKOM & BERGER 1984, RAVELO et al. 1990). Wind-driven coastal upwelling is not important in this region and generally is limited to latitudes south of about 15°S (SHANNON 1985, LUTJEHARMS & MEEUWIS 1987; see also Fig. 1). Nutrients are transported to the region off Angola by the northward-flowing coastal branch of the Benguela Current, which diverts to the subsurface at the Angola-Benguela frontal zone (14-

16°S, SHANNON et al. 1987) and can be traced up to a latitude of about 5°S (VAN BENNEKOM & BERGER 1984).

There is good evidence, based on time series analyses of geochemical and isotopic tracer records in the tropical to subtropical eastern South Atlantic, that the Angola-Benguela Front remained at its present position during the past 200,000 years while an increase in cold water advection from the south, perhaps coupled with a shoaling of the thermocline, caused a higher productivity during glacial stages (SCHNEIDER 1991, RUHLAND et al. 1992, SCHNEIDER et al. 1994). On the basis of this scenario, we infer that the lower ocean-atmosphere pCO_2 difference derived for glacial stages at site GeoB 1016-3 (Fig. 6) is the result of higher biological productivity and an enhanced flux of carbon out of the photic zone. This conclusion is consistent with the results of direct measurements in the North Atlantic (WATSON et al. 1991, ROBERTSON et al. 1993), oceanic modelling (TAYLOR et al. 1991, GARÇON et al. 1992) and a sediment trap experiment off Namibia (MÜLLER et al. 1994), which all show that phytoplankton blooms can cause rapid lowering of surface water PCO_2.

ACKNOWLEDGEMENTS

We would like to thank Birgit Meyer-Schack and Dietmar Grotheer for careful isotopic and geochemical analyses and Monika Segl for taking care of the mass spectrometers. We are also grateful to Walter Hale, Erwin Suess and an anonymous referee for comments which significantly improved the manuscript. This research was funded by the Deutsche Forschungsgemeinschaft (Sonderforschungsbereich 261 at Bremen University, Contribution No. 73) and the Bundesminister für Forschung und Technologie (BMFT), Bonn.

REFERENCES

Andrié C, Oudot C, Genthon C, Merlivat L (1986) CO_2 fluxes in the tropical Atlantic during FOCAL cruises. J Geophys Res 91, C10: 11,741-11,755

Barnola JM, Raynaud D, Korotkevich YS, Lorius C (1987) Vostok ice core provides 160,000-year record of atmospheric CO_2. Nature 329: 408-414

Barnola JM, Pimienta P, Raynaud D, Korotkevich YS (1991) CO_2-climate relationship as deduced from the Vostok ice core: a re-examination based on new measurements and on a re-evaluation of the air dating. Tellus 43B: 83-90

Berger WH (1989) Global maps of ocean productivity. In: Berger WH, Smetacek V, Wefer G (eds) Productivity of the Ocean: Present and Past. J. Wiley & Sons, Chichester, Dahlem Workshop Reports, 429-455

Berger WH, Herguera JC (1992) Reading the sedimentary record of the oceans´s productivity. In: Falkowski PG, Woodhead AD (eds) Primary Productivity and Biogeochemical Cycles in the Sea. Plenum Press, New York, 455-486

Berger WH, Smetacek V, Wefer G (eds) (1989) Productivity of the Ocean: Present and Past. J. Wiley & Sons, Chichester, Dahlem Workshop Reports, 470 pp

Brassell SC, Eglinton G, Marlowe IT, Pflaumann U, Sarnthein M (1986) Molecular stratigraphy: a new tool for climatic assessment. Nature 320: 129-133

Charles CD, Froelich PN, Zibello MA, Mortlock RA, Morley JJ (1991) Biogenic opal in Southern Ocean sediments over the last 450,000 years: Implications for surface water chemistry and circulation. Paleoceanography 6: 697-728

Conte MH, Eglinton G, Madureira LAS (1992) Long-chain alkenones and alkyl alkenoates as palaeotemperature indicators: their production, flux and early sedimentary diagenesis in the Eastern North Atlantic. Org Geochem 19: 287-298

Curry WB, Crowley TJ (1987) The $\delta^{13}C$ of Equatorial Atlantic surface waters: Implications for ice age pCO_2 levels. Paleoceanography 2: 489-517

Degens ET, Guillard RRL, Sackett WM, Hellebust JA (1968) Metabolic fractionation of carbon isotopes in marine plankton - I. Temperature and respiration experiments. Deep-Sea Res 15: 1-9

Deuser WG, Ross EH (1989) Seasonally abundant planktonic foraminifera of the Sargasso Sea: Sucession, deep-water fluxes, isotopic compositions, and paleoceanographic implications. J Foram Res 19: 268-293

Emrich K, Ehhalt DH, Vogel JC (1970) Carbon isotope fractionation during the precipitation of calcium carbonate. Earth Planet Sci Lett 8: 363-371

Fairbanks RG, Sverdlove M, Free R, Wiebe PH, Bé AWE (1982) Vertical distribution and isotopic fractionation of living planktonic foraminifera from the Panama Basin. Nature 298: 841-844

Fontugne MR, Calvert SE (1992) Late Pleistocene variability of the carbon isotopic composition of organic matter in the eastern Mediterranean: Monitor of changes in carbon sources and atmospheric CO_2 concentrations. Paleoceanography 7: 1-20

Fontugne MR, Duplessy J-C (1981) Organic carbon isotopic fractionation by marine plankton in the temperature range 1 to 31°C. Oceanologica Acta 4: 85-90

Freeman KH, Hayes JM (1992) Fractionation of carbon isotopes by phytoplankton and estimates of ancient CO_2 levels. Glob Biogeochem Cycles 6: 185-198

Fry B, Sherr E (1984) $\delta^{13}C$ measurements as indicators of carbon flow in marine and freshwater ecosystems. Contr Mar Sci 27: 13-47

Ganssen G (1983) Dokumentation von küstennahem Auftrieb anhand stabiler Isotope in rezenten Foraminiferen vor Nordwest-Afrika. "Meteor" Forsch Ergebn C37: 1-46

Garçon VC, Martinon L, Andrié C, Andrich P, Minster J-F (1989) Kinematics of CO_2 fluxes in the tropical Atlantic Ocean during the 1983 northern summer. J Geophys Res 94, C1: 855-870

Garçon VC, Thomas F, Wong CS, Minster J-F (1992) Gaining insight into the seasonal variability of CO_2 at ocean station P using an upper ocean model. Deep-Sea Res 39: 921-938

Imbrie J, Hays JD, Martinson DG, McIntyre A, Mix AC, Morley JJ, Pisias NG, Prell WL, Shackleton NJ (1984) The orbital theory of Pleistocene climate: Support from a revised chronology of the marine ^{18}O record. In: Berger AL, Imbrie J, Hays JD, Kukla J, Saltzman J (eds) Milankovitch and Climate, Part I. Reidel, Hingham, Mass, 269-305

Jansen JHF, van Weering TCE, Gieles R, van Iperen J (1984) Late Quaternary oceanography and climatology of the Zaire-Congo Fan and the adjacent eastern Angola Basin. Neth J Sea Res 17: 201-249

Jasper JP, Gagosian RB (1990) The sources and deposition of organic matter in the Late Quaternary Pigmy Basin, Gulf of Mexico. Geochim Cosmochim Acta 54: 1117-1132

Jasper JP, Hayes JM (1990) A carbon isotope record of CO_2 levels during the late Quaternary. Nature 347: 462-464

Jouzel J, Lorius C, Petit JR, Genthon C, Barkov NI, Kotlyakov VM, Petrov M (1987) Vostok ice core: a continuous isotope temperature record over the last climatic cycle (160,000 years). Nature 329: 403-408

Keir RS (1990) Reconstructing the ocean carbon system variation during the last 150,000 years according to the Antarctic nutrient hypothesis. Paleoceanography 5: 253-276

Knox F, McElroy M (1984) Change in atmospheric CO_2: Influence of the marine biota at high latitude. J Geophys Res 89: 4629-4637

Kumar N, Gwiadzda R, Anderson RF, Froelich PN (1993) $^{231}Pa/^{230}Th$ ratios in sediments as a proxy for past changes in Southern Ocean productivity. Nature 362: 45-48

Lutjeharms JRE, Meeuwis JM (1987) The extent and variability of south-east Atlantic upwelling. In: Payne AIL, Gulland JA, Brink KH (eds) The Benguela and Comparable Ecosystems. S Afr J mar Sci 5: 51-62

Lyle M, Murray DW, Finney BP, Dymond J, Robbins JM, Brooksforce K (1988) The record of late Pleistocene biogenic sedimentation in the eastern tropical Pacific Ocean. Paleoceanography 3: 39-60

Martin JH (1990) Glacial-interglacial CO_2 change: the iron hypothesis. Paleoceanography 5: 1-13

McArthur JM, Tyson RV, Thomson J, Matey D (1992) Early diagenesis of marine organic matter: Alteration of the carbon isotopic composition. Mar Geol 105: 51-61

McCabe B (1985) The dynamics of $\delta^{13}C$ in several New Zealand lakes. Thesis, University of Waikato, 278 pp

Mariotti A, Gadel F, Giresse P, Kinga-Mouzeo (1991) Carbon isotope composition and geochemistry of particulate organic matter in the Congo River (Central Africa): Application to the study of Quaternary sediments off the mouth of the river. Chem Geol (Isotope Geoscience Section) 86: 345-357

Metzl N, Beauverger C, Brunet C, Goyet C, Poisson A (1991) Surface water carbon dioxide in the southwest Indian sector of the Southern Ocean: a highly variable CO_2 source/sink region in summer. Mar Chem 35: 85-96

Mix AC (1989) Influence of productivity variations on long-term atmospheric CO_2. Nature 337: 541-544

Mook WG, Bommerson JC, Staverman WH (1974) Carbon isotope fractionation between dissolved bicarbonate and gaseous carbon dioxide. Earth Planet Sci Lett 22: 169-176

Mortlock RA, Charles CD, Froelich PN, Zibello MA, Saltzman J, Hays JD, Burckle LH (1991) Evidence for lower productivity in the Antarctic Ocean during the last glaciation. Nature 351: 220-223

Müller PJ, Erlenkeuser H, von Grafenstein R (1983) Glacial-interglacial cycles in oceanic productivity inferred from organic carbon contents in eastern North Atlantic sediment cores. In: Thiede J, Suess E (eds) Coastal Upwelling, Part B. Plenum Press, New York, 365-398

Müller PJ, Fischer G, Ruhland G, Andersen N, Wefer G (1994) Seasonal $\delta^{13}C_{org}$ variations of sedimenting POM off Namibia: Implications for changes in ocean PCO_2. Limnol Oceanogr, ms submitted

Murphy PP, Feely RA, Gammon RH, Harrison DE, Kelly KC, Waterman LS (1991) Assessment of the air-sea exchange of CO_2 in the South Pacific during austral autumn. J Geophys Res 96, C11: 20,455-20,465

Pedersen TF, Nielsen B, Pickering M (1991) The timing of late Quaternary productivity pulses in the Panama Basin and implications for atmospheric CO_2. Paleoceanography 6: 657-677

Popp BN, Takigiku R, Hayes JM, Louda JW, Baker EW (1989) The post-paleozoic chronology and mechanism of $\delta^{13}C$ depletion in primary marine organic matter. Am J Sci 289: 436-454

Prahl FG, De Lange GJ, Lyle M, Sparrow MA (1989) Post-depositional stability of long-chain alkenones under contrasting redox conditions. Nature 341: 434-437

Prahl FG, Muehlhausen LA, Zahnle DL (1988) Further evaluation of long-chain alkenones as indicators of paleoceanographic conditions. Geochim Cosmochim Acta 52: 2303-2310

Radke M, Sittardt HG, Welte DH (1978) Removal of soluble organic matter from rock samples with a flow-through extraction cell. Anal Chem 50: 663-665

Rau GH, Takahashi T, Des Marais DJ (1989) Latitudinal variations in plankton $\delta^{13}C$: implications for CO_2 and productivity in past oceans. Nature 341: 516-518

Rau GH, Froelich PN, Takahashi T, Des Marais DJ (1991) Does sedimentary organic $\delta^{13}C$ record variations in Quaternary ocean $[CO_2(aq)]$? Paleoceanography 6: 335-347

Ravelo AC, Fairbanks RG, Philander SGH (1990) Reconstructing tropical Atlantic hydrography using planktonic foraminifera and an ocean model. Paleoceanography 5: 409-431

Robertson JE, Watson AJ, Langdon C, Ling RD, Wood JW (1993) Diurnal variation in surface pCO_2 and O_2 at 60°N, 20°W in the North Atlantic. Deep-Sea Res *II* 40: 409-422

Ruhland G, Müller PJ, Schneider R (1992) Late Quaternary variability of alkenone sea-surface temperatures in the eastern Angola Basin. 4th Int Conf Paleoceanog (ICP IV). Kiel, Abstr vol, p 244

Sarnthein M, Winn K, Duplessy J-C, Fontugne M (1988) Global variations of surface ocean productivity in low and mid latitudes: influence on CO_2 reservoirs of the deep ocean and atmosphere during the last 21,000 years. Paleoceanography 3: 361-399

Schneider R (1991) Spätquartäre Produktivitätsänderungen im östlichen Angola-Becken: Reaktion auf Variationen im Passat-Monsun-Windsystem und in der Advektion des Benguela-Küstenstroms. Thesis. Berichte FB Geowissenschaften, Nr. 21, Univ. Bremen, 198 pp

Schneider R, Bickert T, Müller PJ, Wefer G (1993) Comparison of cyclic equatorial and coastal productivity changes in the South Atlantic for the last 300,000 years. EUG VII, Strassbourg. Terra Nova 5, Abstract supplement No. 1, p 741

Schneider R, Müller PJ, Wefer G (1994) Late Quaternary paleoproductivity changes off the Congo deduced from stable carbon isotopes of planktonic foraminifera. Palaeogeogr Palaeoclimatol Palaeoecol, ms submitted

Schwartz D, Mariotti A, Lanfranchi R, Guillet B (1986) $^{13}C/^{12}C$ ratios of soil organic matter as indicators of vegetation changes in the Congo. Geoderma 39: 97-103

Shackleton NJ, Le J, Mix A, Hall MA (1992) Carbon isotope records from Pacific surface waters and atmospheric carbon dioxide. Quat Sci Rev 11: 387-400

Shannon LV (1985) The Benguela ecosystem. Part I. Evalution of the Benguela physical features and processes. Oceanogr Mar Biol Ann Rev 23: 105-182

Shannon LV, Agenbag JJ, Buys MEL (1987) Large- and mesoscale features of the Angola-Benguela Front. In: Payne AIL, Gulland JA, Brink KH (eds) The Benguela and Comparable Ecosystems. S Afr J mar Sci 5: 11-34

Shimmield GB, Mowbray SR, Weedon GP (1990) A 350 ka history of the Indian Southwest Monsoon - Evidence from deep-sea cores, northwest Arabian sea. Trans R Soc Edinb Earth Sci 81: 289-299

Skirrow G (1975) The Dissolved Gases - Carbon Dioxide. In: Riley JP, Skirrow G (eds) Chemical Oceanography, vol 2, 2nd ed. Academic Press, London, 1-192

Tans PP, Fung IY, Takahashi T (1990) Observational constraints on the global atmospheric CO_2 budget. Science 247: 1431-1438

Taylor AH, Watson AJ, Ainsworth M, Robertson JE, Turner DR (1991) A modelling investigation of the role of phytoplankton in the balance of carbon at the surface of the North Atlantic. Glob Biogeochem Cycles 5: 151-171

Thunell RC, Quingmin M, Calvert SE, Pedersen TF (1992) Glacial-Holocene biogenic sedimentation patterns in the South China Sea: productivity variations and surface water pCO_2. Paleoceanography 7: 143-162

Van Bennekom AJ, Berger GW (1984) Hydrography and silica budget of the Angola Basin. Geol Soc Am Mem 143: 134 pp

Van Leeuwen RJW (1989) Sea-floor distribution and Late Quaternary faunal patterns of planktonic and benthic foraminifers in the Angola Basin. Utrecht Micropaleontol Bull 38: 287 pp

Volk T, Hoffert MI (1985) Ocean carbon pumps: Analysis of relative strengths and efficiencies in ocean-driven atmospheric CO_2 changes. In: Sundquist ET, Broecker WS (eds) The Carbon Cycle and Atmospheric CO_2: Natural Variations Archaen to Present. Geophys Monogr Ser, vol. 32. AGU, Washington, D.C., 99-110

Watson AJ, Robinson C, Robinson JE, Williams PJ le B, Fasham MJR (1991) Spatial variability in the sink for atmospheric carbon dioxide in the North Atlantic. Nature 350: 50-53

Wefer G, and cruise participants (1988) Bericht über die METEOR-Fahrt M 6-6, Libreville - Las Palmas, 18.2.1988 - 23.3.1988. Berichte FB Geowissenschaften, Nr. 3, Univ. Bremen, 97 pp

Westerhausen L, Poynter J, Eglinton G, Erlenkeuser H, Sarnthein M (1993) Marine and terrigenous origin of organic matter in modern sediments of the equatorial East Atlantic: the $\delta^{13}C$ and molecular record. Deep-Sea Res 40: 1087-1121

Weiss RF (1974) Carbon dioxide in water and seawater: The solubility of a non-ideal gas. Mar Chem 2: 203-215

Appendix

Table 2 Analytical data for core GeoB 1016-3 (11°46.2´S, 11°40.9´E, 3410 m)

DEPTH (cm)	AGE (ka)	CaCO3 (%)	Corg (%)	Ntot (%)	C/N wt.ratio	$\delta^{13}C$ (‰) *G.ruber*	$\delta^{13}C$ (‰) organic	U^{k}_{37}	SST (°C)
6	1.3	15.6	1.56	0.202	7.72	1.59	-20.47	0.892	25.1
12	2.6	22.3	1.15	0.148	7.77	-	-20.42	-	-
18	3.9	27.2	0.67	0.090	7.44	1.81	-20.93	0.907	25.5
23	4.9	31.9	0.69	0.092	7.50	-	-20.73	-	-
28	6.0	25.6	0.80	0.100	8.00	1.65	-20.36	0.925	26.1
33	6.7	20.5	1.52	0.160	9.50	1.67	-20.36	0.854	24.0
38	7.5	19.9	1.83	0.199	9.20	1.56	-20.37	-	-
43	8.2	18.2	1.77	0.182	9.73	1.23	-20.29	0.861	24.2
48	9.0	18.7	1.58	0.165	9.58	1.06	-19.96	-	-
53	9.7	15.4	1.68	0.166	10.12	0.77	-19.69	0.848	23.8
58	10.4	9.2	2.37	0.236	10.04	1.18	-19.49	-	-
63	11.2	8.8	2.26	0.233	9.70	1.09	-19.51	0.813	22.8
68	11.9	8.9	2.26	0.234	9.66	-	-19.37	-	-
73	12.6	3.7	2.85	0.281	10.14	1.72	-19.48	0.806	22.6
78	13.4	2.9	2.90	0.282	10.28	1.95	-19.39	-	-
83	14.1	1.8	3.24	0.294	11.02	-	-19.44	0.812	22.7
88	14.9	2.8	2.96	0.273	10.84	1.78	-19.60	-	-
93	15.6	2.7	2.66	0.246	10.81	1.68	-19.40	0.800	22.4
98	16.3	3.0	2.58	0.249	10.36	1.27	-19.44	-	-
103	17.1	1.9	2.60	0.232	11.21	1.70	-19.47	-	-
108	17.8	2.4	2.40	0.227	10.57	1.24	-19.55	0.776	21.7
113	18.4	3.1	2.65	0.248	10.69	-	-19.65	-	-
118	19.0	2.2	3.12	0.276	11.30	1.30	-19.63	-	-
123	19.6	2.5	3.06	0.278	11.01	1.30	-19.56	0.771	21.5
128	20.2	4.1	2.80	0.257	10.89	1.80	-19.65	-	-
133	20.8	4.6	2.72	0.261	10.42	1.82	-19.53	-	-
138	21.4	4.3	2.86	0.282	10.14	1.49	-19.57	0.818	22.9
143	22.2	2.6	3.02	0.283	10.67	1.59	-19.76	-	-
148	23.1	3.7	3.24	0.318	10.19	1.82	-19.87	-	-
153	23.9	2.3	3.43	0.357	9.61	2.00	-19.99	0.812	22.7
158	24.7	3.6	3.26	0.340	9.59	1.61	-19.92	-	-
163	25.5	3.7	3.01	0.312	9.65	2.09	-19.86	0.792	22.1
168	26.4	3.9	2.90	0.295	9.83	1.84	-19.72	-	-
173	27.2	3.8	2.42	0.253	9.57	1.77	-19.64	-	-
178	28.0	3.6	2.77	0.280	9.89	1.76	-19.57	0.792	22.1
183	29.4	4.4	2.93	0.301	9.73	1.58	-19.84	-	-
188	30.8	4.2	3.44	0.349	9.86	1.37	-19.81	0.755	21.1
193	32.2	3.3	4.35	0.425	10.24	1.71	-19.77	0.757	21.1
198	33.6	3.7	4.26	0.420	10.14	-	-19.74	-	-
203	34.9	3.6	3.81	0.378	10.08	1.10	-19.76	0.755	21.1
208	36.3	2.5	3.21	0.318	10.09	-	-19.67	0.750	20.9
213	37.7	3.1	3.30	0.323	10.22	1.68	-19.74	-	-
218	39.1	3.0	2.86	0.279	10.25	1.18	-19.75	0.756	21.1
223	40.5	2.2	2.75	0.275	10.00	2.08	-19.65	-	-
228	41.9	3.3	2.88	0.288	10.00	1.92	-19.64	0.771	21.5
233	43.3	5.1	2.75	0.267	10.30	1.42	-19.65	-	-
238	44.7	4.3	3.00	0.293	10.24	1.58	-19.59	0.747	20.8
243	46.1	2.4	2.73	0.268	10.19	1.86	-19.57	-	-

Table 2 (continued)

DEPTH (cm)	AGE (ka)	CaCO3 (%)	Corg (%)	Ntot (%)	C/N wt.ratio	$\delta^{13}C$ (‰) *G.ruber*	$\delta^{13}C$ (‰) organic	U^{k}_{37}	SST (°C)
258	50.2	2.7	2.51	0.250	10.04	1.26	-19.67	0.760	21.2
263	51.6	3.0	2.25	0.219	10.27	1.33	-19.71	-	-
268	53.0	3.8	1.76	0.172	10.23	1.28	-19.80	0.773	21.6
273	53.8	4.1	1.87	0.174	10.75	1.05	-20.01	-	-
278	54.6	4.1	2.27	0.215	10.56	1.08	-19.94	0.767	21.4
283	55.3	4.7	2.37	0.227	10.44	1.04	-19.88	-	-
288	56.1	4.7	2.56	0.254	10.08	1.20	-19.86	0.771	21.5
293	56.9	4.9	2.60	0.267	9.74	1.44	-19.78	-	-
298	57.7	3.2	2.59	0.262	9.89	1.38	-19.82	0.769	21.5
303	58.5	3.8	2.63	0.267	9.85	0.99	-19.81	-	-
308	59.2	3.6	2.64	0.271	9.74	1.09	-19.91	0.798	22.3
313	60.0	4.1	2.55	0.246	10.37	0.97	-19.86	-	-
318	60.8	4.3	2.30	0.226	10.18	1.07	-19.72	-	-
323	61.6	2.7	2.29	0.236	9.70	1.14	-19.41	0.768	21.4
328	62.4	2.2	2.50	0.263	9.51	1.31	-19.37	-	-
333	63.1	0.2	2.90	0.303	9.57	-	-19.48	0.730	20.3
338	63.9	1.5	2.88	0.307	9.38	-	-19.40	-	-
343	64.7	1.8	3.02	0.301	10.03	-	-19.42	0.741	20.6
348	65.5	2.2	3.16	0.303	10.43	-	-19.54	-	-
353	66.2	2.4	2.92	0.290	10.07	-	-19.75	0.774	21.6
358	67.0	5.9	2.80	0.267	10.49	-	-19.96	-	-
363	67.8	4.4	2.98	0.268	11.12	1.14	-20.08	0.769	21.5
368	68.5	8.2	2.46	0.221	11.13	-	-20.26	-	-
373	69.3	4.3	2.49	0.246	10.12	-	-20.42	0.769	21.5
378	70.1	4.1	2.31	0.227	10.18	-	-20.37	-	-
383	70.8	3.7	2.12	0.224	9.46	1.48	-20.25	0.804	22.5
388	71.6	3.0	2.33	0.250	9.32	1.27	-20.25	-	-
393	72.3	2.3	2.27	0.249	9.12	1.51	-20.42	0.810	22.7
398	73.1	2.8	2.41	0.256	9.41	1.70	-20.20	-	-
403	73.9	3.0	2.46	0.266	9.25	1.83	-20.50	0.843	23.6
408	74.6	3.7	2.30	0.252	9.13	1.46	-20.23	-	-
413	75.4	3.7	2.28	0.242	9.42	-	-20.67	0.826	23.1
418	76.2	4.7	2.33	0.265	8.79	1.77	-20.56	-	-
423	76.9	4.3	1.89	0.218	8.67	1.77	-20.81	0.834	23.4
428	77.7	5.4	1.71	0.202	8.47	2.00	-20.57	-	-
433	78.5	6.6	1.58	0.189	8.36	2.04	-20.71	0.849	23.8
438	79.2	8.4	1.47	0.174	8.45	1.95	-20.46	-	-
443	80.0	7.7	1.46	0.174	8.39	1.86	-20.11	0.869	24.4
448	81.0	8.5	1.62	0.181	8.95	1.72	-20.15	-	-
453	82.0	6.8	1.61	0.181	8.90	1.49	-19.90	0.848	23.8
458	83.0	5.0	1.84	0.203	9.06	1.29	-19.86	-	-
463	84.0	4.4	2.18	0.239	9.12	1.41	-19.94	0.832	23.3
468	85.0	4.6	2.69	0.285	9.44	-	-20.10	-	-
473	86.0	3.9	4.34	0.449	9.67	1.62	-20.24	0.893	25.1
478	87.0	3.7	4.72	0.467	10.11	-	-20.48	0.883	24.8
483	88.3	4.1	4.20	0.410	10.24	-	-20.44	-	-
488	89.7	4.2	3.53	0.352	10.03	1.94	-20.58	0.872	24.5
493	91.0	4.9	2.82	0.297	9.49	1.33	-20.68	0.873	24.5
498	92.3	6.0	2.26	0.249	9.08	1.11	-20.57	-	-
503	93.7	6.8	2.02	0.224	9.02	1.29	-20.46	0.889	25.0

Table 2 (continued)

DEPTH (cm)	AGE (ka)	CaCO3 (%)	Corg (%)	Ntot (%)	C/N wt.ratio	$\delta^{13}C$ (‰) *G.ruber*	$\delta^{13}C$ (‰) organic	$U^{k}37$	SST (°C)
518	97.7	13.1	1.45	0.170	8.53	1.65	-20.24	-	-
523	99.0	15.5	1.28	0.157	8.15	1.34	-20.22	0.888	25.0
528	99.8	16.1	1.29	0.159	8.11	1.75	-20.22	0.890	25.0
533	100.6	18.7	1.26	0.160	7.88	1.56	-20.37	0.870	24.4
538	101.4	12.7	1.33	0.156	8.53	1.31	-20.39	-	-
543	102.2	15.2	1.44	0.165	8.73	1.45	-20.61	0.858	24.1
548	103.0	3.9	1.74	0.212	8.21	1.28	-20.49	-	-
553	103.8	5.2	1.93	0.213	9.06	1.59	-20.46	0.854	24.0
558	104.6	2.1	2.41	0.261	9.23	-	-20.60	-	-
563	105.4	1.7	2.84	0.295	9.63	1.75	-20.62	0.849	23.8
568	106.2	2.1	2.63	0.276	9.53	1.46	-20.61	-	-
573	107.0	2.5	2.64	0.274	9.64	1.61	-20.62	0.865	24.3
578	109.5	2.0	2.06	0.225	9.16	1.09	-20.82	0.865	24.3
583	112.0	2.7	1.85	0.210	8.81	1.40	-20.95	0.881	24.8
588	114.5	4.6	1.61	0.181	8.90	1.63	-20.92	0.884	24.9
593	117.0	9.1	1.48	0.189	7.83	1.45	-20.98	0.889	25.0
598	119.5	21.7	1.06	0.126	8.41	1.35	-21.09	0.904	25.4
603	122.0	20.1	1.07	0.124	8.63	1.42	-21.06	0.933	26.3
608	123.8	19.1	1.02	0.120	8.50	1.74	-20.94	0.923	26.0
613	125.5	15.5	1.39	0.145	9.59	1.31	-20.17	0.940	26.5
618	127.3	18.3	1.46	0.149	9.80	1.17	-19.76	0.931	26.2
623	129.0	19.8	1.85	0.179	10.34	0.89	-19.69	0.908	25.6
628	130.8	17.6	2.43	0.215	11.30	0.99	-19.58	0.880	24.7
633	132.5	10.7	2.86	0.255	11.22	1.01	-19.99	0.857	24.1
638	134.3	6.6	3.33	0.316	10.54	0.84	-19.69	0.861	24.2
643	135.7	4.5	3.26	0.298	10.94	0.81	-19.49	0.852	23.9
648	136.9	5.3	2.90	0.264	10.98	1.30	-19.33	-	-
653	138.1	4.5	2.79	0.257	10.86	1.39	-19.26	0.846	23.7
658	139.3	3.5	2.78	0.259	10.73	1.59	-19.23	-	-
663	140.5	2.7	2.65	0.238	11.13	1.32	-18.95	0.856	24.0
668	141.7	2.9	2.60	0.232	11.21	1.35	-19.03	-	-
673	142.9	4.2	2.51	0.226	11.11	1.51	-18.97	0.857	24.1
678	144.1	4.2	2.53	0.229	11.05	1.52	-18.94	-	-
683	145.3	3.1	2.53	0.224	11.29	1.35	-18.74	0.849	23.8
688	146.3	2.7	2.48	0.217	11.43	1.34	-18.71	-	-
693	147.1	3.4	2.36	0.217	10.88	1.50	-18.64	0.854	24.0
698	147.9	5.6	2.37	0.210	11.29	1.12	-18.61	-	-
703	148.7	7.3	2.11	0.185	11.41	1.09	-18.73	0.823	23.1
708	149.4	7.9	2.06	0.180	11.44	1.30	-18.55	-	-
713	150.2	7.2	2.02	0.180	11.22	1.02	-18.61	0.778	21.7
718	151.0	7.2	2.19	0.193	11.35	1.40	-18.83	-	-
723	152.3	6.3	2.68	0.236	11.36	1.47	-19.32	0.801	22.4
728	153.7	5.6	2.97	0.268	11.08	-	-19.25	-	-
733	155.0	8.1	2.31	0.219	10.55	1.13	-19.00	0.816	22.9
738	156.3	17.7	2.36	0.214	11.03	1.41	-19.08	-	-
743	157.7	18.5	2.32	0.205	11.32	1.53	-19.01	0.841	23.6
748	159.0	11.1	2.44	0.216	11.30	1.01	-18.90	-	-
753	160.3	7.1	2.21	0.199	11.11	1.18	-18.60	0.841	23.6
758	161.7	5.2	2.20	0.208	10.58	0.97	-18.52	-	-
763	163.0	4.7	2.26	0.209	10.81	-	-18.45	0.833	23.4

Table 2 (continued)

DEPTH (cm)	AGE (ka)	$CaCO_3$ (%)	Corg (%)	Ntot (%)	C/N wt.ratio	$\delta^{13}C$ (‰) *G.ruber*	$\delta^{13}C$ (‰) organic	U^{k}_{37}	SST (°C)
778	167.0	6.4	1.99	0.190	10.47	1.38	-18.45	-	-
783	168.3	7.5	2.05	0.196	10.46	1.29	-18.30	0.887	24.9
788	169.7	6.7	2.14	0.200	10.70	1.23	-19.03	-	-
793	171.0	5.7	2.17	0.203	10.69	1.17	-19.21	0.881	24.8
798	172.3	5.7	2.02	0.193	10.47	0.95	-19.09	-	-
803	173.7	5.3	1.89	0.183	10.33	1.18	-19.72	0.868	24.4
808	175.0	4.9	2.00	0.191	10.47	0.93	-19.38	-	-
813	176.3	4.4	1.95	0.187	10.43	1.21	-19.50	0.848	23.8
818	177.7	6.7	1.82	0.177	10.28	1.20	-18.99	-	-
823	179.0	6.7	2.01	0.205	9.80	0.98	-19.46	0.865	24.3
828	180.3	8.0	2.16	0.215	10.05	1.17	-19.61	-	-
833	181.7	8.3	2.77	0.266	10.41	1.05	-19.08	0.883	24.8
838	183.0	6.6	3.30	0.303	10.89	1.09	-19.73	-	-
843	183.7	5.4	3.46	0.319	10.85	0.98	-19.70	0.863	24.2
848	184.4	6.1	3.53	0.309	11.42	-	-19.76	-	-
853	185.1	2.2	3.34	0.311	10.74	1.00	-19.92	0.867	24.4
858	185.8	2.7	2.93	0.287	10.21	0.94	-19.85	-	-
863	186.4	1.2	2.88	0.272	10.59	1.69	-20.03	-	-
868	187.1	2.2	2.56	0.251	10.20	1.74	-20.00	0.891	25.1
873	187.8	27.1	0.77	0.107	7.20	1.67	-19.48	-	-
878	188.5	4.0	1.99	0.197	10.10	1.53	-19.98	-	-
883	189.2	6.1	1.85	0.188	9.84	1.35	-19.90	0.922	26.0
888	189.9	4.7	1.71	0.179	9.55	1.00	-20.00	-	-
893	190.6	5.7	1.44	0.158	9.11	1.79	-20.24	0.892	25.1
898	191.2	13.3	0.70	0.076	9.21	1.52	-20.18	-	-
903	191.9	4.2	1.50	0.163	9.20	-	-20.06	0.895	25.2
908	192.6	5.0	1.63	0.183	8.91	1.38	-20.42	-	-
913	193.3	5.2	1.38	0.167	8.26	1.69	-20.59	0.883	24.8
918	194.0	9.4	1.22	0.153	7.97	1.70	-20.57	-	-
923	196.2	11.2	1.09	0.142	7.68	1.74	-20.56	0.910	25.6
928	198.4	8.7	1.27	0.154	8.25	1.69	-20.00	-	-
933	200.6	8.1	1.62	0.184	8.80	1.63	-20.05	-	-
938	202.8	9.1	1.58	0.183	8.63	0.99	-20.03	-	-
943	205.0	4.6	1.81	0.196	9.23	1.09	-20.23	-	-
948	206.8	4.2	2.06	0.216	9.54	1.40	-20.25	-	-
953	208.7	4.7	1.92	0.201	9.55	1.86	-20.16	-	-
958	210.5	5.5	1.84	0.202	9.11	1.82	-20.30	-	-
963	212.3	6.5	1.70	0.190	8.95	1.67	-20.22	-	-
968	214.2	15.3	1.34	0.158	8.48	1.66	-19.53	-	-
973	216.0	2.6	2.07	0.212	9.76	1.60	-19.96	-	-
978	217.1	21.8	0.68	0.100	6.80	1.59	-19.62	-	-
983	218.2	16.5	1.19	0.138	8.62	1.81	-20.23	-	-
988	219.3	18.2	1.17	0.132	8.86	1.25	-19.92	-	-
993	220.4	8.0	1.36	0.144	9.44	1.29	-19.47	-	-
998	221.5	6.2	1.60	0.164	9.76	0.49	-19.62	-	-
1003	222.6	15.5	1.49	0.147	10.14	1.54	-19.83	-	-
1008	223.7	6.2	1.92	0.182	10.55	1.64	-20.01	-	-
1013	224.8	6.0	1.93	0.200	9.65	1.33	-20.02	-	-
1018	225.9	6.7	1.93	0.194	9.95	1.26	-20.18	-	-
1023	227.0	7.7	2.22	0.212	10.47	1.21	-20.21	-	-

Table 2 (continued)

DEPTH (cm)	AGE (ka)	CaCO3 (%)	Corg (%)	Ntot (%)	C/N wt.ratio	$\delta^{13}C$ (‰) *G.ruber*	$\delta^{13}C$ (‰) organic	$U^{k}37$	SST (°C)
1038	230.3	2.0	2.35	0.227	10.35	1.26	-20.32	-	-
1043	231.4	2.1	2.06	0.208	9.90	-	-20.47	-	-
1048	232.5	2.2	1.98	0.201	9.85	1.58	-20.49	-	-
1053	233.6	5.3	1.69	0.186	9.09	1.59	-20.67	-	-
1058	234.7	12.6	1.19	0.141	8.44	1.48	-20.68	-	-
1063	235.8	16.2	1.02	0.131	7.79	-	-20.54	-	-
1068	236.9	15.7	1.11	0.140	7.93	1.23	-20.42	-	-
1073	238.0	8.0	1.69	0.183	9.23	1.05	-20.41	-	-
1078	239.6	12.8	1.89	0.184	10.27	1.65	-20.52	-	-
1083	241.1	18.0	2.69	0.245	10.98	0.77	-20.42	-	-
1088	242.7	12.5	3.20	0.297	10.77	1.33	-20.29	-	-
1093	244.3	7.0	2.91	0.287	10.14	1.34	-20.02	-	-
1098	245.9	7.7	3.94	0.335	11.76	0.84	-20.26	-	-
1103	247.4	5.2	4.47	0.363	12.31	-	-20.37	-	-
1108	249.0	4.3	4.42	0.395	11.19	0.85	-20.68	-	-
1113	251.0	2.0	3.56	0.321	11.09	-	-20.49	-	-
1118	253.0	2.4	3.52	0.314	11.21	-	-20.43	-	-
1123	255.0	1.7	3.48	0.310	11.23	-	-20.57	-	-
1128	257.0	2.7	3.43	0.306	11.21	0.82	-20.45	-	-
1133	259.0	4.5	3.53	0.317	11.14	-	-20.63	-	-
1138	261.0	4.7	3.31	0.298	11.11	1.33	-20.55	-	-
1143	263.0	3.9	3.29	0.293	11.23	-	-20.48	-	-
1148	265.0	2.4	3.59	0.307	11.69	-	-20.59	-	-
1153	267.0	4.3	2.36	0.212	11.13	0.46	-19.83	-	-
1158	269.0	0.9	2.06	0.186	11.08	-	-19.42	-	-
1163	271.6	1.8	2.10	0.199	10.55	0.53	-19.71	-	-
1168	274.1	1.9	3.35	0.287	11.67	-	-20.05	-	-
1173	276.7	2.0	4.22	0.361	11.69	-	-20.48	-	-
1178	279.3	0.2	3.03	0.264	11.48	-	-20.31	-	-
1183	281.9	0.5	3.05	0.281	10.85	-	-20.57	-	-
1188	284.4	3.3	2.43	0.249	9.76	1.59	-20.75	-	-
1193	287.0	8.7	1.62	0.189	8.57	1.68	-20.33	-	-
1198	289.6	16.7	1.56	0.166	9.40	1.46	-20.20	-	-
1203	292.1	9.0	1.86	0.193	9.64	1.28	-19.86	-	-
1208	294.7	4.5	2.30	0.230	10.00	1.06	-20.12	-	-
1213	297.3	3.4	2.04	0.213	9.58	1.43	-20.34	-	-
1218	299.9	4.0	1.50	0.163	9.20	1.48	-20.48	-	-
1223	302.4	4.6	1.54	0.165	9.33	1.09	-20.79	-	-

PCO_2 VARIATIONS OF EQUATORIAL SURFACE WATER OVER THE LAST 330,000 YEARS: THE $\delta^{13}C$ RECORD OF ORGANIC CARBON

Westerhausen, L., M. Sarnthein, U. Struck, H. Erlenkeuser[1], J. Poynter[2]
Geologisch-Paläontologisches Institut und Museum
Christian-Albrechts-Universität Kiel
Olshausenstraße 40-60
D-24118 Kiel, Germany

ABSTRACT. We reconstructed past variations in CO_2 partial pressure (local PCO_2) in the surface waters of the East Atlantic equatorial upwelling zone over the last 330,000 years, based on the $\delta^{13}C$ record of the (marine) organic matter in 'Meteor' core 16772. To deduce the initial $\delta^{13}C_{organic}$ values of plankton and the CO_2 solubility in surface water, the $\delta^{13}C$ record was adjusted for i) past variations in (winter) sea surface temperature, ii) variations in the $\delta^{13}C$ composition of inorganic carbon dissolved in the surface waters, using the $\delta^{13}C$ values of *G. ruber* (white), and iii) isotopic fractionation during the degradation of settling organic matter in the water column and on the sediment surface.

The calculated paleo-PCO_2 variations in the surface waters show a strong signal at the obliquity frequency and are approximately parallel to the VOSTOK ice-core record of atmospheric pCO_2 over the last 140,000 years. Holocene pCO_2 values varied within the range of modern local PCO_2, which is 350-400 ppmv compared to a pre-industrial atmospheric pCO_2 level of 280 ppmv. This positive anomaly demonstrates the persistent CO_2 release from upwelled subsurface water. The glacial-to-interglacial amplitudes of local PCO_2 (at the core site) exceeded those of atmospheric pCO_2 by 20-60%, with values of less than 250 to 300 ppmv during cold isotopic stages, which indicate a decreased net carbon outgassing from the ocean to the atmosphere. The close correlation between high paleo-PCO_2 and low paleo-nutrient contents and paleoproductivity (r=0.7-0.8) suggests that the local PCO_2 variations resulted mainly from CO_2 transfer by phytoplankton production, especially over the last 170,000 years.

---

[1] C14-Labor, Institut für reine und angewandte Kernphysik, Universität Kiel
[2] Organic Geochemistry Unit, School of Chemistry, University of Bristol, UK

NATO ASI Series, Vol. I 17
Carbon Cycling in the Glacial Ocean:
Constraints on the Ocean's Role in Global Change
Edited by R. Zahn et al.

INTRODUCTION

Records of paleo-PCO_2 variations in oceanic surface waters have recently been reconstructed from the $\delta^{13}C$ records of total marine organic carbon (TOC) in deep-sea sediments (Thunell et al., 1992; Fontugne and Calvert, 1992; Schneider et al., this volume; Jasper and Hayes, 1990, employing the $\delta^{13}C$ record of organic compounds) using the recent models of Rau et al. (1989) and Popp et al. (1989). These records partially reflect the global atmospheric CO_2 variations as depicted in ice-core records (Barnola et al., 1987), but leave a number of discrepancies between the two kinds of records unresolved.

As shown by Tans et al. (1990), the modern surface ocean PCO_2 is highly variable on regional and local scales (200-450 ppmv) due to upwelling, downwelling, and the "biological pump", a variability that may complicate and/or dominate the global paleo-PCO_2 signal in local records. Thus, no simple correlation between local and global pCO_2 variations can be expected.

It is the objective of this paper to specify the contributions of regional mechanisms of carbon transfer in a high-resolution paleo-PCO_2 record from the most intensive upwelling region in the eastern equatorial Atlantic, that is the region with the highest annual temperature variation at 50 m depth (Gorshkov, 1979; Figure 1). We compare this record with the potential local sinks and sources of carbon, such as variations in upwelling and outgassing of CO_2, in paleopaleoproductivity, and in paleo-nutrient budgets. In summary, we start from the assumption that the global sum of local paleo-PCO_2 variations has controlled the global variations in atmospheric pCO_2 as depicted in ice-core records. That is, increasing PCO_2 values in the surface ocean have led to a net flux of C from the ocean to the atmosphere, whereas low oceanic PCO_2 values, such as observed in northern high latitudes today, lead to a sink for atmospheric CO_2 (Tans et al., 1990; Quay et al. 1992).

STRATEGY FOR THE RECONSTRUCTION OF PALEO-PCO_2

Experimental and field observations suggest that there is a significant relationship between the $\delta^{13}C$ values of marine TOC in marine sediments and the concentrations of dissolved CO_2 ($CO_{2(aq)}$) in the ambient surface water (Degens et al., 1968; Popp et al., 1989; Rau et al., 1989, 1991, 1992). Rau et al. (1989) defined the following relationship for the South Atlantic and the Southern Ocean (equation 1):

$$\text{plankton } \delta^{13}C = -0.8\ (CO_{2(aq)}) - 12.6 \qquad (1)$$

Thus, $\delta^{13}C$ values of plankton-borne organic matter increase linearly with decreasing ($CO_{2(aq)}$). Below 18-20°C, $\delta^{13}C$ values also reflect variations in sea surface temperature (SST) due to a temperature-dependent fractionation by phytoplankton production (about 0.5 ‰ per 1°C; Fontugne, 1983; Fontugne et al., 1991; Sackett et al., 1974). Above 18-20°C, no such

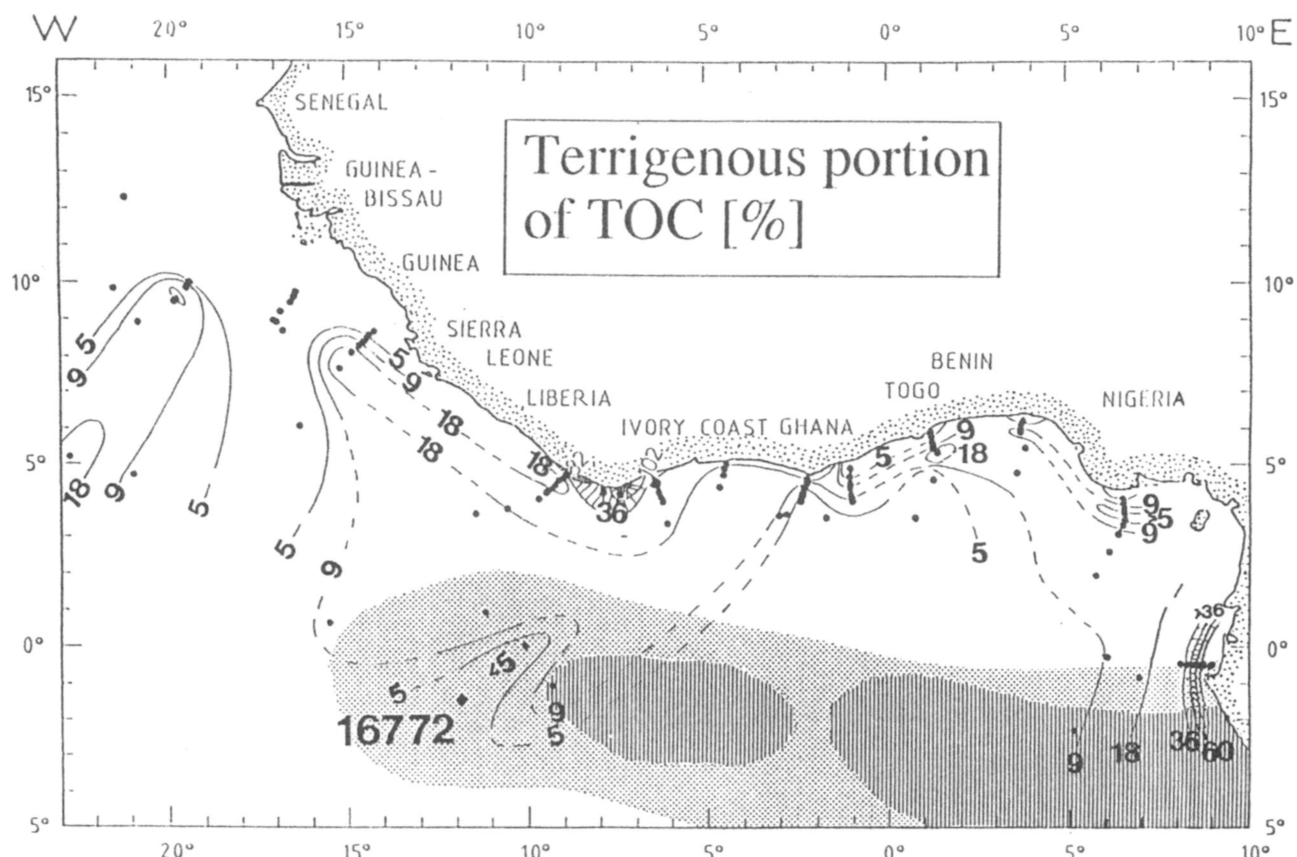

Figure 1: Location of `Meteor` core 16772 at 3912 m water depth in the equatorial East Atlantic. Contour lines mark terrigenous portion of organic carbon in surface sediments (Westerhausen et al., 1993, modified), which is less than 5% at the core location. Vertical hatching (>4°C) and shading (3-4°C) depict ranges of enhanced seasonal temperature variations at 50 m (Gorshkov, 1979), indicating equatorial upwelling.

metabolic temperature effect occurs (Fontugne, 1983), but SSTs control only the solubility and concentrations of CO_2 in the surface water (Weiss, 1974). Hence, the PCO_2 in surface waters above 18°C can be simply estimated from the dissolved $CO_{2(aq)}$ using Henry's law (equation 2) that quantifies the temperature-sensitive CO_2 solubility a:

$$PCO_2 = \alpha\text{-}1\ (CO_{2(aq)}) \qquad (2)$$

Paleosalinity, which also controls α (Weiss, 1974), was normalized to 35.75 ‰, since the impact of potential salinity variations on α is insignificant in the range expected for the equatorial Atlantic. Our $\delta^{13}C_{TOC}$ values were adjusted to SST of the cold season linked to equatorial upwelling, using a local paleotemperature record with temperatures rarely falling below 17°. This record is based on planktonic foraminifera assemblages (modified from Kähler, 1990) using the new modern analogue technique SIMMAX of Pflaumann et al. (1993). The cold season of upwelling represents the main production season of marine organic carbon in the equatorial Atlantic, especially during glacial times (see below). This choice of temperatures of the cold season is supported by first trap data (Fischer, 1991; Fischer and Wefer, 1991), obtained adjacent to our core site 16772, which depict a maximum in total flux and $\delta^{13}C_{TOC}$ during austral winter. For comparison, a U^K_{37}-based paleotemperature record (Figure 2C; Poynter, 1989; Eglinton et al., 1992) of the immediate sea surface was also employed, which generally represents the regime of the upwelling-free summer season (Kähler, 1990).

In harmony with the modern distribution patterns of $\delta^{13}C_{TOC}$ in the equatorial East Atlantic mapped by Westerhausen et al. (1993), the $\delta^{13}C$ values of organic matter were corrected i) for fluctuations in Dissolved Inorganic Carbon (DIC) as recorded by $\delta^{13}C$ deviations from the modern value (2.0 ‰), measured on tests of the surface dwelling planktonic foraminifer *G. ruber* (white) (Fairbanks et al., 1980; Ganssen and Sarnthein, 1983) and ii) the systematic ^{13}C enrichment produced by the degradation of settling particulate organic matter in the water column and on the sediment surface, using equation (3) (Westerhausen et al., 1993; based on plankton hauls, sediment- trap data, and surface-sediment samples):

$$\delta^{13}C_{TOC} = \delta^{13}C + 1.165 * 10^{-3}\, z - 0.141 * 10^{-6}\, z^2 \qquad (3)$$

where the $\delta^{13}C_{TOC}$ value is standardized to the sea surface and z is the water depth in m. Without this correction the Holocene estimates of paleo-PCO_2 decrease by about 95 ppmv, that is far below the values expected (see Figure 3).

We applied equations 1 to 3 to a $\delta^{13}C$ record of sedimentary TOC in the 15 m long piston core M16772 (1°21.0'S; 11°58.4'W; 3913 m water depth; Lutze et al., 1988). At this distal site near the mid-Atlantic Ridge, we assumed that the sedimentary organic matter of the last 330,000 years has been produced almost exclusively by plankton and a $\delta^{13}C$ shift by the input of terrigenous carbon can be ignored (Westerhausen et al., 1993; Figure 1).

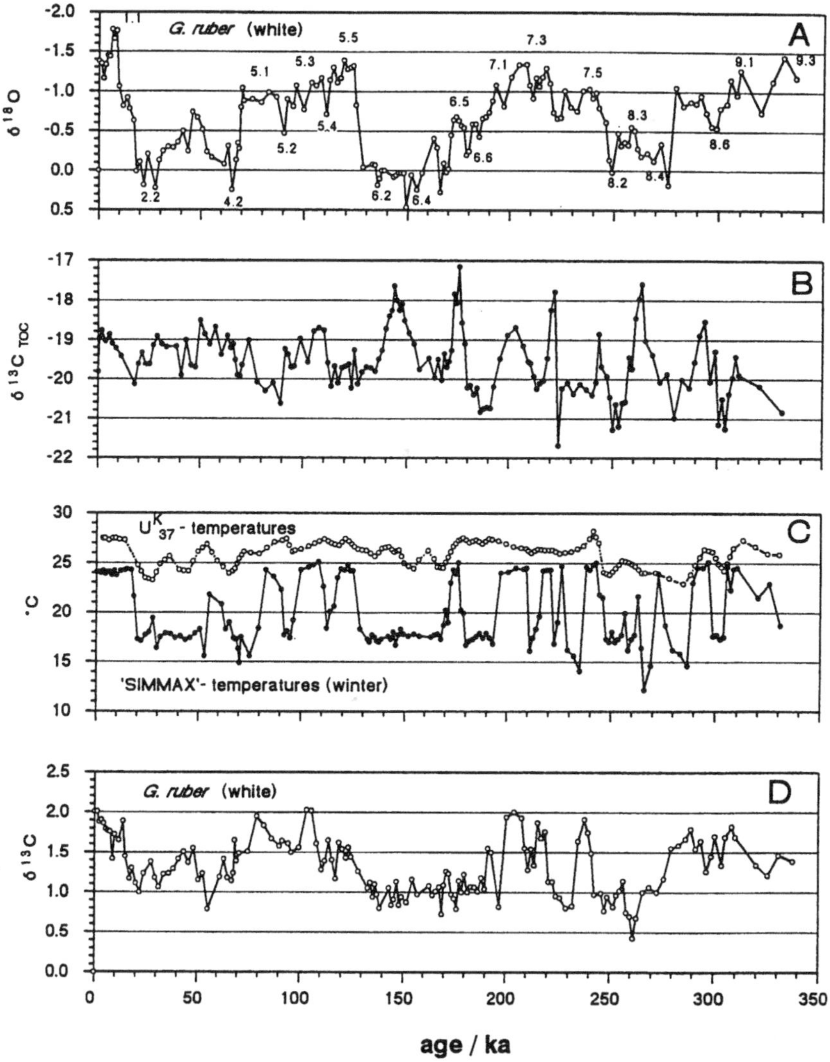

Figure 2: Quaternary isotopic and paleotemperature records in core 16772. A) $\delta^{18}O$-stratigraphy (‰ vs PDB) based on Globigerinoides ruber (white). Numbers mark isotopic events up to event 9.3 (330,000 y B.P.). B) Carbon isotopes of organic sediment fraction (‰ vs PDB). C) Variations in surface water temperature estimated from the U^{K}_{37}-index (open dots; Poynter, 1989; Prahl Wakeham, 1987) and `SIMMAX` winter temperatures (full dots; Kähler, 1990, Pflaumann, 1993). D) d13C values (‰ vs PDB) of plankton carbonate, used to correct $\delta^{13}C_{TOC}$ values for fluctuations in DIC (Fig. 2A).

MATERIALS AND METHODS

Piston core M 16772 was sampled at 10 cm intervals, which correspond to an average of 2200 years per sample (or to a range of 1600-3400 years over the last 130,000 years).

The chronostratigraphy of core 16772 was based on a $\delta^{18}O$ record (Figure 2A) from the planktonic foraminifer *Globigerinoides ruber* (white; 315-400 µm), which also produced the $\delta^{13}C$ record of DIC variations (Figs. 2D). The ages of oxygen-isotope stages and events were adopted from the SPECMAP $\delta^{18}O$ standard curve (Imbrie et al., 1984; Prell et al., 1986; Martinson et al., 1987; and Kähler, 1990). The coherence between the adjusted $\delta^{18}O$ record of *G. ruber* and the SPECMAP stack is excellent (coherence in the 23,000 year band: 0.93; phase angle of -1°; r=0.86). For the last 30,000 years, the key events of the $\delta^{18}O$ record were dated by correlation with AMS-14C dated records, ages that were converted into "calibrated U/Th ages" (Bard et al., 1990) prior to age interpolations, following the technique of Winn et al. (1991). Thus the record can be directly compared with ice core records using a calendar age scale. The surface of the piston core corresponds to about 27.5 cm (composite) depth in a neighbour box core, based on a fit of $\delta^{18}O$ records. Average sedimentation rates range from 3.5 to 8.0 cm/1000 years.

For the analysis of $\delta^{13}C_{TOC}$ values about 0.3 g of dried (at 40°C) and weighed samples were gently treated with dilute HCl (2%) for 1h at 40°C to remove inorganic carbon, collected onto pre-ignited (550°C) glass fibre filters, washed to neutrality with double distilled water and dried again. The flow-line combustion system (Deprie, 1983; Erlenkeuser, unpublished) ensured complete dry oxidation of organic carbon (at 960°C for 10 min.) in an excess of high-purity oxygen. Cryogenic distillation was used to purify and dry the evolved CO_2. Nitrogen oxides were removed on hot copper to avoid interference with the C-isotopes in the mass spectrometer. The pressure of a known volume of CO_2 gas was used to calculate the TOC content with a relative precision better than 2%.

The isotopic composition of carbon dioxide, analyzed with a Finnigan MAT 251 mass spectrometer, was measured relative to a working standard and calibrated against the PDB-reference scale through international isotope standards. The isotope results are expressed in the usual notation $d = (Rs/Rst - 1) * 1000$, with Rs and Rst denoting the $^{13}C/^{12}C$ ratios in the sample CO_2 and in the international PDB standard, respectively. The total standard deviation (1σ) of $\delta^{13}C_{TOC}$ values is 0.04 ‰.

The spectral analysis (Blackman-Tuckey technique; Hays et al., 1976) of the local seawater PCO_2 record was performed using the computer program of the SPECMAP group (kindly provided by J. Imbrie, adjusted to our computer system by U. Pflaumann, 1991). The data were linearly interpolated at 2 kyr intervals.

RESULTS AND DISCUSSION

The $\delta^{13}C$ record of the organic sediment fraction (Table 1, Figure 2B) depicts two different modes of variation over the last 330,000 years. Smooth oscillations between -18.5 and -20.2/-20.7‰ have characterized the last 140,000 years, whereas short-term and high-amplitude (-17.1 to -21.3‰) fluctuations have prevailed from 140,000-330,000 years B.P. Since the peaks of most $\delta^{13}C$ excursions are supported by several data points, we assume that these excursions are real features.

The winter temperatures (Figure 2C) that were employed for deducing the temperature-sensitive CO_2 solubility coefficient (a), varied between 24-25°C during warm stages and 15-18°C during cold stages, except for stage 8, where they fell to 12.5°C. The U^K_{37} -based summer SST record (Figure 2C), which is used for comparison, varied between minimum values of 22.5°-24°C in cold stages 2, 4, 6, and 8 and maximum values of 25.7-28.3°C in warm stages 3, 5, 7, and 9. The upper parts of cold stage 6 are characterized by surprisingly warm surface waters. The $\delta^{13}C$values of the surface dweller *G. ruber* (white) (Figure 2D), which reflect the local fluctuations of DIC, vary by about 1 ‰ over the last 250,000 years and by up to 1.3 ‰ prior to this time, with high values mostly linked to warm stages and low values to cold stages (Figure 2A).

Figure 3A displays the main result of this study, the paleo-PCO_2 record of surface waters in the East Atlantic equatorial upwelling zone (during winter), together with the modern range of annual PCO_2 variations after Oudot et al. (1987). The high outlier value at the very sediment surface can be ignored as a result of incomplete early diagenesis (Westerhausen, 1992). The average range of Holocene PCO_2 values (385 ppmv) matches the values measured in modern sea water. During the Last Glacial Maximum (LGM; and during early stage 3 and stage 4), the CO_2 pressure decreased by about 100 ppmv to 300 ppmv and to even lower values during earlier cold stages 6, 7.4, and 8. PCO_2 maxima reaching up to 440 ppmv marked various warm stages.

Over the last 135,000 years the variations in local PCO_2 basically parallel the atmospheric pCO_2 changes as recorded in ice cores (Figure 3B; Barnola et al., 1987; time scale of ice record adjusted according to Petit et al., 1990). This applies to all parts of the local PCO_2 record except for stage 5.1 (r=0.61; r=0.74 without stage 5.1; Figure 4A). The local PCO_2 record (thin line in Figure 3A) derived from U^K_{37} -based summer SST differs markedly from this record. Probably, the $\delta^{13}C_{TOC}$ signal is largely produced during the cold season, especially during cold climatic stages, when upwelling was strongly enhanced.

Figure 3C depicts the difference between local surface water PCO_2 and atmospheric CO_2, a central question of this study. The glacial-to-interglacial changes in local PCO_2 clearly exceeded the shifts in atmospheric CO_2 by about 20-60 % (during glacial terminations) and therefore contributed significantly to the atmospheric CO_2 balance. The net outgassing

Table 1: Carbon isotopes of organic sediment fraction (‰ vs PDB) versus sediment depth and age.

depth (m)	age (ka)	$\delta^{13}C_{TOC}$ (‰vsPDB)
0.005	0.18	-19.81
0.020	0.73	-18.95
0.050	1.81	-18.77
0.100	3.63	-19.04
0.150	5.44	-18.87
0.200	7.05	-19.10
0.250	8.81	-19.22
0.300	11.30	-19.42
0.445	17.90	-20.13
0.545	19.61	-19.62
0.645	21.47	-19.33
0.745	23.34	-19.63
0.845	25.21	-19.62
0.945	27.07	-19.14
1.045	28.94	-18.91
1.145	31.15	-19.11
1.245	33.50	-19.18
1.445	38.23	-19.17
1.545	40.59	-19.91
1.645	42.95	-19.01
1.745	45.31	-19.65
1.845	47.67	-19.70
1.945	50.02	-18.51
2.045	52.38	-18.85
2.145	54.74	-19.11
2.245	57.56	-18.68
2.345	60.58	-19.37
2.445	63.60	-18.90
2.545	65.26	-19.21
2.645	66.37	-19.11
2.745	67.47	-19.51
2.845	68.57	-19.90
2.945	69.67	-19.92
3.045	70.77	-19.64
3.145	74.04	-19.01
3.245	78.24	-20.08
3.345	82.07	-20.30
3.445	85.73	-20.10
3.545	89.40	-20.62
3.645	91.50	-19.24
3.745	92.92	-19.38
3.845	94.35	-19.69
3.945	95.77	-19.67
4.045	98.75	-18.97
4.145	102.41	-19.57
4.245	105.20	-18.78
4.345	107.64	-18.70
4.445	110.07	-18.76
4.545	111.97	-19.59
4.645	113.64	-20.18
4.745	115.31	-19.67
4.845	116.98	-20.10
4.945	118.66	-19.72
5.045	120.33	-19.67
5.145	122.00	-19.62
5.245	123.38	-20.23
5.345	124.63	-19.26
5.445	126.44	-20.12
5.545	128.50	-19.82
5.645	130.56	-19.69
5.745	132.62	-19.72
5.845	134.68	-19.80
5.945	136.47	-19.49
6.045	138.15	-19.27
6.145	139.82	-18.72
6.245	141.50	-18.40
6.345	142.85	-18.26
6.445	144.07	-17.63
6.545	145.29	-18.00
6.645	146.50	-18.25
6.745	147.72	-18.10
6.845	148.94	-18.51
6.945	151.12	-18.82
7.045	153.72	-19.11
7.145	156.32	-19.75
7.245	161.09	-19.47
7.345	163.97	-19.96
7.445	165.63	-19.50
7.545	167.30	-20.02
7.645	168.62	-19.35
7.745	169.80	-19.69
7.845	170.97	-19.56
7.945	172.15	-19.27
8.045	173.38	-17.84
8.145	174.63	-18.07
8.245	175.88	-17.16
8.345	177.13	-18.57
8.445	178.38	-19.10
8.545	179.63	-20.21
8.645	181.15	-20.16
8.745	182.78	-20.40
8.845	184.42	-20.22
8.945	186.06	-20.83
9.045	187.70	-20.75
9.145	189.33	-20.71
9.245	190.97	-20.74
9.345	192.61	-20.19
9.445	195.73	-19.48
9.545	199.48	-18.88
9.645	203.19	-18.69
9.745	206.89	-19.15
9.875	209.50	-19.56
9.945	210.55	-19.60
10.045	212.05	-19.93
10.145	213.55	-20.25
10.245	215.05	-20.11
10.345	216.73	-20.03
10.445	218.48	-19.47
10.545	220.23	-18.25
10.645	221.98	-17.78
10.745	223.73	-21.68
10.845	225.48	-20.24
10.945	228.10	-20.10
11.045	231.10	-20.37
11.145	234.10	-20.13
11.245	237.10	-20.27
11.345	240.10	-20.41
11.445	241.92	-20.09
11.545	243.24	-18.85
11.645	244.56	-19.68
11.845	247.20	-19.93
11.945	248.62	-20.46
12.045	250.07	-21.28
12.145	251.52	-20.64
12.245	252.97	-21.19
12.345	254.50	-20.61
12.445	256.06	-20.58
12.545	257.63	-19.45
12.645	259.20	-19.74
12.745	260.76	-18.45
12.845	262.33	-17.97
12.945	263.90	-17.59
13.045	265.46	-19.02
13.145	269.00	-19.39
13.245	272.60	-20.08
13.345	276.20	-19.87
13.445	279.80	-20.98
13.545	283.40	-20.01
13.645	287.00	-20.23
13.745	289.40	-19.57
13.845	291.80	-18.89
13.945	294.20	-18.54
14.045	296.60	-20.07
14.145	299.00	-19.29
14.245	300.57	-21.14
14.345	302.14	-20.50
14.445	303.71	-21.26
14.545	305.29	-20.37
14.645	306.86	-19.96
14.745	308.43	-19.43
14.845	310.00	-19.90
14.945	320.00	-20.18
15.145	331.00	-20.83

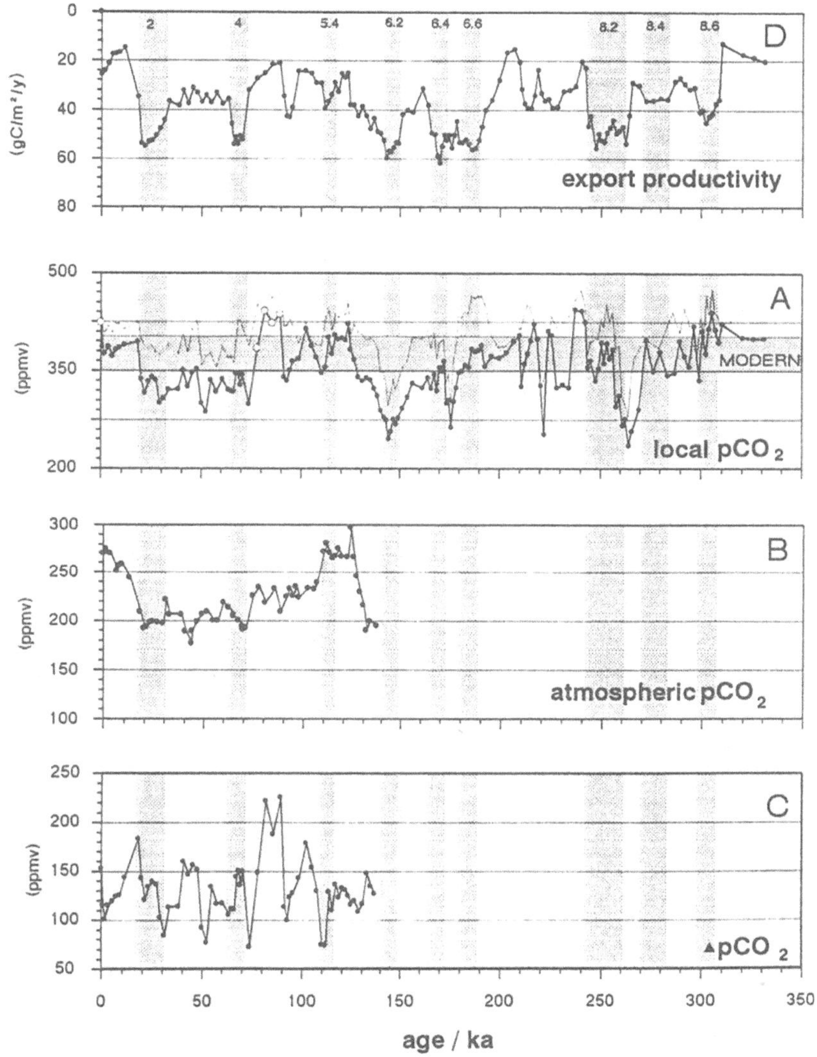

Figure 3: Quaternary fluctuations in surface-water pCO_2 at Site 16772. A) Variations in local pCO2 (ppmv) over the last 330,000 y, calculated from $\delta^{13}C_{TOC}$ normalized to 'SIMMAX' winter temperatures (solid line) and U^{K}_{37} temperatures (thin line), 0 m water depth, and modern DIC (2‰ $\delta^{13}C$). Shaded belt marks range of modern pCO_2 variations in equatorial East Atlantic (Oudot et al., 1987). B) Ice-core record of atmospheric pCO_2 variations (Barnola et al., 1987), chronostratigraphy after Petit et al. (1990). C) pCO_2 difference between local pCO_2 at Site 16772 and global atmospheric pCO_2. D) Export paleoproductivity (on inverse scale!) based on TOC accumulation rates (Sarnthein et al., 1992).

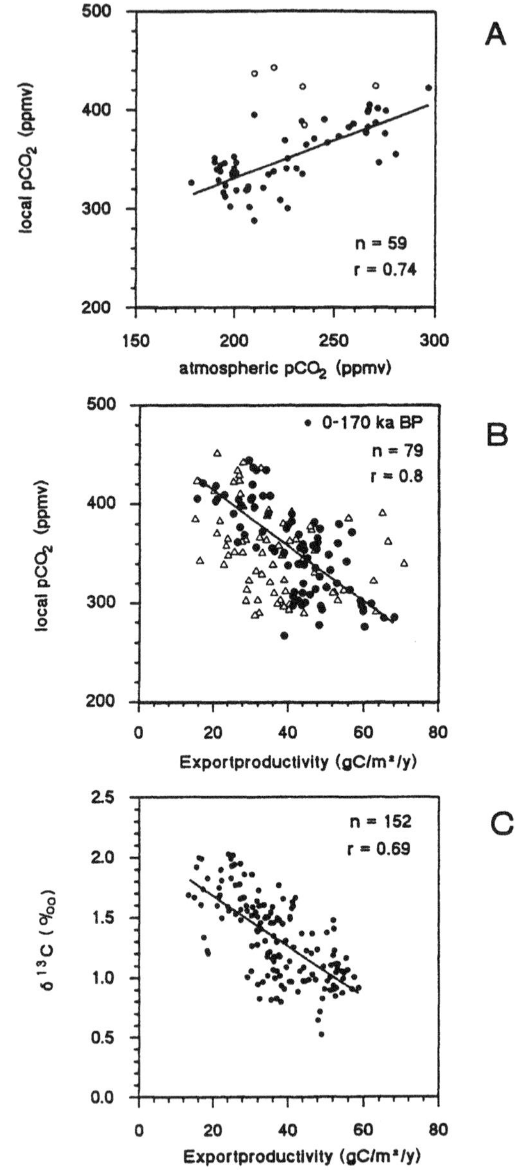

Figure 4: A) Relationship between variations in pCO_2 at the local sea surface and global atmospheric pCO_2 (Barnola et al., 1987). For outlyers (open dots) see Fig. 3A and text. B) Relationship between local pCO_2 and export paleoproductivity over the last 170,000 y (full dots). Correlation coefficient for complete time span (full dots plus open triangles): r=0.7. C) Relationship between export paleoproductivity and $\delta^{13}C$ values of G. ruber (white) recording the nutrient content in surface water 0-330,000 y B.P. (Sarnthein and Winn, 1990). Strictly speaking this use of a parametric statistical technique to test the degree of correlation between two time series is illegitimate. A correct coherency test is published in Struck et al. (1993).

of CO_2 from the equatorial Atlantic ocean to the atmosphere was reduced during peak cold periods (for example, during stages 2, 3, 4, 5.2, 5.4, and 6.4), whereas it was clearly enhanced during deglacial phases such as the onset of terminations 1 and 2, during early stage 3, and especially during early stages 5.1 and 5.3.

In most cases, both the pCO_2 differences and the local PCO_2 itself are maximal during the onset of phases of low plankton production in the equatorial upwelling belt and tend to decrease with increasing productivity (Figure 3D; Sarnthein et al., 1992). A highly significant correlation (r=0.8; parametric correlation legitimized due to a strict in-phase relationship of the two variables at all Milankovitch frequencies; Struck et al., 1993) between the local PCO_2 estimates and paleopaleoproductivity rates over the last 170,000 years (Figure 4B) suggests that indeed the local "biological pump" controls the CO_2 extraction from the surface water, especially the deviations of local PCO_2 from the global atmospheric record. The correlation between local PCO_2 and paleopaleoproductivity is less significant (r=0.7), when calculated for the complete records in core 16772 covering the last 330,000 years (Figure 4B). Thunell et al. (1992) discussed a similar link for the South China Sea; however, they did not test this relationship quantitatively. Our observation is further supported by the persistent lead of the local versus the global atmospheric pCO_2 signal (Figs. 3B and C).

In summary, both the regressions in Figure 4 and the apparent phase relationships described previously suggest that atmospheric pCO_2 changes resulted mainly from a global integration of the various local PCO_2 variations in the low-latitude surface ocean, such as described here and from several other regions (Jasper and Hayes, 1990; Fontugne and Calvert, 1992; Thunell et al., 1992; Schneider et al., this volume). A detailed study of the phase relationships in the frequency domain will be given in Struck et al. (1993). This concept of carbon transfer, which is different from various model calculations (for example, Heinze et al., 1991; Leuenberger and Siegenthaler, 1992; Michel, 1991; Siegenthaler, 1990) applies, at least, for the last 170,000 years.

In principle, the local PCO_2 oscillations in the equatorial surface ocean are a product of both the CO_2 extraction by plankton production (the "biological pump") and the CO_2 outgassing linked to upwelling. Based on our new results, which show that there is a close correlation between local PCO_2 and paleopaleoproductivity, we infer a dominant role of productivity for the carbon transfer between ocean and atmosphere. In a first step, this interpretation is corroborated by a good cross-correlation between high paleopaleoproductivity and low $\delta^{13}C$ values of *G. ruber* (white), which serve as a tracer for enriched nutrients and DIC in the surface waters (r=0.69 for the last 330,000 y; r=0.74 for the last 170,000 y; Figure 4C).

Contrary to many model expectations, however, minimum values of local surface water PCO_2 are confined to times of maximum upwelling and physical outgassing (minimum winter SST; Figure 2C), maximum seasonal temperature contrasts (sensu McIntyre et al., 1989), and to a maximum concentration of upwelling foraminiferal species at Site 16772 (Kähler, 1990).

Thus, CO_2 outgassing may contribute only to a minor extent to changes in the local PCO_2 of the equatorial Atlantic and, hence, its impact on atmospheric pCO_2.

No explanation is available for the observation that the various paleoclimatic records are less well correlated prior to 140,000/170,000 years B.P., when they show generally stronger and more short-term fluctuations (Figs. 2 and 3). Seasonal variations in local upwelling productivity and SST (Figure 1) were possibly different during this time, as may be inferred also from changes in the local seasonal SST records (Figure 3 and McIntyre et al., 1989).

In the spectral domaine (Figure 5) the local PCO_2 values show maximal variance at the narrow frequency bands of 40 ka and 111 ka. Thus the Earth's excentricity and obliquity provide the dominant signals of insolation forcing. On the other hand, precessional forcing, which is characteristic of low latitudes, and which control the trade winds for example (Berger, 1988; McIntyre et al., 1989), can be almost ignored in our record. Only a minor periodicity occurs at 17 ka, which probably represents a slightly aliased signal in the precessional band. The unexpectedly strong 40-ka obliquity signal strongly suggests a forcing of the equatorial paleo-PCO_2 record by high-latitude climate components, a result that is difficult to explain. The 111-ka signal is characteristic of many CO_2 records (Imbrie et al., 1992). A detailed discussion of the various cross-correlations and phase relationships between the paleo-PCO_2 and other proxy

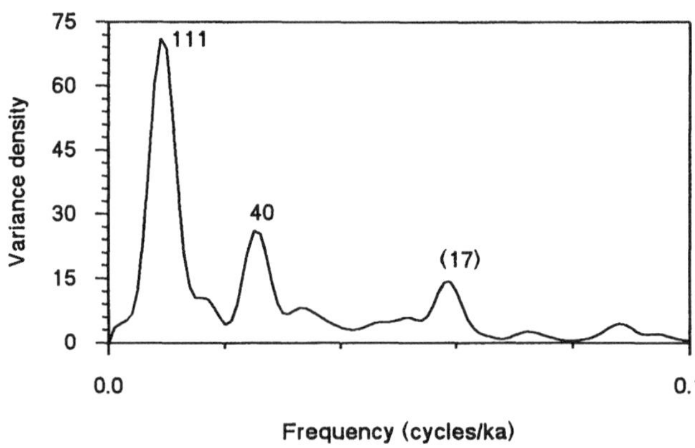

Figure 5: Time spectrum of local PCO_2 variations in the surface water over the last 330,000 years, with dominant cyles indicated in kilo years.

data, such as atmospheric pCO_2, ocean paleopaleoproductivity, nutrients, temperatures, and seasonality in the frequency domain, is given elsewhere (Struck et al., 1993).

CONCLUSIONS

1. The local CO_2 pressure in surface water of the equatorial Atlantic varied between about 230 and 440 ppmv over the last 330,000 years. This range was deduced from $\delta^{13}C$ ratios of sedimentary organic carbon, where the input of terrigenous organic carbon is insignificant (<5%) and the $\delta^{13}C$ ratios were normalized to i) surface water and ii) the DIC as recorded in the $\delta^{13}C$ ratio of the surface dweller G. ruber. The temperature-sensitive CO_2 solubility was assessed using a paleo-SST record for winter, reflecting the upwelling season of maximum plankton production.

2. The local pCO_2 estimates generally paralleled the global atmospheric CO_2 variations. However, the glacial-to-interglacial changes in local pCO_2 exceeded the atmospheric CO_2 shifts by about by 20-60 %. This difference implies a reduced net CO_2 outgassing by the equatorial upwelling belt during peak glacial times, such as during stages 2, 4, 5.4, and 6.4, and an enhanced local CO_2 degassing along with deglacial phases such as during terminations 1 and 2, the earliest stages 3 and 5.1, especially, early stage 5.1.

3. The close correlation between low local pCO_2 and both high local plankton productivity and nutrient concentrations in the surface ocean suggests that the carbon transfer through the "biological pump" dominates over the role of physical outgassing linked to variations in equatorial upwelling over the last 330,000 years. On the contrary, high paleo-pCO_2 values mostly parallel phases of low equatorial upwelling as reflected by SST and seasonality records.

4. Orbital forcing in the narrow frequency bands of excentricity (111 ka) and obliquity (40 ka) clearly dominates the local pCO_2 variations.

5. 330,000 to 170,000 years ago the paleoceanographic regime partly differed from the last 170,000 years, with very short-term fluctuations controlling the CO_2 exchange between the atmosphere and the equatorial Atlantic, a system that is less understood yet.

ACKNOWLEDGMENTS

U. Pflaumann kindly provided his recent unpublished SIMMAX 11 transfer formula for recalculating the paleotemperature record of core 16772 from the planktonic foraminiferal counts of Kähler (1990). The manuscript strongly improved from the careful comments of S. Calvert and M. Fontugne. This study was generously funded by the German National Climate Project (07KF021) and the EPOCH project of the European Community.

REFERENCES

Bard E, Hamelin B, Fairbanks RG, Zindler A,(1990) Calibration of the 14C timescales over the past 30,000 years using mass spectrometric U-Th ages from Barbados corals, Nature, 345, 405-410.

Barnola JM, Raynaud D, Korotkevich YS, Lorius C,(1987) Vostok ice core provides 160,000-year record of atmospheric CO_2, Nature, 329, 408-414.

Berger A, (1988) Milankovich theory and climate, Rev. Geophys. 26, 624-657.

Degens ET, Guillard RRL, Sackett WM, Hellebust JA (1968) Metabolic fractionation of carbon isotopes in marine plankton.-I. Temperature and respiration experiments, Deep-Sea Res., 15, 1-9.

Deprie E, (1983) Aufbau und Erprobung einer Kleinverbrennungs-anlage zur Untersuchung der stabilen Kohlenstoffisotope in organischem Material, Thesis, Univ. Kiel, 85 pp.

Eglinton G, Bradshaw SA, Rosell A, Sarnthein M, Pflaumann U, Tiedemann R, (1992) Molecular record of secular sea surface temperature changes on 100-year timescales for glacial terminations I, II and IV, Nature, 356, 423-426.

Fairbanks RG, Wiebe PH, Bé AWH, (1980) Vertical distribution and isotopic composition of living planktonic foraminifera in the western North Atlantic, Science, 207, 61-63.

Fischer, G (1991) Sedimentation organischen Kohlenstoffs im östlichen Atlantik und die Veränderung des $\delta^{13}C$-Signals in der Wassersäule, Berichte, Fachber. Geowissenschaften, Universität Bremen, 16, 26-29.

Fischer G, Wefer G (1991) Saisonaler Partikelfluß und das Auftriebs geschehen in einigen Auftriebsgebieten des östlichen Atlantiks, Berichte, Fachber. Geowissenschaften, Universität Bremen, 16, 29-31.

Fontugne MR, (1983) Les isotopes stables du carbone organique dans l'océan. Application à la paléoclimatologie, Thèse Doctorat d'Etat Université de Paris XI, 224 pp.

Fontugne MR, Descolas-Gros C, Billy de G, (1991) The dynamics of CO_2-fixation in the Southern Ocean as indicated by carboxylase activities and organic carbon isotopic ratios, Mar. Chem., 35, 371-380.

Fontugne MR Calvert SE, (1992) Late Pleistocene variability of the carbon isotopic composition of organic matter in the eastern Mediterranean: Monitor of changes in carbon sources and atmospheric CO_2 concentrations, Paleoceanography, 7(1), 1-20.

Ganssen G, Sarnthein M (1983) Stable-isotope composition of foraminifers: The surface and bottom water record of coastal upwelling, In E. Suess J. Thiede (eds.): Coastal Upwelling. Its Sediment record, A.-Plenum, New York, 99-121.

Gorshkov SG, (1979) World Ocean Atlas, vol. 2, Atlantic and Indian Oceans, Ministry of Defense USSR Navy, Pergamon, New York.

Hays JD, Imbrie J, Shackleton NJ (1976) Variations in the Earth's orbit: Pace maker of the Ice ages, Science, 194(4270), 1121-1132.

Heinze C, Maier-Reimer E, Winn K, (1991) Glacial reduction by the world ocean-experiments with the Hamburg carbon cycle model, Paleoceanography, 6, 395-430.

Imbrie J, Hays JD, Martinson DG, McIntyre A, Mix AC, Morley JJ, Pisias NG, Prell WL, Shackleton NJ, (1984) The orbital theory of Pleistocene climate: Support from a revised chronology of the marine $\delta^{18}O$ record, In: A.L. Berger et al. (eds.): Milankovitch and Climate, I, D. Reidel Publ. Co., Dordrecht, 269-305.

Imbrie J, Boyle EA, Clemens SC et al. (1992) On the structure and origin of major glaciation cycles. 1. Linear responses to Milankovitch forcing, Paleoceanography 7,701-738.

Jasper JP, Hayes JM, (1990) A carbon isotope record of CO_2 levels during the Late Quaternary, Nature 347, 462-464.

Kähler G, (1990) Oberflächentemperaturen im äquatorialen Atlantik während der letzten 330 000 Jahre (Meteor-Kern 16772), Unpubl. M. Sc. Thesis, Univ of Kiel, 102 pp.

Leuenberger M, Siegenthaler U, (1992) Ice-age atmospheric concentration of nitrous oxide from an Antarctic ice core, Nature, 360, 449-451.

Lutze GF et al., (1988) Bericht über die Meteor-Fahrt 6-5 Dakar-Libreville 15.1.-16.2.1988, Berichte, Geol. Paläont. Inst. Univ. Kiel, 22, 60 S.

McIntyre A, Ruddiman WF, Karlin K, Mix AC, (1989) Surface water response of the equatorial Atlantic Ocean to orbital forcing, Paleoceanography, 4, 19-55.

Martinson DG, Pisias NG, Hays JD, Imbrie J, Moore TC, Shackleton NJ, (1987) Age dating and the orbital theory of the ice ages: development of high resolution 0 to 300,000-year chronostratigraphy, Quatern. Res., 27, 1- 29.

Michel E, (1991) L'océan au dernier maximum glaciaire: La cycle du carbone et la circulation. Contraintes isotopiques et modelisation, Thèse Univ. de Paris-Sud, Centre d'Orsay, 191 pp.

Oudot C, Andrie C, Montel Y, (1987) Evolution du CO_2 océanique et atmos phérique sur la période 1982-1984 dans l'Atlantique tropical, Deep-Sea Res., 34, 1107-1137.

Petit JR, Mounier L, Jouzel J, Korotkevich YS, Kotlyakov VI, Lorius C, (1990) Palaeoclimatological and chronological implications of the Vostok core dust record, Nature, 343, 56-58.

Pflaumann U, (1991) Temperaturreaktionen des nordäquatorialen Atlantiks auf globale Klima-Anfachung während der letzten 750.000 Jahre, Akademie d. Wiss u.d.Liter. Mainz, Abh. Math.-Naturw. Kl. (B Frenzel, Edit.), 177-196.

Pflaumann U, Duprat J, Pujol C, Labeyrie LD, (1993) Simmax, a transfer technique to deduce sea surface temperatures from planktonic foraminifera - the EPOCH approach, Paleoceanography, (in press)

Popp BN, Takigiku JM, Hayes JM, Louda JW, Baker EW, (1989) The post-paleozoic chronology and mechanism of ^{13}C depletion in primary marine organic matter, Am. J. Sci., 289, 436-454.

Poynter J, (1989) The recognition of paleoclimatic signals in organic geochemical data, Ph. D. Thesis, Univ. of Bristol, U.K.

Prahl FG Wakeham SG, (1987) Calibration of unsaturation patterns in long-chain ketone compositions for paleo-temperature assessment, Nature, 330, 367-369.

Prell WL, Imbrie J, Martinson DG, Morley JJ, Pisias NG, Shackleton NJ, Streeter HF, (1986) Graphic correlation of oxygen isotope stratigraphy: Application to the late Quaternary, Paleoceanography, 1, 137-162.

Quay PD, Tilbrook B, Wong CS, (1992) Oceanic uptake of fossil fuel CO_2: Carbon-13 evidence, Science, 256, 74-79.

Rau GH, Takahashi T, Des Marias DJ, (1989) Latitudinal variations in plankton $\delta^{13}C$: implications for CO_2 and productivity in past oceans, Nature, 341, 516-518.

Rau GH, Froelich PN, Takahashi T, Des Marais DJ, (1991) Does sedimantary organic $\delta^{13}C$ record variations in Quaternary ocean ($CO_{2(aq)}$)?, Paleoceanography, 6, 335-347.

Rau GH, Takahashi T, Des Marais DJ, Repeta DJ, Martin JH (1992) The relationship between $\delta^{13}C$ of organic matter and ($CO_{2(aq)}$) in ocean surface water: Data from a JGOFS site in the northeast Atlantic Ocean and a model, Geochim. Cosmochim. Acta. 56, 1413-1419.

Sackett WM, Eadie BJ, Exner ME, (1974) Stable isotope composition of organic carbon in recent Antarctic sediments, In: B. Tissot F. Briemer (eds.): Advances in organic geochemistry. Paris. Technip. (Actes du 6'congrès international de géochimie organique), Rueil-Malmaison, France, 661- 671.

Sarnthein M, Pflaumann U, Ross R, Tiedemann R, Winn K, (1992) Transfer functions to reconstruct ocean paleoproductivity, a comparison, In: Evolution of upwelling systems since the Miocene, Summerhayes CP, Prell WL, Emeis K., editors, Geol. Soc. Spec. Publ., Blackwell, London, 411-427.

Sarnthein M, Winn K (1990) Reconstruction of low and middle latitude export productivity, 30,000 years B.P. to Present: Implications for global carbon reservoirs, In: ME Schlesinger, editor, Climate- Ocean Interaction, Kluwer Acad. Publ., 319-342.

Schneider R, Müller PJ, Ruhland G, (1993) Late Quaternary pCO_2 variations in the Angola current: Evidence from the organic carbon $\delta^{13}C$ and alkenone temperatures, this volume.

Siegenthaler U, (1990) Carbon-14 in the oceans, In: P Fritz JC Fontes, editors, Isotope Geochemistry, Volume 3, Elsevier, 75-138.

Struck U, Sarnthein M, Erlenkeuser H, Westerhausen L, (1993) Ocean-Atmosphere Carbon Exchange: Impact of the 'Biological Pump' in the Atlantic equatorial upwelling belt over the last 330,000 years, Palaeogeogr., Palaeoclimatol., Palaeoecol., 103, 41-56.

Tans PP, Fung IY, Takahashi T, (1990) Observational constraints on the global atmospheric CO_2 budget, Science, 247, 1431-1438.

Thunell RC, Miao Q, Calvert SE, Pedersen TF, (1992) Glacial-Holocene biogenic sedimentation patterns in the South China Sea: productivity variations and surface water pCO_2, Paleoceanography, 7(2), 143-162.

Weiss RF, (1974) Carbon dioxide in water and seawater: The solubility of a non-ideal gas, Mar. Chem., 2, 203-215.

Westerhausen L, (1992) Organische Sedimente im Ostatlantik: Einflüsse von Herkunft, Transportmustern, Diagenese und Klimaschwankungen, Berichte-Reports, Geol.-Paläont. Inst. Univ. Kiel, 48, 109 pp.

Westerhausen L, Poynter J, Eglinton G, Erlenkeuser, K Sarnthein M, (1993) Marine and terrigenous origin of organic matter in modern sediments of the equatorial east Atlantic: The $\delta^{13}C$ and molecular record, Deep- Sea Res., 40(5),

Winn K, Sarnthein M, Erlenkeuser H, (1991) $\delta^{18}O$ stratigraphy and age control of Kiel sediment cores in the East Atlantic, Berichte-Reports, Geol. Paläont. Inst. Univ. Kiel, 45, 1-99.

PART IV:

GEOCHEMICAL AND MICROPALEONTOLOGICAL INDEXES OF PALEOPRODUCTIVITY

PALEOPRODUCTIVITY: FLUX PROXIES VERSUS NUTRIENT PROXIES AND OTHER PROBLEMS CONCERNING THE QUATERNARY PRODUCTIVITY RECORD

W. H. Berger, J. C. Herguera, C. B. Lange, and R. Schneider[1]
Scripps Institution of Oceanography
University of California, San Diego
La Jolla, California 92093-0215

ABSTRACT. There are three fundamentally different proxies recording ocean productivity in sediments: those representing flux (that is, export production), those representing nutrient concentrations (e.g., nitrate or phosphate), and those representing aspects of the trophic structure of the pelagic environment. Flux proxies disagree due to different power-law relationships to the source term. The particular power-laws (expressed in terms of proxy exponents) change as a function of geography, and may change as a function of time. Modification of the accumulation rate by changing preservation on the sea floor (and within the sediment) is an important aspect of this problem.

Combining the use of flux proxies and nutrient proxies makes it possible to estimate the magnitude of physical processes such as mixing and upwelling, and the positive feedback from nutrient buildup in the thermocline. These effects are subsumed into a parameter [v] in the equation Flux = nutrients x [v]. A study of data from the western equatorial Pacific suggests that mixing, upwelling and subsurface nutrient buildup were greatly increased in glacial Stage 6. Diatom abundance does not go parallel to either the flux or the nutrient proxy. Paradoxically, it shows reverse fluctuations. Silicate starvation during glacials may be indicated.

INTRODUCTION: SURPRISE ON ONTONG JAVA PLATEAU

It is generally agreed that productivity in the equatorial Pacific was higher during glacials than during interglacials, as first proposed by Arrhenius (1952). Much of the evidence for this consensus comes from the eastern Pacific (e.g., Pedersen, 1983; Pisias and Prell, 1985; Rea et al., 1986; Arrhenius, 1988; Lyle et al., 1988), but the same pattern prevails in the western Pacific also. An increased rate of glacial carbonate deposition was postulated on the

---

[1] Universität Bremen, Fachbereich Geowissenschaften

NATO ASI Series, Vol. I 17
Carbon Cycling in the Glacial Ocean:
Constraints on the Ocean's Role in Global Change
Edited by R. Zahn et al.

basis of spectral analysis of oxygen isotope records from Ontong Java Plateau, to explain discrepancies between orbital input spectrum and (distorted) output spectra in the sediments (Schiffelbein and Dorman, 1986).

Recently, we have argued that the abundance (or more precisely, the flux) of benthic foraminifers should be a good productivity index (Herguera and Berger, 1991; Herguera, 1992; Burke et al., 1993). These protists depend on the fallout of organic flux which reaches the sea floor from the photic zone (e.g., Berger and Diester-Haass, 1988; Altenbach and Sarnthein, 1989; Loubere, 1991). The accumulation rate of benthic foraminifera in Core ERDC 113P (Ontong Java Plateau) indeed varies in the fashion expected from the hypothesis of high glacial productivity, during the last 250,000 years (Fig. 1a). (For locations of cores see Fig. 2).

Surprisingly, the diatom abundance record does not agree with the late Pleistocene productivity pattern established from the benthic record. From the generally strong positive correlation between biogenic silica flux and productivity (e.g., Barron and Baldauf, 1989), we would expect increased diatom abundance during glacial intervals. A detailed study of diatom abundances and preservation in ODP Hole 806B (Lange and Berger, 1993) shows that diatoms tend to be less plentiful (and more dissolved) during cold periods than during warm ones (Fig. 1b). The range of variation is such (note log scale) that counting errors and interference from fluctuations in sedimentation rates become insignificant. Within the eccentricity band (85 to 115 ky) the correlation between diatom abundance and sealevel is very striking (involving about 50% of the amplitude) (Fig. 1c). Also, there is a small but distinct shift in phase, with the diatom fluctuations leading the changes in sea level. Perhaps the most unexpected and intriguing finding is the presence of layers barren of diatoms, within Stage 6 (Fig. 1b, between 150 and 170 ky, marked B). This pattern is repeated during Stages 14, 16, 18 and 22 further down in the record (Lange and Berger, 1993).

There have been previous indications that diatom abundance fails as a productivity proxy in the western equatorial Pacific, from box cores on Ontong Java Plateau (Mikkelsen, 1978; Herguera, 1992). (However, in the short cores discussed by these authors the possibility arises that diagenesis is a major factor, that is, that glacial-age diatoms have been partly re-dissolved within the sediment, while the younger material is as yet better preserved.) In the eastern equatorial Pacific, also, there are suggestions that opal abundance does not track productivity very well. This may be one reason for the very high carbonate values in glacial-age sediments there (Berger, 1992).

It is difficult to see how diagenesis could produce the patterns in the data under discussion (Fig. 1b). The reversed diatom abundance pattern in late Pleistocene sediments in the western equatorial Pacific (the "opal paradox") sets a sign that respectable productivity proxies may, under certain conditions, give the wrong result when interpreted in conventional ways. This invites us to rethink the relationship between productivity proxies and their target variable, going beyond mere correlations in an attempt to find underlying processes.

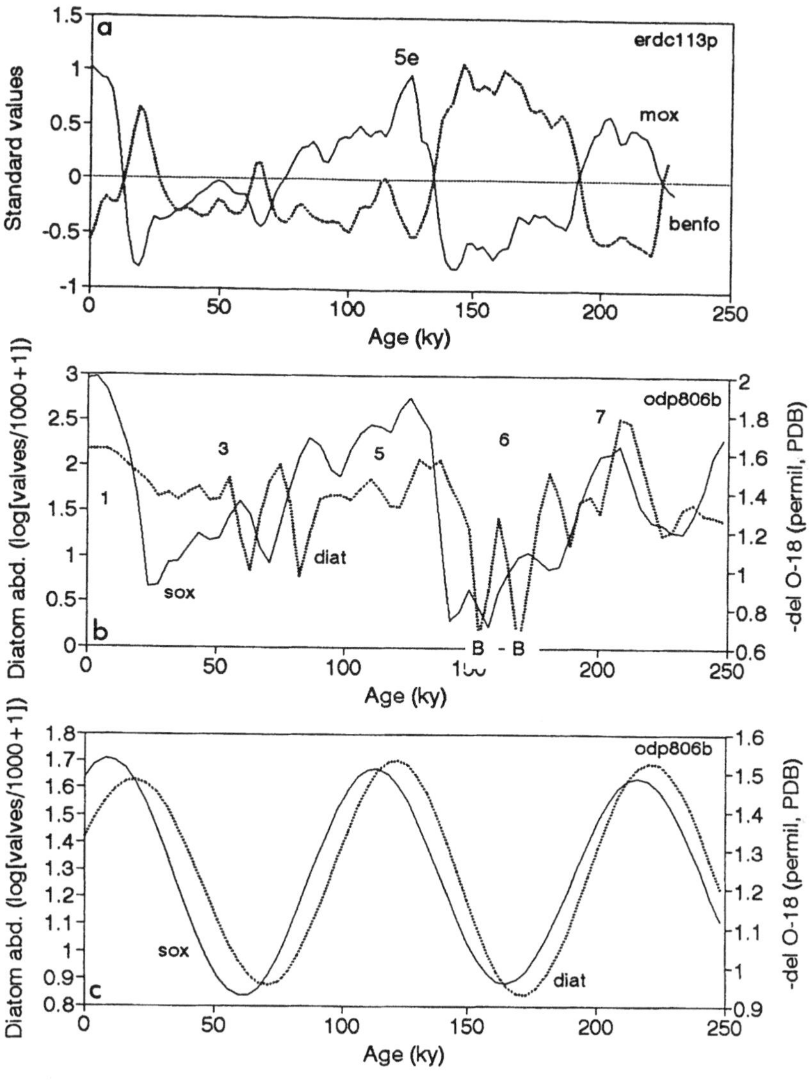

Figure 1 Comparison of proxies of productivity with proxies of sea level. Age scale from setting isotopic substage 5e to 126 ky. a) benfo: rate of accumulation of benthic foraminifers (standardized; i.e., de-meaned and divided by twice the standard deviation); mox: mean values of $\delta^{18}O$ records for *G. sacculifer*, *Pulleniatina*, and *P. wuellerstorfi* (standardized); data from Herguera (this volume). Core ERDC113P, Ontong Java Plateau, 1°38.1'S, 159°13.2'E, 2158 m. b) sox: $\delta^{18}O$ of *G. sacculifer*; diat: diatom abundance; data from Lange and Berger (1993). Note logarithmic scale on diatom abundance spanning more than 2 orders of magnitude (hence no corrections for sedimentation rate). c) Fourier components (85 ky to 115 ky) of sox and diat in Fig. 1b.

Here we review some basic concepts used in the reconstruction of productivity and point out open problems, employing illustrations from recent studies. For a specific exercise in proxy analysis we use examples from the western equatorial Pacific. In this exercise we illustrate a new method for extracting the intensity of physical mixing (that is, a measure of wind speed) from combining the information from two kinds of productivity proxies, one representing flux and the other nutrient concentration.

THE BASIC QUESTIONS

We wish to know how productivity in past oceans was different from today's patterns, and why different proxies give different answers to this question. As far as glacial-to-postglacial contrast (which is the "present" that is widely used as a key to the more distant past), the consensus is as follows (Suess and Thiede, 1983; Thiede and Suess, 1983; Berger et al., 1989a; Bleil and Thiede, 1990; Stein, 1991; Summerhayes et al., 1992):

1. Productivity was increased during glacials in many parts of the ocean margins, that is, above the upper slope, the shelf being exposed (e.g., Prell and Curry, 1981; Sarnthein et al., 1987, 1988; Lapenis et al., 1990). The prime example of increased productivity in such a coastal-ocean setting is the region off NW Africa (Diester-Haass et al., 1973; Müller and Suess, 1979; Sarnthein et al., 1982). There may be striking exceptions to this pattern (e.g., on the Walvis Ridge: Diester-Haass 1985; off Peru: Schrader and Sorknes, 1991).

2. Upwelling was increased along the equator, both in the Pacific and in the Atlantic (e.g., Pedersen, 1983; Stabell, 1986; Pokras, 1987; Lyle, 1988; Lyle et al., 1988; Mix, 1989a,b; Berger and Herguera, 1992). Increased trade winds are responsible, due to increased temperature gradients (e.g., CLIMAP Project Members, 1981; Leinen and Sarnthein, 1989).

3. Productivity in the Southern Ocean was increased (to help explain the decreased glacial pCO_2; Keir, 1990), or decreased (Mortlock et al. 1990), or about the same (see Berger and Wefer, 1991, for review). The main change in the glacial seems to be a shift of the productive zone equator-ward (Mackensen et al., this volume).

4. Productivity in the central gyres was about the same, or perhaps somewhat lowered, due to increased competition for nutrients in the coastal ocean (Sarnthein et al., 1988).

These various notions are derived from a number of different proxies which include the following: (1) flux rates of organic carbon, of opal (diatoms and radiolarians), of carbonate, and of phosphatic materials; (2) concentrations of the above within bulk sediment, or within sediment fractions (when accumulation rates are poorly known); (3) species assemblages of planktonic and benthic remains; (4) carbon and nitrogen stable isotopes in fossils and in organic matter, and elemental ratios such as Ge/Si; (5) biomarkers (ratios of closely related organic compounds); (6) trace metals such as Cd, Ba, and Cu, within calcareous or siliceous shells and within the bulk sediment.

It will be noted that there are in essence two types of proxies: (1) flux proxies that represent themselves an important fraction of the material delivered to the sea floor, and (2) nutrient proxies which contain information about the chemistry of the seawater (i.e., nutrient content) in which production took place. It is clear that fundamentally different things about productivity are being recorded by these two sets of proxies.

Since the several proxies do not measure the same thing, they do not necessarily agree on the amplitude of productivity changes inferred. Disagreement is to be expected and indeed is necessary for study of the dynamics of the system. Proxies may record different things at different times. For example, the organic matter in margin sediments derives both from pelagic sources and from shelf by-passing. The proportions can change through time as a function of sea level, and this change will introduce systematic errors into the estimates of "upwelling" history.

Even if we had perfect proxies, there remains one pervasive difficulty: to calibrate proxies we need to know the present patterns of productivity. In fact, we should be able to define precisely what we mean by "productivity". Neither of these requirements can be met at present. Definitions are somewhat vague. "Carbon fixed per unit area per unit time" is good in principle, but "fixed" by whom and for how long? Does this include the microbial loop? In practice, the rate of uptake of radiocarbon under certain experimental conditions, and the extrapolation to ocean conditions, will have to serve for most purposes (for discussion see Bender et al., 1987).

The global productivity maps available (Koblentz-Mishke et al., 1970; Berger et al., 1987) are similar overall, but deviate considerably from each other in detail (Herzfeld and Berger, 1993). Thus, while the general outlines (Fig. 2) are secure, uncertainties of a factor of 1.5 in any one area should be taken for granted. Geologists use such maps as representing the average conditions of a "present" that is relevant for comparison with surface sediments. These, however, contain mixtures of particles whose ages range over a thousand years or more. In contrast, the productivity maps are based on contouring data from a few specific expeditions in any one region, with all the special conditions of particular seasons in particular years.

In addition, there is no guarantee that present-day calibrations will hold up for the past: Productivity is a result of a complicated mix of different factors (mixed layer thickness, frequency of storms, nutrient concentrations, light conditions, seasonal stratification). It is unlikely that one or two proxies can properly reflect such complexity.

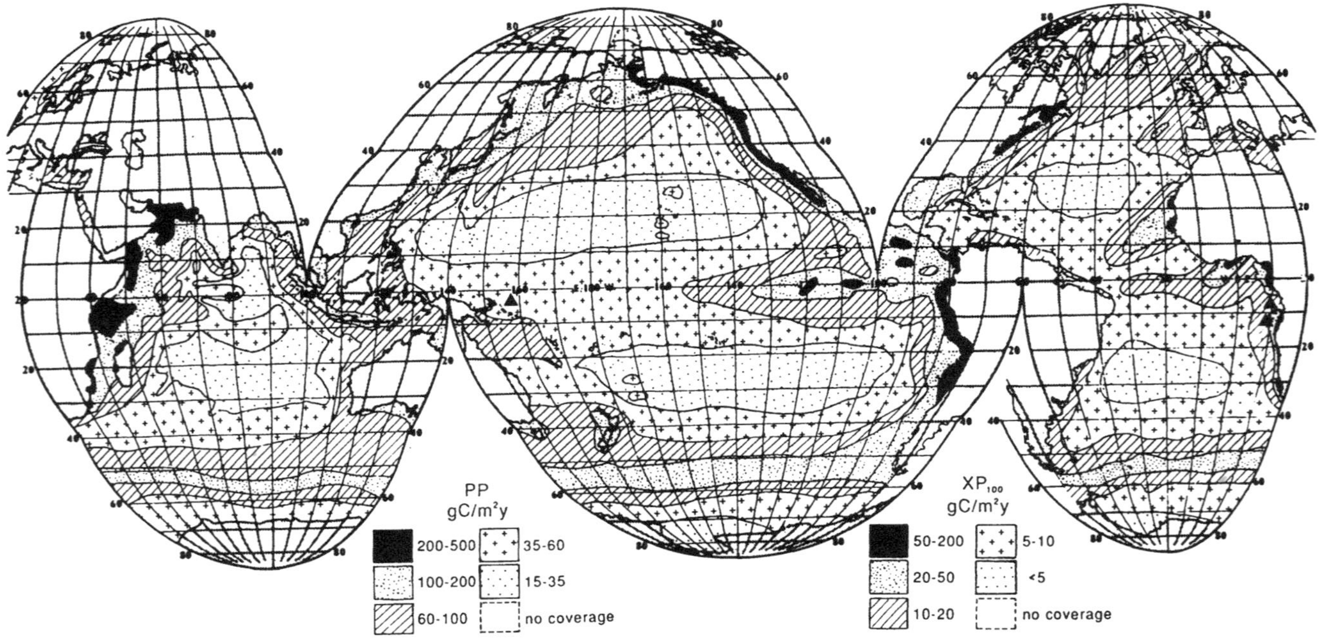

Figure 2 World map of primary productivity (PP) and of export production at 100 m depth (XP_{100}). After Berger et al. (1987), as modified by Berger and Wefer (1991). Position of core locations discussed are shown as black triangles.

EXPORT PRODUCTION

The link from primary production to sedimentary record is through "export production" (Berger et al., 1987). This is the most important productivity-related oceanographic variable (Fig. 2, legend to right). It is the export that contains the materials that will eventually reach the sea floor and hence have a chance to enter the sediment. The concept of export production (that which can be caught in a sediment trap) is closely related to the "new production" of the plankton biologists (the production which results from addition of nitrate from below the photic zone). For steady state, over some sufficiently large area, and at the base of the photic zone, export and new production should balance. This balance is expressed thus:

$$XP = k\,[N] \times [v] \qquad (1)$$

where XP is the rain of organic matter, k is the Redfield ratio, [N] is the concentration of a nutrient, and [v] denotes the rate of advection, mixing, and diffusion to the site of production (loosely referred to as "upwelling" in many geologic reconstructions, and in the following). The export production XP can be reconstructed from flux proxies, [N] from nutrient proxies. The physical mixing parameter then emerges as the ratio of the two.

Exact assignment of XP-values from the sedimentary record is not possible. Because of oxidation, export of organic matter changes with depth of water, according to a function depending on 1/z (Suess, 1980). A simple fit to available data suggests an equation of the form (Berger et al., 1989b):

$$J(z) = 17 \times PP/z + PP/100 \qquad (2)$$

where J(z) is the organic flux at depth z expressed in units of 100 m, and PP is the primary productivity. The equation postulates two types of export components, one decaying with depth, the other resistant to oxidation. The export flux, then, may be taken as 18% of primary production, at 100 m depth, as a first approximation.

However, the proportion of export from surface-near production (the export factor xf) is not constant, but is itself a function of the magnitude of productivity (Eppley and Peterson, 1979). This relationship may be expressed as follows (Berger and Wefer, 1990):

$$xf = 2 \times \sqrt{PP} \qquad (3)$$

where PP is given in gC m^{-2} yr^{-1}. Strong seasonality will yield greater export (as a proportion of total primary production) than will weak seasonality (Platt and Harrison, 1985). An equation relating xf to seasonality (expressed in terms of the number of months it takes to generate one half of the annual production) is given in Berger and Wefer (1990). Strongly pulsed production, according to this formulation, has a percentage output about twice higher than constant production. Pulsing may affect the export of opal more than that of organic carbon, or (with opposite consequences) it may result in the dominance of weakly silicified species whose tests are easily re-dissolved.

FLUX PROXIES

One way to minimize the various problems mentioned is to reconstruct changes in productivity in terms of factors by which the past differs from the present (rather than stating absolute values of flux, which are meaningless unless the base map is known). Another is to use several proxies simultaneously.

A flux proxy much in use is organic matter (Fig. 3a). Another sediment component whose distribution is similar to that of productivity is opal flux (compare Figs. 2 and 3b). Like C_{org}, it is a useful index for productivity in upwelling regions. However, there are important differences between the patterns of the two proxies. For example, C_{org} depends notably on depth, while such a dependency is not in evidence for opal. The most striking feature of the opal map are the large rates of deposition in the Southern Ocean: about one half of all the silica is sequestered in the sediments here. C_{org} does not show such a preference for the Southern Ocean.

There are areas where fluctuations in C_{org} and opal agree very well. For example, there is excellent concordance in the region of the Congo Fan (Fig. 4a). The area is characterized by coastal upwelling, resulting in robust siliceous frustules (*Thalassionema nitzschoides*, *Chaetoceros* spores) with a high preservation potential. A substantial input of riverine silicate is suggested (at least during warm periods) by the abundant occurrence of freshwater diatoms in postglacial sediments. Both proxies indicate an increase in upwelling during glacial periods (except Stage 4), with a strikingly high spike in Stage 2. The amplitude of C_{org} accumulation is typically a factor of somewhat over 2, which translates into a factor of productivity amplitude of 1.5, by the "0.7-rule" (Berger and Herguera, 1992). This rule, derived from comparing organic carbon content and overlying productivity, estimates changes in paleoproductivity as follows:

$$PaP = k \times C_{org}^{0.7} \qquad (4)$$

which reflects the fact that changes in organic carbon content tend to exceed the changes in productivity that generated them.

It will be noted, in Fig. 4a, that the amplitudes of opal variations surpass those of C_{org} for much of the record. Thus, the exponent on the equation for opal would have to be smaller than that for C_{org}, in equation (4), for this particular environment. The variations in opal content, presumably, reflect chiefly the abundance of diatoms, which is known to vary considerably during the late Pleistocene, in the sense described by Mikkelsen (1984).

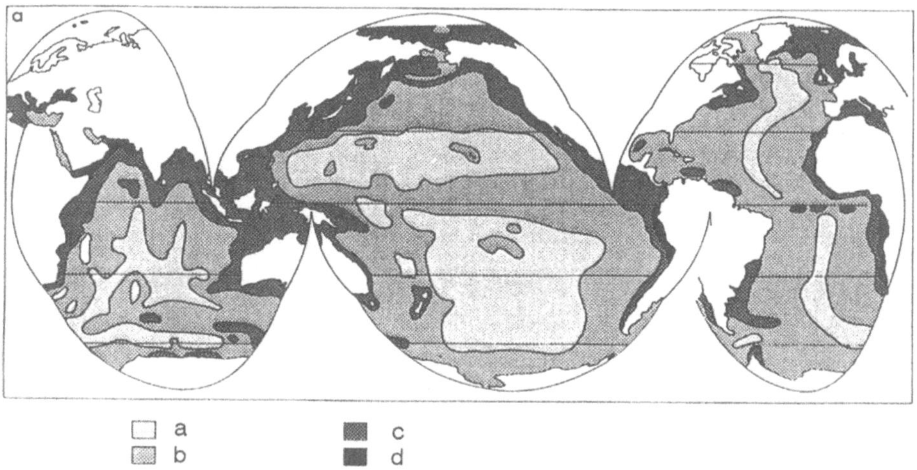

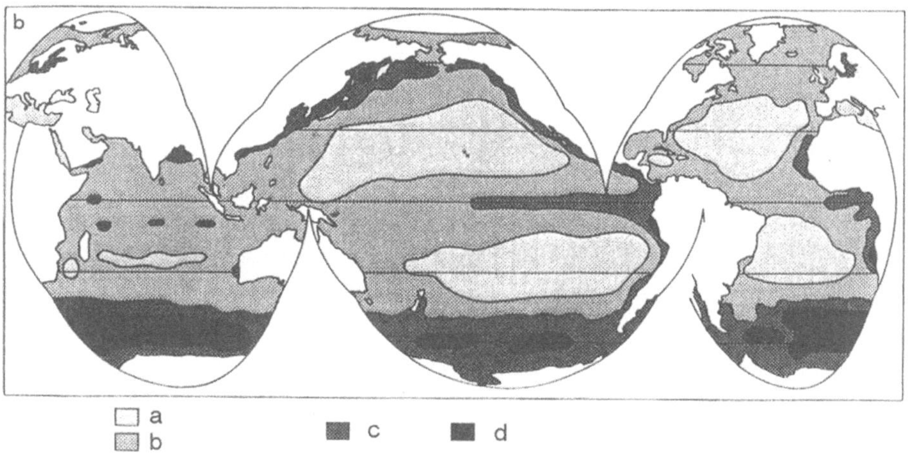

Fig. 3 World maps of productivity proxies. a): C_{org}; categories: a= <0.25%, b= >0.25 and <0.5%, c= >0.5 and <1.0%, d= >1%. b): opal; categories: a= few or no siliceous fossils, b= siliceous microfossils abundant, opal flux < 0.1 g $cm^{-2} ky^{-1}$, c= opal flux > 0.1, d= opal flux > 1 (typically 1 to 2, up to ca. 10 g $cm^{-2} ky^{-1}$). From Berger and Herguera (1992). Prime sources for distributional patterns: Romankevich (1984) and Leinen et al. (1986).

In contrast to the Congo fan area, a core from farther south, off Angola, shows hardly any variation in opal at all. Clearly, productivity in this latter area is not smaller by the same factor that describes the difference in opal accumulation, as is seen when comparing C_{org} fluxes between the two sites. The reason why opal fluxes are so different between these two sites, then, must involve silicate chemistry, that is, silification of diatom frustules and their preservation. We must consider, among other factors, the depths from which upmixing of cold water occurs, and the silica concentrations at those depths (see van Bennekom and Berger [1984] and Jansen and van Iperen [1991], for background).

The observation that a change in proxy flux must be adjusted for exaggeration of a change in productivity in the manner prescribed by equation (4) leads to the definition of a "proxy exponent". We write for "factor", so that

$$PP = F(PP)^{px} \quad (5)$$

where PP is primary productivity, F(PP) is the flux proxy for PP, and px is the proxy exponent. On the basis of deep-sea box core data from western and eastern equatorial Pacific, and from the central Atlantic, Herguera (1992) has compiled a table of factors, which allow us to estimate px for the pelagic environment. The proxy exponent for opal flux comes out near 0.4, and that for benthic foraminifers near 0.5 (Fig. 5), so that

$$PP = F(op)^{0.4} \quad (6)$$

and

$$PP = F(bf)^{0.5} \quad (7)$$

respectively, where F(op) is the accumulation rate of opal and F(bf) that of benthic foraminifers. For a change in the benthic foraminifer flux rate of a factor of two (as for example in the western equatorial Pacific, from glacial to Holocene; Herguera and Berger, 1991), we obtain an estimate for the productivity change of a factor of square root of 2, that is, 1.4 (or slightly greater when using 0.53 for exponent).

Also from the box-core work of Herguera (1992) it emerges that carbonate flux is a poor estimator of productivity change, presumably as a result both of low sensitivity to productivity and of interference from dissolution and winnowing. The line shown in Fig. 5 (corresponding to px=3) indicates that to double carbonate flux we would have to have an eight-fold increase in productivity ($2^{\wedge}xp=2^3$). This line is not based on the data plotted, but the fit has been shifted to take account of the observation that in the regions of lowest productivity (say, 30 gC m^{-2} y^{-1}) the accumulation rate of well preserved calcareous ooze is close to 1 cm ky^{-1}, while in regions of moderately high productivity (say, 150 gC $m^{-2}y^{-1}$) it is near 2 cm ky^{-1} (for an xp between 2 and 3). In the absence of pronounced dissolution, the low sensitivity of carbonate flux to productivity makes it possible to use concentrations of non-carbonate proxies as substitutions for their fluxes (which require knowledge of accumulation rates), for moderate ranges of productivity.

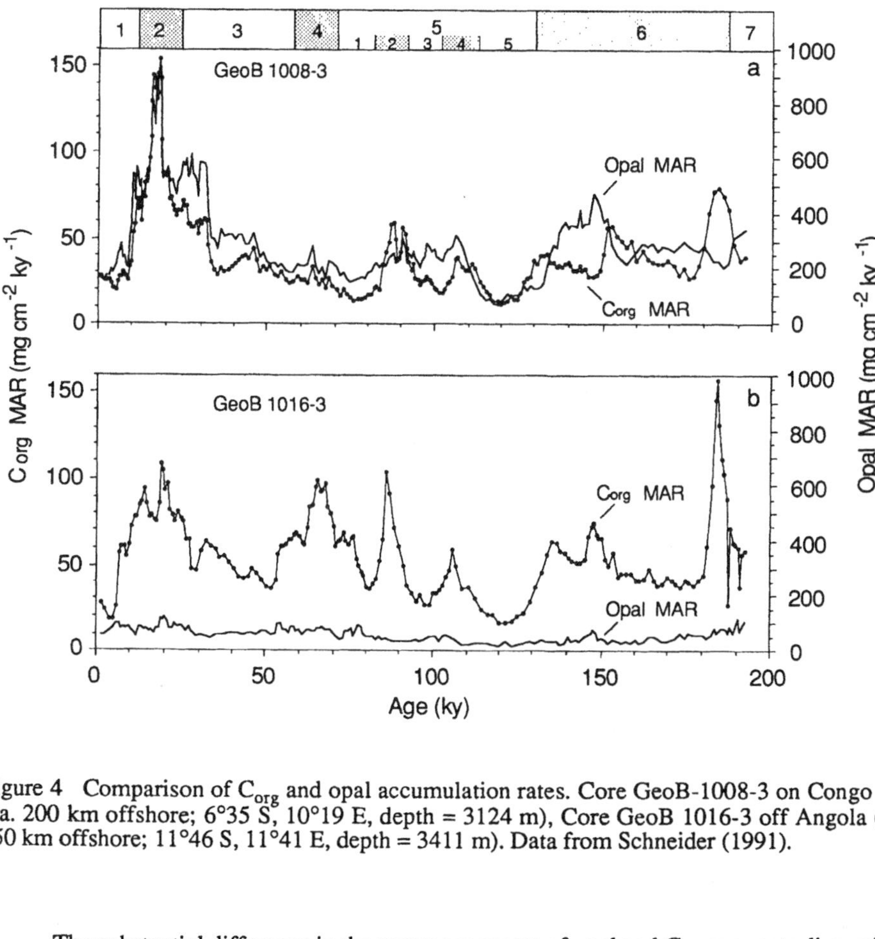

Figure 4 Comparison of C_{org} and opal accumulation rates. Core GeoB-1008-3 on Congo fan (ca. 200 km offshore; 6°35 S, 10°19 E, depth = 3124 m), Core GeoB 1016-3 off Angola (ca. 250 km offshore; 11°46 S, 11°41 E, depth = 3411 m). Data from Schneider (1991).

The substantial difference in the proxy exponents of opal and C_{org} warrants discussion. Apparently this is not a result of differing export factors, as can be ascertained in the trapping data compiled by Milliman and Takahashi (in press) (Fig. 6). Their results show that generally opal export has just a slightly greater sensitivity to productivity change than does C_{org} (exponent of 1.05 on C_{org}). The notable exception is the Antarctic realm, where opal is much more sensitive (exponent of 1.6). Thus, the difference in proxy exponents suggested in Fig. 5 must be due to other effects. We suggest preservation on the sea floor as an important factor. The greater sensitivity of opal, we propose, stems from a greater increase in preservation in opal, during increased delivery, compared with organic carbon. The reason, we submit, is that interstitial waters can reach near-saturation with opal rather close to the sediment-water interface, whereupon dissolution of siliceous tests decreases drastically. Organic matter, on the other hand, has no such limit to decay: after oxygen is used up, reduction of sulfate continues

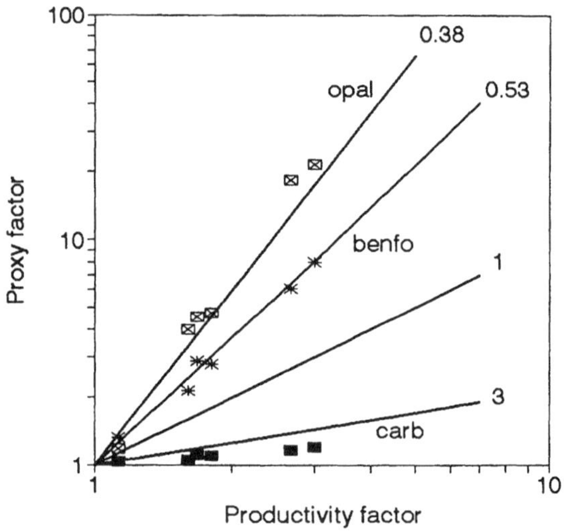

Figure 5 Comparison of proxy factors (the factors by which accumulation rates differ from one site to the next) with productivity factors (for the same sites, in overlying waters). Box-core data from Herguera (1992, Table 2).

the process of destruction (see Bender and Heggie, 1984; Emerson and Hedges, 1988).

In any case, the changing preservation of flux proxies must be counted as a major source of interference with productivity reconstruction. Since so little is preserved of either opal or organic matter, compared with the export, relatively small changes in the efficiency of burial can simulate large changes in the source. For example, a change from one percent to two percent preservation would suggest a doubling of productivity, unless the change is recognized as being due to preservation. These considerations are valid not only for bulk proxies such as opal and C_{org}, but also for shell flux determined by counting abundances of diatoms, radiolarians, or foraminifers. The interference from preservation patterns is another argument for multiple proxy use, since preservation rarely affects different materials in the same fashion.

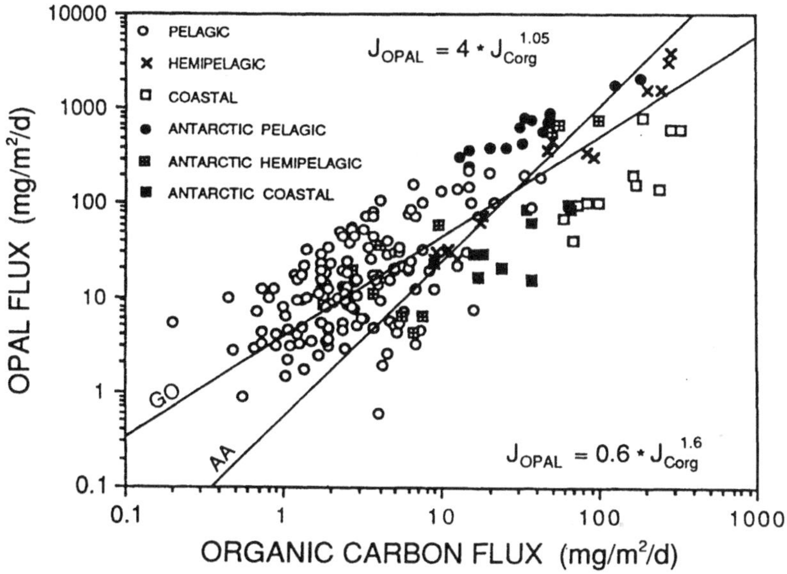

Figure 6 Relationship between opal flux and organic carbon flux in traps from various regions of the ocean, as compiled by Milliman and Takahashi (in press). GO: general ocean; flux of opal increases at a slightly greater factor than flux of organic carbon (exponent of 1.05). AA: Southern Ocean; flux of opal increases by a much greater factor than the flux of organic carbon (although scatter is large). Equations and fit (by eye) added by Berger and Herguera (1992).

The determination of proxy exponents for different geographic settings should prove to be a useful exercise, since it will allow us to find constraints for the different types of productivity and associated preservation patterns. For the past, when productivity is unknown, we can at least assess the differences between the exponents of several proxies.

NUTRIENT PROXIES

The mass flux proxies opal and C_{org} are directly related to export production, being the remnants of that flux within the sediment. However, there are other important proxies which are more closely associated with nutrient concentrations within the productive waters (parameter [N] in equation [1]). Examples, already mentioned, are N/C ratio in C_{org}, $\delta^{13}C$ in foraminifers,

Cd/Ca ratios in calcareous shells, Ge/Si ratios in diatoms, and any other tracer that correlates with phosphate and can be locked into the substance or shell of living organisms.

The best known of these proxies, and the most widely discussed, is the $\Delta\delta^{13}C$ parameter, which measures the difference in $\delta^{13}C$ of surface waters to that of deeper waters. The surface water $\delta^{13}C$ is recorded in shallow-living planktonic foraminifers, while the deep water $\delta^{13}C$ is captured within the shells of benthic foraminifers. The reasoning behind using $\Delta\delta^{13}C$ as a nutrient proxy is as follows (Broecker, 1982): Phosphate concentrations are close to zero in surface waters, but are well above zero in subsurface waters. Thus, the difference in phosphate between surface and deep water is the same as the concentration of phosphate in deep waters. Also, $\delta^{13}C$ at depth is correlated with phosphate: it becomes more negative as phosphate increases in deep waters during oxidation of organic matter (roughly $1^{o}/_{oo}$ for every $1\mu g\ kg^{-1}$; see Kroopnick, 1985). Thus, the difference in $\delta^{13}C$ corresponds to the difference in phosphate (i.e., the concentration in deep water).

Difficulties arise from a number of sources, some having to do with the way foraminifers incorporate the carbon isotopes into their shells, others with oceanographic processes (summarized in Wefer and Berger, 1991). Most planktonic foraminifers have symbionts, which may affect $\delta^{13}C$ of their shells (Spero and DeNiro, 1987). Benthic species show different values, depending on whether they live on or within the sediment (e.g., Zahn et al., 1986). Some phosphate in deep waters (the "preformed" portion not delivered from oxidation at depth) is not correlated with $\delta^{13}C$. Finally, the initial setting of the $\delta^{13}C$ of future deep waters (before the water sinks and feels the effects of adding CO_2 from oxidation) depends on exchange processes with the atmosphere which in turn depend on time of exposure and on temperature (e.g., Charles and Fairbanks, 1990).

The study of productivity variations from $\Delta\delta^{13}C$ in the pelagic record has received much impetus from the discovery of atmospheric pCO_2 fluctuations in polar ice cores (Berner et al., 1980; Delmas et al., 1980; Barnola et al., 1987). A number of mechanisms have been proposed to connect productivity and pCO_2 (see Sundquist and Broecker, 1985; Berger et al., 1989a). They call on changes in the partitioning of total CO_2 between ocean and atmosphere, or on changes in the rates of sequestration of carbon within sediments. One well-known example of the link between productivity and pCO_2 is the parallelism between $\delta^{13}C$ and atmospheric CO_2, here exemplified by data from Ontong Java Plateau (Fig. 7). This correlation was first postulated by Broecker (1982), and shortly after demonstrated by Shackleton et al. (1983).

Initially it was thought that a change in pCO_2 can be calculated directly from a change in $\Delta\delta^{13}C$. This concept, however, is open to doubt for various reasons. There is, for example, the question how representative any one curve (such as the one in Fig. 7) can be for the globally effective difference in $\delta^{13}C$ between surface waters and deep waters. Another question concerns the role of competing controls on pCO_2 that are not reflected in $\delta^{13}C$, especially the changing alkalinity of deep waters (Keir and Berger, 1983; Berger and Spitzy, 1988; Broecker and Peng, 1989; Opdyke and Walker, 1992).

A realistic way to view the $\Delta\delta^{13}C$ curve is to accept it as one of the many productivity proxies that can be extracted from the record. It so happens (for reasons to be found) that such proxies tend to correlate with pCO_2, as do many other proxies of climatic change. An interesting companion proxy to $\delta^{13}C$ is the Cd/Ca ratio in benthic foraminifers (e.g., Boyle and Keigwin, 1987). This parameter closely parallels $\delta^{13}C$ in much of the record studied, but the additional information it offers can reduce uncertainties regarding the initial setting of $\delta^{13}C$ and other related problems. Of course, as a proxy, it also carries its own problems and uncertainties (such as preformed cadmium concentrations).

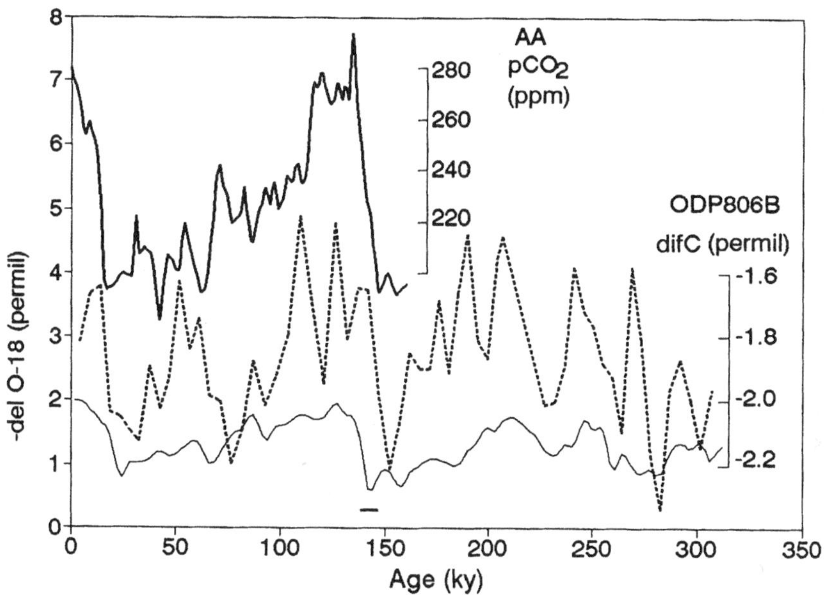

Figure 7 Comparison of atmospheric pCO_2 in ice cores (top, Antarctic ice data Barnola et al., 1987) with the difference between benthic and planktonic $\delta^{13}C$ (difC) in the western equatorial Pacific (Site 806, *P. wuellerstorfi-G. sacculifer*, data Bickert et al., 1993). Bottom: $\delta^{18}O$ curve for guidance (Site 806, *P. wuellerstorfi*, same source). Bar at bottom, to left of 150 ky., indicates the apparent phase shift between $\Delta\delta^{13}C$ and $\delta^{18}O$, with $\delta^{18}O$ lagging the carbon record. From Berger and Wefer (1992).

TROPHIC PROXIES

Besides flux proxies and nutrient proxies there is a third class of estimators for productivity. These are properties of fossil assemblages that can be made to yield information on the environment, including the level of productivity. We call them trophic proxies to distinguish them from the other two sets. Planktonic foraminifers have been used to reconstruct productivity levels in the glacial central Atlantic, yielding evidence for higher productivity of the equatorial system, for example (Mix, 1989a,b). Diatom assemblages change as a function of productivity (Schuette and Schrader, 1979). They were shown to contain clues about greatly increased productivity in the equatorial Atlantic (Pokras, 1987) and in the North Atlantic, during glacial time (Sancetta, 1992), and they have been used to demonstrate interannual fluctuations in upwelling off California (Lange et al., 1990). Schrader (1992), examining the uppermost cores of Hole 680B off Peru, used diatom assemblages to calculate paleoproductivity. He found that his index agrees quite reasonably with C_{org} content.

Species assemblages reflect the total environment of growth of the different organisms represented. Thus, in principle, it should be possible to obtain seasonal information on productivity, and perhaps even on the nature of productivity pulsations. In this connection it is interesting that the living assemblage in the photic zone is not necessarily similar to the fossil one on the sea floor (e.g., Calvert, 1964), which points to the selectivity of the process of information storage (Sancetta, 1989). The use of trophic proxies presupposes a constant relationship, through time, between the particular proxy group (e.g., planktonic foraminifers) and primary producers (dinoflagellates, coccolithophorids, diatoms, etc.). Clearly, there is much room here for "non-analog" situations, as the mix of factors controlling primary productivity changes with changing climatic conditions.

THE TASK OF RECONSTRUCTION

Equation (1) suggests that the reconstruction of changes in productivity involves several tasks: estimation of changes in export production XP, of changes in nutrient concentrations in subsurface waters [N], and of changes in the intensity of physical processes that make these nutrients available [v]. We next turn to an exercise which exemplifies the approach we have in mind.

For basis of reconstruction we use the data for the piston core ERDC113P from Ontong Java Plateau, taken during the Eurydice Expedition (1975) and analyzed by Herguera (this volume). We concentrate on the last 250,000 years. Opal determinations were carried out following the method of Müller and Schneider (1993). The core is well above the lysocline (2158 m). It was previously reported on by Schiffelbein (1984), and by Wu and Berger (1991).

The chief basis for reconstruction are two records: the accumulation rate for benthic foraminifers and the difference between the $\delta^{13}C$ values of *G. sacculifer* and those of *P. wuellerstorfi* ($\Delta\delta^{13}C$[bp]) (Fig. 8). The first is a proxy for the organic matter reaching the sea floor (Herguera and Berger, 1991); it is given in terms of standard values (that is, de-meaned and normalized by 2 standard deviations). The second tracks deep-sea nutrient content, as mentioned; it is shown in terms of scaled values, whereby the $\Delta\delta^{13}C$[bp] for the glacial-Holocene contrast is set to unity. The two records are clearly correlated, as expected if both keep track of aspects of productivity. The large positive excursion between 150,000 and 200,000 years is noteworthy. Also, a phase difference is apparent, with the nutrient proxy leading the flux.

We next scale the benthic record in similar fashion as $\Delta\delta^{13}C$[bp], by setting the glacial-Holocene contrast to a factor of 1.6 (on the conservative side of the range for change in export production shown in Herguera and Berger, 1991). The resulting curve (Fig. 9a) suggests that during Stage 6 export was greater than at present by a factor of about 2. During the same time $\Delta\delta^{13}C$[bp] was increased by a factor of almost 1.25 (log[factor] approaching 0.1) (Fig. 9a). Comparison with the $\delta^{18}O$ record ("mox", that is, mean of values for *G. sacculifer*, *Pulleniatina*, and *P. wuellerstorfi*, standardized) shows a rather nice correspondence of high export (and increased $\Delta\delta^{13}C$[bp]) with glacial conditions.

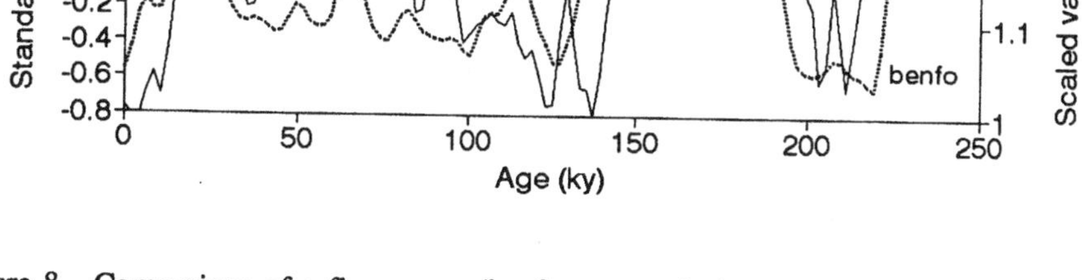

Figure 8 Comparison of a flux proxy (benfo: accumulation rate of benthic foraminifera, standardized) with a nutrient proxy ($\Delta\delta^{13}C$ of *G. sacculifer-P. wuellerstorfi*), scaled to the Holocene-glacial contrast. Core is ERDC113P from Ontong-Java Plateau (see Figs. 1 and 2).

Having in hand the proxies for amplitude of export and nutrients, we can now estimate [v] in equation (1), by forming the ratio between the two (Fig. 9b). This ratio can be read as an index of "upwelling". However, it is not a purely physical parameter: increased upwelling will tend to concentrate nutrients below the surface waters, within a strong oxygen minimum. The nutrient content in this thermocline water is not the same as that reflected in the benthic $\delta^{13}C$ values, so that the $\Delta\delta^{13}C$[bp] is unable to show the effect. The enhancement of productivity through thermocline enrichment, by trapping of nutrients below the upwelling area, should then be contained in the [v] index. Accordingly, the excursion at Stage 6 (150,000 to 200,000 ky) would suggest the presence of a strong oxygen minimum.

To check on the presence of such a minimum, we turn to the isotopic differences between the shallow-living *G. sacculifer* and the deeper living *Pulleniatina* ($\Delta\delta^{18}O$[pp]) (Fig. 9c). In Stage 6 there is a notable minimum in the difference of $\delta^{18}O$ values, with $\Delta\delta^{18}O$[pp] dropping to about one half of the present value (log of factor = -0.3). Excluding effects from differential dissolution at this site (2158 m), this excursion (which suggests that the habitats of *G. sacculifer* and *Pulleniatina* were more similar than now) presumably shows a deepening of the mixed layer. This raises the problem that nutrients would be further removed from the photic zone during the interval. To keep up productivity, therefore (Fig. 9a), higher nutrient content is required within the thermocline. A hint of this is seen in the planktonic $\Delta\delta^{13}C$: it does not follow the downward excursion of the $\Delta\delta^{18}O$[pp]. Clearly, to explain the increased export within Stage 6, upward mixing into the thickened surface layer has to be vigorous (as reflected in the parameter [v]; Fig. 9b). Enhanced upward motions may in fact involve considerable depths along the equator (Semtner and Chervin, 1991). Increased tidal diffusion in the deep sea during glacial time (resulting from the absence of shelves where much tidal energy is otherwise dissipated) could greatly enhance this "deep elevator" transport.

The pelagic record is dominated by cyclic sedimentation with strong orbital forcing (Hays et al., 1976; Imbrie et al., 1984), and the sediments on Ontong Java show this same pattern (Schiffelbein, 1984; Schiffelbein and Dorman, 1986; Berger et al., 1993; Bickert et al., 1993; Schmidt et al., 1993). The obliquity component is especially strong in the carbonate record. To take advantage of this cyclicity in describing the "typical" course of events, we have replotted the productivity-related data therefore as averages within a modulus-41-ky cycle (Fig. 10a). The result is an average 41-ky cycle contained within the record (Fig. 10a). In this plot, for example, a point at 10 ky represents the average value of available data at 10, 51, 92, 133, 174, and 215 ky. (Interpolations were made for steps of 2.05 ky.)

Our "mod-plot" (Fig. 10a) shows that maximum export tends to occur somewhat before the maximum glaciation. It also suggests that the onset of increased export is caused by increased nutrient supply (for example from increased erosion of shelves, as suggested by Broecker, 1982). Subsequently, however, the nutrient content decreases, while the export

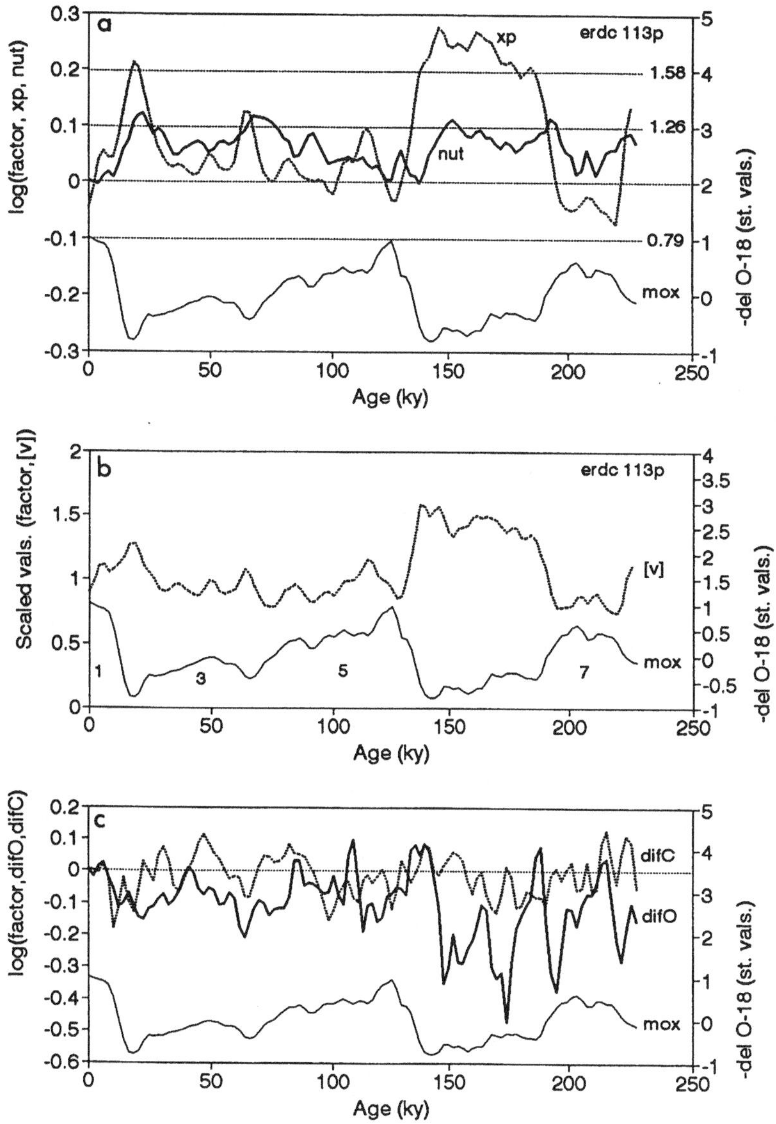

Figure 9 Comparison of productivity proxies in ERDC 113P, Ontong-Java Plateau. a) Proxies for export production xp (benthic foram accumulation) and nutrient content nut ($\Delta\delta^{13}C$[bp] see Fig. 8), both scaled to the Holocene-glacial contrast. mox: proxy for sea level (see Fig. 1). b) Proxy for "upwelling" [v], as the ratio between xp and nut. c) Proxies for mixed layer thickness (difO: $\Delta\delta^{18}O$ *G. sacculifer-Pulleniatina*) and oxygen minimum (difC: $\Delta\delta^{13}C$ of *G. sacculifer-Pulleniatina*).

continues to rise. It appears that nutrients are being "pumped out" (perhaps largely in the margins, due to increased upwelling there: Sarnthein et al., 1989), or they cease to be supplied from the deep Atlantic (due to decreased basin-basin fractionation). The decrease in nutrient supply does not decrease productivity because of enhanced deep mixing and upwelling. The diatom record in the mod-plot does not show much variation. All that can be said from this particular representation is that the diatom abundance adds little or nothing to the productivity story.

A similar plot for the "mixing" variable [v], and for the isotopic contrasts $\Delta\delta^{13}C$[pp] and $\Delta\delta^{18}O$[pp] (Fig. 10b), suggests that nutrients are unusually high in the thermocline during glacierization, while the mixed layer is unusually thick during that time. Maximum upwelling is pretty much centered on maximum glaciation, but decreases well before deglaciation sets in.

RETURN TO THE OPAL PARADOX

Have we now gathered any clues to get a better grasp of the opal paradox? Clearly, opal concentration (i.e., leachable silica, chiefly provided by the test of diatoms and of radiolarians) is neither a good proxy for export production (Fig. 11a) or for nutrient concentration in deep water (Fig. 11b) in the western equatorial Pacific. The latter is especially surprising, because the distribution patterns of phosphate and silicate are highly correlated in the present ocean (Sverdrup et al., 1942; Heath, 1974; Reid, 1965; Calvert, 1983; Chester, 1990). There is, however, a difference between the two that may contain the answer (or one possible answer): silicate precipitates more readily than phosphate, and it is more difficult to remobilize. The first effect produces a drop in silicate-to-phosphate ratios in subsurface waters, going from east to west in the equatorial Pacific (Herzfeld and Berger, 1993). The second leads to vertical separation, with silicate accumulating in excess near the bottom (Berger, 1970). Here it does no good as far as the growth of diatoms, although it can indeed facilitate heavy silicification of deep-living radiolarians.

Increased overall glacial productivity, which is now fairly well established (Sarnthein et al., 1988; Mix, 1989a,b; Berger and Herguera, 1992), would have increased intrabasinal fractionation of silicate from phosphate, mainly by depositing the silica in marginal upwelling areas. The glacial Atlantic, with a reduced production of NADW (Duplessy et al., 1988) and thus not being emptied of silicate as vigorously as is now the case (Redfield et al., 1963), would have efficiently removed silicate in its margins, as well. This would have starved the Pacific basin, which at present relies on interbasinal exchange to maintain its high silicate concentrations. In addition, while phosphate would have been contributed from the shelves during glacial times, the same cannot be assumed regarding silicate: whatever was washed into the ocean would have been rapidly buried, and whatever was dissolved (from diatom frustules

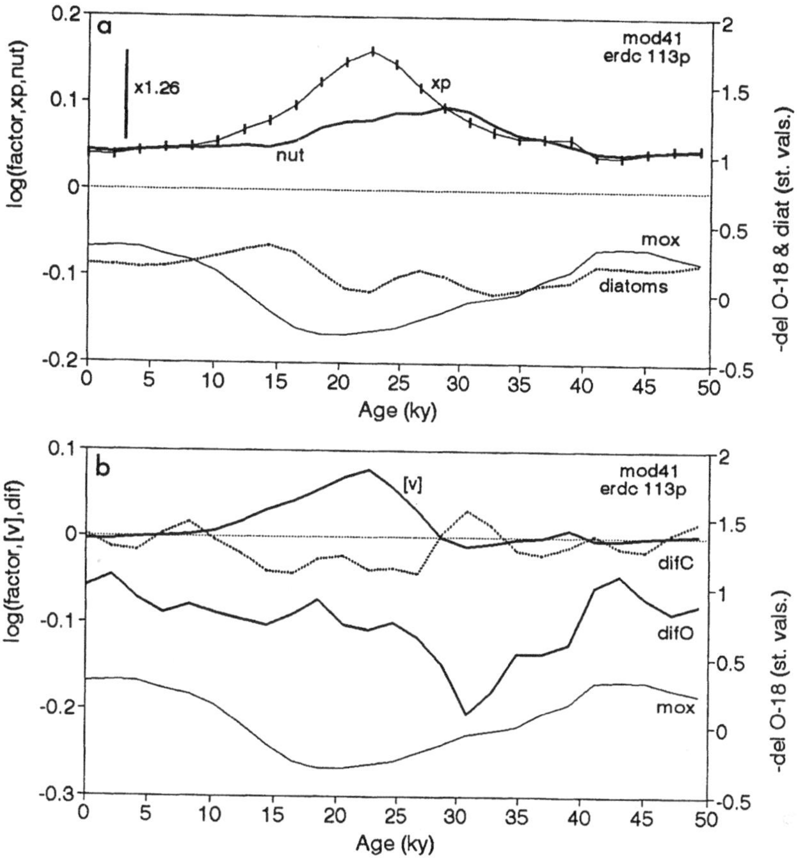

Figure 10 Modular plots of productivity proxies, ERDC 113P, Ontong Java Plateau. See text for construction. Graph shows "typical" obliquity-related cycle. xp, nut, mox, [v], difO, difC, as in Fig. 9; diatoms: log (abundance), see Fig. 1b.

etc.) would have been readily reprecipitated in the coastal zone. The glacial shortage of dissolved silicate in the thermocline waters of the western equatorial Pacific would have led to poor silicification of diatoms during growth, and the resulting less robust frustules would have more readily dissolved at depth. Support for this supposition (that diatom abundances do not correspond to surface productivity under conditions of silicate-starvation) comes from the mapping of abundances in the North Atlantic (Schrader et al., 1992). On Ontong Java Plateau, in addition, the higher degree of winnowing during glacials (as seen in increased sand content; Wu and Berger, 1991) could have played an important role, by removing (the easily resuspended) diatoms from sediments on the upper parts of the plateau.

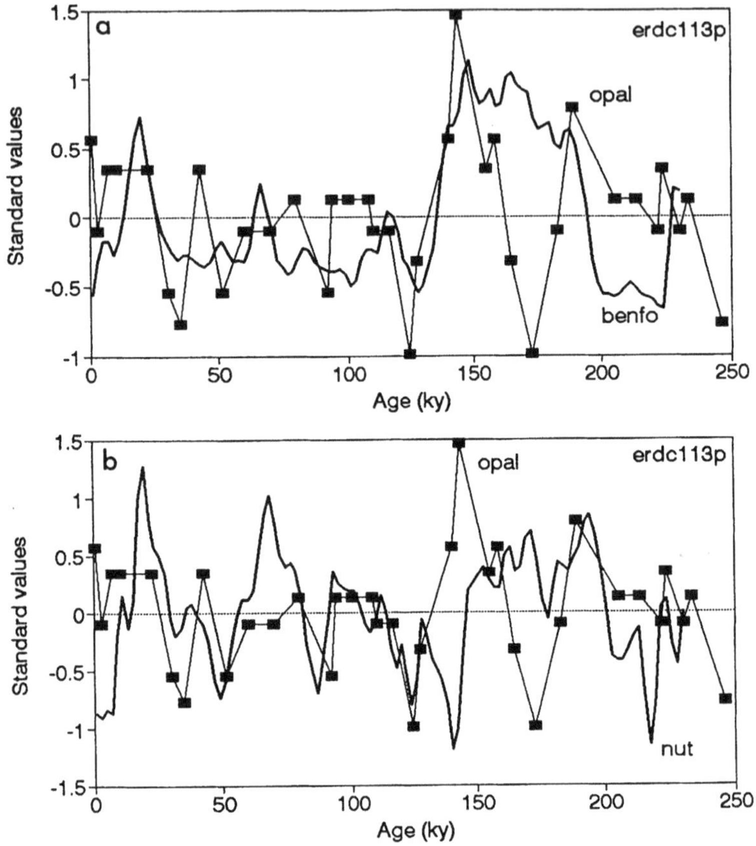

Figure 11 Comparison of opal concentrations with benthic foram flux (a), and with $\Delta\delta^{13}C[bp]$ (b). All values standardized.

SUMMARY AND CONCLUSION

The fact that different productivity proxies yield different answers about fluctuations in productivity during the late Quaternary at a given site points to the necessity of careful evaluation of such proxies. For the past, this can only be done by using several proxies simultaneously. When doing so, it is found that flux proxies disagree because of different power-law relationships to the source term. The particular power-laws (expressed in terms of proxy exponents) change as a function of geography, and may change as a function of time.

Nutrient proxies also cannot be expected to agree in detail. The $\Delta\delta^{13}C$ parameter, a nutrient proxy, cannot be read directly as a measure of atmospheric pCO_2, both because of difficulties in finding a globally representative signal, and also because such a reading neglects changes in alkalinity of deep waters.

Combining the use of flux proxies and nutrient proxies makes it possible to estimate the magnitude of physical processes such as mixing and upwelling, and the positive feedback from nutrient buildup in the thermocline. These effects are subsumed into a parameter [v] in the equation Flux = nutrients x [v]. A study of data from the western equatorial Pacific suggests that mixing, upwelling and nutrient buildup were greatly increased in glacial Stage 6. Also, there is a suggestion that nutrients tend to increase in deep waters during glacierization, ultimately enhancing productivity, and that productivity continues to rise due to upwelling, even though nutrient contents of deep waters decrease during the glacial proper. The opal paradox is explained as the result of decreased supply of dissolved silicate to the deep Pacific, during glacial time, and increased sequestration into the margins.

The change in the overall dynamics of equatorial upwelling which happened during glacial time can perhaps be anticipated from the present-day distribution of productivity (Fig. 12). It is characterized by a profound asymmetry between a highly productive eastern region, and a low-productive western region (Chavez and Barber, 1987; Chisholm and Morel, 1991).

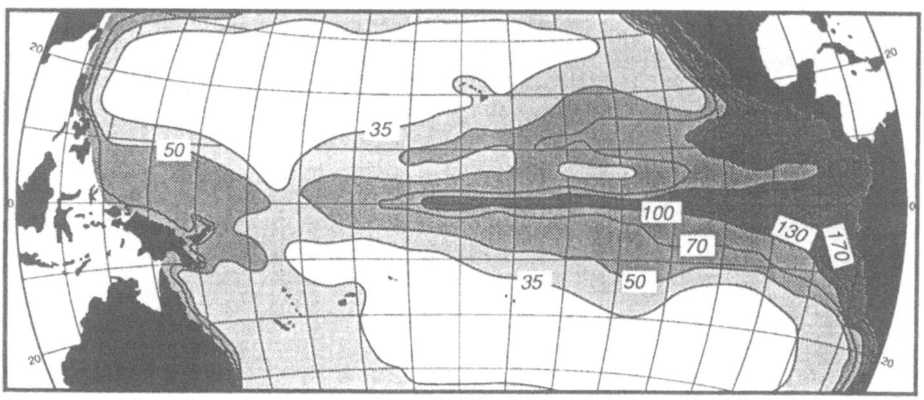

Figure 12 Primary production in the tropical Pacific. Estimates based on phosphate distributions at 100 m depth (Reid et al., 1978, Herzfeld and Berger, 1993), applying the productivity algorithms of Berger et al. (1987), and supplemented by the patterns shown in the world pigment map of NASA Goddard Space Flight Center and University of Miami, Nov 78 through June 81. The patterns suggest that the Ontong Java region is not only (or mainly) influenced by Pacific equatorial upwelling dynamics.

A westward extension of the narrow zone of upwelling reaching into the central Pacific, or an eastward extension of the elevated productivity zone off New Guinea (increased storms?), or (more probably) both, would be able to generate the glacial-time increase in productivity observed.

ACKNOWLEDGEMENTS

The work on productivity-related proxies in the western equatorial Pacific was supported by the U.S. National Science Foundation and the Petroleum Research Fund, through grants to W.H.B. and co-workers. R. Schneider acknowledges support from the Deutsche Forschungsgemeinschaft, Sonderforschungsbereich 261 at the University of Bremen. We thank P. Müller and G. Wefer (Bremen) for continuing interest in these studies and for access to laboratory facilities. We also thank Hans Schrader for a critical review.

REFERENCES

Altenbach AV, Sarnthein M (1989) Productivity record in benthic foraminifera. In: Berger WH, Smetacek VS, Wefer G (eds) Productivity of the Ocean: Present and Past, John Wiley and Sons, Chichester, p 255-269.

Arrhenius GOS (1952) Sediment cores from the east Pacific. Rep Swed Deep Sea Exped (1947-1948) 5:1-288.

Arrhenius GOS (1988) Rate of production, dissolution and accumulation of biogenic solids in the ocean. Palaeogeogr Palaeoclimat Palaeoecol 67:119-146.

Barnola JM, Raynaud D, Korotkevich YS, Lorius C (1987) Vostok ice cores provides 160,000 year record of atmospheric CO_2. Nature 329:408-414.

Barron JA, Baldauf J (1989) Tertiary cooling steps and paleoproductivity as reflected by diatoms and biosiliceous sediments. In: Berger WH, Smetacek VS, Wefer G (eds) Productivity of the Ocean: Present and Past, John Wiley, Chichester, p 341-354

Bender ML, Heggie DT, (1984) Fate of organic carbon reaching the seafloor: a progress report. Geochim Cosmochim Acta 48:977-986.

Bender M, Grande K, Johnson K, Marra J, Williams PJLeB, Sieburth J, Pilson M, Langdon C, Hitchcock G, Orchado J, Hunt C, Donaghay P, Heinemann K (1987) A comparison of four methods for determining planktonic community production. Limnol Oceanogr 32:1085-1098.

Bennekom AJ van, Berger GW (1984) Hydrography and silica budget of the Angola Basin. Neth J Sea Res 17(2-4):149-200.

Berger WH (1970) Biogenous deep-sea sediments: fractionation by deep-sea circulation. Bull Geol Soc Am 81:1385-1402.

Berger WH (1992) Pacific carbonate cycles revisited: arguments for and against productivity control. In: Ishizaki K, Saito T (eds) Centenary of Japanese Micropaleontology, Terra Scientific Publ Co, Tokyo, p 15-25

Berger WH, and Diester-Haass L (1988) Paleoproductivity: the benthic/planktonic ratio in foraminifera as a productivity index. Mar Geol 81:15-25.

Berger WH, Herguera JC (1992) Reading the sedimentary record of the ocean's productivity. In: Falkowski PF, Woodhead AD (eds) Primary Productivity and Biogeochemical Cycles in the Sea, Plenum Press, New York, p 455-486

Berger WH, Spitzy A (1988) History of atmospheric CO_2: constraints from the deep-sea record. Paleoceanography 3:401-411.
Berger WH, Wefer G (1990) Export production: seasonality and intermittency, and paleo-oceanographic implications. Global and Planetary Change 3(3):245-254.
Berger WH, Wefer G (1991) Productivity of the glacial ocean: discussion of the iron hypothesis. Limnol Oceanogr 36:1899-1918.
Berger WH, Wefer G (1992) Klimageschichte aus Tiefseesedimenten - Neues vom Ontong-Java Plateau. Naturwissenschaften 79:541-550.
Berger WH, Fischer K, Lai C, Wu G (1987) Ocean productivity and organic carbon flux. Part I. Overview and maps of primary production and export production. Univ Calif San Diego, SIO Reference 87-30.
Berger WH, Smetacek VS, Wefer G (1989a) Productivity of the Ocean: Present and Past. Dahlem Konferenzen. John Wiley, Chichester.
Berger WH, Smetacek VS, Wefer G (1989b) Ocean productivity and paleoproductivity - an overview. In: Berger WH, Smetacek VS, Wefer G (eds), Productivity of the Ocean: Present and Past, Dahlem Konferenzen. John Wiley, Chichester, p 1-34.
Berger, WH, Bickert T, Schmidt H, Wefer G (1993) Quaternary oxygen isotope record of pelagic foraminifers: Site 806, Ontong Java Plateau. Proc ODP Sci Res 130:381-395.
Berner W, Oeschger H, Stauffer B (1980). Information on the CO_2 cycle from ice core studies. Radiocarbon 22:227-235.
Bickert T, Berger WH, Burke S, Schmidt H, Wefer G (1993) Late Quaternary stable isotope record of benthic foraminifers: Sites 805 and 806, Ontong Java Plateau. Proc ODP Sci Res 130:411-420.
Bleil U, Thiede J (1990) Geological history of the polar oceans: Arctic versus Antarctic. Kluwer, Dordrecht.
Boyle EA, Keigwin LD (1987) North Atlantic thermohaline circulation during the past 20,000 years linked to high-latitude surface temperature. Nature 330:35-40.
Broecker WS (1982) Ocean chemistry during glacial time. Geochim Cosmochim Acta 46:1689-1705.
Broecker WS, Peng T-H (1989) The cause of the glacial to interglacial atmospheric CO_2 change: A polar alkalinity hypothesis. Global Biogeochem Cycles 3:215-239.
Burke SK, Berger WH, Coulbourn WT, Vincent E (1993) Benthic foraminifera in box core ERDC 112, Ontong Java Plateau. J Foram Res 23:19-39.
Calvert SE (1964) Factors affecting distribution of laminated diatomaceous sediments in the Gulf of California. Am Assoc Petrol Geol Mem 3:311-330.
Calvert SE (1983) Sedimentary geochemistry of silicon. In: Aston SR (ed) Silicon Geochemistry and Biogeochemistry, Academic Press, p 143-186.
Charles CD, Fairbanks RG (1990) Glacial to interglacial changes in the isotopic gradients of Southern Ocean surface water. In: Bleil U, Thiede J (eds) Geological History of the Polar Oceans: Arctic versus Antarctic, Kluwer, Dordrecht, p 519-538.
Chavez FP, Barber RT (1987) An estimate of new production in the equatorial Pacific. Deep-Sea Res 34:1229-1243.
Chester R (1990) Marine Geochemistry. Unwin Hyman, London.
Chisholm SW, Morel FM (1991) What controls phytoplankton production in nutrient-rich areas of the open sea? Limnol Oceanogr 36(8):1507-1970 (Special Issue).
CLIMAP Project Members (1981) Seasonal reconstruction of the Earth's surface at the last glacial maximum. Geol Soc Am Map Chart Ser MC36, Boulder, Colorado.
Delmas RJ, Ascencio JM, Legrand M (1980) Polar ice evidence that atmospheric CO_2 20 000 yr BP was 50% of present. Nature 284:155-157.
Diester-Haass L, Schrader HJ, Thiede J (1973) Sedimentological and paleoclimatological investigation of two pelagic ooze cores off Cape Barbas, Northwest Africa. "Meteor" Forsch-Ergebnisse C16:19-66.
Diester-Haass L (1985) Late Quaternary sedimentation on the eastern Walvis Ridge, SE Atlantic (HPC 532 and four piston cores). Mar Geol 65:145-189.
Duplessy JC, Shackleton NJ, Fairbanks RG, Labeyrie L, Oppo D, Kallel N (1988) Deepwater source variations during the last climatic cycle and their impact on the global deepwater circulation. Paleoceanography 3:343-360.

Emerson S, Hedges JI (1988) Processes controlling the organic carbon content of open ocean sediments. Palaeogeogr Palaeoclimat Palaeoecol 3:621-634.
Eppley RW, Peterson BJ (1979) Particulate organic matter flux and planktonic new production in the deep ocean. Nature 282:677-680.
Hays JD, Imbrie J, Shackleton NJ (1976) Variations in the Earth's orbit: pacemaker of the Ice Ages. Science 194:1121-1132.
Heath GR (1974) Dissolved silica and deep-sea sediments. In: Hay W (ed) Studies in Paleo-Oceanography, Soc Econ Paleont Mineral Spec Publ 20:77-93.
Herguera, JC (1992) Deep-sea benthic foraminifera and biogenic opal: glacial to postglacial productivity changes in the western equatorial Pacific. Mar Micropaleont 19:79-98.
Herguera JC, Berger WH (1991) Paleoproductivity: glacial to postglacial change in the western equatorial Pacific, from benthic foraminifera. Geology 19:1173-1176.
Herzfeld, UC, Berger WH (1993) Ocean Productivity and Indicator Variables: Map Comparison for Atlantic and World Oceans. SIO Ref Ser 93-7, La Jolla, California.
Imbrie J, Hays JD, Martinson DG, McIntyre, A, Mix AC, Morley JJ, Pisias NG, Prell WL, Shackleton NJ (1984) The orbital theory of Pleistocene climate: Support from a revised chronology of the marine $\delta^{18}O$ record. In: Berger AL, Imbrie J, Hays JD, Kukla J, Saltzman J (eds) Milankovitch and Climate (Part 1), Reidel, Hingham, Mass, p 269-305.
Jansen JHF, Iperen JM van (1991) A 220,000-year climatic record for the east equatorial Atlantic Ocean and equatorial Africa: evidence from diatoms and opal phytoliths in the Zaire (Congo) deep-sea fan. Paleoceanography 6:573-592.
Keir RS, Berger WH (1983) Atmospheric CO_2 content in the last 120,000 years: the phosphate-extraction model. J Geophys Res 88(C10):6027-6038
Keir RS (1990) Reconstructing the ocean carbon system variation during the last 150,000 years according to the Antarctic nutrient hypothesis. Paleoceanography 5:253-276.
Koblentz-Mishke OI, Volkovinsky VV, Kabanova JG (1970) Plankton primary production of the world ocean. In: Wooster W (ed) Scientific Exploration of the South Pacific, National Academy of Sciences, Washington, D.C.
Kroopnick PM (1985) The distribution of ^{13}C of CO_2 in the world oceans. Deep-Sea Res 32:57-84.
Lange CB, Berger WH (1993) Diatom productivity and preservation in the western equatorial Pacific: the Quaternary record. Proc ODP Sci Res 130:509-523.
Lange CB, Burke SK, Berger WH (1990) Biological Production off Southern California is linked to climatic change. Climatic Change 16:319-329.
Lapenis AG, Os'kina NS, Barash MS, Biyum NS, Vasileva YeV (1990) Late Quaternary variations in the productivity of the ocean's biota. Okeanologiya 30:93 [In Russian].
Leinen M, Sarnthein M (1989) Paleoclimatology and Paleometeorology: Modern and Past Patterns of Global Atmospheric Transport. Kluwer, Dordrecht.
Leinen M, Cwienk D, Heath GR, Biscaye P, Kolla V, Thiede J, Dauphin JP (1986) Distribution of biogenic silica and quartz in recent deep-sea sediments. Geology 14:199-203.
Loubere P (1991) Deep-sea benthic foraminiferal assemblage response to a surface ocean productivity gradient: a test. Paleoceanography 6:193-204.
Lyle M (1988) Climatically forced organic carbon burial in equatorial Atlantic and Pacific oceans. Nature 335:529-532.
Lyle M, Murray DW, Finney BP, Dymond J, Robbins JM, Brooksforce K (1988) The record of late Pleistocene biogenic sedimentation in the eastern tropical Pacific Ocean. Paleoceanography 3:39-59.
Mikkelsen N (1978) Preservation of diatoms in glacial to Holocene deep-sea sediments of the equatorial Pacific. Geology 6:553-555.
Mikkelsen N (1984) Diatoms in the Zaire deep-sea fan and Pleistocene paleoclimatic trends in the Angola Basin and west equatorial Africa. Neth J Sea Res 17(2-4): 280-292.
Milliman JD, Takahashi K (in press) Carbonate and opal production and accumulation in the ocean. In: Hay W, Meybeck M, Usselman T (eds) Global Surficial Geofluxes: Modern to Glacial, National Research Council, Washington, DC.
Mix AC (1989a) Pleistocene paleoproductivity: evidence from organic carbon and foramini-feral species. In: Berger WH, Smetacek VS, Wefer G (eds) Productivity in the ocean: Present and Past, John Wiley, Chichester, p 313-340.

Mix AC (1989b) Influence of productivity variations on long term atmospheric CO_2. Nature, 337:541-544.
Mortlock RA, Charles CD, Froelich PN, Zibello MA, Saltzman J, Hays JD, Burckle LH (1991) Evidence for lower productivity in the Antarctic Ocean during the last glaciation. Nature 351:220-223.
Müller PJ, Schneider R (1993) An automated leaching method for the determination of opal in sediments and particulate matter. Deep-Sea Res 40(3):425-444.
Müller PJ, Suess E (1979) Productivity, sedimentation rate, and sedimentary organic matter in the oceans - I. Organic carbon preservation. Deep-Sea Res 26A:1347-1362.
Opdyke BN, Walker JCG, 1992. Return of the coral reef hypothesis: Basin to shelf partitioning of $CaCO_3$ and its effect on atmospheric CO_2. Geology 20:733-736.
Pedersen TF (1983) Increased productivity in the eastern equatorial Pacific during the last glacial maximum (19,000 to 14,000 yr B.P.). Geology 11:16-19.
Pisias NG, Prell WL (1985) Changes in calcium carbonate accumulation in the equatorial Pacific during the late Cenozoic: evidence from HPC Site 572. In: Sundquist ET, Broecker WS (eds) The Carbon Cycle and Atmospheric CO_2: Natural Variations Archean to Present, AGU Geophys Monogr 32:443-454.
Platt T, Harrison G (1985) Biogenic fluxes of carbon and oxygen in the ocean. Nature 318:55-58.
Pokras EM (1987) Diatom record of late Quaternary climatic change in the eastern equatorial Atlantic and tropical Africa. Paleoceanography 2:273-286.
Prell WL, Curry WB (1981) Faunal and isotopic indices of monsoonal upwelling: Western Arabian Sea. Oceanol Acta 4:91-98.
Rea DK, Chambers LW, Chuey JM, Janecek TR, Leinen M, Pisias NG (1986) A 420,000-year record of cyclicity in oceanic and atmospheric processes from the eastern equatorial Pacific, Paleoceanography 1:577-586.
Redfield AC, Ketchum BH, Richards FA (1963) The influence of organisms on the composition of sea-water. In: Hill MN (ed) The Sea, vol.2, Wiley Interscience, New York, p26-77.
Reid JL (1965) Intermediate Waters of the Pacific. Johns Hopkins, Baltimore, Md.
Reid JL, Brinton E, Fleminger A, Venrick EL, McGowan JA (1978) Ocean circulation and marine life. In: Charnock H, Deacon G (eds) Advances in Oceanography, Plenum, New York, p 65-130.
Romankevich EA (1984) Geochemistry of Organic Matter in the Ocean. Springer, Berlin Heidelberg New York.
Sancetta C (1989) Processes controlling the accumulation of diatoms in sediments: a model derived from British Columbian Fjords. Paleoceanography 4:235-251.
Sancetta C (1992) Primary production in the glacial North Atlantic and North Pacific oceans. Nature 360:249-251.
Sarnthein M, Thiede J, Pflaumann U, Erlenkeuser H, Fütterer D, Koopmann B, Lange H, Seibold E (1982). Atmospheric and oceanic circulation patterns off Northwest Africa during the past 25 million years. In: Rad U von, Hinz K, Sarnthein M, Seibold E (eds) Geology of the Northwest African Continental Margin, Springer, Berlin Heidelberg New York, pp 584-604.
Sarnthein M, Winn K, Zahn R (1987) Paleoproductivity of oceanic upwelling and the effect on atmospheric CO_2 and climatic change during deglaciation times. In: Berger WH, Labeyrie LD (eds) Abrupt Climatic Change - Evidence and Implications, Reidel, Dordrecht, p 311-337.
Sarnthein M, Winn K, Duplessy J-C, Fontugne MR (1988) Global carbon variations of surface ocean productivity in low and mid latitudes: Influence on CO_2 reservoirs of the deep ocean and atmosphere during the last 21,000 years. Paleoceanography 3:361-399.
Semtner A, and Chervin R (1991) A thermohaline conveyor belt in the world ocean. WOCE Notes 3:12-15.
Schiffelbein P (1984) Stable isotope systematics in Pleistocene deep-sea sediment records. PhD-thesis, Univ Calif, San Diego, La Jolla, California.
Schiffelbein P, Dorman L (1986) Spectral effects of time-depth nonlinearities in deep sea sediment records: a deconvolution technique for realigning time and depth scales. J Geophys

Res 91(B3):3821-3835.
Schmidt H, Berger WH, Bickert T, Wefer G (1993) Quaternary carbon isotope record of pelagic foraminifera: Site 806, Ontong Java Plateau. Proc ODP Sci Res 130:397-409.
Schneider R (1991) Spätquartäre Produktivitätsänderungen im östlichen Angola-Becken: Reaktionen auf Variationen im Passat-Monsun-Windsystem und in der Advektion des Benguela Küstenstromes. Berichte Fachbereich Geowissenschaften Univ Bremen 21, Bremen.
Schrader H (1992) Comparison of Quaternary coastal upwelling proxies off central Peru. Mar Micropaleont 19:29-47.
Schrader H, Sorknes R (1991) Peruvian coastal upwelling: Late Quaternary productivity changes revealed by diatoms. Mar Geol 97:233.
Schrader H, Swanberg IL, Burckle LH (1992) Diatoms in recent North Atlantic sediments I: Abundance patterns and what they mean. Hydrobiologia (in press).
Schuette G, Schrader H (1979) Diatom taphocoenoses in the coastal upwelling area off western South America. Nova Hedwigia 64:359-378.
Shackleton NJ, Hall MA, Line J, Shuxi C (1983) Carbon isotope data in Core V19-30 confirm reduced carbon dioxide concentration in the ice age atmosphere. Nature 306:319-322.
Spero HJ, and DeNiro MJ (1987) The influence of symbiont photosynthesis on the $\delta^{18}O$ and $\delta^{13}C$ values of planktonic foraminiferal shell calcite. Symbiosis, 4: 213-228.
Stabell B (1986) Variations of diatom flux in the eastern equatorial Atlantic during the last 400,000 years ("Meteor" cores 13519 and 13521). Mar Geol 72:305-323.
Stein R (1991) Accumulation of Organic Carbon in Marine Sediments. Lecture Notes in Earth Sciences 34, Springer, Berlin.
Suess E (1980) Particulate organic carbon flux in the oceans: surface productivity and oxygen utilization. Nature 288:260-263.
Suess E, Thiede J (1983) Coastal upwelling - Its sediment record. Part A: Responses of the sedimentary regime to present coastal upwelling. Plenum Press, New York and London.
Summerhayes CP, Prell WL, Emeis KC (1992) Upwelling systems: evolution since the early Miocene. Geol Soc London Spec Publ 64.
Sundquist ET, Broecker, WS (1985) The Carbon Cycle and Atmospheric CO_2: Natural Variations Archean to Present. AGU Geophys Monogr 32.
Sverdrup HU, Johnson MW, Fleming RH (1942) The Oceans - Their Physics, Chemistry and General Biology. Prentice Hall, Englewood Cliffs, NJ.
Thiede J, Suess E (1983) Coastal upwelling its sediment record. Part B: Sedimentary records of ancient coastal upwelling. Plenum Press, New York and London.
Wefer G, Berger WH (1991) Isotope paleontology: growth and composition of extant calcareous species. Mar Geol 100:207-248.
Wu G, Berger WH (1991) Pleistocene $\delta^{18}O$ records from Ontong-Java Plateau: effects of winnowing and dissolution. Mar Geol 96:193.
Zahn R, Winn K, Sarnthein M, (1986) Benthic foraminiferal $\delta^{13}C$ and accumulation rates of organic carbon: *Uvigerina peregrina* and *Cibicidoides wuellerstorfi*. Paleoceanography 1:27-42.

FROM MODERN FLUX TO PALEOFLUX: ASSESSMENT FROM SINKING ASSEMBLAGES TO THANATOCOENOSIS

Kozo Takahashi
Department of Marine Sciences and Technology
School of Engineering
Hokkaido Tokai University
Minamisawa, Sapporo 005
Japan

ABSTRACT. Particle fluxes using sediment traps have been seriously studied in the past decade for various microfossil groups including biogenic opal. We have accumulated a substantial amount of information pertinent to properly interpret the past environments. The majority of biogenic opal production is lost due to dissolution mainly at the sea-floor and only a minute fraction can be preserved in the fossil record. Numerical correlations between fluxes and the fossil record are needed to establish a linkage between modern and fossil assemblages based on the minute quantity preserved. A prototype numerical paleoecological method has been constructed. This method is used to quantitatively interpret paleofluxes of diatoms. Paleoflux results from two diatom taxa are presented from the northeastern Pacific: *Denticulopsis seminae* and *Coscinodiscus marginatus*. *Denticulopsis seminae*, a pennate diatom taxon widely occurring in the high latitude subarctic Pacific, is characterized as a productivity taxon in the pelagic realm with high fossil abundance. The other taxon important for winter conditions in the present day subarctic Pacific is *Coscinodiscus marginatus*, one of the best preserved diatoms in the fossil record of the region.

INTRODUCTION

The value of marine microfossils for reconstructing the earth's paleoenvironments has well been recognized among the scientific community. The ocean's environmental parameters can be compared with fossil signals retained in the underlying sediments and regression methods can be developed. Among the many methods, one of the most useful techniques thus far was developed by Imbrie and Kipp (1971). This method has been applied in many areas of the world oceans (e.g., CLIMAP, 1976). Background environmental parameters such as sea surface temperature (SST) or nutrient concentrations and core top faunal/floral assemblages are correlated in the mathematical equations. The resulting equations can be applied to reconstruct past environmental conditions such as SST at 18,000 years ago. Although useful, a major deficiency in it stems from the empirical nature of the method. The empirical relationships without understanding true fossilization mechanisms make wide applications of thin method impractical in a variety of oceanic conditions.

NATO ASI Series, Vol. I 17
Carbon Cycling in the Glacial Ocean:
Constraints on the Ocean's Role in Global Change
Edited by R. Zahn et al.

The most desirable is the integration of well understood fossilization processes beginning with living planktonic assemblages (= biocoenosis) and ending with the sedimentary record (thanatocoenosis). These include, for instance, seasonality of plankton production, detailed settling processes of sinking assemblages (= sidocoenosis: Takahashi, 1993), of plankton remains (or sometime actively sinking living cells), and dissolution processes.

In this paper, sinking processes and quantitative fluxes of several diatom taxa from the northeastern North Pacific will be discussed in the view of fossilization steps. The diatom fluxes are then compared with surficial sedimentary diatom assemblages in order to derive a numerical relationship between the two. The relationship is applied to interpret paleo-fluxes of diatoms.

FATE OF SILICEOUS MICROPLANKTON-FOSSILS: PRODUCTION TO BURIAL

Living shell-bearing planktonic assemblages in the upper water column will sooner or later become sinking assemblages (unless they are dissolved/recycled in situ), regardless of their passive or active sinking mode. One of the conventional views of the vertical transport mode for small sized phytoplankton is via fecal pellets (Honjo, 1976). The phytoplankton cells are ingested by metazoan or other grazers and egested as fecal pellets. Thus, their shells become incorporated in the pellets, often involving pulverization, and the pellets descend toward the sea-floor. More recently, vertical transport in aggregates such as marine snow has been recognized as a major sinking mode for a variety of particle sizes and morphologies (e.g., Takahashi, 1986; Takahashi et al., 1989). Without knowing the transport mode, one cannot assess early fossilization processes such as dissolution in the water column. With the aggregate particles as a major sinking mode in mind, sediment trapping becomes one of the indispensable methods of collecting flux samples (Honjo and Doherty, 1988). Data from the trapping are incorporated in the methods used in this study.

The fate of a given species from production to burial can be numerically traced. This type of approach will help in understanding the magnitude of signals (e.g., temperature, productivity) preserved in individual species of siliceous microfossil groups. The following is an example of the fate of *Denticulopsis seminae**. Taylor and Waters (1982) reported standing stocks of *D. seminae* in waters from 0-150 m depth near Station Papa in March and May 1969. The cumulative standing stocks down to 150 m are 156 x 10^6 cells m^{-2} and 2.81 x 10^6 cells m^{-2} in March and May respectively. These values can be respectively compared with the mean flux values of *D. seminae* during late March through April and late May through June in

---

* *Denticulopsis seminae* (Simonsen and Kanaya) Simonson = *Neodenticula seminae* (Simonsen and Kanaya) Akiba and Yanagisawa.

the flux samples collected during 1982 - 1986, allowing a settling time of approximately three weeks from the mixed layer to the deep trap depth of 3800 m (Figure 1). Flux values are 4.91 x 10^6 valves m^{-2} d^{-1} (or 2.46 x 10^6 cells m^{-2} d^{-1}) for late March - April and 7.71 x 10^6 valves m^{-2} d^{-1} (or 3.86 x 10^6 cells m^{-2} d^{-1}) for mid May -June. Although direct comparisons are difficult because of different years involved, the daily fluxes are the equivalent of about 1.6% and 140% of *D. seminae* standing stocks in the upper layer respectively. It may be that the May standing stock measured by Taylor and Waters (1982) was much lower than a typical value due to a temporal shift of spring high production toward an earlier period in the season in that year. Alternatively, earlier in the season (March) only a small fraction of the standing stock is converted to the sinking flux whereas later in the season (May) much of the standing stock becomes the flux exported to the depth. The daily flux falls via large and rapidly sinking aggregates to the sea floor without losing many frustules (Takahashi, 1986; Takahashi et al., 1989).

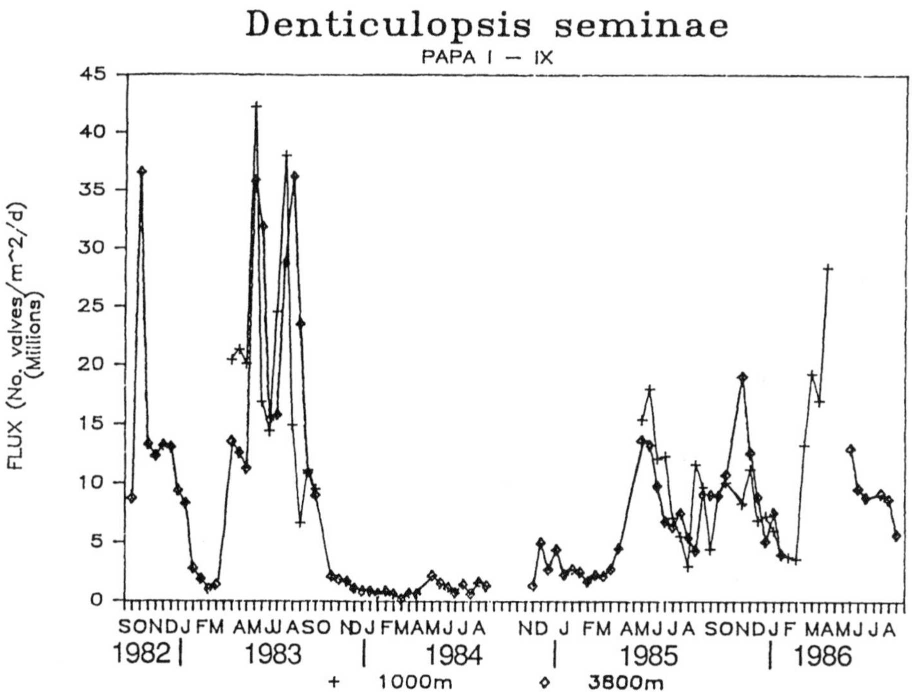

Figure 1. Four year (1982-1986) time-series flux of *D. seminae* at Station Papa in the eastern subarctic Pacific. Note that flux peaks of this taxon are observed during high mass flux periods such as spring and fall.

Approximately 2% of diatom sidocoenosis is preserved in core tops at Station Papa (Takahashi, 1986). In other parts of the world oceans the preservation values are not too far from this value. It has been observed that opal preservation is often a function of opal flux (Milliman and Takahashi, in press). It is important to assess whether thanatocoenosis signals of certain taxa are amplified relative to sidocoenosis signals or vise versa. Such a relationship is a key to reconstruction of paleofluxes.

PALEOFLUX RECONSTRUCTION: THE PRESENT IS A KEY TO THE PAST

One of the best ways to understand paleoenvironments is to start from present day opal flux and then interpret the past. Data from Station Papa (50°N, 145°W) in the eastern subarctic Pacific represents one of the most complete and high resolution time series flux records we ever collected. Therefore, it is pertinent to apply such useful data to paleoceanography in this region. I compared the daily mean percentages of diatom fluxes averaged over four years of sediment trapping experiments with core top samples at two sites near Station Papa (Table 1). Papa core II is a 1.5 m long gravity core which was raised at Station Papa during sediment trap recovery/deployment cruise IV which took place in 1984. The other core RC10-216, listed in Table 1, is Lamont-Doherty core which has been well studied (e.g., Sancetta and Silverstri, 1984) and core top diatom counts were made available (C. Sancetta, pers. comm., 1989). The assumption that the core tops are true representatives of the Holocene is substantiated by the fact that the extinction of *R. curvirostris* occurred well below the top (Sancetta and Silverstri, 1984) and the measured oxygen isotope values (L. Keigwin, pers. comm., 1988).

Table 1. Diatom species percents in sediment trap samples PAPA, 3800 m (average of four year record) and core top samples at Station PAPA and a nearby site.

	PAPA 3800m SED. TRAP	PAPA CORE II	CORE RC10-216
	(This paper)	(Sancetta, pers.comm, 1989)	
Latitude	50° 00' N	50° 00' N	50° 58' N
Longitude	145° 00' W	145° 00' W	151° 10' W
Water Depth (m)	4200	4200	4989
	%	%	%
C. marginatus	**1.1**	**21.4**	**3.9**
T. trifulta group	8.0	11.5	15.6
Small Unidentified	10.0	5.7	-
Coscinodiscids	0.0	0.0	0.0
A. curvatulus	3.4	3.3	11.2
D. seminae	**73.4**	**54.9**	**65.7**
R. alata gracillima	2.0	0.0	0.
R. styliformis	0.2	0.0	0.0
C. atlanticum	0.7	0.0	0.0
C. concavicorne	0.7	0.0	0.0

Percent preservation of diatoms

The two values of estimated diatom percent preservation in core tops, described below, are on the same order of magnitude as previous estimates of diatom (2%: Takahashi, 1986), silicoflagellate (1%: Takahashi, 1987b), and radiolarian (Takahashi, 1987c) preservation in core top samples.

High range (4.9%). A comparison from the four year record of a mean total diatom flux (11.2 x 10^6 valves m^{-2} d^{-1}) with an accumulation rate of 20 x 10^6 valves cm^{-2} 10^3 yr^{-1} in the Station Papa region [using diatom counts in a dry gram of sediments (Jouse et al., 1971) and density of 0.8 g cm^{-3} and sedimentation rate of 1 cm 10^3 yrs^{-1} (Curry and Lohmann, 1986; Takahashi, 1986)] provides a result of 4.9% diatom preservation relative to the mean diatom flux (as 100%).

Low range (0.56%). As an independent estimate of percent preservation, I counted diatoms in Papa Core II 0-1 cm sample using the same method as in my flux study (e.g., Takahashi, 1986). The total diatoms of 2.87 x 10^6 valves dry g^{-1} of sediment was obtained. This is a factor of 9 less than the counts by Jouse et al. (1971). Using the same values of density and accumulation rate as the above, we get 6.29 x 10^4 valves m^{-2} d^{-1} preserved in the core top sample. This is 0.56% of the mean total diatom flux. These two values will be used in the computations described below.

Differential dissolution between taxa

As shown in the Table 1, *Denticulopsis seminae* is the dominant diatom taxon in vertical flux as well as in sediments. The dominance of this taxon in flux is substantiated with four years of the flux record. This species is a productivity indicator in the northeastern North Pacific since its strong co-occurrence with high mass, opal and organic carbon fluxes (Honjo, 1984), especially in spring and fall (Figure 2), has been well established (Takahashi, 1986, 1987a; Takahashi et al., 1989). A relatively large degree of dissolution loss of *D. seminae* in the sediments is apparent compared to *Coscinodiscus marginatus*, a more dissolution resistant taxon whose high production occurs during winter periods (Figure 3) [when the nutrient levels are high (and perhaps other limiting trace elements such as Fe, see Martin et al., 1989) and light levels are low]. Therefore, differential dissolution among different taxa must take place. For the above mentioned 4.9% preservation (high range) of total diatoms in the core top of RC10-216, much higher percentage of *C. marginatus* flux is preserved: 20.7%. For the low range 0.56 % preservation of total diatoms, 2.40% of *C. marginatus* flux is preserved (Table 2). This can be compared with *D. seminae* preservation of 0.76% (high range) and 0.09% (low range) of the measured flux, respectively (Table 2). Therefore, ranges of 79.3 to 97.6% of *C. marginatus* and 99.24 to 99.91% of *D. seminae* fluxes (100%), respectively, were dissolved on or near the sea floor. Such a differential dissolution of thinly silicified and robust taxa needs to be mathematically expressed.

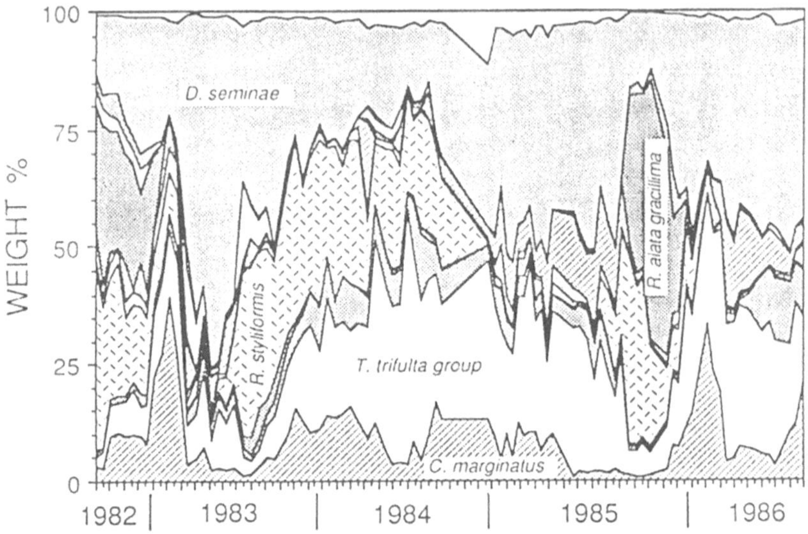

Figure 2. Taxon percent contribution of diatom export carbon production at Station Papa during 1982-1986.

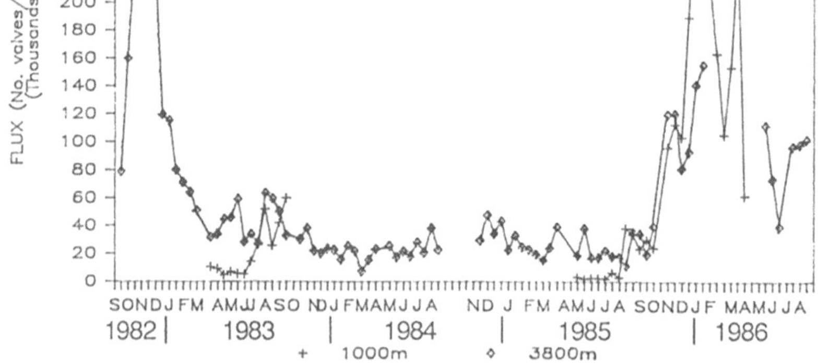

Figure 3. Four year (1982-1986) time-series flux of *C. marginatus* at Station Papa in the eastern subarctic Pacific. Flux of this taxon tends to increase during late fall/early winter periods. The 1983-1984 low fluxes of this taxon reflect the flux suppression linked with the low latitude 1982-1983 El Nino (Takahashi et al., 1989).

Table 2. Specific percent preservation in fossil record (core tops) of sinking flux of major diatom taxa. The preservation percents are expressed in two cases: Case (1) 0.56% total diatom preservation (this study) relative to 100% total diatom flux; and Case (2) 4.9% preservation (Jouse et al., 1971; Takahashi, 1986). Two important and representative taxa are shown in bold face.

Taxon	Core/depth: RC10-216 0-1 cm %	Core/depth: Papa Core II 0-1 cm %	*k*
Case 1: where 0.56 % of total diatom flux (100 %) preserved.			
C. marginatus	**2.4**	**2.4**	**3.74**
T. trifulta group	0.23	0.17	6.39
Small unidentified Coscinodiscids	NA	0.07	7.32
A. curvatulus	0.38	0.11	6.79
D. seminae	**0.10**	**0.09**	**7.05**
R. alata gracillima	0	0	∞
R. styliformis	0	0	∞
C. atlanticum	0	0	∞
C. concavicorne	0	0	∞
weighted mean	0.56	0.56	
Case 2: where 4.9 % of total diatom flux (100%) preserved.			
C. marginatus	**20.7**	**20.7**	**1.58**
T. trifulta group	2.0	1.5	4.23
Small unidentified Coscinodiscids	NA	0.58	5.15
A. curvatulus	0.34	0.98	4.62
D. seminae	**0.91**	**0.76**	**4.88**
R. alata gracillima	0	0	∞
R. styliformis	0	0	∞
C. atlanticum	0	0	∞
C. concavicorne	0	0	∞
weighted mean	4.9	4.9	

Mathematical modeling

Taking a well established approach (Hurd, 1983; Nelson, 1975; Erez, Takahashi and Honjo, 1982), one can express diatom percents due to the dissolution loss using the following equation:

$$dP/dt = -k \cdot P \tag{1}$$

where P is the percent of diatoms and k is a constant. The solutions of this equation are:

$$P_t = P_o e^{-k \cdot t} \tag{2}$$

and:

$$-k = \ln (P_t/P_o) \cdot (1/t) \tag{3}$$

where P_t is the diatom percent after time t and P_o is the initial percent. The computed values for constant k are listed in Table 2. It should be noted that the results of species preservation % values from the two different core tops recovered from nearby Station Papa are similar to one another (Table 2).

Flux to paleoflux computations

Only 0.09% to 0.76% of 100% *D. seminae* which sink toward the sea floor are preserved in Papa Core II top sample (Table 2). Parallel to such an absolute change from flux to accumulation in sediments, *D. seminae* represents 73% of total diatom flux which subsequently becomes 55% in total diatom assemblages in the same core (Papa II) top (Table 1; Figure 4). I interpret that much of the dissolution/preservation processes, including cation

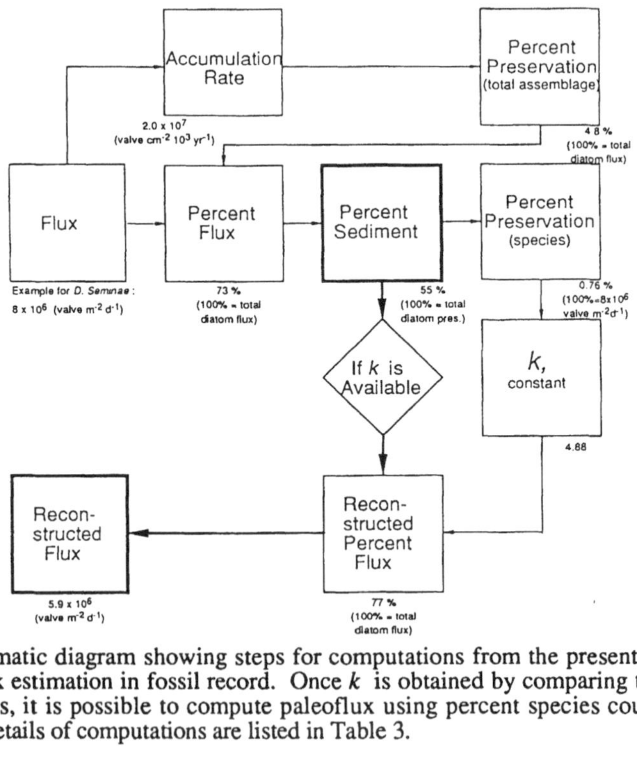

Figure 4. Schematic diagram showing steps for computations from the present day sinking flux to paleoflux estimation in fossil record. Once k is obtained by comparing the flux with core top samples, it is possible to compute paleoflux using percent species counts in fossil assemblages. Details of computations are listed in Table 3.

coating (e.g., Fe, Mn, Al etc.) on or slightly below the sea floor has been completed in the core top sample since the *D. seminae* % values did not decrease with core depth, but increased instead (Figure 5). Alternatively, paleoproductivity between 5 and 20 cm was higher than other time periods, causing high *D. seminae* % values.

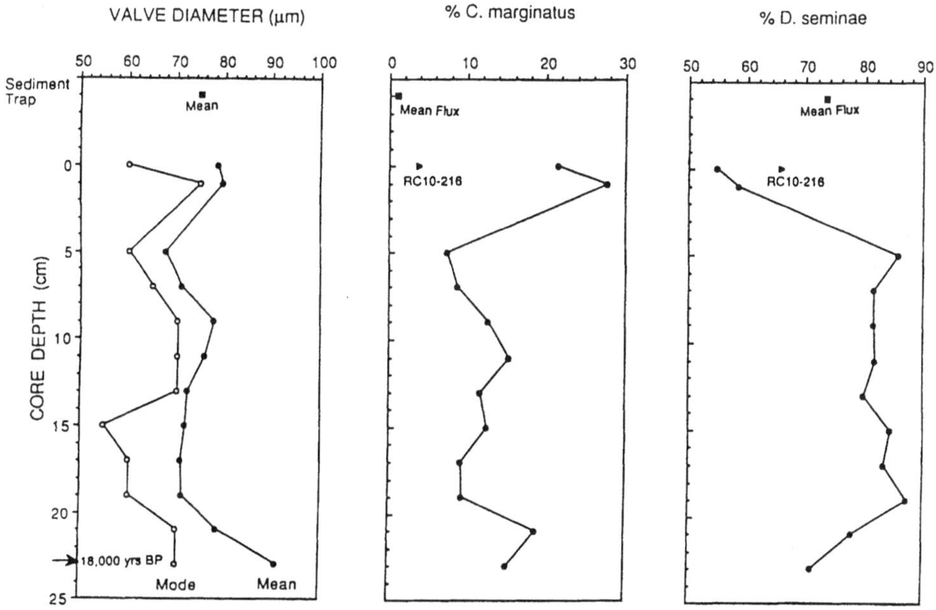

Figure 5. Percents of *D. seminae* and *C. marginatus* in total diatom assemblages and valve size of *C. marginatus* in Papa Core II, Holocene through LGM. Core top percents of the two species from RC10-216 as well as mean sinking flux values from Station Papa sediment traps are also plotted.

Proceeding one step further, using the k values for *D. seminae* from the core top results (k =4.88 for Papa Core II), we can reconstruct (see Figure 4 and Appendix 1 for computation steps involved) paleoflux at Station Papa (Figure 6). The reconstructed percent flux of *D. seminae* at the core top is 77%. Using this value the paleoflux of 5.9 x 10^6 valve $m^{-2}d^{-1}$ at the core top is obtained. High fluxes of *D. seminae,* two to three times higher than today, occurred during 4,000 to 15,000 years BP (Figure 6). Changes in *C. marginatus* paleoflux (Figure 6), %, and size are correlated (Figure 5). This is most likely due to changes in paleoproductivity whose levels are of interest. Admittedly, the k values, which are assumed to be the same throughout the core for a given taxon here, could differ from one sample to the other. The objective of this ecxercise is to demonstrate the above method and not the actual values.

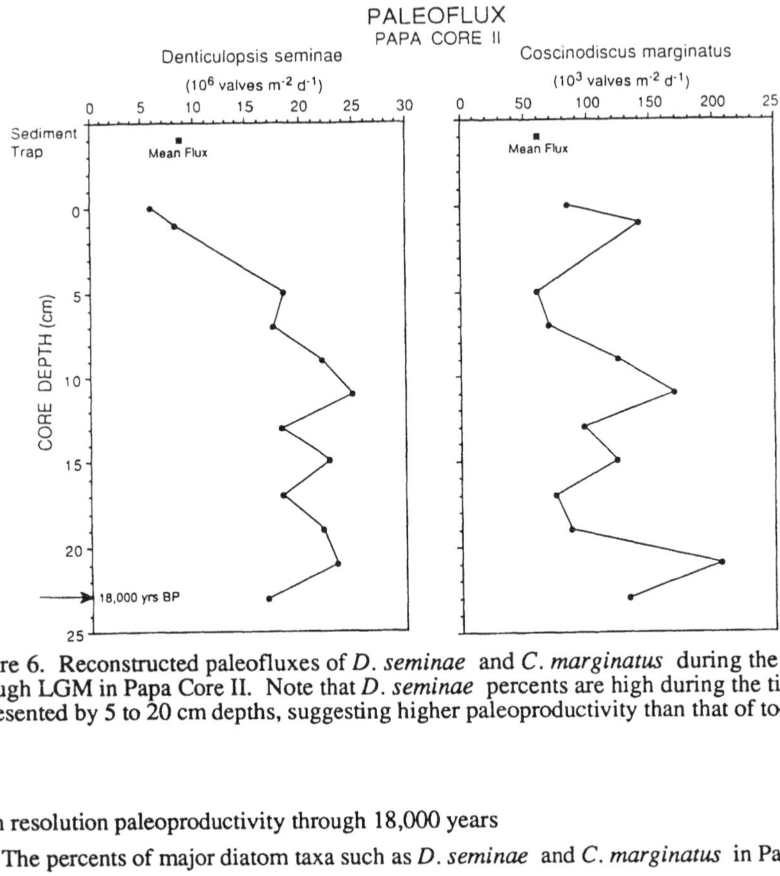

Figure 6. Reconstructed paleofluxes of *D. seminae* and *C. marginatus* during the Holocene through LGM in Papa Core II. Note that *D. seminae* percents are high during the time period represented by 5 to 20 cm depths, suggesting higher paleoproductivity than that of today.

High resolution paleoproductivity through 18,000 years

The percents of major diatom taxa such as *D. seminae* and *C. marginatus* in Papa Core II samples down to the last glacial maximum (LGM: 18,000 yrs BP, 23-24 cm) are illustrated (Figure 5). The LGM has been confirmed with oxygen isotope data by Dr. L. Keigwin (Pers. Comm., 1989). The diatom % data show a co-variation with % *C. marginatus* and valve diameter of *C. marginatus* (Figure 5). Since % *C. marginatus* is in part dependent on *D. seminae*, a mirror image of % *C. marginatus* with % *D. seminae* is not unexpected. Moreover, it is of interest that the frustule diameter of *C. marginatus*, independent from % *D. seminae*, measured throughout the last 18,000 years, correlates well with the percent counts of the above two dominant diatom taxa. According to % *D. seminae* data in Papa Core II and flux data, it is possible that productivity was continuously high during the time between 5 to 20 cm (approximately 4,000 to 15,000 years BP). The size of *C. marginatus* frustules was generally small during this time period. This also suggests high productivity, since the size of *C. marginatus* from the sediment trap samples tends to be smaller during seasonal high production periods than during low production periods (Takahashi, unpublished), although we need additional data to confirm this. In addition, the assumption that the degree of the siliceous microfossil preservation in down-core sediments are the same as the core tops, needs to be closely examined and validated in future work.

ACKNOWLEDGMENT

I am grateful to Dr. Constance Sancetta, Lamont-Doherty Earth Observatory, who kindly provided me with her unpublished diatom counts in the Lamont core RC10-216 top sample: the counts became very useful in this paper. The sediment trap program in the eastern subarctic Pacific was coordinated by Dr. Susumu Honjo, Woods Hole Oceanographic Institution; without his leadership this work could not be materialized. The early version of this paper has been reviewed by Dr. David Lazarus and Dr. Constance Sancetta for whom I am grateful. I also thank Dr. Michael Kaminski and Dr. Rainer Zahn, organizers, for invitation to attend the NATO Advanced Research Workshop on Carbon Cycling in the Glacial Ocean: Constraints on the Ocean's Role in Global Change, which inspired this paper.

REFERENCES

CLIMAP Project Members (1976) The surface of the Iceage Earth. Science 191:1131-1137
Curry WB, Lohmann GP (1986) Late Quaternary carbonate sedimentation at the Sierra Leone Rise (Eastern Equatorial Atlantic Ocean). Mar Geol 70:223-250
Erez J, Takahashi K, Honjo S (1982) In situ dissolution of Radiolaria in the Central North Pacific Ocean. Earth Planet Sci Lett 59:245-254
Honjo S (1976) Coccoliths: production, transportation and sedimentation. Mar Micropaleontol 1:65-79
Honjo S, Doherty KW (1988) Large aperture time-series sediment traps; design objectives, construction and application. Deep-Sea Res 35:133-149
Hurd DC (1983) Physical and chemical properties of siliceous skeletons. In: Aston SR (ed) Silicon Geochemistry and Biogeochemistry, Academic Press, London, p 187-244
Imbrie J, Kipp NG (1971) A new micropaleontological method for quantitative paleoclimatology: application to a late Pleistocene Caribbean Core. In: Turekian KK (ed) The Late Cenozoic Glacial Ages, New Haven, Connecticut, Yale University Press. p 71-181
Jouse A P , Kozlova O G, Mukhina VV (1971) Distribution of diatoms in the surface layer of sediment from the Pacific Ocean. In: Riedel WR, Funnell BM (eds) The Micropaleontology of the Oceans. Cambridge University Press, p 263-269
Martin JH, Gordon RM, Fitzwater S, Broenkow WW (1989) VERTEX: phytoplankton/iron studies in the Gulf of Alaska. Deep-Sea Research 36:649-680
Nelson DM (1975) Uptake and regeneration of silicic acid by marine phytoplankton. Ph.D. Thesis, Univ. Alaska, Fairbanks, Alaska. p 157
Sancetta C, Silverstri S (1984) Diatom stratigraphy of the Late Pleistocene (Brunhes) subarctic Pacific. Mar Micropaleontol 9:263-274
Smetacek VS (1985) Role of sinking in diatom life-history cycles: ecological, evolutionary and geological significance. Mar Biol 84:239-251
Takahashi K (1986) Seasonal fluxes of pelagic diatoms in the subarctic Pacific, 1982-1983. Deep-Sea Res 33:1225-1251
Takahashi K (1987a) Response of subarctic Pacific diatom fluxes to the 1982-1983 El Nino. J Geophys Res 92:4,387-14,392
Takahashi K (1987b) Seasonal Fluxes of Silicoflagellates and Actiniscus in the Subarctic Pacific During 1982-1984. J Mar Res 45:397-425

Takahashi K (1987c) Radiolarian Flux and Seasonality: Climatic and El Nino Response in the Subarctic Pacific, 1982-1984. Global Biogeochemical Cycles. 1: 213-231

Takahashi K (to be published) Sidocoenosis: a sinking assemblage mediating rapid particle transport to deep sea. Mar Micropaleontol

Takahashi K, Honjo S, Tabata S (1989) Siliceous phytoplankton flux: interannual variability and response to hydrographic changes in the northeastern Pacific. In: Peterson D (ed) Aspects of Climate Variability in the Pacific and Western Americas, Geophysical Monograph 55, Amer. Geophys. Union, p 151-160

Taylor FJR, Waters RE (1982) Spring phytoplankton in the subarctic North Pacific Ocean. Mar Biol 67:323-335

Appendix 1. Definitions of the paleoflux computations involved.

Flux	$F_x = C_x \cdot MF_x$
Accumulation Rate	$AR = DC \cdot r \cdot SR$
Percent Preservation (total assemblage)	$TP = AR / (\sum_{x=1}^{n} F_x \cdot 36.5)$
Percent Flux (species)	$PF_x = F_x \cdot 100 / \sum_{x=1}^{n} F_x$
Percent Sediment (species)	$PS_{x,y} = S_{x,y} \cdot 100 / \sum_{x=1}^{n} S_{x,y}$
Percent Preservation (species)	$PP_x = PS_{x,1} \cdot n \cdot TP / [PF_x \cdot \sum_{x=1}^{n} (PS_{x,1} / PF_{x,1})]$
k, constant	$K_x = -1 \cdot \ln(PP_x / 100)$
Reconstructed Percent Flux (species)	$RP_{x,y} = \quad PS_{x,y} \cdot 100 / [e^{-Kx} \cdot \sum_{x=1}^{n} (PS_{x,y} / e^{-Kx})]$
Reconstructed Flux (species)	$RF_{x,y} = 31569 \cdot 10\text{^} (0.03 \cdot RP_{1,y}) \cdot RP_{x,y} / RP_{1,y}$

x = Species : x = 1 = *D. seminae*

y = Core depth : y = 1 = core top

C_x = Species counts of sediment trap samples

MF_x = Multiplication factor

DC = Diatom counts (# valves $\cdot$ dry g^{-1})

r = Density ($g \cdot cm^{-3}$)

SR = Sedimentation rate ($cm \cdot 10^3\ yrs^{-1}$)

S_x = Species counts of core samples

n = Number of species

1 / 36.5 : Conversion factor from ($\# \cdot cm^{-2} \cdot 10^3\ yrs^{-1}$) to ($\# \cdot m^{-2} \cdot d^{-1}$)

LATE QUATERNARY PALEOPRODUCTIVITY VARIATIONS IN THE NE AND EQUATORIAL ATLANTIC: DIATOM AND C_{org} EVIDENCE

Fatima Abrantes, Kyaw Winn[1] and Michael Sarnthein[1]
Instituto Geológico e Mineiro
Rua Academia das Ciências, 19 2º
1200 Lisboa
Portugal

ABSTRACT. We have generated 350 ka records of diatom accumulation rates (DAR) and export production ($P_{exp.}$), based on organic carbon accumulation rates, for Core 16772 in the eastern Equatorial Atlantic (EEA). These records were compared to published records for the coastal upwelling areas off Portugal and NW Africa, in order to obtain a better understanding of NE Atlantic paleoproductivity variations.

Accumulation rates of the various siliceous microfossils correlate well at each location. The DAR shows a good overall correspondence to $P_{exp.}$ at all locations; however, discrepancies between the DAR and $P_{exp.}$ records at the beginning of the interglacial stages are noted for all areas. Assuming that both records are basically a measure of the same process, the primary productivity in the surface ocean, the observed discrepancies can be explained by changes in opal preservation and/or in the plankton community that dominated the equatorial waters during this time. The EEA records exhibit marked oscillations, with maxima occurring during glacial stages. This oscillating character is interpreted as deriving from variations in productivity through time, so that increased primary production has characterized every major glacial period for the last 350 ka in the area. However, only glacial Stages 6 and 4 are characterized by high abundance spikes of the large diatom *Ethmodiscus rex* (Rattray) Hendey.

Comparison of this equatorial upwelling record with the record of the coastal upwelling areas (1326 KS11 off Portugal and 12392-1 off NW Africa) for the last 120 ka, shows that although the maxima are of the same order of magnitude, the times of maximum values are not the same. In the eastern Equatorial Atlantic the most intense DAR and $P_{exp.}$ peaks are observed during Stage 4 but barely recorded in the coastal areas where the most productive period is Stage 2. This high productivity recorded during Stage 2 on coastal areas is also recorded in the EEA by a $P_{exp.}$ maximum of the same magnitude, although DAR values are 4xs lower than the observed in the coastal record. The difference in the timing of maximum siliceous production (DAR) between oceanic and coastal regimes which is not observed for the carbon

---

[1] Geologisch-Paläontologisches Institut und Museum, Universität Kiel, Germany

NATO ASI Series, Vol. I 17
Carbon Cycling in the Glacial Ocean:
Constraints on the Ocean's Role in Global Change
Edited by R. Zahn et al.

production record ($P_{exp.}$), is interpreted as an indication of important changes in the silica concentration of the upwelling source waters (intermediate waters).

INTRODUCTION

During the late Quaternary the earth experienced major changes in climate and in atmospheric CO_2 concentration. For some authors biological productivity is the dominant mechanism controlling atmospheric CO_2 (Sarnthein et al., 1988). Although this is a source of extensive debate, there are no doubts that marine plankton play an important role within the carbon cycle because they mediate the transfer of carbon between the atmosphere and the ocean. Thus, to constrain global models of the carbon cycle, chronologically well-constrained proxy records of primary production are essential.

It is well known that the Equatorial Atlantic is a region of intense high primary production due to westward advection of cool waters from the Benguela Current and Ekman divergence a few degrees south of the equator (Cane, 1979; Philander and Pacanowski, 1980). Model simulations show significant upwelling from deeper than 1000m at the equator both in winter and summer (Philander and Pacanowski, 1986). Upwelling is also widespread along the Portuguese and NW African coasts as a consequence of the interaction of the trade winds and the Canary Current. The upwelling is variable both in space and time: it occurs year round from 20°N to 25°N, being most intensive off Cape Blanc. During the winter it occurs south of Sierra Leone and during the summer (from May to September) it extends further north up to Portugal (Barton et al., 1977; Fiuza, 1983, 1984; Jones and Halpern, 1981; Smith, 1968).

Due to the high productivity (Koblentz-Mishke et al. 1970; Lisitzin, 1972; Berger, 1989), there is an abundant supply of siliceous microfossils to the sea floor and biogenic opal is relatively well preserved in the eastern Equatorial Atlantic sediments (Pokras, 1987; Pokras and Molfino, 1986; Van Bennekom and Berger, 1984). It is the goal of this paper to offer an insight into the paleoproductivity variations in the eastern Equatorial Atlantic (EEA), by examining the temporal changes in diatom accumulation rates (DAR) and export production ($P_{exp.}$) obtained from the C_{org} content of the sediments (Sarnthein et al., 1992), and, to allow a better understanding of NE Atlantic paleoproductivity and nutrient availability by comparing the EEA record to records previously obtained in the NE Atlantic upwelling areas off Portugal and NW Africa (Abrantes, 1990a, b).

METHODS

We focus here on data from piston cores 16772 recovered from the eastern equatorial Atlantic by the R/V Meteor, 1326 KS11 from the Portuguese margin, and 12392 from the NW African margin (Abrantes, 1990a,b) (Figure 1).

For this study the core was sampled at approximately 10 cm intervals, and each sample integrates 2 cm of record. All samples were analyzed for siliceous microfossils (diatoms, silicoflagellates, and radiolaria), diatom assemblage composition, and organic carbon content (C_{org}). Quantitative microfossil slides were prepared using the procedures described by Abrantes (1988). Siliceous microfossil abundances and accumulation rates were also calculated as outlined in Abrantes (1990a). In order to compare samples in a statistically significant way, three subsamples of each sample were counted. These data indicate that as the number of individuals present in the sample increases, so does the sample range, however in no case is the sample range larger than the difference between two consecutive samples.

All marine diatoms were counted as whole valves except for *Ethmodiscus rex* (Rattray) Hendey. *Ethmodiscus rex* , one of the largest diatoms in marine plankton (Wiseman and Hendey, 1953), has been described as an important component of the diatom assemblage found in tropical sediments, often forming *Ethmodiscus* oozes (Gardner and Burckle, 1975, Mikkelsen, 1977; Schrader, 1974; Stabell, 1986). This species was never observed complete; counts numbers refer to fragments as large or larger than 50µm.

Export production (P_{exp}) was calculated by the empirical formula proposed by Sarnthein et al. (1988) and modified by the same author in 1992 (Sarnthein et al., 1992). The organic carbon concentration was determined by calculating the difference between total carbon and carbonate carbon. The total carbon content was measured by both mass spectrometry and a Carlo Elba elemental analyzer.

Chronostratigraphy is based on oxygen isotope analyses made on the planktic foraminifera *G. sacculifer*. Correlation of the $\delta^{18}O$ record with the curve of Martinson et al. (1987) provides the chronologic framework prior to 30 ka. For the last 30 kyr, age control is based on interpolation of U/Th age (Bard et al., 1990), converted from Accelerated Mass Spectrometry (AMS) ^{14}C ages following the technique of Winn et al. (1991). All the data is available upon request.

To analyze the relationships between the productivity proxies utilized and the isotopic (climatic) signal a cross-correlation procedure was applied (Davis, 1986). The $\delta^{18}O$ record was cross correlated with: (1) DAR, (2) $P_{exp.}$, (3) radiolarian accumulation rates (AR), and (4) silicoflagellate AR for Core 16772. The DAR record was also cross correlated with the accumulation rates observed for the other siliceous microfossils and P_{exp} of Cores 16772 and 12392. Prior to this analyze, all the data were linearly interpolated to 2 kyr intervals.

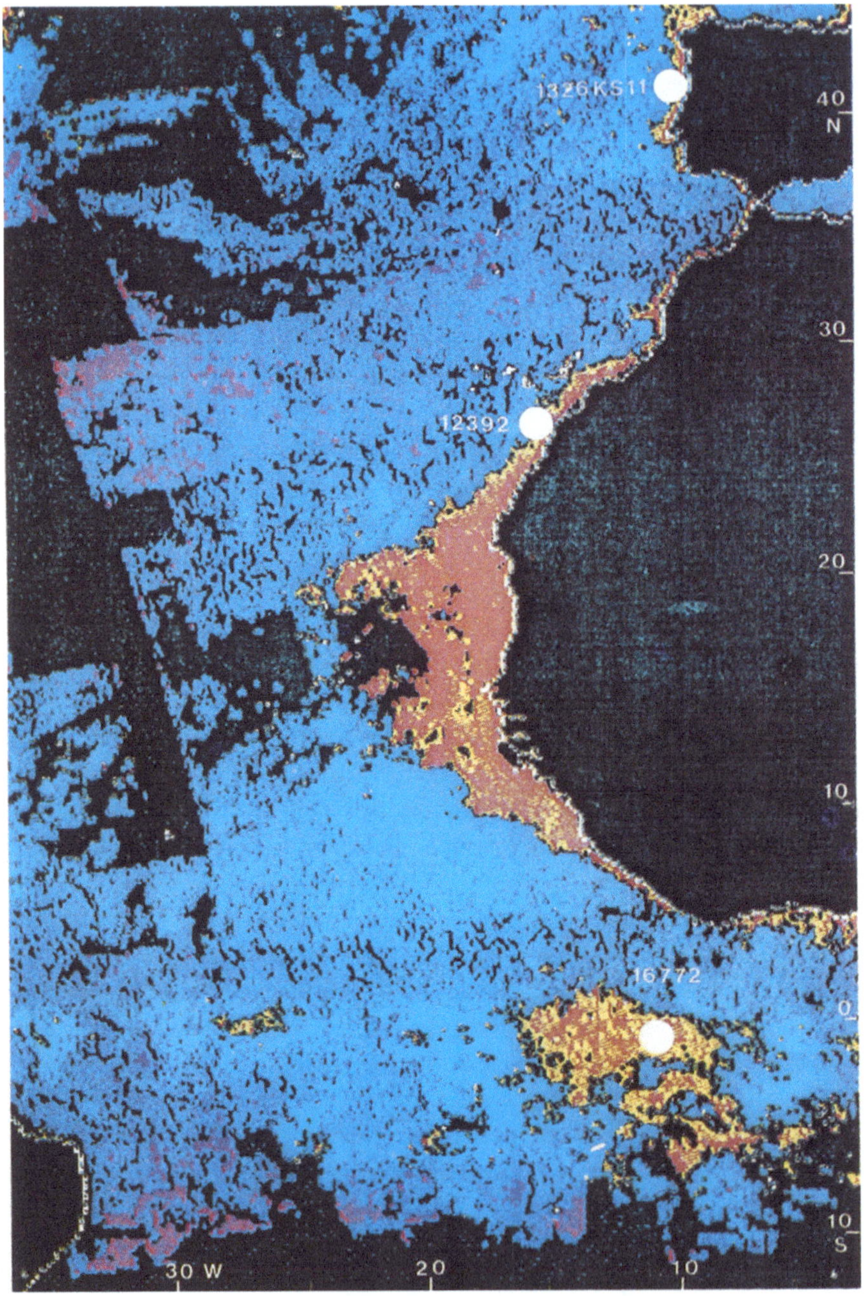

Figure 1. Satellite-derived image estimating estimating distribution of phytoplankton, taken in September 1979, and where the red, orange and yellow colors correspond to high chlorophyll concentrations. Solid dots represent cores location. Cores position is as follows: 16772 1°21' S and 11° 58' W; 12392-1 25°10' N and 16°51' W; 1326 KS11 40° 41'N and 10°12' W

RESULTS

Core 16772 is a 15.07 m long sequence of undisturbed pelagic sediments. The sedimentation rate at this site varies from 2 to 9 cm/kyr. During interglacials the sedimentation rates average 4.5 cm/kyr, increasing up to 6 cm/kyr during glacials.

The DAR record for Core 16772 (Figure 2) shows an oscillating pattern with relative maxima values in glacials and glacial terminations and relative minima during interglacials, although, somewhat high values are also observed for substages 7.3, 5.5 and 5.3. Note that during Stage 2 DAR values are four times lower than the observed at other glacial stages and of the same order of magnitude of the observed at the 2/1 boundary. The described pattern is also observable in the accumulation rate records of the diatom fragments, silicoflagellates, and radiolaria (Figure 2).

The diatom genus *Ethmodiscus* is present throughout the core, showing an increase in AR by 1.5 to 2 orders of magnitude during glacial Stages 6 and 4 (Figure 2).

The $P_{exp.}$ record, calculated from Sarnthein et al. (1992) (Figure 3) varies cyclically and relatively regularly over the last 350 ka. Maximum concentrations occur in glacial Stages 8, 6, 4 and 2, while relative minima characterize the interglacial periods, implying that production was higher during glacials.

The results of the cross correlation procedure are presented in Table 1. By this analysis, two time series can be compared with each other in order to determine positions of pronounced correspondence. Two items of information emerge from such a comparison: The strength of the relationship between the two series (R), and the lag or offset in time between them at their position of maximum equivalence. The position of greatest correlation ($r \geq 0.6$) between DAR and the AR of any other siliceous microfossil considered occurs at lag 0, when both series are in phase. The highest correspondence between DAR and P_{exp} occurs also at lag 0 and is positive, however, there is a somewhat higher correlation for core 12392 ($r=0.52$) than for core 16772 ($r=0.42$). The oxygen isotopic record shows no correlation with all the siliceous microfossils, but has a positive correlation ($r=0.6$) at zero lag with the P_{exp}.

DISCUSSION

The DAR record of the EEA shows maximum concentrations during glacial periods while relative minima characterize the interglacial periods. In Stages 8, 6 and 2, DAR values are also typically high near each termination. Note also the decline in DAR younger than 50 ka, and, the amplitude of the Stage 2 maximum relative to the maxima and mean values observed from 350 to 50 ka. Do these variations reflect the history of paleoproductivity in the EEA for the last 350 kyr?

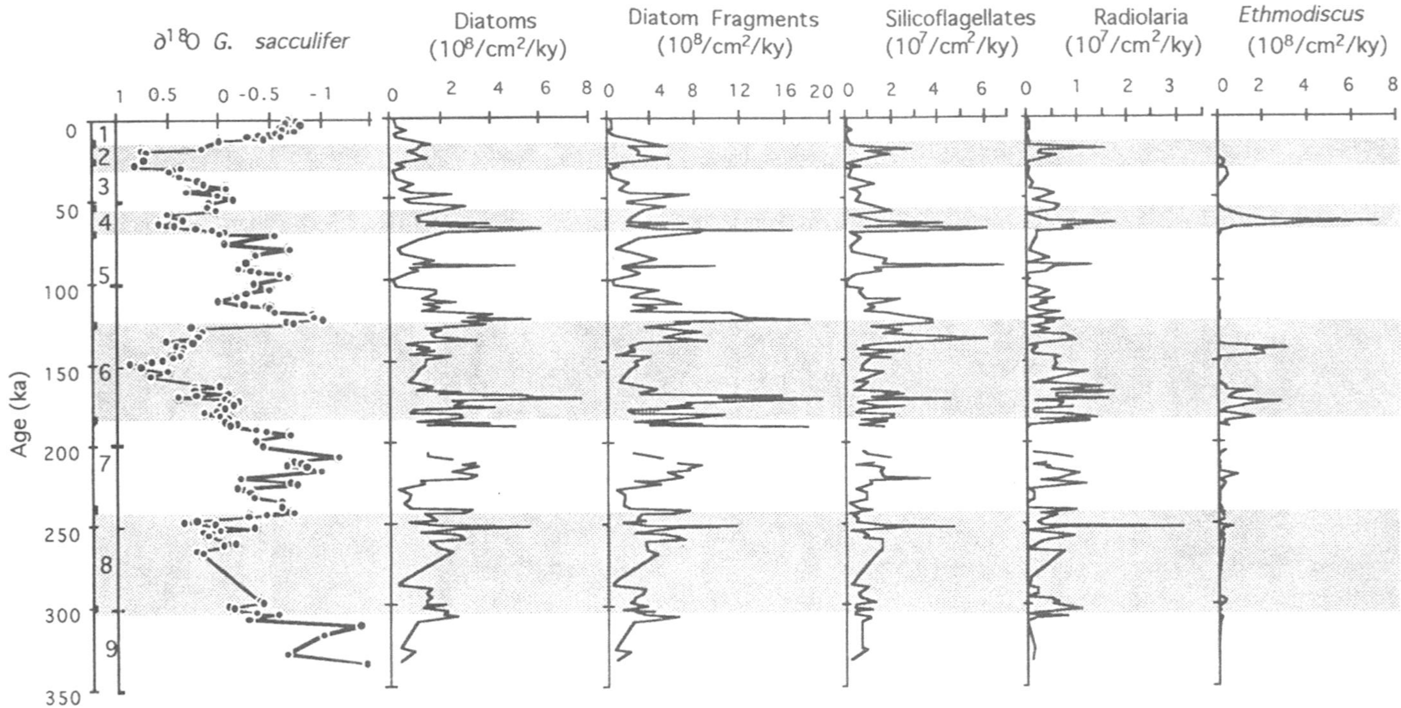

Figure 2. $\delta^{18}O$ and accumulation rates (AR) for diatoms, diatom fragments, silicoflagellates, radiolarians and *Ethmodiscus*. Shading indicates glacial periods.

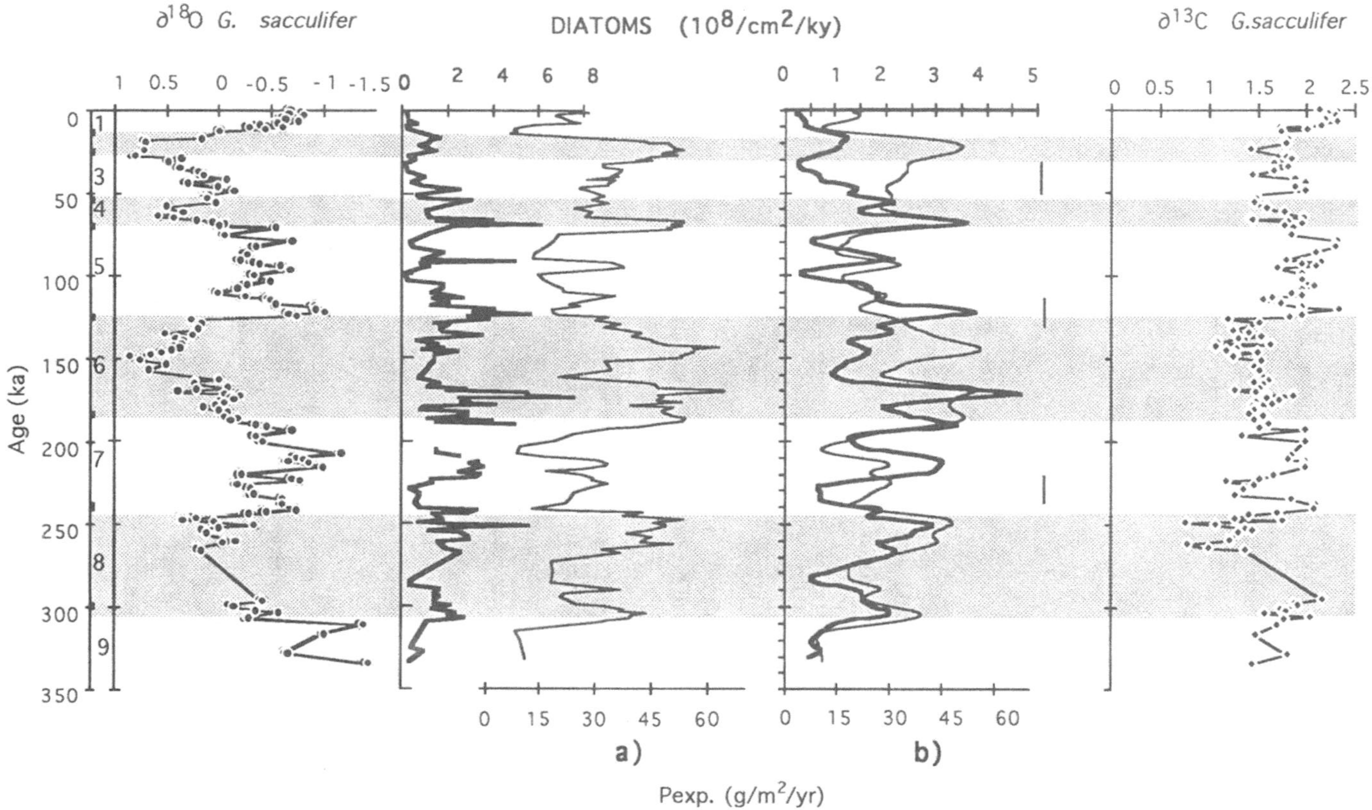

Figure 3. $\delta^{18}O$, Diatom Accumulation Rate (DAR - heavy line) and organic carbon derived export Production (P_{exp}, thin line) for Core 16772; a) calculated values; b) smoothed curves; c) $\delta^{13}C$ record of *G. sacculifer*, from Westerhausen (1991). Shading indicates glacial periods.

Table 1. Results of Cross-Correlation Analysis for Cores 16772 and 12392. DAR- diatom accumulation rate; $P_{exp.}$- export production derived from the C_{org} content; RadAR- radiolaria accumulation rates; SilAR- silicoflagellates accumulation rates; FragInd- fragmentation index.

	R	Lag (ky)
16772 - EEA		
DAR/Pexp	0.42	0
DAR/RadAR	0.60	0
DAR/SilAR	0.73	0
$\delta^{18}O$/DAR	0.23	42
$\delta^{18}O$/Pexp	0.62	0
$\delta^{18}O$/RadAR	0.33	0.5
$\delta^{18}O$/SilAR	-0.20	-27
DAR/FragInd	0.21	38
12392 - NW Africa		
DAR/Pexp	0.52	0
DAR72/DAR92	0.36	-37

Broecker and Peng (1982) argue that production and preservation of opaline tests in the sediments are closely correlated, thus, periods of higher productivity should be reflected in higher accumulation rates of biogenic silica. Abrantes (1992) has shown that the variation observed in the diatom record of NE Atlantic coastal upwelling areas agreed with $P_{exp.}$ estimates as well as with the bottom-pore water difference in $\delta^{13}C$ ($\Delta\delta^{13}C$). More recently, however, Archer et al., (1993) suggest that opal accumulation rates reflect a significant artifact of dilution by other phases. If so, we need to determine whether the higher glacial DAR are due to increased surface productivity, better preservation, or additional supply of other sediment components, that is, increased sedimentation rate. There is evidence that sedimentation rates increased during glacials. As such, it is possible that the apparent higher DAR are a result of higher bulk sedimentation. If that is the case, a high correlation rate should exist between the diatom abundance and the sedimentation rate. However, diatom abundances both as nº/cc or nº/g are not correlated to sedimentation rate (r^2 =0.02 and 0.01 respectively), and the r^2 obtained between DAR and the sedimentation rate is 0.31. When we plot diatom abundance (nº/g), and DAR the same pattern emerges (Figure 4), and the r^2 between these two variables equals 0.55. On the basis of these results, we conclude that even though an increase in sedimentation rate may have led to a better preservation of the siliceous material as argued by Broecker and Peng (1982), the DAR variations observed are not just a function of the sedimentation rate. The possibility of differential dissolution was also considered and a first check was achieved by means of a cross correlation between the silicoflagellate record and diatom and radiolaria, which are progressively more resistant to dissolution (Schrader, 1972). The results show strong statistical coherence and zero phase lag (Table 1) which provide some evidence to minimize the effect of dissolution on the record. The fragmentation index, usually calculated as the ratio between diatom fragments and whole diatoms, is another way of evaluating the preservation state of the samples (Abrantes, 1990a, b). Figure 5 shows the distribution pattern of such index for Core 16772. In order to facilitate the comparison, we have generated a smoothed curve

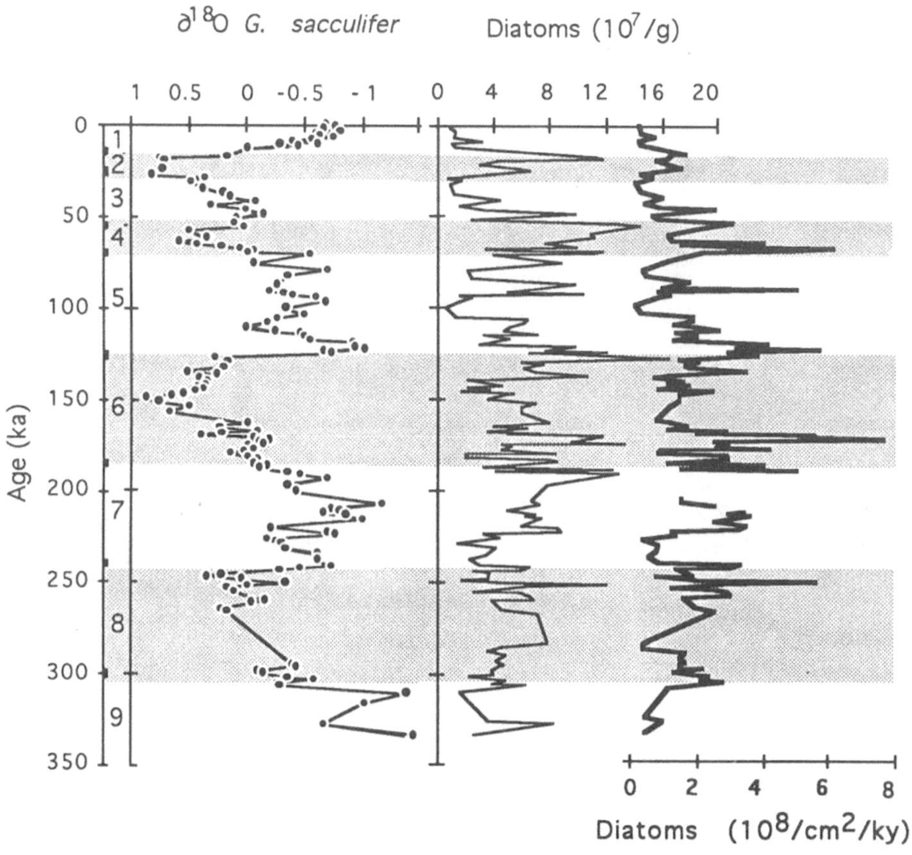

Figure 4. Distribution pattern of diatoms as number per gram of dry sediment (thin line), and as accumulation rate (bold line). Shading indicates glacial periods.

of both the FragIndex and the DAR records (Figure 5). Visual comparison of the two curves shows a generally good agreement of higher fragmentation rates at times of higher DAR, the opposite of what would be expected if dissolution were controlling the diatom content of the samples. Exceptions to this general trend are observed for the period between 336 to 250 Ka where the fragmentation rate appears to be more or less constant and, two intervals, 250-220 Ka and 105-50 Ka where the two curves have a mirror type image indicating that the low DAR values may be a reflection of increased fragmentation. A cross correlation of the DAR and FragIndex shows no correlation (r=0.21) indicating that the variation in DAR is not a simple matter of preservation.

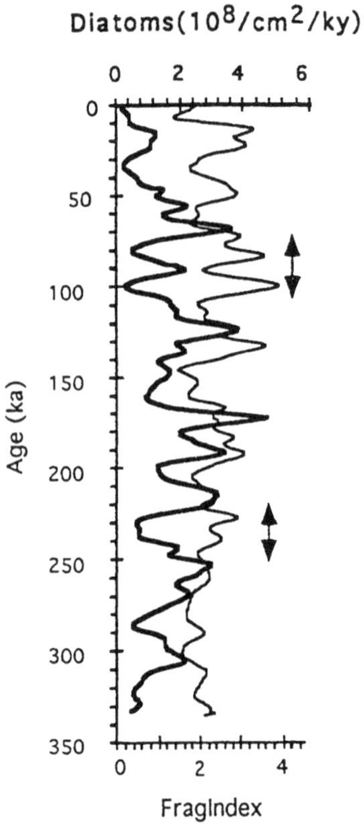

Figure 5. Distribution pattern of Diatom accumulation rate (DAR - heavy line) and the fragmentation index (FragIndex = nº fragments/nº complete diatoms). Both data sets were smoothed with a three-point running average.

Both analyses agree and indicate that the major peaks of the DAR record are not just artifacts of sedimentation rate or dissolution. We therefore believe that they are representative of true increases in primary production and flux to the seafloor. Even though we can offer no absolute paleoproductivity values based on the DAR record, we can estimate the changes relative to the late Holocene values. In this way we estimate an increase in siliceous productivity (DAR) in the EEA by a factor of 6 for the last transition and the last glacial, and by one order of magnitude for the maxima during glacial Stages 4,6 and 8.

Intuitively it seems reasonable that relationships should exist among the DAR and the $P_{exp.}$ records. Diatoms are the phytoplankters that transform excess nutrients into new production

(Margalef, 1978, 1979, 1985), and the $P_{exp.}$ empirical relationship was calibrated by Sarnthein et al., (1988, 1992) having in mind the concept of new production. How closely tied are these two records in the EEA?

The comparison of the DAR and $P_{exp.}$ records for the EEA Core 16772 (Figure 3) indicates that the two profiles are similar in overall pattern but show important differences. The two records differ at the base of interglacial Stages 7 (235-225 ka), 5 (130-115 ka) and all of Stage 3 (50-30 ka). For Stages 7 and 3 the DAR record decreases while the $P_{exp.}$ curve shows a continuous increase. At the Stage 5 DAR shows a maximum while $P_{exp.}$ has a minimum.

Several authors have focused their attention in the problem of the C_{org} record. Their arguments are discussed in detail by Calvert (1987) and Pedersen and Calvert (1990), who concluded that carbon accumulation in marine sediments is largely controlled by production and not by bottom water oxygenation. Here we propose that changes in productivity are the major cause of the observed DAR variations. The results of the cross correlation analysis indicate a correlation of 0.52 between the DAR and P_{exp} records. These results provide some indication that both proxies react in the same way to productivity variations. Although, a check to these paleoproductivity indicators, by means of an independent productivity proxy would better support such indications. According to Broecker and Peng (1982), a decrease in nutrients in the surface waters parallels a decrease in the $^{13}C/^{12}C$ ratio of the bicarbonate in the surface water. Planktonic foraminifera species, such as *Globigerinoides sacculifer*, which dwell in the surface mixed layer, may provide a $\delta^{13}C$ record of the nutrient content in the surface ocean (Deuser et al., 1981; Deuser and Ross, 1980). The observation of the *Globigerinoides sacculifer* $\delta^{13}C$ record for core 16772 from Westerhausen (1991), (Figure 3) clearly demonstrates that glacial surface waters were ^{13}C depleted, that is nutrient enriched, relatively to Holocene values by about 1‰ during Stages 2 and 4 and by 1.5‰ during Stage 6, confirming the DAR and $P_{exp.}$ indications.

Therefore, the deviations observed between the DAR and $P_{exp.}$ records may be an important indication of changes in the preservation potential of opal or in the type of particles exported to the sediment. The first hypothesis involves variations in post-depositional dissolution of opal. An inversion of the productivity signal due to opal dissolution in the sediments is unlikely, since opal preservation is enhanced by an increase in production and/or bulk sediment accumulation (Broecker and Peng, 1982). Opal preservation can also be a function of bottom water chemistry, yet, deep waters are equally undersaturated with respect to Si everywhere in the ocean, a situation that according to Mortlock et al. (1991) is not likely to have changed during the late Quaternary. The decrease in DAR values observed for Stages 7 and 3 do not correlate with a sedimentation rate decrease, though high values of the fragmentation index at the base of Stage 7 are indicative of poor preservation. However, no dissolution spike is observed for the Stage 3 (50-30 ka) discrepancy, a feature also observed in Cores 12392 (off NW Africa) and 1326 KS11 (off Portugal) (Figure 6). Besides, the Stage 5 (130-115 ka) interval shows an increase in DAR while $P_{exp.}$ decreases. The lack of evidence for dissolution in Stage 3 as well as the reverse trend observed for the DAR curve in early Stage 5, when similar thermohaline conditions are likely to have occurred during 5 and 3, based on the work of Oppo and Fairbanks (1990), weaken the case for bottom water chemistry changes. We therefore suggest, but cannot prove at this point, that a decrease in opal production at the surface as a result of a decrease in upwelling (increasing the water column stability) and/or in the PO_4/NO_3 versus SiO_2 ratio, and consequent change in the plankton community may be a better explanation for the observed discrepancies. In fact, the available data on planktonic foraminifera (Figure 6 of Sarnthein et al., 1992) reveal the occurrence of important maxima at the time intervals at which DAR decreases, the 235-225 and 50-30 ka intervals, and a lower magnitude foram maximum at the base of Stage 5 (130-115 ka), when DAR is observed to increase.

COASTAL VERSUS EQUATORIAL ATLANTIC

A comparison of the coastal upwelling records (Abrantes, 1990a, b) and the EEA (equatorial upwelling) record, for the last 120 ka (Figure 6) reveals that they differ in two important aspects:

a) Maximum DAR production in coastal areas occurred during Stage 2 while in the EEA maximum production occurred during Stage 4.

b) Higher productivity in coastal areas is recorded during Stage 2 in both DAR and $P_{exp.}$ by maxima of the same magnitude, while in the EEA the two records differ substantially in magnitude.

Changes in productivity in the NE Atlantic coastal upwelling areas have been hypothesized to have a direct climatic link, and increased production during Stage 2 has been interpreted as a result of the intensification of coastal upwelling due to the intensification of the meridional trade winds (Abrantes, 1992; Sarnthein et al, 1982).

In the EEA, during Stage 2 low DAR values are observed, while both the $P_{exp.}$ record as well as the planktonic foraminiferal $P_{exp.}$ support the hypothesis of high productivity (Sarnthein et al., 1992). A nutrient depletion of the intermediate waters (oceanic upwelling source) through this coastal upwelling intensification (Herguera, 1992), or by a high latitude stripping of the intermediate waters (Labeyrie, 1987) may be possible explanations for the Stage 2 data, as well as for the reverse situation that occurs in Stage 4. However, it is difficult to explain the maxima of the same magnitude that occur in both coastal and oceanic areas during Stage 6, a time at which the thermohaline circulation of the north Atlantic was similar to that of Stages 2 and 4 (Oppo and Fairbanks, 1990). Another consideration is the important contribution of *Ethmodiscus rex* (Rattray) Hendey to the diatom assemblage of the EEA during

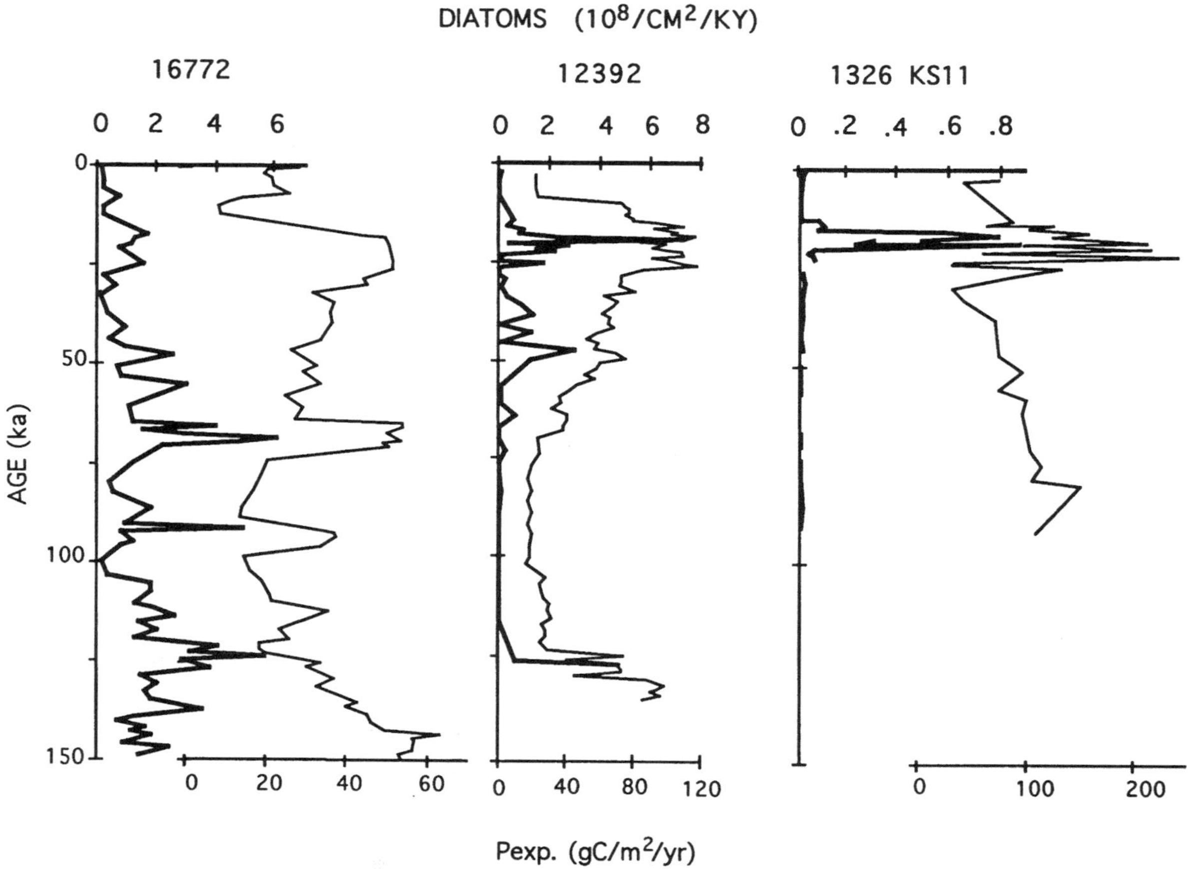

Figure 6. Distribution pattern of the DAR (bold line) and P_{exp} (thin line) for the last 120 ka in Cores 16772, 12392 and 1236-KS11.

glacial Stages 6 and 4 but not in Stage 2 (Figure4). The formation of *E. rex* oozes has been attributed to high primary productivity by Gardner and Burckle (1975), by Stabell (1986), and, supported by Pokras (1987). The occurrence in such large amounts of one of the largest diatoms of the marine plankton would require an important supply of Si (Gardner and Burckle, 1975). It seems though, that besides changes in the Si cycling and consequent changes in the silica concentration of the upwelling source waters through time, we may also need to consider changes in the supply of silica to the ocean. A two-fold higher dissolved silica river flux into the oceans during glacial periods, as a result of continental weathering intensification, has in fact been proposed by Froelich et al. (1992). This would support the higher siliceous productivity observed during glacials in general. However, to explain the high siliceous production observed in both coastal and equatorial regions during Stage 6, as opposed to high coastal production for Stage 2, or high equatorial siliceous productivity for Stage 4, we may need to invoke stronger continental weathering associated to a more intense glaciation for Stage 6.

CONCLUSIONS

The analysis of the DAR record from Core 16772 reveals that productivity in the eastern Equatorial Atlantic fluctuates between glacials and interglacials, being relatively higher in glacials, glacial terminations and substages 7.3, 5.5 and 5.3, and lower during interglacials. An overall agreement between the DAR and $P_{exp.}$ records is observed, however, there are significant discrepancies during certain periods between the two records. The reasons for these discrepancies are not entirely clear, however, we believe that they reflect shifts in the plankton community driven by changes in the water column stability (mixing/upwelling) and/or changes in the intermediate waters PO_4/NO_3 versus Si ratio.

The intercomparison of the records from the EEA and the NE Atlantic upwelling areas off Portugal and NW Africa, reveals:

- an overall correspondence between the DAR and $P_{exp.}$ records, in each location, supporting the idea that both indicators primarily reflect export production.
- a lack of response of the siliceous record in the EEA during Stage 2, time of maximum DAR values in the coastal upwelling areas suggesting that the origin and Si content of the upwelling source waters (intermediate depths) may be a major factor determining production.

ACKNOWLEGMENTS

Funding has been provided by JNICT grant PMCT/C/MAR698/90 and the German National Program of Climate Research. The authors thank C. Sancetta and J. Herguera for critical reviews and comments, H. Monteiro for reviewing an early version of this manuscript, and A. Inês for laboratory collaboration.

REFERENCES

Abrantes F (1988) Diatom productivity peak and increased circulation during latest Quaternary deglaciation: Western Mediterranean, Mar. Micropaleontol. 13: 76-96.

Abrantes F (1990a) Increased upwelling off Portugal during the last glaciation: Diatom evidence, Mar. Micropaleontol. 17: 285-310.

Abrantes F (1990b) Variability of upwelling off NW Africa during the latest Quaternary: Diatom evidence,Paleoceanography 6: 431-460.

Abrantes F (1992) Paleoproductivity oscillations during the last 130Ky along the Portuguese and NW African margins, In: C. Summerhays and W. Prell (Eds) Evolution of the upwelling systems from the Miocene to the Present, Geolog. Soc. Spec. Pub. 64,. London. 499-510.

Archer D, Lyle M, Rodgers K, Froelich P (1993)What controls Opal Preservation in Tropical Deep-Sea Sediments?, Paleoceanography 8: 7-22.

Bard E, Hamelin R, Fairbanks R, Zindler A (1990) Calibration of the 14C timescale over the past 30,000 years using mass spectrometric U-Th ages from Barbados corals, Nature 345: 405-410.

Barton E, Huyer A, Smith R (1977) Temporal variation observed in the hydrographic regime near Cabo Corveiro in the NW African upwelling region, February to April 1974, Deep Sea Res. 24: 7-23.

Berger WH (1989) Global Maps of Ocean Productivity. In: W. H. Berger, V. S. Smetacek and G. Wefer (ed.), Productivity of the Ocean: Present and Past. John Wiley & Sons, p. 429-455.

Broecker WS , Peng TH (1982) Tracers in the Sea. Eldigio Press.

Calvert SE (1987) Oceanographic controls on the accumulation of organic matter in marine sediments. In: J. B. a. A. J. Fleet (ed.), Marine Petroleum Source Rocks. Blackwell, p. 137-151.

Cane MA (1979) The response of an equatorial ocean to simple wind stress patterns: I, Model formation and analytic results, J. Mar. Res. 37: 233-252.

Davis JC (1986) Statistics and data analysis in Geology, John Wiley & Sons. 646 pp.

Deuser WG , Ross EH (1980) Seasonal change in the flux of organic carbon to the deep Sargasso Sea, Nature 238: 364-365.

Deuser WG, Ross EH, Hemleben C , Spindler M (1981) Seasonal changes in species composition, numbers, mass, size and isotopic composition of planktonic foraminifera settling in the deep Sargasso Sea, Palaeogeography, Palaeoclimatology, Palaecology 33: 103-127.

Fiuza A (1983) Upwelling patterns off Portugal, In: E. Suess and J. Thiede (Ed.) Coastal Upwelling its Sediment Record. Part A: Responses of the sedimentary regime to Present Coastal Upwelling. Plenum, New York, 85-98.

Fiuza A (1984) Hidrologia e Dinamica das Aguas Costeiras de Portugal, Universidade de Lisboa.

Froelich P, Blanc V, Mortlock R, Chillrud S, Dunstan W, Udomkit A, Peng T-H (1992) River fluxes of Dissolved Silica to the Ocean Were Higher During Glacials: Ge/Si in Diatoms, Rivers and Oceans, Paleoceanography 7: 739-768.

Gardner JV, Burckle LH (1975) Upper Pleistocene *Ethmodiscus rex* oozes from the eastern equatorial Atlantic, Microp. 21: 236-242.
Herguera JC (1992) Deep-sea benthic foraminifera and biogenic opal: Glacial to postglacial productivity changes in the western equatorial Pacific, Mar. Microp. 19: 79-98.
Koblentz-Mishke O J, Volkovinsky VV, Kabanova JG (1970) Plankton productivity of the world ocean, National Academy of Sciences.
Labeyrie L, Duplessy JC, Blanc PL (1987) Variations in the mode of formation and temperature of oceanic deep waters over the past 125,000 years, Nature 327: 477-482.
Lisitzin AP (1972) Sedimentation in the world ocean, Spec. Pub. Soc. Econ. Paleontol. Mineral. 17: 218pp.
Margalef R (1978) Life-forms of phytoplankton as survival alternatives in an unstable environment, Oceanol. Acta 1: 493-509.
Margalef R (1979) Phytoplankton communities in upwelling areas. The example of NW Africa, Oecol. Aquat. 3: 97-132.
Margalef R (1985) Primary production in upwelling areas. Energy, global ecology and resources, Proceedings of the Simposio International Sobre las areas de Afloramiento mas importantes del oeste Africano (Cabo Blanco Y Buenguela)., Instituto de Investigaciones Pesqueras, Barcelona., 225-232.
Martinson DG, Pisias NG, Hays J D, Imbrie J, Jr Moore, Cand T, Shackleton NJ (1987) Age dating and the Orbital theory of the Ice Ages: Development of a High-Resolution 0 to 300,000-Year Chronostratigraphy, Quat. Res. 27: 1-29.
Mikkelsen N (1977)On the origin of *Ethmodiscus* ooze, Mar. Microp. 2: 35-46.
Mortlock R A , Charles CD, Froelich PN, Zibello MA, Saltzman J, Hays JD, Burckle LH (1991) Evidence for lower productivity in the Antarctic Ocean during the last glaciation, Nature 351: 220-223.
Oppo DW, Fairbanks RG (1990) Atlantic Ocean Thermohaline circulation of the last 150,000 years: Relationship to Climate and Atmospheric CO_2, Paleoceanography 5: 277-288.
Pedersen TF, Calvert SE (1990) Anoxia vs. productivity: What controls the formation of organic-carbon-rich sediments and sedimentary rocks?, Am. Assoc. Pet. Geol. Bull. 74: 454-466.
Philander SGH, Pacanowski R (1980) The generation of equatorial currents, J. Geophys. Res. 85: 1123-1136.
Philander SGH, Pacanowski RC (1986) The mass and heat budget in a model of the Tropical Atlantic ocean, J. Geophys. Res. 91: 14212-14220.
Pokras E (1987) Diatom record of late Quaternary climatic change in the eastern equatorial Atlantic and Tropical Africa, Paleoceanography 2: 273-286.
Pokras E, Molfino B (1986) Oceanographic control of diatom abundances and species distributions in surface sediments of the Tropical and southeast Atlantic, Mar. Microp. 10: 165-188.
Sarnthein M, Thiede J, Pflaumann U, Erlenkeuser H, Fütterer D, Koopmann B, Lange H, Seibold E (1982) Atmospheric and Oceanic Circulation patterns off Northwest Africa during the past 25 million years, p. 584-604. In: U. V. Rad, K. Hinz, M. Sarnthein and E. Seibold (ed.), Geology of the Northwest African Continental Margin., Springer-Verlag.
Sarnthein M, Winn K, Duplessy JC, Fontugue MR (1988) Global variations of surface ocean productivity, Paleoceanography 3: 361-399.
Sarnthein M , Pflaumann U, Ross R, Tiedemann R, Winn K (1992) Transfer functions to reconstruct ocean palaeoproductivity: a comparison, In: C. Sumerhays and W. Prell (Eds) Evolution of the upwelling systems from the Miocene to the Present, Geol. Soc. Spec. Pub. 64,. London. 411-428.
Schrader H (1972) Kieselsäure-Skelette in Sedimenten des ibero-marokkanischen Kontinentalrandes und angrezender Tiefsee-Ebenen, Meteor-Forsch.-Ergebnisse Reihe C: 10-36.
Schrader HJ (1974) Cenozoic marine planktonic diatom stratigraphy of the tropical Indian Ocean, Deep Sea Drill. Proj. Initial Rep. 24: 887-967.
Smith RL (1968) Upwelling, Oceanogr. Mar. Biol. 6: 11-46.
Stabell B (1986) Variations of diatom flux in the eastern equatorial Atlantic during the last 400,000 years ("Meteor" cores 13519 and 13521), Mar. Geology 72: 305-323.

Van Bennekom AJ, Berger GW (1984) Hydrography and silica budget of the Angola Bassin, Neth. J. Sea Res. 17: 149-200.
Westerhausen L (1992) Organic sediments of the equatorial east Atlantic: Effects of origin, transport, diagenesis, and climate,Dissertation, Christian-Albrechts-Universitat Kiel.
Winn K, Sarnthein M, Erlenkeuser H (1991) $\delta^{18}O$ Stratigraphy and Chronology of Kiel Sediment cores from the East Atlantic, Berichte - Reports 45: 99.
Wiseman JDH, Hendey NI (1953). The significance and diatom content of a deep-sea floor sample from the neighborhood of the greatest oceanic depth, Deep-Sea Res. 1: 47-59.

GLACIAL-HOLOCENE PALEOPRODUCTIVITY OFF WESTERN AUSTRALIA: A COMPARISON OF PROXY RECORDS

D. C. McCorkle, H. H. Veeh[1], and D. T. Heggie[2]
Department of Geology and Geophysics
Woods Hole Oceanographic Institution
Woods Hole, MA 02543
USA

ABSTRACT. We compare paleoproductivity proxy records from a set of gravity cores from the Exmouth Plateau (≈ 19°S, 113°E, 950 to 2250 m) and the Perth Basin (≈ 27°S, 111°E, 2750 m) in the southeastern Indian Ocean. In general, these proxies indicate higher surface-ocean productivity in this region at the Last Glacial Maximum (LGM, isotope Stage 2). LGM sediment accumulation rates and the accumulation rates of biogenic sediment components ($CaCO_3$ and organic carbon) are a factor of 1.5 to 2 higher than Holocene values. Benthic foraminiferal abundances and accumulation rates are both higher in glacial sediments, as are the concentrations and accumulation rates of authigenic uranium in the sediments. These benthic foraminiferal abundance and authigenic uranium data suggest higher surface ocean productivity during the glacial, but we cannot yet relate them to carbon flux quantitatively. In contrast to these three approaches, a productivity proxy based on paired-species benthic foraminiferal $\delta^{13}C$ differences shows little glacial-Holocene change. Possible explanations for this disagreement are discussed.

Together, the data suggest that the glacial productivity off Western Australia was elevated relative to Holocene values, and support the hypothesis that a north-flowing west Australian current led to coastal upwelling and enhanced primary productivity off western Australia during the Last Glacial Maximum. However, glacial productivity was high only relative to the low productivity characteristic of this region in the modern ocean. We see no evidence of strong upwelling similar to that observed in the modern ocean off the southwestern coasts of Africa and South America.

---

[1] School of Earth Sciences, Flinders University, GPO Box 2100, Adelaide, 5001 Australia
[2] Australian Geological Survey Organization, Program in Marine Geosciences and Petroleum Geology, GPO Box 378, Canberra, 2601, Australia

NATO ASI Series, Vol. I 17
Carbon Cycling in the Glacial Ocean:
Constraints on the Ocean's Role in Global Change
Edited by R. Zahn et al.

INTRODUCTION

Primary productivity and the accompanying flux of organic matter and skeletal material ($CaCO_3$ and SiO_2) from the surface ocean exert a strong influence on the geochemical cycles and oceanic distributions of carbon, carbon isotopes, and many trace elements. Several recent studies suggest that variations in surface ocean productivity may provide a mechanism for glacial-interglacial (G-I) changes in atmospheric pCO_2 (e.g., Broecker, 1982; Knox and McElroy, 1984; Sarmiento and Toggweiler, 1984; Siegenthaler and Wenk, 1984; Dymond and Lyle, 1985; Broecker and Peng, 1987, 1989; Boyle, 1988a, b; Keir, 1989). Thus, we need to understand the history of productivity variations both in order to interpret paleo-water mass tracer distributions correctly, and in order to evaluate the role of surface ocean productivity in setting atmospheric pCO_2.

A wide variety of chemical and micropaleontological approaches have been used to interpret sediment properties in terms of paleoproducivity (e.g., Müller and Suess, 1979; Müller *et al.*, 1983; Pedersen, 1983; Sarnthein *et al.*, 1987; Sarnthein *et al.*, 1988; Lyle *et al.*, 1988; Pedersen *et al.*, 1988; Mix, 1989; Berger *et al.*, 1989). Although regions of lower glacial productivity have been identified (e.g., Lyle *et al.*, 1992), a general tendency of higher glacial productivity is observed. This LGM productivity increase has been attributed to stronger glacial trade winds, which drive increased upwelling and/or supply nutrients from the continents.

In this paper we compare a set of paleoproductivity proxies, and use them to assess G-I productivity variations in the Indian Ocean off Western Australia. The proxy comparison is motivated by the uncertainties associated with the most widely-used paleoproductivity proxy, the accumulation rate of sediment organic carbon (C_{org}). In addition to determining $CaCO_3$ and C_{org} accumulation rates, we consider three proxy records which are independent of the preservation of biogenic material from the upper ocean: the abundance and accumulation rate of benthic foraminifera, the accumulation of authigenic uranium in the sediments, and the $\delta^{13}C$ difference between two species of benthic foraminifera.

Our interest in the productivity history of this particular region comes from the CLIMAP LGM sea surface temperature (SST) reconstructions for the Indian Ocean (CLIMAP, 1976). In contrast to the small SST changes calculated for most of the tropical and subtropical Indian Ocean, the CLIMAP SST maps show a zone of substantially cooler water off the west coast of Australia during the LGM (CLIMAP, 1976; Prell *et al.*, 1980; Webster and Streten, 1978, 1982; Prell, 1982; Wells, 1992). Recent reevaluation of the CLIMAP faunal data using a different statistical approach (the modern analog technique) reduces, but does not eliminate, this regional SST depression (Prell, 1985). The CLIMAP studies suggested that this SST lowering was the result of equatorward transport of cold water due to the more northerly position of the Glacial West Wind Drift, a portion of which is suggested to have been diverted to the north after impinging on Australia. We have analyzed a set of gravity cores to test the hypothesis that this Glacial West Australian current, not observed in the modern circulation of the region, may have led to coastal upwelling off western Australia as the north-flowing water was turned offshore by Coriolis acceleration.

STUDY AREA

Core locations are presented in Table 1 and Figure 1. The bathymetry of the Exmouth Plateau displays relatively gentle slopes from just under 1000 meters to abyssal depths, and includes several broad shoulders. The Exmouth Plateau is isolated from the continental shelf by a slight trough; as a result, the Quaternary sediments of the Plateau are predominantly marine carbonates (e.g., Kolla and Biscaye, 1977; Haq *et al* 1990). There is no bathymetric barrier between the Perth Basin study area (57GC-15 and 57GC-19) and the Australian continent.

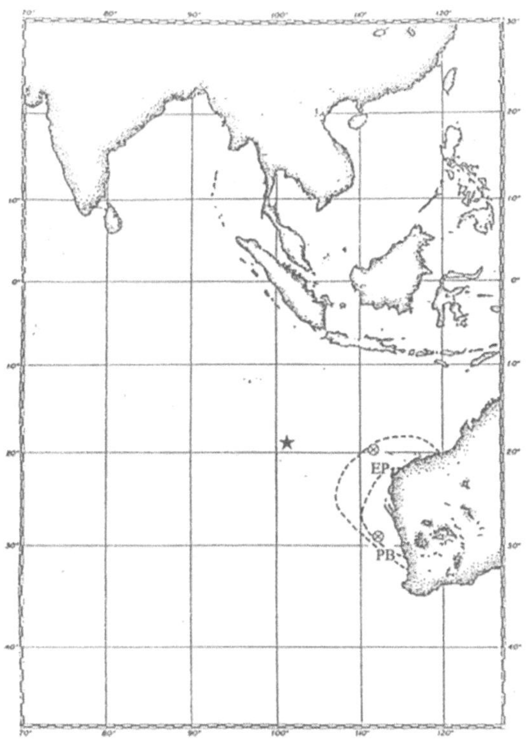

Figure 1. Locations of the Exmouth Plateau cores (EP) and Perth Basin (PB) cores. GEOSECS station 438 is indicated by the star. The approximate extent of LGM sea surface cooling off western Australia (more than -4°C relative to modern for August and February) is indicated by the dashed lines (Prell *et al.*, 1980; Prell, 1985).

METHODS

The gravity cores (9 cm diameter) were collected during cruises 53 and 57 of the research vessel "Rig Seismic" (Choi *et al.*, 1987), split on board the ship, and archived at the Australian Bureau of Mineral Resources (BMR, now the Australian Geologic Survey Organization (AGSO)).

Dry bulk density was measured by determining the weight after drying of a known volume of wet sediment. The data have not been corrected for salt content.

$CaCO_3$ and C_{org} concentrations were determined at Flinders University, by coulometric titration of the CO_2 released upon acidification with phosphoric acid ($CaCO_3$), and upon oxidation with dichromate/sulfuric acid (C_{org}), using an adaptation of the method of Weliky *et al.*, (1983). Sediment Ca contents were also measured by X-ray fluorescence at the AGSO (Cruikshank and Pyke, 1993), and percent $CaCO_3$ was calculated assuming that all Ca was present as $CaCO_3$. The two $CaCO_3$ methods show good agreement, and data from both labs are discussed below.

Foraminiferal data were obtained at Woods Hole Oceanographic Institution (WHOI). Stable isotope analyses were carried out in Dr. Lloyd Keigwin's lab at WHOI, using a VG Micromass 602E mass spectrometer, upgraded with triple collectors and a high-sensitivity source. Uncrushed foraminiferal samples were roasted under a vacuum at 370°C, and reacted with 100% phosphoric acid at 90°C. Planktonic samples consisted of 10 to 15 individuals, and benthic analyses used 2 to 5 individuals. All stable isotopic data are reported relative to the PDB standard. Instrument precision is ± 0.1‰, and the average precision of replicate analyses of planktonic foraminifera is ± 0.13‰.

The uranium-series data were obtained by alpha spectrometry at Flinders University. Details of the U-series methods will be presented in a separate report (Veeh *et al.*, in prep.). Briefly, the samples were dissolved and a calibrated $^{232}U/^{232}Th$ tracer was added. Uranium and thorium isotopes were then separated by a series of ion exchange and solvent extraction steps, followed by electrodeposition onto stainless steel disks. The one-sigma counting statistics yield precisions of ± 3 % and ±7 % for the uranium and thorium concentrations, respectively.

Four radiocarbon dates were obtained on bulk $CaCO_3$ samples submitted to the AMS facility at the Institute of Geological and Nuclear Studies, Ltd., Lower Hutt, New Zealand.

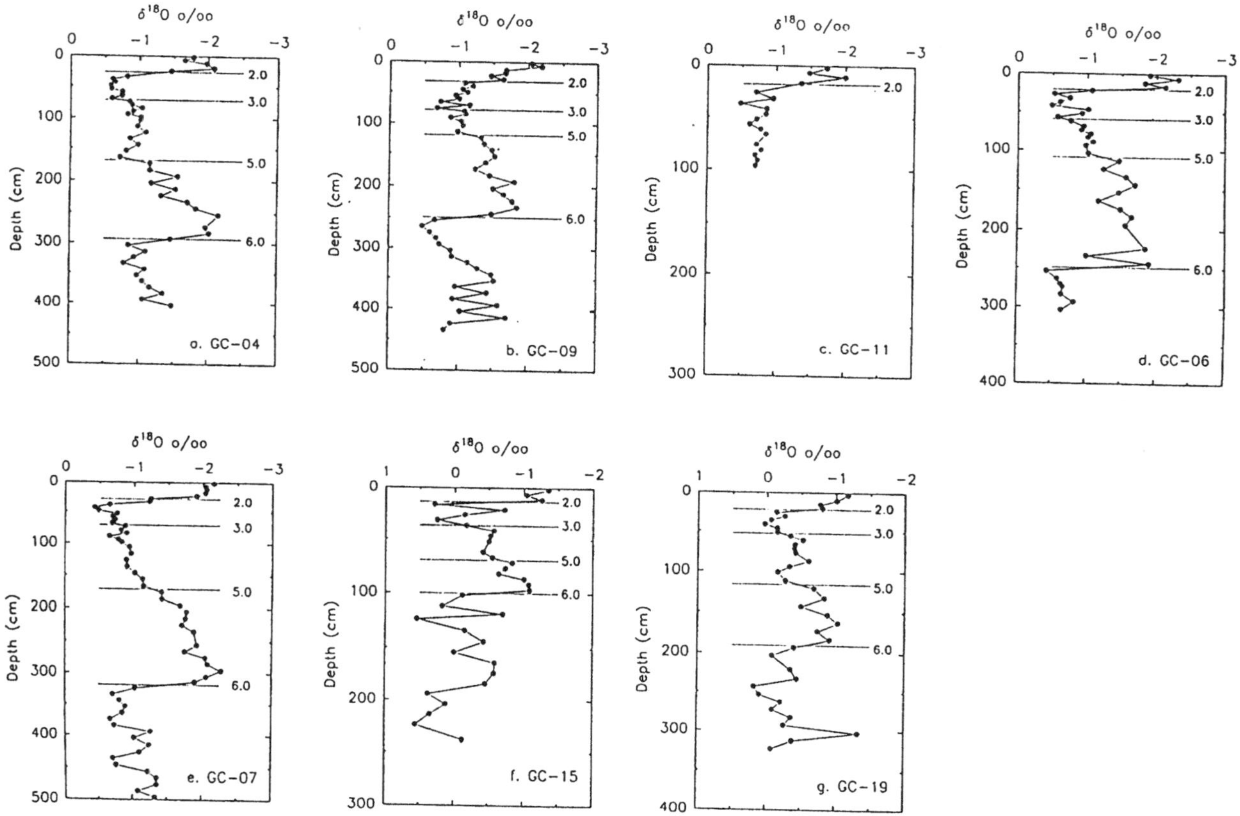

Figure 2. Planktonic foraminiferal $\delta^{18}O$ (*G. sacculifer*, 250-300 m, w/o sac). Estimated depths for isotopic stage boundaries are indicated by dotted lines. Event 2.0 = 12.05 ky, 3.0 = 24.11 ky, 5.0 = 73.91 ky, 6.0 = 129.84 ky (Pisias *et al.*, 1984; Martinson *et al.*, 1987). Where ambiguities exist in picking these $\delta^{18}O$ events, we have used benthic foraminiferal $\delta^{18}O$ data to help define the Stage 2/1 boundary (event 2.0), and comparison of percent $CaCO_3$ data with adjacent cores to help position the Stage 6/5 boundary (event 6.0). Data are available in tabular form from the first author.

RESULTS

Stratigraphy

$\delta^{18}O$ records from the planktonic foraminifera *Globigerinoides sacculifer* (250-300 m, w/o sac) are shown in 2. Although bulk sediment accumulation rates are low at these sites, the $\delta^{18}O$ profiles show the characteristic G-I pattern, with LGM-to-Holocene $\delta^{18}O$ amplitudes ranging from 0.95 to 1.50‰. Core-top planktonic $\delta^{18}O$ values in 53GC-11 are a few tenths of a per mil higher than the other Exmouth Plateau cores, suggesting that we may not have recovered a full Holocene section at this site. Oxygen isotope events 2.0, 3.0, 5.0, and 6.0 are indicated by the dashed lines on each plot in Figure 2. We have assigned ages to these boundaries based on correlation with the SPECMAP stacked $\delta^{18}O$ record (Pisias *et al.*, 1984; Martinson *et al.*, 1987). Where ambiguities exist in this correlation, we have used benthic foraminiferal $\delta^{18}O$ data to help constrain the Stage 2/1 boundary, and percent $CaCO_3$ data to help position the Stage 6/5 boundary. Sedimentation rates based on these age assignments are presented in Table 2a and Figure 3. The long-term average sediment accumulation rates derived from ^{230}Th data for cores 53GC-04, 53GC-07, 57GC-15, and 57GC-19 show good agreement with these $\delta^{18}O$ age models (Table 2b) (Veeh *et al.*, in prep.).

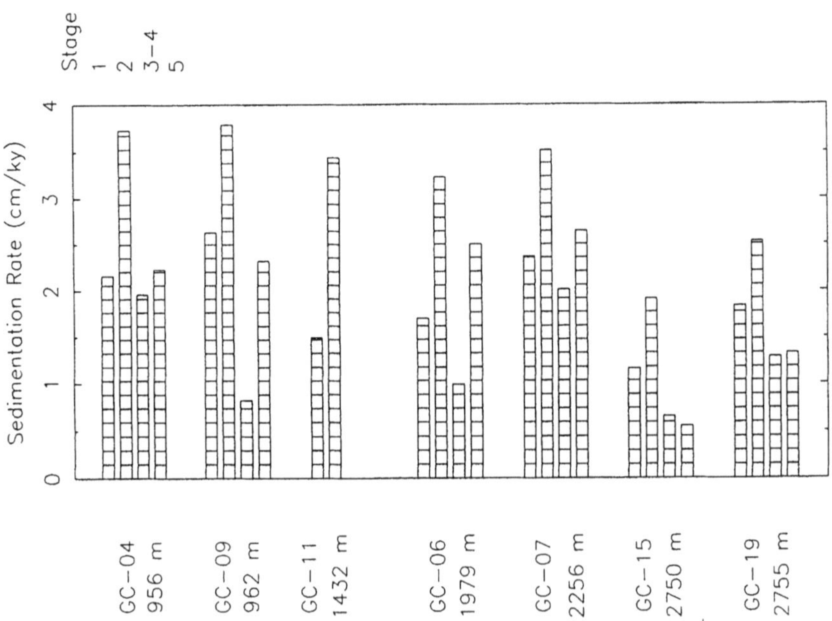

Figure 3. A comparison of $\delta^{18}O$-derived sediment accumulation rates. Average rates for isotopic stages 1, 2, 3-4, and 5 were calculated based on the age assignments shown in Figure 2. Stage 2 sedimentation rates are a factor of 1.5 to 2.3 times higher than Holocene (stage 1) rates at each site. As noted in the text, the Holocene sedimentation rate in 53GC-11 is a lower limit, due to the possibility of a missing core top at this site.

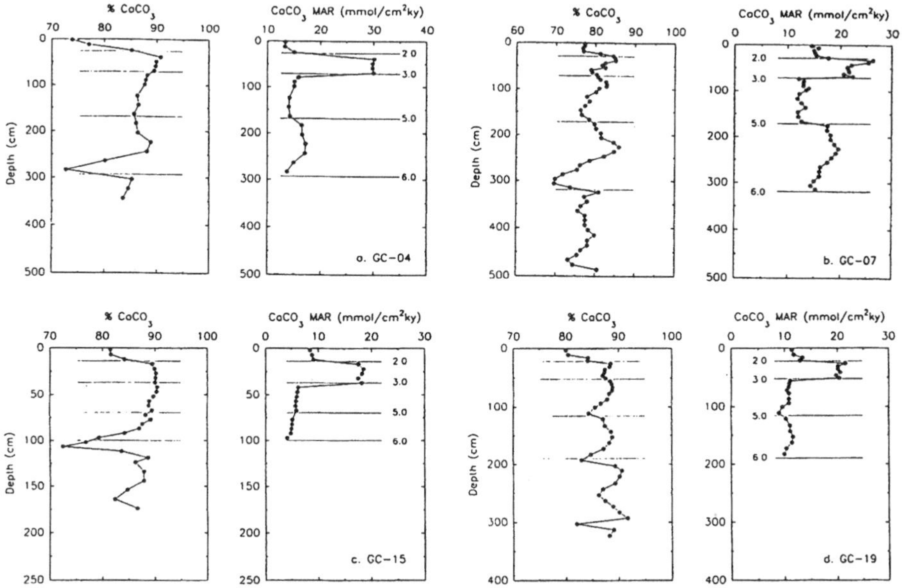

Figure 4. Percent $CaCO_3$ profiles and $CaCO_3$ mass accumulation rates. The positions of isotopic stage boundaries are indicated by dotted lines. The percent $CaCO_3$ records show low values during the Holocene in each core, during isotope Stage 5e in 53GC-04 and 53GC-07 (Exmouth Plateau, 956 and 2256 m), and at the 6/5 boundary in cores 57GC-15 and 57GC-19 (Perth Basin, 2750 and 2755 m). The abrupt shifts in MAR at the 2/1 and 3/2 boundaries are a consequence of the calculation method, which uses measured percent $CaCO_3$ and dry bulk density values, and the average sedimentation rate for each interval (see text). Stage 2 $CaCO_3$ MARs in these cores are a factor of 1.4 to 2.1 higher than Holocene rates.

The LGM-Holocene transition is well-defined in these cores, but the Holocene sections only range from 14 to 30 cm thick. As a result, Holocene accumulation rates are sensitive to either core-top sediment loss or mispositioning of event 2.0 (the Stage 2/1 boundary). Holocene accumulation rates on the Exmouth Plateau (1.5 to 2.5 cm/ky) are slightly higher than rates in the two Perth Basin cores (1.2 to 1.8 cm/ky).

Sediment accumulation rate estimates for isotope Stage 2 depend on correct identification of event 2.0 and event 3.0 (the Stage 3/2 boundary) (Pisias *et al.*,1984). In cores 53GC-04, 57GC-15, and 57GC-19, the event 3.0 pick is relatively clear, but in both 53GC-06 and 53GC-07 there are two plausible depths for this event (Figure 2). In both cores, choice of the deeper position yields a Stage 2/Stage 1 accumulation rate ratio which agrees with cores

53GC-04, 57GC-15, and 57GC-19. Four bulk $CaCO_3$ radiocarbon dates from 53GC-07 and 57GC-19 are also consistent with these $\delta^{18}O$ age models (Table 2c), though additional radiocarbon determinations would improve our confidence in identification of the event 3.0 boundary. Based on these 3/2 boundaries, Stage 2 sedimentation rates are a factor of 1.5 to 2 higher than Holocene rates at each site. These bulk sedimentation rate differences set the pattern for the Glacial-Holocene decrease in the accumulation rates of biogenic sediment components, as discussed below.

Events 5.0 and 6.0 are fairly clear in those cores with long $\delta^{18}O$ records. In 57GC-15, event 6.0 was chosen to keep the prominent $CaCO_3$ minimum close to the 6/5 boundary, as is observed in the other cores (Figure 4). Because the Stage 3-4 and Stage 5 sections are thicker, the sedimentation rate estimates are less sensitive to the exact picks of events 5.0 and 6.0. The Stage 3-4 and Stage 5 sedimentation rates are generally less than or equal to Holocene rates.

Biogenic Component Accumulation Rates

$CaCO_3$ concentrations and $CaCO_3$ mass accumulation rate estimates for the four cores with $CaCO_3$ records are plotted in Figure 4. Low carbonate concentrations are observed in the Holocene sections of all four cores. Cores 53GC-04 and 53GC-07 also display a strong $CaCO_3$ minimum just above event 6.0 (i.e., during isotope Stage 5e). In the deeper cores from the Perth Basin, 57GC-15 and 57GC-19, there is a strong $CaCO_3$ minimum which is centered on the 6/5 transition (event 6.0), rather than during Stage 5e.

The glacial-interglacial decrease in bulk sediment accumulation rate accentuates these percent $CaCO_3$ differences when we calculate $CaCO_3$ mass accumulation rates ($CaCO_3$ MAR (mmol/cm^{2}ky) = MAR $\times$ %$CaCO_3$ $\div$ 10, where the total mass accumulation rate, MAR, is the dry bulk density times the sediment accumulation rate). The abrupt changes in MAR at the stage boundaries are clearly an artifact of our use of average sediment accumulation rates for each stage. Glacial (Stage 2) $CaCO_3$ MARs are a factor of 1.4 to 2.1 higher than Holocene values, reflecting Stage 2 maxima in both percent $CaCO_3$ and MAR (Table 2a).

C_{org} concentrations and C_{org} mass accumulation rates are plotted in Figure 5. The core top samples in three of the four cores show elevated C_{org} concentrations, but even in the shallowest core (53GC-04) these "high" values are just 0.5%. Below the Holocene sections, C_{org} contents are low and fairly constant, ranging from 0.1 to 0.2 % in 57GC-15 and 57GC-19, and from 0.3 to 0.4 % in 53GC-04 and 53GC-07. We do not have an independent estimate of the relative proportions of terrestrial and marine C_{org} in these cores.

The higher core-top (Stage 1) C_{org} contents work against the G-I drop in sediment accumulation rate when we calculate C_{org} mass accumulation rates (C_{org} MAR (mmol/cm^{2}ky) = MAR $\times$ %C_{org} $\div$ 1.2). Average isotope Stage 2 organic carbon accumulation rates are higher

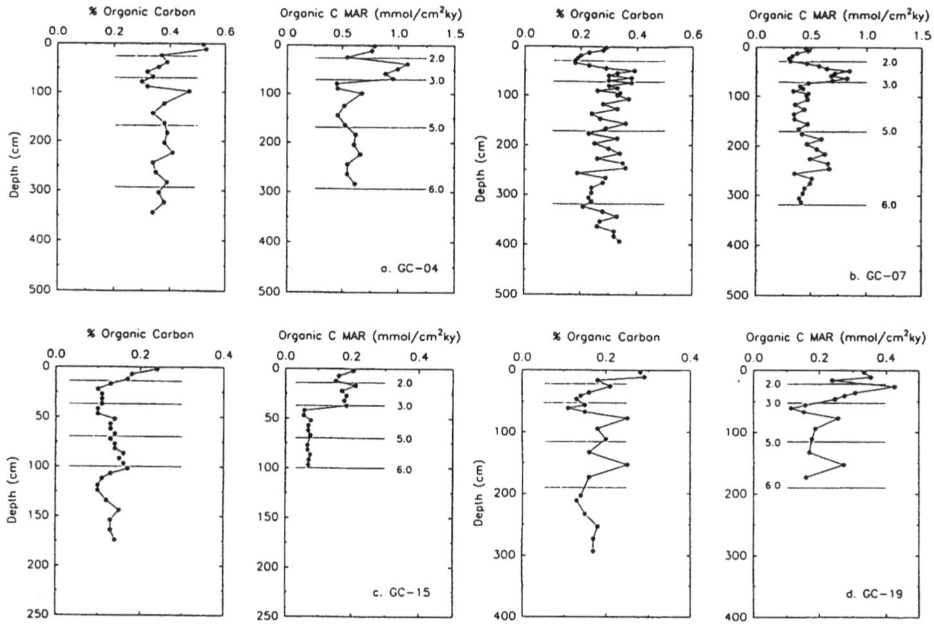

Figure 5. Percent C_{org} profiles and C_{org} mass accumulation rates. The positions of isotopic stage boundaries are indicated by dotted lines. C_{org} contents are low in all four cores. The percent C_{org} profiles display relatively high core top values, with lower concentrations down core. 53GC-07 is the only core with a Stage 2 increase in percent C_{org}. Because Stage 2 sediment accumulation rates are high in each core, the calculated C_{org} MARs are also elevated during Stage 2 (see text). As discussed in the text, the Stage 2 maxima in C_{org} MAR in 53GC-04 and 53GC-07, and the relatively high Stage 2 MARs in 57GC-15 and 57GC-19 (comparable to Holocene rates) are driven by changes in sediment accumulation rate, not by increases in percent C_{org} in the sediment.

than Holocene values in cores 53GC-04 and 53GC-07, and average LGM and Holocene C_{org} MARs are roughly equal in cores 57GC-15 and 57GC-19 (Table 2a and Figure 5).

Benthic Foraminiferal Abundances

Cibicidoides wuellerstorfi and *Uvigerina peregrina* are the two most abundant benthic foraminiferal species in the >250 m fraction of these cores - we also observe *Pyrgo spp.*, and occasional individuals of other *Cibicidoides* and *Uvigerina* species. Abundances of *C. wuellerstorfi* and *U. peregrina* show strong G-I variations in each core, with highest abundances of both taxa occurring during glacial periods (Figure 6). The total number/gram of *Cibicidoides* spp. and *Uvigerina* spp. in the >250 m fraction are used to calculate benthic foraminiferal accumulation rates (BFARs)(Figure 7 and Table 2a). These accumulation rates should be less sensitive to carbonate dissolution or sediment winnowing than absolute

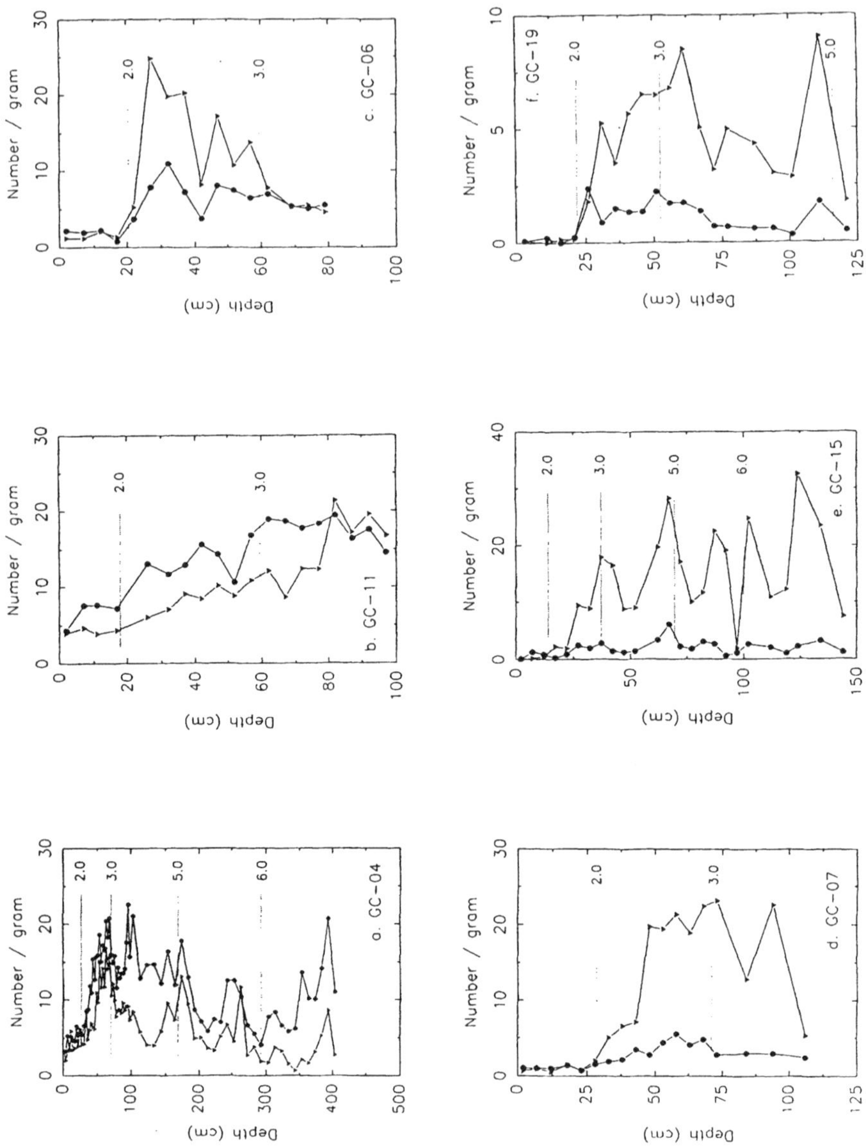

Figure 6. The abundance (number of individuals greater than 250 m per gram of dry sediment) of the benthic foraminiferal genera *Cibicidoides* (circles) and *Uvigerina* (triangles). In each core, *C. wuellerstorfi* is the most abundant *Cibicidoides* species, and *U. peregrina* is the most abundant *Uvigerina* species. In 53GC-04 and 53GC-07, only the abundance of *C. wuellerstorfi* is plotted. Holocene abundances are low at all sites, and there is a dramatic increase in abundance during Stage 2 in each core except 53GC-11. (The dashed lines show the isotopic stage boundaries from the planktonic $\delta^{18}O$ records.)

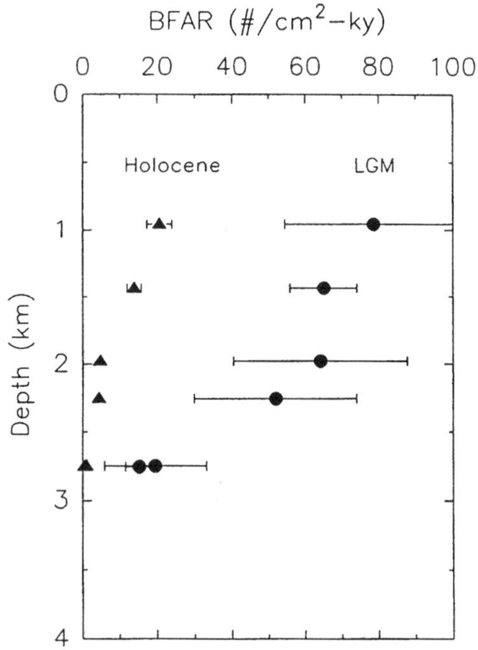

Figure 7. Average Holocene and Last Glacial Maximum (Stage 2) benthic foraminiferal accumulation rates for each core, plotted as a function of water depth. The BFARs are calculated from the abundance (#/g) of *Cibicidoides spp.* and *Uvigerina spp.* greater than 250 m, measured dry bulk density values, and the average sedimentation rate for each interval. BFARs decrease with increasing water depth during both the Holocene and LGM. LGM BFARs are from 4 to 40 times Holocene rates.

foraminiferal abundances (Herguera and Berger, 1991). Stage 2 BFARs range from 4 to 40 times the Holocene rates.

Authigenic Uranium.

Authigenic uranium concentrations for cores 53GC-04, 53GC-07, and 57GC-19 are plotted in Figure 8. The U-series data (including uranium and thorium concentrations and isotopic results) will be presented in detail in a separate report (Veeh *et al.*, in prep.). The isotopic composition ($^{234}U/^{238}U$) of the sediment uranium suggests that a significant portion of this uranium is derived from sea water. Authigenic uranium concentrations were estimated from the measured total U and Th concentrations, assuming that all the Th is derived from detrital phases, and that the U/Th (w/w) of detrital material is 0.27 (Anderson, 1982; Anderson *et al.*, 1989):

$$\text{authigenic uranium} = \text{U(total)} - \text{U/Th(detrital)} \times \text{Th}$$

Estimation of authigenic uranium based on the percent $CaCO_3$ data and an assumed U concentration of 2.7 ppm in the non-carbonate fraction of pelagic sediments (Colley *et al.*, 1985; Thomson *et al.*, 1990) yields virtually identical values.

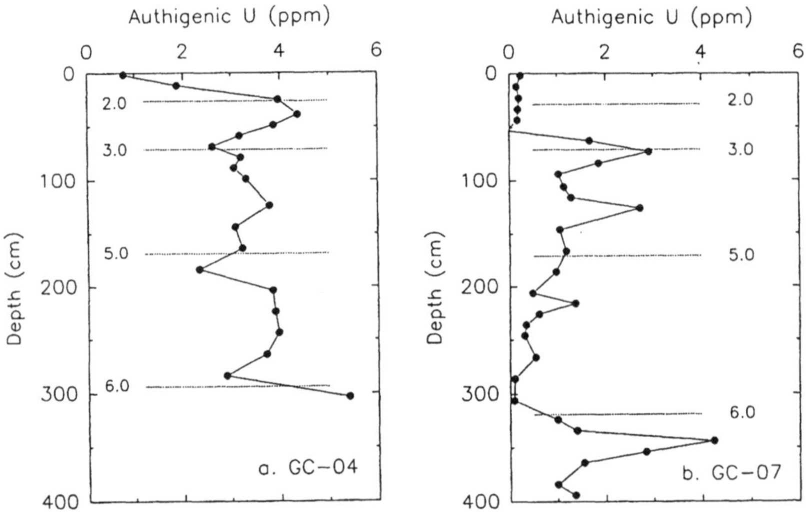

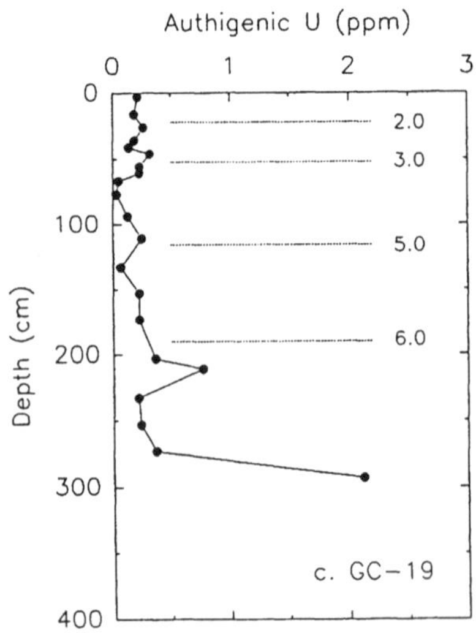

Figure 8. Authigenic uranium contents for cores 53GC-04, 53GC-07 and 57GC-19. Authigenic uranium values were calculated from measured U and Th concentrations, and an assumed detrital U/Th ratio of 0.27. Note the change of scale for 57GC-19. Zones of high authigenic uranium are thought to result from high C_{org} fluxes to the sea floor, and thus to mark times of high surface ocean productivity. However, these authigenic uranium peaks will be located below the position of the sediment-water interface at the time of their formation, as discussed in the text.

Core top authigenic U concentrations are low in all three cores. In 53GC- 04, authigenic uranium concentrations rise to a maximum at the LGM, and stay high throughout the length of this core, with local minima observed at $\delta^{18}O$ event 3.0, and in early and late isotope Stage 5. Authigenic uranium reaches its highest value in glacial isotope Stage 6. In 53GC-07, authigenic uranium concentrations display maxima at the base of isotope Stage 2, in the middle of Stage 3, and during Stage 6. Low values, comparable to Holocene concentrations, are observed during the last interglacial (Stage 5e). Authigenic uranium concentrations stay low through most of 57GC-19, except for a small maximum during isotope Stage 6. The single high values at the bottom of 53GC-04 and 57GC-19 may also indicate authigenic uranium maxima, but additional measurements at greater depths will be required to confirm this.

Paired-species benthic foraminiferal $\Delta\delta^{13}C$ values.

Holocene and isotope Stage 2 benthic foraminiferal stable isotope data are presented in Table 3 and Figure 9. All results are for *C. wuellerstorfi* and *U. peregrina* from the >250 m fraction. $\delta^{13}C$ values for the two species are plotted as a function of depth in the sediments in the left-hand panel of each plot. Event 2.0 (Termination 1), as determined from the planktonic $\delta^{18}O$ profiles, is indicated by the dashed line on each profile. Because sediment accumulation rates and the abundances of both benthic species are low in the Holocene sections of these cores, bioturbation is a potential problem, and we cannot simply make a sample-by-sample comparison of $\delta^{13}C$ values from the two benthic species. Instead, we plot the stable isotope data in $\delta^{13}C$ - $\delta^{18}O$ space to help identify LGM (high $\delta^{18}O$) and Holocene (low $\delta^{18}O$) specimens (right-hand panels in Figure 9). In these plots, the filled symbols to the right identify the Holocene *C. wuellerstorfi* (circles) and *U. peregrina* (triangles), and the filled symbols to the left identify the LGM specimens. Holocene and LGM interspecies $\delta^{13}C$ differences ($\Delta\delta^{13}C$ values) are indicated by the vertical bars to the right and left of each plot. These $\Delta\delta^{13}C$ results are summarized in Table 2a and Figure 10. Perth Basin core 57GC-15 did not contain Holocene *Uvigerina*, and 57GC-19 did not contain Holocene *Cibicidoides*, so we could not calculate $\delta^{13}C$ differences at these sites. There is no significant G-I change in the paired-species $\Delta\delta^{13}C$ values in any of these cores.

DISCUSSION

Biogenic Sediment Components

The observed patterns of $CaCO_3$ and C_{org} accumulation rates (i.e., generally higher during isotope Stage 2) are consistent with higher productivity during the LGM. Sediment trap studies show that primary production in the surface ocean drives the flux of calcium carbonate

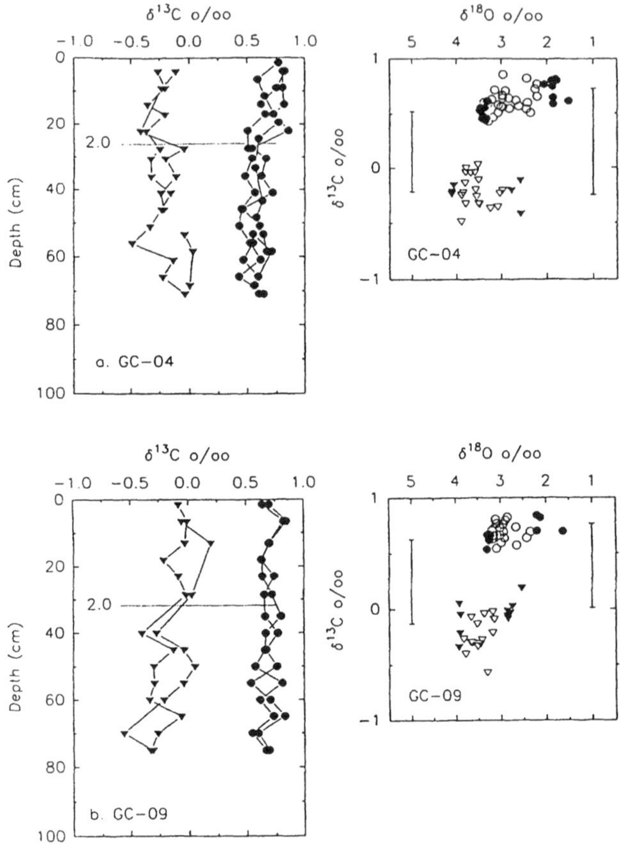

Figure 9. *C. wuellerstorfi* (circles) and *U. peregrina* (triangles) stable isotope results. The left-hand graphs show Holocene and Stage 2 $\delta^{13}C$ data for both species plotted as a function of depth in the sediments. There are no significant differences between replicate profiles run on different days (lines in left-hand graphs). The event 2.0 boundary, derived from planktonic foraminiferal $\delta^{18}O$ data, is shown by the dotted line on each graph. Because sediment accumulation rates and Holocene benthic foraminiferal abundances are low, we cannot rely on stratigraphic position to identify pairs of Holocene and LGM individuals. Instead, we plot the $\delta^{13}C$ of each sample as a function of its $\delta^{18}O$, and use the maximum and minimum $\delta^{18}O$ values to pick full LGM and full Holocene samples for each species. These samples are indicated by the solid symbols in the right-hand graphs. The LGM and Holocene *C. wuellerstorfi* - *U. peregrina* $\delta^{13}C$ differences ($\Delta\delta^{13}C$values) are shown by the vertical bars to the left and right of the $\delta^{13}C$-$\delta^{18}O$ graphs. Figure 9 continued on next page.

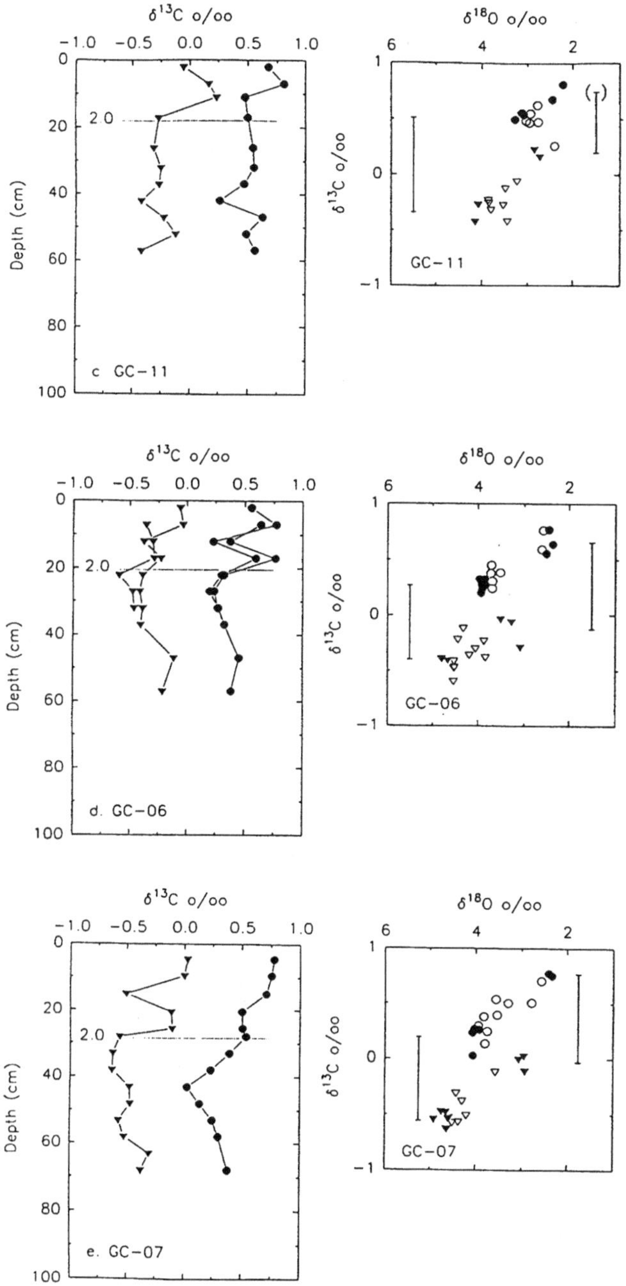
δ13C o/oo
Depth (cm)
2.0
c GC-11
δ18O o/oo
δ13C o/oo
GC-11
d. GC-06
GC-06
e. GC-07
GC-07

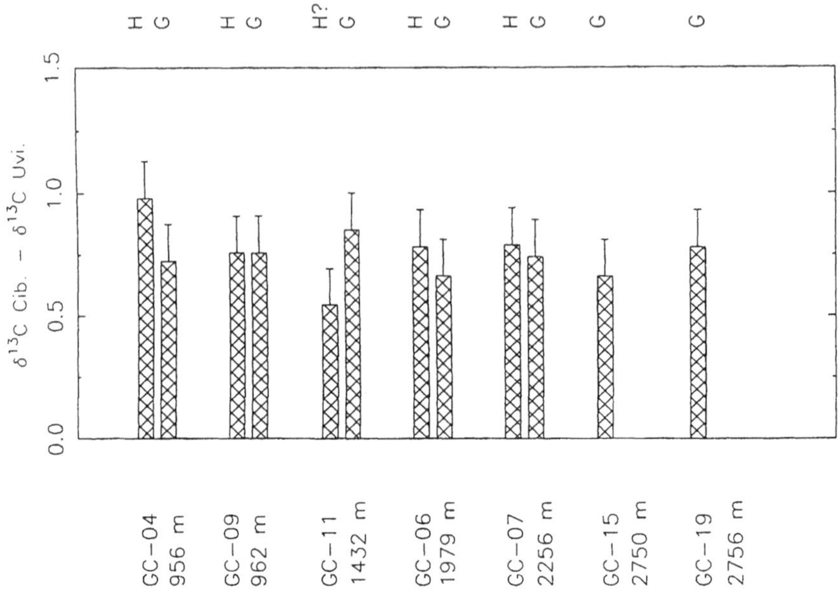

Figure 10. Average Holocene and LGM paired-species benthic foraminiferal $\delta^{13}C$ values at each site, arranged by water depth. There are no Holocene values for cores 57GC-15 and 57GC-19 because there were no full Holocene isotopic samples for one species in each core. There is no significant LGM-Holocene change in $\delta^{13}C$ in any of these cores.

and C_{org} to the sea floor (e.g., Honjo, 1980; Martin *et al.*, 1987; Lyle *et al.*, 1988). The relationship between total production and export ("new") production varies, and the Corg/$CaCO_3$ export ratio is not constant, but there is in general a positive correlation between surface ocean productivity and the flux of these two biogenic phases to the sea floor. However, sea floor diagenetic reactions can have an important impact on the accumulation of both C_{org} and $CaCO_3$ in the sediments (e.g., Martin and Sayles, 1991, and references therein).

Calcium carbonate.

Carbonate dissolution in the oceans varies as a function of water depth, bottom water chemistry, and carbonate flux; $CaCO_3$ preservation can range from high values (presumably approaching 100%) to essentially zero below the carbonate compensation depth (e.g. Broecker and Takahashi, 1978). There is also considerable evidence that organic matter decomposition in the sediments can drive $CaCO_3$ dissolution in sediments located above the saturation horizon (e.g. Emerson and Bender, 1981; Emerson and Archer, 1990). Glacial-interglacial changes in $CaCO_3$ accumulation rates have been interpreted in terms of variations in both dissolution intensity and $CaCO_3$ supply (e.g., Peterson and Prell, 1985a, b; Curry and Lohmann, 1985; Lyle *et al.*, 1988; Farrell and Prell, 1989; Archer, 1991).

The low percent $CaCO_3$ during interglacials in our Indian Ocean cores (Figure 4) is similar to the "Pacific" carbonate pattern (Arrhenius, 1952; Berger, 1973). Peterson and Prell (1985a,b) observed evidence of enhanced dissolution centered on interglacials and interglacial-to-glacial transitions in a set of deep water cores (2.9 - 4.4 km) from the Ninetyeast Ridge in the eastern equatorial Indian Ocean (90°E, 6°S). The Exmouth Plateau and Perth Basin core locations are all shallower than the shallowest cores studied by Peterson and Prell, and they are all located well above the modern lysocline and calcite saturation horizon (≈ 3700 m) on the Ninetyeast Ridge. Core-top $CaCO_3$ contents increase slightly with increasing water depth, while LGM percent $CaCO_3$ values show no clear trend with depth (Table 4). To assess the possibility of Holocene $CaCO_3$ dissolution on these cores, we have determined Holocene and LGM values for both the percent coarse fraction, and the percent fragments of planktonic foraminifera (Table 4) (Berger *et al.*, 1982; Peterson and Prell, 1985a,b). There are no consistent G-I variations in either parameter. We cannot rule out the possibility that the low-carbonate Holocene sections reflect dissolution, rather than times of low carbonate supply, but based on the absence of a clear G-I signal in either percent coarse or percent fragments, we conclude that changes in the rate of $CaCO_3$ supply drive the G-I decrease in $CaCO_3$ MAR.

Organic Carbon (C_{org})

In most of the ocean, only a small fraction of the C_{org} flux to the sea floor is buried. As a result, variations in C_{org} preservation efficiency may have a strong influence on sediment C_{org} concentrations (e.g., Emerson, 1985; Curry and Lohmann, 1985; Emerson and Hedges, 1988; Martin and Sayles, 1991). Despite this possible complication, the areal distribution of the burial flux of organic C is well correlated with surface ocean productivity in the modern ocean (e.g., Berger *et al.*, 1987). In light of this correlation, it is frequently assumed that changes through time in the C_{org} accumulation rate at a given site are driven by changes in the C_{org} flux to the sediments. Based on this assumption, elevated C_{org} concentrations in Glacial sediment have been interpreted as evidence of higher Glacial surface ocean productivity (e.g., Müller and Suess, 1979; Müller *et al.*, 1983; Pedersen, 1983; Sarnthein *et al*, 1988; Lyle *et al.*, 1988; Herguera and Berger, 1991; Berger and Herguera, 1992). Agreement between carbon preservation and independent sedimentary proxies of carbon flux lend support to the link between productivity and sediment carbon concentration (e.g., Lyle *et al.*, 1988; Pedersen *et al.*, 1988), although poor agreement between C_{org} accumulation rate-based and faunal transfer function-derived paleoproductivity proxies has also been reported (Mix, 1989).

The C_{org} MAR variations we observe are dominated by variations in the bulk sediment accumulation rates at each site. Only in 53GC-07 are the percent C_{org} values higher in Stage 2 than in the Holocene; in the other three cores, the LGM maximum in C_{org} MAR is entirely due to the higher Stage 2 sediment accumulation rates. This is a different, and substantially less compelling, situation than previous C_{org}-based paleoproductivity studies which observe a Stage 2 maximum in the C_{org} content of the sediments (e.g., Pedersen *et al.*, 1988).

As noted above, the C_{org} contents of our cores are low, and at present we have no information on the source of this C_{org}. There are no large rivers on the Australian coast in this region, but we cannot rule out the possibility that a significant fraction of the C_{org} remaining in these cores is refractory terrestrial material, unrelated to phytoplankton production in the overlying surface ocean. This is particularly true in downcore sections with only 0.1 or 0.2 % C_{org}.

The biogenic material data ($CaCO_3$ and C_{org} MARs) suggest an increase in the glacial productivity of this region. LGM accumulation rates range up to a factor of 1.5 to 2.0 higher than Holocene values (Figures 4 and 5), and these accumulation rate changes may be attenuated relative to the corresponding increases in surface ocean primary production (Herguera and Berger, 1991). The covariation of $CaCO_3$ and C_{org} MARs in 53GC-04 and 53GC-07 indicates that the G-I C_{org} variations are not simply an artifact of $CaCO_3$ dissolution. This covariation is consistent with a common productivity control for both proxies. The observation that there is little G-I change in the C_{org} MAR in 57GC-15 and 57GC-19 may be a consequence of the very low absolute values for the C_{org} flux at these sites - essentially all of the C_{org} falling to the sea floor is decomposed during both glacial and interglacial periods.

Benthic Foraminiferal Abundances and Accumulation Rates

We observe a dramatic glacial increase in benthic foraminiferal abundances and accumulation rates (Figures 6 and 7). Recent studies have summarized the evidence linking benthic foraminiferal biomass and benthic foraminiferal accumulation rates (BFARs) to the flux of C_{org} to the sea floor (e.g., Pedersen *et al.*, 1988; Altenbach and Sarnthein, 1989; Herguera and Berger, 1991; Altenbach, 1992). These studies conclude that food supply is a primary control on the overall abundance of benthic foraminifera at the sea floor. The organic matter flux to the sediments - the food supply to the benthos - is driven by surface ocean productivity. These studies also suggest that different benthic foraminiferal taxa may be adapted to different rates of food supply, and that these adaptations may be responsible for the microhabitat preferences of different taxa (Corliss, 1985; Mackensen and Douglas, 1989; Corliss and Emerson, 1990; McCorkle *et al.*, 1990). These adaptations may in turn lead to a correlation between benthic foraminiferal faunal assemblages and surface ocean productivity (Loubere, 1991).

Ecological and environmental factors other than food supply may influence the abundance of a given foraminiferal species. *C. wuellerstorfi* and *U. peregrina*, the two most abundant species in these cores, have been suggested to occupy different benthic microhabitats, with *C. wuellerstorfi* characterized as epibenthic or "elevated" epibenthic, and *U. peregrina* characterized as a near-interface infaunal taxon (Corliss, 1985; Lutze and Thiel, 1987; Corliss and Emerson, 1990). The observed covariation of the abundances of these two species (with different microhabitat preferences) is consistent with high productivity during the glacial leading to a general increase in benthic foraminiferal abundances. We do not simply observe a shift in dominance, but a strong, across-the-board increase in benthic foraminiferal abundances. The rate data (below), and the lack of G-I changes in percent coarse fraction or percent fragments (Table 4) suggest that the abundance variations are not an artifact of winnowing or carbonate dissolution.

We observe Stage 2 benthic foraminifera accumulation rates in the greater than 250 m fraction roughly an order of magnitude larger than the corresponding stage 1 values (Figure 7). Herguera and Berger (1991) and Herguera (1992) recently derived an empirical relationship between benthic foraminiferal accumulation rate (in the greater than 150 m size range) and surface ocean productivity. They used the Berger *et al.*, (1987) productivity maps to estimate modern primary productivity, and included terms to account for the relationship between primary productivity and the carbon flux from the surface ocean (Eppley and Peterson, 1979) and for the decrease in carbon flux with depth (Suess, 1980). This results in a linear relationship between BFAR and the estimated carbon flux to the sea floor, and a non-linear relationship between BFAR and primary productivity - BFAR increases roughly 1.5 times as fast as primary productivity. Our results are not directly comparable with those of Herguera and Berger (1991), because of the different size fractions used in each study. Pedersen *et al.*, (1988) reported a positive correlation between carbon flux and the average test size of *Uvigerina*, which suggests that the BFAR in the >250 m fraction will be more sensitive to productivity variations than the >150 m BFAR. This sensitivity is apparent in the dramatic increase in the G/I BFAR ratio with increasing water depth. The factor of 20-40 increase in BFAR in 57GC-15 and 57GC-19, much larger than any plausible increase in productivity, suggests a nonlinear response of >250 m BFAR. This sensitivity may itself be a function of carbon flux, and would therefore require independent calibration. Overall, however, the glacial increase in >250 m benthic foraminiferal abundances and accumulation rates is strong evidence for a glacial increase in carbon flux, presumably driven by higher surface ocean productivity.

Authigenic Uranium

Authigenic uranium concentrations are highest in the shallowest core, 53GC-04, and decrease with increasing water depth (Figure 8). The high authigenic uranium values do not appear to be due to sedimentary apatite, which was not detected in X-ray diffraction spot checks, and they are not a consequence of high sediment C_{org} levels, since there is little correlation between sediment C_{org} content and authigenic uranium in these cores. The most likely explanation for the observed enrichment is diffusion of uranium from bottom water into the sediments (i.e., down a pore water uranium concentration gradient) to the level where reducing conditions lead to uranium uptake (e.g., Colley and Thomson, 1985; Santschi *et al.*, 1988; Anderson *et al.*, 1989; Barnes and Cochran, 1990; Thomson *et al.*, 1990; Klinkhammer and Palmer, 1991). This decomposition-driven uranium uptake should vary with the C_{org} flux to the sea floor, and thus would be expected to show the observed inverse correlation with water depth. The same mechanism may also enable authigenic uranium concentrations to serve as a paleoproductivity proxy. Authigenic uranium should accumulate more rapidly during times when a relatively high carbon flux decreases the thickness of the oxic zone at the top of the sediment column. Changes in the uranium precipitation depth will not be a linear function of paleoproductivity (since carbon flux is a non-linear function of productivity, and pore water oxygen penetration is a non-linear function of carbon flux), but high authigenic uranium concentrations will mark times of high C_{org} flux. We observe such zones of high authigenic uranium in the Stage 2 section of core 53GC-04, at the 2/3 boundary in 53GC-07, and in the isotope Stage 6 sections of cores 53GC-07, 57GC-19, and possibly 53GC-04 (Figure 8).

Elevated levels of authigenic uranium observed in the glacial sections of a set of Eastern Atlantic Ocean cores (including a sharp maximum just below the G-I transition) have been interpreted as a consequence of uranium uptake from the pore water in the suboxic portion of the sediment column (Colley and Thomson, 1985; Wilson *et al.*, 1986; Thomson *et al.*, 1990). These studies suggest that the position of the uranium precipitation zone remained essentially constant due to a G-I decrease in sediment accumulation rate. The sedimentation rate changes in these Atlantic cores are much larger than we observe on the Exmouth Plateau/Perth Basin, but a G-I decrease in carbon flux would have a similar effect, even without major sediment accumulation rate changes.

The pore water redox front which determines the position of authigenic uranium precipitation horizon will, at steady state, move up in pace with the sediment-water interface. However, the depth to this zone will decrease during times of high C_{org} flux, and will tend to deepen again when surface productivity is reduced. Thus the zone of authigenic uranium precipitation would be relatively close to the sediment-water interface during high-carbon-flux glacial periods, steepening the pore water uranium gradient, and resulting in high authigenic uranium precipitation rates. When productivity decreased during interglacials, the oxic zone would tend to deepen and the authigenic uranium accumulation rate would decrease. The depth of oxygen penetration could be limited by C_{org} in the glacial sediments, which would tend to fix the position of the precipitation zone (Jahnke *et al.*, 1989), and result in continued authigenic uranium accumulation at the top of the glacial authigenic uranium peak. A subsequent return to glacial (high-flux) conditions would decrease the oxic zone thickness again. Authigenic uranium would begin to accumulate at this new horizon, located above a zone of low-authigenic uranium interglacial sediments.

It is difficult to assign a chronology to authigenic uranium in the sediments because of the offset between the sediment-water interface and the zone of authigenic uranium precipitation. For instance, the low core-top authigenic uranium concentrations in 53GC-04 and 53GC-07 need not imply that authigenic uranium is not accumulating today. Low core-top concentrations would be predicted if the zone of authigenic uranium precipitation is a few tens of centimeters below the sediment-water interface at these sites. The top of the authigenic uranium maxima at ≈25 cm in 53GC-04 and at ≈70 cm in 53GC-07 may mark the depth at which modern authigenic uranium is precipitating. A rough calculation suggests that the observed authigenic uranium concentrations at these depths could be supported by a diffusive flux of uranium from the bottom water.

If the shallowest peak of uranium is accumulating today, the accumulation rate can be estimated as the product of the uranium concentration and the modern (Holocene) bulk sediment accumulation rate.

For 53GC-04 $\quad UMAR \approx 4\ g/g * 1.75\ g/cm^2ky \approx 7\ g/cm^2ky$

For 53GC-07 $\quad UMAR \approx 2.5\ g/g * 1.97\ g/cm^2ky \approx 5\ g/cm^2ky$

The diffusive flux of uranium to this (assumed) zone of precipitation can be estimated as the product of the sediment porosity (ϕ), the effective diffusivity (D) of U in the sediments, and the concentration gradient:

$$Uflux \approx \phi * D(sed) * \delta[U]/\partial z$$

We assume a porosity of 0.85 and a diffusivity of $2.7*10^{-6} cm^2/s$ (Barnes and Cochran, 1990). The change in concentration is set equal to the bottom water uranium concentration (i.e., we assume that pore water [U] goes to zero at the depth of precipitation), and the depth is the distance from the sediment water interface to the top of the solid-phase uranium peak.

For 53GC-04 $\quad Uflux \approx 0.85 * 2.7*10^{-6} * (1.4*10^{-5}\ mol/cm^3/25cm) \approx 9.6\ g/cm^2ky$

For 53GC-07 $\quad Uflux \approx 0.85 * 2.7*10^{-6} * (1.4*10^{-5}\ mol/cm^3/70cm) \approx 3.4\ g/cm^2ky$

The modern burial fluxes and potential pore water fluxes are comparable at each site. We do not have pore water data with which to test this hypothesis, but this calculation indicates that quantitative estimates of paleo-accumulation rates of authigenic uranium will be complicated, since both the authigenic uranium content and the sediment accumulation rates used in these calculations must be those appropriate for the time of authigenic uranium precipitation.

The maxima in authigenic uranium observed in Stage 6 (53GC-07, 57GC-19, and possibly 53GC-04) and Stage 2/3 (53GC-07 and 53GC-04), and the authigenic uranium minima observed during Stage 5e (53GC-07, and possibly 53GC-04) are consistent with high productivity during full glacial periods, and low productivity during interglacials. The low levels of authigenic uranium observed in core 57GC-19, and the lack of a strong G-I signal in authigenic uranium in this core, suggest both a generally lower carbon flux at this site, and perhaps less G-I variation in this flux.

Paired-Species $\delta^{13}C$ Differences

Several studies have suggested that the $\delta^{13}C$ difference ($\Delta\delta^{13}C$) between the benthic foraminifera *C. wuellerstorfi* and *U. peregrina* may reflect the productivity of the overlying surface waters (Woodruff and Savin, 1985; Zahn *et al.*, 1986; McCorkle *et al.*, 1990). A proposed mechanism for this relationship is that the $\delta^{13}C$ of *U. peregrina* records productivity-driven changes in the $\delta^{13}C$ of pore water just below the sediment-water interface. The principal assumptions required by this proposed mechanism seem to be supported by the available data (McCorkle *et al.*, 1990). The flux of C_{org} to the sea floor increases as surface-ocean productivity increases, although the flux/productivity ratio is not constant. The pore water $\delta^{13}C$ gradient at the sediment water interface is largely controlled by the flux of organic matter to the sea floor (McCorkle *et al.*, 1985, McCorkle, 1987). *C. wuellerstorfi* $\delta^{13}C$ values seem to faithfully record the carbon isotopic composition of bottom water (e.g. Woodruff *et al.*, 1980; Belanger *et al.*, 1981; Graham *et al.*, 1981; Duplessy *et al.*, 1988; McCorkle and Keigwin, in press). The vertical distribution of living (stained) *Uvigerina* spp. within the sediments is consistent with a shallow infaunal habitat for these species (Corliss, 1985; Lutze and Thiel, 1987; Corliss and Emerson, 1990). Finally, a strong correlation is observed between the $\delta^{13}C$ of living (stained) *Uvigerina* spp. and the $\delta^{13}C$ of pore water dissolved inorganic carbon in the top half centimeter of the sediments (Figure 11) (McCorkle *et al.*, 1990). (Note that living *C. wuellerstorfi* were not observed in these cores, preventing a direct *Cibicidoides* - *Uvigerina* comparison.).

The $\delta^{13}C$ difference ($\Delta\delta^{13}C$) between *C. wuellerstorfi* and *U. peregrina* shows no significant G-I change in any of the Exmouth Plateau cores. This result is clearly at odds with the evidence for higher LGM productivity, derived from biogenic sediment MARs, benthic foraminiferal accumulation rates, and authigenic uranium concentrations. The glacial productivity increase implied by the changes in bulk sediment MAR and $CaCO_3$ MAR (i.e., a factor of 1.5 to 2.0 higher productivity during Stage 2) should result in a roughly 0.3 to 0.7‰ increase in glacial $\Delta\delta^{13}C$ values, relative to the Holocene. The C_{org} MAR data suggest little

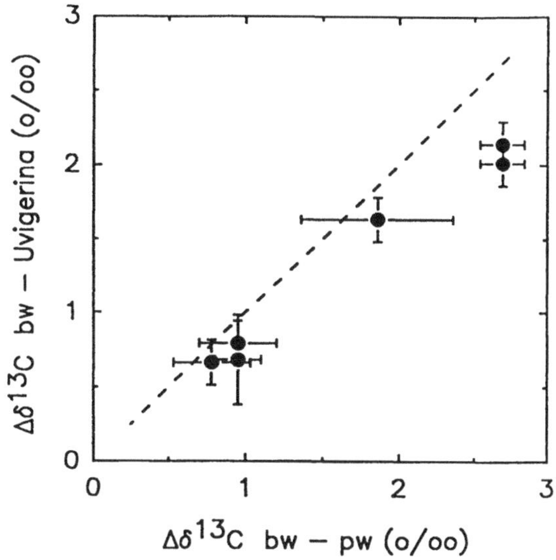

Figure 11. The relationship between the carbon isotopic composition of living (stained) *Uvigerina* and that of pore water dissolved inorganic carbon (DIC) (McCorkle *et al.*, 1990, and McCorkle, unpublished data). The $\delta^{13}C$ difference between bottom water DIC and stained *Uvigerina* is plotted as a function of the $\delta^{13}C$ difference between bottom water DIC and pore water in the 0 to 0.5 centimeter sample interval, in order to account for the site-to-site differences in the $\delta^{13}C$ of bottom water in these Atlantic and Pacific Ocean cores. The dashed line has a slope of 1. The pore water $\delta^{13}C$ gradient (bottom axis) is driven by the flux of C_{org} to the sediments. The roughly 1:1 relationship suggests a similar carbon flux (productivity) control on *Uvigerina* $\delta^{13}C$. Average C_{org} mass accumulation rates for stages 1, 2, 3-4, and 5 are shown by the vertical bars in the right-hand graphs

productivity change in 57GC-15 and 57GC-19, but would still predict a 0.5‰ increase in glacial $\Delta\delta^{13}C$ values in 53GC-04 and 53GC-07.

We do not believe we have incorrectly estimated the LGM and Holocene $\delta^{13}C$ values for the two species (Figure 9). Neither including nor excluding samples from the averages (i.e., by including a wider range of $\delta^{18}O$ values in the LGM and Holocene averages, or by using only the single sample with the most extreme $\delta^{18}O$ value instead of an average of several samples) results in a systematic change in calculated $\Delta\delta^{13}C$ values. However, we note that the Holocene *C. wuellerstorfi* $\delta^{13}C$ values are consistently ≈0.5‰ higher than $\delta^{13}C$-DIC values from the nearest GEOSECS site, Station 438 (McCorkle *et al.*, 1992). If this difference reflects a real offset from local bottom water, and if this offset has changed through time, then the paired-species $\Delta\delta^{13}C$ values would reflect changes in this offset, rather than changes in productivity.

Another possibility is that the foraminiferal $\delta^{13}C$ values are biased in a way which makes them insensitive to carbon flux at the low absolute carbon fluxes typical of this region. One way this could occur is if the Holocene *U. peregrina* specimens in fact lived during brief higher-flux episodes. Perhaps when *C. wuellerstorfi* and *U. peregrina* coexist in low-carbon flux environments, *U. peregrina* only becomes abundant (at least in the >250 m fraction) when the carbon flux reaches some threshold value. In this case the*U. peregrina* $\delta^{13}C$ values in a sample would reflect episodes of elevated carbon flux, rather than typical conditions. There would be no $\Delta\delta^{13}C$ record of the long low-flux periods, because the *Uvigerina* were not present at these times. Instead, all the observed $\Delta\delta^{13}C$ values would hover around the $\Delta\delta^{13}C$associated with the minimum carbon flux needed for *Uvigerina* to thrive.

We observe average $\Delta\delta^{13}C$ values of 0.77 and 0.74‰ for Holocene and LGM foraminifera. These differences are slightly smaller than the interspecies $\Delta\delta^{13}C$ offset (=0.90‰) noted in earlier studies (e.g., Shackleton, 1977). These values, and the data from living *Uvigerina* shown in Figure 11, suggest that the hypothesized threshold carbon flux results in a bottom water - pore water $\Delta\delta^{13}C$ of about 0.7‰. The smallest $\Delta\delta^{13}C$ values on Figure 11 are about 0.7‰ for both the bottom water - pore water difference and the bottom water - *Uvigerina* difference. We have measured pore water profiles with smaller $\Delta\delta^{13}C$ values in the Central Equatorial Pacific (McCorkle *et al.*, 1985), but did not find living (stained) *Uvigerina* at these low-flux sites. As a result, the correlation shown in Figure 11 does not include any data from low carbon flux sites.

A $\Delta\delta^{13}C$ value of 0.7‰ corresponds to an C_{org} flux to the sea floor of approximately 600 mgC/m^{2}ky (McCorkle, 1987). The modern carbon flux in this region can be estimated by combining the surface primary productivity (=60 to 90 gC/m^{2}y (Berger *et al.*, 1987)) with an expression relating the benthic carbon flux (J) to primary productivity (PP) and water depth (Z) (e.g., Berger and Wefer, (1990):

$$J = 2 \times PP^{0.5} \div Z + 0.5 \times PP \div Z^{0.5}$$

This calculation predicts a modern flux of 150 to 300 mgC/m^{2}ky at the depths of the Exmouth Plateau cores. These values are between one quarter and one half the hypothesized threshold for the *Cibicidoides* - *Uvigerina* $\Delta\delta^{13}C$ proxy. This threshold for *Uvigerina* must be tested in the modern ocean before we can accept this explanation for the discrepancy between paired-species $\Delta\delta^{13}C$ values and the other paleoproductivity proxies we have measured.

CONCLUSIONS

We have determined biogenic sediment accumulation rates, benthic foraminiferal accumulation rates, authigenic uranium contents, and paired-species benthic foraminiferal $\Delta\delta^{13}C$

values on a set of cores from the Indian Ocean off western Australia. These data enable us to compare several proxy records of ocean productivity, and to estimate glacial-interglacial changes in paleoproductivity in this region.

Biogenic sediment accumulation rates are probably the most widely used proxies for paleoproductivity. In this low-flux region, C_{org} concentrations quickly drop to very low values down core, so that C_{org} accumulation rates do not provide an independent productivity record - changes in C_{org} MAR are controlled entirely by changes in the bulk accumulation rate of the sediments. However, strong G-I changes in the concentration and accumulation rates of $CaCO_3$ are observed, which do not appear to be controlled by processes such as carbonate dissolution or winnowing. Together, these biogenic sediment MAR changes suggest that surface ocean productivity in this region was at least a factor of 1.5 to 2.0 higher during the Last Glacial Maximum (isotope Stage 2) than during the Holocene.

The overall abundance of benthic foraminifera in modern sediments is thought to be controlled by the rate of food supply to the sea floor. This food supply rate is clearly linked to the C_{org} flux out of the surface ocean, and thus to primary production. Strong G-I variations in benthic foraminiferal accumulation rates were observed in all of the Exmouth Plateau and Perth Basin cores. Glacial BFARs range from 4 to 40 times Holocene rates. Because foraminiferal abundance is not a linear function of carbon flux, especially for the large (>250 m) individuals used in this study, we cannot translate these rates into quantitative productivity estimates. However, the carbon flux to the sea floor, and by implication the productivity of the surface ocean, was clearly higher during Stage 2 than during the Holocene.

The accumulation of authigenic uranium in the sediments may provide a record of changes in the C_{org} flux to the sea floor. The rate of authigenic uranium precipitation is limited by the rate at which dissolved uranium can diffuse from bottom water into the oxygen-free zone of the sediments. Higher carbon fluxes decrease the thickness of the oxic zone of the sediments, and thereby increase the uranium flux and the authigenic uranium precipitation rate. Quantitative application of this proxy is complicated by the difficulty of fixing a chronology for authigenic uranium, which may be deposited tens of centimeters below the sediment-water interface. However, glacial maxima and interglacial minima in authigenic uranium are observed in these cores, suggesting that productivity in this region was higher during glacial periods. The data also show that uranium removal into the suboxic zone of continental margin sediments has varied on G-I time scales.

We do not observe any G-I change in the $\delta^{13}C$ difference ($\Delta\delta^{13}C$) between *C. wuellerstorfi* and *U. peregrina*. At present, we do not have a firm explanation for the disagreement between this result and the lines of evidence for higher glacial productivity discussed above. One possibility is that this paired-species $\Delta\delta^{13}C$ proxy is inherently biased against recording low-productivity periods because of a carbon flux (i.e., food supply) threshold for the *Uvigerina* species.

This proxy comparison shows generally good agreement between three of these four very different paleoproductivity indicators. Each of the three non-traditional proxies (benthic foraminiferal accumulation rates, authigenic uranium content, and paired-species $\Delta\delta^{13}C$ values) has its own uncertainties, and each requires further calibration in the modern ocean before it can become a quantitative tool. Given the uncertainty of any single proxy, including biogenic sediment accumulation rate estimates, simultaneous application of more than one tool can substantially increase our confidence in paleoproductivity determinations.

Together, the data suggest that the glacial productivity off Western Australia was elevated relative to Holocene values, and support the hypothesis that a north-flowing west Australian current led to coastal upwelling and enhanced primary productivity off western Australia at the Last Glacial Maximum. However, glacial productivity was high only relative to the low productivity characteristic of this region in the modern ocean. We see no evidence of strong upwelling similar to that observed in the modern ocean off the southwestern coasts of Africa and South America.

ACKNOWLEDGMENTS

We thank the officers, crew, and scientists of *R/V Rig Seismic* Survey 53 (H. Stagg, Chief Scientist) and Survey 57 (J. Marshall, Chief Scientist) for collecting the cores which made this study possible. The cores were collected as part of the Continental Margins Program of the Australian Geological Survey Organization.We also thank P. Davis, C. Tindall, and J. Lane (AGSO), E. Barnett (Flinders University) and E. Franks, S. O'Connor-Lough, E. Roosen, and L.-P. Zhou (WHOI) for assistance in the lab. J. Pyke conducted the XRF analyses at AGSO. R. Francois and L. Keigwin provided helpful reviews of an early draft of this manuscript. H. Veeh received support from the Australian Research Council, and D. McCorkle was supported by NSF grant OCE-8817620. This is WHOI contribution 8251.

REFERENCES

Altenbach, A.V., 1992. Short term processes and patterns in the foraminiferal response to organic flux rates. Mar. Micropaleontol., vol. 19, pp. 119-129.

Altenbach, A.V., and M. Sarnthein, 1989. Productivity record in benthic foraminifera, in, W.H. Berger, V.S. Smetacek, and G. Wefer (eds.), Productivity of the Ocean: Present and Past. Wiley, New York, pp. 255-269.

Anderson, R.F., 1982. Concentration, vertical flux, and remineralization of particulate uranium in seawater. Geochim. Cosmochim. Acta, vol. 46, pp. 1293-1299.

Anderson, R.F., A.P. Lehuray, M.Q. Fleisher, and J.W. Murray, 1989. Uranium deposition in Saanich Inlet sediments, Vancouver Island. Geochim. Cosmochim. Acta, vol. 53, pp. 2205-2213.

Arrhenius, G., 1952. Sediment cores from the east Pacific, Rep. Swed. Deep Sea Exped. 1947-1948, no. 5, 228 p.
Barnes, C.E. and J.K. Cochran, 1990. Uranium removal in oceanic sediments and the oceanic U balance. Earth Planet. Sci. Lett., vol. 97, pp. 94101.
Belanger, P.E., W.B. Curry, and R.K. Matthews, 1981. Core-top evaluation of benthic foraminiferal isotopic ratios for paleoceanographic interpretations. Palaeogeogr., Palaeoclimatol., Palaeoecol., vol. 33, pp. 205-220.
Berger, W.H., 1973. Deep-sea carbonates: Pleistocene dissolution cycles. Jour. Foram. Res., vol. 3, pp. 187-195.
Berger, W.H., M.-C. Bonneau, and F.L. Parker, 1982. Foraminifera on the deep-sea floor: lysocline and dissolution rate. Oceanolog. Acta, vol. 5, pp. 249-258.
Berger, W.H., K. Fischer, C. Lai, and G. Wu, 1987. Ocean productivity and organic carbon flux. Part I. Overview and maps of primary production and export production. Scripps Institution of Oceanography Ref. 87-30:1.
Berger, W.H., V.S. Smetacek, and G. Wefer, 1989. Ocean productivity and paleoproductivity - an overview. in Productivity of the Ocean: Present and Past, Berger, W.H., V.S. Smetacek, and G. Wefer, eds., John Wiley and Sons, pp. 1-34.
Berger, W.H., and G. Wefer, 1990. Export production: seasonality and intermittency, and paleoceanographic implications. Paleogeogr. Paleoclimatol. Paleoecol., 89, 245-254.
Berger, W.H. and J.C. Herguera, 1992. Reading the sedimentary record of the ocean's productivity. in Primary Productivity and Biogeochemical Cycles in the Sea, P.G. Falkowski and A.D. Woodhead, eds., Plenum Press, pp. 455-486.
Boyle, E.A., 1988a. Vertical oceanic nutrient fractionation and glacial/interglacial CO_2 cycles. Nature, vol. 331, pp. 55-56.
Boyle, E.A., 1988b. The role of vertical chemical fractionation in controlling Late Quaternary atmospheric carbon dioxide. J. Geophys. Res., vol 93, no. C12, pp. 15701-15714.
Broecker, W.S. (1982) Ocean chemistry during glacial time. Geochim. Cosmochim. Acta, 46, pp.1689-1705.
Broecker, W.S., and T. Takahashi, 1978. The relationship between lysocline depth and *in situ* carbonate ion concentration, Deep Sea Res., vol. 25, pp. 65-95.
Broecker, W.S., and T-.H. Peng, 1987. The role of $CaCO_3$ compensation in the glacial to interglacial atmosphere CO_2 change, Global Biogeochemical Cycles, vol. 1, no. 1, pp. 15-29.
Broecker, W.S., and T.-H. Peng, 1989. The cause of the Glacial to Interglacial atmospheric CO_2 change: A polar alkalinity hypothesis, Global Biogeochemical Cycles, v. 3, n. 3, pp. 215-239.
Choi, D.R., H.M.J. Stagg, and others, 1987. Rig Seismic research cruise 6: northern Australia heat flow - post-cruise report. Australian Bureau of Mineral Resources, Canberra, Report 274, 40 pp.
CLIMAP, 1976. The surface of the Ice Age Earth, Science, vol. 191, pp. 1131-1137.
Colley, S. and J. Thomson, 1985. Recurrent uranium relocations in distal turbidites emplaced in pelagic conditions. Geochim. Cosmochim. Acta, vol. 49, pp. 2339-2348.
Corliss, B.H., 1985. Microhabitats of benthic foraminifera within deep-sea sediments, Nature, vol. 314, pp. 435-438.
Cruikshank, B.I. and J.G. Pyke, 1993. Analytical methods used in the Minerals and Land Use Program's Geochemical Laboratory. Australian Geological Survey Organisation Record 1993/26.
Corliss, B.H., and S. Emerson, 1990. Distribution of rose bengal stained deep-sea benthic foraminifera from the Nova Scotian continental margin and Gulf of Maine. Deep-Sea Res., vol. 37, n. 3, pp. 381-400.
Curry, W.B., and G.P. Lohmann, 1985. Carbon deposition rates and deep water residence time in the equatorial Atlantic Ocean throughout the last 160,000 years, in, T. E. Sundquist., and S. W. Broecker. (eds.), The Carbon Cycle and Atmospheric CO_2: Natural Variations Archean to Present. American Geophysical Union, Washington DC, pp. 285-301.

Duplessy, J.C., N.J. Shackleton, R.G. Fairbanks, L. Labeyrie, D. Oppo, and N. Kallel, 1988. Deep water source variations during the last climatic cycle and their impact on the global deep water circulation. Paleoceeanography, vol. 3, no. 3, pp. 343-360.
Dymond, J., and M. Lyle, 1985. Flux comparisons between sediments and sediment traps in the eastern tropic Pacific: Implications for atmospheric CO_2 variations during the Pleistocene, Limnol. Oceanogr., vol. 30 no. 4, pp. 699-712.
Emerson, S., and M. Bender, 1981. Carbon fluxes at the sediment water interface of the deep sea: calcium carbonate preservation. J. Mar. Res., vol. 39, no. 1, pp. 139-162.
Emerson, S., 1985. Organic carbon preservation in marine sediments, in, E.T. Sundquist, and W.S. Broecker (eds.), The Carbon Cycle and Atmospheric CO_2: Natural Variations Archean to Present. American Geophysical Union, Washington DC, pp. 78-87.
Emerson, S., and J.I. Hedges, 1988. Processes controlling the organic carbon content of open ocean sediments, Paleoceanography, vol. 3, no. 5, pp. 621-634.
Emerson, S.R., and D. Archer, 1990. Calcium carbonate preservation in the ocean. Phil. Trans. R. Soc. Lond., vol. A331, pp. 29-40.
Farrell, J.W., and W.L. Prell, 1989. Climatic change and $CaCO_3$ preservation; an 800,000 year bathymetric reconstruction from the Central Equatorial Pacific Ocean. Paleoceanography, vol. 4, no. 4, pp. 447-466.
Haq, B.U., U. von Rad, and ODP Leg 122 Shipboard Scientific Party, 1990. Proc. ODP, Init. Repts. 122. Ocean Drilling Program, College Station, TX, pp.9-15.
Keir, R.S., 1989. Paleoproduction and atmospheric CO_2 based on ocean modeling, in, W.H. Berger, V.S. Smetacek, and G. Wefer (eds.), Productivity of the Ocean: Present and Past. Wiley, New York, pp. 377-394.
Klinkhammer, G.P. and M.R. Palmer, 1991. Uranium in the oceans: where it goes and why. Geochim. Cosmochim. Acta, vol. 55, pp. 1799-1806.
Knox, F., and M.B. McElroy, 1984. Changes in atmospheric CO_2: influence of the marine biota at high latitude, J. Geophys. Res., vol. 89, pp. 4629-4637.
Kolla, V., and P.E. Biscaye, 1977. Distribution and origin of quartz in the sediments of the Indian Ocean, J. Sed. Pet., vol. 47 n. 2, pp. 642-649.
Lutze, G.-F. and H. Thiel, 1987. *Cibicidoides wuellerstorfi* and *Planulina ariminensis*, elevated epibenthic foraminifera. Univ. of Kiel, Ber. Sonderforschungsbereich 313, vol. 6, pp. 17-30.
Lyle, M., D.W. Murray, B.P. Finney, J. Dymond, J.M. Robbins, and K. Brooksforce, 1987. The record of Late Pleistocene biogenic sedimentation in the Eastern Tropical Pacific Ocean. Paleoceanography, vol. 5, pp. 15-41.
Lyle, M., R. Zahn, F. Prahl, J. Dymond, R. Collier, N. Pisias, and E. Suess, 1992. Paleoproductivity and carbon burial across the California Current: the Multitracers transect, 42°N. Paleoceanography, vol. 7, pp. 251-272.
McCorkle, D.C., 1987. Stable carbon isotopes in deep sea pore waters: modern geochemistry and paleoceanographic applications. Ph. D. Dissertation, University of Washington, Seattle, WA, 209 pp.
McCorkle, D.C., S.R. Emerson, and P.D. Quay, 1985. Stable carbon isotopes in marine porewaters. Earth Planet. Sci. Lett., 74, 13-26.
McCorkle, D.C., L.D. Keigwin, B.H. Corliss, and S.R. Emerson, 1990. The influence of microhabitats on the carbon isotopic composition of deepsea benthic foraminifera. Paleoceanography, vol. 5, pp. 161-185.
McCorkle, D.C. and L.D. Keigwin. Depth profiles of $\delta^{13}C$ in bottom water and core-top *C. wuellerstorfi* on the Ontong-Java Plateau and Emperor Seamounts. in press, Paleoceanography.
McCorkle, D.C., H.H. Veeh, and D.T. Heggie, 1992. LGM and Holocene $\delta^{13}C$ profiles from the Indian Ocean off Western Australia. Fourth International Conference on Paleoceanography (ICP-IV), pp. 194.
Mix, A.C., 1989. Pleistocene paleoproductivity: evidence from organic carbon and foraminiferal species, in, W.H. Berger, V.S. Smetacek,and G. Wefer (eds.), Productivity of the Ocean: Present and Past. Wiley, New York, pp. 313-340.

Müller, P.J., and E. Suess, 1979. Productivity, sedimentation rate, and sedimentary organic matter in the oceans - I. Organic carbon preservation, Deep Sea Res., vol. 26A, pp. 1347-1362.

Müller, P.J., H. Erlenkeuser, and R. von Grafenstein, 1983. Glacial-interglacial cycles in oceanic productivity inferred from organic carbon contents in eastern North Atlantic sediment cores, in, J. Thiede, and E. Suess (eds.), Coastal Upwelling, its sediment record. Plenum Presss, New York, pp. 365-398.

Pedersen, T.F., 1983. Increased productivity in the eastern equatorial Pacific during the Last Glacial Maximum (19,000 - 14,000 yr. B.P.), Geology, vol. 11, pp. 16-19.

Pedersen, T.F., M. Pickering, J.S. Vogel, J.N. Southon, and D.E. Nelson, 1988. The response of benthic foraminifera to productivity cycles in the Eastern Equatorial Pacific: faunal and geochemical constraints on glacial bottom-water oxygen levels, Paleoceanography, v. 3, pp. 157-168.

Peterson, L.C., and W. Prell, 1985a. Carbonate dissolution in Recent sediments of the eastern equatorial Indian Ocean : preservation patterns and carbonate loss above the lysocline, Mar. Geol., vol. 64, pp. 259-290.

Peterson, L.C., and W.L. Prell, 1985b. Carbonate preservation and rates of climatic change: an 800 kyr record from the Indian Ocean, in, T. E. Sundquist., and S. W. Broecker. (eds.), The Carbon Cycle and Atmospheric CO_2: Natural Variations Archean to Present. American Geophysical Union, Washington DC, pp. 251-269.

Prell, W.L., W.H. Hutson, D.F. Williams, A.W.H. Be, K. Geitzenauer, and B. Molfino, 1980. Surface circulation of the Indian Ocean during the Last Glacial Maximum, approximately 18000 yr BP, Quat. Res., vol. 14, pp. 309-336.

Prell, W.L., 1982. Reply to comments by P.J. Webster and N.A. Streten regarding "Surface circulation of the Indian Ocean during the Last Glacial Maximum, approximately 18000 YBP", Quat. Res., vol. 17, pp. 128-131.

Prell, W.L. 1985. The stability of low-latitude sea-surface temperatures: An evaluation of the CLIMAP reconstructoin with emphasis on the positive SST anomalies. D.O.E. Tech. Report, 60 pp.

Santschi, P.H., C. Bajo, M. Mantovani, D. Orciuolo, R.E. Cranston, and J. Bruno, 1988. Uranium in pore waters from North Atlantic (GME and Southern Nares Abyssal Plain) sediments. Nature, vol. 331, pp. 155-157.

Sarmiento, J.L., and J.R. Toggweiler, 1984. A new model for the role of the oceans in determining atmospheric pCO_2, Nature, vol. 308, pp. 621-623.

Sarnthein, M., K. Winn, and R. Zahn, 1987. Paleoproductivity of oceanic upwelling and the effect on atmospheric CO_2 and climatic change during deglaciation times. in Abrupt Climate Change, Proceedings of the NATO/NSF A.R.W. Symposium at Biviers/Grenoble, 1985. W.H.Berger and L.D. Labeyrie, eds. pp. 311-337.

Sarnthein, M., K. Winn, J.-C. Duplessy, and M.R. Fontugne, 1988. Global variations of surface ocean productivity in low and mid latitudes: Influence on CO_2 reservoirs of the deep ocean and atmosphere during the last 21,000 years. Paleoceanography, vol. 3, pp. 361-399.

Shackleton, N.J. 1977. Carbon-13 in *Uvigerina*: tropical rainforest history and the Equatorial Pacific carbonate dissolution cycles, in The Fate of Fossil Fuel CO_2 in the Oceans, Andersen, N.R., and Malahoff, A. (Eds.). Plenum, New York. p. 401-427.

Siegenthaler, U., and T. Wenk, 1984. Rapid atmospheric CO_2 variations and ocean circulation, Nature, vol. 308, pp. 624-625.

Thomson, J., H.E. Wallace, S. Colley, and J. Toole, 1990. Authigenic uranium in Atlantic sediments of the last glacial stage - a diagenetic phenomenon. Earth Planet. Sci. Lett., vol. 98, pp. 222-232.

Webster, P.J., and N.A. Streten, 1978. Late Quaternary ice age climates of tropical Australasia: interpretations and reconstructions, Quat. Res., vol. 10, pp. 279-309.

Webster, P.J., and N.A. Streten, 1982. Comments on : "Surface circulation of the Indian Ocean during the Last Glacial Maximum approximately 18000 yr BP" by Warren L Prell et al, Quat. Res., vol. 16, pp. 421-423.

Weliky, K., E. Suess, C.A. Ungerer, P.J. Müller, and K. Fischer, 1983. Problems with accurate carbon measurements in marine sediments and particulate matter in seawater: A new approach. Limnol. Oceanogr., vol. 28, pp. 1252-1259.

Wilson, T.R.S., J. Thomson, D.J. Hydes, S. Colley, F. Culkin, and J. Sorensen, 1986. Oxidation fronts in pelagic sediments: diagenetic formation of metal-rich layers, Science, vol. 232, pp. 972-975.

Woodruff, F., S.M. Savin, and R.G. Douglas, 1981. Miocene stable isotope record: a detailed deep Pacific Ocean study and its paleoclimatic implications. Science, vol. 212, pp. 665-668.

Woodruff, F., and S.M. Savin, 1985. $\delta^{13}C$ values of Miocene Pacific benthic foraminifera: correlations with seal level and biological productivity, Geology, vol. 13, pp. 119-122.

Zahn, R., K. Winn, and M. Sarnthein, 1986. Benthic foraminiferal $\delta^{13}C$ and accumulation rates of organic carbon: *Uvigerina peregrina* group and *Cibicidoides wuellerstorfi*, Paleoceanography, vol. 1, 27-42.

Table 1. Locations of Exmouth Plateau and Perth Basin cores.

Core	Latitude	Longitude	Depth
Exmouth Plateau:			
53GC-04	19°35.1'S	113°32.1'E	956
53GC-09	20° 0.3'S	112°55.9'E	962
53GC-11	20°53.7'S	112°20.0'E	1432
53GC-06	19° 3.2'S	112°45.1'E	1979
53GC-07	18°54.5'S	112°37.9'E	2256
Perth Basin:			
57GC-15	29°22.9'S	113°13.0'E	2750
57GC-19	27°19.2'S	111°37.6'E	2755

Table 2a. Isotopic stage boundaries, bulk sedimentation rates, $CaCO_3$ and organic C accumulation rates, benthic foraminiferal abundances and accumulation rates, and paired-species benthic ^{13}C differences.

Core	Stage	Depth interval cm	Bulk sed. rate cm/ky	$CaCO_3$ MAR mmol/ cm^2ky	Organic C MAR mmol/ cm^2ky	Total Cib.+Uvi #/g	Benthic foram. MAR #/cm^2ky	Cib.-Uvi. ^{13}C
53GC-04	1	0 - 26	2.16	13.93	0.70	11.79	20.63	0.98
(956 m)	2	26 - 71	3.73	29.95	0.98	23.96	78.65	0.72
	3-4	71 - 169	1.96	14.79	0.52	30.19	51.48	
	5	169 - 294	2.23	16.04	0.60			
53GC-09	1	0 - 32	2.63					0.76
(962 m)	2	32 - 78	3.79					0.76
	3-4	78 - 119	0.82					
	5	119 - 249	2.32					
53GC-11	1	0 - 32	1.49			10.85	13.74	0.55
(1432 m)	2	32 - 78	3.44			22.22	64.97	0.85
53GC-06	1	0 - 21	1.70			3.18	4.60	0.78
(1979 m)	2	21 - 60	3.23			21.91	60.15	0.67
	3-4	60 - 109	0.99					
	5	109 - 249	2.50					
53GC-07	1	0 - 29	2.37	15.67	0.37	2.16	4.25	0.79
(2256 m)	2	29 - 71	3.52	22.81	0.68	18.67	51.92	0.74
	3-4	71 - 171	2.01	12.88	0.42			
	5	171 - 319	2.65	17.38	0.50			
57GC-15	1	0 - 14	1.16	8.77	0.17	0.90	0.96	
(2750 m)	2	14 - 37	1.91	18.04	0.19	9.68	19.41	0.66
	3-4	37 - 70	0.65	5.93	0.07	19.12	12.68	
	5	70 - 100	0.55	4.58	0.06	15.55	8.55	
57GC-19	1	0 - 22	1.83	12.39	0.31	0.24	0.36	
(2755 m)	2	22 - 53	2.53	20.41	0.31	6.49	15.11	0.78
	3-4	53 - 116	1.28	10.49	0.17	6.37	7.66	
	5	116 -190	1.32	10.83	0.20			

Table 2b. Long-term (stages 1-5) ^{230}Th sediment accumulation rates [Veeh et al., in prep.]

Core	cm/ky
53GC-04	2.27
53GC-07	2.23
57GC-15	0.67
57GC-19	1.46

Table 2c. Bulk $CaCO_3$ radiocarbon dates.

Core	Depth (cm)	Age (yr)
53GC-07	58	21,930 ± 210
57GC-19	3	3946 ± 60
	16	10,204 ± 93
	36	19,600 ± 170

Table 3a. 53GC-04 Holocene and Stage 2 benthic isotope data.

Sample Interval	Depth		*C. wuellerstorfi*				*U. peregrina*		
		^{18}O	^{13}C	^{18}O	^{13}C	^{18}O	^{13}C	^{18}O	^{13}C
0 - 3	1.5	2.07	0.77						
3 - 5	4.0	2.45	0.82	1.91	0.81	3.55	-0.26	2.58	-0.12
5 - 8	6.5	1.87	0.59						
8 - 10	9.0	1.89	0.75	1.80	0.80	2.78	-0.20	3.04	-0.23
10 - 13	11.5	1.87	0.65						
13 - 15	14.0	1.53	0.62	1.82	0.82	3.09	-0.35		
16 - 18	17.0	2.22	0.66	2.26	0.73	2.98	-0.21		
18 - 21	19.5	2.22	0.77						
21 - 23	22.0	2.36	0.51	2.97	0.86	3.25	-0.36	2.57	-0.41
23 - 26	24.5	2.43	0.60						
26 - 29	27.5	3.07	0.51	2.45	0.54	3.89	-0.25	3.69	-0.05
29 - 32	30.5	2.82	0.55	2.98	0.67	3.80	-0.33	3.57	-0.20
32 - 35	33.5	2.62	0.57						
35 - 37	36.0	3.42	0.48	2.68	0.62	3.51	-0.32	3.50	-0.11
40 - 42	41.0	3.01	0.56	3.15	0.72	4.06	-0.16	4.11	-0.24
42 - 45	43.5	3.22	0.63						
45 - 47	46.0	3.43	0.45	3.33	0.46	4.10	-0.23	4.10	-0.22
47 - 50	48.5	3.07	0.58						
50 - 52	51.0	3.35	0.43	3.32	0.61	3.49	-0.33		
52 - 55	53.5	3.37	0.55	2.84	0.64	3.80	-0.04		
55 - 57	56.0	3.48	0.53	3.47	0.55	3.89	-0.49		
57 - 60	58.5	2.91	0.71	2.98	0.67	3.52	0.03		
60 - 62	61.0	3.31	0.61	3.19	0.47	3.81	-0.14		
62 - 65	63.5								
65 - 67	66.0	3.30	0.43	3.29	0.59	3.85	-0.23		
67 - 70	68.5	2.97	0.56			3.79	0.00		
70 - 72	71.0	2.98	0.64	3.39	0.60	3.59	-0.04		

Table 3b. 53GC-09 Holocene and Stage 2 benthic isotope data.

Sample Interval	Depth	*C. wuellerstorfi* ^{18}O	^{13}C	^{18}O	^{13}C	*U. peregrina* ^{18}O	^{13}C	^{18}O	^{13}C
0 - 3	1.5	1.63	0.70	2.92	0.64	2.84	-0.08		
5 - 8	6.5	2.20	0.84	2.13	0.82	2.86	-0.01	2.86	-0.06
11.5 - 14.5	13.0	2.20	0.70	2.35	0.69	2.79	-0.03	2.54	0.19
16.5 - 19.5	18.0	3.17	0.63	3.19	0.63	3.18	-0.21		
21.5 - 24.5	23.0	2.42	0.64	2.66	0.74	3.15	-0.09		
27.0 - 30.0	28.5	3.15	0.72	3.14	0.66	3.18	-0.02	2.74	0.03
33.5 - 36.5	35.0	2.89	0.80	3.17	0.66				
38.5 - 41.5	40.0	3.30	0.67	3.09	0.77	3.78	-0.40	3.42	-0.27
43.5 - 46.5	45.0	3.23	0.66	3.27	0.67	3.52	-0.13	3.37	-0.04
48.5 - 51.5	50.0	2.65	0.57	2.94	0.76	3.63	-0.30	3.92	0.05
53.5 - 56.5	55.0	3.11	0.81	3.30	0.54	3.67	-0.29	3.90	-0.04
58.5 - 61.5	60.0	3.26	0.62	3.20	0.70	3.94	-0.34	3.90	-0.21
63.5 - 66.5	65.0	3.01	0.73	2.85	0.83	3.66	-0.07		
68.5 - 71.5	70.0	3.10	0.55	2.99	0.60	3.83	-0.26	3.29	-0.56
73.5 - 76.5	75.0	3.02	0.67	2.94	0.69	3.51	-0.33	3.45	-0.30

Table 3c. 53GC-11 Holocene and Stage 2 benthic isotope data.

Sample Interval	Depth	*C. wuellerstorfi* ^{18}O	^{13}C	*U. peregrina* ^{18}O	^{13}C
1 - 3	2.0	2.45	0.68	3.22	-0.05
6 - 8	7.0	2.22	0.81	2.73	0.17
10 - 12	11.0	2.76	0.48	2.85	0.23
16 - 18	17.0	3.27	0.50	3.52	-0.27
25 - 27	26.0	3.08	0.54	3.79	-0.31
31 - 33	32.0	2.93	0.55	3.84	-0.25
36 - 38	37.0	2.96	0.47	4.07	-0.26
41 - 43	42.0	2.40	0.26	4.14	-0.42
46 - 48	47.0	2.78	0.63	3.85	-0.22
51 - 53	52.0	3.03	0.49	3.48	-0.12
56 - 58	57.0	3.13	0.56	3.43	-0.42
61 - 63	62.0	3.22	0.44	3.47	-0.20
66 - 68	67.0	3.15	0.52	3.49	-0.40
71 - 73	72.0	2.95	0.49	3.88	-0.41
76 - 78	77.0	2.89	0.48	3.68	-0.24

Table 3d. 53GC-06 Holocene and Stage 2 benthic isotope data.

Sample Interval	Depth	*C. wuellerstorfi* ^{18}O	^{13}C	^{18}O	^{13}C	*U. peregrina* ^{18}O	^{13}C	^{18}O	^{13}C
1 - 3	2.0	2.51	0.56			3.28	-0.06		
6 - 8	7.0	2.45	0.77	2.37	0.64	3.51	-0.03	4.19	-0.36
11 - 13	12.0	3.91	0.23	3.72	0.38	3.84	-0.38	4.06	-0.30
16 - 18	17.0	2.61	0.59	2.58	0.77	3.87	-0.23	3.08	-0.29
21 - 23	22.0	3.68	0.30	3.85	0.32	4.79	-0.39	4.54	-0.59
26 - 28	27.0	3.93	0.20	3.69	0.24	4.53	-0.41	4.53	-0.47
31 - 33	32.0	3.92	0.27	3.83	0.27	4.81	-0.39	4.52	-0.46
36 - 38	37.0	3.96	0.32			4.67	-0.40		
41 - 43	42.0								
46 - 48	47.0	3.70	0.45			4.32	-0.11		
51 - 53	52.0								
56 - 58	57.0	3.51	0.38			4.44	-0.21		

Table 3e. 53GC-07 Holocene and Stage 2 benthic isotope data.

Sample Interval	Depth	*C. wuellerstorfi* ^{18}O	^{13}C	^{18}O	^{13}C	*U. peregrina* ^{18}O	^{13}C	^{18}O	^{13}C
0.5 - 3.5	2.0								
3.5 - 5.5	4.5	2.41	0.78			2.95	0.03		
5.5 - 8.5	7.0								
8.5 - 10.5	9.5	2.34	0.75			3.06	-0.00		
10.5 - 13.5	12.0								
13.5 - 16.5	15.0	2.57	0.71			4.20	-0.50		
16.5 - 19.5	18.0								
19.5 - 21.5	20.5	3.29	0.50			3.57	-0.12		
21.5 - 24.5	23.0								
24.5 - 26.5	25.5	2.79	0.51			2.93	-0.11		
26.5 - 29.5	28.0	3.56	0.54			4.37	-0.56		
31.5 - 34.5	33.0	3.54	0.40			4.63	-0.62		
36.5 - 39.5	38.0	4.07	0.23	3.93	0.26	4.63	-0.63	4.60	-0.53
41.5 - 44.5	43.0	4.06	0.03	4.03	0.27	4.65	-0.47	4.91	-0.54
46.5 - 49.5	48.0	3.80	0.14			4.75	-0.47		
51.5 - 54.5	53.0	3.75	0.24			4.51	-0.57		
56.5 - 59.5	58.0	3.94	0.30			4.58	-0.52		
61.5 - 64.5	63.0					4.42	-0.30		
66.5 - 69.5	68.0	3.82	0.38			4.29	-0.38		

Table 3f. 57GC-15 Holocene and Stage 2 benthic isotope data.

Sample Interval	Depth	*C. wuellerstorfi* ^{18}O	*C. wuellerstorfi* ^{13}C	*U. peregrina* ^{18}O	*U. peregrina* ^{13}C
1 - 3	2.0	2.53	0.55		
6 - 8	7.0	2.40	0.48	4.77	-0.50
11 - 13	12.0	3.06	0.26	4.30	-0.41
16 - 18	17.0	4.26	0.20	4.49	-0.49
21 - 23	22.0	3.76	0.18	3.94	-0.30
26 - 28	27.0	3.78	0.12	3.80	-0.23
31 - 33	32.0	3.66	-0.12	4.11	-0.34
36 - 38	37.0	3.39	0.30	3.78	-0.61

Table 3g. 57GC-19 Holocene and Stage 2 benthic isotope data.

Sample Interval	Depth	*C. wuellerstorfi* ^{18}O	*C. wuellerstorfi* ^{13}C	*U. peregrina* ^{18}O	*U. peregrina* ^{13}C
1.5 - 4.5	3.0	3.70	0.13	2.36	0.05
9.5 - 12.5	11.0				
14.5 - 17.5	16.0				
19.5 - 22.5	21.0			4.33	-0.40
24.5 - 27.5	26.0	3.91	0.17	4.51	-0.55
29.5 - 32.5	31.0	4.17	0.06	4.70	-0.63
34.5 - 37.5	36.0	4.12	0.13	4.79	-0.74
39.5 - 42.5	41.0	3.84	0.29	4.60	-0.63
44.5 - 47.5	46.0	3.51	0.14	4.46	-0.56
49.5 - 52.5	51.0	3.83	0.21	4.38	-0.51

Table 4. Holocene-LGM comparisons of percent $CaCO_3$, percent planktonic fragments, and percent coarse fraction (>63 m).

		$CaCO_3$				Fragments			
		Holocene		LGM		Holocene		LGM	
Core	Depth	%	SD	%	SD	%	SD	%	SD
53GC-04	956	78.67	4.80	89.98	0.44	8.0	5.7	3.8	2.4
53GC-09	962	77.72	5.83	86.44	1.53				
53GC-11	1432								
53GC-06	1979								
53GC-07	2256	79.60	2.80	82.41	2.69	4.0	3.0	6.2	3.2
57GC-15	2750	82.43	1.25	89.94	0.24				
57GC-19	2755	82.23	2.01	87.70	0.54	0.5	1.0	0.8	0.9

		Coarse			
		Holocene		LGM	
Core	Depth	%	SD	%	SD
53GC-04	956	29.3	3.2	32.2	2.6
53GC-09	962	25.0	6.4	28.3	6.7
53GC-11	1432	38.6	2.7	49.1	1.8
53GC-06	1979	37.3	8.5	48.1	11.7
53GC-07	2256	24.1	1.2	29.5	1.7
57GC-15	2750	35.5	13.0	50.2	8.6
57GC-19	2755	23.1	1.2	16.4	1.6

NUTRIENT, MIXING AND EXPORT INDICES: A 250 KYR PALEOPRODUCTIVITY RECORD FROM THE WESTERN EQUATORIAL PACIFIC

J. C. Herguera
Centro de Investigación Científica y
de Enseñanza Superior de Ensenada
División de Oceanología
Carretera Transpeninsular Km. 107
Ensenada, Baja California
México

ABSTRACT. The seminal proposition by Arrhenius (1952) that colder periods favor higher productivity oceans than warmer ones, together with the relatively recent hypothesis linking the surface ocean export productivity levels and atmospheric pCO_2 have motivated a renewed interest in the reconstruction of the ocean's paleoproductivity from deep-sea sediments. In this paper we present a record of the relative magnitude and timing of the paleoexport of organic matter to the sea-floor for the last 250 Kyr for the western equatorial Pacific region. The benthic foraminifera preserved in deep-sea sediments are used to estimate the organic matter flux to the sea floor and indirectly they are also a measure of the organic carbon export of from the photic layer. This paleoexport record shows a remarkable coherence with the ice record, as indicated by the $\delta^{18}O$ of benthic foraminifera. We further estimate the relative importance of changing nutrients and upper ocean mixing rates to explain the observed paleoproductivity fluctuations. The record of changes in nutrient concentrations is based on the interpretation of the $\Delta\delta^{13}C_{PB}$ between planktics and benthics as a paleonutrient tracer. To estimate the magnitude of the physical forcing, or the upper ocean mixing rates, we use a simple box model, together with a paleotemperature record based on the interpretation of the $\Delta\delta^{18}O_{GsPo}$ between two planktic foraminifera living at different depths within the mixed layer. Paleoexport, the nutrient content, and the ocean mixing rates records are compared to estimate the relative importance of each mechanism in explaining the observed fluctuations of the paleoexport record.

NATO ASI Series, Vol. I 17
Carbon Cycling in the Glacial Ocean:
Constraints on the Ocean's Role in Global Change
Edited by R. Zahn et al.

INTRODUCTION

Increasing evidence indicates that the paleoproductivity of the ocean and climate are closely knit on glacial to interglacial timescales. The paleoproductivity record has received renewed attention since the discovery by Berner et al. (1980), Delmas et al. (1980), and Barnola et al. (1987) that the atmospheric carbon content has varied on glacial to interglacial time scales. The relative size of the marine carbon reservoir (approx. 60 atmospheric carbon masses, ACM) has led to the attribution of the observed fluctuations in atmospheric pCO_2 to changes in ocean productivity (Broecker, 1982; Sundquist and Broecker, 1985; and references therein; Boyle, 1988b; Keir, 1988; Martin, 1990).

The climate-productivity relationship was first proposed by Arrhenius (1952) in his interpretation of eastern equatorial Pacific sediment accumulation patterns and has been subsequently confirmed by various authors working in the same region (Pedersen, 1983; Pisias and Prell, 1985; Arrhenius, 1988; Lyle et al., 1988; Pedersen et al., 1991). Basically it proposes that colder periods favor higher productivity oceans than warmer ones. Just recently this same pattern has been reported from the western equatorial Pacific based on the interpretation of the accumulation rates of benthic foraminifera as an index for the organic matter flux to the sea-floor (Herguera and Berger, 1991; Herguera, 1992).

Here we extend the paleoproductivity reconstruction, derived from benthic foraminifera accumulation rates, through the last 250 Kyr in western equatorial Pacific deep-sea sediments. Our purpose is to estimate the relative importance to paleoproductivity of changes in the oceanic nutrient concentration, and the relative magnitude of the physical forcing, i. e. the rate of mixing of deep and surface waters.

To estimate the change in productivity that results from variations in mixing intensity, we make use of the $\delta^{18}O$ record of two planktic foraminifera. Changing the mixing rates of deep and surface waters should in principle affect the temperature structure within the mixed layer. These temperatures are recorded in the $\delta^{18}O$ of planktic foraminiferal shells living at different depths within the mixed layer column. The relative difference in $^{18}O/^{16}O$ ratios between shallow and deeper living species should therefore be a function of the difference in temperature of the waters where they grew. To estimate the changes in mixing of upper ocean waters we will make use of a relatively simple heat balance box model, that relates relative differences in temperature within the mixed layer to a mixing index.

Basically what we exploit here are the gradient in temperature between the base and the top of the mixed layer. When this difference is relatively high we interpret it as a warm, stratified surface layer characterized by low mixing rates; by increasing these rates the surface layer will be cooled, stratification will be weakened, and the temperature difference will decrease accordingly.

We derive variations in the nutrient content of deep waters from the difference between the planktic and benthic foraminifera $\delta^{13}C$ records ($\Delta\delta^{13}C_{PB}$). We compare the surface ocean record, thought of as stripped of nutrients, with the deeper one which records the overall deep water composition. The relative $\delta^{13}C$ difference is a function of the ocean's nutrient concentration and is not affected by other processes that change the $\delta^{13}C$ of the ocean carbon reservoir (e.g. carbon fluxes between reservoirs, circulation changes). We recover this difference for the past by measuring the $\Delta\delta^{13}C$ between planktic and benthic foraminifera from deep-sea sediments.

This interpretation of the $\Delta\delta^{13}C$ is based on the observed parallelism between the $^{13}C/^{12}C$ ratio, nutrients, and the apparent oxygen utilization in the ocean (Broecker and Peng, 1982, Kroopnick, 1985, Berger and Vincent, 1986). When a deep water mass, rich in nutrients, reaches the surface ocean it enhances the biological productivity in the sunlit zone. During photosynthesis, nutrients and isotopically light carbon are assimilated into the tissues of phytoplankton; the sinking of this organic matter and regeneration at depth, the so called "biological pump", enriches the deeper ocean in the lighter carbon by depleting the surface reservoir. This removal by export depends on the nutrient content of the upmixed water; assuming a complete utilization of nutrients in the surface layer the magnitude of the $\Delta\delta^{13}C_{PB}$ in the foraminifera can be taken as a measure of nutrient content of deep waters. For this reconstruction we chose the planktic foraminifera record of *Globigerinoides sacculifer*, thought to represent the nutrient-depleted surface waters, and the benthic foraminifera record of *Planulina wüllerstorfi*, assumed to represent the average value for Pacific deep water masses (Herguera et al., 1992).

From the comparison between the paleoexport record derived from the benthic foraminifera, and the calculated paleoexport -the combined product of the nutrient concentration and mixing intensity-, we should learn about the relative importance of both processes in driving the observed paleoproductivity fluctuations in the western equatorial Pacific. From their disagreement we shall gain some clues on when and how other mechanisms may overrule our simple representation of the paleoproductivity record.

METHODS AND DATA BASE

We used core ERDC 113P (1°38' S, 159°13'E; 2158 m) located well above the regional lysocline, to minimize dissolution-related problems. To verify that the foraminiferal assemblages in the samples studied are not affected by dissolution processes we constructed a preservation stratigraphy for this core. The preservation index is based on the amount of fragmentation present in a narrow size fraction (125 to 150 μm). We found this size fraction to be very sensitive to dissolution processes (Berger et al., 1982; Herguera, 1991). At depths close to the lysocline, which for the Ontong Java Plateau is located close to 3.3 km depth (Wu and Berger, 1989), the preservation index shows values around 50% fragmentation, which makes it especially sensitive to changes of the lysocline depth and in turn of the carbonate saturation levels back through time. The core was sampled on average every 2 cm, which translates on average into a 2.5 to 3 Kyr interval between samples (Table 1).

Table 1. Depth column is in cm. $\delta^{18}O$ *P. obl.* and *G. sacc.* correspond to the $\delta^{18}O$ of the planktic foraminifera *Pulleniatina obliquiloculata* and *Globigerinoides sacculifer*. $\delta^{13}C$ *P. wüll.* corresponds to the $\delta^{13}C$ of the benthic foraminifer *Planulina wüllerstorfi*, and $\delta^{13}C$ *G. sacc.* corresponds to the $\delta^{13}C$ of the planktic foraminifer *Globigerinoides sacculifer*. L stands for the laboratory where the determination was made (LJ, SIO at La Jolla; C, OSU at Corvallis; B, UB at Bergen). BF/g corresponds to the concentration or the number of specimens of benthic foraminifera per gram of dry sediment.

Depth cm.	$\delta^{18}O$ *P. obl.*	L	$\delta^{18}O$ *P. obl.*	L	$\delta^{18}O$ *G.sacc.*	$\delta^{13}C$ *G.sacc.*	L	$\delta^{13}C$ *P.wüll.*	L	$\delta^{13}C$ *P.wüll.*	L	BF/g
0.5			-1.33	C	-2.08	1.72	LJ	0.31	B			
0.5			-1.43	C	-2	1.84	LJ					
0.5			-1.42	C								
1	-1.48	LJ			-2.03	1.87	LJ	0.37	B	0.37	C	61.70
1	-1.28	LJ			-1.96	1.89	LJ	0.25	LJ			
3	-1.59	LJ			-2.17	1.82	LJ	0.34	LJ	0.38	C	
3	-1.44	LJ			-1.99	1.87	LJ					
5	-1.47	LJ			-2.16	1.66	LJ	0.22	LJ	0.16	C	83.50
5					-2.31	1.9	LJ			0.26	C	
7	-1.37	LJ			-2.18	1.68	LJ	0.36	B			
7	-1.21	LJ			-2	1.98	LJ					
9	-1.41	LJ			-1.97	1.76	LJ	0.33	LJ	0.31	C	
9	-1.21	LJ				1.94	LJ					

Table 1, continued

Depth	$\delta^{18}O$		$\delta^{18}O$		$\delta^{18}O$	$\delta^{13}C$		$\delta^{13}C$		$\delta^{13}C$		
cm.	*P. obl.*	L	*P. obl.*	L	*G.sacc.*	*G.sacc.*	L	*P.wüll.*	L	*P.wüll.*	L	BF/g
10.5	-1.29	LJ			-1.85	1.75	LJ			0.22	C	86.96
10.5	-1.43	LJ			-1.83	1.91	LJ					
11.5	-1.32	LJ	-1.36	C	-1.84	1.82	LJ	0.38	B	0.33	C	78.54
13	-1.43	LJ			-1.97	1.71	LJ	0.18	LJ			68.30
13					-1.88	1.87	LJ					
15	-1.32	LJ			-1.85	1.73	LJ	0.08	LJ	0.22	C	62.50
15	-1.12	LJ						0.15	LJ			
16	-1.28	LJ										
17	-1.16	LJ			-1.81	1.72	LJ	0.17	B			78.15
17	-1.22	LJ			-1.85	1.87	LJ	0.24	LJ			
17								-0.02	LJ			
19	-0.64	LJ			-1.78	1.32	LJ	0.11	B			
19	-0.77	LJ			-1.75	1.58	LJ					
21	-0.76	LJ	-0.74	C	-1.18	1.7	LJ	-0.02	LJ	-0.02	C	88.30
21	-0.66	LJ			-1.17	1.56	LJ	0.09	LJ			
23	-0.34	LJ	-0.66	C	-1.07	1.57	LJ	-0.11	LJ			
25	-0.25	LJ			-0.59	1.66	LJ	-0.09	LJ	0.06	C	103.50
25					-1.25	1.54	LJ					
27	-0.27	LJ	-0.25	C	-0.97	2.09	LJ	-0.27	LJ	0.02	C	136.50
27	-0.15	LJ	0.14	C	-0.65	1.81	LJ	-0.22	LJ	0.06	C	
27					-0.73	1.95	LJ	-0.08	LJ			
27								-0.08	B			
29	-0.36	LJ			-0.66	1.66	LJ					
29	-0.28	LJ			-1.07	1.8	LJ					
31	-0.18	LJ	-0.21	C	-0.49	2.05	LJ	-0.27	LJ			133.40
31	-0.28	LJ			-1.05	1.91	LJ	-0.07	LJ			
33	-0.19	LJ	-0.45	C	-0.59	1.82	LJ	-0.09	LJ			
33	-0.4	LJ			-1.08	1.73	LJ					
35	-0.26	LJ			-0.78	1.82	LJ	-0.08	LJ	-0.03	C	128.20
35	-0.33	LJ			-1	1.66	LJ	-0.1	LJ			
35	-0.36	LJ			-.8	1.73	LJ					
35	-0.21	LJ										
37	-0.6	LJ			-1.26	1.79	LJ	0.03	LJ			
37	-0.73	LJ			-0.95	1.94	LJ	-0.13	LJ			

Table 1, continued

Depth	$\delta^{18}O$		$\delta^{18}O$		$\delta^{18}O$	$\delta^{13}C$		$\delta^{13}C$		$\delta^{13}C$		
cm.	*P. obl.*	L	*P. obl.*	L	*G.sacc.*	*G.sacc.*	L	*P.wüll.*	L	*P.wüll.*	L	BF/g
37					-0.91	1.85	LJ					
39	-0.47	LJ			-1.12	1.61	LJ			0.08	C	104.89
39	-0.61	LJ			-0.92	1.89	LJ					
41	-0.62	LJ			-1.02	2.07	LJ	0.18	LJ			102.00
41	-0.38	LJ			-1.1	1.87	LJ	0	LJ			
43	-0.5	LJ	-0.6	C	-1.13	1.79	LJ	0.12	LJ	0.15	C	
43	-0.66	LJ			-0.93	2.13	LJ					
45.5			-0.44	C	-0.99	1.93	LJ	0.26	B	0.17	C	96.00
47.5			-0.58	C	-1.07	1.99	LJ			0.2	C	
51	-0.54	LJ	-0.67	C	-1.2	2.06	LJ	0.25	LJ	0.2	C	92.76
51			-0.56	C	-1	1.94	LJ					
53	-0.63	LJ	-0.62	C	-1.14	1.77	LJ			0.16	C	
53			-0.72	C								
55.5	-0.67	LJ	-0.62	C	-1.21	1.85	LJ	0.11	B			101.90
57.5			-0.61	C	-1.27	1.83	LJ			0.14	C	
61	-0.66	LJ	-0.72	C	-1.23	1.79	LJ	-0.17	LJ	0.22	C	90.41
61			-0.62	C	-1.51	1.91	LJ	-0.1	LJ			
63	-0.67	LJ	-0.75	C	-1.29	2.01	LJ			0.22	C	
63					-1.32	1.72	LJ					
65.5	-0.82	LJ	-0.78	C	-1.34	1.96	LJ			0.18	C	102.70
67.5			-0.78	C	-1.38	1.85	LJ			0.18	C	117.51
71	-0.68	LJ	-0.82	C	-1.26	1.57	LJ	-0.04	LJ	0.06	C	104.08
71					-1.32	1.82	LJ	-0.1	LJ			
73	-0.65	LJ	-0.89	C	-1.17	1.84	LJ			0.04	C	
75			-0.63	C	-1.24	1.7	LJ			0.02	C	
77.5			-0.7	C	-1.23	1.77	LJ			0.06	C	100.00
81	-0.65	LJ			-1.23	1.9	LJ	-0.12	LJ	-0.05	C	97.16
81					-1.29	1.7	LJ	-0.22	LJ			
83	-0.67	LJ			-1.11	1.81	LJ	-0.04	LJ	0.05	C	
83	-0.61	LJ			-0.75	1.67	LJ					
84.5			-0.51	C	-0.96	1.75	LJ			-0.12	C	143.10
85			-0.51	C						-0.11	C	
87.5			-0.5	C	-0.99	1.81	LJ			-0.1	C	
91	-0.53	LJ			-1.11	1.83	LJ	-0.12	LJ	-0.02	C	123.58

Table 1, continued

Depth cm.	$\delta^{18}O$ *P. obl.*	L	$\delta^{18}O$ *P. obl.*	L	$\delta^{18}O$ *G.sacc.*	$\delta^{13}C$ *G.sacc.*	L	$\delta^{13}C$ *P.wüll.*	L	$\delta^{13}C$ *P.wüll.*	L	BF/g
91					-0.83	1.96	LJ	-0.08	LJ			
93	-0.69	LJ			-0.94	2.08	LJ			0.11	C	
93	-0.71	LJ			-1.27	1.92	LJ					
93					-1.4	2	LJ					
95.5			-0.88	C	-1.39	2.1	LJ			0.1	C	90.20
97.5			-0.71	C	-1.35	2.01	LJ			0.21	C	
101	-1.01	LJ	-0.96	C	-1.54	1.9	LJ	0.19	LJ	0.17	C	107.68
101					-1.28	2.1	LJ					
103	-0.93	LJ			-1.45	2.04	LJ			0.26	C	
105.5			-0.95	C	-1.52	1.97	LJ			0.22	C	96.90
107.5			-0.72	C	-1.59	1.82	LJ			0.25	C	
111			-1.03	C	-1.46	2.11	LJ	0.2	LJ	0.17	C	91.27
111					-1.5	1.71	LJ	0.28	LJ			
113	-0.88	LJ			-1.37	2.02	LJ			0.21	C	
115.5			-0.83	C	-1.43	2.1	LJ			0.18	C	81.52
117.5			-0.79	C	-1.51	1.99	LJ			0.21	C	
117.5			-1.03	C								
121	-0.99	LJ	-1	C	-1.55	1.91	LJ	0.29	LJ	0.22	C	93.56
123	-1.04	LJ			-1.49	1.86	LJ			0.25	C	
125.5			-1.04	C	-1.63	1.82	LJ			0.22	C	77.50
127.5			-0.93	C	-1.73	1.77	LJ			0.19	C	
131	-1.2		-1.17	C	-1.4	1.76	LJ	0.13	LJ	0.07	C	95.42
131					-1.6	1.72	LJ					
133	-1.08	LJ			-1.57	1.61	LJ			0.12	C	
133					-1.77	1.76	LJ					
135.5			-0.87	C	-1.73	1.75	LJ			0.08	C	90.80
137.5			-0.96	C	-1.77	1.72	LJ			0.13	C	
141	-1.18	LJ	-1.22	C	-1.5	1.84	LJ	-0.05	LJ	0.12	C	111.71
141					-1.78	1.54	LJ					
143	-1.21	LJ			-1.59	1.73	LJ			0.1	C	
145.5			-1.12	C	-1.68	1.7	LJ			0.13	C	122.50
147.5			-1.3	C	-1.76	1.66	LJ			0.09	C	
151	-1.44	LJ	-1.4	C	-1.77	1.7	LJ	-0.08	LJ	0.05	C	89.18
151					-1.93	1.42	LJ	0.07	LJ			

Table 1, continued

Depth cm.	$\delta^{18}O$ *P. obl.*	L	$\delta^{18}O$ *P. obl.*	L	$\delta^{18}O$ *G.sacc.*	$\delta^{13}C$ *G.sacc.*	L	$\delta^{13}C$ *P.wüll.*	L	$\delta^{13}C$ *P.wüll.*	L	BF/g
153	-1.39	LJ			-1.88	1.67	LJ			0.27	C	
155.5			-1.35	C	-2	1.69	LJ			0.17	C	80.20
157.5	-1.71	LJ	-1.24	C	-2.12	1.71	LJ			0.11	C	
161	-1.42	LJ	-1.28	C	-1.79	1.45	LJ	-0.23	LJ	-0.05	C	67.14
161					-1.99	1.57	LJ	-0.1	LJ			
163	-1.28	LJ			-1.8	1.51	LJ			-0.11	C	
165.5			-1.14	C	-1.66	1.47	LJ			-0.07	C	83.10
165.5			-1	C								
167.5			-0.61	C	-1.51	1.44	LJ			-0.01	C	
171	-0.34	LJ	-0.34	C	-1.33	1.37	LJ	-0.21	LJ	-0.14	C	116.11
171					-1.25	1.36	LJ	-0.1	LJ			
173	-0.46	LJ			-1.27	1.4	LJ			0.05	C	
175.5			0	C	-1.2	1.35	LJ			-0.05	C	126.80
175.5			-0.3	C								
177.5			0.04	C	-0.76	1.47	LJ			-0.05	C	
181	0.05	LJ	-0.05	C	-0.63	1.64	LJ	-0.26	LJ	-0.18	C	137.57
181					-0.68	1.4	LJ					
183	-0.28	LJ			-0.74	1.55	LJ			-0.18	C	
185.5			-0.14	C	-0.81	1.61	LJ			-0.2	C	145.90
185.5			-0.32	C								
187.5			-0.19	C	-0.77	1.58	LJ			-0.13	C	
187.5			-0.37	C								
191	-0.65	LJ	-0.59	C	-0.93	1.61	LJ	-0.31	LJ			141.02
191					-0.6	1.63	LJ					
191					-0.77	1.54	LJ					
193	-0.35	LJ			-0.73	1.62	LJ			-0.23	C	
195.5			-0.28	C						-0.26	C	142.20
197.5			-0.42	C	-0.79	1.61	LJ			-0.28	C	
201	-0.46	LJ	-0.4	C	-0.77	1.72	LJ	-0.29	LJ			128.72
201					-0.73	1.64	LJ					
203	-0.13	LJ	-0.51	C	-0.63	1.55	LJ			-0.28	C	
203					-0.56	1.31	LJ					
205.5			-0.31	C	-0.55	1.37	LJ			-0.26	C	145.50
207.5			-0.27	C	-0.61	1.45	LJ			-0.26	C	

Table 1, continued

Depth	$\delta^{18}O$		$\delta^{18}O$		$\delta^{18}O$	$\delta^{13}C$		$\delta^{13}C$		$\delta^{13}C$		
cm.	*P. obl.*	L	*P. obl.*	L	*G.sacc.*	*G.sacc.*	L	*P.wüll.*	L	*P.wüll.*	L	BF/g
211			-0.36	C	-0.63	1.61	LJ	-0.34	LJ			149.89
213	-0.37	LJ	-0.33	C	-0.87	1.6	LJ			-0.23	C	
215.5			-0.35	C	-0.82	1.55	LJ			-0.25	C	147.50
217.5			-0.37	C	-0.94	1.64	LJ			-0.23	C	
221	-0.81	LJ	-0.57	C	-1.03	1.6	LJ	-0.31	LJ			148.09
221					-1.04	1.42	LJ					
222.5	-0.53	LJ	-0.81	C	-0.79	1.53	LJ			-0.24	C	
222.5	-0.77	LJ										
225.5			-0.51	C	-0.95	1.44	LJ			-0.21	C	125.80
227.5			-0.69	C	-0.88	1.46	LJ			-0.24	C	
230.5	-0.53	LJ	-0.51	C	-0.93	1.38	LJ			-0.25	C	123.18
232.5	-0.77	LJ	-0.69	C	-0.9	1.69	LJ			-0.27	C	
235.5			-0.48	C	-0.98	1.44	LJ			-0.28	C	123.10
237.5			-0.58	C	-1.07	1.4	LJ			-0.27	C	
240.5	-0.52	LJ	-0.46	C	-1.02	1.55	LJ			-0.23	C	119.31
242.5	-0.47	LJ			-0.98	1.44	LJ			-0.25	C	
245.5			-0.48	C	-0.96	1.49	LJ			-0.26	C	119.80
247.5			-0.4	C	-0.95	1.53	LJ			-0.22	C	
250.5	-0.33	LJ	-0.22	C	-0.9	1.57	LJ			-0.17	C	125.32
252.5	-0.46	LJ	-0.49	C	-1.1	1.62	LJ			-0.18	C	
255.5			-0.46	C	-1.16	1.69	LJ			-0.13	C	131.20
257.5			-1	C	-1.52	1.77	LJ			-0.05	C	
260.5			-0.95	C	-1.32	1.85	LJ			-0.13	C	81.45
260.5					-1.51	1.87	LJ					
262.5	-1.32	LJ	-1.12	C	-1.47	2.06	LJ			0.1	C	
265.5			-1.02	C	-1.51	2.1	LJ			0.18	C	73.60
267.5			-1.09	C	-1.61	1.82	LJ			0.19	C	
270.5			-1.1	C	-1.76	1.54	LJ			0.14	C	73.56
272.5	-1.58	LJ	-0.84	C	-1.58	1.89	LJ			0.14	C	
275.5			-0.91	C	-1.52	1.71	LJ			0.14	C	71.30
277.5			-0.97	C	-1.47	1.75	LJ			0.01	C	
280.5			-1	C	-1.36	1.83	LJ			0.21	C	74.22
280.5					-1.56	1.85	LJ					
282.5	-1.28	LJ	-1.02	C	-1.66	1.7	LJ			0.08	C	

Table 1, continued

Depth cm.	$\delta^{18}O$ *P. obl.*	L	$\delta^{18}O$ *P. obl.*	L	$\delta^{18}O$ *G.sacc.*	$\delta^{13}C$ *G.sacc.*	L	$\delta^{13}C$ *P.wüll.*	L	$\delta^{13}C$ *P.wüll.*	L	BF/g
285.5			-1.12	C	-1.59	1.86	LJ			0.24	C	62.40
287.5			-0.89	C	-1.69	1.78	LJ			0.14	C	
290.5			-0.94	C	-1.49	1.68	LJ			0	C	63.62
290.5					-1.57	1.71	LJ					
292.5	-1.11	LJ	-0.91	C	-1.58	1.5	LJ			-0.05	C	
295.5			-0.78	C	-1.39	1.62	LJ			-0.04	C	62.40
297.5			-0.98	C	-1.31	1.75	LJ			-0.13	C	
300.5			-0.63	C	-1.25	1.67	LJ			-0.14	C	120.20
302.5	-0.71	LJ			-1.16	1.62	LJ					
302.5					-1.42	1.52	LJ					
305.5			-0.76	C	-1.21	1.59	LJ			-0.11	C	
307.5					-1.13	1.56	LJ					
310.5			-0.6	C	-1.09	1.19	LJ			-0.12	C	132.10
312.5	-0.77	LJ	-0.78	C	-1.06	1.76	LJ					
315.5					-1.44	1.74	LJ			0.04	C	
317.5					-1.54	1.73	LJ					
320.5			-1.12	C	-1.63	1.9	LJ			0.04	C	102.00
320.5					-1.61	1.5	LJ					
322.5	-1.19	LJ			-1.6	1.81	LJ					
322.5					-1.65	1.49	LJ					
325.5			-1.2	C	-1.7	1.53	LJ	0.02	B			
327.5					-1.59	1.41	LJ					
330.5			-0.88	C	-1.56	1.53	LJ			0.02	C	81.30
330.5			-1.04	C	-1.52	1.46	LJ					
332.5	-0.95	LJ			-1.44	1.31	LJ					
335.5			-1.1	C	-1.37	1.38	LJ			-0.05	C	
337.5					-0.9	1.47	LJ					
340.5			-0.38	C	-0.81	1.57	LJ			-0.09	C	95.70
340.5					-0.78	1.55	LJ					

For the sample preparation procedures readers are referred to Herguera (1992). Specimens of *Globigerinoides sacculifer* and *Pulleniatina obliquiloculata* were picked, usually 40 to 50 specimens, from the narrow size fraction between 350 to 425µm for isotopic determinations.

This procedure was shown by Schiffelbein and Hills (1984) to minimize the isotopic noise that derives from intraspecies compositional variability. For the benthic foraminifera 6 to 12 specimens of *Planulina wüllerstorfi* greater then 250µm were used for each isotopic determination.

Isotopic analyses were performed by the author at two laboratories: at S.I.O. (La Jolla), and at Oregon State University (Corvallis, Oregon) in the laboratory of A. Mix. A few samples were also analyzed at the University of Bergen in the laboratory of E. Jansen. Isotopic data for the planktic foraminifera *G. sacculifer* and *P. obliquiloculata* and the benthic foraminifer *Planulina wüllerstorfi* are listed in Table 1. The origin of the isotopic data is marked in both tables (LJ, for La Jolla; C, for Corvallis; B, for Bergen).

At S.I.O., measurements were made in the conventional fashion, by reacting carbonate shells with 100% phosphoric acid at 50°C in an on-line vacuum system connected to the mass spectrometer (VG Micromass 602C) (Berger and Killingley, 1977). The resulting CO_2 was analyzed isotopically relative to a known CO_2 reference gas, and the oxygen and carbon isotope values calculated with respect to PDB, following the correction for the ^{17}O effect (Craig, 1957). The standard gas was calibrated against NBS-20 limestone, and treated identically to the samples. The average difference for NBS-20 during the period the samples were run was 0.14‰ for $\delta^{18}O$ and 0.11 ‰ for $\delta^{13}C$. For the samples analyzed at Bergen we refer the reader to Herguera et al. (1992).

At Oregon State University analyses were performed on a Finnegan/MAT251 automated mass spectrometer, equipped with an Autoprep Systems carbonate preparation device. Analytical precision for the local carbonate standard (Wiley marble) during the period samples were analyzed was ±0.06 for $\delta^{18}O$ and ±0.03 for $\delta^{13}C$. The precision of foraminiferal analyses is slightly worse than this, due to the internal variability in the foraminifera, benthic mixing and sample size.

Since we are combining data from the two laboratories we further checked if the differences in the analyses between laboratories are within the error margins of the individual laboratories. In Table 2 we show the results of the calculation of the mean difference between replicates (row M) and standard deviation around the mean (row SD) of the same species within the same laboratory ($\Delta\delta^{18}O$ Sp./lab), and for replicates between both laboratories ($\Delta\delta^{18}O$ Sp./lab-lab), while "n" denotes the number of samples considered for the calculation. As can be seen from Table 2 mean differences between laboratories fall within the same range as differences within each laboratory. For the La Jolla laboratory they also fall within the reported average difference with the NBS-20 standard. They seem to be slightly higher than the reported spread around the standard for the OSU laboratory, which may be due to the low number of replicates made, and in part may be a sign of the internal variability within the foraminiferal species. There is also a slightly higher standard deviation for the interlaboratory calibration of *P. obliquiloculata* $\delta^{18}O$ mainly due to three values located at 157.5, 203 and 272.5 cm. From this exercise we conclude that both laboratories were producing comparable results and that the combination of both data sets is feasible for the type of reconstruction presented in this work.

The age model for ERDC113P was derived from the *P. obliquilocuta* $\delta^{18}O$ record using the age assignments of the SPECMAP age model (Imbrie et al., 1984). The transitions or stage boundaries were preferentially used over the peaks and troughs of the record. The ages used and depths in the core can be found in Table 3, the top of the core was assigned a 0 age. Ages were then assigned by linear interpolation between the selected levels.

Table 2. Interlaboratory comparison of the isotopic determinations. M and SD stand for the mean and the standard deviation, n is the number of samples, $\Delta\delta^{18}O$Po/LJ, are the $\delta^{18}O$ replicates of *P. obliquiloculata* determined at the La Jolla laboratory, $\Delta\delta^{18}O$Po/C, are the $\delta^{18}O$ replicates of *P. obliquiloculata* for the Corvallis laboratory, and $\Delta\delta^{18}O$Po/LJ&C are the $\delta^{18}O$ replicates of *P. obliquiloculata* determined at the La Jolla and Corvallis laboratories. This nomenclature also applies to the rest of the columns with the a$\Delta\delta$ition of Gs, which stands for *G. sacculifer*, and Pw, for *P. wüllerstorfi*.

	$\Delta\delta^{18}O$Po/LJ	$\Delta\delta^{18}O$Po/C	$\Delta\delta^{18}O$Po/LJ&C
M	0.13	0.15	0.13
SD	0.06	0.12	0.15
	n=[22]	n=[8]	n=[39]

	$\Delta\delta^{18}O$Gs/LJ	$\Delta\delta^{13}C$Gs/LJ
M	0.16	0.16
SD	0.13	0.10
	n=[53]	n=[53]

	$\Delta\delta^{13}C$Pw/LJ	$\Delta\delta^{13}C$Pw/C	$\Delta\delta^{13}C$Pw/LJ&C
M	0.12	0.07	0.10
SD	0.06	0.04	0.09
	n=[16]	n=[2]	n=[26]

Sedimentation rates were calculated for each time interval, and assumed to be constant between the selected levels (Table 3, SR). Sediment accumulation rates (SAR), were calculated from the product of the sedimentation rates times the weight of dry mass per cm^3, which was empirically derived from the carbonate content of the core. The benthic foraminiferal accumulation rates (BFAR) are taken as the product of SAR and the abundance of benthic foraminifera per gram of dry sediment (BF/g), and are reported as number of specimens per cm^2 per Kyr (Table 1).

Table 3. Isotopic stages identified in the core and the selected depths of the isotopic events used. The SPECMAP ages were taken from Imbrie et al. (1984). SR, sedimentation rates in cm/Kyr.

Isotopic Stage	Isotopic Event	Depth in cm.	SPECMAP Age	SR
1				1.65
2	2.0	19	11	1.57
3	3.0	41	25	1.18
4	4.0	81	59	1.21
5	5.0	95.5	71	1.23
6	6.0	165.5	128	1.59
7	7.0	257.5	186	1.36
	8.0	337.5	245	

RESULTS

The Paleoexport Record

We have already reported (Herguera and Berger, 1991; Herguera, 1992) how the flux of organic matter to the sea-floor is strongly correlated with the benthic foraminifera accumulation rates (BFAR), and in turn how the BFAR is related to the productivity of the surface waters. The main assumption is that the number of benthic foraminifera produced per unit area per unit time depends on the supply of organic carbon delivered to the sea-floor, and this supply is a reflection of the exported productivity in the photic zone. The assumption appears to be reasonable in light of earlier studies on deep-sea benthos, including studies of foraminifera (Murray and Hjort, 1912; Sverdrup et al., 1942; Phleger, 1960; Murray, 1973;

Douglas and Woodruff, 1981; Rowe, 1983; Thiel, 1983; Caralp, 1984; Lutze and Coulbourn, 1984; Corliss et al., 1986; Lutze et al., 1986; Altenbach, 1988; Berger and Diester-Haass, 1988; Gooday, 1988; Loubere, 1991).

We have already reported how the abundance and the accumulation rates of benthic foraminifera in the surface sediments decreases with increasing depth, a trend that parallels the diminishing food supply (Herguera and Berger, 1991; Herguera, 1992). Their abundance patterns are little if at all affected by dissolution processes at depths above the lysocline. Below this depth we noted an increase in the abundance of benthic foraminifera in the sediments, which we interpreted as a sign of the increasing importance of dissolution in shaping the concentration of benthics. This effect is due to the differential removal by dissolution of nannofossils, planktic foraminifera, and fragments of both, leading to an enrichment of benthics. This enrichment is compensated by the lower sediment accumulation rates at depths below the lysocline, which explains the conservative behavior of the benthic foraminifera accumulation trend at depths below the CCD in the Equatorial Pacific deep-sea sediments.

To verify that the paleoclimatic record in core ERDC113P is minimally affected by dissolution related processes we constructed a preservation stratigraphy for this core (Figure 1). The preservation index used is based on the amount of fragmentation present in a narrow size fraction (125 to 150 μm), which we found is very sensitive to dissolution processes (Berger et al., 1982; Herguera, 1991). At depths close to the lysocline, which for the Ontong Java Plateau is located close to 3.3 km depth (Wu and Berger, 1989), this index shows fragmentation values around 10 to 15% at depths above the lysocline, increasing to 50% fragmentation at 3.3 to 3.5 km depth -the regional lysoclyne-, reaching more than 90% at a depth of 4 km, which makes it especially sensitive to changes of the lysocline depth and in turn of carbonate saturation levels back through time. The amount of fragmentation for the core in this study, which fluctuates between 6 to 12%, leads us to think that the amount of dissolution has been low and that it has not significantly affected the carbonate record we are interpreting (Figure 1).

From the figure it is also apparent how interglacial to glacial transitions are invariably punctuated by an increase in fragmentation, while interglacials do show on average relatively higher fragmentation values than glacial periods. This, together with the fact that the core has a well characterized and fairly complete planktic stratigraphy (Figure 2), in spite of its relatively low sedimentation rate (1.4 cm/Kyr, on the average), leads us to believe that core ERDC113P preserves the major features and most of the details seen in records with higher sedimentation rates and that dissolution has minimally affected the carbonate record in the core.

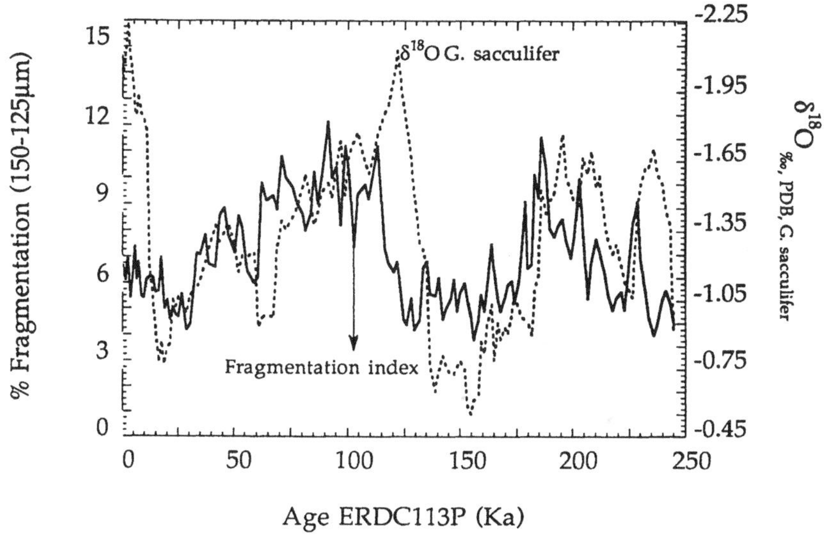

Figure 1. $\delta^{18}O$ of *G. sacculifer* (dashed line) and the fragmentation index (continuous line) used in this work. Note that the highest fragmentation recorded does not go over 11% for this narrow size fraction, and the fact that interglacials show relatively higher values than glacials.

A feature that consistently stands out from the comparison between the $\delta^{18}O$ of *G. sacculifer* and the benthic foraminifera accumulation rate (BFAR) record (Figure 3), is that glacial periods show higher accumulation of benthic foraminifera relative to interglacials, as we would expect from previous work (Herguera and Berger, 1991; Herguera, 1992). The range of values fluctuate between 80 to 220 specimens/cm^2 Kyr, a factor of 2.7 between the lowest, characteristic of interglacials, and the highest values which only occur during glacials. As a further check to assure that this recurrent and prominent feature is not an artifact of dissolution we have compared the benthic foraminifera abundance (BF/g, concentration of benthics in the sediment) with the fragmentation record (Figure 4). Something that stands out from this comparison is that periods with a higher concentration of benthics accompany relatively lower fragmentation, with the exception of stage 4 and stage 5.4, marked in the figure, which show anyway a low benthic foraminifera accumulation rate (Figure 3). From this comparison we can safely say that dissolution has played little or no role in the observed benthic abundance fluctuations in the core.

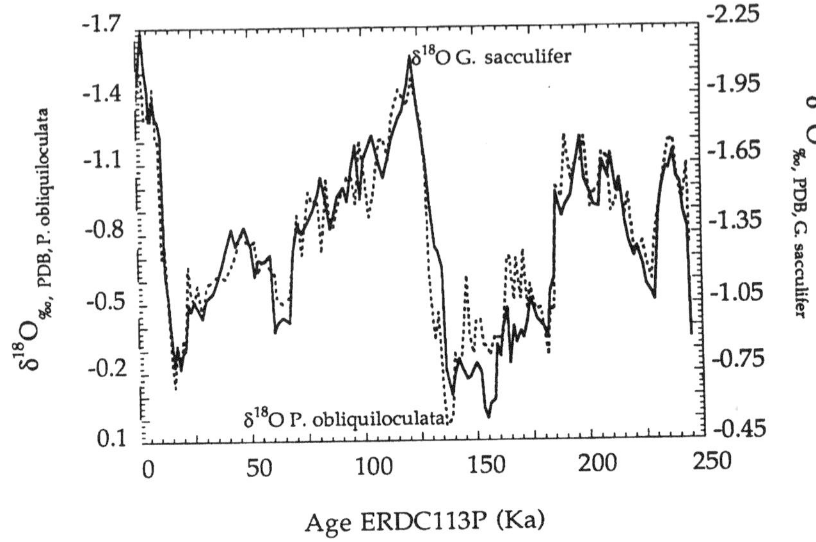

Figure 2. $\delta^{18}O$ stratigraphies of *G. sacculifer* (continuous line) and *P. obliquiloculata* (dashed line) for core ERDC113P. Note how the amplitude of *G. sacculifer* during glacials, but stage 2, seems higher than for *P. obliquiloculata.*

Paleoexport (FPex) was calculated, relative to modern values, with the aid of the equation discussed by Martin et al. (1987). We could also arrive at the same result by using the power law reconstruction discussed in Herguera (1992) and Berger and Herguera (1992), where $FPex = FBFAR^{0.7}$ (F is the factor of change and BFAR are the benthic foraminifera accumulation rates). In this work we will use paleoexport values rather than paleoproductivity for two reasons: first the export production of the photic zone in the ocean is directly related to the flux of nutrients into the mixed layer, a valid assumption to make on geological timescales; and second the flux of organic matter reaching the sediment is modulated by the export from the photic zone to the deep ocean.

The immediate observation from the comparison between the $\delta^{18}O$ of *G. sacculifer* and the paleoexport record (Figure 5) is that glacial periods show higher export productivity levels relative to interglacials, and that transitions beween relatively higher paleoexport periods to the lower ones are relatively fast. A finding that strengthens Arrhenius' (1952) hypothesis on

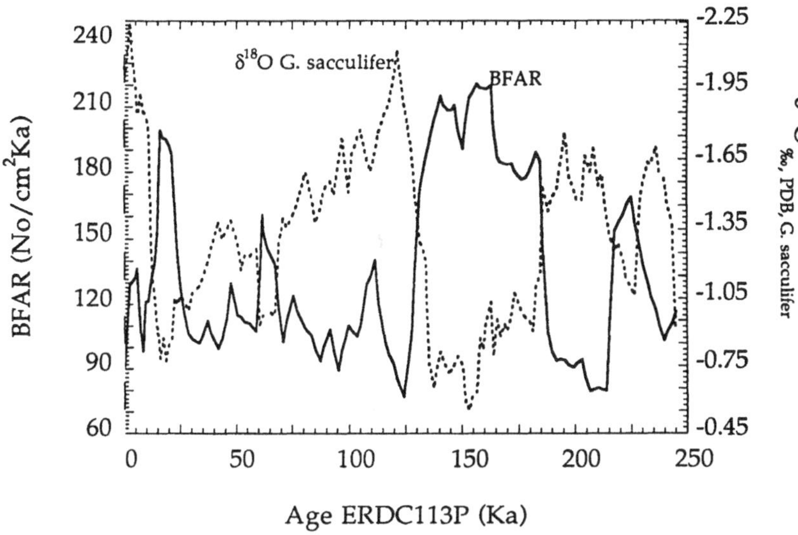

Figure 3. Comparison of the $\delta^{18}O$ of *G. sacculifer* (dashed line) and the benthic foraminifera accumulation rate (BFAR) record (continuous line). Observe how glacials show consistently higher BFAR values than interglacials, and that the timing of the fluctuations follows very closely the ice volume proxy.

the changing productivity levels in the equatorial Pacific on Glacial to Interglacial timescales to explain the observed sediment accumulation patterns. From this reconstruction we can also observe the high degree of coherence between the ice volume and the paleoexport record, a sign of the high sensitivity of the processes controlling paleoproductivity to climate changes at least for the last 250 Kyr (Figure 5).

Also worth mentioning is the relative magnitude of the paleoexport changes. An upper limit of 1.9 occurs during stage 6, and a lower value of 0.9 relative to modern values is seen in isotope stage 5e (Figure 5). Intermediate values are rare, more than half (55%) of the time period under consideration shows export values close to modern levels.

The derived changès in the paleoexport record could be explained, at least in principle, in terms of changes in the ocean's nutrient concentration and the rate of mixing of these waters in the equatorial region (Herguera and Berger, submitted). If we assume a mixed layer in

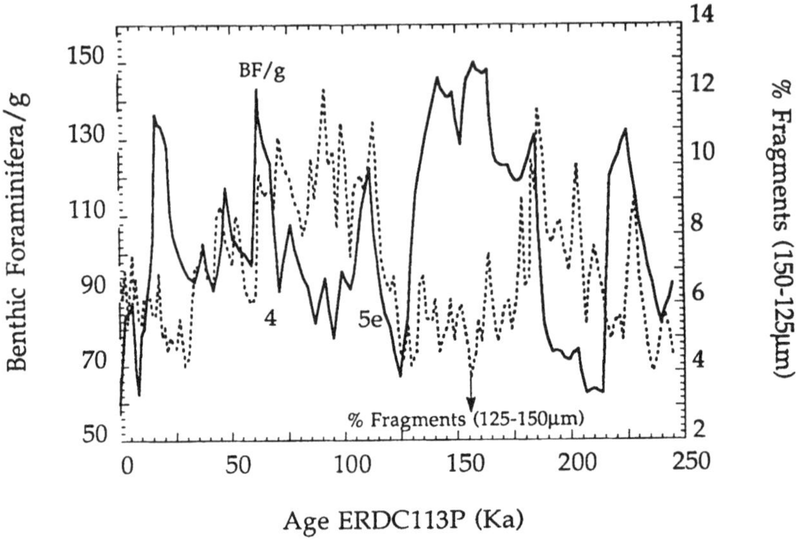

Figure 4. Changes in the concentration of the benthic foraminifera in the sediment (continuous line) and the fragmentation index (dashed line). Note the lack of correlation between both indices, and the fact that high fragmentation periods do not show a concomitant increase in the concentration of benthics. Note also the similarity of the BF/g record and the BFAR.

steady state where the amount of exported production equals the flux of nutrients into this layer then

$$Pex = (R * Nut) * F \quad (1)$$

where Pex is paleoexport production, a concept analogous to new production (Eppley and Peterson, 1979), R is the nutrient-specific Redfield ratio, Nut is the nutrient concentration, and F is the flux of water brought into the mixed later either by advection, upwelling, and/or turbulent diffusion from the deeper waters. We can divide each of the parameters in this expression by its late Holocene values,

$$(Pex/Pex_H) = ((R * Nut)/(R * Nut_H) * (F/F_H))$$

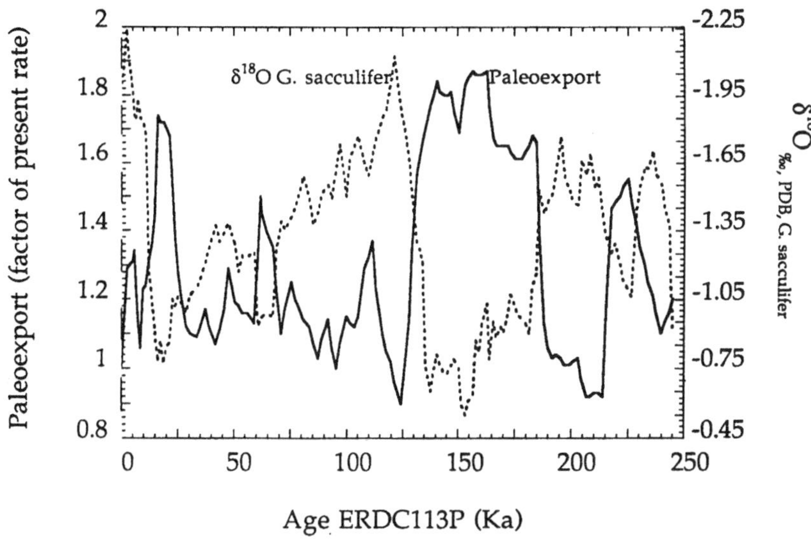

Figure 5. Derived paleoexport record (continuous line) in factors of change relative to Recent values. *G. sacculifer* $\delta^{18}O$ (dashed line) shown for correlation.

so that we can rewrite it in terms of factors (F) relative to present or late Holocene values

$$FPex = F\,Nut * F\,F \tag{2}$$

Note that by taking factors the constant R (Redfield ratio) cancels. We can now compare the paleoexport factor calculated for the last 250 Kyr, based on the abundance changes in benthic foraminifera, with the nutrient and mixing factors to estimate how much of the record can be explained by both mechanisms.

Paleonutrients Record

As discussed earlier we use the $\Delta\delta^{13}C_{PB}$, that is the difference between planktic and benthic foraminifera $\delta^{13}C$, to derive the paleonutrient record. If the biological productivity would strip the surface ocean totally of its phosphate content, assuming a mean phosphate (PO_4) concentration of $2.2*10^{-6}$ µmol/kg and a Redfield ratio of phosphate to carbon of 1:106 (Redfield et al., 1963), then $2.33*10^{-4}$ µmol/kg of organic carbon would be produced, which is roughly 10% of the ΣCO_2 dissolved in the ocean (Broecker and Peng, 1982). If we take a typical fractionation value for marine organic matter of -20‰, that would mean that the surface ocean would show a $\delta^{13}C$ enrichment of 2‰ relative to the deep ocean. This value comes close to the $\delta^{13}C$ difference actually recorded between the shells of benthic and planktic foraminifera.

The observed similarity between modern oceanic phosphate and the $\Delta\delta^{13}C_{PB}$, and the remarkable coherence between the latter and the paleo-pCO_2 record in Antarctic ice cores (Barnola et al., 1987) have led to the proposal that the $\Delta\delta^{13}C_{PB}$ could also be used as a paleo-pCO_2 tracer when properly scaled (Shackleton et al., 1983; Shackleton and Pisias, 1985). The idea behind it is that a nutrient-rich ocean, with higher productivity and consequently higher export levels, will have lower pCO_2 surface in waters than an ocean with lower nutrients (Broecker, 1982). This simple representation of the ocean is complicated by other factors which control the pCO_2 of ocean waters, including changes in the vertical distribution of alkalinity (Keir and Berger, 1983; Boyle, 1988b; Sarmiento et al., 1988; Broecker and Peng, 1989), which do not leave a trace in the $\Delta\delta^{13}C_{PB}$. Another important difference is that while CO_2 cycles through the atmosphere and oceans, nutrients like phosphate do not. The amount of fractionation between atmospheric CO_2 and surface ocean ΣCO_2 is temperature dependent introducing an aΔδitional complication to the correlation between $\delta^{13}C$ and the PO_4 concentration (Keir, 1991; Broecker and Maier-Reimer, 1992).

For the paleonutrient reconstruction we chose the planktic foraminifera record of *Globigerinoides sacculifer* (without the final sac-like chamber), to represent the nutrient-depleted warm surface waters, where it has been reported to thrive best and to show its highest abundance in the equatorial regions of the Atlantic, Indian and Pacific Ocean (Bradshaw, 1959; Bé and Tolderlund, 1971; Parker, 1971; Parker and Berger, 1971; Bé and Hutson, 1977). Maximum export of *G. sacculifer* shells to the sea-floor occurs during the summer months in the Atlantic ocean (Deuser and Ross, 1989) as well as in the equatorial Pacific (Thunell et al., 1983). This species also seems to record best the isotopic composition of the warm, nutrient-depleted, surface mixed-layer (Shackleton and Vincent, 1978; Fairbanks et al., 1982; Curry et al., 1983; Ravelo and Fairbanks, 1992). We are also aware of the offset from equilibrium values induced by other "vital effects"; however for the reconstruction here, we assume constancy throughout the core.

Benthic foraminifera are also tagged by the $\delta^{13}C$ of the ΣCO_2 dissolved in the waters where they grew their shells. Foraminifera that grow within the sediments may record a lower $^{13}C/^{12}C$ ratio than the overlying seawater, the departure from ambient sea water values being controlled by the amount of organic carbon in the sediment and its oxidation rate (McCorkle et al., 1985; McCorkle et al., 1990). To avoid this problem we chose the benthic foram *Planulina wüllerstorfi*, which lives on the surface of the sediment, or above the surface where suitable protruding objects exist (Corliss, 1985; Lutze and Thiel, 1989; Corliss and Emerson, 1990). Such an epifaunal habitat minimizes any effects from changes in $\delta^{13}C$ occurring at the sediment-water interface and within the sediment from the oxidation of organic matter. There appears to be no relationship between the organic matter content in sediments and the $\delta^{13}C$ record of *P. wüllerstorfi* (Zahn et al., 1986).

One of the most remarkable observations from the $\delta^{13}C$ record of *P. wüllerstorfi* (Figure 6) is that it seems to be locked in to the $\delta^{18}O$ record, which confirms the idea of the close ties between climate and the deep-ocean carbon cycle. In contrast the $\delta^{13}C$ record of the planktic foram *G. sacculifer* is punctuated by prominent minima during the late glacial and early transition stages (Figure 6), while for the rest of the record it follows overall the benthic $\delta^{13}C$ record as expected. This puzzling observation has been reported elsewhere in tropical deep-sea cores (Shackleton et al., 1983; Curry and Crowley, 1987; Oppo and Fairbanks, 1989).

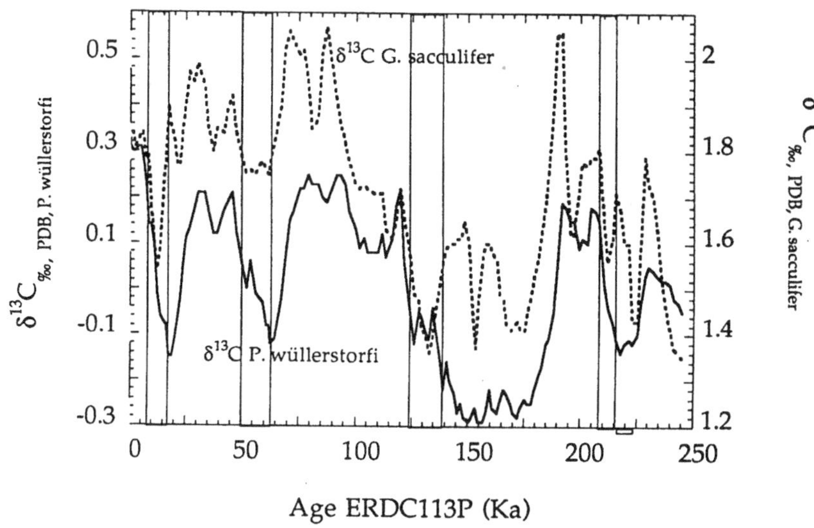

Figure 6. Comparison of the $\delta^{13}C$ record of *P. wüllerstorfi* (continuous line) and *G. sacculifer* (dashed line). This last record has been smoothed to show more clearly the minima of the planktic $\delta^{13}C$ (enclosed in boxes) during the transition from glacial to postglacial values recorded in benthic foraminifera. Note also that the scales have been shifted by 1.5‰.

From the comparison with our Paleoexport reconstruction the planktic signal changes towards more negative values with the onset of deglaciation, in a parallel fashion to the decrease in paleoexport values from glacial to postglacials (Figure 7). This change is already evident during late glacial stages and very early during deglaciation, which shows the high sensitivity of the processes controlling paleoproductivity to climate changes during late stages of glacial periods. We interpret this minimum in the planktic carbon isotope record in terms of the control that productivity, and in turn nutrients, exerts on the $^{13}C/^{12}C$ ratio of the surface ocean. Our evidence shows how a decrease in productivity during the late glacial stages, coupled with the growth of the continental biosphere during the transitions, shape the planktic carbon record of the surface ocean during glacial to postglacial cycles.

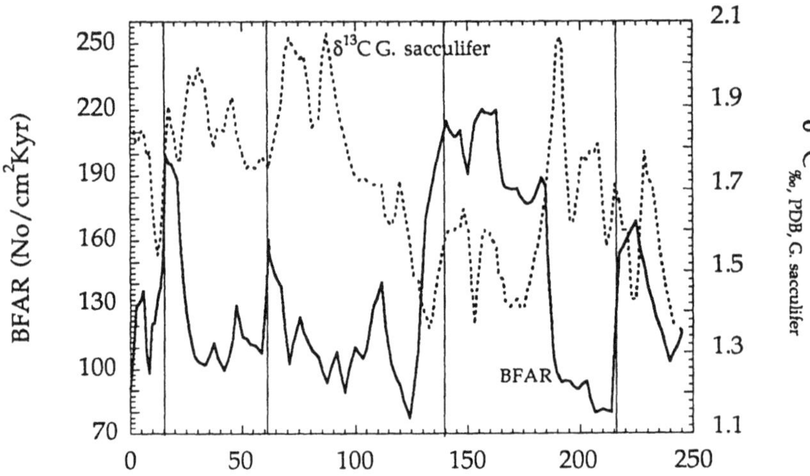

Figure 7. Comparison between the BFAR record (continuous line) and the $\delta^{13}C$ of *G. sacculifer* (dashed line). Marked with lines are the periods of rapid decrease in the BFAR and the planktic $\delta^{13}C$ signals, which are always located at the end of the glacial periods.

For the nutrient reconstruction we use the observed linear correlation between the $\delta^{13}C$ of ΣCO_2 and the nutrient concentration in deep waters (Bainbridge, 1981; Broecker et al., 1982; Weiss et al., 1983; Kroopnick, 1985), to estimate the difference in nutrient concentrations of shallow and deep water masses, that is

$$\Delta\delta^{13}C_{PB} = -k\,\Delta Nut_{SD} \qquad (3)$$

where $\Delta Nut_{SD} = Nut_{Surface} - Nut_{Deep}$. A necessary condition for this model is that shallow water nutrient content is negligible, a condition that is met in the modern western equatorial Pacific; consequently we get $\Delta Nut_{SD} = -Nut_{Deep}$, which allows us to use $F(\Delta\delta^{13}C_{PB})$ to estimate F(Nut) from equation 2. The $\Delta\delta^{13}C_{PB}$ record shows a variability that closely matches the glacial to interglacial climatic fluctuations although there seems to be a 5 to 10 Kyr lag of the $\delta^{18}O$ relative to the increase in the carbon isotope difference especially during transitions to glacial stages (Figure 8), while there is no apparent lag from glacials to interglacials with the exception during transition 6/5. It is yet unresolved whether this feature is just an expression of some sedimentological peculiarity of the core brought about by sediment mixing from older (glacial) levels of the core, or by redeposition processes of older sediments from shallower depths in the Plateau, or whether it is a record of changes in the oceans' $\delta^{13}C$ that preceeded the ice waning. If further data from other cores verifies this lag of the $\delta^{18}O$ record relative to the $\delta^{13}C_{PB}$ it would confirm the importance of the paleoproductivity fluctuations to drive climate change.

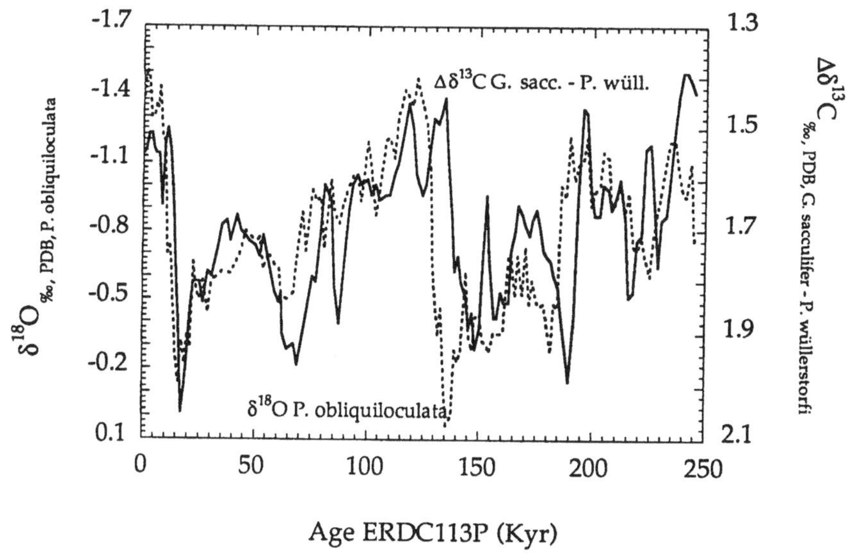

Figure 8. $\Delta\delta^{13}C_{PB}$ record between the planktic and benthic foraminifera (continuous line) plotted on top of the $\delta^{18}O$ of *P. obliquiloculata* (dashed line) for guidance. Note that the $\delta^{13}C_{PB}$ scale has been inverted to compare it better with the ice volume proxy recorded in the $^{18}O/^{16}O$ of the planktic foramifer.

In general the record shows higher values or a stronger gradient during glacials than interglacials (note that in Figure 8 the $\Delta\delta^{13}C$ PB axis has been inverted for a better view of the timing of both records). We take the present carbon isotope difference as 1.6‰, which we use as the normalizing factor for the rest of the record, and compare it with the last glacial value of 2‰. The glacial to postglacial value of F($\Delta\delta^{13}C$ PB), and therefore the F(Nut), is the ratio of 2 to 1.6, that is 1.25. Something that stands out from the comparison between the paleoexport changes and the nutrient factors (FNutrients) (Figure 9) is the relatively minor contribution of nutrients to explain the paleoexport record. On the average only 20 to 40% of the observed increase during glacials can be explained by the aΔδition of nutrients.

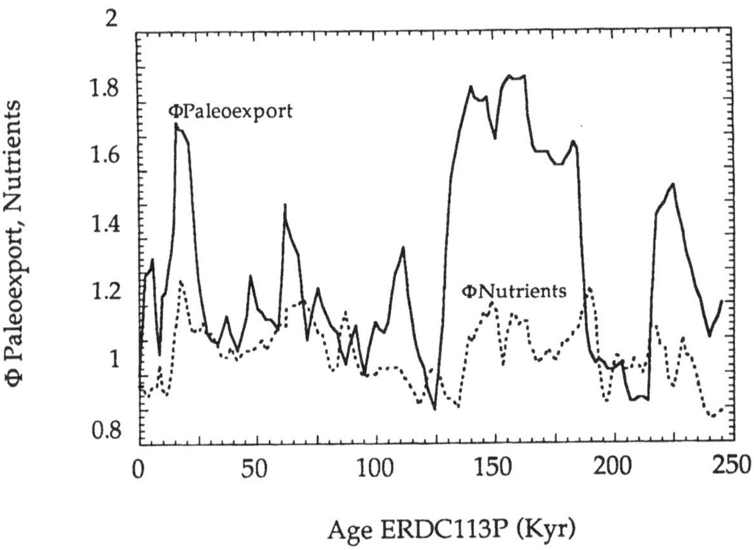

Figure 9. Downcore comparison of the Paleoexport changes (continuous line) with the nutrients factor FNutrients (dashed line).

Mixing Record

As mentioned earlier the paleoexport of organic carbon to the deep ocean is mainly controlled by the flux of nutrients into the mixed layer, and the flux is determined by the concentration and the volume of water transported or mixed. We now then turn to the second part of equation (1), the relative importance of mixing for the export production. By changing the fluxes of water into the mixed layer we are also changing the temperature structure within this layer. For equatorial ocean regions we can simplify the thermal structure of the mixed layer as a balance between the rates of solar heating from above and of mixing of cooler waters from below. If the rate of upward flow of deeper cooler waters increases relative to the rate of solar heating, the temperature gradient in the mixed layer decreases, because of cooling from below, favoring a well-mixed surface ocean. Conversely a relative low upward flux of cooler waters relative to the intensity of insolation allows for a stronger stratification, because heating from above becomes more important, and it is manifested by greater temperature difference between the top and the base of the mixed layer. Thus, relative differences between mixing rates and solar heating should leave an imprint in the temperature structure of the mixed layer and in turn in the $^{18}O/^{16}O$ ratios of planktic foraminifera living at different depths.

To estimate the relative changes in mixing of upper ocean waters we make use of a relatively simple heat balance box model, that relates relative differences in temperature within the mixed layer to a mixing index. We derive the paleotemperature record from the interpretation of the $\Delta\delta^{18}O$GsPo between the shallow dwelling *Globigerinoides sacculifer* and *Pulleniatina obliquiloculata,* which has been reported as living in a relatively deeper habitat, usually in the top of the thermocline and the associated deep chlorophyll maximum (Fairbanks et al., 1982; Ravelo and Fairbanks, 1992). Their relative difference in $^{18}O/^{16}O$ ratios should be a function of the difference in temperature of the waters where they grew, other factors being equal, i.e. assuming that the major control on the isotopic difference is temperature and that salinity and water mass effects are the same for both species. We note that it is difficult to narrow down the precise depth habitat where the planktic foram grew most of its test, due to the lack of data on the life and depth habitat of these species. We assume that this habitat effect, like other "vital" effects, does not change through time.

We used a simple model to relate the temperature changes that would result from changing fluxes of cooler waters into the mixed layer of the western equatorial Pacific (Figure 10). We are interested in the variability of the surface temperatures (To) in the box that result from changes in the input flux of cooler deeper waters (JmTm). Solar heating is assumed constant (Ts), which may be true during glacials, unless there is an important change in the cloud cover over equatorial regions during cooler periods. The temperature of the laterally advected waters into the box (Ti) is also held constant, a condition that may not have been true during

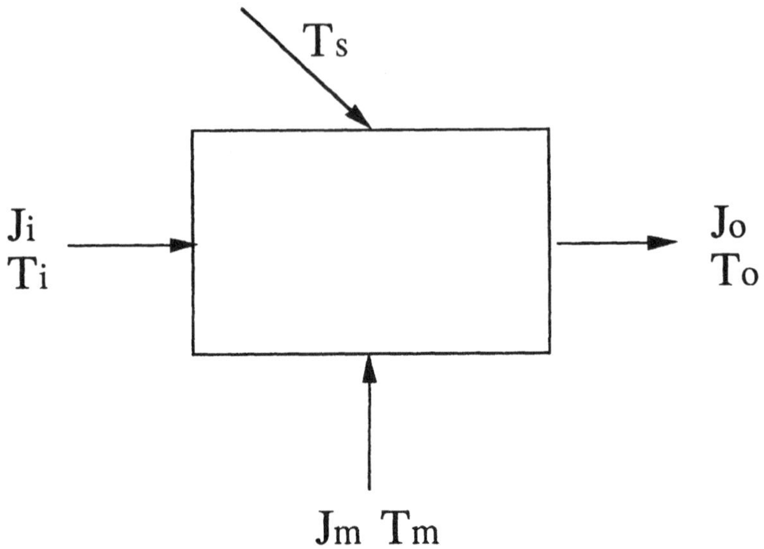

Figure 10. Representation of the box model used to represent the surface ocean and the specified fluxes into and out of the box, see text for further explanations.

glacials. From the model (Figure 10) we solve for the upmixed flux Jm

$$Jm = Ji * (Ti - To) / (To - Tm) \qquad (4)$$

where Ji is the flux of water into the box, To is the temperature of the water as it comes out of the box, Tm is the temperature of the cooler upmixed water. Fluxes in and out of the box are equal to maintain a constant volume. Tm was taken as 19°C, a mean value for the thermocline waters taken from McPhaden and Hayes (1991), and since we are in the warmest pool of water in the Pacific the temperature of the waters advected into the box Ti is set at 30°C, a value commonly found in the western Pacific tropical warm water pool (McPhaden and Hayes, 1991; Lukas and Lindstrom, 1991). The ingoing flux Ji was set at the constant value of 10 Sv derived from Philander et al. (1986) for the modern western equatorial Pacific, and the outgoing temperature varied between 23 and 28°C, a characteristic temperature range between the western and central Equatorial Pacific (Philander, 1990).

The results are plotted on Figure 11, where we relate the changes in mixing -as a factor of change (F Mixing) relative to present values- with the drop in temperature of the surface layer or outgoing flux. The starting temperature value is 29°C which is the average reported for the western equatorial Pacific pool of water (Philander, 1990). We now use these results to convert the temperature difference deduced from the $\Delta\delta^{18}$OGP to mixing rates.

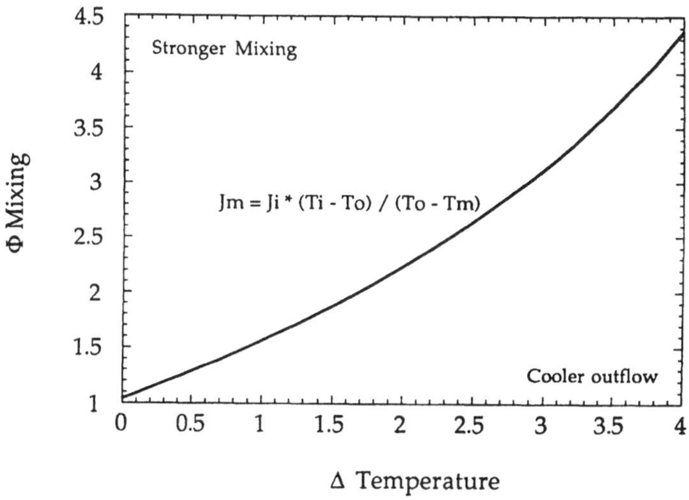

Figure 11. Change in mixing rate (FMixing, factor relative to present values) as a function of the temperature drop (ΔTemperature, starting value of 29°C) obtained from the model for the western equatorial Pacific surface ocean, assuming constant solar heating and lateral advection of warm waters.

To do this we find the difference between the δ^{18}O of *G. sacculifer* and *P. obliquiloculata*, and convert this difference to temperature by taking a ratio of 0.25‰/1°C. The temperature difference between both planktic foraminifera is 2.75°, based on an average δ^{18}O difference of 0.7‰, a result derived from the ERDC collection of core-top values for both species. This temperature difference agrees well with the observed gradient between the upper part of the mixed layer and the top of the thermocline in western equatorial Pacific waters (Lukas and Lindstrom, 1991). Since we want to know temperature differences ΔT with time relative to today, we then subtract 2.75° from the all of the downcore temperature estimates, where

$$\Delta T = \Delta\delta^{18}\mathrm{OGsPo‰} * (1°\mathrm{C}/0.25‰) - 2.75°\mathrm{C} \qquad (5)$$

We then convert ΔT to an upper ocean mixing factor (F Mixing) using the relationship shown in Figure 11. The result of this reconstruction is plotted in Figure 12 where we compare the changes in mixing rates (F Mixing) with the paleoexport record (F Paleoexport).

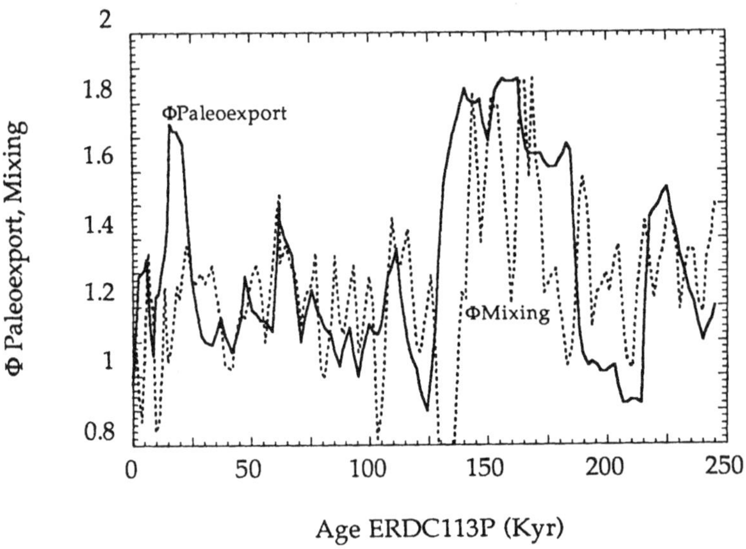

Figure 12. Downcore comparison of both indices expressed as factors of change relative to modern values. Changes in Paleoexport (FPaleoexport, continuous line) and in mixing rates (FMixing, dotted line).

A greater role for mixing than for nutrients is apparent from this reconstruction for most of the record. Note also the higher frequency of the mixing record relative to the other two, which may be an artifact of the small differences in isotopic composition we are comparing, when small errors appear as high frequency fluctuations. The calculated changes in mixing rates from the $\Delta\delta^{18}$OGsPo record can explain by themselves most of the observed changes in paleoexport for the last 250 Kyrs. During glacials export production levels were higher by a factor of 1.7 to 1.9 than during interglacials. Mixing rates show comparable increases between 1.4 to 1.8 during the same intervals, evidence for the important role of mixing modulating the export productivity to the deep ocean in western equatorial Pacific waters.

There are short and intriguing periods, early and late during stage 6 and late during stage 2, when neither the mixing nor the nutrients seem to explain the observed higher paleoexport values. It is yet unresolved whether this short-term features are just an expression of some sedimentological peculiarities of this core (sediment mixing, redeposition), or whether they are a record of changes in upper ocean processes, in which case they could be interpreted as periods of a higher nutrient utilization efficiency and consequently higher export to the deep ocean. There also seems to be a cooling during late stage 7 which is not accompanied by an increase in paleoexport. We interpret this as a possible signal of lateral advection of cooler waters into the region.

DISCUSSION

One of the assumptions for this reconstruction is that changes in deep water $\delta^{13}C$ are representative of corresponding changes in thermocline waters, which are in fact the source of the nutrients. We are basically representing the ocean as a two-box model of the type proposed by Broecker (1982) to derive Equations (1) through (3). The necessary conditions for the model to work are that: a) nutrients in the nutrient-rich thermocline change by the same factor as in deep waters, that is, $FNut_{Deep} = FNut_{Thermocline}$; and b) the relationship between $F(\Delta\delta^{13}C_{PB})$ and F(Nut) is the same for deep water and for the thermocline.

There is reason to believe that neither of these assumptions is strictly correct. In the first case, there is evidence from the Atlantic Ocean of important changes in deep sea circulation during glacials, a reduction of North Atlantic water production (Duplessy et al., 1980; Shackleton et al., 1983; Boyle and Keigwin, 1985) and a vertical reorganization of nutrients and alkalinity (Boyle, 1988b) with a consequent change in the $\delta^{13}C$ structure of deep waters (Duplessy et al., 1988; Kallel et al., 1988). A similar change in the $\delta^{13}C$ of Pacific deep waters was also found, although with a smaller gradient than in the Atlantic (Berger, 1987; Mix et al., 1991; Herguera et al., 1991). From the available limited evidence on the $\delta^{13}C$ structure of mid-depth water masses in the Pacific, we chose *P. wüllerstorfi* from core ERDC113P located at 2.1 km depth, because it seems to record the average $\delta^{13}C$ of Pacific mid-depth and deep waters (Herguera et al., 1992). If new evidence shows that the $\delta^{13}C$ gradient in Pacific mid-depth waters was much stronger during glacials, and that in effect the waters feeding the thermocline in the western equatorial region were isotopically heavier than the average deep waters we used here, than we will have overestimated the nutrient contribution by an amount equivalent to the difference with our ERDC113P *P. wüllerstorfi* record.

As for the second case, if mixing was indeed increased, then the nutrient enrichment which currently exists in the thermocline below the equator could have been greater than now; this would make $FNut_{Thermocline}$ greater than $FNut_{Deep}$. We should see such an effect during glacials as an increased thermocline to mixed layer $F(\Delta\delta^{13}C_{GsPo})$; that is, the $\delta^{13}C$

difference between *G. sacculifer* and *P. obliquiloculata* (thermocline to mixed layer), which exceeds the $FNut_{Deep}$. Our data suggest that $\Delta\delta^{13}C_{GsPo}$ does increase slightly from glacial to postglacial time (Herguera and Berger, in prep), by roughly 10%, ruling out a factor $F(\Delta\delta^{13}C_{GsPo})$ greater than one during glacial stages. Thus our assumption that $FNut_{Deep}=FNut_{Thermocline}$, appears plausible.

Another potential source of error could affect the postulated constant relationship between $\delta^{13}C$ and nutrients. It is true that these two parameters have been observed to follow each other within deep waters, both being tied to the amount of apparent oxygen utilization (AOU) within aging water masses (Kroopnick, 1985). However, in regions where the thermocline crops out, decoupling of the signals is possible, such that $\delta^{13}C$ is reset by exchange with the atmosphere, independently of changes in nutrient content (Charles and Fairbanks, 1990; Keir, 1991; Broecker and Maier-Reimer, 1992). Judging from the observed changes in $\delta^{18}O$ between *G. sacculifer* and *P. obliquiloculata* the changes in $\delta^{13}C$ go in the direction opposite from that expected for a cooling of the upper thermocline, which would lead us to underestimate the nutrient concentrations.

There are still a number of explanations for the differences between the records. One of them has to do with the different sources of error associated with the different analytical procedures and assumptions used in this exercise, e. g. precision of isotopic determinations, intraspecific variability, vital effects. Errors in the calibration between benthic foraminifera abundances and the paleoexport levels, errors in the age assignments and their propagation into sedimentation rates also introduce uncertainty. A further concern is that changes in species composition downcore could introduce spurious effects in the observed gradient between the two foraminifera as pointed out by Hutson (1980) and Schiffelbein and Hills (1984). Although we have no record of the relative abundances of both species downcore, their high abundance in all of the samples studied leads us to suppose a minor role for this effect in this study.

The mixing index and the use of a very simple heat balance model also carry with them assumptions that by themselves are important sources of error. In our model we gave little importance to the lateral advection of cooler waters, which would be felt as a cooling with no accompanying paleoexport increase. Advection of warmer waters was also kept constant: if they decreased during glacials we will have overestimated the amount of mixing. A regional reconstruction along the equator would help to clarify these questions. There is already one paleotemperature record based on the measurement of the undersaturation of alkenone biomarkers ($U^K{}_{37}$) from central equatorial Pacific waters: an increase in upwelling of 50 to 100% was estimated (Lyle et al., 1992), which is higher than the one we obtain.

There are several ways of showing that these patterns are not unique to this particular core, and that they do record real fluctuations in export productivity, nutrients and temperature at least on a regional level. One of them would be to repeat the same measurements on another core of the same region, which would reduce the size of certain errors. The use of other proxies for nutrients like the Cd/Ca ratio (Boyle, 1988a), and for temperature, the measurement of the degree of unsaturation of alkenone biomarkers like $U^K{}_{37}$, would also help to further constrain some of the paleoexport changes here derived.

The enhanced mixing proposed during glacials to explain the higher export production levels could have been driven by stronger pressure gradients in cooler periods, fueled by the much stronger planetary temperature gradients. There is evidence from the CLIMAP (1981) reconstruction of sea surface temperatures, for the last 18 Kyrs, that upwelling was stronger during the last glacial. In low latitudes, equatorial and coastal margin regions, upwelling areas are known to play a key role in the ocean's biological pump, which in turn affects the atmospheric pCO_2, although there seems to be no agreement as to the degree of this effect (Sarnthein et al., 1988; Keir, 1988; Boyle, 1988b; Pedersen et al., 1991; Berger and Wefer, 1991). In a reconstruction of trade wind strength based on the northwestern African margin sediments, Sarnthein et al. (1981) showed that the end of the last two glaciations is marked by a decrease in the upwelling productivity of NW Africa, even before the rise in sea-level. A similar timing of events was observed in the winter trade wind speed near the Peru coast (Molina-Cruz, 1977).

Although the observations presented here confirm the close ties between climate and the export production from the surface ocean, there is some evidence, as yet inconclusive, of periods showing enhanced paleoexport values that are not explained by higher mixing rates nor nutrients. One explanation for this enhanced paleoexport could be that the aΔδition of Fe, as proposed by Martin (1990), and/or other micronutrients to the surface ocean from dust brought about by winds, would have forced a much greater episodicity in the biological productivity of the photic layer, thus favoring a higher export to the deep ocean (Berger and Wefer, 1990). There is evidence from the polar ice cores of increased dust levels during glacials, which carries with it the implication of the greater extension of the source areas, and an intensification of the atmospheric circulation (Angelis et al., 1987; Grousset et al., 1992), while their global nature and timing suggest a link with the Earth's orbital parameters (Petit et al., 1990). This increase in the paleoexport could have led to a further depletion of nutrients in the upper ocean above the permanent thermocline, and an enrichment at depth, that would have resulted in the deepening of the nutricline separating nutrient rich abyssal waters from the more depleted mid-depth ones during glacials. There is evidence for the existence of an upper nutrient-depleted water mass, which in equatorial regions can reach depths of 2 km during the last glacial (Boyle, 1988b; Duplessy et al., 1988; Kallel et al., 1988; Mix et al., 1991; Herguera et al., 1992), although the role of Fe and/or other micronutrients in its development is highly speculative at present.

The decrease in paleoexport appears to happen late during glacials or very early during deglaciation. This observation asserts to the idea that changes in ocean productivity help drive climate changes (Shackleton et al., 1983; Shackleton and Pisias, 1985). There are several explanations as to the connections. The early waning of the ice sheets probably led to the relaxation of the hemispheric temperature gradients and to changes in the patterns of atmospheric circulation. One of the consequences of these changes was the abatement of wind strength which in turn slowed down the ocean mixing rates, thus lowering the paleoexport of organic carbon to the deep ocean. Another possible effect of the changing patterns of atmospheric circulation were the changes in precipitation and evaporation patterns over the continents, that reduced the areal extent of arid areas inducing a lower atmospheric dust load. There is evidence that changes in the dust content of polar ice cores preceded the observed temperature change at every transitional period (Petit et al., 1990). This abatement in the flux of dust to the ocean would have also decreased the contribution of Fe and/or other micronutrients to the surface ocean thus further lowering the export production from the photic zone. Besides the much debated changes in pCO_2 to drive climate, ocean albedo may also have played an important role. By lowering productivity at the end of glacial periods the oceans' albedo would have decreased, leading to an increase solar absorption and thus to further heating.

CONCLUSIONS

From this reconstruction it is apparent that glacial periods show higher accumulation of benthic foraminifera relative to interglacials at least for the last 250 Kyr. The range of values fluctuate between 80 to 220 specimens/cm^2 Kyr, a factor of 2.7 between the lowest, characteristic of interglacials, and the highest values which only happen during glacials. There is also evidence showing that dissolution has played little or no role in the observed fluctuations in benthic abundance .

There is a remarkable degree of coherence between the ice volume and the paleoexport record, a sign of the high sensitivity of the processes controlling paleoproductivity to climate changes at least for the last 250 Kyr. We can also constrain the relative magnitude of the paleoexport changes with an upper limit around a factor of 2 (during stage 6), and a lower limit close to 0.9 relative to modern values during the last interglacial.

We further interpret the minimum in the planktic carbon isotope record in terms of the control that productivity, and in turn nutrients, exerts on the $^{13}C/^{12}C$ ratio of the surface ocean. Our evidence shows how a decrease in productivity during the late glacial stages, coupled with the growth of the continental biosphere during the transitions, shape the tropical surface ocean's planktic carbon record during glacial to postglacial cycles.

From our reconstruction of the nutrient flux to the mixed layer we deduce that this mechanism has a minor role in explaining the observed changes in the paleoexport record. On the average only 20 to 40% of the observed increase during glacials can be explained by the addition of nutrients.

A greater role for mixing during glacials than for nutrients is apparent from the reconstruction here presented. The calculated changes in mixing rates from the $\Delta\delta^{18}$OGsPo record can explain by themselves most of the observed changes in paleoexport for the last 250 Kyrs. During glacials export production levels were higher by a factor of 1.7 to 1.9 than during interglacials. Mixing rates show comparable increases between 1.4 to 1.8 during the same intervals, evidence for the important role of mixing in modulating the export productivity to the deep ocean in the western equatorial Pacific. Although the findings presented here attest to the close ties between climate and paleoexport through upper ocean mixing, they do not presently preclude the role of other mechanisms to explain the observed changes in the paleoexport record. Further work on other cores and indices should help to clarify this latter point.

ACKNOWLEDGMENTS

I would like to express my gratitude and appreciation to Wolf Berger for being a never-ending source of inspiration, and to Robin Keir for his insights and support. I also would like to extend my thanks to Bill Rough whose invaluable help at the OSU mass-spec made this work possible, to Sandra Luz-Correa who helped me with the laborious task of foraminiferal picking, to Alan Mix and Joe Ortiz for their support, discussions and aid during the realization of this work at Corvallis. Thoughtful and thorough reviews by T. Pedersen and an anonymous reviewer have helped to enhance and focus this paper in important ways. This work was partly funded by a postdoctoral fellowship from the Spanish Ministery of Education, and by a grant of the Petroleum Research Fund from the American Chemical Society.

REFERENCES

Arrhenius G (1952) Sediment cores from the East Pacific, Repts Swedish Deep Sea Expedition, 1947-1948, Göteborg, Elander, 227 p

Arrhenius G (1988) Rate of production, dissolution and accumulation of biogenic solids in the ocean: Palaeogeogr Palaeoclimatol Palaeoecol: 67, 119-146

Bainbridge AE (1981) GEOSECS Atlantic Expedition, Hydrographic Data., Washington, D. C., U.S. Govt Printing Office

Barnola JM, Raynaud D, Korotkevich YS, Lorius C, (1987) Vostok ice core provides 160,000-year record of atmospheric CO_2. Nature329: 408-414

Bé AWH, Hutson WH (1977) Ecology of planktonic foraminifera and biogeographic patterns of life and fossil assemblages in the Indian Ocean. Micropaleontology 23: 369-414

Bé AWH, Tolderlund DS (1971) Distribution and ecology of living planktonic foraminifera in surface waters of the Atlantic and Indian oceans. In: Riedel WR, Funnell BM (eds), The Micropaleontology of Oceans, Cambridge University Press, London, 105-149

Berger WH (1987) Ocean ventilation during the last 12,000 years: hypothesis of counterpoint deep water production. Marine Geology 78: 1-10

Berger WH, Bonneau MC, Parker FL(1982) Foraminifera on the deep-sea floor: lysocline and dissolution rate. Oceanol Acta 5: 249-258

Berger WH, Herguera JC (1992) Reading the sedimentary record of the ocean's productivity. In: Falkowski PG, Woodhead AD (eds) Primary Productivity and Biogeochemical Cycles in the Sea, Plenum Press, New York, 455-486

Berger WH, Killingley JS (1977) Glacial-Holocene transition in deep-sea carbonates: selective dissolution and the stable isotope signal. Science 197: 563-566

Berger WH, Vincent E (1986) Deep-sea carbonates: Reading the carbon-isotope signal. Geolog Rundschau 75: 249-270

Berger WH, Wefer G (1990) Export production: seasonality and intermittency, and paleoceanographic implications. Palaeogeog Palaeoclimatol Palaeoecol 89: 245-254

Berner W H, Oeschger H, Stauffer B (1980) Information on the CO_2 cycle from ice core studies. Radiocarbon 22: 227-235

Boyle EA (1988a) Cadmium: chemical tracer of deepwater paleoceanography. Paleoceanography 3: 471-489

Boyle EA (1988b) The role of vertical chemical fractionation in controlling late Quaternary atmospheric carbon dioxide. Jour of Geophys. Res. 93: 15701-15714

Boyle EA, Keigwin L D (1986) Comparison of Atlantic and Pacific paleochemical records for the last 215,000 years: Changes in deep ocean circulation and chemical inventories. Earth Planet Sci Lett 76: 135-150

Bradshaw JS (1959) Ecology of living planktonic foraminifera in the north and equatorial Pacific Ocean. Contrib Cushman Foundation Foram Res 10, 25-64

Broecker WS (1982) Ocean chemistry during glacial time. Geochim Cosmochim Acta 46: 1689-1705

Broecker WS, Maier-Reimer E (1992) The influence of air and sea exchange on the carbon isotope distribution in the sea. Glob Biogeochem Cycles 6: 315-320

Broecker WS, Peng,T-H (1982) Tracers in the Sea, Eldigio Press, New York, 690 p

Broecker WS, Peng T-H (1989) The cause of the glacial to interglacial atmospheric CO_2 change: a polar alkalinity hypothesis. Glob Biogeochem Cycles 3: 215-239

Broecker WS, Spencer DW, Craig H (1982) GEOSECS Pacific Expedition, Hydrographic Data, U.S. Govt Printing Office, Washington, D. C.,

Charles CD, Fairbanks RG (1990) Glacial to interglacial changes in the isotopic gradients of Southern Ocean surface waters. In: Bleil U, Thiede J (eds) Geological History of Polar Oceans: Arctic versus Antarctic, Kluwer Academic, Dordrecht, 519-538

CLIMAP Project Members (1981) Seasonal reconstructions of the Earth's surface at the last Glacial maximum. GSA Map and Chart Series, MC-36

Corliss BH (1985) Microhabitats of benthic foraminifera within deep-sea sediments: Nature 314: 435-438

Corliss BH, Emerson S (1990) Distribution of Rose Bengal stained deep-sea benthic foraminifera from the Nova Scotian continental margin and Gulf of Maine. Deep-Sea Res 37: 381-400

Craig H (1957) Isotopic standards for carbon and oxygen and correction factors for mass-spectrometric analysis of carbon dioxide. Geochim Cosmochim Acta 12: 133-149

Curry WB, Crowley TJ (1987) $\delta^{13}C$ in equatorial Atlantic surface waters: Implications for Ice Age pCO_2 levels. Paleoceanography 2: 489-517

Curry WB, Thunell RC, Honjo S (1983) Seasonal changes in the isotopic composition of planktonic foraminifera collected in Panama Basin sediment traps. Earth Planet Sci Lett 64: 33-43

Delmas RJ, Ascencio JM, Legrand M (1980) Polar ice evidence that atmospheric CO_2 20,000 BP was 50% of present. Nature 284: 155-157

Deuser WG, Ross EH (1989) Seasonally abundant planktonic foraminifera of the Sargasso Sea: succession, deep-water fluxes, isotopic comparisons, and paleoceanographic implications. Jour Foram Res 19: 268-293

Duplessy J-C, Moyes J, Pujol C (1980) Deep water formation in the North Atlantic Ocean during the last ice age. Nature 286: 479-482

Duplessy J-C, Shackleton NJ, Fairbanks R, Labeyrie L, Oppo D, Kallel N (1988) Deepwater source variations during the last climatic cycle and their impact on the global deepwater circulation. Paleoceanography 3: 343-360

Eppley RW, Peterson BJ (1979) Particulate organic matter flux and planktonic new production in the deep ocean. Nature 282: 677-680

Fairbanks, RG, Sverdlove M, Free R, Wiebe PH, Bé AWH (1982) Vertical distribution and isotopic composition of living planktonic foraminifera from the Panama Basin. Nature 298: 841-844

Grousset FE, Biscaye PE, Revel M, Petit J-R, Pye K, Joussaume S, Jouzel J (1992) Antarctic (Dome C) ice-core dust at 18 k.y. B.P.:Isotopic constraints on origins. Earth Planet Sci Lett 111: 175-182

Herguera JC (1991) Paleoproductivity and Deep-Sea Circulation Changes During the late Quaternary: Evidence from the Western Equatorial Pacific, PhD thesis, Univ California, San Diego

Herguera JC (1992) Deep-sea benthic foraminifera and biogenic opal: glacial to postglacial productivity changes in the western equatorial Pacific. Mar Micropalaeo 19: 79-98

Herguera JC, Berger W H (1991) Paleoproductivity from benthic foraminifera abundance: glacial to postglacial change in the west-equatorial Pacific. Geology 19: 1173-1176

Herguera JC, Jansen E, Berger WH (1992) Evidence for a bathyal front at 2000 m depth in the glacial Pacific, based on a depth transect on Ontong Java Plateau. Paleoceanography 7: 273-288

Herguera JC, Stott LD, Berger WH (1991) Glacial deep-water properties in the west-equatorial Pacific: bathyal thermocline near a depth of 2000 m. Mar Geol 100: 201-206

Herguera JC, Berger WH (submitted) Glacial to postglacial drop of productivity in the western equatorial Pacific: mixing rates versus nutrient concentrations

Hutson WH (1980) Bioturbation of deep-sea sediments: Oxygen isotopes and stratigraphic uncertainty: Geology 8: 127-130

Imbrie J, Hays JD, Martinson DG, McIntyre A, Mix AC, Morley JJ, Pisias NG, Prell WL, Shackleton NJ (1984) The orbital theory of Pleistocene climate: support from a revised chronology of the marine $\delta^{18}O$ record. In: Berger AL, Imbrie J (eds) Milankovitch and Climate. Part 1, D. Reidel, Hingham, Mass., 269-305

Kallel N, Labeyrie LD, Juillet-Leclerc A, Duplessy J-C (1988) A deep hydrological front between intermediate and deep-water masses in the glacial Indian Ocean. Nature 333: 651-655

Keir R (1991) The effect of vertical nutrient redistribution on surface ocean $\delta^{13}C$. Glob Biogeochem Cycles 5: 351-358

Keir RS (1988) On the late Pleistocene ocean geochemistry and circulation: Paleoceanography 3: 413-445

Keir RS, Berger WH (1983) Atmospheric CO_2 content in the last 120,000 years: the phosphate extraction model. Jour Geophys Res 88: 6027-6038

Kroopnick PM (1985) The distribution of $\delta^{13}C$ of ΣCO_2 in the world oceans. Deep-Sea Res 32: 57-84

Lukas R, Lindstrom E (1991) The mixed layer of the western Equatorial Pacific Ocean. Jour of Geophys Res 96: 3343-3357

Lutze GF, Thiel H (1989) *Cibicidoides wüllerstorfi* and *Planulina ariminensis*, elevated epibenthic foraminifera. Jour Foram Res 19: 153-158

Lyle M, Murray DW, Finney BP, Dymond J, Robbins JM, Brooksforce K (1988) The record of Late Pleistocene biogenic sedimentation in the eastern tropical Pacific Ocean. Paleoceanography 3: 39-59

Lyle M, Prahl FG, Sparrow MA (1992) Upwelling and productivity changes inferred from a temperature record in the central equatorial Pacific. Nature 355: 812-815

Martin JH (1990) Glacial-interglacial CO_2 change: The iron hypothesis. Paleoceanography 5: 1-13

Martin JH, Knauer GA, Karl DM, Broenkow WW (1987) VERTEX: carbon cycling in the northeast Pacific. Deep-Sea Res 34: 267-285

McCorkle DC, Emerson SR, Quay PD (1985) Stable carbon isotopes in marine porewaters. Earth Planet Sci Lett 74: 13-26

McCorkle DC, Keigwin LD, Corliss BH , Emerson SR (1990) The influence of microhabitats on the carbon isotopic composition of deep-sea benthic foraminifera. Paleoceanography 5: 161-185

McPhaden MJ, Hayes SP (1991) On the variability of winds sea surface temperature surface layer heat content in the western Equatorial Pacific: Jour Geophys Res 96: 3331-3342

Mix AC, Pisias NG, Zahn R, Rugh W, Lopez C, Nelson K (1991) Carbon 13 in Pacific deep intermediate waters 0-370 ka: implications for ocean circulation and Pleistocene CO_2. Paleoceanography 6: 205-226

Molina-Cruz A (1977) The relation of the southern trade winds to upwelling processes during the last 75,000 years: sedimentological evidence. Quat Res 8: 324-338.

Oppo DW, Fairbanks RG (1989) Carbon isotope composition of tropical surface water during the past 22,000 years. Paleocanography 4: 333-351

Parker FL (1971) Distribution of planktonic foraminifera in recent deep-sea sediment *in* Funnell BM, Riedel WR (eds) The Micropaleontology of Oceans, Cambridge University Press, Cambridge, 289-307

Parker FL, Berger WH (1971) Faunal solution patterns of planktonic Foraminifera in surface sediments of the South Pacific: Deep-Sea Res 18: 73-107

Pedersen TF (1983) Increased productivity in the eastern equatorial Pacific during the last glacial maximum (19,000 to 14,000 yr B.P.). Geology 11: 16-19.

Pedersen TF, Nielsen B, Pickering M (1991) Timing of Late Quaternary productivity pulses in the Panama Basin implications for atmospheric CO_2. Paleoceanography 6: 657-677

Petit JR, Mounier L, Jouzel J, Korotkevich YS, Kotlyako VI, Lorius C (1990) Palaeoclimatological and chronological implications of the Vostok core dust record Nature 343: 56-58

Philander G (1990) El Niño, La Niña and the Southern Oscillation. International Geophysics Series, Academic Press, San Diego, 293 p

Philander SG, Hurlin WJ, Seigel AD (1986) Simulation of the seasonal cycle of the tropical Pacific Ocean: Jour Phys Oceanogr 17: 1986-2002

Pisias NG, Prell WL(1985) Changes in calcium carbonate accumulation in the equatorial Pacific during the late Cenozoic: evidence from HPC Site 572. In Sundquist ET , Broecker WS (eds) The Carbon Cycle and Atmospheric CO_2: Natural Variations Archean to Present, Geophys Monog 32, AGU, Washington D.C., 443-454

Ravelo AC, Fairbanks RG (1992) Oxygen isotopic composition of multiple species of planktonic foraminifera: recorders of the modern photic zone temperature gradient: Paleoceanography 7: 815-831

Redfield AC, Ketchum BH, Richards FA (1963) The influence of organisms on the composition of sea-water . In: Hill MN (ed) The Sea 2: New York Wiley Interscience 26-77

Sarmiento JL, Toggweiler JR, Najjar R (1988) Ocean carbon-cycle dynamics and atmospheric pCO_2. In: Charnock H, Lovelock JE, Liss PS, Whitfield M (eds) Tracers in the Ocean, Princeton University Press, Princeton, New Jersey, 3-19

Sarnthein M, Tetzlaff G, Koopmann B, Wolter K, Pflaumann U (1981) Glacial and interglacial wind regimes over the eastern subtropical Atlantic and North-West Africa. Nature 293: 193-196

Sarnthein M, Winn K, Duplessy J-C, Fontugne MR (1988) Global variations of surface ocean productivity in low mid latitudes: influence on CO_2 reservoirs of the deep ocean atmosphere during the last 21,000 years. Paleoceanography 3: 361-399

Schiffelbein P, Hills S (1984) Direct assesment of stable isotope variability in planktonic foraminifera populations. Paleogeog Paleoecol Paleoclimatol 48: 197-213

Shackleton NJ, Hall MA, Line J, Shuxi C (1983) Carbon isotope data from core V19-30 confirm reduced carbon dioxide concentration in the ice-age atmosphere. Nature 306: 319 -322

Shackleton NJ, Imbrie J, Hall MA (1983) Oxygen carbon isotope record of East Pacific core V19-30: Implications for the formation of deep water in the late Pleistocene north Atlantic. Earth Planet Sci Lett 65: 233-244

Shackleton NJ, Pisias NG (1985) Atmospheric carbon dioxide, orbital forcing, and climate. In: Sundquist ET, Broecker WS (eds) The Carbon Cycle and Atmospheric CO_2: Natural Variations Archean to Recent, Geophys Monogr Ser 32, AGU, Washington D.C., 303-317

Shackleton NJ, Vincent E (1978) Oxygen carbon isotope studies in recent foraminifera from the southwest Indian Ocean. Mar Micropaleo 3: 1-13

Sundquist ET, Broecker WS (1985),The Carbon Cycle and Atmospheric CO_2: Natural Variations Archean to Present, Geophys Monograph Series 32, AGU, Washington D.C., 627 p

Thunell RC, Curry WB, Honjo S (1983) Seasonal variation in the flux of planktonic foraminifera: time series sediment trap results from the Panama Basin. Earth Planet Sci Lett 64: 44-55

Weiss RF, Broecker WS, Craig H (1983) GEOSECS Indian Ocean Expedition, Hydrographic Data, U S Government Printing Office, Washington D.C.

Wu G, Berger W.H. (1989) Planktonic foraminifera: differential dissolution and the Quaternary stable isotope record in the west Equatorial Pacific. Paleoceanography 4: 181-198

Zahn R, Winn K, Sarnthein M (1986) Benthic foraminiferal $\delta^{13}C$ accumulation rates of organic carbon: *Uvigerina peregrina* group and *Cibicidoides wuellerstorfi*. Paleoceanography 1: 27-42

DINOFLAGELLATE CYSTS AS PALEOPRODUCTIVITY INDICATORS: STATE OF THE ART, POTENTIAL, AND LIMITS

B. Dale and A. Fjellså
Department of Geology
University of Oslo
P.O. Box 1047 Blindern
N-0316 Oslo 3
Norway

ABSTRACT. Dinoflagellate cysts are useful marine paleoenvironmental indicators of: sea surface temperatures, salinity, and coastal/oceanic water mass boundaries. Most are coastal/neritic; very few oceanic cysts are produced, and deep sea assemblages probably often include many that are transported long distances from coastal environments. This factor must be considered in attempts to apply standard transfer functions in paleoceanographic interpretations from deep sea sediments. Cysts are potentially useful for productivity studies as indicators of upwelling and eutrophication (especially where mineralized microfossils are absent through dissolution).

INTRODUCTION

The best possible estimates of marine paleoproductivity are important to investigations attempting to understand the relationship between the marine carbon cycle and climatic change. On the geological time-scale, we need to investigate how productivity has varied with changes in atmospheric levels of CO_2, and with known drastic global climatic changes. We also need to be able to estimate productivity changes over tens to hundreds of years, in order to investigate the possible effects of human activities.

Two main lines of investigating marine paleoproductivity have been developed to date: geochemical (see other papers in this volume) and micropaleontological. The micropaleontological contribution to paleoceanography, including paleoproductivity, has come mainly from only a few microfossil groups, notably foraminifera, calcareous nannoplankton, diatoms, and radiolaria. Two main factors limit the usefulness of these groups for investigations of the marine carbon cycle and paleoproductivity: 1) all are mineralized (calcareous or siliceous) and therefore may be absent from the record due to dissolution, and 2)

NATO ASI Series, Vol. I 17
Carbon Cycling in the Glacial Ocean:
Constraints on the Ocean's Role in Global Change
Edited by R. Zahn et al.

with the exception of some diatoms and benthic foraminifera, most are oceanic and therefore of little use in the coastal and neritic zones, which today account for about 50% of primary productivity, and which receive a great deal of pollution.

These limitations, together with a need for paleoceanography to include the greatest possible diversity of biological signals, suggest the need to assess other microfossil groups as paleoproductivity indicators. This paper summarizes the state of the art for one of these , the dinoflagellate cysts (referred to as cysts, here, in the interest of brevity). This group is best known for its extensive use in biostratigraphy (e.g. in oil exploration), but attempts are being made to use these microfossils increasingly in paleoceanography (Turon, 1980; Mudie et al., 1990). In contrast to many other groups, most fossilizeable cysts have tests containing sporopollenin-like material (comparable with other palynomorphs such as pollen and spores) and are thus not affected by dissolution of calcium carbonate or silica; also, they are well represented in coastal and neritic environments. It is therefore important to evaluate their potential , and possible factors which may limit their use in paleoceanography, especially with regard to paleoproductivity.

DINOFLAGELLATE CYSTS AS ENVIRONMENTAL INDICATORS

Dinoflagellates are one of the major groups of phytoplankton, occurring in almost all aquatic environments. The main features of dinoflagellates and their resting cysts, which produce the fossil record for the group, have been summarized elsewhere (Dale, 1983; Taylor, 1987).

Several comprehensive studies of the distribution of recent dinoflagellate cysts suggest that the group is particularly useful for marine paleoenvironmental interpretations (Wall et al., 1977; Reid and Harland, 1977; Dale, 1983). These and other studies have identified three main environmental signals in the cyst data:

1) *the surface water temperature signal* - some cysts are restricted to latitudinal (temperature controlled) zones, often corresponding to the standard ecological zones established from the distributions of other marine organisms (e.g. fish, molluscs, and seaweeds) (Dale, 1983).

2) *the coastal/oceanic signal* - one of the most marked ecological boundaries of the marine realm is that between the generally nutrient-rich coastal waters and the relatively nutrient-poor oceanic waters. Most cyst-forming dinoflagellates seem to be coastal/neritic, but several oceanic types have been identified that almost never occur shorewards of the shelf break. Presence or absence of these respective types allows recognition of this boundary in paleontological data (Wall et al., 1977).

3) *the salinity signal* - some main characteristics of both brackish and hypersaline assemblages have been documented, allowing recognition of these paleoenvironments from cyst studies (Wall & Dale, 1973; Wall et al., 1977).

Recent distribution studies thus suggest that cysts may provide paleoenvironmental information at least on a par with the other main microfossil groups. However, the fact that most cyst-forming dinoflagellates seem to be coastal or neritic has to be taken into account when considering their distribution patterns. This may be illustrated by comparing living cyst distributions from coastal regions of the North Atlantic (Dale, 1983, Fig. 46) with their deep sea thanatocoenosis distributions (Harland, 1983). Several cyst types that are clearly restricted to cold temperate zones in coastal samples are also present in deep sea samples from the tropical zone (e.g. *Bitectatodinium tepikiense* Wilson, and *Spiniferites elongatus* Reid). Though pointing out limitations in his data, and cautioning against over interpretation of these, Harland (1983) interpreted the deep sea distributions as representative of cyst production from overlying surface waters, and therefore classified these cysts as cosmopolitan types. However, biologically, it seems most unlikely that a species would inhabit two such very different environments as cold temperate coastal waters and tropical oceanic waters, without also living in either warmer coastal, or colder oceanic waters, as would be expected for true cosmopolitans.

An alternative explanation for these biologically anomalous distributions is that cysts from cold temperate coastal regions may be transported long distances into the deep sea. Supporting evidence for this is emerging from deep sea sediment trap studies (Dale & Dale, 1992). The contribution of transported coastal cysts is probably a main cause of the generally good correspondence between the total concentrations of cysts and terrestrial palynomorphs in some deep sea records (e.g. Zippi, 1992). Further indirect evidence for the long distance transport of cysts is provided by similar transport of other microfossils. Needham et al. (1969) demonstrated transport of reworked Carboniferous palynomorphs by the Western Boundary Current from the St. Lawrence region of SE. Canada as far as offshore Florida. Johnson et al. (1977) demonstrated transport of antarctic diatoms by Antarctic Bottom Water as far north as northern Brazil. Both these examples serve as models for the use of cold temperate cysts as tracers of deep water transport in the North Atlantic.

Even in abyssal sediments, assemblages may be dominated by transported "coastal" cysts, since it is known that relatively few oceanic cysts are produced (Dale, 1986; Dale & Dale, 1992). This factor must be taken into account in attempts to apply transfer functions to cyst assemblages. Other microfossil groups have been used routinely to reconstruct oceanic surface water paleotemperatures, based on transfer equations from their present day distributions in deep sea bottom sediments (CLIMAP, 1976, and many references since). However, these groups are more oceanic, and their distribution in bottom sediments presumably reflects more direct sedimentation from overlying surface waters. It is by no means clear that deep sea cyst assemblages, probably representing a mixture of relatively large amounts of variously transported coastal types and fewer truly oceanic types, may be used in the same way, though attempts to do so are being made (Mudie et al., 1990; Edwards et al., 1991).

Harland (1983) presented a "broader based scheme" for the ecological classification of cysts than that of Wall et al., (1977). He considered his classification possibly better for interpreting Quaternary sequences, since: "the constraints discussed earlier are directly at work here in smudging some of these rather arbitrary ecological divisions" (Harland, 1983, p.376). In the same discussion, he also pointed out that data from living cysts would improve this classification. Our data from living cysts suggest a rather precise ecological zonation in coastal/neritic environments; it is the smudging factors in the deep sea record, such as current transport, sedimentation rates, and differences in sampling and preparation that are more arbitrary. Thus, the "broader" ecological classification of extant cysts suggested by Harland (1983, Table 1) is probably not an ecological classification, since the thanatocoenosis data on which it is based differ so widely from the biocoenosis data.

DINOFLAGELLATE CYSTS AS PALEOPRODUCTIVITY INDICATORS

There have been very few studies using cysts as paleoproductivity indicators. Nevertheless, the data so far (including our latest, unpublished work) suggest both probable applications and probable limits to their usefulness in such studies. They suggest that cyst analysis could contribute to at least two main aspects of paleoproductivity: coastal upwelling and eutrophication.

The present day coastal upwelling signal: previous work

Increased nutrients associated with the various forms of upwelling are extremely important sources of biological production in the present day oceans. Therefore, identifying past upwelling from the sedimentary record has long been recognized as an important aspect of paleoproductivity studies (Thiede & Suess, 1983; Suess, von Huene, et al., 1990). Diatoms are by far the most common primary producers in upwelling areas, and they, together with foraminifera, silicoflagellates, and radiolaria, have proved useful indicators of paleo-upwelling (Schrader & Sorknes, 1990; Oberhänsli et al., 1990).

The few investigations of cysts as possible upwelling indicators were reviewed by Powell et al. (1992). One of the first of these was a study of cyst distribution in bottom sediments from Peru and Namibia/South Africa by Wall et al. (1977). The "upwelling signals" obtained from these two major centers of intensified upwelling were very different from each other. Assemblages from Peru were heavily dominated by species of the genus *Protoperidinium*, mainly unidentified spherical brown cysts, but also including significant amounts of *P. americanum* (Gran et Braarud) Balech, *P. conicum* (Gran) Balech), and several types not found elsewhere. The protoperidinioids represent heterotrophic species that probably feed on other phytoplankton cells, including diatoms (Schnepf and Elbrächter, 1992). Their dominance in the cyst record off Peru probably reflects: 1) optimum feeding conditions for heterotrophic species due to the extremely rich phytoplankton of these waters, and 2) relatively fewer cysts from autotrophic species that are largely out-competed by diatoms.

Only two of the 10 samples representing two transects off Namibia and South Africa contained a cyst assemblage similar to those off Peru. The others contained assemblages heavily dominated by the two cosmopolitan species most commonly found in offshore sediments from many other parts of the world. These species, *Operculodinium centrocarpum* (Deflandre et Cookson) Wall and *Spiniferites ramosus* (Ehrenberg) Mantel, are both autotrophic. The samples which contained more protoperidinioid cysts were those nearest to shore in one of the transects. From this limited number of samples, it is therefore not clear whether the reported cyst assemblages from Namibia/South Africa in fact represent an upwelling signal, and further studies are needed to resolve the issue. Several possible explanations may account for the published data: 1) the two assemblages containing relatively larger amounts of protoperidinioids may represent a similar upwelling signal to that from Peru, here restricted to the nearshore intensified upwelling zone ; 2) the other assemblages may represent a signal from older upwelling waters carried further offshore, where autotrophic species may compete more successfully; 3) the data represent a different type of upwelling system that allows the main cosmopolitan autotrophic dinoflagellates to compete more successfully.

New results

In the past few years, we have investigated cyst assemblages from several other regions of coastal upwelling, as part of a world-wide study of recent cyst distributions (Dale and co-workers, unpublished data). Results of this study will be reported fully elsewhere, but some general observations relevant to the present discussion are summarized here, since they suggest modifications to the published data. New data have been collected from the major upwelling of offshore Mauritania, and from the following coastal regions notably affected by upwelling: southern Chile, southern California, SW. Iceland, and the Faeroe Islands.

The main conclusions confirm the importance of protoperidinioids as the dominant signal for upwelling. Many of the samples from Chile, and some from other regions also confirm the importance of the unidentified spherical brown cysts as a main component of this signal (as previously suggested particularly from Peru). However, the new results suggest that several other species are more important than previously suggested. Particularly surprising were the generally high percentages of *Protoperidinium conicum* in many regions (17-44% off Mauritania, 20-30% in Chile, 19-40% in SW. Iceland, and 14-53% in The Faeroes). This species is probably the most cosmopolitan protoperidinioid, occuring in low amounts (usually <5%) in tropical to cold temperate waters. It has never been recorded in such high percentages elsewhere in our data base of over 400 samples worldwide. The much higher values in these upwelling regions probably reflect some particular competitive advantage in these special environments for the most cosmopolitan, and therefore most adaptable, heterotrophic species.

Another important species from the new data is *Protoperidinium americanum* (6-16% off Mauritania, 15-34% in a few samples from Chile, and 10-27% in The Faeroes). Previously, this species was often recorded in relatively low percentages in coastal assemblages from cold temperate coastal zones in both the northern and southern hemispheres (Dale, 1983; Wall et al., 1977; Dale, unpublished records). The new records from southern Chile and The Faeroes support this, although the percentages are higher than most previous records. However, the significant occurrences off Mauritania strongly suggest that this is a cold temperate species that has also adapted to upwelling regions well outside the main centers of its coastal distribution. At least some of the heterotrophic species seem to have very specific food requirements, for example, preying on one particular species of phytoplankton (Schnepf and Elbrächter, 1992); *P. americanum* may represent an example where its distribution follows that of its main food source, rather than reflecting a special, direct adaptation to the upwelling environment.

A third species, *Lingulodinium machaerophorum* (Deflandre et Cookson) Wall (the paleontological name for the cyst of *Gonyaulax polyedra* Stein), seems to be associated with somewhat more diffuse upwelling effects on the continental shelves of NW. Africa and southern California. This species also forms an important part of the eutrophication signal, discussed more fully below.

Cyst distribution off southeastern Australia

In a study of recent and late Quaternary cysts from the continental shelf and slope of southeastern Australia, McMinn (1992) suggested some limitations to the upwelling signal discussed here. Though his ten samples from the shelf of southern New South Wales yielded few cysts, they included proportionally large amounts of protoperidioids (33-82%). These species seldom account for more than 5% of estuarine assemblages in the region, and their occurrence in such large percentages was therefore particularly characteristic for the shelf. The author noted that assemblages with proportionally large amounts of protoperidiniods were previously reported from major upwelling areas, or occasionally in nutrient enriched areas near river mouths, but concluded that the Australian assemblages probably did not represent an upwelling signal, for the following reasons: 1) most of the protoperidinioid species were different from the spherical brown cysts previously reported from upwelling regions; 2) cyst concentration was so low; and 3) while there is evidence suggesting periodic nutrient enrichment, strong upwelling is unknown in the area.

An alternative explanation was suggested by McMinn (1992, p. 21), based on work cited from Dodge & Harland (1991). He proposed that the high relative abundance of protoperidinioids, there, reflected a distribution pattern found in oligotrophic environments of many Northern Hemisphere shelves. As yet, there is no documentation of high relative abundances of protoperidinioids suggesting an oligotrophic shelf signal. However, the extent to which any alternative signal (e.g. that suggested by McMinn) was indistinguishable from the upwelling signal could seriously diminish the usefulness of the latter. It is therefore important to reconsider McMinn´s reasons for not interpreting his results as an upwelling signal, particularly in light of the more recent work reported here:

1) As discussed above, the upwelling signal is not restricted to proportionally large amounts of spherical brown cysts, but is now known to include several other species of *Protoperidinium*. Different upwelling regions have now been shown to be characterized by different prominent species, and McMinn´s results could be interpreted as further documentation of a new region with some additional important species. Indeed, two of his four main species, *P. conicum* and *P. pentagonum* (Gran) Balech, are now recorded from our new work on other upwelling regions.

2) The fact that he recorded low cyst concentrations, may not be inconsistent with an interpretation involving upwelling. Even in comparable samples associated with major upwelling, the assemblages recording the upwelling signal contain relatively low cyst concentrations (e.g. 500-900 cysts/g in coastal samples, and around 2000 cysts/g in shelf/slope samples from Peru; 700-1000 cysts/g in shelf samples from Namibia; Dale, unpublished records). The more turbulent waters of upwelling environments basically favor the "sinking strategies" of diatoms, whereas the "swiming strategy" of dinoflagellates is clearly suited to more stratified systems (Dale, 1986). The increased phytoplankton production in upwelling systems is therefore overwhelmingly from diatoms; the cyst-forming protoperidinioids which produce the upwelling signal are almost certainly present in far smaller amounts than found for the largely autotrophic cyst-forming dinoflagellates in many other coastal environments. Relatively low cyst productivity is therefore probably a general characteristic of the upwelling signal.

3) The possibility for periodic nutrient enrichment, rather than stronger upwelling, noted by McMinn, could also result in even less cyst production. We know very little about the various degrees of upwelling and nutrient enrichment in different hydrographic regimes, and the effects of these on the cyst signal. However, it is reasonable to suppose that there may be a wide gra-

dation of nutrient enrichment, from occassional small-scale to massive upwelling, producing representative cyst signals. The SE. Australian shelf, and others mentioned by McMinn (1992) may represent the weaker end of this scale, though his alternative suggestion cannot be ruled out.

Cysts as paleo-upwelling indicators

Two main attempts have been made so far to use cysts as paleo-upwelling indicators. Bujak (1984) investigated the cyst record of Cenozoic sediments in several boreholes from the Bering Sea and the northern North Pacific Ocean. He found a major shift in assemblage dominance from autotrophic to heterotrophic species (mostly spherical brown protoperidinioids) that correlated approximately with a massive increase in diatoms in Late Miocene to recent sediments. He interpreted this as evidence for the onset of large-scale upwelling, producing large amounts of diatoms, which in turn produced large amounts of dissolved organic matter that the heterotrophic dinoflagellates were presumed to have utilized as food. An analogy was made between this and the present day upwelling off Peru reported by Wall et al. (1977), also characterized by dominance of spherical brown protoperidinioid cysts.

The basic importance of protoperidinioid cysts as a main element of the upwelling signal, noted by Bujak, has been confirmed by subsequent studies (discussed above). However, our improved undertanding of cyst distributions since Bujak's work suggests that he may have been recording more than just a simple upwelling signal. This issue is not yet resolved, but two main points should be considered: 1) most of the cysts in these deep sea sediments may be transported in from coastal regions, and therefore not represent a direct signal from the frontal upwelling overlying the sample sites; and 2) *Protoperidinium conicoides* (Paulsen) Balech accounted for most of the identified cysts, a species that is common in cold temperate to Arctic coastal assemblages associated with sea ice at present (Dale, 1985). While this does not exclude the possibility that this species could also dominate in upwelling areas (as interpreted by Bujak), it opens up the possibility that the assemblage shift recorded by Bujak (1984) mostly reflected a shift to more sea ice as climate cooled in the Miocene.

An important question, here, is whether upwelling and sea ice produce separate distinctive signals. On the one hand, there seems to be a possibility that *P. conicoides* feeds on phytoplankton associated with sea ice, and is therefore a good indicator of sea ice (this includes shallow coastal bays, remote from typical upwelling). However, melting sea ice may produce enrichment of surface waters in many ways similar to upwelling by: 1) releasing nutrients directly from the ice, and 2) making available to phytoplankton waters containing higher "winter" levels of nutrients (that are then subject to wind-driven mixing of stormy high latitude waters). Presumably, this could select for similar plankton that could therefore represent a more typical "upwelling" signal.

The other main attempt to use cysts as paleo-upwelling indicators was by Lewis et al. (1990). These authors studied Quaternary sediments from the upwelling region off Peru. The relatively widely spaced sampling intervals (representing 20,000-50,000 years between samples) preclude the possibilities for documenting detailed changes in upwelling. Furthermore, since full cyst count data are not presented, the distribution of almost all of the protoperidinioid species and related forms of particular interest, here, cannot be evaluated. Nevertheless, data for the various cyst categories presented show major percentage shifts between the autotrophic (mainly species of the genus *Gonyaulax*) and the heterotrophic types (mainly, though not restricted to, protoperidinioids). Comparing these data with inferred paleo-temperature data, the authors were able to conclude that colder waters, interpreted as stronger upwelling, were represented by higher proportions of peridiniacean (mostly protoperidinioid) cysts.

Lewis et al. (1990) thus suggest that relatively high proportions of peridiniacean cysts probably represent one of the most useful paleo-upwelling signals off Peru. Unfortunately, these authors and Powell et al. (1992) present this in terms which may confuse subsequent workers. They refer to the ratio between P-cysts and G-cysts (P/G ratio), categories of cysts first described by Evitt (1985). This is confusing, however, because Evitt's definition of peridinioid cysts (P-cysts) would not include some of those included in this category by Lewis et al. (1990) and Powell et al. (1992), for example, several species of the genus *Diplopsalis*. In addition, several cyst types included in their P-cysts are clearly autotrophic (*Peridinium faeroense*, and *Scrippsiella* spp. undiff.), while other probably heterotrophic protoperidinioids (spiny forms) are excluded from this category. This mixing of morphologic, taxonomic, and trophic categories thus confuses the basis for this useful signal, which is almost certainly a simple relative increase in heterotrophic species.

It is important to note that P-cysts, as defined by Evitt (1985), are a morphologic group that includes two major living genera, *Peridinium* and *Protoperidinium*. Morphologically, these are differentiated by relatively small features of the motile stages. In most cases, these features cannot be recognized from their cysts, which therefore are collectively referred to as P-cysts (a necessary compromise imposed by the constraints of the fossil record). Ecologically, however, these genera are very different, today; *Protoperidinium* is probably restricted to marine heterotrophic species, and *Peridinium* almost entirely to freshwater species. The evidence, to date, strongly suggests that it is the marked proportional increase in heterotrophic species (mainly protoperidinioids, but including species from other genera such as *Diplopsalis*) that provide the best cyst signal for upwelling.

Probably all of the important cysts of heterotrophic species ocurring in recent and Quaternary sediments can be recognized or inferred by relating them to their respective living motile stages. To the extent that a collective term for these is needed, they could be referred to as H-cysts. Some of these types, or closely related forms, range down into Neogene sediments, where they also may be useful paleo-upwelling indicators. Earlier than this, it will probably be increasingly difficult to differentiate heterotrophic species, and P-cysts may represent the closest possible approximation. In which case, it should be noted that the abundant and varied fossil record of P-cysts (from at least Aptian time) probably includes marine peridinioids that may have been autotrophic, as their freshwater relatives are today.

DINOFLAGELLATE CYSTS AS EUTROPHICATION INDICATORS

In considerations of environmental pollution and climatic change, the one critical parameter that often is most difficult to assess is background levels prior to pollution. There is an obvious need to determine background levels of actual pollutants, and to establish levels of natural variation, but there is also a need to quantify if possible biological signals representing the state of the environment at any given time. This last mentioned factor is important not only for monitoring the effects of pollution, but, not least, for monitoring subsequent environmental recovery. However, in many cases, the relevant environmental parameters have not been measured beforehand, and the question is first addressed only after it is realized that a critical environmental problem exists. This is an aspect of environmental studies where micropaleontological methods, applied to interpret the sedimentary archive, have proved useful. These methods have been applied also to investigations of paleoproductivity and climatic change. For example, Lange et al. (1987; 1990) considered possible effects of climatic changes on coastal primary production, by studying diatoms, radiolaria, and planktonic foraminifera in the sedimentary record off Southern California.

Eutrophication, excessive biological production due to anthropogenic contamination of the world´s oceans, is increasingly the cause of environmental problems. It has recently been suggested that these include increased frequency and intensity of toxic phytoplankton blooms, amounting to a global epidemic (Smayda, 1990). However, it is difficult to assess the extent of this problem, due to a lack of data for toxic phytoplankton concentrations prior to anthropogenic impact on the environment, and to the lack of well developed signals for tracing eutrophication. Since dinoflagellates are one of the major groups of phytoplankton, and they also include many of the toxic species, their cysts would seem to be potentially useful indicators of eutrophication and the history of toxic blooms.

We have recently carried out the first studies of prehistoric and historic "blooms" of one toxic species, from cored sediments in the Kattegat-Skagerrak region (Dale et al., 1993). The results showed evidence of probable massive blooms prior to significant human influence on the environment, which we suggest were caused primarily by the effects of climatic change (Dale & Nordberg, 1993). This demonstrated the usefulness of cyst analysis for investigating blooms, at least for the few toxic species known to produce fossilizeable cysts. However, we have also begun to explore the possible use of cysts as eutrophication indicators, and this is even more pertinent to the present discussion of paleoproductivity.

Cysts as eutrophication indicators - new results from the Oslofjord, Norway

The Oslofjord (Figure 1) has a known history of eutrophication, with increased nutrient levels in the water resulting from human population growth in the Oslo region. This started in the mid - 1800's with the country's first large industrial development growing along the Aker river, and developed further with the introduction of the water closet in the early 1900's, and its widespread use after about 1930. Phytoplankton production is believed to have increased (particularly in the inner fjord), and water quality to have deteriorated through this time due to increased organic matter, resulting in extremely reduced oxygen levels in bottom waters.

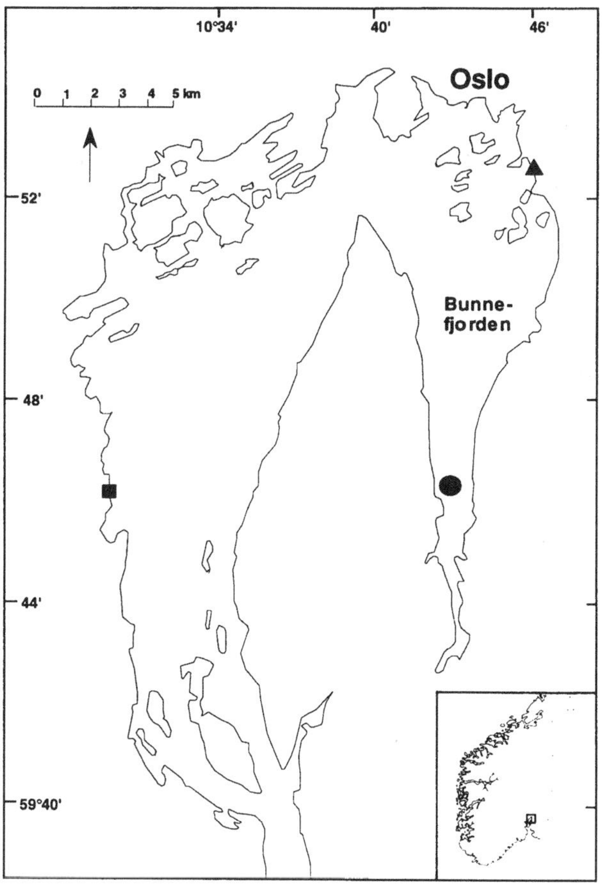

Figure 1. Map of the inner Oslofjord showing the core location (circle) and the sewage treatment plants (triangle = the older plant; square = the newer plant established in the 1880's).

An initial study was carried out on sediments from the most eutrophic part of the inner fjord, to see if cysts showed any changes that could be used elsewhere as a signal to trace the development of eutrophication (Madsen, 1990). For this, a 58cm - long core was taken in anoxic sediments from the deepest part (150m water depth) of the inner basin of Bunnefjord, at the head of the 100km-long Oslofjord (Figure 1). Samples from 2cm - intervals were analyzed for cysts, both quantitatively (cysts/g dry sediment) and qualitatively (percentage species composition of the assemblages). A similar sediment core from the same basin was dated (^{210}Pb method), and these dates were used to infer approximate ages for the cyst assemblages recovered.

The results were very encouraging, and will be reported fully elsewhere. In the meantime, the main conclusions are summarized here to illustrate the potential for this type of application. We believe the results (Figure 2) almost certainly reflect a eutrophication signal, though more such studies are needed to confirm this, both from other eutrophicated and non-eutrophicated environments. Quantitatively, this included a dramatic increase from < 10,000 to almost 160,000 cysts/g, one of the highest cyst concentrations recorded so far. Qualitatively, there was a pronounced shift in species dominance due to a massive increase in *L. machaerophorum* (the paleontological name for cysts of *Gonyaulax polyedra*). The details included the following main points:

1) The lower 15 cm of the core (58 - 43cm, representing from about 1850-1880) contains relatively low cyst concentrations, and a cyst assemblage similar to that found in the outer fjord today. This is thought to represent the pre-eutrophication situation, with lower phytoplankton production, including cyst-forming dinoflagellates, and relatively good water quality comparable with that found in the outer fjord today.

2) From about 45-6 cm (approximately 1880-1980), the record is characterized by a large, general increase in cyst concentration, thought to represent increased production due to eutrophication. The massive increase of *L. machaerophorum* seems to have occurred around the 1950´s, at a time when the water quality is known to have been at about it´s worst. The increase and eventual dominance of this species thus probably corresponds to the establishment of relatively high nutrient levels, with increased eutrophication.

3) The upper 4-6 cm (from about the 1980´s) showed a decline in cyst concentrations towards the pre-eutrophic levels. This may represent a response to improved water quality in the inner fjord, resulting from a new, improved, sewage treatment system that became operational during that time.

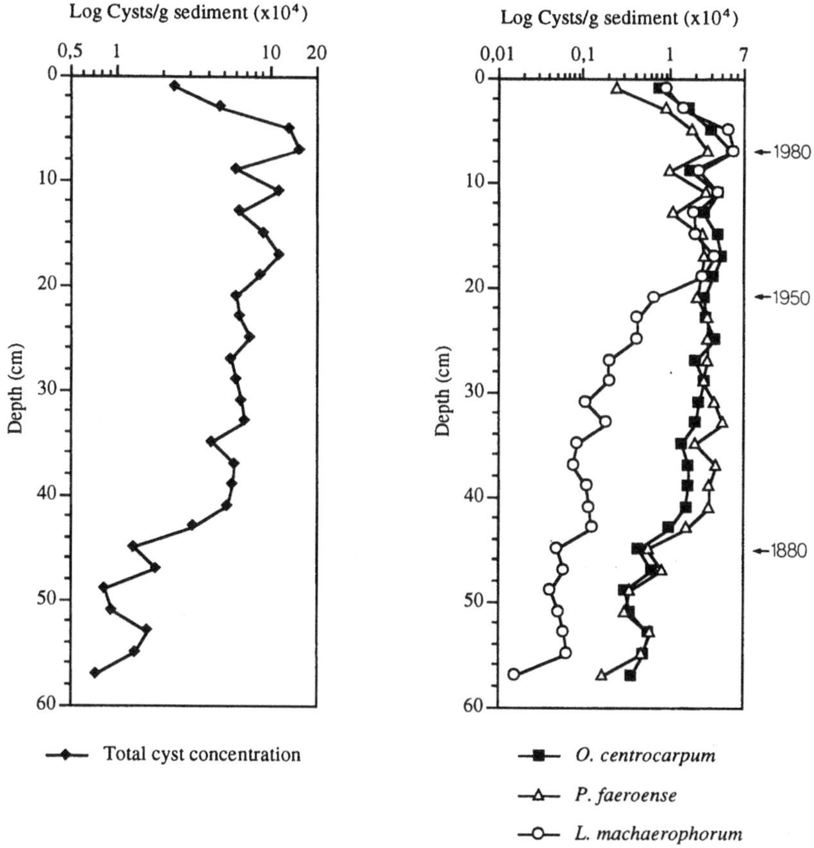

Figure 2. Total cyst concentrations, and concentrations of selected cyst species from the inner Oslofjord core (suggested ages are included on the far right).

The fact that eutrophication was accompanied by increasing cyst concentrations was not surprising. However, the importance of *L. machaerophorum* to the postulated eutrophication signal was more unexpected. Previously, the recorded distribution of this cyst species included relatively high concentrations in river estuaries, and it has therefore most often been interpreted as an indicator of lower salinity (Morzadec-Kerfourn, 1977). There is no evidence of reduced salinity in the Oslofjord record, which in this case strongly suggests higher levels of nutrients as the main determining cause of increased concentrations of this species. Supporting evidence for associating this species with increased nutrients is also reported from the Californian coast, where it has long been recognized as a main component of upwelling plankton (Eppley & Harrison, 1974). This, in turn, suggests the possibility that it also may have been responding to increased nutrients, rather than to lower salinities, in previous estuarine data. The status of *L.*

machaerophorum, established in the literature as a lower salinity indicator, will have to be reevaluated in light of these new findings.

SUMMARY

1) Cysts are good paleoenvironmental indicators, especially in coastal and neritic environments.

2) Relatively few oceanic cysts are formed, and deep sea assemblages may be dominated by coastal/neritic cysts potentially resulting from long distance transport. This will have to be taken into account in attempts to apply transfer functions based on modern deep sea cyst distributions.

3) In paleoproductivity studies, cysts may prove to be particularly useful in more coastal environments where the typically oceanic microfossils (planktonic foraminifera, calcareous nannoplankton, and radiolaria) are poorly represented, and in sediments where mineralized microfossils are absent due to dissolution.

4) Cysts show an upwelling signal characterized by a relatively large component of heterotrophic species (mainly in, but not restricted to, the genus *Protoperidinium*). These species are thought to prey on other phytoplankton, including diatoms, and their importance in the upwelling signal probably reflects: a) the availability of food for heterotrophic species, rather than any special environmental adaptations; b) the relative paucity of autotrophic dinoflagellates which are generally out-competed by diatoms in upwelling environments. It should be noted that overall cyst concentrations in upwelling environments may be relatively low; there is no indication of significantly elevated concentrations (e.g. as found in the eutrophication signal).

5) Different upwelling systems may be characterized by different cyst species: spherical brown protoperidinioids are important off Peru, Chile, and possibly southern Africa; *P. conicum* and *P. americanum* off NW. Africa, Iceland, and the Faeroes; and *L. machaerophorum* on the continental shelves of NW. Africa and California. Thus, there may be a potential for recognizing different types of upwelling from the fossil record.

6) Assemblages dominated by heterotrophic species have also been reported from: a) sea-ice dominated environments (mainly *P. conicoides*) - it is not yet clear whether this represents a distinctive ice signal or an "upwelling effect" from increased nutrients released by melting ice;

and b) oligotrophic shelves, which may nevertheless experience periodic nutrient enrichment - this may represent the weakest form of "upwelling" signal.

7) Attempts to describe the upwelling signal as a ratio between peridiniacean (P-cysts) and gonyaulacacean cysts (G-cysts) may be unnecessarily confusing, since: a) autotrophic species other than gonyaulacacean forms are usually present, sometimes in relatively large concentrations; and b) P-cysts include a mixture of taxonomically, ecologically, and trophically different species. If a collective term is needed, the cysts of heterotrophic species could be called H-cysts, and the signal described as the relative proportion of these in the total assemblage.

8) First attempts to use cysts as eutrophication indicators suggest a eutrophication signal from Oslofjord sediments, characterized by: a) a large increase in total cyst concentration (cysts/g dry sediment); and b) progressively increasing proportions of *L. machaerophorum* at higher degrees of eutrophication.

9) *L. machaerophorum* is considered to be an indicator of higher nutrient levels; earlier reports of increased concentrations in estuaries probably reflect increased nutrients rather than reduced salinities. Its established status in the literature as a low salinity indicator species should be reevaluated.

ACKNOWLEDGEMENTS

We thank BP international Ltd., U.K., and BP Petroleum Development, Norway, for financial support for the world-wide cyst distribution study referred to here. We thank Amy L. Dale for reviewing an earlier draft of the manuscript, and we thank reviewers Geoff Eaton and Rex Harland for good suggestions that improved the final version.

REFERENCES

Bujak JP (1984) Cenozoic dinoflagellate cysts and acritarchs from the Bering Sea and northern North Pacific, DSDP Leg 19. Micropaleontology 30:180-212

CLIMAP Project Members (1976) The surface of the iceage Earth. Science 191:1131-1137

Dale B (1983) Dinoflagellate resting cysts: "benthic plankton". In: Fryxell GA (ed) Survival strategies of the algae. Cambridge University Press Cambridge: 69-136

Dale B (1985) Dinoflagellate cyst analysis of Upper Quaternary sediments in core GIK 15530-4 from the Skagerrak. Norsk Geologisk Tidsskrift 65:97-102

Dale B (1986) Life cycle strategies of oceanic dinoflagellates. Unesco technical papers in marine science 49:65-72

Dale B, Dale A (1992) Dinoflagellate contributions to the deep Sea. Ocean Biocoenosis Series

5, Woods Hole Oceanographic Institution Woods Hole 76 pp
Dale B, Nordberg K (1993) Possible environmental factors regulating prehistoric and historic "blooms" of the toxic dinoflagellate *Gymnodinium catenatum* in the Kattegat-Skagerrak region of Scandinavia. In: Smayda TJ, Shimizu Y (eds) Toxic phytoplankton blooms in the sea. Proceedings of the Fifth International Conference on Toxic Marine Phytoplankton Newport Rhode Island USA 28 October - 1 November 1991. Elsevier Science Publishers Amsterdam
Dale B, Madsen A, Nordberg K, Thorsen TA (1993) Evidence for prehistoric and historic "blooms" of the toxic dinoflagellate *Gymnodinium catenatum* in the Kattegat-Skagerrak region of Scandinavia. In: Smayda TJ, Shimizu Y (eds) Toxic phytoplankton blooms in the sea. Proceedings of the Fifth International Conference on Toxic Marine Phytoplankton Newport Rhode Island USA 28 October - 1 November 1991. Elsevier Science Publishers Amsterdam
Dodge J D, Harland R (1991) The distribution of planktonic dinoflagellates and their cysts in the eastern and northeastern Atlantic Ocean. New Phytologist 118:593-603
Edwards LE, Mudie PJ, de Vernal A (1991) Pliocene paleoclimatic rèconstruction using dinoflagellate cysts: comparison of methods. Quaternary Science Reviews 10:259-274
Eppley RW, Harrison WG (1974) Physiological ecology of *Gonyaulax polyedra*: A red water dinoflagellate of Southern California. In: Locicero V R (ed) The First International Conference on Toxic Dinoflagellate Blooms. The Massachusetts Science and Technology Foundation Wakefield Massachusetts : 11-22
Evitt WR (1985) Sporopollenin dinoflagellate cysts - their morphology and interpretation. American Association of Stratigraphic Palynologists Foundation Austin 333 pp
Harland R (1983) Distribution maps of Recent dinoflagellate cysts in bottom sediments from the North Atlantic Ocean and adjacent seas. Palaeontology 26:321-387
Johnson D A, Ledbetter M, Burkle L H (1977) Vema Channel paleoceanography: Pleistocene dissolution cycles and episodic bottom water flow. Marine Geology 23:1-33
Lange CB, Berger, WH, Burke SK, Casey RE, Schimmelmann A, Soutar A, Weinheimer AL (1987) El Nino in Santa Barbara Basin: diatom, radiolarian and foraminiferan responses to the "1983 El Nino" event. Mar Geol 78:153-160
Lange CB, Burke SK, Berger WH (1990) Biological production off southern California is linked to climatic change. Climatic Change 16:319-329
Lewis J, Dodge JD, Powell AJ (1990) Quaternary dinoflagellate cysts from the upwelling system offshore Peru, Hole 686B, ODP Leg 112. In: Suess E, von Huene R, *et al* Proceedings of the Ocean Drilling Program Scientific Results. Ocean Drilling Program College Station Texas 112:323-328
Madsen A (1990) Dinoflagellate cysts in sediment as eutrophication indicators in the Bunnefjord, inner Oslofjord. Unpublished Cand. Scient thesis University of Oslo Norway
McMinn A (1992) Recent and Late Quaternary dinoflagellate cyst distribution on the continental shelf and slope of southeastern Australia. Palynology 16:13-24
Mudie PJ, De Vernal A, Head MJ (1990) Neogene to Recent palynostratigraphy of circum-arctic basins: results of ODP Leg 104, Norwegian, Sea Leg 105, Baffin Bay, and DSDP Site 611, Irminger Sea. In: Bleil U, Theide J (eds) Geological History of the Polar Oceans: Arctic versus Antarctic. Kluwer The Hague 609-646
Morzadec-Kerfourn M-T (1977) Les Kystes de dinoflagellés dans les sédiments récents le long des côtes Bretonnes. Revue de Micropaléontologie 20:157-166
Needham HD, Habib D, Heezen BC (1969) Upper Carboniferous palynomorphs as a tracer of red sediment dispersal patterns in the Northwest Atlantic. Journal of Geology 77:113- 120
Oberhänsli H, Heinze P, Diester-Haass L, Wefer G (1990) Upwelling off Peru during the last 430, 000 yr and its relationship to the bottom-water environment, as deduced from coarse grain-size distributions and analyses of benthic foraminifers at Holes 679D, 680B, and 681B, Leg 112. In: Suess E, von Huene R, *et al* Proceedings of the Ocean Drilling Program Scientific Results. Ocean Drilling Program College Station Texas 112:369-390
Powell AJ, Lewis J, Dodge JD (1992) The palynological expressions of post-Palaeogene upwelling: a review. In: Summerhayes CP, Prell WL, Emeis KC (eds) Upwelling Systems: evolution since the Early Miocene. Geological Society Special Publication 64: 215-226
Reid PC, Harland R (1977) Studies of Quaternary dinoflagellate cysts from the North Atlantic.

American Association of Stratigraphic Palynologists Contribution Series 5A:147-169
Schnepf E, Elbrächter M (1992) Nutritional strategies in dinoflagellates: a review with emphasis on cell biological aspects. Europ J Protistol 28: 3-24
Schrader H, Sorknes R (1990) Spatial and temporal variation of Peruvian coastal upwelling during the latest Quaternary. In: Suess E, von Huene R, *et al* Proceedings of the Ocean Drilling Program, Scientific Results. Ocean Drilling Program College Station Texas 112:391-406
Smayda TJ (1990) Novel and nuisance phytoplankton blooms in the sea: Evidence for a global epidemic. In: Graneli E, Sundström B, Edler L, Anderson DA (eds) Toxic Marine Phytoplankton. Elsevier New York 29-40
Suess E, von Huene R, *et al* (1990) Proceedings of the Ocean Drilling Program Scientific Results. Ocean drilling Program College station Texas 112
Taylor FJR (ed) (1987) The biology of dinoflagellates. Botanical Monographs Volume 21. Blackwell Scientific Publications Oxford 785 pp
Thiede J, Suess E (eds) (1983) Coastal upwelling, its sediment record; pt. B. Plenum Press New York 610 pp
Turon J-L (1980) Dinoflagellés et environnement climatique. Les kystes de dinoflagellés dans les sédiments récents de l'Atlantique nord-oriental et leurs relations avecl'environnement océanique. Application aux dépots Holocènes du Chenal de Rockall. Memoir Musée Naturelle d'Histoire Paris B27:269-282
Wall D, Dale B (1973) Paleosalinity relationships of dinoflagellates in the Late Quaternary of the Black Sea - a summary. Geoscience and Man 7:95-102
Wall D, Dale B, Lohmann GP, Smith WK (1977) The environmental and climatic distribution of dinoflagellate cysts in modern marine sediments from regions in the North and South Atlantic Oceans and adjacent seas. Marine Micropaleontology 2:121-200
Zippi PA (1992) Dinoflagellate cyst stratigraphy and climate fluctuations in the eastern North Atlantic during the last 150, 000 years. In: Head MJ, Wrenn JH (eds) Neogene and Quaternary Dinoflagellate Cysts and Acritarchs. American Association of Stratigraphic Palynologists Foundation Dallas 55-68

DEEP-SEA BENTHIC FORAMINIFERS: FOOD AND BOTTOM WATER MASSES

Detmar Schnitker
Department of Oceanography
University of Maine
Orono, ME 04469, U.S.A.

ABSTRACT. Benthic foraminifers are an abundant, nearly cosmopolitan and easily preserved component of the deep-sea meiofauna. The benthic foraminiferal faunas are very strongly structured on regional scales. In the western North and South Atlantic the distribution of benthic assemblages follows the distribution of named "water masses," leading to the expectation that fossil benthic foraminifers from sediment cores allow a reconstruction of past deep-water circulation patterns. However, in many other areas the faunal composition is strongly correlated with the productivity of the overlying surface waters. Such results demonstrate that the benthic meiofauna responds to the integral of the deep environment and it is the limiting or overwhelming influence that determines the success or failure of individual species and thus the composition of the fauna. Analysis of available data clearly indicates that benthic foraminifers are unequivocal indicators of productivity in areas where productivity is high. In areas of low or very uniform productivity the composition of the benthic fauna clearly carries the imprint of deep water mass structure as the dominant feature. Faunal sequences from late Quaternary cores from the Atlantic and Pacific oceans show very strong differentiations. Not only is the glacial/postglacial transition shown, but also much superimposed fine structure. Faunal composition is as sensitive to deep-water environmental change as are geochemical indicators. The faunal composition allows at least some appreciation of the cause for the major changes, but a quantitative differentiation between productivity or water mass changes is not yet possible. If, however, an independent assessment of productivity change is available, then the benthic foraminiferal faunas are exquisite indicators of changes in deep water circulation.

INTRODUCTION

Benthic foraminifers are an abundant, diverse, and easily preserved component of the deep-sea meiofauna and therefore have potential utility as monitors and recorders of living conditions on the deep ocean floor. According to the principles of ecology and paleoecology,

NATO ASI Series, Vol. I 17
Carbon Cycling in the Glacial Ocean:
Constraints on the Ocean's Role in Global Change
Edited by R. Zahn et al.

such diverse faunas, composed of hundreds of species, should respond qualitatively and quantitatively to even minor differences and change in their environment.

Early work, such as that of Brady (1884), the papers by Cushman (1910-1917, 1918-1931), or the more recent work by Phleger, Parker and Peirson (1953) demonstrated the diversity of the fauna and its cosmopolitan distribution. Attempts to utilize benthic foraminifers as depth indicators met with only partial success (Bandy and Chierici, 1966). It was only with the onset of quantitative work that the distribution of benthic foraminifers in deep sea sediments began to yield information that could be used for ecological and paleoceanographic purposes (see summaries by Schnitker, 1980, and Douglas and Woodruff, 1981). This work showed that, at least on regional scales, the benthic foraminiferal faunas were very strongly structured. Initially, it was demonstrated that in the North and South Atlantic the distribution of benthic assemblages followed the distribution of named "water masses," which led to the expectation that fossil benthic foraminifers from sediment cores would allow a reconstruction of past deep-water circulation patterns.

These results encouraged many workers in several countries to start investigations of this fascinating group. Controversy arose when it became clear that the most important outcome of those new studies was a demonstration that faunal composition was strongly correlated with the productivity of the overlying surface waters and the delivery of this organic matter to the seafloor (e.g. Lutze and Coulbourn, 1984). Because of this controversy, the paleoceanographic community has come to distrust benthic foraminifers as indicators of past deep oceanic conditions, relying on organic and inorganic geochemical and isotopic indicators instead.

The following narrative constitutes an attempt to summarize the current state of understanding of the ecology of deep water benthic foraminifers. While it seems not yet possible to arrive at an understanding of how to discriminate unequivocally among the two principal determinants of faunal composition, under certain oceanographic settings, and with minimal auxiliary knowledge from other proxies, benthic foraminifers can be used with reasonable assurance as paleoceanographic indicators.

FORAMINIFERS AND WATER MASSES

Studies that demonstrate a correlation between the occurrence of distinct foraminiferal faunas and the distribution of deep water masses have been carried out in all major ocean basins. The earliest of these, which also served as models for subsequent studies elsewhere, were those of Streeter (1973) and Schnitker (1974) on faunas from the North Atlantic. While Streeter (1973) re-analyzed the faunal data of Phleger et al. (1953) by means of a Q-mode factor analysis, Schnitker (1974) used counts from newly acquired bottom samples. Both studies arrived at very similar conclusions: 1) three distinct faunal associations exhibit a very coherent geographic distribution, and 2) these distributions correlate well with the distribution of the three principal water masses in the North Atlantic (Figure 1).

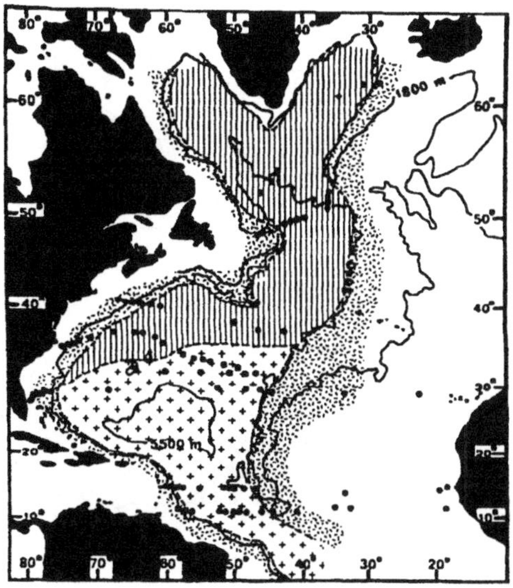

Figure 1. Distribution of three principal faunal associations in the western North Atlantic. A *Uvigerina* and *Hoeglundina* dominated fauna occurs along the basin margin and the Mid-Atlantic Ridge (stipples). The distribution of this association is congruent with the sea-floor intercept of the Upper North Atlantic Deep Water. An *Epistominella exigua* dominated fauna occurs in the deep northern North Atlantic, being roughly coincident with the lower North Atlantic Deep Water (vertical rulings). In the deep southern North Atlantic *Nuttallides umbonifer* dominates, congruent with the occurrence of Antarctic Bottom Water (crosses). After Schnitker, 1974.

Whatever discrepancies existed between the faunal distributions of these two studies were later found to be due to the use of different sieve sizes by the two authors (150 µm by Streeter and 125 µm by Schnitker). Examining the faunas of the Atlantic continental slope of northern North America, Streeter and Lavery (1982) found a conspicuous depth stratification which most strongly followed the depth distribution of the deep water structure.

Lohmann (1978) (Figure 2) and Gofas (1978), studying the faunal distribution along depth transects on the Rio Grande Rise and the Walvis Ridge in the South Atlantic, found a remarkably clear depth zonation of distinct faunas, which correlated accurately with the principal water masses of that area. Most recently Mackensen et al. (in press) found that the distribution of potentially fossilizable benthic foraminifers of the eastern South Atlantic mimic the regional hydrography. They succeeded in delineating faunas associated with a Northern Component Deep Water and the Southern Component Bottom Water.

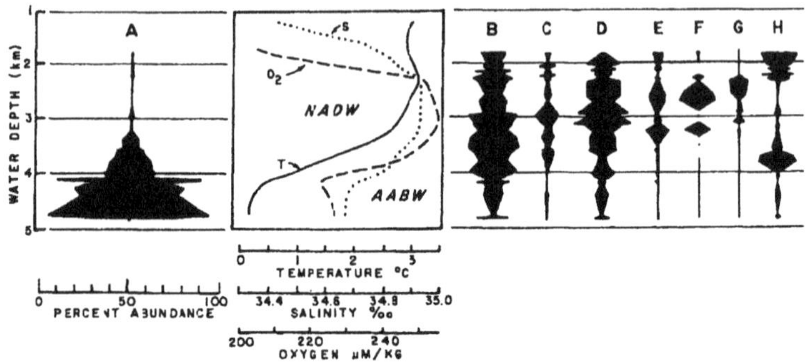

Figure 2. Distribution of benthic foraminifers and deep water hydrography on the Rio Grande Rise. *Nuttallides umbonifer* (A) is closely associated with AABW; *Cibicidoides wuellerstorfi* (B), *C. kullenbergi* (C), and *Hoeglundina elegans* (G) reach their maxima in NADW; while *Uvigerina peregrina* (H) reaches its greatest abundance just above and below the oxygen maximum of the NADW. (after Lohmann, 1978).

The strong vertical differentiation of deep water characteristics in the equatorial Pacific Ocean at the Ontong Java Plateau is reflected in a marked vertical hierarchy of benthic foraminiferal species and faunas (Burke, 1981). Burke found that the faunas underlying the oxygen-poor Pacific Deep Water are dominated by species which elsewhere have also been recognized in similar environments (i.e., *Siphouvigerina, Astrononion*), but, most importantly, that the faunas associated with Pacific Bottom Water are dominated by *Nuttallides umbonifer* and *Epistominella exigua*. Nearly identical faunal/oceanographic relationships are reported by Nienstedt and Arnold (1988) from the East Pacific Rise, at about 13°20' N, suggesting that they are valid for the entire Pacific Ocean basin and that they can serve as monitors of past deepwater conditions. Gofas (1978) found that the two deep water masses occurring in his sampling area of the Samoan Passage (central Pacific) are very strongly reflected in the foraminiferal distribution.

Such findings were summarized by Schnitker (1980), who extracted from the diverse data sets a tentative list of index species that might be used to infer the presence of particular water masses (Table 1). Attempts to resolve what properties of a particular water mass are the determinant entities were only partially successful, and subsequent studies have not advanced the problem much beyond some very clear-cut observations. On a global scale, the most influential water mass property that determines foraminiferal distribution patterns is carbonate saturation. The undersaturation that is a common feature of the coldest and deepest ocean water results in the elimination of calcareous specimens on large areas of the ocean basins. Calcareous benthic foraminifers continue to secrete shells against an increasing dissolution gradient to some distance below the CCD (Schnitker, unpublished observations), but these shells become dissolved soon after their death (or reproduction), leaving behind dissolution resistant agglutinated forms. Bremer and Lohmann (1982) demonstrated that in the Atlantic Ocean it is the increasing degree of carbonate undersaturation with depth, and the generally undersaturated nature of Antarctic Bottom Water that are responsible for the dominance of *Nuttallides umbonifer* in such settings. On an extraregional scale, Streeter and Shackleton (1979) demonstrated that the distribution of *Uvigerina peregrina* descended from less than 2,000 meters in the North Atlantic to over 4,000 meters in the South Atlantic. The only water property that they could identify as having a similar distribution was their relatively low oxygen contents.

The study of Lohmann (1978) best illustrates the promises and problems of determining controlling factors: the correlations between foraminiferal faunas and various physical and chemical attributes of the 'water masses' are very tight and allow the establishment of 'ecological equations' (Figure 3). However, many of the physical and chemical attributes are strongly covariant, and it is not clear which one is the controlling one in the case under investigation. These 'ecological equations' have validity only for the region and for the water characteristics from which they were derived. However, the potential utility of this methodology has not yet been tested. The covariance of the many water mass attributes does not remain the same along the pathways of the thermohaline circulation. The persistence of distinct associations between water masses and faunas in those oceanic settings that have hitherto been investigated, holds open the possibility that some of the controlling properties may be isolated from analysis of a global data set.

Table 1. Relationship between dominant species and water masses.

Water Mass	Western N. Atlantic	Eastern N. Atlantic	Western S. Atlantic	Eastern S. Atlantic	Central Pacific	Southern Indian Ocean
AABW	*N.umbonifer*	*N.umbonifer*	*N.umbonifer*	*N.umbonifer*	*N.umbonifer*	*N.umbonifer*
ABW	*E. exigua*	—	—	—	—	—
lwr NADW	*E. exigua*	*C.wuellerst.*	*C.wuellerst.*	*C.wuellerst.*	—	—
upr NADW	*U. peregrina*	*U. peregrina*	—	—	—	—
CPW	—	—	*U. peregrina*	*U. peregrina*	—	—
PDW	—	—	—	—	*C.wuellerst.*	—
IBW	—	—	—	—	—	*U.peregrina*

AABW = Antarctic Bottom Water
NADW = North Atlantic Deep Water
PDW = Pacific Deep Water
ABW = Arctic Bottom Water
CPW = Circum Polar Water
IBW = Indian Bottom Water

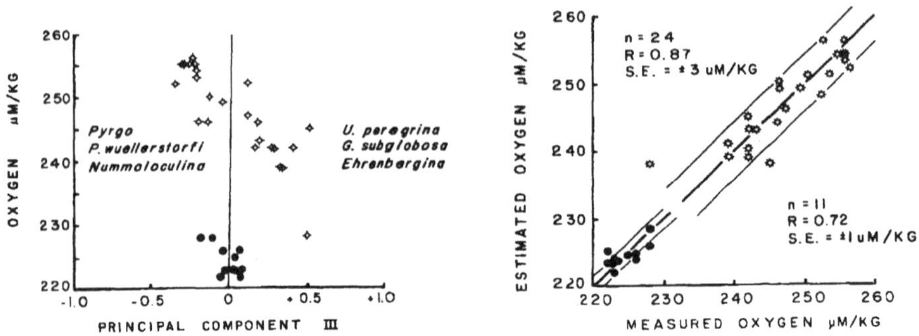

Figure 3. (left panel) Correlation of faunal associations, expressed as a Principal Component, with the oxygen content of the bottom water. Star symbols represent samples from NADW covered areas, solid dots samples from under AABW. (right panel) Estimated oxygen contents (derived from "Ecological Equation") vs. observed oxygen contents. After Lohmann (1978).

FORAMINIFERS AND FOOD SUPPLY

Implicit in the studies cited above is the assumption that on vast regional and oceanic scales the primary productivity, and hence food supply to the ocean floor, would play a negligible role. This assumption arose perhaps fortuitously, because most of these studies were not carried out in areas of strong productivity contrasts. The first large-scale study to demonstrate that surface productivity does play a significant, even dominant, role in the distribution of benthic foraminifers was that of Lutze and Coulbourn (1984). The benthic faunas from beneath and adjacent to the upwelling regions off NW Africa are very strongly differentiated according to the amount of organic carbon contained in the sediment, itself a close proxy of the productivity of the overlying surface waters. The imprint of productivity completely obliterates any reflection of faunal differentiation according to water mass structure. Similarly, Mackensen et al. (1985) and Mackensen (1987) correlated specific faunal associations of the Norwegian and Greenland Seas with the flux of particulate organic matter to the seafloor. More recently, Berger & Herguera (1992) demonstrated that in the western equatorial Pacific the production of benthic foraminiferal shells is directly proportional to the flux of organic matter to the seafloor. (They did not attempt to correlate the *composition* of the benthic faunas to the organic matter flux.) In order to rule out the effects of water mass upon faunal composition, Loubere (1991) employed the elegant strategy of studying samples from about equal depths along the equatorial East Pacific Rise, an area of little bottom water differentiation but with a strong gradient of surface productivity. The succession of species dominance across the productivity gradient allowed Loubere (1991) to derive an 'ecological equation' that seems to be very successful in estimating surface ocean productivity from faunal compositions (Figure 4). In his study of the benthic faunas of the eastern South Atlantic Mackensen (in press) found that in addition to the two faunal groupings that indicated the northern and southern component water masses two faunal components occurred that were uniquely associated with the area of high surface productivity between the Polar Front and the winter sea ice limit.

PALEOECOLOGY AND PALEOCEANOGRAPHIC INTERPRETATIONS

The apparent success in correlating faunal associations with water masses led Streeter (1973) and Schnitker (1974, 1979, 1980) to interpret the distribution of fossil benthic faunas contained in piston cores from the North Atlantic. They determined that during the last glacial maximum the North Atlantic deep circulation had drastically rearranged itself, concluding that the North Atlantic was filled with an 'aged' water mass, containing less oxygen than the present

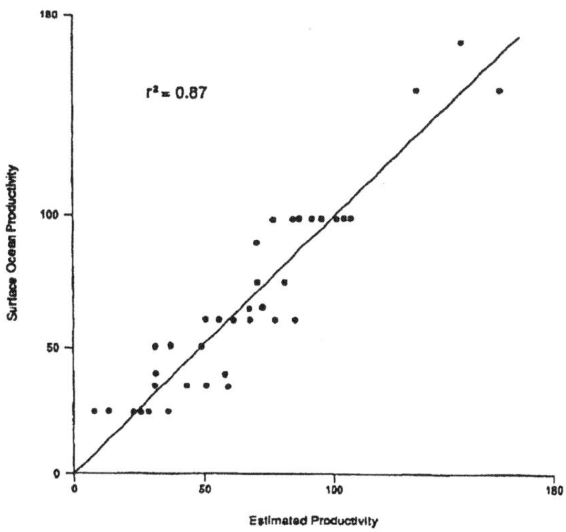

Figure 4. "Ecological Equation" - derived surface productivity estimates vs. observed surface productivity (in g C/m^2/yr). After Loubere (1991).

waters, similar to modern intermediate waters. The North Atlantic apparently had ceased to be a source of well oxygenated NADW. These conclusions were later corroborated by independent geochemical and isotope studies (Boyle and Keigwin, 1982, 1987; Labeyrie et al., 1992). For a much earlier time, and on a truly global scale, Woodruff and Savin (1989) isolated part of their Miocene benthic faunas as tracers of water masses, clearly retracing the indications of stable isotope evidence. On the other hand, interpretations of faunal sequences contained in late Quaternary piston cores were pointing toward drastic changes in past productivities, rather than deep circulation changes. For example, Pedersen et al. (1988), using geochemical tracers, found a pronounced increase in productivity during isotope stage 2 in a piston core from the eastern equatorial Pacific. This increase in productivity is accompanied by a faunal shift, most notably an increase in the abundance of *Uvigerina* spp. (Figure 5). Similarly, Caralp (1984, 1989) documented that strong changes of northeastern North Atlantic surface productivity were paralleled by equally strong changes in the composition of the benthic foraminiferal faunas. She succeeded to derive a link between the abundance of two key species and the abundance and quality of the organic matter contained in the sediment (Figure 6). One of the faunal components isolated by Woodruff and Savin (1989) from their global distribution of Miocene benthic faunas seemed to trace the history of oceanic surface productivity. The fossil distributions thus showed a dichotomy in their environmental linkages, as did their modern counterparts.

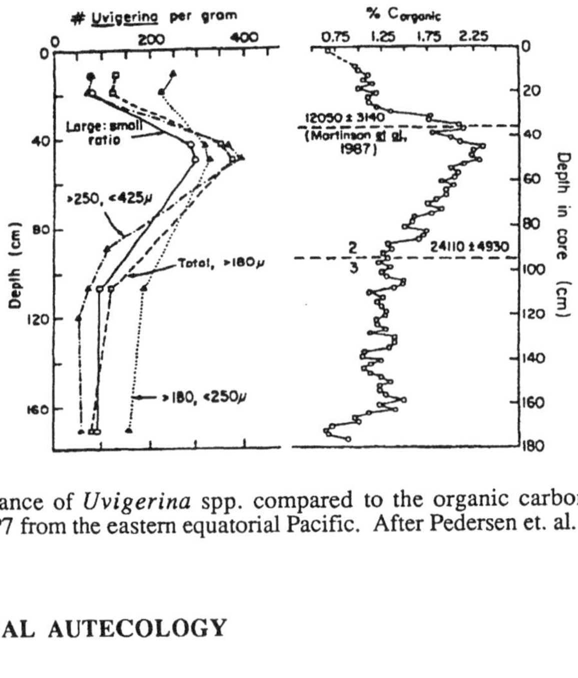

Figure 5. Abundance of *Uvigerina* spp. compared to the organic carbon content of the sediment in Core P7 from the eastern equatorial Pacific. After Pedersen et. al. (1988).

FORAMINIFERAL AUTECOLOGY

Most of the work cited above was done without much regard to, or knowledge of, the mode of life of deep water benthic foraminifers. Compared to the extensive knowledge that has accumulated on the taxonomy and geological record of foraminifers, mostly compiled by geologists, very little is known of the biology of foraminifers in general, and deep water forms specifically. Only recently were foraminifers shown to constitute a major part of the deep sea benthic biomass and to occupy a very important role in the deep-sea benthic food chain (Snider et al., 1984; Smith, 1992; Gooday et al., 1992). In particular, several species of the deep water

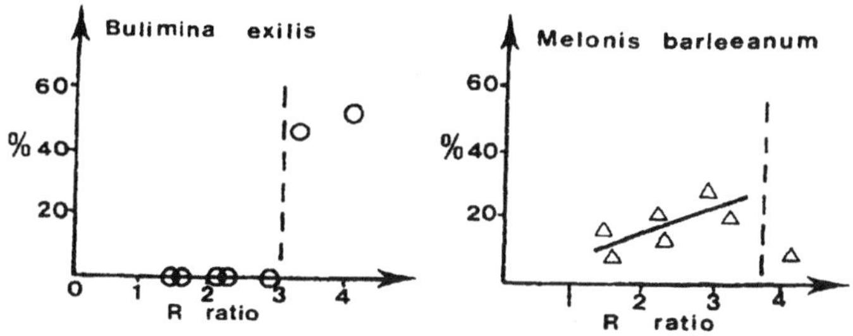

$$R = \frac{\text{Pheopigments } (\mu g/g)}{\text{Glucids } (\mu g/g)} \times 100$$

Figure 6. Abundance of two species known to occur in organic-rich sediments in relation to the quality of the organic matter, expressed as R ratio. After Caralp (1989).

benthic foraminiferal fauna have the ability to respond quickly to sudden, seasonal, fluxes of phytodetritus that arrive on the ocean floor in mid- and high latitudes. Certain species, notably among them *Epistominella exigua* and *Cibicidoides wuellerstorfi*, are capable to quickly invade the mass of phytodetritus and for feeding and eventual reproduction (Gooday, 1988). Linke (1992) demonstrated that several species of deep water benthic foraminifers have developed very specific feeding strategies and metabolic adaptations. *Rupertina stabilis,* which is known to be a suspension-feeder (Lutze and Altenbach, 1988), appeared to be metabolically adapted to a fairly steady supply of food particles obtained from bottom currents, an ecological niche that can also be exploited by *Cibicidoides wuellerstorfi* (Lutze and Thiel, 1987). On the other hand, *Rhabdammina abyssorum*, a species living on the sediment surface, appeared to be well adapted to endure long periods of starvation and to be capable of quickly taking advantage of sudden food inputs (Gooday et al., 1992; Graf and Linke, 1992; Linke, 1992). Using transmission electron microscopy, Heeger (1990) examined the food vacuoles of several species of foraminifers and noted that those of infaunal species (*Melonis zaandami*, *Pyrgo murrhina*) contained little, and only strongly degraded material. However, soon after a fall of phytodetritus, the food vacuoles of epifaunal species were filled with fresh food particles, such as diatoms and dinoflagellates.

Corliss (1985) investigated the depth distribution of living (Rose Bengal stained) foraminiferal species in the top 15 cm of ocean floor sediments. He found recurring hierarchical distributions of species with depth. Furthermore, he found a general correlation between test shape, test wall pore-structure, and depth habitat: plano-convex species with few and coarse pores, usually one one side only, were found at or near the sediment surface; rounded planispiral and cylindrical or ovoid specimens with evenly distributed fine pores were found living within the sediment, some at considerable depths. Corliss and Emerson (1990) related the foraminiferal depth distribution to the flux of organic carbon to the seafloor: infaunal species are abundant where food flux is high, while epifaunal species are best suited to environments of low food fluxes. The actual depth range over which infaunal specimens are encountered was found to be a secondary function of food flux: high fluxes result in a shallow oxic layer; low fluxes allow deeper penetration of oxygen. Corliss and Chen (1988) tested these ideas through an analysis of the Mackensen et al. (1985) foraminiferal data from the Norwegian Sea and found that infaunal species dominated near 1000m depth in organic-rich sediments, and epifaunal species below 1,500 meters where sediments had lower organic matter contents.

It thus appears that several different and independent lines of inquiry into the ecological meaning of deep water benthic foraminifers are converging towards an understanding that food supply, and thus surface productivity, is the determining factor of faunal abundance and especially faunal composition. In fact, many paleoceanographers seem to accept this notion as being firmly established.

ARGUMENTS FOR A RECONSIDERATION

Ecological principles hold that organisms respond to the integral of their environment, and that it is the limiting or overwhelming influence that determines the success or failure of individual species and thus the composition of the fauna. It is therefore not surprising that benthic foraminifers are unequivocal indicators of productivity in areas where productivity is high. However, as Miller and Lohmann (1982) remark, the correlation of a species (*Uvigerina peregrina* in their case) with a particular aspect of the deep environment is not unique: high organic matter loadings within the sediment are usually accompanied with lowered interstitial oxygen levels. It is therefore conceivable that *Uvigerina peregrina* is not discriminating between the possible origins of lowered oxygen levels, and that it could be sensitive to the amount of oxygen in the bottom water, as was argued by Schnitker (1980). For example, Streeter and Shackleton (1979) traced the occurrence of *Uvigerina peregrina* for the entire length of the Atlantic Ocean, crossing areas of strongly contrasting productivities of the overlying waters (Figure 7). Notwithstanding the productivity contrasts, the distribution of this species remained congruent with the distribution of waters with lowered (<5 ml/l) oxygen contents. Another demonstration of this principle is the observation by Mackensen et al. (in press) that on a regional (South Atlantic) scale the faunal composition clearly follows the distribution of water masses, but is strongly modified (overprinted) below a well-defined area of high productivity. But even below the high-productivity zone, the benthic faunas according to the water masses that cover the sampling sites. Bernhard (1992) found evidence that the within-sediment depth distribution of benthic foraminiferal species is dependent on of one of the water mass properties - its oxygen content. On the continental margin off California she found evidence for the influence of organic carbon loading as well as bottom water oxygen content upon the species composition of foraminiferal faunas. While she found that the depth distribution of species was organized similar to the depth distributions described by Corliss and Emerson (1990), bottom water oxygenation strongly overprinted these distributions. The lower the oxygen content of the bottom water, the scarcer the epifaunal component, accompanied by a migration of the deep interstitial fauna towards the surface of the sediment.

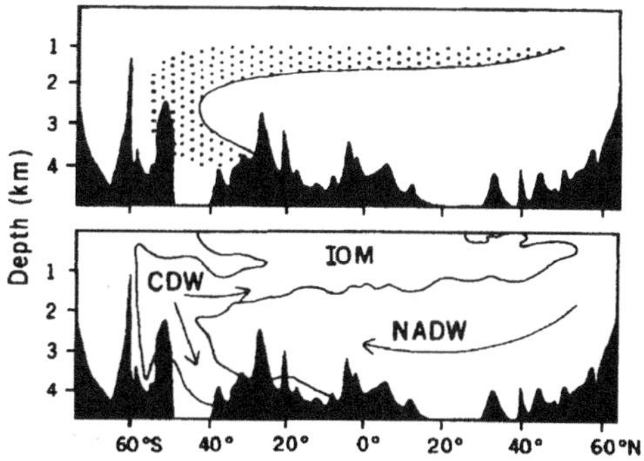

Figure 7 *Uvigerina peregrina* (stippled area: 5 to 30% of total foraminiferal fauna) in the Atlantic Ocean. The species is depth-independent, but occurs in association with low-oxygen (<5 ml/l) deep and intermediate waters. This transect, west of the mid-ocean ridge, passes through several zones of strongly contrasting surface productivity. After Streeter and Shackleton (1979).

In contrast to Pederson et al.'s (1988) finding of higher late glacial productivity in an equatorial Pacific setting, the organic geochemical and stable isotope measurements on core CH84-7P from the western North Atlantic indicates that food flux to the ocean floor was low during the last glacial stage. Such findings corroborate Mix's (1989) determination of just such lowered surface productivity along much of the North American continental margin, as derived by transfer-function from CLIMAP planktonic foraminiferal data. As shown in Figure 8, the composition of the benthic fauna, as exemplified by its key species, clearly does not key in on the productivity signal. Just the opposite: the epifaunal *Hoeglundina elegans* and *Cibicidoides wuellerstorfi* (and also *Epistominella exigua*) are abundant in the high-productivity postglacial, while the infaunal *Uvigerina peregrina*, *Globobulimina auriculata*, and *Cassidulina neocarinata* dominate the low-productivity glacial section of the core. The only possible interpretation of these faunal data is the conclusion that the late glacial bottom water had diminished oxygen contents, while the $\Delta\delta^{13}C$ differential between the epifaunal and infaunal foraminifers leads to the same conclusion, using the arguments of McCorkle and Emerson (1988). Such lowered oxygen content had been postulated by Schnitker (1979, 1980) and later by Boyle and Keigwin (1982, 1987) and Labeyrie et al. (1992).

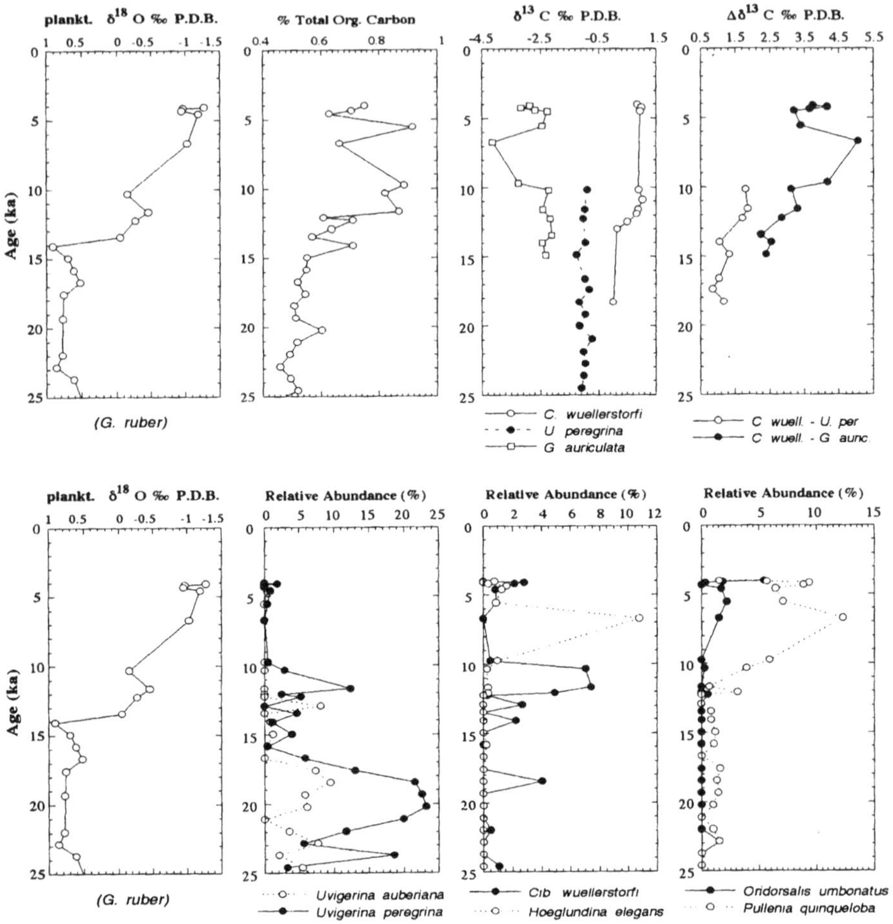

Figure 8. Core CH84-7P from Blake Outer Ridge. Time scale based on correlation of isotope curve to SPECMAP time scale. Top panel - indices of low late-glacial, and of high postglacial productivity are reflected in $\delta^{13}C$ gradients between epifaunal (*Cibicidoides wuellerstorfi*), shallow infaunal (*Uvigerina perigrina*), and deep infaunal (*Globobulimina auriculata*) benthic foraminifers. Bottom panel - *Uvigerina* spp. occur during low-productivity interval, suggesting lowered oxygen as cause, *Cibicidoides wuellerstorfi* and *Hoeglundina elegans* suggest well oxygenated water during postglacial interval, *Oridorsalis umbonatus* and *Pullenia quinqueloba* key in on the high postglacial food input.

AN OUTLOOK

From the examples cited above, it is clear that our understanding of foraminiferal response to their environment is still far from complete. Only when an essentially global set of foraminiferal distribution and their associated physical, chemical and biotic environment has been analyzed will we see the response of foraminifers to the widest possible range of combinations of environmental variables and perhaps grasp its complexity. But even then is there no assurance that paleoceanographic interpretations might not be compromised through no-analog situations. At the present, foraminiferal data have to be accompanied by geochemical or stable isotope information before unequivocal paleoceanographic deductions can attain a useful level of significance. Not all foraminiferal data points can be duplicated by geochemical or stable-isotope analyses, it is only necessary that the principal observed modes be verified.

Why proceed with foraminiferal analyses when they seem to be beset with so many apparent difficulties? As shown in Figure 9, foraminifers are still the finest, most sensitive monitor the deep environment, they are just not yet unequivocally calibrated. Core GPC-5 and its companion core EN120-GGC1 from the western North Atlantic have been analyzed for their foraminiferal and trace element content (Schnitker, 1979; Keigwin et al., 1991). The Cd/Ca ratios of core EN120-GGC1 are accepted as measures of the nutrient content of the deep waters, and hence of the intensity of North Atlantic deep water ventilation. The composition of the benthic fauna, expressed as Principal Components, parallels the ventilation history, even showing differentiations that are not evident in the geochemical record, such as a two-fold division of the record between 12 and 11 ka (radiocarbon years).

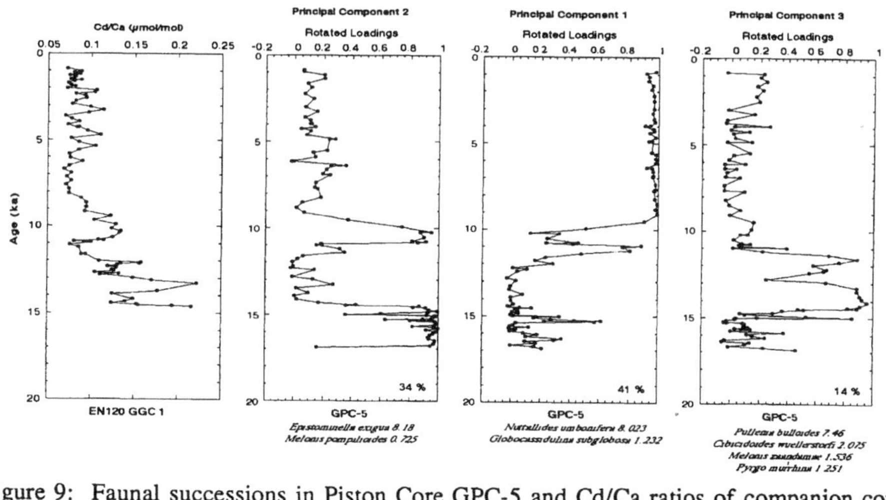

Figure 9: Faunal successions in Piston Core GPC-5 and Cd/Ca ratios of companion core EN120 GGC-1, from 4500m in the western North Atlantic. Cd/Ca ratios are considered to be indicators of bottom water ventilation (data from Keigwin et al., 1991). Note the synchrony of events.

ACKNOWLEDGEMENTS

The conveners of the NATO ASI workshop are thanked for the opportunity to present the information and thoughts contained in this paper. This manuscript benefited from thoughtful reviews by A. Gooday and J. Murray. The author's portions of the research discussed above were supported by various grants from the National Science Foundation.

REFERENCES

Bandy OL, Chierici MA (1966) Depth-temperature evaluation of selected California and Mediterranean bathyal foraminifera. Mar Geol 4:259-271

Berger WH, Herguera JC (1992) Reading the sedimentary record of the ocean's productivity. In: Falkowski PG, Woodhead AD (eds) Primary Productivity and Biogeochemical Cycles in the Sea. Plenum Press, New York, p 455-486

Bernhard JM (1992) Benthic foraminiferal distribution and biomass related to porewater oxygen content: central California continental slope and rise. Deep-Sea Res 39:585-605

Boyle E, Keigwin LD (1982) Deep circulation of the North Atlantic over the last 200,000 years: Geochemical evidence. Science 218:784-787

Boyle E, Keigwin L D (1987) North Atlantic thermohaline circulation during the past 20,000 years linked to high-latitude surface temperature. Nature 330:35-40

Brady HB (1884) Report on the foraminifera dredged by H.M.S. Challenger, during the years 1873-1876. Rep Sci Res Voy Challenger, V 9 (Zool) p 814

Bremer M, Lohmann GP (1982) Evidence for primary control of the distribution of certain Atlantic Ocean benthonic foraminifera by degree of carbonate saturation. Deep-Sea Res 29:987-998

Burke SC (1981) Recent benthic foraminifera of the Ontong Java Plateau. J Foram Res 11:1-19

Caralp MH (1984) Impact de la matière organique dans des zones de forte productivité sur certains foraminifères benthiques. Oceanologica Acta 7:509-515

Caralp MH (1989) Abundance of *Bulimina exilis* and *Melonis barleeanum*: relationship to the quality of marine organic matter. Geo-Mar Lett 9:37-43

Corliss BH (1985) Microhabitats of benthic foraminifera within deep-sea sediments. Nature 314:435-438

Corliss BH, Chen C (1988) Morphotype patterns of Norwegian Sea deep-sea benthic foraminifera and ecological implications. Geology 16:716-719

Corliss BH, Emerson S (1990) Distribution of Rose Bengal stained deep-sea benthic foraminifera from the Nova Scotian continental margin and Gulf of Maine. Deep-Sea Res 37:381-400

Cushman JA (1910-1917) Foraminifera of the Pacific Ocean. U.S. Nat Mus Bull 71

Cushman JA (1918-1931) The Foraminifera of the Atlantic Ocean. U.S. Nat Mus Bull 104

Douglas R, Woodruff F (1981) Deep sea benthic foraminifera. In: Emiliani C (ed.) The Oceanic Lithosphere. The Sea V 7:1233-1327. Wiley-Interscience, New York

Gofas S (1978) Une approche du paléoenvironnement océanique: les foraminifères benthiques calcaires, traçeurs de la circulation abyssale. Thèse doct. Univ. de Bretagne Occidentale, Brest, p 149

Gooday A (1988) A response by benthic foraminifera to the deposition of phytodetritus in the deep sea. Nature 332:70-73

Gooday AJ, Levin LA, Linke P, Heeger T (1992) The role of benthic foraminifera in deep-sea food webs and carbon cycling. In: Rowe GT, Pariente V (eds) Deep-Sea Food Chains and the Global Carbon Cycle. Kluwer Academic Publishers, Dordrecht Boston London, p 63-91

Graf G, Linke P, (1992) Adenosine nucleotides as indicators of deep-sea benthic metabolism.

In: Rowe GT, Pariente V (eds) Deep-Sea Foodchains and the Global Carbon Cycle. Kluwer Academic Publishers, Dordrecht Boston London, p 237-243

Heeger T (1990) Elektronenmikroskopische Untersuchungen zur Ernährungsbiologie benthischer Foraminiferen. Berichte aus dem Sonderforschungsbereich 313. Universität Kiel, No 31, p 1-139

Labeyrie LD, Duplessy JC, Duprat J, Juillet-Leclerc A, Moyes J, Michel E, Kallel N, Shackleton NJ (1992) Changes in the vertical structure of the North Atlantic Ocean between glacial and modern times. Quat Sci Rev 11:401-413

Linke P (1992) Metabolic adaptations of deep-sea benthic foraminifera to seasonally varying food input. Mar Ecol Progr Ser 81:51-63

Lohmann GP (1978) Abyssal benthonic foraminifera as hydrographic indicators in the western South Atlantic Ocean. J Foram Res 7:7-34

Loubere P (1991) Deep-sea benthic foraminiferal assemblage response to a surface ocean productivity gradient: a test. Paleoceanogr 6:193-204

Lutze GF, Altenbach A (1988) *Rupertina stabilis* (Wallich), a highly adapted, suspension feeding foraminifer. Meyniana 40:55-69

Lutze GF, Coulbourn WT (1984) Recent benthic foraminifera from the continental margin of northwest Africa: Community structure and distribution. Mar Micropal 8:361-401

Lutze GF, Thiel H (1987) *Cibicidoides wuellerstorfi* and *Planulina ariminensis*, elevated epibenthic foraminifera. Univ. Kiel, Ber. Sonderforschungsbereich 313 6:17-30

Mackensen A (1987) Benthische Foraminiferen auf dem Island-Schottland Rücken: Umwelt-Anzeiger an der Grenze zweier ozeanischer Räume. Paläont Z 61:149-179

Mackensen A, Fütterer DK, Grobe H, Schmiedl G (in press). Benthic foraminiferal assemblages from the eastern South Atlantic Polar Front region between 35 and 57°S: distribution, ecology and fossilization potential. Mar Micropal

Mackensen A, Sejrup H, Jansen E (1985) The distribution of living benthic foraminifera on the continental slope and rise off southwest Norway. Mar Micropal 9:275-306

McCorkle DC, Emerson SR (1988) The relationship between pore water carbon isotopic composition and bottom water oxygen concentration. Geochim Cosmochim Acta 52:1169-1178

Miller K, Lohmann GP (1982) Environmental distribution of Recent benthic foraminifera on the northeast United States continental slope. Geol Soc America Bull 93:200-206

Mix A (1989) Pleistocene paleoproductivity: evidence from organic carbon and foraminiferal species. In: Berger WH, Smetacek VS, Wefer G (eds) Productivity of the Ocean: Present and Past. Wiley-Interscience, Chichester New York Brisbane Toronto Singapore, p 313-340

Nienstedt JC, Arnold AJ (1988) The distribution of benthic foraminifera on seamounts near the East Pacific Rise. J Foram Res 18:237-249

Pedersen T, Pickering M, et al. (1988) The response of benthic foraminifera to productivity cycles in the eastern equatorial Pacific: faunal and geochemical constraints on glacial bottom water oxygen levels. Paleoceanogr 3:157-168

Phleger FB, Parker FL, and Peirson JF (1953) North Atlantic Foraminifera. Repts Swedish Deep-Sea Expedition, V 7, N 1, p 3-121

Smith KL, Jr. (1992) Benthic boundary layer communities and carbon cycling at abyssal depths in the central North Pacific. Limnol Oceanogr 37:1034-1056

Schnitker D (1974) West Atlantic abyssal circulation during the past 120,000 years. Nature 247:385-387

Schnitker D (1979) The deep waters of the western North Atlantic during the past 24,000 years, and the re-initiation of the Western Boundary Undercurrent. Mar Micropal 4:265-280

Schnitker D (1980) Quaternary deep-sea benthic foraminifers and bottom water masses. In: Donath FE (ed) Annual Review of Earth and Planetary Sciences 8:343-370

Snider LJ, Burnett BR, Hessler RR (1984) The composition and distribution of meiofauna and nanobiota in a central North Pacific deep-sea area. Deep-Sea Res 31:1225-1249

Streeter SS (1973) Bottom water and benthonic foraminifera in the North Atlantic – glacial-interglacial contrasts. Quat Res 3:131-141

Streeter SS, Lavery SA (1982) Holocene and latest glacial benthic foraminifera from the slope

and rise off eastern North America. Geol Soc America Bull 93:190-199
Streeter SS, Shackleton NJ (1979) Paleocirculation in the deep north Atlantic: 150,000 yr. record of benthic foraminifera and oxygen-18. Science 203:168-170
Woodruff F, Savin SM (1989) Miocene deepwater oceanography. Paleoceanogr 4:87-140

THE HISTORY OF BARIUM, BIOGENIC SILICA AND ORGANIC CARBON ACCUMULATION IN THE WEDDELL SEA AND ANTARCTIC OCEAN OVER THE LAST 150,000 YEARS

Graham Shimmield , Shirley Derrick, Andreas Mackensen[1] , Hannes Grobe[1] and Carol Pudsey[2]
Marine Geosciences Unit
Department of Geology & Geophysics
University of Edinburgh
West Mains Road
Edinburgh, EH9 3JW
Scotland, UK

ABSTRACT. Over the last decade many regional palaeoceanographic studies have found evidence for enhanced primary productivity during glacial episodes, particularly in the equatorial Pacific and off northwest Africa. These studies have given rise to the "glacial productivity hypothesis" which has suggested that enhanced oceanic biomass during glacial times supported an efficient removal of organic carbon from the euphotic zone, contributing to lower atmospheric CO_2 levels recorded in ice cores. Recently, studies from the Southern Ocean have shown that south of the Antarctic Polar Front glacial palaeoproductivity was *lower* than during interglacial times. Here we present further evidence for enhanced interglacial productivity in the Southern Ocean using a transect of cores collected from the Scotia Sea and Weddell Sea. Biogenic silica, organic carbon and barium are presented as proxy indicators of past productivity. In order to establish such palaeoproductivity records for this region, where an absence of foraminifera precludes the standard use of $\delta^{18}O$ stratigraphy and carbonate ^{14}C dating, we have developed a method based on the synchronous removal of barium to the sea floor by scavenging and the formation of barite within the frustules of marine diatoms. The barium record is calibrated to a $\delta^{18}O$ Specmap time scale from a single core at 68° 45'S, 5° 53'W (PS1506) where planktonic and benthic foraminifera are found. One core from the Weddell Sea has been studied using the $^{230}Th_{xs}$ dating method to confirm the barium stratigraphy. Using the age models developed here, we identify important increases in palaeoproductivity during isotope stage 5e, and during

[1] Alfred Wegener Institute for Polar and Marine Research, Columbusstrasse, 2850 Bremerhaven, Germany
[2] British Antarctic Survey, High Cross, Madingley Road, Cambridge, CB3 0ET, UK

NATO ASI Series, Vol. I 17
Carbon Cycling in the Glacial Ocean:
Constraints on the Ocean's Role in Global Change
Edited by R. Zahn et al.

the Holocene (Stage 1). Some evidence for a decrease in productivity during a cooling event between 11,000 and 12,000 years BP is observed. Highest palaeoproductivity, defined by biogenic opal accumulation, occurs in the vicinity of the Scotia Arc, just south of the present-day Antarctic Polar Front. Glacial productivity (18 to 72 ky BP) was weaker and displaced slightly to the north. The extent of glacial sea ice is considered to be of primary importance in governing the budget of biogenic detritus to these Southern Ocean sediments.

INTRODUCTION

The importance of the highly productive waters of the Southern Ocean to the productivitycycle of the world's oceans is well known. For instance, DeMaster (1981) and Ledford-Hoffman et al. (1986) have demonstrated that two-thirds of the biogenic silica (opal) removal from the oceans occurs in the Antarctic Ocean, amounting to some 6.9 ± 1.4 Tmol Si year^{-1}. Much of this high productivity is fuelled by the upwelling supply of nutrients associated with the Circumpolar Deep Water (CPW) . The sediments underlying the upwelling frontal zone, just to the south of the Antarctic Polar Front (APF), are often very rich in diatom detritus (75% by weight) forming the so-called "diatom ooze belt" (DeMaster, 1981; Ledford-Hoffman et al., 1986). Recent studies (Treguer and Van Bennekom, 1991; Queguiner et al., 1991) have concentrated on refining the seasonal and local fluxes of opal in the Antarctic Ocean to test the earlier silica budget estimations. Such studies have suggested that the mean ratio of net opal accumulation on the sea bed to net production of opal in surface waters is about 15% (Treguer and Van Bennekom, 1991). Considering the variability in this estimate (primarily due to dissolution and terrigenous sediment dilution), it is clear that the bottom sediments reflect the distribution of primary productivity in the surface waters.

Studies of the distribution of other sedimentary biogenic components in the Antarctic Ocean are rather scarce. Despite the obvious link between the quantity of organic carbon (C_{org}) in the sediment and primary productivity (discussed in Pedersen and Calvert, 1990), few studies have been undertaken in this area. Similarly, several workers have commented on the spatial distribution of high sedimentary barium contents under areas of high productivity (Goldberg and Arrhenius, 1958; Church, 1970; Schmitz, 1987; Shimmield, 1992). Studies within the water column have related the quantity of suspended particulate barium to plankton activity (Dehairs et al., 1980). Recent studies in the Southern Ocean (Dehairs et al., 1991; Stroobants et al., 1991) have confirmed this relationship by matching the water column distribution of barium to the intensity of oxygen and nitrate depletion by seasonal new production. One recent study (Dymond et al., 1992) has presented a cogent case for the use of sedimentary barium as a proxy for primary productivity, using quantitative algorithms developed from analyses of sediment trap material. They argue that the strong correlation between C_{org} and Ba fluxes indicate a link between upper ocean biology and the barium flux to the sediment.

Over the past few years the argument for the forcing of global climate by changes in the biogeochemical cycle of carbon has received a great deal of attention. Large changes in ocean biological productivity and the efficient use of upwelled nutrients, particularly at high latitudes, are thought to have impacted the atmospheric carbon dioxide level through photosynthetic control (Siegenthaler and Wenk, 1984; Kier, 1990). Variation in global climate through the late Pleistocene is known to complement changes in atmospheric CO_2, deduced from the measurement of ice cores (Barnola et al., 1987). One important consequence of this hypothesis is that the sedimentary record should contain traces of the history of surface ocean productivity if the "biological pump" mechanism was effective in lowering atmospheric CO_2 (Martin, 1990) during glacial episodes. Alternatively, Broecker and Peng (1989) have suggested an "alkalinity model" whereby lower atmospheric CO_2 levels occur through increasing the alkalinity of polar surface waters whilst maintaining surface nutrient concentrations. Changes in North Atlantic Deep Water (NADW) thermohaline flow are important in this context.

If such hypotheses are correct then the record of palaeoproductivity from the Southern Ocean over the last glacial cycle should record an inverse relationship between atmospheric CO_2 and enhanced productivity. Recent work by Mortlock et al. (1991) and Charles et al. (1991) has shown that the record from the southern high latitudes is complex. North of the APF the sedimentary records of opal accumulation rate and abundance confirm the high glacial productivity hypothesis, but south of the APF (where present-day productivity reaches a maximum) the sedimentary record suggests *lower* glacial productivity was prevalent. They conclude that models of climate change and atmospheric CO_2 linked to an increase in glacial productivity in the Antarctic are *not* supported by records of opal accumulation from the Southern Ocean. Similarly, through the use of sediment $\delta^{15}N$ measurements on the organic fraction, Francois et al. (1992), have shown that there was relatively little change in the magnitude of the $\delta^{15}N$ shift between nutrient-rich and nutrient-poor waters either side of the APF. They suggest that this result implies that Southern Ocean polar water did not suffer major nitrate uptake during glacial times and therefore productivity was not a major influence on lowering glacial atmospheric CO_2.

Of crucial importance to the study of palaeoproductivity in the Southern Ocean is the establishment of an accurate age model for the sediment cores recovered. Most palaeoceanographic work over the last decade has relied on measurements of the $\delta^{18}O$ content of planktonic and benthic foraminifera. Such time-series measurements can be compared with the globally-averaged SPECMAP stack in order to construct an age-depth relationship, often in parallel with appropriate spectral tuning techniques (Martinson et al, 1987). However, in the Southern Ocean to the south of the APF most of the water column is highly undersaturated with respect to calcite and there is a dearth of both planktonic and benthic foraminifera necessary for stable oxygen isotope measurements. For many years this has hindered palaeoceanographic studies at these high latitudes, save for some very general observations based on geomagnetic chronology (e.g. Pudsey et al., 1988). In contrast, detailed studies by Mortlock et al. (1991) and Charles et al. (1991) were carried out using carbonate-containing cores from north of the APF and extrapolation to more southerly cores (as far south as 54° 23'S). In 1988 workers from the Alfred Wegener Polar Institute cored a carbonate-rich sedimentary province off the eastern Weddell Sea continental margin near Atka Bay (70° S, 10° W) where a reliable $\delta^{18}O$ chronology (Mackensen et al., 1989; Grobe and Mackensen, 1992; Mackensen et al., this volume) could be obtained. However, the important region of the Weddell Sea and Scotia Sea has provided few dated palaeoceanographic records with which to examine glacial to interglacial trends in palaeoproductivity to the south of the Antarctic Polar Front zone.

In this paper we present geochemical data (biogenic silica, organic carbon, barium) to address changes in palaeoproductivity at 68°S in the eastern Weddell Sea. $\delta^{18}O$ measurements on the planktonic foraminiferan *Neogloboquadrina pachyderma* and the epibenthic foraminiferan *Epistominella exigua* are used to provide a reliable age model (Mackensen et al., this volume). A transect of four cores across the Weddell Sea to the Scotia Sea is examined using the same proxy palaeoproductivity indicators. Similarities in the history of barium accumulation provide a means of establishing an approximate chronology. The age model has been corroborated in one core by using the excess ^{230}Th ($^{230}Th_{xs}$) method. The results indicate that glacial productivity was indeed lower south of the APF, confirming the observations of Mortlock et al., (1991) and Charles et al., (1991). This study extends our knowledge of Southern Ocean sediment geochemistry and indicates that significant palaeoceanographic records addressing the glacial productivity hypothesis can be recovered from the deep Scotia Sea and Weddell Sea.

MATERIALS AND METHODS

Core PS1506 was raised from a water depth of 2550 m off the eastern Weddell Sea continental margin (see Figure 1) by the Alfred Wegener Institute, Bremerhaven, Germany. 450 cm of this piston core have been studied in detail at 5 cm intervals using X-ray fluorescence (XRF) analysis for major and minor element bulk chemistry. A rhodium target X-ray tube was used, followed by appropriate mass absorption correction algorithms (see Shimmield, 1985 for further details). By running international certified reference materials we estimate an r.m.s.d. of better than 5%. The precision of XRF analysis is very good; for the trace elements reported here the analyses are better than 2%. Organic carbon was measured on a Carlo-Erba elemental analyser following $CaCO_3$ removal with hot sulphurous acid and freeze drying. The precision of this technique is estimated at 3% 1s r.m.s.d. for samples low in carbonate (<30%). For core PS1506 an estimate of biogenic silica (opal) was obtained by calculating the excess silica present, over and above that found in average shale. This method will have a substantial error if large amounts of quartz are present.

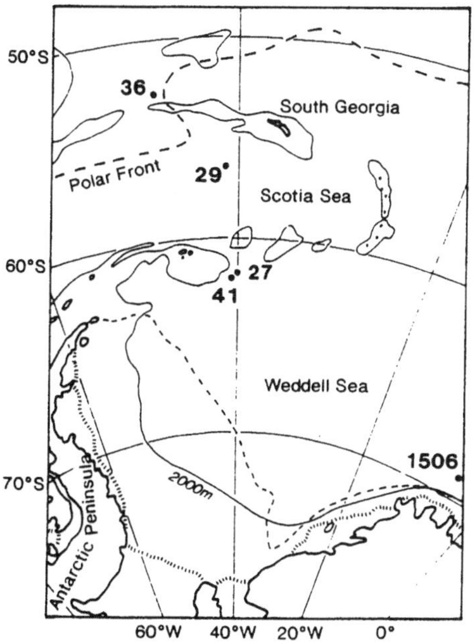

Figure 1. Map of the Weddell Sea and Scotia Sea indicating the positions of cores 36, 29, 27, 41 and PS1506. The short dotted line marks the modern summer sea ice extent.

The $\delta^{18}O$ data presented here for core 1506 were measured on hand-picked samples of *Neogloboquadrina pachyderma* and *Epistominella exigua* at the Alfred Wegener Institute. In core 36, the upper 140 cm contained a sufficient abundance of *N. pachyderma* for isotopic analysis. The ratio of ^{16}O to ^{18}O is given in standard δ notation with reference to the PDB standard. The external precision for the method was better than 0.08 ‰ based on replicate foraminiferan samples. The age model used in this study was derived from comparison of the obtained isotope records (for both planktonic and benthic species) against the SPECMAP stacked record.

The other cores discussed below (Figure 1) are from the British Antarctic Survey collection and have been described in Pudsey et al. (1988) and Pudsey (1992). Core 36 is located just to the north of the present day Antarctic Polar Front in 3552 m of water. Cores 29, 27 and 41 are from the Scotia Sea in 3369, 3470 and 3310 m of water respectively, and have been analysed at 10 cm intervals. Piston cores tend to suffer from core-top loss to some degree; consequently, the trigger cores have been analysed and patched into the record. For these cores opal was analysed by a wet colourimetric method following leaching of the opal in sodium carbonate solution. This technique, developed at Edinburgh, also quantifies the leached aluminium present and corrects for any lithogenic silica in solution by applying an average lithogenic Si/Al ratio. Comparison with results obtained by this method with the XRF Si/Al ratio method applied to core PS1506 suggests that the quartz contribution to these sediments is rather low, and that the estimated opal content for this core is a reasonable proxy for siliceous productivity.

The age control available for the BAS cores comes from (i) radiolarian biostratigraphy, namely the uppermost abundance maximum of *Cycladophora davisiana* (Jordan and Pudsey, 1992); (ii) relative abundance stratigraphy of diatoms, particularly *Eucampia antarctica* (Jordan and Pudsey, 1992), (iii) tentative identification of geomagnetic events (short intervals of low or reversed inclination within the Bruhnes normal chron (Pudsey, 1992); and (iv) excess ^{230}Th dating using the assumption of constant flux with time (see below for further discussion). The ^{230}Th method was performed by isotope dilution a-particle spectrometry using techniques described elsewhere (Shimmield et al, 1986). No down-core, stable isotope measurements on foraminiferal remains were possible due to the absence of calcareous microfossils in these sediments except for the upper 140 cm in core 36 (see below).

RESULTS

*Barium as a palaeo*productivity *indicator*

Sedimentary barium is derived from two main sources. A biogenic component is produced by the precipitation of barite ($BaSO_4$) within reducing microenvironments during the settling of diatom frustules through the water column (Dehairs et al., 1980, 1990; Bishop, 1988; Dymond et al., 1992). The second, minor, component is Ba associated with the lithogenic mineral fraction, usually feldspars, with a typical Ba/Al weight ratio of 0.0075. However, the precise mechanism for barite precipitation in the water column deserves greater attention. The dissolved Ba in the water column displays a nutrient-like distribution that is similar to both alkalinity and silica. Indeed, Lea and Boyle (1990) have argued that the Ba/Ca ratio of benthic foraminifera provides a faithful proxy for deep-water alkalinity over glacial to interglacial timescales. Recently, Dymond et al. (1992) have shown conclusively that particulate Ba flux observed in sediment traps results from the accumulation of dissolved Ba from seawater and displays a linear increase with depth. In their hypothesis the observed depth increase in particulate Ba allows them to formulate an algorithm for estimating past palaeoproductivity. This relies on the observation that barite is very stable and slow to dissolve in deep-sea sediments as long as sulphate is present in the pore waters (Shimmield, 1992). Thus, it may be suggested that

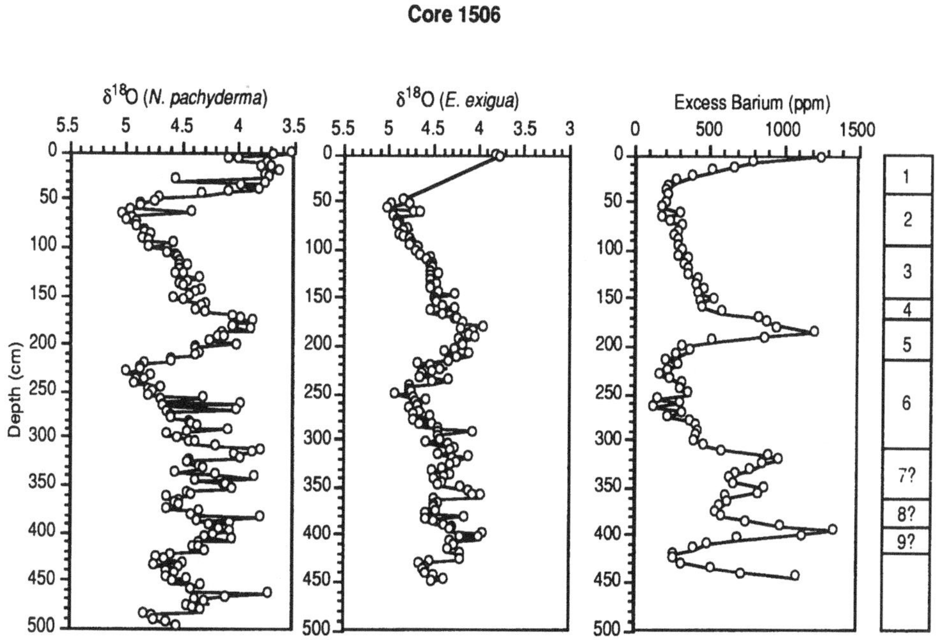

Figure 2. The correlation between excess Ba (weight ratio) and the $\delta^{18}O$ stratigraphy of core PS1506 measured on foraminifera *N. pachyderma* (planktonic) and *E. exigua* (benthic) (Mackensen et al. this volume). The isotope stage boundaries are given in the right hand column using the chronology of Martinson et al. (1987).

initial (enhanced) primary productivity provides suitable substrates, namely diatom frustules, for barite precipitation to occur. In the Southern Ocean this is well documented in the recent work of Dehairs et al. (1991) and Stroobants et al. (1991). Upon sinking, these frustules support continuous precipitation of barite from the water column, probably through some form of dynamic equilibrium between the dissolved and particulate phase in a manner analogous to ^{230}Th (Anderson et al., 1983a and b). Consequently, the historical record of Ba accumulation reflects both initial primary productivity, and the standing stock of dissolved Ba in the water column. If this hypothesis is correct then inter-ocean variation in dissolved Ba, reflecting the intermediate and deep-water nutrient and alkalinity distributions, will be reflected in the sediment composition, over which the local palaeoproductivity signal will be superimposed. One consequence of such a hypothesis should be broadly synchronous increases in sedimentary biogenic barium within ocean basins that contain waters of similar chemistry at intermediate and deeper depths.

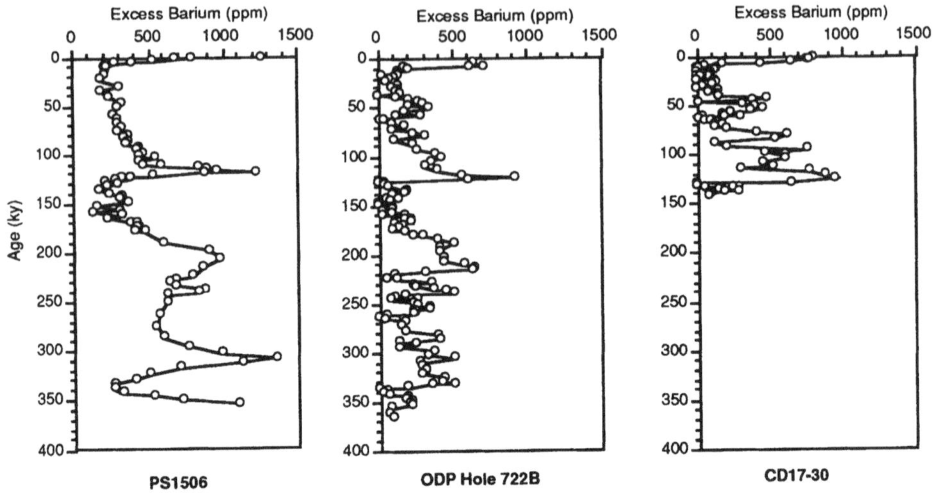

Figure 3. Downcore distributions of excess Ba from the Antarctic continental margin (PS1506) and the Arabian Sea (CD17-30 and ODP Hole 722) showing the excellent synchroneity in Ba content between these areas (age control from $\delta^{18}O$ stratigraphy given in sidereal years).

The eastern Weddell Sea continental margin

In Figure 2 the content of excess Ba in sediments from core PS1506 is presented. The estimation of excess (biogenic) Ba uses a normative calculation based on a typical Ba/Al ratio found in the lithogenic fraction of sediments (Dymond et al., 1992)* . Because Ba and Al are equally diluted by other excess phases (opal, biogenic calcite, quartz), the excess quantity is dilution independent. This record is compared to the $\delta^{18}O$ record obtained from the foraminifera *N. pachyderma* and *E. exigua* (Mackensen et al., this volume). There is a good correspondence between the two records, and furthermore, the $\delta^{18}O$ record provides a means of establishing the chronology of the core by matching to the standard SPECMAP stacked record (Martinson et al., 1987). Graphical correlation was completed using the computer program of Paillard et al. (1992) to produce the age model used here. The high excess Ba ratio observed in the Holocene (Stage 1) sediment and Stages 5 and 7 is very similar to that reported for the north west Arabian Sea (Shimmield and Mowbray, 1991; Clemens et al., 1991; Shimmield, 1992; Figure 3). We may therefore suggest that both the Antarctic Ocean and north west Indian Ocean were characterised by high palaeoproductivity during the major interglacial episodes, and that intermediate and deep

* Excess Ba (ppm) = (Total barium ppm - (Al wt.% * 75)); where 75 is the lithogenic Ba/Al ratio (ppm/wt.%)

water Ba chemistry was somewhat similar. At present, Antarctic Intermediate Water (AAIW) and Antarctic Bottom Water (AABW) are both enriched in dissolved Ba in comparison to North Atlantic Deep Water (NADW), as shown by GEOSECS data. Mixing of NADW and AABW/AAIW results in northward flow into the cul-de-sac of the Arabian Sea. Such similarity in the biogenic Ba records suggests that Ba may also be a first-order proxy for the $\delta^{18}O$ record obtained from foraminifera. Intra-ocean variability under mesotrophic productivity conditions may be low, offering the potential for cross-core correlation between sedimentary Ba records in a similar fashion to the foraminiferal stable oxygen and carbon isotope records.

Opal, C_{org} and carbonate have also been used as proxy indicators of past changes in ocean palaeoproductivity (see Charles et al., 1991 for detailed opal records for the Southern Ocean). In Figure 4 we compare the down-core distribution of these productivity indicators with the excess Ba content discussed above. (Note that the opal is estimated from total Si and Al contents with a correction for lithogenic Si having a ratio of 3.2:1, respectively. Ca/Al ratio is used to represent the $CaCO_3$ record.) Examination of the estimated opal record suggests a reasonable correspondence with the excess Ba signal, although the phasing is different, with opal leading the excess Ba. This is particularly noticeable across Termination I (25 ky to present). Of course, both opal and carbonate are subject to water column and seafloor dissolution, and therefore the records observed in PS1506 are a product of dilution by terrigenous input, dissolution and biological productivity.

Nevertheless, this suite of proxy records confirms the general conclusion that primary productivity was probably enhanced during interglacial periods when sea ice cover was reduced. This confirms the hypothesis that interglacial palaeoproductivity was greater south of the APF (Mortlock et al., 1991; Charles et al., 1991; Francois et al., 1992; Kumar et al., 1993). The C_{org} record suggests some variation with time over the late Quaternary, although the dominant control on the abundance measured is probably microbial degradation, giving rise to the classic log normal decrease with depth.

The Weddell Sea and Scotia Sea

The cores collected by the British Antarctic Survey span the Scotia Sea and northern Weddell Sea, extending from just north of the present day Antarctic Polar Front position to well south of it (Figure 1). Such a transect offers the potential for recording and testing the glacial to interglacial productivity hypothesis and for examining the fluxes of biogenic components over the late Pleistocene. However, as outlined above, the lack of biogenic carbonate throughout the core records severely hampers the ability to use stable isotope stratigraphy as a means of defining the chronology of sedimentation.

We propose the use of barium as a chronostratigraphic tool for correlation of glacial to interglacial change across the Weddell and Scotia Sea. Figure 5 displays the depth distribution of excess Ba in the four BAS cores arranged on the north-south transect. By using the excess Ba to

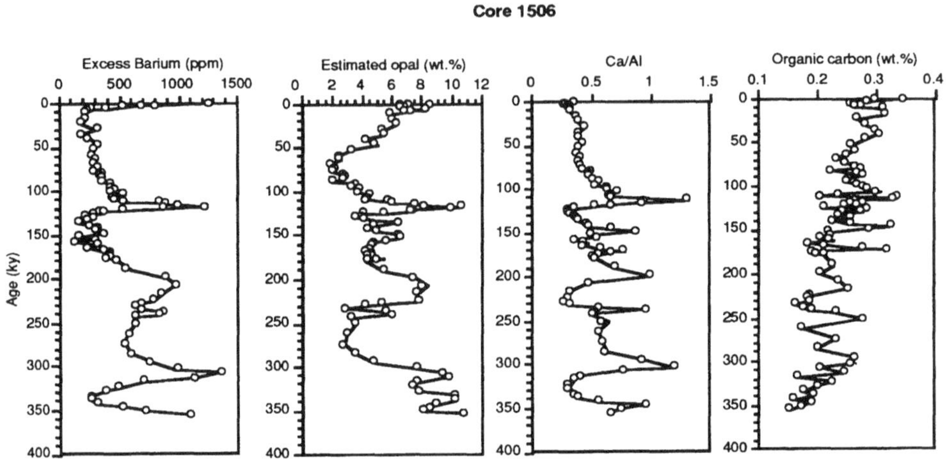

Figure 4. The chronology of excess Ba, opal (estimated from total Si minus an average lithogenic component), C_{org} and Ca/Al in core PS1506. Note the close correspondence of excess Ba and opal, but Ca/Al (an indicator of the amount of calcite present) is out of phase due to selective modification by dissolution. C_{org} decreases with depth due to microbial degradation.

$\delta^{18}O$ relationship observed in core PS1506 and in the northern Arabian Sea, tentative stratigraphic horizons are indicated. Key features include the Stage 5e maxima in excess Ba, and the glacial to Holocene increase. Additional support for this proposed chronology comes from measurements of the relative abundance of the radiolarian *Cycladophora davisiana* (Jordan and Pudsey, 1992). Hays et al (1976) found that the first subsurface peak of this radiolarian in the Southern Ocean may be correlated with the Last Glacial Maximum at 18 ky BP. Jordan and Pudsey (1992) identify this peak at 1.3 m depth in core 27 and 2.5 m in core 29 (see Figure 5). In addition, the abundance of the diatom *Eucampia antarctica*, has been shown to correlate with the radiolarian record (Burckle and Burak, 1988). In most Weddell Sea and Scotia Sea cores, diatom abundance (tracked by the total biogenic opal content) is restricted to the Holocene (Stage 1), ie 2.3 m in core 29 (Jordan and Pudsey, 1992: Unit 1 comprising *Rhizosolenia* spp. and *Chaetoceros* spores and *Thalassiothrix* spp., common *Distephanus speculum* and common to abundant *Nitzschia kerguelensis*, with relatively little *Eucampia antarctica*.) Further clues to the chronology of sedimentation has been presented for core 41 by Pudsey (1992) who describes two magnetic reversal events suggested to be the Lake Mungo at 28-31 ky BP and the Laschamp at 35-42 ky BP. These events are indicated on Figure 5.

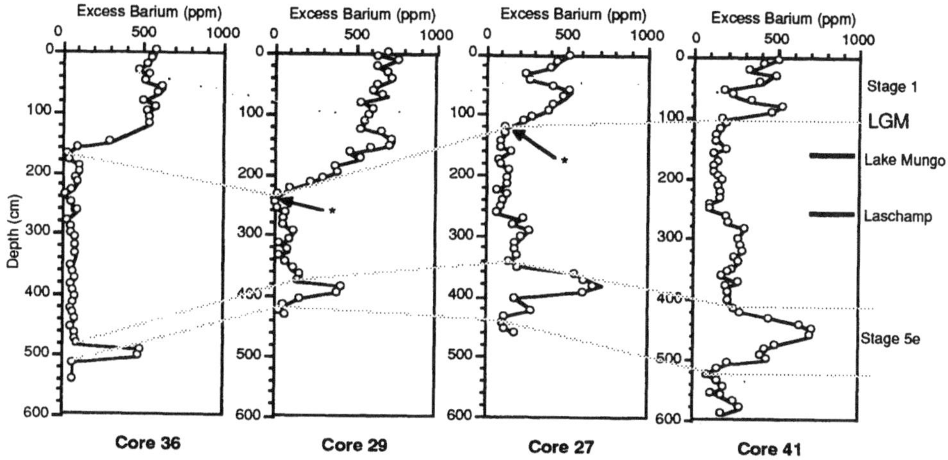

Figure 5. Using the excess Ba chronology from core PS1506, a correlation for cores 36, 29, 27 and 41 may be performed. The Last Glacial Maximum (LGM) time line is based on the disappearance of opal (Unit I: Jordan and Pudsey, 1992) and excess Ba, followed by the first downcore abundance peak of *C. davisiana*. This is marked by the * symbol and occurs at 2.5 m and 1.3 m depth in cores 29 and 27, respectively. No abundance peaks are found in core 36, and no radiolarians are present in core 41. Magnetic events found in core 41 are assigned to the Lake Mungo (28-31 ky BP) and Laschamp (35-42 ky BP) (Jordan and Pudsey 1992). Elevated excess Ba in Stage 5e is characteristic of the Indian Ocean (Shimmield, 1992). The light dashed line at 12 k ^{14}C yr BP is uncertain; it may reflect a cooling event described by Labracherie et al. (1989).

Recent studies by Labracherie et al. (1989) on the isotopic and micropalaeontological deglacial records from three cores taken between 44° to 55°S (Indian Ocean sector) have shown that a cool "oscillation" interrupted the warming phase between 11,000 and 12,000 years BP (event dating by ^{14}C AMS measurements on planktonic foraminifera). This event appears to precede the northern hemisphere Younger Dryas event by about 1,000 years, although Southern Ocean warming appears to be complete by 10,000 years BP. In Figure 6 we compare the Holocene $\delta^{18}O$ signal of both *G. bulloides* and *N. pachyderma* with the excess Ba and $CaCO_3$ records for core 36. These $\delta^{18}O$ records depend upon salinity and temperature variations of the surface waters which is complex in the Southern Ocean (Labeyrie, et al. 1986). *G. bulloides* carries a lighter $\delta^{18}O$ signal, being a planktonic foraminiferan that prefers temperate to sub-polar waters compared to the deeper-dwelling, polar *N. pachyderma* (Kellog et al., 1978). At about 50 cm depth the heaviest $\delta^{18}O$ value for *G. bulloides* was measured. We believe that this event is the synchronous with that measured by Labracherie et al. (1989) in the Indian Ocean sector, and that the decrease in excess Ba, lower $CaCO_3$ content and heavy $\delta^{18}O$ signal are all evidence of a cooling event, expanding sea ice extent and lower palaeoproductivity. We have extrapolated tentatively the Holocene excess Ba decrease in cores 29, 27 and 41 as being chronostratigraphically synchronous, and with a ^{14}C age of 12,000 years BP.

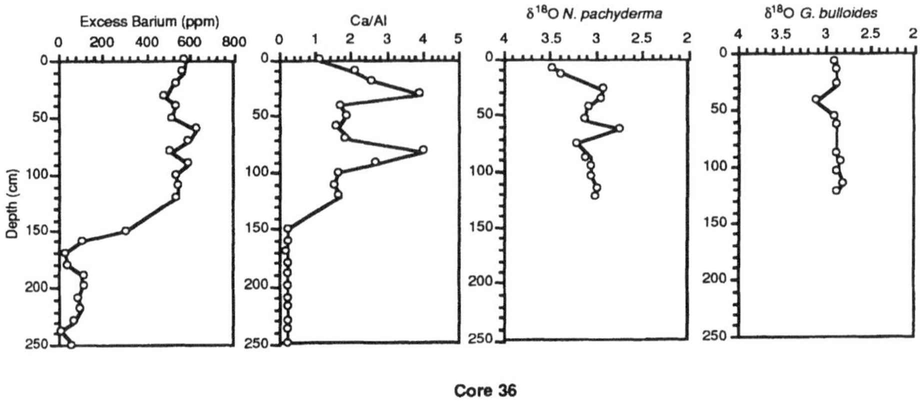

Figure 6. Detail of the correlation between excess Ba (ppm), Ca/Al, and in *N. pachyderma* and *G. bulloides*. The decrease in Ca/Al centered on 50 cm depth is reflected in a decrease in excess Ba and a shift towards heavier $\delta^{18}O$. This event may be synchronous with the cooling episode between 11 and 12 ky BP described by Labracherie et al. (1989) from the Indian Ocean sector of the Southern Ocean.

Figure 7 shows the results of measurements of $^{230}Th_{xs}$ (i.e. ^{230}Th-^{234}U) in core 27. ^{230}Th is an extremely particle reactive radionuclide that is quantitatively scavenged from seawater by settling particles, with a residence time that is short (10-40 years) relative to lateral mixing in the deep sea (Francois et al, 1990; Lao et al., 1992). Consequently, the flux of $^{230}Th_{xs}$ from the water column roughly balances local production from parent ^{234}U decay (Francois et al, 1990; Kumar et al., 1993). Recently, Lao et al. (1992) have undertaken a study to examine the effect ocean margin boundary scavenging has had on glacial to interglacial fluxes of $^{230}Th_{xs}$. They show that over periods of a few thousand years changes in particle flux and particle composition may have an impact on $^{230}Th_{xs}$ (and $^{231}Pa_{xs}$) by changing the nature and intensity of boundary scavenging. This phenomenon has been exploited by Kumar et al. (1993) who suggested that $^{231}Pa_{xs}/^{230}Th_{xs}$ ratios may be used to evaluate past changes in productivity. However, Lao et al. (1992) point out that, "If detailed profiles of ^{230}Th and ^{231}Pa covering a sufficiently long time period are available, one can correctly interpret the results as containing short-term cyclic variability superimposed on a long-term average sedimentation rate." In this context, long-term defines an entire glacial-interglacial cycle of ~100,000 years, with episodes such as Termination 1 representing rapid climate and oceanic change on a short time scale.

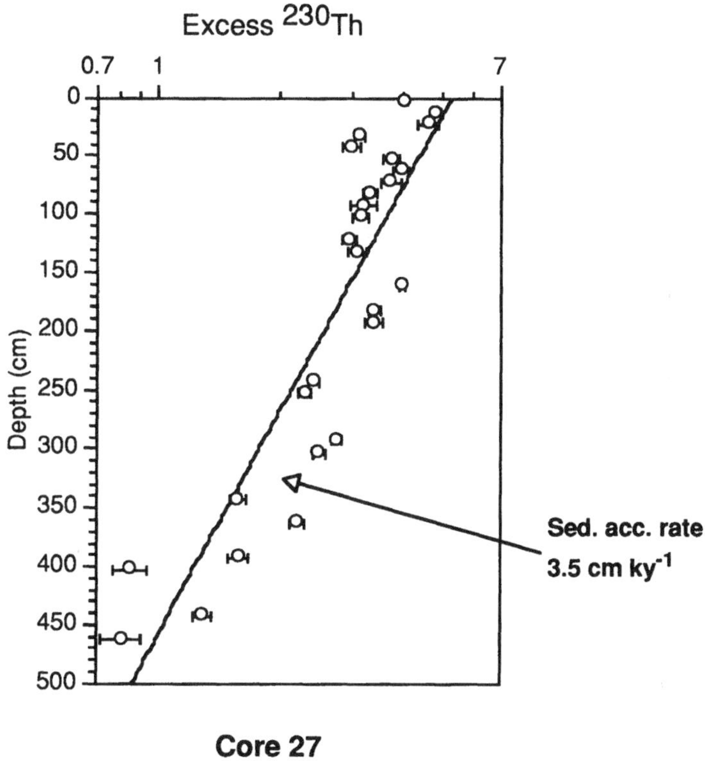

Figure 7. Excess ^{230}Th (^{230}Th-^{234}U) plotted against depth in core 27. A weighted least-squares weighted regression gives an overall sedimentation rate of 3.5 cm ky^{-1}, helping to confirm the proposed Ba stratigraphy, particularly the stage 5e subsurface maximum. See text for further details on the use of excess ^{230}Th for estimating sedimentation rates in this context.

Consequently, we examine here $^{230}Th_{xs}$ data to address, primarily, the overall sedimentation rate for core 27 in order to provide independent confirmation of the excess Ba chronostratigraphic model. This assessment suggests that an overall average sedimentation rate of ~3.5 cm yr^{-1} matches the observations, providing that a constant input of $^{230}Th_{xs}$ is assumed. This approach indicates that the obvious subsurface maximum in excess Ba at 420 cm depth represents Stage 5 rather than Stage 3. Detailed measurements in the upper 200 cm of the core reveal reasonable coherence between $^{230}Th_{xs}$ and the excess Ba record suggesting that the assumption of constant thorium input flux is subject to some qualification, and that "short-term" changes in particle flux arising from fluctuation in new production may affect scavenging of this radionuclide, as suggested by Lao et al. (1992).

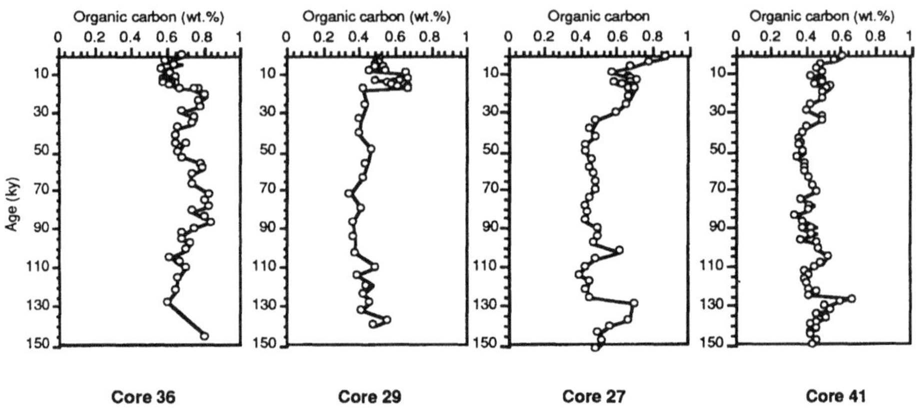

Figure 8. The chronology of C_{org} (weight %) in cores 36, 29, 27 and 41 using the chronology established from the excess Ba records and matching to the timescale of Martinson et al. (1987).

The distribution of organic carbon and opal (as wt.%) along the Weddell Sea transect using the age models developed above is presented in Figures 8 and 9. Despite uncertainties regarding the age of the core top (50 to 100 cm were probably lost from core 27) there appears to be a regionally coherent pattern in the distribution of these biogenic components. C_{org} varies between 0.3 and 0.9 wt.%, with highest contents in the Holocene of the northern Weddell Sea (cores 27 and 41), and in Stage 5. There is very good downcore coherence in the records from these two cores, adding to the confidence in the tentative age model. Further north, in the Scotia Sea, C_{org} appears to be slightly higher in the glacial Stage 2 sediments, and is generally greater in abundance in sediments underlying the present day APF (core 36). The highest opal contents are found near the APF, with core 36 showing little major variation over its 130 ky history. In contrast, core 29 displays pronounced high opal contents in Stage 1 and Stage 5. Further south the opal content is dramatically lower, but still exhibits increased abundance during the major interglacial periods. These results support and extend the work of Mortlock et al. (1991) and Charles et al. (1991) into the Weddell Sea. The decrease in glacial sedimentary opal contents reflects both terrigenous dilution and silica dissolution, together with a substantial decrease in siliceous primary productivity. It should be noted that recently Archer et al. (1993) completed modelling studies on the parameters controlling opal preservation in tropical and subtropical sediments. They concluded that changes in downcore sedimentary opal cannot simply be interpreted as reflecting changes in ocean productivity. However, in the Southern Ocean, where the burial flux of opal represents a significant proportion of the total biogenic opal production, the kinetic competition with dissolution may allow the preservation of sedimentary records that better reflect primary productivity (Archer et al., 1993). Perhaps this leads to the observation of Charles et al. (1991) that sedimentary opal in the Southern Ocean is a useful tracer of productivity over glacial cycles.

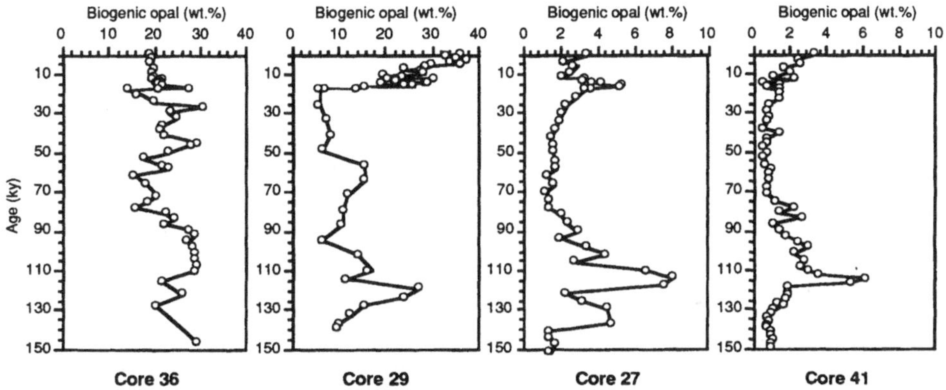

Figure 9. The chronology of biogenic opal (weight %) in cores 36, 29, 27 and 41.

DISCUSSION

The relatively high flux of warm NADW to the Southern Ocean plays an important role in governing the properties of Circumpolar Deep Water and surface water (Corliss, 1982; Charles and Fairbanks, 1990) as well as the extent of sea ice formation. In the high-nutrient waters lying between the Polar Front and the Antarctic Divergence the extent of sea ice, and the resulting inhibition of phytoplankton photosynthesis, is the key variable affecting primary production and resulting impact on CO_2 drawdown. Consequently, climatic cycles which impact on both NADW flux and the extent of sea ice exert a strong control on Southern Ocean productivity. Over the past few years many workers have established that Antarctic productivity was lower during glacial episodes south of the Polar Front (Mackensen et al., 1989; Grobe et al., 1990; Mortlock et al., 1991; Charles et al., 1991; Francois et al., 1992; Kumar et al., 1993). Jordan and Pudsey (1992) used diatom abundance and stratigraphy for the same BAS cores studied here to produce a map illustrating the hypothesised extent of maximum and minimum glacial sea ice in Scotia and Weddell Sea. In this paper we present geochemical data (excess Ba, biogenic opal, organic carbon and Ca/Al) that indicate the effect the sea ice boundary has on palaeoproductivity.

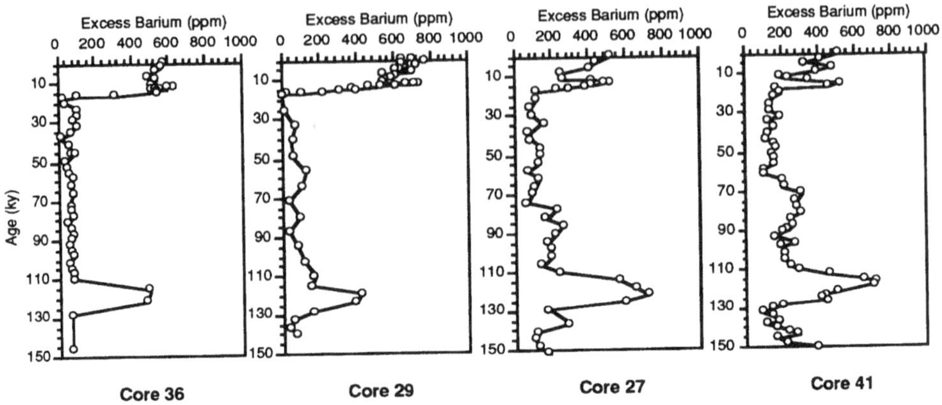

Figure. 10. The chronology of excess Ba (ppm) in cores 36, 29, 27 and 41.

*The synchroneity of Southern Ocean palaeo*productivity *signals*

Across the Weddell and Scotia Sea transect there is a marked variation in the level of biogenic opal recorded in the sediments, despite the fact that accumulation rates do not change dramatically (see Figure 5). In core 36 (the most northerly) the Ca/Al ratio in Holocene (Stage 1) sediment is high, with a clear decrease into glacial sediment that is coincident with the level of excess Ba. However, the quantity of biogenic opal remains high throughout the history of the core. Further south, this changes dramatically with a virtual absence of biogenic Ca, and a lack of opal during glacial episodes (Stages 2-4). The quantity of excess Ba (Figure 10) displays a strong interglacial to glacial contrast, but with little variation in the overall content within Stage 1 sediment along the transect, in contrast to biogenic opal. Jordan and Pudsey (1992) point out that the upper diatomaceous interval in the BAS cores (their Unit I) is synchronous with Stage 1, with the first *C. davisiana* peak lying immediately below (see Figure 2). In order to resolve this ambiguity it is suggested that a marked change in sedimentation rate exists between glacial Stage 2 sediment and Stage 1. The *C. davisiana* first subsurface maximum becomes practically synchronous with the stage boundary due to the lack of sedimentation during the ice-covered glacial stage.

In core PS1506 the increase in excess Ba appears to occur somewhat later when compared to the $\delta^{18}O$ signal of *N. pachyderma* (Figure 2). It is possible that the excess Ba record reflects the degree of ice cover, which lags somewhat the $\delta^{18}O$ isotopic signature of the water mass (although local variations due to salinity gradients are unquantified). In Figure 4 it is clear that the interglacial climatic optima are characterised by high contents of excess Ba and opal, but that the biogenic carbonate (shown by Ca/Al) appears to lag the primary maxima. Mackensen et al. (this volume) show that these maxima are low in benthic foraminiferal numbers. Rather than reflecting low carbonate productivity at these times, it is suggested that intensive carbonate dissolution has occurred due to the increased flux of biogenic detritus to the sea floor.

Climatic amelioration during Stages 1 and 5

Morley (1989) showed that southward migrations of the Polar Front to ~55°S in the Indian Ocean sector occurred during Stage 1 and early Stage 5. In ODP hole 704A (46°53'S, 7°25'E) Gard and Crux (1991) describe the presence of the subtropical coccolith *Calcidiscus leptoporus* during Stage 5e. Hodell (1993) used high carbonate contents in Hole 704a (46°52.8'S, 7°25.3'E) to identify episodes of climatic amelioration during stages 7, 9 and 11. We believe that excess Ba can be used to trace the timing of such events to higher latitudes (where biogenic carbonate is absent) than has previously been possible. Hodell (1993) used $\delta^{13}C$ record of benthic foraminifera in Hole 704A to estimate the relative contribution of NADW in influencing the isotopic record. Enhanced NADW flux is recorded during stages 7, 9 and 11, probably greater than during stages 1 and 5. If the extent of excess Ba in the sediment record faithfully reflects palaeoproductivity associated with NADW flow (either directly as suggested by the records from the northern Arabian Sea (Figure 3), or indirectly through the sensible heat transfer to CDW and melting of sea ice and resulting primary productivity), then an independent chronology of Scotia Sea and Weddell Sea sediments (AMS ^{14}C dating of organic carbon) would provide a valuable insight into the effects and timing of NADW flow to high latitudes.

SUMMARY AND CONCLUSIONS

1. Through the use of inorganic geochemical tracers and measurements of the opal and organic carbon content of piston cores, it has been possible to examine the palaeoceanography of the last 150 ky in the Weddell Sea and Antarctic Ocean. This study confirms previously published work (Mackensen et al. 1988; Mortlock et al. 1991; Charles et al. 1991; Grobe and Mackensen, 1992; Francois et al, 1992; Kumar et al., 1993) that *interglacial* productivity was higher immediately south of the present-day Antarctic Polar Front position, and contrasts with many other ocean settings where glacial productivity is higher (e.g. the equatorial Pacific and off northwest Africa).

2. Barium contents within the sediments follow a clear trend with time, believed to result from variation in palaeoproductivity and water column scavenging, rather than detrital input and/or diagenetic remobilisation. Comparison with a $\delta^{18}O$ record from off Akta Bay (PS1506), establishes a barium chronology which is consistent with those obtained from the Arabian Sea. From core PS1506 a speculative chronology is erected for cores

spanning a transect of the Weddell and Scotia Seas. $^{230}Th_{xs}$ dating from one core (27) provides additional confirmation of this age model.

3. The major control on the distribution of biogenic detritus is considered to be sea ice extent. Glacial conditions effectively quenched productivity further south than 60°S. Interestingly, both organic carbon and barium productivity appears to be less influenced by the ice extent, maintaining moderate fluxes south of the opal shut-down. Dissolution of opal within the water column is severe and may account for this observed distribution.

ACKNOWLEDGEMENTS. The authors wish to thank the marine staff of the British Antarctic Survey and Alfred Wegener Institute for their help in obtaining the cores described here. Special thanks to Rainer Zahn for organising the NATO workshop, and to Tom Pedersen and two anonymous reviewers for comments on an earlier draft of this paper. Dr Caroline Bertram and Prof Nick Shackleton (Cambridge University) carried out the oxygen isotope measurements. This study was carried out with the financial support from a British Antarctic Survey Special Topic Award (GST/02/451) to GBS and SD. This is SFB Contribution 72 and publication number 649 of the Alfred Wegener Institute for Polar and Marine Research.

REFERENCES

Anderson RF, Bacon MP, Brewer PG (1983a) Removal of ^{230}Th and ^{231}Pa from the open ocean. *Earth Planet Sci Lett* **62**: 7-23

Anderson RF, Bacon MP, Brewer PG (1983b) Removal of ^{230}Th and ^{231}Pa at ocean margins. *Earth Planet Sci Lett* **66**: 73-90

Archer D, Lyle M, Rodgers K, Froelich P (1993) What controls opal preservation in tropical deep-sea sediments? *Paleoceanography* **8**: 7-22

Barnola JM, Raynaud D, Korokevitch YS, Lorius C (1987) Vostok ice core provides 160,000-year record of atmospheric CO_2. *Nature* **329**: 408-414

Bishop JKB (1988) The barite-opal-organic carbon association in oceanic particulate matter. *Nature* **331**: 341-343

Broecker WS, Peng T-H (1989) The cause of glacial to interglacial atmospheric CO_2 change: A polar alkalinity hypothesis. *Global Biogeochemical Cycles* **3**: 215-239

Burckle LH, Burak RW (1988) Fluctuations in Late Quaternary diatom abundances: stratigraphic and paleoclimatic implications from subantarctic deep sea cores. *Palaeogeogr Palaeoclimatol Palaeoecol* **67**: 147-156

Charles CD, Fairbanks RG (1990) Glacial to interglacial changes in isotopic gradients of Southern Ocean surface water. in *Geological History of Polar Oceans: Arctic versus Antarctic*, eds U Bleil and J Thiede, NATO/ASI Series C: Dordrecht, Netherlands, 729-760

Charles CD, Froelich PN, Zibello MA, Mortlock RA, Morley JJ (1991) Biogenic opal in Southern Ocean sediments over the last 450,000 years: Implications for surface water chemistry and circulation. *Palaeoceanography* **6**: 697-728

Church TM (1970) Marine barite, Ph.D. dissertation, 100 pp., Univ. Calif., San Diego.

Clemens S, Prell W, Murray D, Shimmield GB, Weedon GP (1991) Forcing mechanisms of the Indian Ocean monsoon. *Nature* **353**: 720-725

Corliss B (1982) Linkage of North Atlantic and Southern Ocean deep-water circulation during glacial intervals. *Nature* **298:** 458-460

Dehairs F, Chesselet R, Jedwab J (1980) Discrete suspended particles of barite and the barium cycle in the open ocean. *Earth Planet Sci Lett* **49:** 528-550

Dehairs F, Stroobants N, Goeyens L (1991) Suspended barite as a tracer of biological activity in the Southern Ocean. *Marine Chemistry* **35:** 399-410

DeMaster DJ (1981) The supply and accumulation of silica in the marine environment. *Geochim Cosmochim Acta* **45:** 1715-1732

Dymond J, Suess E, Lyle M (1992) Barium in the deep-sea sediment: A geochemical proxy for paleoproductivity. *Palaeoceanography* **7**:163-181

Francois R, Bacon MP, Suman DO (1990) Thorium 230 profiling in deep-sea sediments: High-resolution records of flux and dissolution of carbonate in the equatorial Atlantic during the last 24,000 years. *Paleoceanography* **5**: 761-788

Francois R, Altabet MA, Burckle LH (1992) Glacial to interglacial changes in surface nitrate utilization in the Indian sector of the Southern Ocean as recorded by sediment $\delta^{15}N$. *Paleoceanography* **7:** 589-606

Gard G, Crux J (1991) Preliminary results from Hole 704A: Arctic-Antarctic correlation through nannofossil biochronology. *Proc Ocean Drilling Program Science Results* **114**: 193-200

Goldberg ED, Arrhenius GOS (1958) Chemistry of Pacific pelagic sediments.*Geochim Cosmochim Acta* **13:** 153-212

Grobe H, Mackensen A (1992) Late Quaternary climatic cycles as recordd in sediments from the Antarctic continental margin. in *The Antarctic Paleoenvironment: A Perspective on Global Change Antarctic Research Series* **56:** 349-376

Hays JD, Lozano JA, Shackleton NJ, Irving G (1976) Reconstruction of the Atlantic and western Indian Ocean sectors of the 18,000 BP Antarctic Ocean. In: *Investigation of the Late Quaternary Palaeoceanography and Paleoclimatology*, eds RM Cline, JD Hays, *Geological Society of America Memoir*, **145**: 337-372

Hodell DA (1993) Late Pleistocene paleoceanography of theSouth Atlantic sector of the Southern Ocean: Ocean Drilling Program Hole 704A. *Paleoceanography* **8**: 47-67

Jordan RW, Pudsey CJ (1992) High-resolution diatom stratigraphy of Quaternary sediments from the Scotia Sea. *Mar Micropalaeo* **19**: 201-237

Kellog TB, Shackleton NJ, Duplessey J-C (1978) Planktonic foraminiferal and oxygen isotopic stratigraphy and paleoclimatology of the Norwegian sea deep sea cores. *Boreas* **7**: 61-73

Kier RS (1990) Reconstructing the ocean carbon system variation during the last 150,000 years according to the Antarctic nutrient hypothesis. *Palaeoceanography* **5**: 253-276

Kumar N, Gwiazda R, Anderson RF, Froelich PN (1993) $^{231}Pa/^{230}Th$ ratios in sediments as a proxy for past changes in Southern Ocean productivity. *Nature* **362**: 45-48

Labeyrie LD, Pichon J-J, Labracherie M, Ippolito P, Duprat J, Duplessey J-C (1986) Melting history of Antarctica during the past 60,000 years *Nature* **322**: 701-706

Labracherie M, Labeyrie LD, Duprat J, Bard E, Arnold M, Pichon J-J, Duplessey J-C (1989) The last deglaciation in the Southern Ocean. *Paleoceanography* **6:** 629-638

Lao Y, Anderson RF, Broecker WS (1992) Boundary scavenging and deep-sea sediment dating: Constraints from excess ^{230}Th and ^{231}Pa. *Paleoceanography* **6**: 783-798

Lea DW, Boyle EA (1990) Foraminiferal reconstruction of barium distributions in water masses of the glacial oceans. *Paleoceanography* **5**: 719-742

Ledford-Hoffman PA, Demaster DJ, Nittrouer CC (1986) Biogenic silica accumulation in the Ross Sea and the importance of the Antarctic continental shelf deposits in the marine silica budget. *Geochim Cosmochim Acta* **50**: 2099-2110

Martin JH (1990) Glacial-interglacial CO_2 change: the iron hypothesis *Palaeoceanography* **5**: 1-13

Mackensen A, Grobe H, Hubberten H-W, Spiess V, Futterer, DK (1989) Stable isotope stratigraphy from the Antarctic continental margin during the last one million years. *Mar Geol* **87:** 315-321

Martinson DG, Pisias NG, Hays JD, Imbrie J, Moore TC, Shackleton NJ (1987) Age dating and the orbital theory of the ice ages: Development of a high-resolution 0 to 300,000-year chronostratigraphy. *Quat Res* **27**: 1-29

Morley JJ (1989) Variations in high-latitude oceanographic fronts in the southern Indian Ocean: An estimation based on faunal changes *Paleoceanography* **4**: 547-554

Mortlock RA, Charles CD, Froelich PN, Zibello MA, Saltzman J, Hays JD, Burckle LH (1991) Evidence for lower productivity in the Antarctic Ocean during the last glaciation. *Nature* **351**: 220-222

Paillard D, Lecoat F, Chol E, Jodet, L (1992) Analyseries. Computer Program for graphical correlation on a Macintosh. Preliminary version

Pedersen TF, Calvert SE (1990) Anoxia vs productivity: What controls the formation of organic-carbon-rich sediments and sedimentary rocks *Am Assoc Petrol Geol Bull* **74**: 454-466

Pudsey CJ, Barker PF, Hamilton N (1988) Weddell Sea abyssal sediments: A record of Antarctic Bottom Water flow. *Mar Geol* **81**: 289-314

Pudsey CJ (1992) Late Quaternary changes in Antarctic Bottom Water velocity inferred from sediment grain size in the northern Weddell Sea. *Mar Geol* **107**: 9-33

Queguiner B, Treguer P, Nelson DM (1991) The production of biogenic silica in the Weddell and Scotia Seas. *Mar Chem* **35**: 449-459

Schmitz B (1987) The TiO_2/Al_2O_3 ratio in the Cenozoic Bengal abyssal fan sediments and its use as a paleostream energy indicator. *Mar Geol* **76**: 195-206

Siegenthaler U, Wenk T (1984) Rapid atmospheric CO_2 variations and ocean circulation. *Nature* **308**: 624-626

Shimmield GB (1985) The geochemistry and mineralogy of East Pacific sediments, Baja California, Mexico, Unpub PhD Thesis University of Edinburgh

Shimmield GB (1992) Can sediment geochemistry record changes in coastal upwelling palaeoproductivity? Evidence from northwest Africa and the Arabian Sea. *In* Summerhayes, C, Prell W, Emeis K (eds) *Upwelling Systems Since the Early Miocene*, Geol Soc Spec Publ **64**: 29-46

Shimmield GB, Mowbray SR (1991) The inorganic geochemical record of the northwest Arabian Sea: A history of productivity variation over the last 400 ka from Sites 722 and 724. *In*: Prell, WL and Niitsuma N et al *Proceedings of the Ocean Drilling Program, Science Results*, **117**: College Station, Texas, 409-429

Shimmield GB, Murray JW, Thomson J, Bacon MP, Anderson RF, Price NB (1986) The distribution and behaviour of ^{230}Th and ^{231}Pa at an ocean margin, Baja California, Mexico. Geochim Cosmochim Acta **50**: 2499-2507

Siegenthaler U, Wenk T (1984) Rapid atmospheric CO_2 variations and ocean circulation. *Nature* **308**: 624-626

Stroobants N, Dehairs F, Goeyens L, Vanderheijden N, Van Grieken R (1991) Barite formation in the Southern Ocean water column. *Mar Chem* **35**: 411-421

Treguer P, Van Bennekom AJ (1991) The annual producton of biogenic silica in the Antarctic Ocean. *Mar Chem* **35**: 477-487

SUBJECT INDEX

Springer-Verlag and the Environment

We at Springer-Verlag firmly believe that an international science publisher has a special obligation to the environment, and our corporate policies consistently reflect this conviction.

We also expect our business partners – paper mills, printers, packaging manufacturers, etc. – to commit themselves to using environmentally friendly materials and production processes.

The paper in this book is made from low- or no-chlorine pulp and is acid free, in conformance with international standards for paper permanency.